AF776913

Plant Genomics

Plant Genomics

Editor

Frank M. You

MDPI • Basel • Beijing • Wuhan • Barcelona • Belgrade • Manchester • Tokyo • Cluj • Tianjin

Editor
Frank M. You
Ottawa Research and
Development Centre
Agriculture and Agri-Food Canada
Ottawa
Canada

Editorial Office
MDPI
St. Alban-Anlage 66
4052 Basel, Switzerland

This is a reprint of articles from the Special Issue published online in the open access journal *International Journal of Molecular Sciences* (ISSN 1422-0067) (available at: www.mdpi.com/journal/ijms/special_issues/plant_genomics).

For citation purposes, cite each article independently as indicated on the article page online and as indicated below:

LastName, A.A.; LastName, B.B.; LastName, C.C. Article Title. *Journal Name* **Year**, *Volume Number*, Page Range.

ISBN 978-3-0365-7227-7 (Hbk)
ISBN 978-3-0365-7226-0 (PDF)

© 2023 by the authors. Articles in this book are Open Access and distributed under the Creative Commons Attribution (CC BY) license, which allows users to download, copy and build upon published articles, as long as the author and publisher are properly credited, which ensures maximum dissemination and a wider impact of our publications.
The book as a whole is distributed by MDPI under the terms and conditions of the Creative Commons license CC BY-NC-ND.

Contents

About the Editor

Frank M. You

Dr. Frank You is a highly accomplished senior research scientist in bioinformatics and genomics at the Agriculture and Agri-Food Canada (AAFC) Ottawa Research and Development Centre. He is also an adjunct professor in the Department of Plant Science at the University of Manitoba and a guest professor at Nanjing Agricultural University in China. Dr. You received his Ph.D. in plant genetics and breeding with a specialization in statistical genetics in 1989, and holds two Bachelor's degrees, one in agronomy in 1982 and another in computer science in 1999.

Dr. You is an expert in computational biology and bioinformatics, statistical genetics, and plant genetics and breeding. He has a wealth of experience in plant comparative and statistical genomics, quantitative genetics, genome assembly and annotation of complex genomes, gene expression and microarray data analysis, physical mapping and data analysis, high-throughput molecular marker design and development, and bioinformatics software development.

Dr. You's recent research projects focus on genome sequencing and annotation, QTL mapping, and identification and characterization of genes associated with seed yield and disease resistance in flax and cereal crops. Additionally, he is actively involved in breeding database development, gene identification, and marker development for wheat disease and pest resistance. With his impressive expertise and extensive research accomplishments, Dr. You continues to contribute to the advancement of the field of plant genomics and genetics.

Preface to "Plant Genomics"

Plant genomics is a rapidly growing field that has revolutionized our understanding of the genetic makeup and biological processes of plant species. The advancements in genomic technologies have enabled researchers to study the entire genetic composition of plants, including the function and regulation of genes, genome structure, and evolution. The impact of plant genomics is far-reaching and has significant implications for agriculture, environment, and human health.

In this Special Issue Reprint "Plant Genomics", we are pleased to present 28 papers that showcase the latest research in plant genomics. These papers cover a wide range of topics, including genome-wide association study and genomic prediction, genome sequencing, gene expression in response to abiotic stresses, genome-wide screening, and characterization of various gene families. Moreover, they involve more than 20 different plant species, highlighting the diversity of plants and the importance of studying their genomes.

The papers in this Special Issue have been contributed by leading experts in the field, who have used cutting-edge techniques and approaches to provide insights into various aspects of plant genomics. We hope that these papers will stimulate further research and inspire new collaborations that will enhance our understanding of plant biology and contribute to the development of sustainable agriculture.

We would like to express our gratitude to all the authors who have contributed their work to this Special Issue, as well as the reviewers who have provided valuable feedback and ensured the quality of the papers. We also thank the editorial team for their support and dedication in bringing this Special Issue to fruition.

We hope that this Special Issue will serve as a valuable resource for researchers, students, and professionals interested in plant genomics and related fields.

Frank M. You
Editor

International Journal of
Molecular Sciences

MDPI

Article

Genome-Wide Association Study and Selection Signatures Detect Genomic Regions Associated with Seed Yield and Oil Quality in Flax

Frank M. You [1,2,*], Jin Xiao [1,3], Pingchuan Li [1], Zhen Yao [2], Gaofeng Jia [1,4], Liqiang He [1], Santosh Kumar [5], Braulio Soto-Cerda [6,7], Scott D. Duguid [2], Helen M. Booker [4], Khalid Y. Rashid [2] and Sylvie Cloutier [1,6,*]

1 Ottawa Research and Development Centre, Agriculture and Agri-Food Canada, Ottawa, ON K1A 0C6, Canada; xiaojin@njau.edu.cn (J.X.); lipingchuan@gmail.com (P.L.); gaofeng.jia@usask.ca (G.J.); liqiang.he@canada.ca (L.H.)
2 Morden Research and Development Centre, Agriculture and Agri-Food Canada, Morden, MB R6M 1Y5, Canada; zhen.yao@canada.ca (Z.Y.); scott.duguid@agr.gc.ca (S.D.D.); khalid.rashid@agr.gc.ca (K.Y.R.)
3 Department of Agronomy, Nanjing Agricultural University, Nanjing 210095, China
4 Crop Development Centre, University of Saskatchewan, Saskatoon, SK S7N 5A8, Canada; helen.booker@usask.ca
5 Brandon Research and Development Centre, Agriculture and Agri-Food Canada, Brandon, MB R7A 5Y3, Canada; Santosh.kumar@agr.gc.ca
6 Department of Plant Science, University of Manitoba, Winnipeg, MB R3T 2N2, Canada; braulio.soto@cgna.cl
7 Agriaquaculture Nutritional Genomic Center, CGNA, Temuco 4871158, Chile
* Correspondence: frank.you@agr.gc.ca (F.M.Y.); sylvie.cloutuer@agr.gc.ca (S.C.); Tel.: +1-613-759-1539 (F.M.Y.); +1-613-759-1744 (S.C.)

Received: 20 July 2018; Accepted: 3 August 2018; Published: 6 August 2018

Abstract: A genome-wide association study (GWAS) was performed on a set of 260 lines which belong to three different bi-parental flax mapping populations. These lines were sequenced to an averaged genome coverage of 19× using the Illumina Hi-Seq platform. Phenotypic data for 11 seed yield and oil quality traits were collected in eight year/location environments. A total of 17,288 single nucleotide polymorphisms were identified, which explained more than 80% of the phenotypic variation for days to maturity (DTM), iodine value (IOD), palmitic (PAL), stearic, linoleic (LIO) and linolenic (LIN) acid contents. Twenty-three unique genomic regions associated with 33 quantitative trait loci (QTL) for the studied traits were detected, thereby validating four genomic regions previously identified. The 33 QTL explained 48–73% of the phenotypic variation for oil content, IOD, PAL, LIO and LIN but only 8–14% for plant height, DTM and seed yield. A genome-wide selective sweep scan for selection signatures detected 114 genomic regions that accounted for 7.82% of the flax pseudomolecule and overlapped with the 11 GWAS-detected genomic regions associated with 18 QTL for 11 traits. The results demonstrate the utility of GWAS combined with selection signatures for dissection of the genetic structure of traits and for pinpointing genomic regions for breeding improvement.

Keywords: flax; genome-wide association study (GWAS); selective sweep; genotyping by sequencing (GBS); bi-parental population; single nucleotide polymorphism (SNP); seed yield; plant height; maturity; fatty acid composition

1. Introduction

Flax (*Linum usitatissimum* L., $2n = 2x = 30$) is a self-pollinating annual crop from the Linaceae family. It is a dual-purpose crop grown for its seed oil or stem fiber, resulting in two morphotypes:

linseed and fiber. The linseed or flaxseed morphotype is rich in oil (40–50%) containing five main fatty acids: palmitic (PAL, C16:0, ~6%), stearic (STE, C18:0, ~2.5%), oleic (OLE, C18:1$^{\Delta 9}$, ~19%), linoleic (LIO, C18:2$^{\Delta 9, 12}$, ~13%), and linolenic (LIN, C18:3$^{\Delta 9, 12, 15}$, ~55%) [1,2]. Because of its high LIN content, linseed is the richest plant source of omega-3 fatty acid which is beneficial for reducing blood cholesterol levels and mitigating heart diseases in humans [3,4]. The same attributes make it ideal as industrial oil for use in paints, linoleum flooring, inks, soaps and varnishes [4].

Linseed breeding has focused on high seed yield (YLD), high oil content (OIL), and either high or low LIN content. Low LIN (2–4%) and high LIO (65–70%) lines have been developed through mutation breeding. NuLin™ 50 with 67.8% LIN (http://www.viterra.ca) and Omégalin with 65.8% (http://www.terredelin.com) are examples of high LIN linseed cultivars currently registered. Extremely low LIN lines such as Linola™ or Solin™ improve oxidative stability, making such cultivars suitable for the fabrication of margarine [3]. Since 1910, a total of 82 flax cultivars have been released in Canada [5]. These cultivars and elite breeding lines provide diverse genetic materials for dissecting the genetic architecture of oil biosynthesis and yield related traits in linseed.

Several methods can be used to dissect the genetic architecture of crop traits. QTL or linkage mapping uses bi-parental populations to identify loci responsible for trait variation between parents based on a recombination-based genetic linkage map [6]. Bi-parental populations, such as F_2, recombinant inbred line (RIL), doubled haploid (DH) and backcross (BC) populations, are the most widely used genetic resources for mapping QTL for traits of interest in self-fertilizing crops, including flax [7–12]. While bi-parental populations are easy to develop and have power for QTL detection, only the a limited number of alleles from the parental genotypes are analyzed in a single population, resulting in a narrow genetic base and low representation of allelic diversity [13]. In addition, genetic recombination is limited in these populations [14]. To increase the QTL dissection power, attempts have been made to expand the genetic diversity through other multiple-parent population types, such as nested association mapping (NAM) populations [15–17] and multi-parent advanced generation intercross (MAGIC) populations [18–25], while retaining the advantages of association mapping and bi-parental populations. However, the development of such populations requires careful planning and time. Natural populations that possess tremendous phenotypic diversity can be advantageous in genome-wide association study (GWAS) with various molecular markers in plants and animals [26–31]. Association mapping using a diverse germplasm panel overcomes the phenotypic diversity limitation of bi-parental populations, thereby increasing the QTL mapping power [32] but is impeded by low detection power of association of rare alleles. GWAS usually uses a natural population to investigate wider phenotypic variation for complex traits by taking advantage of ancient genetic recombination events in populations [33].

GWAS may be complemented by performing genome-wide selective sweep scan (GW3S) which identifies selection signatures that are beneficial for plant adaptation. A selective sweep is the reduction or elimination of variation among the nucleotides near a new beneficial mutation. Following strong positive natural selection or artificial selection during domestication or breeding, selective sweeps affect nearby linked alleles [34]. Ancient selective sweeps are relevant to natural evolution and domestication of crop species that are subjected to natural and artificial selective pressures and forced to adapt rapidly to new environments and thus drive speciation [35]. Breeding selects favorable alleles and retains them in new cultivars. These signatures of selection can be detected by a cross-population comparison approach [34]. Recent studies demonstrated that genomic regions that exhibit selection signatures are also enriched for genes associated with biologically important traits [36–40]. Thus, detection of selection signatures is emerging as an additional approach to identify and validate novel gene-trait associations [41].

Genetic regions associated with storage oil biosynthesis in flax have been studied based on QTL mapping using bi-parental populations. Several QTL responsible for oil content and fatty acid composition have been mapped in independent studies including the three populations used herein. The first population (BM) of 243 $F_{2:6}$ recombinant inbred lines (RILs) from a cross between the Canadian linseed varieties CDC Bethune and Macbeth was used for a linkage mapping study and detected three QTL each for OLE and STE, two each for LIO and IOD, and one each for PAL, LIN and OIL with several QTL co-locating at the same locus [8]. The second population (EV) was a cross between E1747 and Viking. The third population (SU) was a cross between SP2047 (a yellow-seeded Solin^{TM} line with 2–4% LIN) and UGG5-5 (a brown-seeded flax line with 63–66% LIN) and comprised of 78 lines generated through DH method. It was used in a linkage mapping study using simple sequence repeat (SSR) markers which identified QTL for LIO, LIN and iodine value (IOD) co-locating on LG7 and LG16, and a QTL for PAL on LG9 [7]. The linkage-based studies from these populations provided numerous QTL for important traits but the QTL were generally far from the markers and poorly delimited because of the low resolution of the genetic maps [18,19,42].The three bi-parental populations were also used to construct a consensus genetic map [43], and to perform genomic selection [44] primarily using SSR markers. Because the three populations have been simultaneously phenotyped for several common agronomic and seed oil quality traits in the same environments (years/locations), we designed the present study to test the efficiency of the combined bi-parental population approach for GWAS and GW3S to detect genomic regions associated with seed yield and seed oil quality traits using genotyping by sequencing (GBS).

2. Results

2.1. Re-Sequencing and Genome-Wide SNPs

In the present study, a set of 260 genotypes (97 from the recombinant inbreeding line (RIL) population from a cross between CDC Bethune and Macbeth (BM), 91 from the RIL population from a cross between E1747 and Viking (EV) and 72 from the doubled haploid population from a cross between SP2047 and UGG5-5 (SU) along with the 5 of 6 parents except for the reference CDC Bethune) were re-sequenced using GBS to identify genome-wide single nucleotide polymorphism (SNP) markers on the chromosome-based flax pseudomolecules [45]. An average of ~57.7 million paired end reads were generated for each individual, corresponding to 5754 Mb sequences or 19.2× genome equivalents of the reference scaffolds (~302 Mb) [46] (Table S1). Paired-end reads of each genotype were aligned to the flax scaffolds [46], resulting in a total of 536,186 SNPs. After filtering off SNPs with minor allele frequency (MAF) <0.05 and genotyping rate <60% [47,48], 17,288 SNPs were retained on the flax pseudomolecules [45] (Table S2). Out of these, 15,284 segregated in BM, 15,397 in EV and 7568 in SU. The SNPs were mostly uniformly distributed across all 15 chromosomes (chr), ranging from 601 on chr11 to 1572 on chr13 (Figure 1, Table S2). Approximately 71.1% of all SNPs were located in intergenic regions, 16.2% were in introns and 12.7% were in exons (Table S2). These SNPs were used for further population structure analysis, GWAS and GW3S.

2.2. Whole-Genome Pattern of LD

The LD and LD decay rates were analyzed for each population separately as well as the merged population using the filtered SNP data. The physical distances of pair-wise SNPs at which the LD r^2 dropped to half were 1242, 223, 728 and 272 kb for BM, EV, SU and merged populations respectively. This indicated substantial variation in LD decay rate across populations (Figure 2). The average LD r^2 of BM, EV, SU, and merged populations were 0.37, 0.26, 0.28 and 0.30, respectively, with the number of haplotype blocks for each population estimated at 599, 648, 206 and 1205, respectively (Table S3).

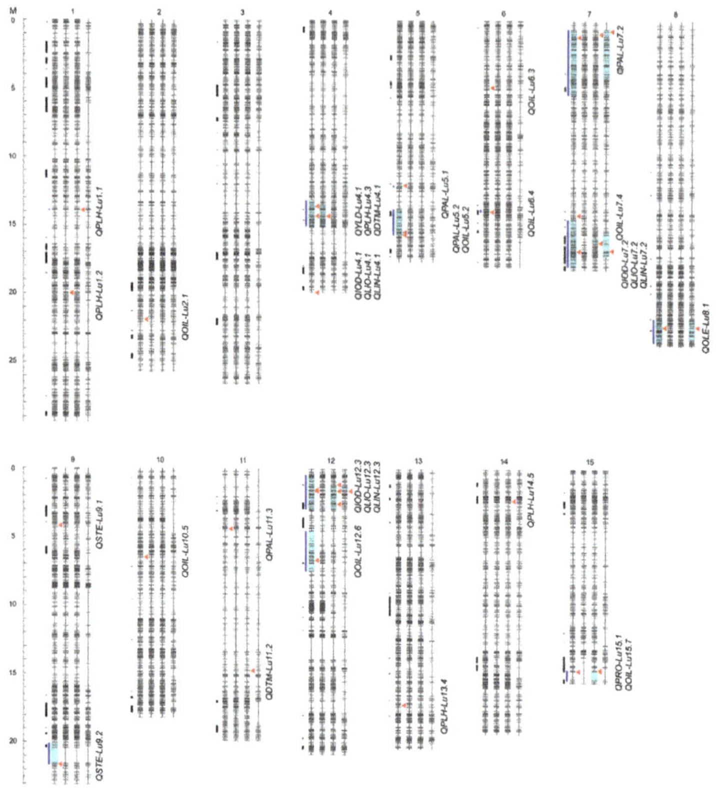

Figure 1. Distribution of 17,288 SNPs, 114 selective sweeps and 33 QTL on the 15 chromosomes of flax for each of three bi-parental populations BM, EV and SU and, for the merged population (BM + EV + SU). Four vertical bars from left to right for each chromosome represent the BM + EV + SU, BM, EV and SU populations, respectively. Short horizontal lines on bars represent SNPs. QTL regions are highlighted in cyan and by vertical blue lines. Red triangles identify QTL's peak SNP. Selective sweeps are represented by short vertical black lines.

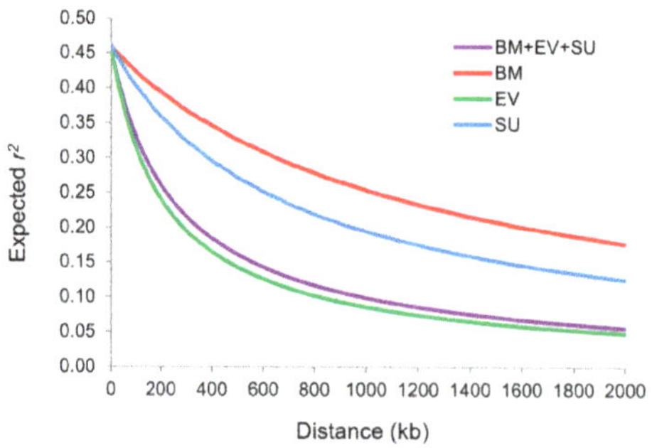

Figure 2. Intra-chromosome LD (r^2) decay of SNP pairs over the entire flax genome as a function of physical distances (kb) of pair-wise SNPs for the three individual and merged populations. The curves are drawn based on a fitted non-linear model (see Section 4.2).

2.3. Genetic Diversity and Population Structure

Nucleotide diversity (π) was estimated at 41.52, 38.26 and 3.95 for the BM, EV and SU populations, respectively (Table 1), and was consistent with the number of SNPs identified from the three populations. A strong population-differentiation (F_{st}) was observed at 0.44 between BM and SU and 0.48 between EV and SU. However, F_{st} was weaker at 0.04 between the BM and EV (Table 1).

Table 1. Genetic differentiation (F_{st}) between three bi-parental (upper triangle elements) and nucleotide diversity (π) within these populations (diagonal elements).

Population	BM	EV	SU
BM	41.52	0.04	0.44
EV		38.26	0.48
SU			3.95

BM: CDC Bethune/Macbeth; EV: E1747/Viking; SU: SP2047/UGG5-5.

The genetic structure within the merged population was assessed based on the 17,288 SNP loci from the 260 individuals using two methods: principal component analysis (PCA) and discriminant analysis for principal components (DAPC). Bi-plots of the first three principal components of the PCA showed five distinct clusters (Figure 3a,b). The BM (BM1 and BM2) and EV (EV1 and EV2) populations each contained two sub-populations, while SU produced a single cluster. DAPC corroborated the same five clusters (Figure 3c,d). Therefore, a DAPC Q matrix based on the five clusters was generated and used as covariates to assess the population stratification in GWAS and phenotypic variation explained by the SNPs.

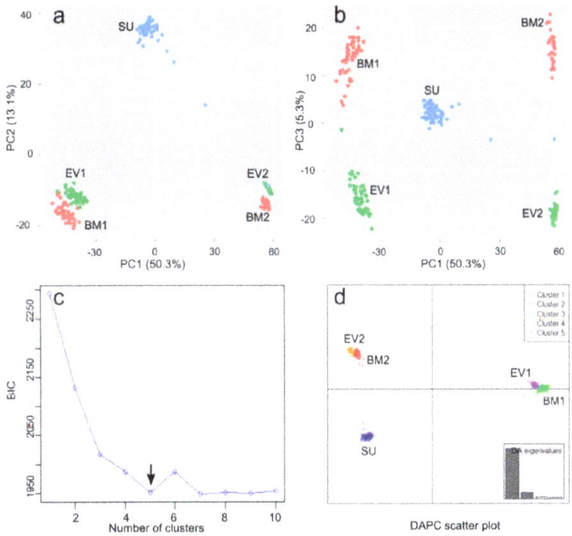

Figure 3. Principal component analysis (PCA) and discriminant analysis of principal components (DAPC) of the 260 individuals in three bi-parental populations (BM, EV and SU) based on 17,288 SNPs. (**a**) Bi-plot of the first and second principal components (PCs); (**b**) Bi-plot of the first and third PCs; (**c**) k-means clustering analysis based on 100 chosen PCs shows that the optimal number of clusters (k) is 5, that is where the Bayesian information criterion (BIC) is lowest (arrow); (**d**) DAPC scatter plot. Percentages in parentheses in the axis titles of (**a**) and (**b**) represent the variance explained by the respective PCs. Individuals from the BM and EV populations grouped into two subpopulations each, BM1 and BM2, and EV1 and EV2, respectively.

2.4. h^2_{SNP}

Phenotypic variation of traits was largely explained by SNPs in the three individual and the merged populations (Table 2). The average h^2_{SNP} for all 11 traits was 0.51. The largest h^2_{SNP} values among the four populations ranged from 0.45 (YLD) to 0.90 (PAL). More than 80% of the phenotypic variation in one of the populations was explained by identified SNPs for days to maturity (DTM), IOD, PAL, STE, LIO and LIN. The h^2_{SNP} varied from one population to another depending on the genetic variation between the two parents. For SU, little or no phenotypic variation was explained by SNPs for DTM, plant height (PLH) and STE. For EV, a relatively low phenotypic variation ($h^2_{SNP} < 0.1$) was explained by SNPs for STE and OLE.

Table 2. Phenotypic variation explained by all SNPs (h^2_{SNP}) and identified QTL (h^2_{GWAS}) for 11 traits in different populations.

Trait	Population	$h^2_{SNP} \pm s$	No. QTL	$h^2_{GWAS} \pm s$
YLD	BM + EV + SU	0.43 ± 0.12	1	0.14 ± 0.09 §
	BM	0.22 ± 0.25		
	EV	0.15 ± 0.24		
	SU	0.45 ± 0.21		
PLH	BM + EV + SU	0.53 ± 0.12	1	0.08 ± 0.11
	BM	0.76 ± 0.12	2	0.21 ± 0.15
	EV	0.76 ± 0.14	2	0.22 ± 0.18
	SU	0.06 ± 0.20		
DTM	BM + EV + SU	0.43 ± 0.13	1	0.10 ± 0.07
	BM	0.81 ± 0.11	1	0.18 ± 0.13
	EV	0.36 ± 0.24	1	0.18 ± 0.22
	SU	0.00 ± 0.20		
PRO	BM + EV + SU	0.51 ± 0.11	1	0.12 ± 0.16
	BM	0.52 ± 0.20		
	EV	0.34 ± 0.23	1	0.09 ± 0.12
	SU	0.58 ± 0.19		
OIL	BM + EV + SU	0.66 ± 0.09	7	0.62 ± 0.14
	BM	0.46 ± 0.22		
	EV	0.39 ± 0.21	1	0.08 ± 0.08
	SU	0.70 ± 0.15		
IOD	BM + EV + SU	0.80 ± 0.06	3	0.57 ± 0.10
	BM	0.49 ± 0.19		
	EV	0.78 ± 0.12	2	0.51 ± 0.14
	SU	0.66 ± 0.17	2	0.35 ± 0.18
PAL	BM + EV + SU	0.79 ± 0.06	4	0.48 ± 0.11
	BM	0.12 ± 0.26		
	EV	0.55 ± 0.20	1	0.09 ± 0.11
	SU	0.90 ± 0.07	1	0.56 ± 0.18
STE	BM + EV + SU	0.21 ± 0.15	2	0.41 ± 0.19
	BM	0.85 ± 0.09		
	EV	0.02 ± 0.14		
	SU	0.00 ± 0.22	1	
OLE	BM + EV + SU	0.55 ± 0.10	1	0.16 ± 0.13
	BM	0.36 ± 0.22		
	EV	0.09 ± 0.25		
	SU	0.72 ± 0.16	1	0.20 ± 0.19

Table 2. *Cont.*

Trait	Population	$h^2_{SNP} \pm s$	No. QTL	$h^2_{GWAS} \pm s$
LIO	BM + EV + SU	0.80 ± 0.06	3	0.73 ± 0.07
	BM	0.54 ± 0.20		
	EV	0.75 ± 0.13	2	0.54 ± 0.14
	SU	0.66 ± 0.17	2	0.36 ± 0.18
LIN	BM + EV + SU	0.80 ± 0.06	3	0.56 ± 0.09
	BM	0.49 ± 0.19		
	EV	0.76 ± 0.13	2	0.55 ± 0.14
	SU	0.66 ± 0.17	2	0.36 ± 0.18

YLD: seed yield; PLH: plant height; DTM: days to maturity; PRO: protein content; OIL: oil content; IOD: iodine value; PAL: palmitic acid content; STE: stearic acid content; OLE: oleic acid content; LIO: linoleic acid content; LIN: linolenic acid content; BM: CDC Bethune/Macbeth; EV: E1747/Viking; SU: SP2047/UGG5-5. § h^2_{GWAS} of YLD was estimated based on the phenotypes in a single environment (Morden/2012). For all other traits, h^2_{GWAS} was estimated based on the BLUP estimates of phenotypes.

2.5. QTL Identified from 11 Traits

Using the best linear unbiased prediction (BLUP) values of phenotyping data collected from six to eight year/location environments with both generalized linear model (GLM) and mixed linear model (MLM), we identified a total of 33 QTL for 11 traits, one for YLD, eight for OIL, five for PLH, four for PAL, three each for IOD, LIO, and LIN, two each for DTM and STE, and one each for protein content (PRO) and OLE (Table 3, Figure 1, Figures S1 and S2). Thirty-one of the 33 QTL were detected using GLM and 13 with MLM (Tables S4 and S5). Of these latter 13, two QTL (QTL 18 for IOD and QTL 31 for LIN) were detected only by MLM, while the remaining 11 were identified by both MLM and GLM (Table S4).

Out of 33 QTL identified, 12, 6, 3 and 27 were from EV, SU, BM and merged population, respectively. Only six QTL were detected exclusively from two individual populations, including four (QTL 2 and 6 for PLH, QTL 8 for DTM and QTL 17 for OIL) from EV and two (QTL 3 and 4 for PLH) from BM. Eighteen were identified exclusively from the merged population. Ten QTL were detected simultaneously from the merged population and one or more individual populations (Tables S4 and S5).

QTL for YLD (QTL 1) was identified only in two environments (2010/Morden and 2012/Saskatoon) (Figure S2) but not in other environments or using BLUP estimates over the six year/location environments. We also performed GWAS for all other traits with phenotypic data from individual environments and obtained similar results with the QTL identified using BLUP values over multiple environments (Table S6).

2.6. QTL Effect Significance

To validate the QTL, we tested statistical significance of difference of phenotypes between two contrasting haplotype pairs for each QTL in the merged and individual populations and in different year/location environments. QTL effect differences between two contrasting haplotype pairs for all 33 QTL were significant (Figure 4, Table S7). We also assessed relationship of the number of pyramiding positive-effect QTL in individuals with trait phenotypes. Significant linear relations for all eight traits which had two or more QTL identified in this study were observed, showing primarily additive or accumulative QTL effects (Figure 5).

Table 3. QTL and associated gene candidates.

Trait	QTL No.	QTL Name	Chr.	Start Position (bp)	End Position (bp)	XP-CLR Score	Known QTL or Marker	Candidate Gene IDs	Candidate Gene Location (bp)	Candidate Gene Name	Gene Annotation
YLD	1	*QYLD-Lu4.1*	4	13,594,936	14,968,389	12.54	*QYld.BM.crc-LG4* [a]				
PLH	2	*QPLH-Lu1.1*	1	13,887,715	13,930,292						
	3	*QPLH-Lu1.2*	1	20,012,490	20,012,490			Lus10020835	19,610,837	*BRI1* [49]	Leucine-rich receptor-like protein kinase family protein
								Lus10020865	19,790,777	*GA2* [49]	Terpenoid cyclases/Protein prenyltransferases superfamily protein
	4	*QPLH-Lu4.3*	4	14,305,982	15,042,104	12.54		Lus10034358	14,006,288	*BIM2* [49]	BES1-interacting Myc-like protein 2
								Lus10041435	14,157,752	*MYB62* [49]	Myb domain protein 62
								Lus10041481	14,398,338	*LMCO4* [49]	Laccase/Diphenol oxidase family protein
								Lus10041794	15,920,170	*ROT3* [49]	Cytochrome P450 superfamily protein
								Lus10041801	15,948,434	*WAT1* [49]	Walls Are Thin 1
	5	*QPLH-Lu13.4*	13	17,243,884	17,243,884			Lus10030567	18,680,474	*GA2OX8* [49]	Gibberellin 2-oxidase 8
	6	*QPLH-Lu13.5*	14	2,320,469	2,320,469	40.61		Lus10021395	3,647,029	*HCT* [49]	Hydroxycinnamoyl-CoA shikimate/quinate hydroxycinnamoyl transferase
DTM	7	*QDTM-Lu4.1*	4	13,171,757	15,042,104	12.54	*QDm.BM.crc-LG4* [a]	Lus10015766	13,094,864	*FLC* [50]	K-box region and MADS-box transcription factor family protein
								Lus10034461	13,434,121	*CDF3* [50]	Cycling DOF factor 3
								Lus10034370	13,933,421	*AP1* [50]	K-box region and MADS-box transcription factor family protein
								Lus10041483	14,411,103	*PFT1* [50]	Phytochrome and flowering time regulatory protein (PFT1)
								Lus10041500	14,512,085	*ATAN11* [50]	Transducin/WD40 repeat-like superfamily protein
								Lus10041540	14,716,950	*RGL1* [50]	RGA-like 1
								Lus10041595	14,966,739	*AP2* [50]	Integrase-type DNA-binding superfamily protein
	8	*QDTM-Lu11.2*	11	14,768,686	14,768,686						
PRO	9	*QPRO-Lu15.1*	15	14,746,288	14,746,310	8.50		Lus10030671	22,732,660	*WRI* [50]	Integrase-type DNA-binding superfamily protein

Table 3. *Cont.*

Trait	QTL No.	QTL Name	Chr.	Start Position (bp)	End Position (bp)	XP-CLR Score	Known QTL or Marker	Candidate Gene IDs	Candidate Gene Location (bp)	Candidate Gene Name	Gene Annotation
OIL	10	*QOIL-Lu2.1*	2	21,913,720	21,913,720						
	11	*QOIL-Lu5.2*	5	15,704,607	15,705,039						
	12	*QOIL-Lu6.3*	6	4,879,632	4,879,632						
	13	*QOIL-Lu6.4*	6	13,799,180	13,970,951	50.58					
	14	*QOIL-Lu7.4*	7	14,209,179	14,209,179						
	15	*QOIL-Lu10.5*	10	6,517,448	6,517,448						
	16	*QOIL-Lu12.6*	12	4,591,214	7,491,405	27.77					
	17	*QOIL-Lu15.7*	15	14,665,900	15,429,055	8.89		Lus10039906	19,833,852	*KCS14-2* [51]	3-ketoacyl-CoA synthase
IOD	18	*QIOD-Lu4.1*	4	19,909,467	19,909,467			Lus10039906	19,833,852	*KCS14-2* [51]	3-ketoacyl-CoA synthase
	19	*QIOD-Lu7.2*	7	15,346,458	17,977,459	45.70	*QIOD.crc-LG7* [b]	Lus10038321	16,089,922	*FAD3a* [52]	Fatty acid desaturase
	20	*QIOD-Lu12.3*	12	489,561	2,981,642	106.22	*QIOD.crc-LG16* [b]	Lus10036184	1,035,336	*FAD3b* [52]	Fatty acid desaturase
								Lus10023359	1,729,292	*FAH1* [50]	Fatty acid hydroxylase 1
PAL	21	*QPAL-Lu5.1*	5	12,062,376	12,182,441			Lus10029880	12,062,376	*KCS12-3* [51]	3-ketoacyl-CoA synthase
	22	*QPAL-Lu5.2*	5	13,797,851	15,668,995	12.14					
	23	*QPAL-Lu7.3*	7	624,461	5,423,691	17.74	*QPal.BM.crc-LG7* [a] *QPAL.crc-LG9* [b] *c79-s540_Lu2534* [c]	Lus10001814	79,471	*KAS Ic-1* [51]	3-ketoacyl-acyl carrier protein synthase I
								Lus10028925	1,085,389	*KAS IIIb-2* [51]	3-ketoacyl-acyl carrier protein synthase III
								Lus10028885	1,262,079	*SUN1* [50]	SAD1/UNC-84 domain protein 1
	24	*QPAL-Lu11.4*	11	4,417,685	4,429,424			Lus10026345	4,333,672	*KCS7-1* [51]	3-ketoacyl-CoA synthase
OLE	25	*QOLE-Lu8.1*	8	21,782,841	23,527,563	12.64		Lus10006636	22,165,534	*KCS9-1* [51]	3-ketoacyl-CoA synthase
								Lus10006637	22,174,324	*KCS1-1* [51]	3-ketoacyl-CoA synthase
								Lus10018485	23,111,453	*DES-1-LIKE* [50]	Fatty acid desaturase family protein
STE	26	*QSTE-Lu9.1*	9	4,229,230	4,229,230			Lus10040333	4,275,842	*KCS18-2* [51]	3-ketoacyl-CoA synthase
	27	*QSTE-Lu9.2*	9	20,080,531	21,636,823	27.55		Lus10011877	20,059,127	*SAD1* [51]	Stearoyl acyl carrier protein desaturase
								Lus10011839	20,227,416	*FatA2-2* [51]	FatA acyl-ACP thioesterase
LIO	28	*QLIO-Lu4.1*	4	19,909,467	19,909,467			Lus10039906	19,833,852	*KCS14-2* [51]	3-ketoacyl-CoA synthase
	29	*QLIO-Lu7.2*	7	14,540,706	17,977,459	45.70	*QLIO.crc-LG7* [b] *c281-s185_ Lu566* [c]	Lus10038321	16,089,922	*FAD3a* [52]	Fatty acid desaturase
	30	*QLIO-Lu12.3*	12	489,561	2,981,642	106.22	*QLIO.crc-LG16* [b] *Llio-LG12.3* [c]	Lus10036184	1,035,336	*FAD3b* [52]	Fatty acid desaturase
LIN	31	*QLIN-Lu4.1*	4	19,909,467	19,909,467			Lus10039906	19,833,852	*KCS14-2* [51]	3-ketoacyl-CoA synthase
	32	*QLIN-Lu7.2*	7	14,540,719	17,977,459	45.70	*QLIN.crc-LG7* [b] *c281-s185_ Lu566* [c]	Lus10038321	16,089,922	*FAD3a* [52]	Fatty acid desaturase
	33	*QLIN-Lu12.3*	12	489,561	2,981,642	106.22	*QLIN.crc-LG16* [b] *Llin-LG12.3* [c]	Lus10036184	1,035,336	*FAD3b* [52]	Fatty acid desaturase
								Lus10023359	1,729,292	*FAH1* [50]	Fatty acid hydroxylase 1

[a] QTL identified in [8]; [b] QTL identified in [7]; [c] QTL identified in [53]. All candidate genes are labelled by references.

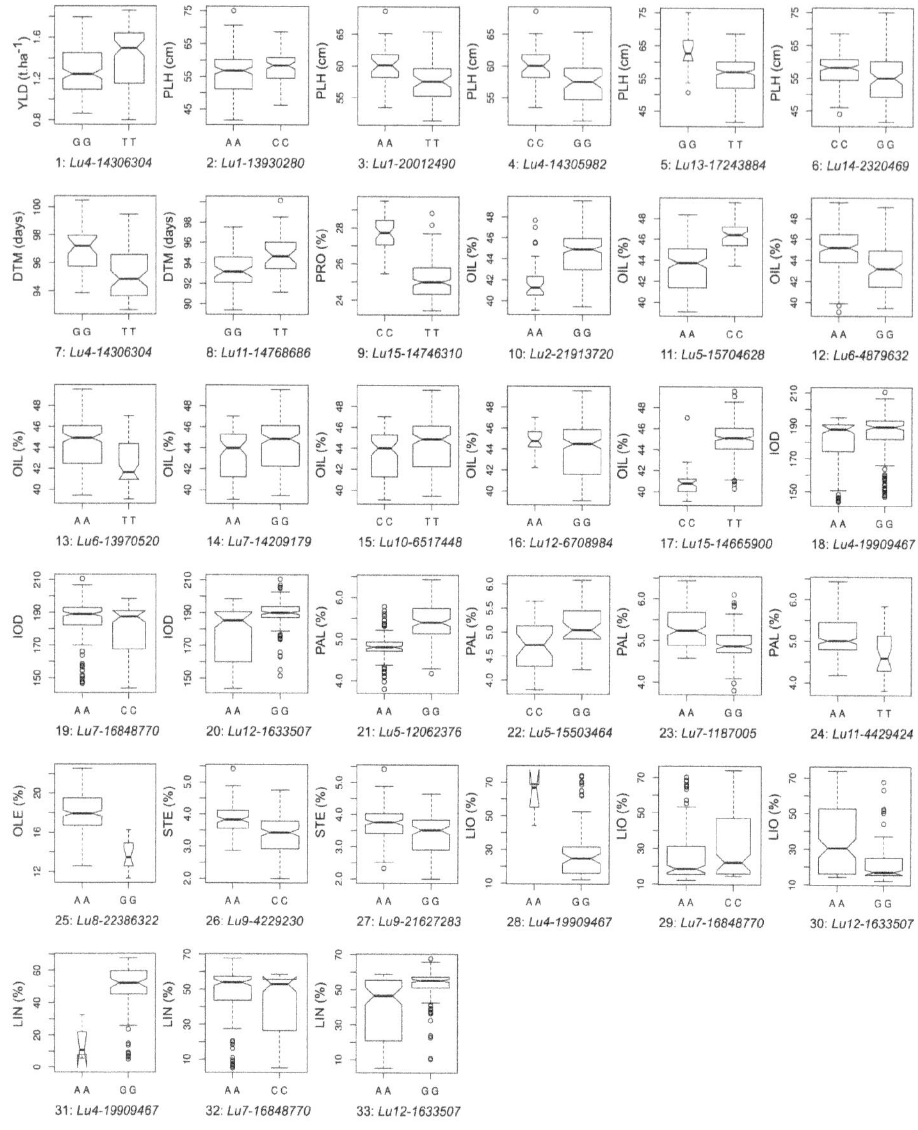

Figure 4. Trait performance of two contrasting haplotype pairs for each of 33 QTL identified from 11 traits. A QTL is represented by the peak SNP identified in the association study. The numbers of QTL correspond to QTL No in Table 3. The BLUP values of the 11 traits in the merged population were used except for PLH/QTL 3 and DTM/QTL 7 for which BM population was used, DTM/QTL 8 for which EV population was used, and PAL/QTL 22, LIO/QTL 28 and LIN/QTL 31 for which SU population was used. The box width is proportional to the size of the subpopulations. Phenotype differences between two contrasting haplotype pairs for each QTL are shown by boxes' notches. For any given QTL, boxes' notches that do not overlap indicate significant median differences at 95% confidence level.

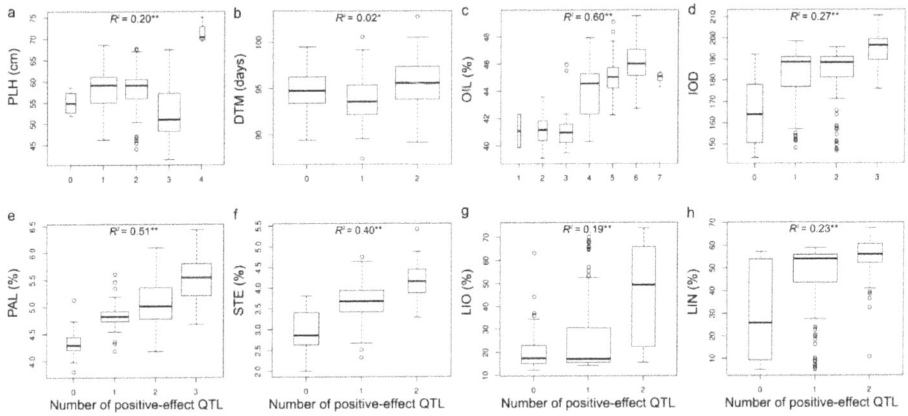

Figure 5. The relationship of phenotypes with the number of positive-effect QTL in individuals. Eight traits with two or more QTL identified were analyzed: (**a**) plant height (PLH), (**b**) days to maturity (DTM), (**c**) oil content (OIL), (**d**) iodine value (IOD), (**e**) palmitic acid content (PAL), (**f**) steric acid content (STE), (**g**) linoleic acid content (LIO), and (**h**) linolenic acid content (LIN). The BLUP values of the eight traits in the merged population were used. The correlation of phenotypes with the number of positive-effect QTL was calculated. * and ** represent statistical significance at 0.05 and 0.01 probability level.

2.7. Pleiotropy of QTL

Sixteen of the 33 QTL co-located at six genomic regions concerning nine traits (Figures 1 and 6, Table S8). QTL for PLH, DTH and YLD co-located on chr4. QTL for IOD, LIO and LIN co-located on chr4, 7 and 12. Chromosome 15 harbored QTL for OIL and PRO while chr5 had QTL for OIL and PAL.

2.8. Phenotypic Variation Explained by QTL

Phenotypic variations explained by individual QTL (h^2_{QTL}) were estimated (Table S4). Overall, the QTL explained 4 to 66% of the total phenotypic variation, with an average of 32.5% which is more than half of the average h^2_{SNP} (51%). For five traits (IOD, LIO, LIN, PAL and OIL), QTL explained an average of 61% of the variation (Table 2 and Table S4) We also estimated the phenotypic variation explained by all QTL for a trait (h^2_{GWAS}) (Table 2). In the merged population, the QTL explained 48–73% of the phenotypic variation for OIL, IOD, PAL, LIO and LIN but only 8–14% for PLH, DTM and YLD.

2.9. Candidate Genes Underlying QTL

Based on the GWAS results, we investigated the genes annotated in the flax genome [54] in an attempt to predict candidate genes from loci significantly associated with each trait. The genomic locations of SNP markers at the peaks of the QTL were scanned within a 500 Kb window in either direction to constitute a subset of genes from which we deduced a candidate gene list based on *a priori* knowledge of their function(s). Candidate genes were identified for every QTL except for the YLD QTL (Table 3). We discovered seven candidate genes underlying QTL for DTM on chr4. The QTL for PLH harbors five candidate genes of completely different function. The genes underlying QTL for fatty acid composition include *KCS14-2*, *FAD3a*, and *FAD3b* for IOD/LIN/LIO, *KCS12-3* and *KAS Ic-1* for PAL, *KCS9-1* and *KCS1-1* for OLE, and *KCS18-2* and *SAD1* for STE.

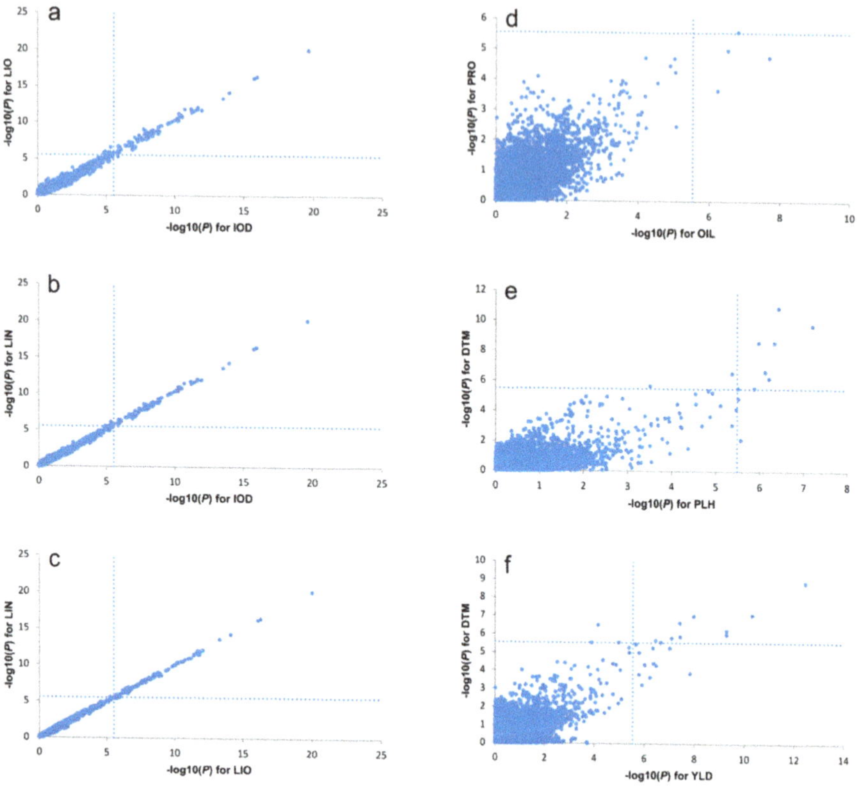

Figure 6. Relations of $-\log_{10}(P)$ values of SNP markers between two traits showing pleiotropy or linkage relationship of SNP markers in different pairs of traits. (**a**) IOD vs. LIN; (**b**) IOD vs. LIO; (**c**) LIN vs. LIO; (**d**) OIL vs. PRO; (**e**) PLH vs. DTM; (**f**) DTM vs. YLD. Results of the GWAS using a GLM and data from the BM + EV + SU population for IOD, LIO, and LIN (**a–c**), the EV population for OIL and PRO (**d**), the BM population for PLH and DTM (**e**) and the BM + EV + SU population for DTM and YLD (**f**) are shown. The vertical and horizontal dashed lines show the cut-off value of significant SNP markers associated with a trait. YLD: seed yield (t ha^{-1}); DTM: days to maturity; OIL: oil content (%); PRO: protein content (%); IOD: iodine value; LIO: linoleic acid content (%); LIN: linolenic acid content (%).

2.10. Selection Signatures in Bi-Parental Populations

A GW3S was performed to identify potential selection signatures during breeding improvement using XP-CLR [34]. Due to the high genetic diversity in BM and EV (Table 1) and large phenotypic differences between them (Table S9), GW3S between BM and EV was conducted. A total of 114 selection signatures with an average size of 226.3 kb were identified (Figures 1 and 7, Table S10), accounting for 7.82% of the flax pseudomolecules (~316 Mb). These putative selection signatures overlapped with 11 GWAS-detected genomic regions associated with 18 QTL (Figures 1 and 7).

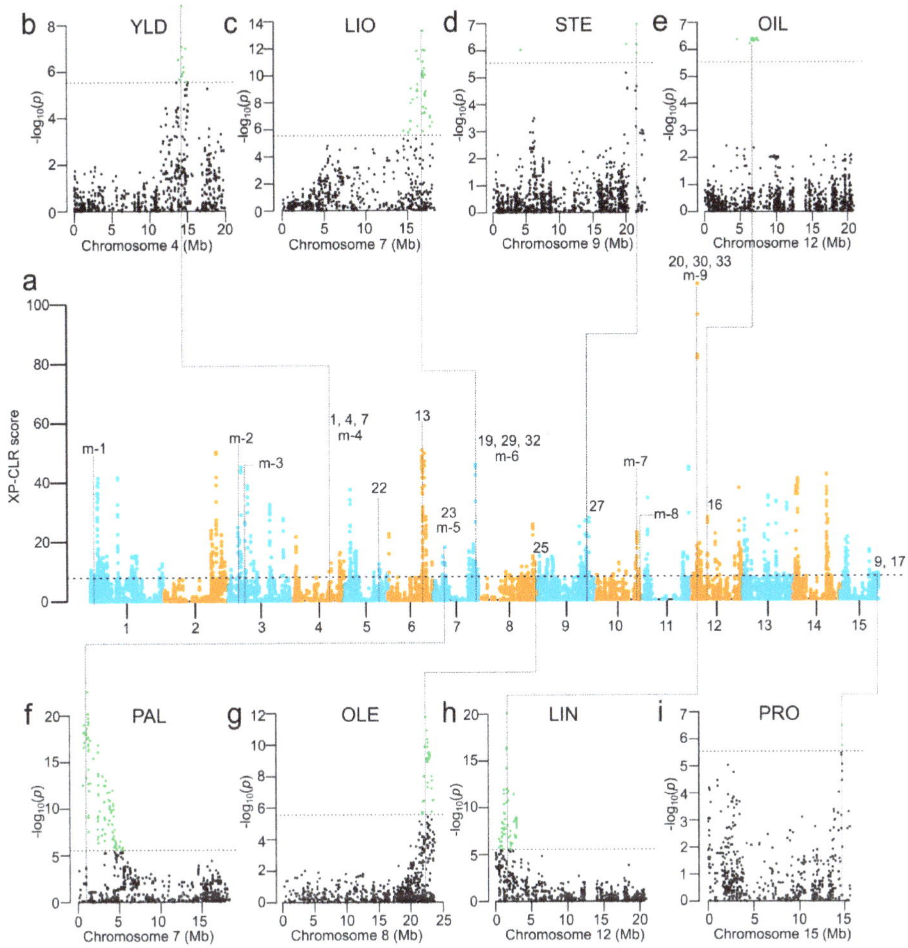

Figure 7. Genome-wide selective sweep scan using XP-CLR between BM and EV (**a**), and Manhattan plots of QTL overlapping with selective sweeps for (**b**) seed yield (YLD), (**c**) linoleic acid content (LIO), (**d**) steric acid content (STE), (**e**) oil content (OIL), (**f**) palmitic acid content (PAL), (**g**) oleic acid content (OLE), (**h**) linolenic acid content (LIN), and (**i**) protein content (PRO). QTL associated with selective sweeps are also labeled on peaks of selective sweeps. The numbers represent the QTL numbers listed in Table 3. Multiple numbers on the same peak represent genomic regions co-located with more than one trait. The labels 'm-#' represent the genomic regions associated with QTL previously identified and listed in Table S11. The green dots on Manhattan plots represent significant SNPs.

Some selection signatures were also associated with previously identified QTL (Table S11). For example, the selection signatures were associated with 10 previously reported QTL (Figure 7). The signatures at position 2.45–2.46 Mb on chr1 overlapped with SNP marker *Lu1_2670961* linked to QTL *QSte.BM.crc-LG1* for STE; the ones at 4.74–4.77 Mb on chr3 overlapped with *Lu3_5950394*, a SNP linked to QTL *QOle.BM.crc-LG3-1*/*QLio.BM.crc-LG3* for OLE and LIO; signatures at 7.24–7.25 Mb on chr3 overlapped with SNP *Lu3_8415336* linked to QTL *QSte.BM.crc-LG3* for STE [8]; position 16.80–16.81 Mb on chr10 harbors signatures that overlap with SSR *Lu2262* linked to an unnamed QTL for OIL; finally, position 17.52–17.53 Mb on chr10 has selection signatures that coincide with SSR *Lu2746* linked to an unnamed QTL for LIN/IOD [53].

3. Discussion

3.1. QTL Associated with Seed Yield and Seed Oil Quality Traits

Thirty-three QTL were identified in the current study. Of which, nine QTL were identified in previous studies [7,8] for the same traits, including seed yield and seed oil quality traits. Cloutier et al. [7] detected six major QTL for LIO, LIN and IOD in SU population. These six QTL correspond to the two underlying genes, *FAD3a* and *FAD3b*. Some of these QTL were in close proximity on the same chromosome. We identified the same QTL by association mapping that were previously detected by linkage mapping [7] using the same phenotype and SNP genotype data in the SU population (Table 3). The refinement of flax pseudomolecule [45] between the linkage study and our current association study allowed reassignment of chr12 for LIO, LIN and IOD QTL which were previously assigned to LG16 [8]. In addition, the same QTL were also detected in the EV population as well as the merged population. Our association study also validated three QTL for YLD, DTM and PAL which were previously identified using linkage mapping using SSRs and SNPs [8,9] and from the association mapping using a flax core collection population with SSR markers [53] (Table 3). These verified QTL for fatty acid composition, seed yield and maturity demonstrate the feasibility of the association mapping method to detect QTL in a bi-parental population as well as a multi-parent population.

An additional 24 novel QTL were detected in our current study which were not discovered in previous studies using individual BM or SU populations. These new QTL were detected using the merged population which greatly increased the population size, thereby enhancing the association power and resolution for QTL detection. We noted that only two QTL were discovered from the BM population alone. This is likely the result of significantly reduced representation of lines re-sequenced from BM population [8]. The discovery of new QTL demonstrates that GWAS using multiple bi-parental populations is equally or more efficient for QTL detection than QTL mapping using single bi-parental populations alone.

We tested the statistical significance of QTL effects for all 33 QTL identified for the 11 traits and found that all effect differences were significant. We also observed significant positive correlation between the number of positive-effect QTL and corresponding trait phenotypes in individuals for eight traits from which had two or more QTL were identified (Figures 4 and 5, Table S7). These results not only corroborate the significance of the QTL but also demonstrate that effects of QTL in an individual performed additively, suggesting that marker-assisted selection (MAS) for these QTL would be effective in breeding. Thus, we listed the flanking sequences of these QTL in Table S12 for MAS purpose.

3.2. Pleotropic QTL Associated with Seed Yield and Quality Traits

Six genomic regions associated with more than one trait were identified. QTL for IOD, LIO, and LIN were concurrent on chromosomes 4, 7 and 12; QTL for YLD, PLH, and DTM co-located on chr4; QTL for PRO and OIL were on chr15 and QTL for PAL and OIL were on chr5 (Figures 1 and 6, Table S8).

IOD is a measure of the degree of unsaturation of the oil that is calculated from the GC-derived fatty acid composition. Thus, breeding lines with high LIN normally show high IOD [7] due to the high correlation between IOD, LIO, and LIN [44] (Table S13). QTL co-located at the same genomic regions indicate that the traits may be controlled by the same gene or tightly linked genes. The two genomic regions on chromosomes 7 and 12 harbor the two fatty acid desaturase genes, *FAD3a* and *FAD3b*. These genes are responsible for linoleic and linolenic acid composition [52,55].

PLH and DTM are complex traits that considerably impact the adaptability, biomass, and economic yield of agricultural crops [56,57]. In soybean, one QTL that strongly associated with both PLH and DTM traits was identified with an SNP at 45.0 Mb position on chromosome 19 and it harbors the candidate gene *DT1*, which is homolog to *Arabidopsis terminal flower 1* (*TFL-1*, AT5G03840) [56]. Based on in silico gene annotation, the *DT1* homolog are located on chromosomes 6 and 8 in flax but no QTL for either PLH or DTM were identified on these two chromosomes. This could be due to the lack of functional polymorphism(s) at those loci among the parents of our three populations. However, a different genomic region on chr4 harbours five candidate genes for PLH and seven for DTM, raising the possibility that PLH and DTM are controlled by tightly linked genes in flax. The same genomic region was also associated with YLD. Because plant height and maturity affect seed yield, it is not surprising that QTL for PLH, DTM and YLD were mapped to the same locus. This pleiotropic relationship between YLD and DTM was previously validated [8] (Table 3).

Inheritance of seed oil content is complicated due to its quantitative nature. The seed oil content was directly affected by fatty acid composition traits, such as PAL, STE, OLE, LIO, and LIN, or indirectly by several major agronomic traits, such as seed yield and protein content [58]. Significant correlations of OIL were observed with PAL (-0.57; $p = 0$) and PRO (-0.70; $p = 0$) (Table S13). OIL is also usually negatively correlated with PRO in oilseed crops [59]. Of the eight QTL associated with oil content, two co-located with QTL for PAL on chr5 and for PRO on chr15, respectively.

3.3. Phenotypic Variation Explained by SNPs and QTL

SNP heritability (h^2_{SNP}) for a trait is the total proportion of phenotypic variance explained by additive contributions from genome-wide SNPs. A method for estimating h^2_{SNP} for a complex trait was initially proposed in 2011 [60,61] and implemented in GCTA (Genome-wide Complex Trait Analysis) software [61]. Since then, the method has been applied to many quantitative traits largely in human and animal genetic studies [62,63]. The method was also used to estimate phenotypic variance explained by a subset of SNPs selected by p-values from GWAS in an independent sample [64]. However the estimate of variance explained by the SNP subsets ascertained by the p-values from GWAS in the same sample may be inflated due to positive correlation between true SNP effects and estimation errors (personal communication to the GCTA author, Jian Yang). However, as the GCTA-based heritability estimation method includes the population structure effect in the linear model and also considers heritability estimates to be irrelevant to the number of SNPs used [60,61], the accuracy of estimates should be higher than those obtained simply using the simple multivariate regression adopted in most GWAS of plant traits. In the current study, for the first time we applied this method to estimate h^2_{SNP} for 11 agronomic and seed quality traits in three bi-parental populations and a merged population. As the number of SNPs identified from a population depends on its genetic variation for the traits, the trait-associated h^2_{SNP} estimates vary across populations and traits. Overall, seed yield had a lower h^2_{SNP} than seed quality traits as expected considering the extent of genetic complexity of the former (Table 2). We also used the same method to estimate phenotypic variation explained by individual QTL (h^2_{QTL}) and by all QTL for a specific trait (h^2_{GWAS}). h^2_{GWAS} measures the extent of the phenotypic variation explained by QTL compared to that of all SNPs. This comparison led to the conclusion that many QTL for PLH, DTM and YLD were not detected in our study but the QTL for seed quality traits identified herein likely represent major genetic regions or genes controlling these traits.

3.4. Selection Signatures Associated with Seed Yield and Seed Quality Traits

GW3S has been used for screening putative genomic regions under selection pressure caused by domestication or artificial selection [36,38]. Usually, contrasting genetic populations are compared (such as wild accessions vs. cultivated accessions, landraces vs. breeding lines) to identify the allele frequency differentiation between different populations. In this study, we alternatively used two contrasting bi-parental mapping populations and identified 114 selection signatures with an average size of 226.3 kb. Some of these selection signatures support nearly 50% of the 23 GWAS-detected genomic regions associated with 33 QTL. Some of the QTL identified by GWAS have no overlapping selection signatures, partially because the regions of QTL had XP-CLR (Cross Population Composite Likelihood Ratio) scores less than the predetermined cut-off values. On the other hand, many selection signatures have high XP-CLR scores but no associated QTL (Figure 7). These significant selection signatures may be associated with QTL for traits not included in this study. This is suggested by the fact that five previously identified genomic regions related to seven QTL overlapped with the selection signatures identified in our current study comparing BM and EV (Table S10). These putative selection signatures provide useful candidates for further QTL-trait association study. Our results combined with previous studies demonstrate that GW3S combined with GWAS is a powerful approach for dissecting genetic structure of breeding populations and for the identification of underlying genomic regions for breeding improvement. Using GWAS with bi-parental populations and selection signatures allowed the cross validation of QTL previously identified by other mapping methods and established the foundation for genomic assisted breeding in flax.

4. Materials and Methods

4.1. Plant Materials

Three bi-parental mapping populations of different genetic backgrounds served as genotype panel for the association study. The first population (BM) consisted of 243 F_6-derived RILs generated by single seed descent from a cross between CDC Bethune and Macbeth. Its two parents are Canadian high-yielding conventional linseed cultivars with 55–57% LIN [65,66]. The second population (EV) contained 90 F_6-derived RILs from a cross between E1747, an ethyl methanesulfonate (EMS)-induced low LIN breeding line [67], and Viking, a French fiber flax cultivar with ~55% LIN that was grown extensively in the 2000's but deregistered in 2012. The third population (SU) is an F_1-derived DH population of 78 lines obtained from a cross between the breeding line SP2047, from which a yellow-seeded Solin™ 2047 with only 2–3% LIN has been derived, and breeding line UGG5-5, which is a high LIN line with 63–66% LIN [7,55]. BM was designed to study yield-related traits while EV and SU were intended for QTL identification for fatty acid composition and fiber traits.

4.2. Whole Genome Resequencing, SNP Calling, SNP Imputation and LD Analysis

Three populations consisting of 97 randomly selected lines from BM, 91 from EV, 72 from SU including five parents (one parent is the reference genome) were grown in growth cabinets with a 16-h light and 8-h dark cycle at 20/18 °C. DNA was extracted from young leaf tissue using the DNeasy 96 Plant kit (Qiagen, Mississauga, ON, Canada) according to the manufacturer's instructions. The DNA was subsequently restricted, size-selected and quantified prior to the construction of the reduced representation libraries used for Illumina sequencing as previously described [47]. Reduced representation libraries from a total of 260 individuals of the three populations, i.e., 96 randomly selected from BM, 89 from EV, 70 from SU, and five parents (One parent CDC Bethune of BM is used as a reference genome) were re-sequenced by the Michael Smith Genome Sciences Centre of the BC Cancer Agency, Genome British Columbia (Vancouver, BC, Canada) using 100-bp paired-end reads on an Illumina HiSeq 2000 platform (Illumina Inc., San Diego, CA, USA).

SNP calling was performed using the revised AGSNP pipeline [47,48,68]. The flax scaffold sequences of cultivar CDC Bethune [46] were used as reference for read mapping. Then SNPs were called using SAMtools [69] and further filtered using a set of criteria such as mapped read depth, consensus base ratio, mapping quality score and homopolymers with a validation rate of 96.8% for the called SNPs as described in detail [47]. Finally SNPs with a MAF < 0.05 and a genotyping rate <60% were removed for further analysis. The coordinates of all SNPs were extracted from the chromosome-based flax pseudomolecules v2.0 [45]. Missing data for these filtered SNPs were imputed using Beagle v.4.2 [70].

Intra-chromosome LD (r^2) was calculated using plink ver. 1.9 [71] with the parameters "-r2 -ld-window-kb 2000 -ld-window-r2 0". Before LD calculation, SNP data were pruned using the parameter "–indep-pairwise 2000 50 0.9" to remove SNPs with high r^2 (>0.9) in a 2000 kb window with step size of 50 SNPs. Pair-wise r^2 values were plotted against the base pair distance, and curves of LD decay with distances of paired SNPs were fitted using a non-linear regression model [72] as follows:

$$r^2 = \frac{10+cd}{(2+cd)(11+cd)} \times \left\{1+\frac{(3+cd)\left(12+12cd+(cd)^2\right)}{n(2+cd)(11+cd)}\right\}, \quad (1)$$

where c is the coefficient to be estimated, d is the distance between pair-wise SNPs, and n is the number of total gametes, corresponding to twice the number of individuals in a population. The R function *nls* was used to fit the model. The rate of LD decay for each population was determined from the fitted model at the half of the maximum LD (r^2). Haplotype blocks were calculated using plink with the parameters "–blocks no-pheno-req –blocks-max-kb 2000".

4.3. Differentiation and Stratification

Nucleotide diversity (π) of three bi-parental populations and genetic differentiation (F_{st}) between the populations were estimated using the R package "PopGenome" [73]. The genetic structures of the three separate inbreeding populations and the combined population were assessed using both PCA and DAPC [74]. Analyses with DAPC included several steps: (1) PCA was conducted using the imputed SNPs. According to the curve of accumulative variances against the number of principle components (PCs), the optimum number of PCs was chosen at which the cumulative variance had no obvious increase; (2) *k*-means clustering analysis was performed based on the chosen PCs. To identify the optimal number of clusters, *k*-means was run sequentially with increasing *k* values and the Bayesian information criterion (BIC) was calculated for each *k*. The optimum *k* corresponded to the lowest BIC; (3) Discriminant analysis was conducted based on the chosen clusters and individuals were reassigned to the different clusters. The posterior probability of cluster membership was calculated based on the retained discrimination functions and used as the Q matrix for GWAS and heritability estimation. PCA was performed using the function implemented in TASSEL while DAPC was conducted using the R package "adegenet" 2.0 [75].

4.4. Phenotyping of Bi-Parental Populations

Individuals from the three populations were evaluated in field trials over four years (2009–2012) at two sites, Morden Research and Development Centre, Manitoba (MD) and Kernen Crop Research Farm near Saskatoon, Saskatchewan (SAS) in Canada. A type-2 modified augmented design (MAD) [76] was used for the field experiments from which phenotypic data were collected. The detailed experimental design was previously described [44,77]. All 243 individuals of the BM population were phenotyped in four years (2009–2012) and two sites (MD and SAS), while 86 individuals of the EV population and 72 individuals of the SU population were evaluated in three years (2010–2012) and two sites (MD and SAS).

Eleven common traits were evaluated in the three populations, including YLD, PLH, DTM, PRO, OIL, IOD and five fatty acid composition traits (OLE, PAL, STE, LIO, and LIN). PLH was measured from ground to the uppermost part of the plant at maturity. DTM was recorded from sowing to 95% of capsule maturity (seeds rattling in the capsules or bolls). Seed yield data were measured by harvesting two 0.5-m sections from rows located in the central part of each subplot (0.2 m^2). A total of 1 g of seed from each line at each environment was sampled for OIL measurement and fatty acid composition. Methyl esters of fatty acids were prepared according to the American Oil Chemists' Society (AOCS) Official Method Ce 2-66 [78] and fatty acid composition was measured by gas chromatography (GC) following AOCS Official Method Ce 1e-91. OIL was determined by nuclear magnetic resonance calibrated against the FOSFA extraction reference method. PRO was measured using near-infrared spectroscopy calibrated against the combustion analysis reference method and expressed on an N $\times$ 6.25 dry basis. Phenotyping of these seed quality traits has been previously described [53]. All phenotypic data from the field experiments and laboratory measurements were adjusted for soil heterogeneity as previously described based on the MAD pipeline [77]. The BLUP values over multiple environmental phenotypes estimated using R package "lme4" [79] were used for further association study analyses. The Shapiro-Wilk normality test was performed for all traits using the R function "shapiro.test". All 11 traits followed approximately a normal or mixed normal distribution (Figure S3). Simple correlations among 11 traits were calculated using the function "rcorr" of the R package "Hmisc".

4.5. Phenotypic Variation Explained by All SNPs

The phenotypic variation explained by all SNPs, denoted as h^2_{SNP}, was estimated for all traits based on the following mixed linear model [60] implemented in the GCTA software [61]:

$$\boldsymbol{y} = \boldsymbol{X\beta} + \boldsymbol{g} + \boldsymbol{\varepsilon} \text{ with its variance } \boldsymbol{V} = \boldsymbol{A}\sigma^2_g + \boldsymbol{I}\sigma^2_\varepsilon \tag{2}$$

where $\boldsymbol{y}$ is an $n \times 1$ vector of phenotypes with n individuals in a population, $\boldsymbol{X}$ is the $n \times n$.

structure matrix, $\boldsymbol{\beta}$ is a vector of fixed effects of population structure, including posterior probabilities of an individual assigning to a cluster in DAPC, $\boldsymbol{g}$ is an $n \times 1$ vector of the total genetic effects of the individuals with $\boldsymbol{g} \sim N\,(\boldsymbol{0}, \boldsymbol{A}\sigma^2_g)$, and $\boldsymbol{\varepsilon}$ is a vector of residual effects with $\boldsymbol{\varepsilon} \sim N\,(\boldsymbol{0}, \boldsymbol{I}\sigma^2_\varepsilon)$. $\boldsymbol{A}$ is interpreted as the genetic relationship matrix (GRM) between individuals and estimated from SNPs. σ^2_g is estimated using the restricted maximum likelihood (REML) method based on the GRM estimated from all SNPs. Thus, SNP heritability h^2_{SNP} was estimated as

$$h^2_{SNP} = \frac{\sigma^2_g}{\sigma^2_g + \sigma^2_\varepsilon} \tag{3}$$

4.6. Genome-Wide Association Study

GWAS was performed with the GLM and compressed MLM [80,81] implemented in TASSEL (v5.2) [82], which employs the EMMA and P3D algorithms to reduce computing time. For MLM, the kinship matrices for the merged population and the three single populations were calculated using TASSEL (v5.2) [82]. Manhattan plots and quantile-quantile (Q-Q) plots of GWAS were obtained using the R package "qqman" [83].

SNP markers for candidate QTL were determined based on the p-value for each marker estimated in the GLM or MLM analysis. The p-values were adjusted by the Bonferroni correction, being α (0.05)/No. of SNPs used in the analyses. Allele effects of significant markers were calculated as the difference between the average phenotypic values of homozygous alleles which were obtained directly from the TASSEL outputs. Candidate QTL were defined based on peaks of SNPs exceeded the significance threshold for the trait. The genomic region for a QTL was defined as a genome block spanning all significant SNPs.

The amount of phenotypic variation explained by significant QTL was estimated for all SNP markers within the QTL regions using the same method as described above [61], denoted as h^2_{QTL}. We similarly estimated phenotypic variation explained by all significant QTL for a single trait and denoted it h^2_{GWAS}.

4.7. Candidate Gene Mining

Genome-wide gene scan along chromosomes for significant QTL was performed to characterize the underlying genomic regions and identify candidate genes. First, all orthologous genes of the model species *Arabidopsis thaliana* were mapped to the flax genome using BLASTP of flax protein sequences against *A. thaliana* protein sequences at an E-value of 1.0×10^{-10}. A total of 15,323 unique *A. thaliana* genes were mapped. A list of known flax or *A. thaliana* genes associated with our studied traits and their associations was drawn based on literature and database searches [49,51,84]. We investigated candidate genes within QTL regions or within a 500 kb window upstream and downstream of the peaks depending on the LD decay estimates. In addition, previously identified QTL (SSR markers) in flax [7,8,53] were mapped to the flax pseudomolecules to validate the QTL results from this study.

4.8. QTL Validation

Three approaches were applied to validate QTL identified by GWAS. The first approach was to compare our QTL with previously identified QTL as described above. The same QTL was inferred if two QTL were mapped to the same recombination block or haplotype block. The second approach tested the significance of difference of phenotypes between two contrasting haplotype pairs of a QTL in the populations. Statistically significant differences served to validate significant QTL. Both *t* and Wilcox non-parametric tests were performed using the R functions "t.test" and "wilcox.test" for each QTL in the merged and individual populations and in different year/location environments. To test the positive correlations of a trait upon pyramiding of QTL, a simple regression of the number of positive-effect QTL on phenotypic values of a trait was calculated. A positive-effect QTL in an individual meant that this individual possessed a positive effect allele for the QTL. The last approach was to perform genome-wide selective sweep scans to confirm QTL associated genomic regions as described below.

4.9. Genome-Wide Selective Sweep Scan

A WG3S was performed along chromosomes across two populations using the program XP-CLR [34]. Comparisons between BM and EV using XP-CLR were conducted. The genetic distances (cM) between SNPs were estimated using the integrated flax consensus genetic map [43], assuming uniform recombination between SNPs. For each chromosome, XP-CLR was executed with the parameters "-w1 0.005 100 100 1 -p1 0.7" to estimate XP-CLR scores for 100-bp windows. Each chromosome was then divided into 10-kb segments and the highest XP-CLR score from windows with at least one SNP were assigned to each 10-kb segment ($x_{max,i}$). If the XP-CLR scores ($x_{max,\ i}$ and $x_{max,\ i+1}$) of two adjacent 10-kb segments were greater than the 80th percentile ($x_{max,80th}$) of the genome-wide scores of all 10-kb fragments, then they were grouped as a single putative selective sweep. In addition, putative selective sweeps were also merged if they were separated by no more than one low score ($<x_{max,80th}$) segment. Merged selective sweeps were assigned the highest score from their merged 10-kb segments. These merged segments were further combined into a larger region if these segments belonged to the same peak in the genome-wide selective sweep plot (Figure 5a). Finally, the combined regions falling in the highest 10th percentile of all putative selective sweeps were considered differentially selected regions or selection signatures.

The selection signatures were compared to both our detected QTL and previously reported QTL on the genetic loci to find associations between them. Positions where the QTL corresponding markers were located were extended by 100 kb on both sides and then compared with the position of the selection signatures. The QTL and selection signatures were considered associated when they overlapped.

Supplementary Materials: Supplementary materials can be found at http://www.mdpi.com/1422-0067/19/8/2303/s1.

Author Contributions: S.C., F.M.Y., S.D.D., H.M.B. and K.Y.R. conceived and designed the study. S.C. performed sequencing. S.D.D., H.M.B. and K.Y.R. performed the phenotyping. F.M.Y., J.X., P.L., Z.Y., G.J., L.H., S.K. and B.S.-C. analyzed the data. F.M.Y., J.X., S.K. and S.C. wrote the manuscript. All authors reviewed and edited the manuscript.

Funding: This research was funded by Genome Canada and other industrial stakeholders for the Total Utilization Flax GENomics (TUFGEN) project, by Agriculture and Agri-Food Canada for an A-base project, and by Western Grain Research Foundation (WGRF) and the Saskatchewan Flax Development Commission (SFDC) for the flax breeding database project.

Conflicts of Interest: The authors declare no conflict of interest. The funders had no role in the design of the study; in the collection, analyses, or interpretation of data; in the writing of the manuscript, and in the decision to publish the results.

Abbreviations

DH	doubled haploid
GBS	genotyping by sequencing
GW3S	genome-wide selective sweep scan
GWAS	genome-wide association study
IOD	iodine value
LD	linkage disequilibrium
LIN	linolenic acid
LIO	linoleic acid
MAF	minor allele frequency
MAGIC	multi-parent advanced generation intercross
NAM	nested association mapping
OIL	oil content
OLE	oleic acid
PAL	palmitic acid
QTL	quantitative trait loci
RIL	recombinant inbred line
SNP	single nucleotide polymorphism
SSR	simple sequence repeat
STE	stearic acid
YLD	seed yield

References

1. Westcott, N.D.; Muir, A.D. Chemical studies on the constituents of Linum spp. In *Flax, the Genus Linum*; Muir, A.D., Westcott, N.D., Eds.; Taylor and Francis: New York, NY, USA, 2003; pp. 55–73.
2. Diederichsen, A.; Kusters, P.M.; Kessler, D.; Bainas, Z.; Gugel, R.K. Assembling a core collection from the flax world collection maintained by Plant Gene resources of Canada. *Genet. Resour. Crop Evol.* **2013**, *60*, 1479–1485. [CrossRef]
3. Green, A.G.; Chen, Y.; Singh, S.P.; Dribnenki, J.C.P. Flax. In *Compendium Transgenic Crop Plants: Transgenic Oilseed Crops*; Kole, C., Hall, T.C., Eds.; Blackwell Publishing Ltd.: Oxford, UK, 2008; pp. 199–226.
4. Tolkachev, O.N.; Zhuchenko, A.A. Biologically active substances of flax: Medicinal and nutritional properties. *Pharm. Chem. J.* **2000**, *34*, 360–367. [CrossRef]
5. You, F.M.; Duguid, S.D.; Lam, I.; Cloutier, S.; Rashid, K.Y.; Booker, H. Pedigrees and genetic base of the flax varieties registered in Canada. *Can. J. Plant Sci.* **2016**, *96*, 837–852. [CrossRef]
6. Price, A.H. Believe it or not, QTLs are accurate! *Trends Plant Sci.* **2006**, *11*, 213–216. [CrossRef] [PubMed]
7. Cloutier, S.; Ragupathy, R.; Niu, Z.; Duguid, S.D. SSR-based linkage map of flax (*Linum usitatissimum* L.) and mapping of QTLs underlying fatty acid composition traits. *Mol. Breed.* **2011**, *28*, 437–451. [CrossRef]
8. Kumar, S.; You, F.M.; Duguid, S.; Booker, H.; Rowland, G.; Cloutier, S. QTL for fatty acid composition and yield in linseed (*Linum usitatissimum* L.). *Theor. Appl. Genet.* **2015**, *128*, 965–984. [CrossRef] [PubMed]
9. Asgarinia, P.; Cloutier, S.; Duguid, S.; Rashid, K.; Mirlohi, A.; Banik, M.; Saeidi, G. Mapping quantitative trait loci for powdery mildew resistance in flax (*Linum usitatissimum* L.). *Crop Sci.* **2013**, *53*, 2462–2472. [CrossRef]
10. Fu, Y.-B. Genetic evidence for early flax domestication with capsular dehiscence. *Genet. Resour. Crop Evol.* **2011**, *58*, 1119–1128. [CrossRef]
11. Soto-Cerda, B.J.; Maureira-Butler, I.; Muñoz, G.; Rupayan, A.; Cloutier, S. SSR-based population structure, molecular diversity and linkage disequilibrium analysis of a collection of flax (*Linum usitatissimum* L.) varying for mucilage seed-coat content. *Mol. Breed.* **2012**, *30*, 875–888. [CrossRef]
12. Wiesnerova, D.; Wiesner, I. ISSR-based clustering of cultivated flax germplasm is statistically correlated to thousand seed mass. *Mol. Biotechnol.* **2004**, *26*, 207–213. [CrossRef]
13. McMullen, M.D.; Kresovich, S.; Villeda, H.S.; Bradbury, P.; Li, H.; Sun, Q.; Flint-Garcia, S.; Thornsberry, J.; Acharya, C.; Bottoms, C. Genetic properties of the maize nested association mapping population. *Science* **2009**, *325*, 737–740. [CrossRef] [PubMed]

14. Bandillo, N.; Raghavan, C.; Muyco, P.A.; Sevilla, M.A.L.; Lobina, I.T.; Dilla-Ermita, C.J.; Tung, C.-W.; McCouch, S.; Thomson, M.; Mauleon, R.; et al. Multi-parent advanced generation inter-cross (MAGIC) populations in rice: Progress and potential for genetics research and breeding. *Rice* **2013**, *6*, 11. [CrossRef] [PubMed]
15. Yu, J.; Holland, J.B.; McMullen, M.D.; Buckler, E.S. Genetic design and statistical power of nested association mapping in maize. *Genetics* **2008**, *178*, 539–551. [CrossRef] [PubMed]
16. Monir, M.M.; Zhu, J. Dominance and epistasis interactions revealed as important variants for leaf traits of maize NAM population. *Front. Plant Sci.* **2018**, *9*, 627. [CrossRef] [PubMed]
17. Ren, D.; Fang, X.; Jiang, P.; Zhang, G.; Hu, J.; Wang, X.; Meng, Q.; Cui, W.; Lan, S.; Ma, X.; et al. Genetic architecture of nitrogen-deficiency tolerance in wheat seedlings based on a nested association mapping (NAM) population. *Front. Plant Sci.* **2018**, *9*, 845. [CrossRef] [PubMed]
18. Mackay, I.; Powell, W. Methods for linkage disequilibrium mapping in crops. *Trends Plant Sci.* **2007**, *12*, 57–63. [CrossRef] [PubMed]
19. Cavanagh, C.; Morell, M.; Mackay, I.; Powell, W. From mutations to MAGIC: Resources for gene discovery, validation and delivery in crop plants. *Curr. Opin. Plant Biol.* **2008**, *11*, 215–221. [CrossRef] [PubMed]
20. Mathew, B.; Leon, J.; Sannemann, W.; Sillanpaa, M.J. Detection of epistasis for flowering time using Bayesian multilocus estimation in a barley MAGIC population. *Genetics* **2018**, *208*, 525–536. [CrossRef] [PubMed]
21. Camargo, A.V.; Mackay, I.; Mott, R.; Han, J.; Doonan, J.H.; Askew, K.; Corke, F.; Williams, K.; Bentley, A.R. Functional mapping of quantitative trait loci (QTLs) associated with plant performance in a wheat magic mapping population. *Front. Plant Sci.* **2018**, *9*, 887. [CrossRef] [PubMed]
22. Ongom, P.O.; Ejeta, G. Mating design and genetic structure of a multi-parent advanced generation intercross (magic) population of sorghum (*Sorghum bicolor* (L.) Moench). *G3 (Bethesda)* **2018**, *8*, 331–341. [CrossRef] [PubMed]
23. Huynh, B.L.; Ehlers, J.D.; Huang, B.E.; Munoz-Amatriain, M.; Lonardi, S.; Santos, J.R.P.; Ndeve, A.; Batieno, B.J.; Boukar, O.; Cisse, N.; et al. A multi-parent advanced generation inter-cross (MAGIC) population for genetic analysis and improvement of cowpea (*Vigna unguiculata* L. Walp.). *Plant J.* **2018**, *93*, 1129–1142. [CrossRef] [PubMed]
24. Ponce, K.S.; Ye, G.; Zhao, X. Qtl identification for cooking and eating quality in indica rice using multi-parent advanced generation intercross (MAGIC) population. *Front. Plant Sci.* **2018**, *9*, 868. [CrossRef] [PubMed]
25. Huang, C.; Shen, C.; Wen, T.; Gao, B.; Zhu, D.; Li, X.; Ahmed, M.M.; Li, D.; Lin, Z. SSR-based association mapping of fiber quality in upland cotton using an eight-way MAGIC population. *Mol. Genet. Genom.* **2018**, *293*, 793–805. [CrossRef] [PubMed]
26. Garrido-Cardenas, J.A.; Mesa-Valle, C.; Manzano-Agugliaro, F. Trends in plant research using molecular markers. *Planta* **2018**, *247*, 543–557. [CrossRef] [PubMed]
27. Pena, R.N.; Ros-Freixedes, R.; Tor, M.; Estany, J. Genetic marker discovery in complex traits: A field example on fat content and composition in pigs. *Int. J. Mol. Sci.* **2016**, *17*. [CrossRef] [PubMed]
28. Zhu, X.M.; Shao, X.Y.; Pei, Y.H.; Guo, X.M.; Li, J.; Song, X.Y.; Zhao, M.A. Genetic diversity and genome-wide association study of major ear quantitative traits using high-density SNPs in maize. *Front. Plant Sci.* **2018**, *9*, 966. [CrossRef] [PubMed]
29. Chen, L.; Wan, H.; Qian, J.; Guo, J.; Sun, C.; Wen, J.; Yi, B.; Ma, C.; Tu, J.; Song, L., et al. Genome-wide association study of cadmium accumulation at the seedling stage in rapeseed (*Brassica napus* L.). *Front. Plant Sci.* **2018**, *9*, 375. [CrossRef] [PubMed]
30. MacGregor, S.; Ong, J.S.; An, J.; Han, X.; Zhou, T.; Siggs, O.M.; Law, M.H.; Souzeau, E.; Sharma, S.; Lynn, D.J.; et al. Genome-wide association study of intraocular pressure uncovers new pathways to glaucoma. *Nat. Genet.* **2018**, *50*, 1067–1071. [CrossRef] [PubMed]
31. Huang, X.; Wei, X.; Sang, T.; Zhao, Q.; Feng, Q.; Zhao, Y.; Li, C.; Zhu, C.; Lu, T.; Zhang, Z.; et al. Genome-wide association studies of 14 agronomic traits in rice landraces. *Nat. Genet.* **2010**, *42*, 961–967. [CrossRef] [PubMed]
32. Meng, L.; Zhao, X.; Ponce, K.; Ye, G.; Leung, H. QTL mapping for agronomic traits using multi-parent advanced generation inter-cross (MAGIC) populations derived from diverse elite indica rice lines. *Field Crops Res.* **2016**, *189*, 19–42. [CrossRef]
33. Huang, X.; Han, B. Natural variations and genome-wide association studies in crop plants. *Annu. Rev. Plant Biol.* **2014**, *65*, 531–551. [CrossRef] [PubMed]

34. Chen, H.; Patterson, N.; Reich, D. Population differentiation as a test for selective sweeps. *Genome Res.* **2010**, *20*, 393–402. [CrossRef] [PubMed]
35. Gore, M.A.; Chia, J.M.; Elshire, R.J.; Sun, Q.; Ersoz, E.S.; Hurwitz, B.L.; Peiffer, J.A.; McMullen, M.D.; Grills, G.S.; Ross-Ibarra, J.; et al. A first-generation haplotype map of maize. *Science* **2009**, *326*, 1115–1117. [CrossRef] [PubMed]
36. Xie, W.; Wang, G.; Yuan, M.; Yao, W.; Lyu, K.; Zhao, H.; Yang, M.; Li, P.; Zhang, X.; Yuan, J.; et al. Breeding signatures of rice improvement revealed by a genomic variation map from a large germplasm collection. *Proc. Natl. Acad. Sci. USA* **2015**, *112*, E5411–5419. [CrossRef] [PubMed]
37. Wen, Z.; Boyse, J.F.; Song, Q.; Cregan, P.B.; Wang, D. Genomic consequences of selection and genome-wide association mapping in soybean. *BMC Genom.* **2015**, *16*, 671. [CrossRef] [PubMed]
38. Zhou, Z.; Jiang, Y.; Wang, Z.; Gou, Z.; Lyu, J.; Li, W.; Yu, Y.; Shu, L.; Zhao, Y.; Ma, Y.; et al. Resequencing 302 wild and cultivated accessions identifies genes related to domestication and improvement in soybean. *Nat. Biotechnol.* **2015**, *33*, 408–414. [CrossRef] [PubMed]
39. He, C.; Fu, J.; Zhang, J.; Li, Y.; Zheng, J.; Zhang, H.; Yang, X.; Wang, J.; Wang, G. A gene-oriented haplotype comparison reveals recently selected genomic regions in temperate and tropical maize germplasm. *PLoS ONE* **2016**, *12*, e0169806. [CrossRef] [PubMed]
40. Jordan, K.W.; Wang, S.; Lun, Y.; Gardiner, L.J.; MacLachlan, R.; Hucl, P.; Wiebe, K.; Wong, D.; Forrest, K.L.; Sharpe, A.G.; et al. A haplotype map of allohexaploid wheat reveals distinct patterns of selection on homoeologous genomes. *Genome Biol.* **2015**, *16*, 48. [CrossRef] [PubMed]
41. Cadzow, M.; Boocock, J.; Nguyen, H.T.; Wilcox, P.; Merriman, T.R.; Black, M.A. A bioinformatics workflow for detecting signatures of selection in genomic data. *Front. Genet.* **2014**, *5*, 293. [CrossRef] [PubMed]
42. Korte, A.; Farlow, A. The advantages and limitations of trait analysis with GWAS: A review. *Plant Methods* **2013**, *9*, 29. [CrossRef] [PubMed]
43. Cloutier, S.; Ragupathy, R.; Miranda, E.; Radovanovic, N.; Reimer, E.; Walichnowski, A.; Ward, K.; Rowland, G.; Duguid, S.; Banik, M. Integrated consensus genetic and physical maps of flax (*Linum usitatissimum* L.). *Theor. Appl. Genet.* **2012**, *125*, 1783–1795. [CrossRef] [PubMed]
44. You, F.M.; Booker, M.H.; Duguid, D.S.; Jia, G.; Cloutier, S. Accuracy of genomic selection in biparental populations of flax (*Linum usitatissimum* L.). *Crop J.* **2016**, *4*, 290–303. [CrossRef]
45. You, F.M.; Xiao, J.; Li, P.; Yao, Z.; Jia, G.; He, L.; Zhu, T.; Luo, M.C.; Wang, X.; Deyholos, M.K.; et al. Chromosome-scale pseudomolecules refined by optical, physical and genetic maps in flax. *Plant J.* **2018**, *95*, 371–384. [CrossRef] [PubMed]
46. Wang, Z.; Hobson, N.; Galindo, L.; Zhu, S.; Shi, D.; McDill, J.; Yang, L.; Hawkins, S.; Neutelings, G.; Datla, R.; et al. The genome of flax (*Linum usitatissimum*) assembled de novo from short shotgun sequence reads. *Plant J.* **2012**, *72*, 461–473. [CrossRef] [PubMed]
47. Kumar, S.; You, F.M.; Cloutier, S. Genome wide SNP discovery in flax through next generation sequencing of reduced representation libraries. *BMC Genom.* **2012**, *13*, 684. [CrossRef] [PubMed]
48. You, F.M.; Deal, K.R.; Wang, J.; Britton, M.T.; Fass, J.N.; Lin, D.; Dandekar, A.M.; Leslie, C.A.; Aradhya, M.; Luo, M.C.; et al. Genome-wide SNP discovery in walnut with an AGSNP pipeline updated for SNP discovery in allogamous organisms. *BMC Genom.* **2012**, *13*, 354. [CrossRef] [PubMed]
49. Sun, C.; Wang, B.; Yan, L.; Hu, K.; Liu, S.; Zhou, Y.; Guan, C.; Zhang, Z.; Li, J.; Zhang, J.; et al. Genome-wide association study provides insight into the genetic control of plant height in rapeseed (*Brassica napus* L.). *Front. Plant Sci.* **2016**, *7*, 1102. [CrossRef] [PubMed]
50. Lamesch, P.; Berardini, T.Z.; Li, D.; Swarbreck, D.; Wilks, C.; Sasidharan, R.; Muller, R.; Dreher, K.; Alexander, D.L.; Garcia-Hernandez, M.; et al. The Arabidopsis Information Resource (TAIR): Improved gene annotation and new tools. *Nucleic Acids Res.* **2012**, *40*, D1202–D1210. [CrossRef] [PubMed]
51. You, F.M.; Li, P.; Kumar, S.; Ragupathy, R.; Li, Z.; Fu, Y.-B.; Cloutier, S. Genome-wide identification and characterization of the gene families controlling fatty acid biosynthesis in flax (*Linum usitatissimum* L.). *J. Proteom. Bioinf.* **2014**, *7*, 310–326.
52. Vrinten, P.; Hu, Z.; Munchinsky, M.A.; Rowland, G.; Qiu, X. Two FAD3 desaturase genes control the level of linolenic acid in flax seed. *Plant Physiol.* **2005**, *139*, 79–87. [CrossRef] [PubMed]
53. Soto-Cerda, B.J.; Duguid, S.; Booker, H.; Rowland, G.; Diederichsen, A.; Cloutier, S. Association mapping of seed quality traits using the Canadian flax (*Linum usitatissimum* L.) core collection. *Theor. Appl. Genet.* **2014**, *127*, 881–896. [CrossRef] [PubMed]

54. You, F.M.; Li, P.; Ragupathy, R.; Kumar, S.; Zhu, T.; Luo, M.-C.; Duguid, S.D.; Rashid, K.Y.; Booker, H.M.; Deyholos, M.K.; et al. The Draft Flax Genome Pseudomolecules. In Proceedings of the 66th Flax Institute of the United States, Fargo, ND, USA, 31 March–1 April 2016; pp. 17–24.
55. Banik, M.; Duguid, S.; Cloutier, S. Transcript profiling and gene characterization of three fatty acid desaturase genes in high, moderate, and low linolenic acid genotypes of flax (*Linum usitatissimum* L.) and their role in linolenic acid accumulation. *Genome* **2011**, *54*, 471–483. [CrossRef] [PubMed]
56. Zhang, J.; Song, Q.; Cregan, P.B.; Nelson, R.L.; Wang, X.; Wu, J.; Jiang, G.L. Genome-wide association study for flowering time, maturity dates and plant height in early maturing soybean (*Glycine max*) germplasm. *BMC Genom.* **2015**, *16*, 217. [CrossRef] [PubMed]
57. Zhang, W.K.; Wang, Y.J.; Luo, G.Z.; Zhang, J.S.; He, C.Y.; Wu, X.L.; Gai, J.Y.; Chen, S.Y. QTL mapping of ten agronomic traits on the soybean (*Glycine max* L. Merr.) genetic map and their association with EST markers. *Theor. Appl. Genet.* **2004**, *108*, 1131–1139. [CrossRef] [PubMed]
58. Eskandari, M.; Cober, E.R.; Rajcan, I. Genetic control of soybean seed oil: Ii. QTL and genes that increase oil concentration without decreasing protein or with increased seed yield. *Theor. Appl. Genet.* **2013**, *126*, 1677–1687. [CrossRef] [PubMed]
59. Hwang, E.Y.; Song, Q.; Jia, G.; Specht, J.E.; Hyten, D.L.; Costa, J.; Cregan, P.B. A genome-wide association study of seed protein and oil content in soybean. *BMC Genom.* **2014**, *15*, 1. [CrossRef] [PubMed]
60. Yang, J.; Benyamin, B.; McEvoy, B.P.; Gordon, S.; Henders, A.K.; Nyholt, D.R.; Madden, P.A.; Heath, A.C.; Martin, N.G.; Montgomery, G.W.; et al. Common SNPs explain a large proportion of the heritability for human height. *Nat. Genet.* **2010**, *42*, 565–569. [CrossRef] [PubMed]
61. Yang, J.; Lee, S.H.; Goddard, M.E.; Visscher, P.M. GCTA: A tool for genome-wide complex trait analysis. *Am. J. Hum. Genet.* **2011**, *88*, 76–82. [CrossRef] [PubMed]
62. Yang, J.; Bakshi, A.; Zhu, Z.; Hemani, G.; Vinkhuyzen, A.A.E.; Lee, S.H.; Robinson, M.R.; Perry, J.R.B.; Nolte, I.M.; van Vliet-Ostaptchouk, J.V.; et al. Genetic variance estimation with imputed variants finds negligible missing heritability for human height and body mass index. *Nat. Genet.* **2015**, *47*, 1114–1120. [CrossRef] [PubMed]
63. Yang, J.; Manolio, T.A.; Pasquale, L.R.; Boerwinkle, E.; Caporaso, N.; Cunningham, J.M.; de Andrade, M.; Feenstra, B.; Feingold, E.; Hayes, M.G.; et al. Genome partitioning of genetic variation for complex traits using common SNPs. *Nat. Genet.* **2011**, *43*, 519–525. [CrossRef] [PubMed]
64. Wood, A.R.; Esko, T.; Yang, J.; Vedantam, S.; Pers, T.H.; Gustafsson, S.; Chu, A.Y.; Estrada, K.; Luan, J.; Kutalik, Z.; et al. Defining the role of common variation in the genomic and biological architecture of adult human height. *Nat. Genet.* **2014**, *46*, 1173–1186. [CrossRef] [PubMed]
65. Duguid, S.D.; Kenaschuk, E.O.; Rashid, K.Y. Macbeth flax. *Can. J. Plant Sci.* **2003**, *83*, 803–805. [CrossRef]
66. Rowland, G.G.; Hormis, Y.A.; Rashid, K.Y. CDC bethune flax. *Can. J. Plant Sci.* **2002**, *82*, 101–102. [CrossRef]
67. Rowland, G.G.; Bhatty, R.S. Ethyl meththane-sulphonate induced fatty acid mutations in flax. *J. Am. Oil Chem. Soc.* **1990**, *67*, 213–214. [CrossRef]
68. You, F.M.; Huo, N.; Deal, K.R.; Gu, Y.Q.; Luo, M.C.; McGuire, P.E.; Dvorak, J.; Anderson, O.D. Annotation-based genome-wide SNP discovery in the large and complex *Aegilops tauschii* genome using next-generation sequencing without a reference genome sequence. *BMC Genom.* **2011**, *12*, 59. [CrossRef] [PubMed]
69. Li, H.; Handsaker, B.; Wysoker, A.; Fennell, T.; Ruan, J.; Homer, N.; Marth, G.; Abecasis, G.; Durbin, R. The sequence alignment/map format and SAMtools. *Bioinformatics* **2009**, *25*, 2078–2079. [CrossRef] [PubMed]
70. Browning, S.R.; Browning, B.L. Rapid and accurate haplotype phasing and missing-data inference for whole-genome association studies by use of localized haplotype clustering. *Am. J. Hum. Genet.* **2007**, *81*, 1084–1097. [CrossRef] [PubMed]
71. Purcell, S.; Neale, B.; Todd-Brown, K.; Thomas, L.; Ferreira, M.A.; Bender, D.; Maller, J.; Sklar, P.; de Bakker, P.I.; Daly, M.J.; et al. Plink: A tool set for whole-genome association and population-based linkage analyses. *Am. J. Hum. Genet.* **2007**, *81*, 559–575. [CrossRef] [PubMed]
72. Hill, W.G.; Weir, B.S. Variances and covariances of squared linkage disequilibria in finite populations. *Theor. Popul. Biol.* **1988**, *33*, 54–78. [CrossRef]
73. Pfeifer, B.; Wittelsburger, U.; Ramos-Onsins, S.E.; Lercher, M.J. Popgenome: An efficient swiss army knife for population genomic analyses in R. *Mol. Biol. Evol.* **2014**, *31*, 1929–1936. [CrossRef] [PubMed]

74. Jombart, T.; Devillard, S.; Balloux, F. Discriminant analysis of principal components: A new method for the analysis of genetically structured populations. *BMC Genet.* **2010**, *11*, 94. [CrossRef] [PubMed]
75. Jombart, T. Adegenet: A R package for the multivariate analysis of genetic markers. *Bioinformatics* **2008**, *24*, 1403–1405. [CrossRef] [PubMed]
76. Lin, C.S.; Poushinsky, G. A modified augmented design (type 2) for rectangular plots. *Can. J. Plant Sci.* **1985**, *65*, 743–749. [CrossRef]
77. You, F.M.; Duguid, S.D.; Thambugala, D.; Cloutier, S. Statistical analysis and field evaluation of the type 2 modified augmented design (MAD) in phenotyping of flax (*Linum usitatissimum*) germplasms in multiple environments. *Aust. J. Crop Sci.* **2013**, *7*, 1789–1800.
78. Association of Official Analytical Chemists. Fat (total, saturated and unsaturated) in foods: Hydrolytic extraction gas chromatographic method. In *Official Methods of Analysis of AOAC International*, 18th ed.; Horwitz, W., Ed.; AOAC International: Gaithersburg, MD, USA, 2001.
79. Bates, D.; Maechler, M.; Bolker, B.; Walker, S. Fitting linear mixed-effects models using lme4. *J. Stat. Softw.* **2015**, *67*, 1–48. [CrossRef]
80. Zhang, Z.; Ersoz, E.; Lai, C.Q.; Todhunter, R.J.; Tiwari, H.K.; Gore, M.A.; Bradbury, P.J.; Yu, J.; Arnett, D.K.; Ordovas, J.M.; et al. Mixed linear model approach adapted for genome-wide association studies. *Nat. Genet.* **2010**, *42*, 355–360. [CrossRef] [PubMed]
81. Yu, J.; Pressoir, G.; Briggs, W.H.; Vroh Bi, I.; Yamasaki, M.; Doebley, J.F.; McMullen, M.D.; Gaut, B.S.; Nielsen, D.M.; Holland, J.B.; et al. A unified mixed-model method for association mapping that accounts for multiple levels of relatedness. *Nat. Genet.* **2006**, *38*, 203–208. [CrossRef] [PubMed]
82. Bradbury, P.J.; Zhang, Z.; Kroon, D.E.; Casstevens, T.M.; Ramdoss, Y.; Buckler, E.S. Tassel: Software for association mapping of complex traits in diverse samples. *Bioinformatics* **2007**, *23*, 2633–2635. [CrossRef] [PubMed]
83. Turner, S.D. Qqman: An R package for visualizing GWAS results using Q-Q and manhattan plots. *biorXiv* **2014**. [CrossRef]
84. Thambugala, D.; Duguid, S.; Loewen, E.; Rowland, G.; Booker, H.; You, F.M.; Cloutier, S. Genetic variation of six desaturase genes in flax and their impact on fatty acid composition. *Theor. Appl. Genet.* **2013**, *126*, 2627–2641. [CrossRef] [PubMed]

© 2018 by the authors. Licensee MDPI, Basel, Switzerland. This article is an open access article distributed under the terms and conditions of the Creative Commons Attribution (CC BY) license (http://creativecommons.org/licenses/by/4.0/).

International Journal of
Molecular Sciences

MDPI

Article

Evaluation of Genomic Prediction for Pasmo Resistance in Flax

Liqiang He [1,2], Jin Xiao [2], Khalid Y. Rashid [3], Gaofeng Jia [4], Pingchuan Li [3], Zhen Yao [3], Xiue Wang [2], Sylvie Cloutier [1,*] and Frank M. You [1,2,*]

[1] Ottawa Research and Development Centre, Agriculture and Agri-Food Canada, Ottawa, ON K1A 0C6, Canada; liqiang.he@canada.ca

[2] State Key Laboratory of Crop Genetics and Germplasm Enhancement, College of Agriculture, Nanjing Agricultural University/JiangSu Collaborative Innovation Center for Modern Crop Production, Nanjing 210095, China; xiaojin@njau.edu.cn (J.X.); xiuew@njau.edu.cn (X.W.)

[3] Morden Research and Development Centre, Agriculture and Agri-Food Canada, Morden, MB R6M 1Y5, Canada; khalid.rashid@canada.ca (K.Y.R.); lipingchuan@gmail.com (P.L.); zhen.yao@canada.ca (Z.Y.)

[4] Crop Development Centre, University of Saskatchewan, Saskatoon, SK S7N 5A8, Canada; gaofeng.jia@usask.ca

* Correspondence: sylvie.cloutier@canada.ca (S.C.); frank.you@canada.ca (F.M.Y.); Tel.: +1-613-759-1744 (S.C.); +1-613-759-1539 (F.M.Y.)

Received: 28 November 2018; Accepted: 11 January 2019; Published: 16 January 2019

Abstract: Pasmo (*Septoria linicola*) is a fungal disease causing major losses in seed yield and quality and stem fibre quality in flax. Pasmo resistance (PR) is quantitative and has low heritability. To improve PR breeding efficiency, the accuracy of genomic prediction (GP) was evaluated using a diverse worldwide core collection of 370 accessions. Four marker sets, including three defined by 500, 134 and 67 previously identified quantitative trait loci (QTL) and one of 52,347 PR-correlated genome-wide single nucleotide polymorphisms, were used to build ridge regression best linear unbiased prediction (RR-BLUP) models using pasmo severity (PS) data collected from field experiments performed during five consecutive years. With five-fold random cross-validation, GP accuracy as high as 0.92 was obtained from the models using the 500 QTL when the average PS was used as the training dataset. GP accuracy increased with training population size, reaching values >0.9 with training population size greater than 185. Linear regression of the observed PS with the number of positive-effect QTL in accessions provided an alternative GP approach with an accuracy of 0.86. The results demonstrate the GP models based on marker information from all identified QTL and the 5-year PS average is highly effective for PR prediction.

Keywords: genomic selection; genomic prediction; genotyping by sequencing; pasmo resistance; pasmo severity; quantitative trait loci; single nucleotide polymorphism; *Septoria linicola*; flax

1. Introduction

Flax (*Linum usitatissimum* L.) is an important food and fibre crop cultivated and grown in cooler regions of the world, such as Canada [1]. Pasmo, elicited by the fungus *Septoria linicola*, is one of the most widespread diseases of flax, causing reductions in seed and oil yield, as well as fibre quality and durability [2]. Developing resistant cultivars is the most viable and effective option to control this disease that has become widespread in all flax production areas of North America and other parts of the world. Resistance to pasmo has a low heritability [3] and is quantitatively inherited [4]. Large variations in pasmo disease severity were observed in the flax core collection, which can be capitalized upon to develop resistant cultivars [3]. Phenotypic recurrent selection is typically used to develop cultivars with improved resistance and selection is usually carried out based on phenotypic assessments of resistance in field conditions [5]. However, field assessment of pasmo severity (PS) in germplasm and breeding lines is costly and, is heavily influenced by the environments due to strong genotype × environment (G × E) interactions [3,4].

With the advancements in molecular marker development over the last decade, efforts to use marker-assisted breeding strategies have been pursued. One such strategy involves identifying quantitative trait loci (QTL) in biparental mapping populations and using markers to efficiently backcross QTL into elite breeding materials [6]. This so-called marker-assisted recurrent selection (MARS) or simply marker-assisted selection (MAS) characterizes many breeding programs that employ molecular markers to select non-phenotyped individuals for crossing and downstream selection of segregating populations [7]. This method is suitable for the selection of monogenic or oligo-genic architectures but has limited use for quantitative traits controlled by many genes of smaller effects [8]. Genomic selection (GS) or prediction (GP) is an alternative marker-assisted breeding strategy better suited to polygenic quantitative traits, especially those with low heritability, because it makes use of all marker effects across the entire genome to calculate genomic estimated breeding values (GEBVs) [9] for individual plant selection [9,10].

In GP, a training population (TP) is genotyped with genome-wide markers and phenotyped for the trait(s) under selection; statistical models that best predict the breeding values from the marker data are then applied to select non-phenotyped germplasm. GP has been used to select for disease resistance in several crops such as *Fusarium* head blight (FHB) in wheat, a typically quantitatively inherited trait with predominantly additive genetic variation, where GP had a significantly higher accuracy than pedigree-based information alone [11]. GP feasibility has also been studied for selection of wheat rust resistance and was found particularly effective when validation lines had at least one which is close to the reference lines [12]. The implementation of GP on northern leaf blight, a complex genetic architecture trait in maize, resulted in superior gains and reduced breeding cycle time to ≤80% of the phenotypic cycle [13]. Despite the many successful examples, the use of GP to improve disease resistance in crops has been challenging for two reasons: (i) selection for major resistance genes can be ephemeral due to changes in pathogen races; and (ii) breeding for minor resistance genes with small effects may face the remarkable complexities encountered in GP [14].

The fast-evolving genotyping platforms have been a game-changer in the implementation of GP, allowing the production of large numbers of genome-wide markers, whereas progresses in phenotyping were not associated with similar cost reduction or quantum leaps in throughput. Given the number of markers (p) and sample size (n) in a given population, there are many more p effects to be estimated than the n, leading to an infinite number of possible marker effect estimates [15], that is, the so-called "large p, small n problem" ($p >> n$) when applying markers to predict phenotypes [11]. Several GP statistical models have been proposed to address this issue [16]. For example, the ridge-regression best linear unbiased prediction (RR-BLUP) is a mixed linear model that considers markers as random effects. Covariance between markers is considered to be zero and the marker variance is assumed to be the total genetic variance divided by the number of markers. The variance is assumed to be equal for all markers, allowing many more marker effects to be estimated than there are phenotypic records [17]. Unlike RR-BLUP, the Bayesian LASSO (BL) assumes markers to have unequal variances and, performs continuous shrinkage and variable selection simultaneously, with small-effect markers shrinking more severely than larger-effect loci. In the $p >> n$ setting, LASSO will select at most $n - 1$ variables and set the effects of the remaining predictors at zero [18]. Although the problem is solved statistically in these models, improving the accuracy and efficiency of GP by reducing the number of genome-wide markers would be advantageous because any increment in the TP size comes at a cost [19–22]. Genome-wide association study (GWAS) is an approach to identify genome-wide markers linked to QTL, resulting in a limited number of favourable genetic loci responsible for traits of interest [23]. For example, GP of crown rust resistance in *Lolium perenne* demonstrated GWAS's ability to identify and rank markers, which enabled the identification of a small subset of single nucleotide polymorphisms (SNPs) that could achieve predictive abilities close to that attained using the complete marker set [24]. Utilization of GWAS removes a large proportion of unrelated markers and in the construction of prediction models.

The only GP empirical study published to date in flax, which used bi-parental populations for yield, oil content and fatty acid composition traits, indicated that GP could increase genetic gain per unit time in linseed breeding. The GP results significantly exceeded those from direct phenotypic selection, especially for traits with low broad-sense heritability [25]. Resistance to flax pasmo is polygenic. Our previous study reported 500 non-redundant QTL for PR from 370 diverse flax accessions of a core collection based on five-year pasmo field assessments; of those, 134 QTL were statistically stable in all five years and 67 had relatively stable and large effects [4].

The objective of this study was to evaluate the potential of QTL markers in GP and compare the GP efficiency affected by different markers, including genome-wide SNPs and QTL markers, to provide a realistic and highly accurate model for germplasm evaluation and parent selection in pasmo resistance breeding.

2. Results

2.1. Evaluation of Pasmo Resistance

PS ratings at green boll stage or maturity across five consecutive years were similar but on average PS ratings in 2014 and 2016 were higher than those in other years (Table 1). They had single peak distributions but skewed towards high PS ratings except for those in 2014 (Figure 1). Scatter plots of PS ratings between years indicated strong genotype × year interaction even though statistically significant correlations of PS ratings between years were observed (Figure 1), as shown in the variance analysis results in the previous study [4]. However, the Pearson correlations of 5-year averages of PS ratings (PS-mean) with those in individual years (r = 0.72–0.83) were much higher than the Pearson correlations between individual years (r = 0.31–0.62) (Figure 1), implying that the mean PS ratings over multiple years or environments were a more suitable data set than individual year's data sets for model construction of genomic prediction.

Table 1. Pasmo severity of 370 flax accessions across five years in the field condition.

Data Set	$\bar{x} \pm s$	Range	CV (%)
PS-2012	5.57 ± 1.86	1.00–9.00	32.76
PS-2013	5.69 ± 1.91	2.00–9.00	33.20
PS-2014	6.86 ± 2.07	1.00–9.00	29.41
PS-2015	6.11 ± 1.55	1.00–9.00	25.44
PS-2016	6.72 ± 1.37	2.00–9.00	20.39
PS-mean	6.22 ± 1.32	1.80–9.00	21.27

$\bar{x}$: average pasmo severity across five years; s: standard deviation; CV: coefficient of variation.

2.2. Evaluation of Marker Sets Used in Genomic Prediction

Four marker sets were used for GP of pasmo resistance. The first marker set contained 52,347 genome-wide SNPs (SNP-52347) that were correlated to the five-year average PS and the PS of the five individual years at a 10^{-5} probability level [4]. The other three marker sets were the 500 unique QTL (SNP-500QTL), the 134 QTL statistically stable over five consecutive years (SNP-134QTL) and the 67 stable and relatively large-effect QTL (SNP-67QTL) sets previously identified [4]. The SNP-500QTL dataset comprises markers for all small- or large-effects, including QTL stable across environments and environment-specific QTL identified using three single-locus and seven multi-locus statistical models and all six phenotypic datasets (Figure 2). The SNP-134QTL dataset is a subset of the SNP-500QTL dataset whereas SNP-67QTL is a subset of the former; all SNP-500QTL markers were included in SNP-52347. These four marker sets explained 54%, 72%, 27% and 29% of the phenotypic variation of the five-year PS average (PS-mean), respectively; these values exceeded those of the individual year PS data (Table 2). Although SNP-500QTL was a subset of SNP-52347, this marker set explained a greater percentage of the phenotypic variation for PS than SNP-52347 for all datasets.

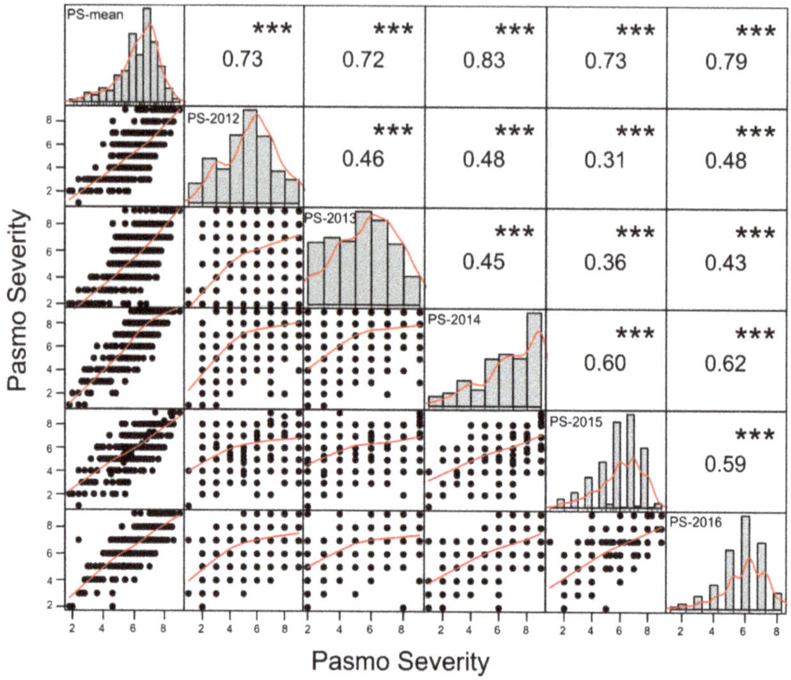

Figure 1. Dot plots (lower triangle), histograms (diagonal) and Pearson correlations (upper triangle) between six pasmo severity datasets. Best curves are fitted in dot plots and histograms. *** represents significance at the <0.001 probability level.

Table 2. Phenotypic variation of pasmo severity (PS) ($h^2 \pm s$) explained by the four marker sets.

PS Dataset	Marker Set			
	SNP-500QTL	SNP-134QTL	SNP-67QTL	SNP-52347
PS-mean	0.72 ± 0.04	0.27 ± 0.05	0.29 ± 0.05	0.54 ± 0.07
PS-2012	0.64 ± 0.06	0.18 ± 0.05	0.16 ± 0.04	0.43 ± 0.08
PS-2013	0.63 ± 0.06	0.12 ± 0.04	0.12 ± 0.04	0.38 ± 0.08
PS-2014	0.65 ± 0.06	0.23 ± 0.05	0.20 ± 0.05	0.45 ± 0.08
PS-2015	0.56 ± 0.06	0.20 ± 0.05	0.17 ± 0.04	0.44 ± 0.09
PS-2016	0.53 ± 0.06	0.18 ± 0.05	0.18 ± 0.05	0.38 ± 0.07

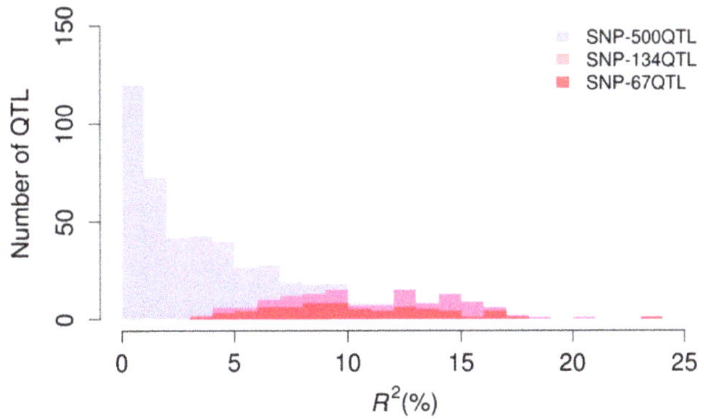

Figure 2. Distribution of R^2 (%) (phenotypic variation explained by individual QTL) in the three QTL marker sets.

2.3. Accuracy of Genomic Prediction in Relation to Marker Sets and Pasmo Severity Datasets

Genomic prediction models were constructed using RR-BLUP with pairwise combinations of the four marker sets and the six PS datasets. Statistical models for the 24 combinations were generated and evaluated for their accuracy (*r*) and relative efficiency (*RE*) using a five-fold random cross-validation scheme (Table 3). *RE* represents the relative efficiency of GP over direct phenotypic selection which depends on the heritability of a selective trait. Direct phenotypic selection for a trait was considered to have a baseline efficiency of 1. Thus, *RE* values greater than 1 indicate GP models more efficient than direct phenotypic selection in one selection cycle [25–27]. Analysis of variance (ANOVA) (Table S1) indicated that *r* and *RE* both significantly differed among the four marker sets and the six PS datasets; there was also a significant interaction effect between marker sets and PS datasets (Table S1). Owing to the significant marker × phenotype dataset interaction, multiple comparisons of the 24 combinations were performed. For all marker sets, the PS-mean models significantly outperformed those based on individual year datasets (Table 3). The SNP-500QTL marker set models generated significantly higher *r* and *RE* values than any other marker sets (Figure 3). Interestingly, the SNP-67QTL derived models produced slightly but significantly higher values of *r* and *RE* than SNP-134QTL models. The highest *r* and *RE* values were obtained for models combining the SNP-500QTL and PS-mean datasets (Table 3, Figure 3). Intriguingly, the SNP-52347 models yielded the lowest *r* and *RE* values despite including all QTL markers (Table 3, Figure 3); both BL and Bayesian ridge regression (BRR) corroborated this finding (Figure S1). No significant differences in *r* and *RE* values were observed among the three statistical models: RR-BLUP, BL and BRR (Figure S1).

2.4. Sample Size of Training Populations versus Genomic Prediction Accuracy

To find an optimal size for the TP, the relationship between TP size and prediction accuracy was analysed. TPs of various sizes from 18 to 351, corresponding to 5% to 95% of the total 370 accessions, were used to build models with the SNP-500QTL marker set and the PS-mean phenotypic dataset. The prediction accuracy significantly increased for TP sizes up to 100, followed by smaller accuracy gains with every additional TP size increments (Figure 4). A GP accuracy >0.9 was obtained once the TP size reached 185 (Figure 4).

Table 3. Accuracy (*r*) and relative efficiency (*RE*) values of the 24 combinations representing the four marker sets and six pasmo severity (PS) datasets using RR-BLUP obtained using a random five-fold cross-validation.

Marker Set	PS Dataset	r ($\bar{x} \pm s$) [1]	RE ($\bar{x} \pm s$) [1]
SNP-500QTL	PS-mean	0.92 ± 0.02a	1.84 ± 0.04a
	PS-2012	0.84 ± 0.03b	1.68 ± 0.06b
	PS-2013	0.81 ± 0.04c	1.62 ± 0.07c
	PS-2014	0.82 ± 0.04c	1.63 ± 0.07c
	PS-2015	0.76 ± 0.05d	1.52 ± 0.09d
	PS-2016	0.76 ± 0.05d	1.52 ± 0.11d
SNP-134QTL	PS-mean	0.75 ± 0.06e	1.49 ± 0.11e
	PS-2012	0.68 ± 0.06f	1.36 ± 0.11f
	PS-2013	0.60 ± 0.07ij	1.19 ± 0.14ij
	PS-2014	0.60 ± 0.07i	1.21 ± 0.14i
	PS-2015	0.47 ± 0.09o	0.94 ± 0.18o
	PS-2016	0.56 ± 0.09l	1.12 ± 0.17l
SNP-67QTL	PS-mean	0.76 ± 0.05d	1.53 ± 0.1d
	PS-2012	0.67 ± 0.06g	1.35 ± 0.11g
	PS-2013	0.60 ± 0.07ij	1.20 ± 0.14ij
	PS-2014	0.60 ± 0.07ij	1.20 ± 0.14ij
	PS-2015	0.50 ± 0.09n	1.00 ± 0.17n
	PS-2016	0.59 ± 0.08k	1.17 ± 0.17k
SNP-52347	PS-mean	0.67 ± 0.07g	1.33 ± 0.14g
	PS-2012	0.63 ± 0.06h	1.27 ± 0.12h
	PS-2013	0.59 ± 0.07jk	1.19 ± 0.14jk
	PS-2014	0.53 ± 0.08m	1.06 ± 0.17m
	PS-2015	0.38 ± 0.09q	0.77 ± 0.17q
	PS-2016	0.46 ± 0.09p	0.93 ± 0.18p

[1] Different letters represent multiple test significance among the 24 combinations at the 0.05 probability level.

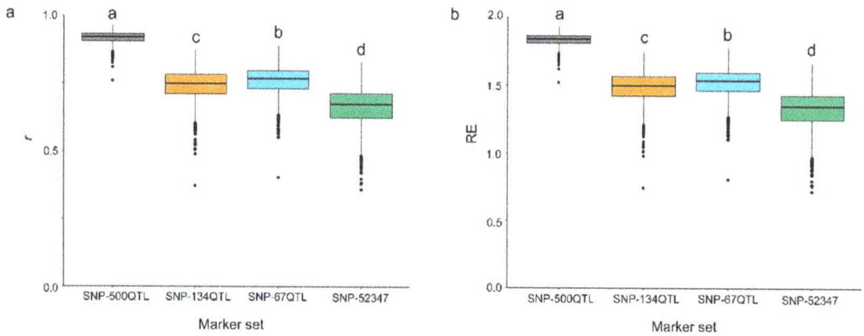

Figure 3. Accuracy (*r*) (**a**) and relative efficiency (*RE*) (**b**) of RR-BLUP prediction models built with combinations of four marker sets using the five-year average PS dataset (PS-mean) and random five-fold cross-validations. Letters above box plots indicated statistical significance ($p < 0.05$) for *r* and *RE* among marker sets.

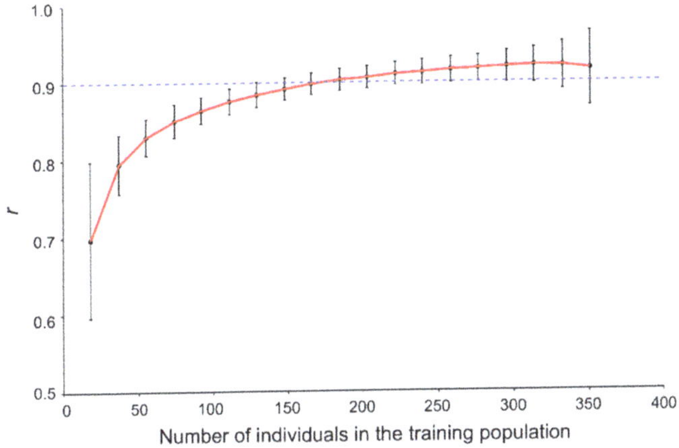

Figure 4. Relationship between the genomic prediction accuracy (*r*) and the size of the training population based on the SNP-500QTL marker set, the PS-mean dataset and the RR-BLUP models. The dash line represents a prediction accuracy of 0.9.

2.5. Prediction Models of Pasmo Resistance

All 370 accessions were used as a training population to build a prediction model using the SNP-500QTL genotypic dataset and the PS-mean phenotypic dataset because this combination outperformed all other models. The model was then employed to predict PS in each year (Table 4). Prediction accuracies (*r*) ranging from 0.71 to 0.81 and *RE* values of 1.42 to 1.62 were obtained when predicting PS for individual years (Table 4).

A prediction accuracy as high as 0.98 and a *RE* value of 1.96 were obtained when the model was used to predict PS-means of the 370 accessions (Table 4). A linear relationship was observed between the observed (y) and predicted PS (x): $y = 1.0522x - 0.3267$ ($R^2 = 0.96$) (Figure 5a). Based on this equation, the average prediction interval between the two red dashed lines, representing the 95% confidence interval, was only less than 1 (an average of 0.97) on the PS ratings (Figure 5a).

NPQTL in the 370 accessions for the 500 QTL set was tallied. Significant linear correlation between PS-mean and NPQTL ($r = 0.86$ or $R^2 = 0.73$) was observed (Figure 5b). This correlation was less than but close to the accuracy of the GP model with SNP-500QTL and higher than the GP models using other marker sets (Table 3). However, the single linear regression equation ($y = -0.0262x + 11.934$) of the observed PS (y) to NPQTL (x) had a large standard deviation for each prediction value, with an average prediction interval width of 2.70, nearly three times the average prediction interval width of the GP model; that is, the NPQTL model had a higher prediction error than the GP model.

Table 4. Accuracy (*r*) and relative efficiency (*RE*) of genomic prediction for pasmo severity in different years using the RR-BLUP model built with the SNP-500QTL marker set and the PS-mean phenotypic data using all 370 accessions as training data set.

PS Dataset for Prediction	*r*	*RE*
PS-mean	0.98	1.96
PS-2012	0.73	1.46
PS-2013	0.71	1.42
PS-2014	0.81	1.62
PS-2015	0.71	1.43
PS-2016	0.77	1.55

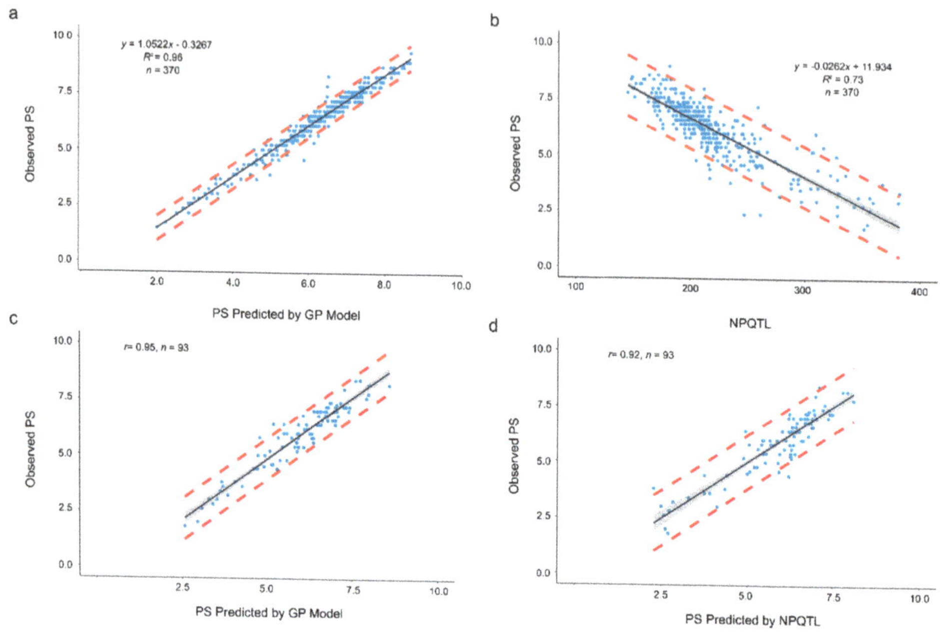

Figure 5. Relationship of observed pasmo severity (PS) with PS predicted by a GP model (**a**,**c**) or with PS predicted by the number of QTL with positive-effect alleles (NPQTL) (**b**,**d**). (**a**) Linear regression of observed PS (y) to predicted PS (x) using the genomic prediction model built with the PS-mean dataset and the SNP-500QTL marker set of all 370 accessions as training data set. (**b**) Linear regression of observed PS (y) to NPQTL (x) in the 370 flax accessions. (**c**) Relationship of observed PS of 93 randomly chosen accessions with the PS predicted by the genomic model constructed with the SNP-500QTL marker set and PS-mean dataset when a random subset of 277 accessions was used as training population. (**d**) Relationship of observed PS of 93 randomly chosen accessions with the PS predicted by NPQTL (Figure S2) The red dashed lines represent upper and lower boundaries of the 95% prediction intervals, that is, it is expected that the value of a sample lies within that prediction interval in 95% of the samples. The grey band represents the 95% confidence interval, that is, 95% of those intervals include the true value of the population mean.

2.6. A Case Study of Genomic Prediction

To assess GP prediction accuracy, a training-testing partition was generated with random assignment of breeding lines to either training or testing subsets. Considering the different improvement status of accessions in the population (cultivars, breeding lines, landraces or unknown types) and different levels of resistance, we randomly chose 20% of the 370 accessions in the population, that is, 93 accessions (52 cultivars, 21 breeding lines, 3 landraces and 17 unknown types) as validation dataset, that is, a five-fold random cross-validation set. To predict the PS of these 93 accessions, a RR-BLUP model using the SNP-500QTL set and the PS-mean of the remaining 277 accessions as TP set was built to predict PS. The predicted results are shown in Figure 5c and Table S2. The prediction accuracy was as high as 0.95 (r between observed and predicted PS). Similarly, a linear regression model of observed PS (y) to NPQTL (x) of the 277 accessions (the same TP as GP) produced $y = -0.026x + 11.902$ (Figure S2), which was similar to the regression equation previously obtained with the complete accession set (Figure 5b). Using this prediction model, predicted PS and intervals were calculated (Figure 5d, Table S2). The prediction accuracy of 0.92 for NPQTL was slightly inferior to that of the GP model. The observed PS values all fell within prediction intervals (Table S2).

3. Discussion

Cross-validation remains the most popular method to evaluate GP accuracy [14,28]. Our RR-BLUP model prediction accuracy of 0.92 for PR is the highest of all published GP models for plant disease resistance traits [14]. This model is especially valuable because PR has low heritability and high inheritance complexity [3,4]. The QTL markers, multi-year phenotypic data and the genetic diversity and size of the population likely contributed positively to this high prediction accuracy [29].

3.1. All Detected QTL Used as Markers in Genomic Prediction

Three sets of QTL markers (SNP-500QTL, SNP-134QTL and SNP-67QTL) and a genome-wide SNP marker set (SNP-52347) were evaluated here. GP models built using SNP-500QTL consistently outperformed models derived with any of the other three marker sets (Table 3, Figure S1), lending credence to the robustness and reliability of the QTL identified using multiple single-locus and multi-locus GWAS statistical methods [4]. Most GWAS aim to detect large-effect QTL, such as the SNP-67QTL set. While potentially useful in MAS, these tend to explain a reduced portion of the phenotypic variation compared to more comprehensive models (Table 2). Consequently, the GP models built with such marker sets have lower GP accuracies. Therefore, using all potential QTL associated with the selective trait to build GP models is advantageous because it greatly improves prediction accuracy. Prediction accuracies of models obtained with SNP-134QTL and SNP-67QTL data sets were comparable (Table 3, Figure S1) and they explained a similar proportion of the phenotypic variation for PS (Table 2), confirming the redundancy or overlap between the two datasets. Removal of redundant QTL from SNP-134QTL to produce SNP-67QTL produced slightly higher accuracy models (Figure 3). Simplifying GP models by removal of redundant and unrelated markers will ease the practical implementation of GP in breeding programs.

3.2. Superior Performance of Genomic Prediction Combined with GWAS

Surprisingly, the GP models built using SNP-52347 generated a lower prediction accuracy than the models with SNP-500QTL (Table 3, Figure S1), regardless of the statistical methods (Figure S1). Similarly, SNP-52347 explained a lower percentage of the phenotypic variation for PS than SNP-500QTL (Table 2). Besides interaction between SNPs, introduction of noise from genome-wide markers [30], the low prediction accuracy may also be owing to some of the erroneously called SNPs and imputation of missing SNP data. SNP-500QTL includes all or nearly all QTL potentially associated with PS; additional markers, not only failed to increase but actually reduced the prediction accuracy, further emphasizing the effectiveness of the QTL identification methodology adopted in our previously published GWAS study [4]. Similar findings were found for FHB in wheat where deoxynivalenol (DON) concentration QTL-linked markers significantly improve prediction accuracy compared to random genome-wide markers [30]. Markers linked to QTL underlying important traits are deemed more useful for prediction strategies because genome-wide markers may introduce noise, thereby reducing accuracy [30]. Using QTL for GP models may be beneficial to balance genetic backgrounds along with maximum gain of breeding value [31]. Genome-wide prediction models based on ~5000 SNPs from de novo GWAS for tropical rice improvement were as effective for prediction as the full marker set of 108,005 SNPs, indicating that the relationship between marker number and prediction accuracy is neither strict nor linear [32]. To sum up, combined applications of the QTL discovered via GWAS and the accelerated breeding cycles through GP facilitate the full use of genome-wide markers in crop disease resistance breeding [10,33]. Removal of redundant markers has the potential to alleviate the effect of the "large p, small n" issue.

3.3. Accuracy of GP Modelling by Environment, Training Population and Statistical Methods

G × E interactions, which affects the accuracy of trait assessment, are common for plant traits. A strong G × E interaction was observed in flax PR [4]. As a consequence, different PS QTL were identified for individual years and for the 5-year average [4]; similarly, GP efficiencies differed when individual yearly and average PS data sets were used as training sets (Table 3). The highest accuracies were obtained when the 5-year mean phenotypic data was used as training data (Table 4), suggesting that the average phenotypic data across multiple environments should be used for GP model construction. Because phenotypic values of genotypes in each year had one replication, the average phenotypic data across multiple years is actually equivalent to the best linear unbiased prediction values (BLUPs) or the best linear unbiased estimators (BLUEs). Therefore, the means across multiple environments estimate or reflect the true breeding values of a trait.

Some studies report that prediction accuracy of GP is highly affected by the size of the TP. In general, the prediction accuracy increases with TP size [21,28,29,34–36]. In the GP of seed weight in soybean, for example, prediction accuracy was sensitive to changes in TP size, which may have led to changes of relatedness between training and validation sets [21]. Lorenzana and Bernardo observed that, in an Arabidopsis family, prediction accuracy improved by 0.10 when TP size increased from 48 to 96, by an additional 0.07 when TP size was increased to 192 and by a further 0.05 with a TP size of 332 [37]. Here GP accuracy >0.9 was observed when the TP size reached 185 which slightly increased to 0.921 with a TP size of 314 (Figure 4). Large TPs provide the statistical power needed to improve prediction accuracy [38], especially for traits with low heritability [34,39]. When TP size is sufficiently large, even low heritability traits can be accurately predicted [28,40], including the low heritability PS studied therein. Diversity of the population also affect prediction accuracy [21,29,34,41–43]. A diverse TP may contain more QTL associated with selective traits and increase the correlation of the TP with validation populations (VPs) or test/prediction populations (PPs), resulting in a subsequent increase in prediction accuracy. Although some breeding lines [11,30,44] and bi-parental derived lines [25,41,45,46] are used for TPs, many studies have opted for a more diverse TP germplasm [29,41–43]. Our core collection TP preserves the variation present in the world collection of 3378 accessions maintained by Plant Gene Resources of Canada (PGRC) and represents a broad range of geographical origins, different improvement statuses (landraces, historical and modern cultivars, breeding lines) and two morphotypes (linseed and fibre types) [1,3]. This collection also contains most parents of modern Canadian flax cultivars [25]. Therefore, diverse phenotypic and genetic variabilities within the flax core collection render it useful as a resource for breeding and as a TP for GP model construction.

A variety of statistical methods have been proposed to estimate marker effects for GP. In general, GP methods are based on additive genetic models and their accuracies may vary depending on genetic architecture of target traits. According to the assumptions for statistical distributions of the marker effects, two groups of GP models have been proposed. The first group of models, such as RR-BLUP, genomic BLUP (GBLUP) and BRR, assume that all markers have some effects on the target trait and the same variance, that is, all makers contribute to the variation of the trait. The second group of models, including BayesA, BayesB, BayesC and BL, assume a specific variance for each marker. Some of these models such as BayesB, BayesC and BL, also allow variable (marker) selection when some of markers have very small or no effects. Based on these assumptions, the first group of models are expected to be useful for complex quantitative traits that have a polygenic architecture, while the second group of models are suitable for traits that controlled by a small number of genes or QTL with large effects. Several studies have shown better performance of BayesB for traits controlled by a few of genes with large effect [47–50]. Some simulation studies have also shown that BayesB outperformed GBLUP that is equivalent to RR-BLUP, when the number of QTL underlying a trait are small [47,51]. However, BayesB, RR-BLUP and other models had a similar prediction accuracy under the infinitesimal model [51] or for some complex traits [19,49]. In this study, no difference among RR-BLUP, BRR and BL was observed (Figure S1), primarily because flax pasmo resistance is a complex and polygenic trait and most of QTL associated with it had similar and small effects (Figure 2). RR-BLUP is most commonly used because of some superior features [11,14,42,52–54]. For example, RR-BLUP successfully recognized complex patterns with additive effects and delivered good GP in wheat disease resistance [55]. RR-BLUP also has a clear-cut computational efficiency compared with any other statistical models [11,54,56,57]. Here the RR-BLUP model with the 500 QTL markers and the 5-year mean PS produced high prediction accuracy and is therefore recommended for the prediction of PR in flax.

3.4. Pasmo Severity Prediction Using Number of Positive-Effect QTL

A highly significant correlation (r = 0.86 or R^2 = 0.73) between NPQTL and PS (Figure 5b) provides an alternative approach to directly predict PS phenotypes. The prediction accuracy using the linear regression equation of PS to NPQTL was inferior to the GP model (Figure 5) because the QTL effects were variable (Figure 2), whereas the linear regression equation considered only the number of QTL but not their individual effects. However, NPQTL is advantageous because it can be readily calculated based on the genotyping by sequencing (GBS) or other genotyping data for the QTL markers [14] and the prediction accuracy based on the NPQTL is comparable to most GP models. Thus, the NPQTL-based prediction equation provides a simple alternative model for PS prediction.

3.5. Breeding Application of Genomic Prediction

Plant breeding is to pyramid favourite alleles from distinct parents using different approaches such as conventional crossing, mutation or transgenic methods to develop new varieties. However, most traits of agronomic importance are genetically controlled by polygenes and have a low heritability such as seed yield and horizontal resistance to diseases. Conventional phenotype selection for these traits is usually inefficient because assessment for them must be performed in multiple environments to obtain breeding values of individuals and thus it is very costly, time consuming and inaccurate; and also because of difficulty of evaluation in fields, greenhouses or laboratories. GS or GP provides an efficient approach to increase selection efficiency by not only increasing selection accuracy but also shortening breeding cycles [58]. In this study, we demonstrate a good example of GP for flax pasmo resistance that is environment-sensitive, costly and difficult for field evaluation. As high as 0.92 of prediction accuracy was obtained for PR, corresponding to 1.84 of relative efficiency over the direct phenotypic selection (Table 3), demonstrating efficiency of GP for low heritability traits. Because the training population underlying the GP models is a diverse germplasm collection that contains more than 90 breeding lines and 245 varieties from different breeding programs [3], the GP models developed in this study are expected to be used for germplasm evaluation, parent selection and individual selection of segregation populations for PR.

4. Materials and Methods

4.1. Population

A total of 370 diverse flax accessions from the core collection [1] were used to evaluate different GP models. This subset of the core collection collected from 38 countries in 12 geographic regions has been used to identify the QTL associated with PS used in our PS models [4].

4.2. Pasmo Resistance Data

All flax accessions were assessed for PS in the same pasmo nursery from 2012 to 2016 at the Morden Research and Development Centre, Agriculture and Agri-Food Canada (AAFC), Morden, Manitoba, Canada [4]. A type-2 modified augmented design (MAD2) [59,60] was used for the field trials [3]. Accessions were seeded during the second or third week of May every year. Approximately 200 g of pasmo-infested chopped straw from the previous growing season was spread between rows as inoculum when plants were approximately 30-cm tall. A misting system was operated for 5 min every half hour for 4 weeks, except on rainy days, to ensure conidia dispersal and disease infection and development. Field assessments were conducted at the early (P1) and late flowering stages (P2, 7–10 days after P1), the green boll stage (P3, 7–10 days after P2) and the early brown boll stage (P4, 7–10 days after P3). In 2014 and 2015, only the first three field assessments were conducted because early maturity of the plants did not allow for a fourth rating. The PS observed at green boll stage or maturity was used for GP as previously described [4]. PS was assessed on leaves and stems of all plants in a single row plot using a 0–9 scale (0 = no sign of infection and 9 = > 90% leaf and stem area infected) [4]. Six sets of PS, including five individual year datasets and the 5-year average, were used for GP modelling. The function "chart.Correlation" of the R package PerformanceAnalytics (v1.5.2, https://cran.r-project.org/web/packages/PerformanceAnalytics/index.html) was used to analyse correlations between different PS datasets and draw histograms and scatter plots.

4.3. Genomic Data

A total of 258,873 SNPs were obtained from the 370 accessions after pruning by removing redundant SNPs [4]. The missing data of SNPs (on average 14.13% of a missing data rate) were imputed using Beagle v.4.2 with default parameters [61]. Our previous GWAS analyses of PS in flax were conducted separately for combinations of the five individual year and the 5-year average datasets with ten statistical methods [4]. The statistical methods for GWAS included three single locus models (GLM [62], MLM [63] and GEMMA [64]) and seven multi-locus models (FarmCPU [65], mrMLM [66], FASTmrEMMA [67], ISIS EM-BLASSO [68], pLARmEB [69], pKWmEB [70], FASTmrMLM [71]). For GLM, MLM and FarmCPU, the first six principal components (PCs), accounting for 33.04% of the total variation, were chosen as covariates to measure population structure, while Frappe (http://med.stanford.edu/tanglab/software/frappe.html) was used to estimate the population structure of the 370 accessions for other six multi-locus models. GEMMA does not require a Q matrix. The threshold of significant associations for all three single-locus methods (GLM, MLM and GEMMA) and the multi-locus method FarmCPU was determined by a critical p value ($\alpha = 0.05$) subjected to Bonferroni correction, that is, the corrected p value = 1.93×10^{-7} (0.05/258,873 SNPs), while a log of odds (LOD) score of three was used to detect robust association signals for the remaining six multi-locus models. The R package MVP (https://github.com/XiaoleiLiuBio/MVP) was used for GWAS analyses for the GLM, MLM and FarmCPU, the GEMMA software (https://github.com/genetics-statistics/GEMMA) for GEMMA and the R package mrMLM (https://cran.r-project.org/web/packages/mrMLM/index.html) for the additional six multi-locus models. The details of GWAS analyses were described in Reference [4]. A total of 500 non-redundant QTL for PS were identified from 370 diverse flax accessions, including 134 QTL that statistically stable in all five years and 67 QTL with relatively stable and large effects [4]. These three QTL datasets (500 unique QTL, 134 statistically stable QTL and 67 stable and large-effect QTL) were used for GP model construction. In addition, we performed Pearson's χ^2 test with Yate's continuity correction to detect all SNPs significantly associated with PS using a 10^{-5} probability level. The three QTL sets and the genome-wide SNP set were used to construct the GP models. Thus, GP models with the 24 combinations of the four marker sets and the six phenotypic datasets were built and compared.

4.4. Genomic Prediction Models

Three statistical methods RR-BLUP [9,17,20], Bayesian LASSO (BL) [20,25,33] and Bayesian ridge regression (BRR) [25,72] were used to build GP models for PS. These predictive models estimate marker effects by modelling markers as random effects. No fixed effects were fitted in the models. The statistical models and their computation procedures are described in detail elsewhere [40,73]. The R package rrBLUP [56] was used to fit the RR-BLUP model and the R package BLR [74] was used to fit the BL and BRR models. The parameters used to fit BL and BRR were determined based on suggestions of de los Campos et al. [74]. Broad-sense heritability (0.25) of PS estimated in the population [3] was used. When preparing QTL marker data for model construction, the positive-effect allele of the tag SNP of a QTL was coded '1' and the alternative allele '−1'. Similarly for the SNP marker set, the reference allele of an SNP was coded '1' and the alternative allele '−1'. Missing data were coded '0'. The EM algorithm implemented in the R package rrBLUP [56] was used to impute the missing marker data because missing marker data were not allowed in the model construction.

4.5. Evaluation of Prediction Models

Two validation methods were used to evaluate prediction models generated from combinations of statistical models, marker sets and PS datasets. The first method was a five-fold random cross-validation. The 370 flax accessions were randomly partitioned into five subsets. For a given partition, each subset was in turn used as validation or test data and the remaining four subsets made the training dataset. This partitioning was repeated 500 times. In this manner, a total of 2500 training data sets were created to build GP models and estimate marker effects. These were used to predict the breeding values of the individuals in the corresponding 2500 test/validation datasets. The accuracy of the genomic predictions (r) was defined by the Pearson's simple correlation coefficient between the genetic values predicted by GP and the observed phenotypic values. The relative efficiency of genomic prediction over phenotypic selection (RE) was estimated using $|r|/H^2$ [26,27], where H^2 refers to the broad-sense heritability of PS, estimated to be 0.25 [3]. RE was used as a criterion to compare the response to one cycle of genome-wide selection versus one cycle of phenotypic selection. Means of r and RE of the 500 samplings for each marker set, GP model and PS dataset were used to describe the prediction accuracy of GP and the efficiency of one GP cycle relative to one phenotypic selection cycle, respectively. To compare different marker and PS datasets, a joint analysis of variance with Tukey multiple pairwise-comparisons was performed to test the statistical significance of differences in r and RE using R. As a case study, we randomly selected 20% of all 370 accessions as validation dataset and used the remaining 277 accessions as training dataset to build a GP model for genomic prediction of unknown germplasm.

The second cross-validation approach involved comparisons across different PS datasets, that is, each of the six complete PS phenotypic datasets were used as training datasets to build GP models that were applied to itself and to the other five phenotypic datasets. The same set of markers for all 370 accessions was used for training and validation. This method tests the relevance of models built based on single year phenotypic data to predict phenotypes measured in different years.

4.6. Phenotypic Variation Explained by Markers

The phenotypic variation explained by all markers in various marker sets, denoted h^2_{SNP}, was estimated for all PS datasets based on the mixed linear model [75] implemented in the GCTA software [76]. The detailed calculation is described in Reference [77].

5. Conclusions

Using a diverse worldwide flax core collection of 370 accessions as a training and test population with 500 QTL identified by GWAS, the 5-year average PS data and the RR-BLUP statistical model, we developed a highly effective GP model with a prediction accuracy as high as 0.92 for pasmo, a low heritability and high inheritance complexity trait. This is the highest reported accuracy value of all GP models for plant disease resistance traits and comparable with previously published results. As an alternative, we developed a linear regression prediction model based on NPQTL that also produced a high prediction accuracy of 0.86. The GP model and the NPQTL-based regression equation were validated and deemed to be applicable to the evaluation of flax germplasm including parent selection for PR. The use of all potential QTL associated with a target trait would be beneficial because the exclusion of a large proportion of unrelated markers would facilitate the construction of highly accurate GP models.

Supplementary Materials: Supplementary Materials can be found at http://www.mdpi.com/1422-0067/20/2/359/s1.

Author Contributions: Conceptualization, F.M.Y. and S.C.; Methodology, F.M.Y.; Software, F.M.Y.; Formal Analysis, F.M.Y., G.J., P.L. and L.H.; Resources, K.Y.R. (field phenotypic data), S.C. and F.M.Y.; Data Curation, F.M.Y., K.Y.R. and Z.Y.; Writing-Original Draft Preparation, F.M.Y., L.H. and J.X.; Writing-Review & Editing, F.M.Y., S.C. and X.W.; Visualization, Z.Y.; Supervision, F.M.Y., S.C. and X.W.; Funding Acquisition, S.C., K.R. and F.M.Y.

Funding: This work was part of the Total Utilization Flax GENomics (TUFGEN) project funded by Genome Canada and other stakeholders, the A-base project (J-001004) funded by Agriculture and Agri-Food Canada and the flax cluster project funded by the Western Grains Research Foundation (WGRF) and the Canada-China science and technology and innovation action plan (2017ZJGH0106002).

Acknowledgments: We thank the China Scholarship Council for their financial support of L.H. for his research at Agriculture and Agri-Food Canada (AAFC).

Conflicts of Interest: The authors declare no conflict of interest. The funders had no role in the design of the study; in the collection, analysis or interpretation of the data; in the writing of the manuscript; and in the decision to publish the results.

Abbreviations

ANOVA	Analysis of variance
BL	Bayesian LASSO
BRR	Bayesian ridge regression
DON	Deoxynivalenol
FHB	*Fusarium* head blight
G × E	Genotype by environment interaction
GBS	Genotyping by sequencing
GEBV	Genomic estimated breeding value
GP	Genomic prediction
GS	Genomic selection
GWAS	Genome-wide association study
MARS	Marker-assisted recurrent selection
MAS	Marker-assisted selection
NPQTL	Number of QTL with positive-effect alleles
PGRC	Plant Gene Resources of Canada
PP	Test/prediction population
PR	Pasmo resistance
PS	Pasmo severity
QTL	Quantitative trait locus/loci
RE	Relative efficiency
RR-BLUP	Ridge regression best linear unbiased prediction
SNPs	Single nucleotide polymorphisms
TP	Training population
VP	Validation population

References

1. Diederichsen, A.; Kusters, P.M.; Kessler, D.; Bainas, Z.; Gugel, R.K. Assembling a core collection from the flax world collection maintained by Plant Gene Resources of Canada. *Genet. Resour. Crop Evol.* **2012**, *60*, 1479–1485. [CrossRef]
2. Vera, C.L.; Irvine, R.B.; Duguid, S.D.; Rashid, K.Y.; Clarke, F.R.; Slaski, J.J. Pasmo disease and lodging in flax as affected by pyraclostrobin fungicide, N fertility and year. *Can. J. Plant Sci.* **2014**, *94*, 119–126. [CrossRef]
3. You, F.M.; Jia, G.; Xiao, J.; Duguid, S.D.; Rashid, K.Y.; Booker, H.M.; Cloutier, S. Genetic variability of 27 traits in a core collection of flax (*Linum usitatissimum* L.). *Front. Plant Sci.* **2017**, *8*, 1636. [CrossRef]
4. He, L.; Xiao, J.; Rashid, K.Y.; Yao, Z.; Li, P.; Jia, G.; Wang, X.; Cloutier, S.; You, F.M. Genome-wide association studies for pasmo resistance in flax (*Linum usitatissimum* L.). *Front. Plant Sci.* **2019**, *9*, 1982. [CrossRef]
5. Diederichsen, A.; Rozhmina, T.A.; Kudrjavceva, L.P. Variation patterns within 153 flax (*Linum usitatissimum* L.) genebank accessions based on evaluation for resistance to *fusarium* wilt, anthracnose and pasmo. *Plant Genet. Resour.* **2008**, *6*, 22–32. [CrossRef]
6. Collard, B.C.Y.; Mackill, D.J. Marker-assisted selection: An approach for precision plant breeding in the twenty-first century. *Philos. Trans. R. Soc. Lond. B Biol. Sci.* **2008**, *363*, 557–572. [CrossRef]
7. Heslot, N.; Jannink, J.L.; Sorrells, M.E. Perspectives for genomic selection applications and research in plants. *Crop Sci.* **2015**, *55*, 1–12. [CrossRef]
8. Xu, Y.; Crouch, J.H. Marker-assisted selection in plant breeding: From publications to practice. *Crop Sci.* **2008**, *48*, 391–407. [CrossRef]

9. Meuwissen, T.H.; Hayes, B.J.; Goddard, M.E. Prediction of total genetic value using genome-wide dense marker maps. *Genetics* **2001**, *157*, 1819–1829.
10. Lipka, A.E.; Kandianis, C.B.; Hudson, M.E.; Yu, J.M.; Drnevich, J.; Bradbury, P.J.; Gore, M.A. From association to prediction: Statistical methods for the dissection and selection of complex traits in plants. *Curr. Opin. Plant Biol.* **2015**, *24*, 110–118. [CrossRef]
11. Arruda, M.P.; Brown, P.J.; Lipka, A.E.; Krill, A.M.; Thurber, C.; Kolb, F.L. Genomic selection for predicting *Fusarium* head blight resistance in a wheat breeding program. *Plant Genome* **2015**, *8*. [CrossRef]
12. Daetwyler, H.D.; Bansal, U.K.; Bariana, H.S.; Hayden, M.J.; Hayes, B.J. Genomic prediction for rust resistance in diverse wheat landraces. *Theor. Appl. Genet.* **2014**, *127*, 1795–1803. [CrossRef]
13. Technow, F.; Burger, A.; Melchinger, A.E. Genomic prediction of northern corn leaf blight resistance in maize with combined or separated training sets for heterotic groups. *G3 Genes Genomes Genet.* **2013**, *3*, 197–203. [CrossRef]
14. Poland, J.; Rutkoski, J. Advances and challenges in genomic selection for disease resistance. *Annu. Rev. Phytopathol.* **2016**, *54*, 79–98. [CrossRef]
15. Gianola, D. Priors in whole-genome regression: The Bayesian alphabet returns. *Genetics* **2013**, *194*, 573–596. [CrossRef]
16. Desta, Z.A.; Ortiz, R. Genomic selection: Genome-wide prediction in plant improvement. *Trends Plant Sci.* **2014**, *19*, 592–601. [CrossRef]
17. Whittaker, J.C.; Thompson, R.; Denham, M.C. Marker-assisted selection using ridge regression. *Genet. Res.* **2000**, *75*, 249–252. [CrossRef]
18. Tibshirani, R. Regression shrinkage and selection via the Lasso. *J. R. Stat. Soc. Ser. B Methodol.* **1996**, *58*, 267–288. [CrossRef]
19. Jiang, Y.; Zhao, Y.; Rodemann, B.; Plieske, J.; Kollers, S.; Korzun, V.; Ebmeyer, E.; Argillier, O.; Hinze, M.; Ling, J.; et al. Potential and limits to unravel the genetic architecture and predict the variation of *Fusarium* head blight resistance in European winter wheat (*Triticum aestivum* L.). *Heredity* **2015**, *114*, 318–326. [CrossRef]
20. Spindel, J.; Begum, H.; Akdemir, D.; Virk, P.; Collard, B.; Redona, E.; Atlin, G.; Jannink, J.L.; McCouch, S.R. Genomic selection and association mapping in rice (*Oryza sativa*): Effect of trait genetic architecture, training population composition, marker number and statistical model on accuracy of rice genomic selection in elite, tropical rice breeding lines. *PLoS Genet.* **2015**, *11*, e1004982.
21. Zhang, J.; Song, Q.; Cregan, P.B.; Jiang, G.L. Genome-wide association study, genomic prediction and marker-assisted selection for seed weight in soybean (*Glycine max*). *Theor. Appl. Genet.* **2016**, *129*, 117–130. [CrossRef]
22. Li, Y.; Ruperao, P.; Batley, J.; Edwards, D.; Khan, T.; Colmer, T.D.; Pang, J.; Siddique, K.H.M.; Sutton, T. Investigating drought tolerance in chickpea using genome-wide association mapping and genomic selection based on whole-genome resequencing data. *Front. Plant Sci.* **2018**, *9*, 190. [CrossRef]
23. Yu, J.; Buckler, E.S. Genetic association mapping and genome organization of maize. *Curr. Opin. Biotechnol.* **2006**, *17*, 155–160. [CrossRef]
24. Arojju, S.K.; Conaghan, P.; Barth, S.; Milbourne, D.; Casler, M.D.; Hodkinson, T.R.; Michel, T.; Byrne, S.L. Genomic prediction of crown rust resistance in *Lolium perenne*. *BMC Genet.* **2018**, *19*, 35. [CrossRef]
25. You, F.M.; Booker, H.M.; Duguid, S.D.; Jia, G.; Cloutier, S. Accuracy of genomic selection in biparental populations of flax (*Linum usitatissimum* L.). *Crop J.* **2016**, *4*, 290–303. [CrossRef]
26. Dekkers, J.C. Prediction of response to marker-assisted and genomic selection using selection index theory. *J. Anim. Breed. Genet.* **2007**, *124*, 331–341. [CrossRef]
27. Ziyomo, C.; Bernardo, R. Drought tolerance in maize: Indirect selection through secondary traits versus genomewide selection. *Crop Sci.* **2013**, *53*, 1269–1275. [CrossRef]
28. Crossa, J.; Perez-Rodriguez, P.; Cuevas, J.; Montesinos-Lopez, O.; Jarquin, D.; de Los Campos, G.; Burgueno, J.; Gonzalez-Camacho, J.M.; Perez-Elizalde, S.; Beyene, Y.; et al. Genomic selection in plant breeding: Methods, models, and perspectives. *Trends. Plant Sci.* **2017**, *22*, 961–975. [CrossRef]
29. Gowda, M.; Das, B.; Makumbi, D.; Babu, R.; Semagn, K.; Mahuku, G.; Olsen, M.S.; Bright, J.M.; Beyene, Y.; Prasanna, B.M. Genome-wide association and genomic prediction of resistance to maize lethal necrosis disease in tropical maize germplasm. *Theor. Appl. Genet.* **2015**, *128*, 1957–1968. [CrossRef]
30. Rutkoski, J.; Benson, J.; Jia, Y.; Brown-Guedira, G.; Jannink, J.-L.; Sorrells, M. Evaluation of genomic prediction methods for Fusarium head blight resistance in wheat. *Plant Genome* **2012**, *5*, 51–61. [CrossRef]

31. Deshmukh, R.; Sonah, H.; Patil, G.; Chen, W.; Prince, S.; Mutava, R.; Vuong, T.; Valliyodan, B.; Nguyen, H.T. Integrating omic approaches for abiotic stress tolerance in soybean. *Front. Plant Sci.* **2014**, *5*, 244. [CrossRef]
32. Spindel, J.E.; Begum, H.; Akdemir, D.; Collard, B.; Redona, E.; Jannink, J.L.; McCouch, S. Genome-wide prediction models that incorporate de novo GWAS are a powerful new tool for tropical rice improvement. *Heredity* **2016**, *116*, 395–408. [CrossRef]
33. Kayondo, S.I.; Pino del Carpio, D.; Lozano, R.; Ozimati, A.; Wolfe, M.; Baguma, Y.; Gracen, V.; Offei, S.; Ferguson, M.; Kawuki, R.; et al. Genome-wide association mapping and genomic prediction for CBSD resistance in *Manihot esculenta*. *Sci. Rep.* **2018**, *8*, 1549. [CrossRef]
34. Wang, X.; Xu, Y.; Hu, Z.L.; Xu, C.W. Genomic selection methods for crop improvement: Current status and prospects. *Crop J.* **2018**, *6*, 330–340. [CrossRef]
35. Jarquin, D.; Kocak, K.; Posadas, L.; Hyma, K.; Jedlicka, J.; Graef, G.; Lorenz, A. Genotyping by sequencing for genomic prediction in a soybean breeding population. *BMC Genom.* **2014**, *15*, 740. [CrossRef]
36. Asoro, F.G.; Newell, M.A.; Beavis, W.D.; Scott, M.P.; Jannink, J.-L. Accuracy and training population design for genomic selection on quantitative traits in elite North American oats. *Plant Genome* **2011**, *4*, 132–144. [CrossRef]
37. Lorenzana, R.E.; Bernardo, R. Accuracy of genotypic value predictions for marker-based selection in biparental plant populations. *Theor. Appl. Genet.* **2009**, *120*, 151–161. [CrossRef]
38. Goddard, M. Genomic selection: Prediction of accuracy and maximisation of long term response. *Genetica* **2009**, *136*, 245–257. [CrossRef]
39. Nielsen, H.M.; Sonesson, A.K.; Yazdi, H.; Meuwissen, T.H.E. Comparison of accuracy of genome-wide and BLUP breeding value estimates in sib based aquaculture breeding schemes. *Aquaculture* **2009**, *289*, 259–264. [CrossRef]
40. Lorenz, A.J.; Chao, S.; Asoro, F.G.; Heffner, E.L.; Hayashi, T.; Iwata, H.; Smith, K.P.; Sorrells, M.E.; Jannink, J.L. Genomic selection in plant breeding. *Adv. Agron.* **2011**, *110*, 77–123.
41. Cuevas, J.; Crossa, J.; Montesinos-Lopez, O.A.; Burgueno, J.; Perez-Rodriguez, P.; de los Campos, G. Bayesian genomic prediction with genotype x environment interaction kernel models. *G3 Genes Genomes Genet.* **2017**, *7*, 41–53.
42. Dong, H.; Wang, R.; Yuan, Y.; Anderson, J.; Pumphrey, M.; Zhang, Z.; Chen, J. Evaluation of the potential for genomic selection to improve spring wheat resistance to Fusarium head blight in the Pacific Northwest. *Front. Plant Sci.* **2018**, *9*, 911. [CrossRef]
43. Isidro, J.; Jannink, J.L.; Akdemir, D.; Poland, J.; Heslot, N.; Sorrells, M.E. Training set optimization under population structure in genomic selection. *Theor. Appl. Genet.* **2015**, *128*, 145–158. [CrossRef] [PubMed]
44. Rutkoski, J.E.; Poland, J.A.; Singh, R.P.; Huerta-Espino, J.; Bhavani, S.; Barbier, H.; Rouse, M.N.; Jannink, J.-L.; Sorrells, M.E. Genomic selection for quantitative adult plant stem rust resistance in wheat. *Plant Genome* **2014**, *7*. [CrossRef]
45. McElroy, M.S.; Navarro, A.J.R.; Mustiga, G.; Stack, C.; Gezan, S.; Pena, G.; Sarabia, W.; Saquicela, D.; Sotomayor, I.; Douglas, G.M.; et al. Prediction of cacao (*Theobroma cacao*) resistance to Moniliophthora spp. diseases via genome-wide association analysis and genomic selection. *Front. Plant Sci.* **2018**, *9*, 343. [CrossRef]
46. Enciso-Rodriguez, F.; Douches, D.; Lopez-Cruz, M.; Coombs, J.; de Los Campos, G. Genomic selection for late blight and common scab resistance in tetraploid potato (*Solanum tuberosum*). *G3 Genes Genomes Genet.* **2018**, *8*, 2471–2481. [CrossRef]
47. Daetwyler, H.D.; Pong-Wong, R.; Villanueva, B.; Woolliams, J.A. The impact of genetic architecture on genome-wide evaluation methods. *Genetics* **2010**, *185*, 1021–1031. [CrossRef]
48. Jannink, J.L.; Lorenz, A.J.; Iwata, H. Genomic selection in plant breeding: From theory to practice. *Brief Funct. Genom.* **2010**, *9*, 166–177. [CrossRef]
49. Thavamanikumar, S.; Dolferus, R.; Thumma, B.R. Comparison of genomic selection models to predict flowering time and spike grain number in two hexaploid wheat doubled haploid populations. *G3 Genes Genomes Genet.* **2015**, *5*, 1991–1998. [CrossRef]
50. VanRaden, P.M.; Van Tassell, C.P.; Wiggans, G.R.; Sonstegard, T.S.; Schnabel, R.D.; Taylor, J.F.; Schenkel, F.S. Invited review: Reliability of genomic predictions for North American Holstein bulls. *J. Dairy Sci.* **2009**, *92*, 16–24. [CrossRef]

51. Clark, S.A.; Hickey, J.M.; van der Werf, J.H. Different models of genetic variation and their effect on genomic evaluation. *Genet. Sel. Evol.* **2011**, *43*, 18. [CrossRef]
52. Rutkoski, J.; Singh, R.P.; Huerta-Espino, J.; Bhavani, S.; Poland, J.; Jannink, J.L.; Sorrells, M.E. Genetic gain from phenotypic and genomic selection for quantitative resistance to stem rust of wheat. *Plant Genome* **2015**, *8*. [CrossRef]
53. Gonzalez-Camacho, J.M.; Ornella, L.; Perez-Rodriguez, P.; Gianola, D.; Dreisigacker, S.; Crossa, J. Applications of machine learning methods to genomic selection in breeding wheat for rust resistance. *Plant Genome* **2018**, *11*. [CrossRef]
54. Liabeuf, D.; Sim, S.C.; Francis, D.M. Comparison of marker-based genomic estimated breeding values and phenotypic evaluation for selection of bacterial spot resistance in tomato. *Phytopathology* **2018**, *108*, 392–401. [CrossRef]
55. Ornella, L.; Singh, S.; Perez, P.; Burgueño, J.; Singh, R.; Tapia, E.; Bhavani, S.; Dreisigacker, S.; Braun, H.-J.; Mathews, K.; et al. Genomic prediction of genetic values for resistance to wheat rusts. *Plant Genome* **2012**, 5. [CrossRef]
56. Endelman, J.B. Ridge regression and other kernels for genomic selection with R package rrBLUP. *Plant Genome* **2011**, *4*, 250–255. [CrossRef]
57. Piepho, H.P. Ridge regression and extensions for genomewide selection in maize. *Crop Sci.* **2009**, *49*, 1165–1176. [CrossRef]
58. Bassi, F.M.; Bentley, A.R.; Charmet, G.; Ortiz, R.; Crossa, J. Breeding schemes for the implementation of genomic selection in wheat (*Triticum* spp.). *Plant Sci.* **2016**, *242*, 23–36. [CrossRef]
59. Lin, C.S.; Poushinsky, G. A modified augmented design (type 2) for rectangular plots. *Can. J. Plant Sci.* **1985**, *65*, 743–749. [CrossRef]
60. You, F.M.; Duguid, S.D.; Thambugala, D.; Cloutier, S. Statistical analysis and field evaluation of the type 2 modified augmented design (MAD) in phenotyping of flax (*Linum usitatissimum*) germplasms in multiple environments. *Aust. J. Crop Sci.* **2013**, *7*, 1789–1800.
61. Browning, S.R.; Browning, B.L. Rapid and accurate haplotype phasing and missing-data inference for whole-genome association studies by use of localized haplotype clustering. *Am. J. Hum. Genet.* **2007**, *81*, 1084–1097. [CrossRef] [PubMed]
62. Price, A.L.; Patterson, N.J.; Plenge, R.M.; Weinblatt, M.E.; Shadick, N.A.; Reich, D. Principal components analysis corrects for stratification in genome-wide association studies. *Nat. Genet.* **2006**, *38*, 904–909. [CrossRef] [PubMed]
63. Yu, J.; Pressoir, G.; Briggs, W.H.; Vroh Bi, I.; Yamasaki, M.; Doebley, J.F.; McMullen, M.D.; Gaut, B.S.; Nielsen, D.M.; Holland, J.B.; et al. A unified mixed-model method for association mapping that accounts for multiple levels of relatedness. *Nat. Genet.* **2006**, *38*, 203–208. [CrossRef]
64. Zhou, X.; Stephens, M. Genome-wide efficient mixed-model analysis for association studies. *Nat. Genet.* **2012**, *44*, 821–824. [CrossRef] [PubMed]
65. Liu, X.; Huang, M.; Fan, B.; Buckler, E.S.; Zhang, Z. Iterative usage of fixed and random effect models for powerful and efficient genome-wide association studies. *PLoS Genet.* **2016**, *12*, e1005767. [CrossRef] [PubMed]
66. Wang, S.B.; Feng, J.Y.; Ren, W.L.; Huang, B.; Zhou, L.; Wen, Y.J.; Zhang, J.; Dunwell, J.M.; Xu, S.; Zhang, Y.M. Improving power and accuracy of genome-wide association studies via a multi-locus mixed linear model methodology. *Sci. Rep.* **2016**, *6*, 19444. [CrossRef] [PubMed]
67. Wen, Y.J.; Zhang, H.; Ni, Y.L.; Huang, B.; Zhang, J.; Feng, J.Y.; Wang, S.B.; Dunwell, J.M.; Zhang, Y.M.; Wu, R. Methodological implementation of mixed linear models in multi-locus genome-wide association studies. *Brief. Bioinform.* **2017**, *19*, 700–712. [CrossRef] [PubMed]
68. Tamba, C.L.; Ni, Y.L.; Zhang, Y.M. Iterative sure independence screening EM-Bayesian LASSO algorithm for multi-locus genome-wide association studies. *PLoS Comput. Biol.* **2017**, *13*, e1005357. [CrossRef] [PubMed]
69. Zhang, J.; Feng, J.Y.; Ni, Y.L.; Wen, Y.J.; Niu, Y.; Tamba, C.L.; Yue, C.; Song, Q.; Zhang, Y.M. pLARmEB: Integration of least angle regression with empirical Bayes for multilocus genome-wide association studies. *Heredity* **2017**, *118*, 517–524. [CrossRef]
70. Ren, W.L.; Wen, Y.J.; Dunwell, J.M.; Zhang, Y.M. pKWmEB: Integration of Kruskal-Wallis test with empirical Bayes under polygenic background control for multi-locus genome-wide association study. *Heredity* **2017**, *120*, 208–218. [CrossRef]

71. mrMLM. Available online: https://cran.r-project.org/web/packages/mrMLM/index.html (accessed on 25 August 2018).
72. De los Campos, G.; Naya, H.; Gianola, D.; Crossa, J.; Legarra, A.; Manfredi, E.; Weigel, K.; Cotes, J.M. Predicting quantitative traits with regression models for dense molecular markers and pedigree. *Genetics* **2009**, *182*, 375–385. [CrossRef] [PubMed]
73. De los Campos, G.; Hickey, J.M.; Pong-Wong, R.; Daetwyler, H.D.; Calus, M.P. Whole-genome regression and prediction methods applied to plant and animal breeding. *Genetics* **2013**, *193*, 327–345. [CrossRef] [PubMed]
74. de Los Campos, G.; Perez, P.; Vazquez, A.I.; Crossa, J. Genome-enabled prediction using the BLR (Bayesian Linear Regression) R-package. *Methods Mol. Biol.* **2013**, *1019*, 299–320. [PubMed]
75. Yang, J.; Benyamin, B.; McEvoy, B.P.; Gordon, S.; Henders, A.K.; Nyholt, D.R.; Madden, P.A.; Heath, A.C.; Martin, N.G.; Montgomery, G.W.; et al. Common SNPs explain a large proportion of the heritability for human height. *Nat. Genet.* **2010**, *42*, 565–569. [CrossRef] [PubMed]
76. Yang, J.; Lee, S.H.; Goddard, M.E.; Visscher, P.M. GCTA: A tool for genome-wide complex trait analysis. *Am. J. Hum. Genet.* **2011**, *88*, 76–82. [CrossRef]
77. You, F.M.; Xiao, J.; Li, P.; Yao, Z.; Jia, G.; He, L.; Kumar, S.; Soto-Cerda, B.; Duguid, S.D.; Booker, H.M.; et al. Genome-wide association study and selection signatures detect genomic regions associated with seed yield and oil quality in flax. *Int. J. Mol. Sci.* **2018**, *19*, 2303. [CrossRef]

© 2019 by the authors. Licensee MDPI, Basel, Switzerland. This article is an open access article distributed under the terms and conditions of the Creative Commons Attribution (CC BY) license (http://creativecommons.org/licenses/by/4.0/).

International Journal of
Molecular Sciences

MDPI

Article

Genome-Wide Association Analysis of Mucilage and Hull Content in Flax (*Linum usitatissimum* L.) Seeds

Braulio J. Soto-Cerda [1,*], Sylvie Cloutier [2], Rocío Quian [1], Humberto A. Gajardo [1], Marcos Olivos [1] and Frank M. You [2,3]

1 Agriaquaculture Nutritional Genomic Center (CGNA), Las Heras 350, Temuco 4781158, Chile; rocio.quian@cgna.cl (R.Q.); humberto.gajardo@cgna.cl (H.A.G.); marcos.olivos@cgna.cl (M.O.)
2 Ottawa Research and Development Centre, Agriculture and Agri-Food Canada, Ottawa, ON K1A 0C6, Canada; sylvie.j.cloutier@agr.gc.ca (S.C.); frank.you@agr.gc.ca (F.M.Y.)
3 Morden Research and Development Centre, Agriculture and Agri-Food Canada, Morden, MB R6M 1Y5, Canada
* Correspondence: braulio.soto@cgna.cl; Tel.: +56-45-2740412

Received: 14 August 2018; Accepted: 18 September 2018; Published: 21 September 2018

Abstract: New flaxseed cultivars differing in seed mucilage content (MC) with low hull content (HC) represent an attractive option to simultaneously target the food and feed markets. Here, a genome-wide association study (GWAS) was conducted for MC and HC in 200 diverse flaxseed accessions genotyped with 1.7 million single nucleotide polymorphism (SNP) markers. The data obtained for MC and HC indicated a broad phenotypic variation and high (~70%) and a moderate (~49%) narrow sense heritability, respectively. MC and HC did not differ statistically between fiber and oil morphotypes, but yellow-seeded accessions had 2.7% less HC than brown-seeded ones. The genome-wide linkage disequilibrium (LD) decayed to $r^2 = 0.1$ at a physical distance of ~100 kb. Seven and four quantitative trait loci (QTL) were identified for MC and HC, respectively. Promising candidate genes identified include *Linum usitatissimum* orthologs of the *Arabidopsis thaliana* genes *TRANSPARENT TESTA 8, SUBTILISIN-LIKE SERINE PROTEASE, GALACTUROSYL TRANSFERASE-LIKE 5, MUCILAGE-MODIFIED 4, AGAMOUS-LIKE MADS-BOX PROTEIN AGL62, GLYCOSYL HYDROLASE FAMILY 17,* and *UDP-GLUCOSE FLAVONOL 3-O-GLUCOSYLTRANSFERASE*. These genes have been shown to play a role in mucilage synthesis and release, seed coat development and anthocyanin biosynthesis in *A. thaliana*. The favorable alleles will be useful in flaxseed breeding towards the goal of achieving the ideal MC and HC composition for food and feed by genomic-based breeding.

Keywords: flaxseed; *Linum usitatissimum*; GWAS; seed mucilage content; seed hull content; single nucleotide polymorphism (SNP)

1. Introduction

Flaxseed (*Linum usitatissimum* L.), one of the oldest crops, has been used as human food and animal feed since ancient times [1]. The two main morphotypes of cultivated *L. usitatissimum* are oil morphotype (flaxseed) and fiber morphotype (fiber flax). Flaxseed plants are shorter, more branched, and larger seeded, and branches cover a greater proportion of the main stem compared to fiber flax. Flaxseed currently enjoys new prospects in the functional food market because of growing consumer interest in food with health benefits [1]. Flaxseed is rich in bioactive compounds, such as α-linolenic acid (omega-3) that have cardiovascular benefits, lignans with anticancer properties, and insoluble and soluble fiber (mucilage) that is capable of lowering cholesterol and insulin [2].

Flaxseed mucilage is a heterogeneous polysaccharide composed of xylose, arabinose, glucose, galactose, rhamnose, and fructose [3] that can be purified into neutral and acidic polymers. Mucilage abounds in the seed coat, where it makes up to 8–10% of the seed weight [4]. Mucilage synthesis is tightly linked to seed coat development [5] and both tissues form the seed hull, a structure representing 37–48% of the seed weight [6,7]. These two fractions, rich in polysaccharides, are components of the flaxseed meal, primarily used as a protein rich livestock and poultry feed [6,8]. Absorption of flaxseed meal's advantageous 31–45% protein content [9] may be hindered by mucilage and cell wall polysaccharides. This is due to the swelling capacity of polysaccharides in the digestive tract of monogastric animals that causes concomitant growth depression and reduced feed efficiency [7,10]. In that context, reduction of mucilage (MC) and hull (HC) contents in flaxseed meal is desirable to achieve improved feeding value for livestock and poultry. Studies of flaxseed mucilage degradation are focused on chemical retting, enzyme retting, and steam explosion [11]. Reduction of the hull content in flaxseed and rapeseed meal has been achieved through dehulling methods [12] and the use of yellow-seeded genotypes [7,13]. Food and feed markets demand flaxseed cultivars differing in mucilage and hull content. It is, therefore, crucial to decipher the genetic factors underlying these complex traits in order to accelerate the development of market-specific flaxseed cultivars.

In the model plant *Arabidopsis thaliana*, the genes necessary for the synthesis, modification, and release of mucilage, as well as seed coat development, are well understood [5,14]. Putative flax orthologs of the *RHAMNOSE SYNTHASE* (*AtRHM1*), *GALACTURONOSYLTRANSFERASE-LIKE 3* (*GATL3*), *GALACTURONOSYLTRANSFERASE 11* (*GAUT11*), *XYLOGLUCAN ENDOTRANSGLUCOSYLASE/HYDROLASE 3*(*XTH3*), and *ALPHA-XYLOSIDASE-1* (*AtBXL1*), involved in mucilage production, have been identified using gene expression analysis during seed development [15]. Similarly, putative flax orthologs of the *TRANSPARENT TESTA 3, 4, 5,* and *7* (*TT3, TT4, TT5,* and *TT7*), *FLAVONOL SYNTHASE* (*FLS*) and *BANYULS* (*BAN*), involved in flavonoids synthesis during seed coat development, have also been identified [15].

Genetic variation for MC and HC in flaxseed has been assessed [4,7,16–18] but no quantitative trait loci (QTL) have been reported so far. QTL for *Fusarium* wilt resistance [19], powdery mildew [20], iodine value, palmitic, linoleic, and linolenic acids [21,22], and seed and flower color [23], were reported. QTL for seed protein, cell wall, straw weight, yield-related traits, and phenological traits, have also been reported using bi-parental mapping and association mapping [22,24,25]. Recently, genome-wide association studies (GWAS) have been conducted for agronomic and seed quality traits using thousands of single nucleotide polymorphism (SNP) loci [26,27]. GWAS mines the natural sequence diversity within a species and captures historical recombination events. It is therefore a suitable approach to discover loci that control complex traits, leading to a higher mapping resolution to facilitate the identification of candidate genes [28]. Thus, the suite of genomic tools available for flaxseed genetic studies [21,22,26,27,29–32] make genomic evaluation of MC and HC feasible.

The objective of this research was to identify QTL and candidate genes contributing to mucilage content and hull content that could be capitalized upon to assist in breeding flaxseed cultivars with different mucilage content and with reduced hull content. Improving these traits will increase seed value of this important cash crop.

2. Results

2.1. Phenotypic Evaluation

Two hundred flax accessions from the Canadian flax core collection were planted in two environments. Evaluation of MC and HC showed a normal distribution across the two environments according to the Shapiro–Wilk normality test and normality plots (Table S1, Figure S1). Variance component analysis indicated significant effects of genotype, environment, and genotype × environment interaction, according to the Wald statistic ($p < 0.001$). The phenotypic variation for MC in Vilcún location 2014 (CAR2014) ranged from 23.52 to 103.57 mg g^{-1} with an average of 58.67 mg g^{-1}. A lower variation was observed for MC in Huichahue location 2015 (HU2015), which ranged from

18.88 to 91.90 mg g^{-1} with an average of 55.04. mg g^{-1} HC variation ranged from 35.56 to 48.59% in CAR2014 and from 35.73 to 48.59% in HU2015 (Figure 1, Table S1). MC and HC were significantly positively correlated in CAR2014 and HU2015 with coefficients of 0.28 and 0.29, respectively. Narrow sense heritability (h^2) for MC attained 70.7 and 73.8% in CAR2014 and HU2015, respectively. Lower h^2 of 51.4 and 46.2% for HC at CAR2014 and HU2015 were observed. MC did not differ statistically between flax morphotypes nor seed color classes, according to the Kruskal–Wallis non-parametric test. The average MC was 55.33 and 56.63 mg g^{-1} for the fiber and oil morphotypes, respectively ($p = 0.651$) (Figure S2a). The average MC registered values of 56.63 and 59.22 ($p = 0.517$) for the brown and yellow seeded classes, correspondingly. The average HC did not differ statistically between flax morphotypes (fiber = 43.41%, oil = 42.79%; $p = 0.373$). On the other hand, yellow-seeded genotypes averaged 2.66% less HC than brown seeded accessions ($p = 3.2 \times 10^{-5}$) (Figure S2b).

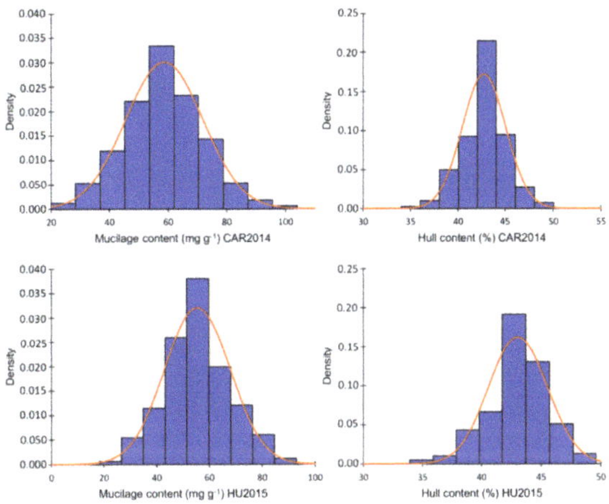

Figure 1. Mucilage (MC) and hull (HC) contents distribution in the association panel in two environments: CAR2014 = Vilcún location 2014, HU2015 = Huichahue location 2015. Values represent the mean of three biological replicates for each trait.

2.2. Population Structure and Linkage Disequilibrium

The dendrogram based on 771,914 SNPs and the STRUCTURE plot grouped the 200 individuals into two major clusters arbitrarily assigned as "red" and "blue" (Figure 2a,b). In the K against ΔK plot, a break in the slope was clearly observed at $K = 2$ (Figure 2b). The red cluster comprised almost exclusively genotypes belonging to the oil morphotype, while the blue cluster included both flax morphotypes. The coefficient of population differentiation ($F_{ST} = 0.08$) indicated a weak population structure between the two clusters.

The genome-wide linkage disequilibrium (LD) decayed to $r^2 = 0.1$ at a physical distance of ~100 kb (Figure 2c). Intrachromosomal LD decayed to $r^2 = 0.1$ at a distance between marker pairs ranging from ~40 kb on chromosome 8 to ~210 kb on chromosome 13 (Figure S3). A highly significant positive correlation ($r = 0.75$, $p = 0.0012$) between marker density and the intrachromosomal LD blocks was observed (Figure 3). For example, chromosomes 4 and 8 with the smallest number of markers, and chromosomes 6 and 13 with the largest, displayed the fastest and slowest LD decays, respectively (Figure 3 and Figure S3). The fast LD decays observed in this association panel are indicative of its advantageous potential for reducing QTL intervals and fine mapping of candidate genes for MC and HC.

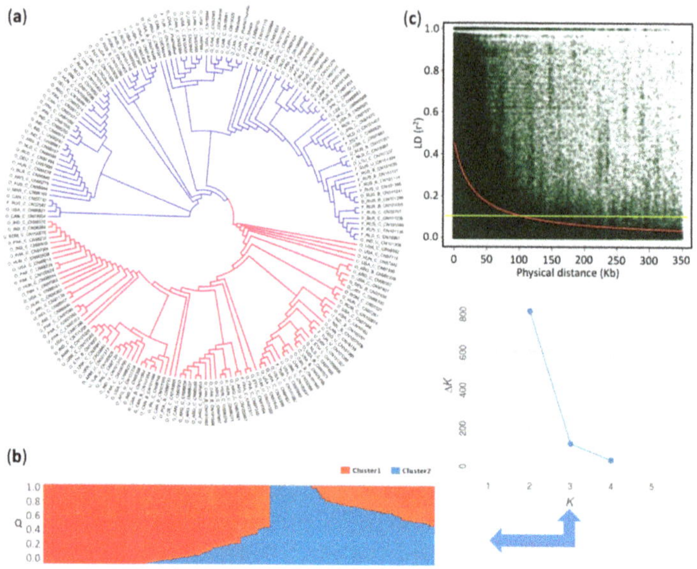

Figure 2. Population structure and genome-wide linkage disequilibrium decay. (**a**) Neighbor-joining (NJ) tree for 200 flax accessions based on 779,914 single nucleotide polymorphisms (SNPs); (**b**) Model-based population structure of 200 flax accessions belonging to two clusters predefined by the STRUCTURE software. Each accession is represented by a vertical bar. The color subsections within each vertical bar indicate membership coefficient (Q) to different clusters; (**c**) Genome-wide linkage disequilibrium decay of r^2 values (red line), against physical distance (kb) using the Hill and Weir (1988) function in *L. usitatissimum*. Yellow line indicates the cutoff value (r^2 = 0.1) used to determine the genome-wide linkage disequilibrium (LD) block size.

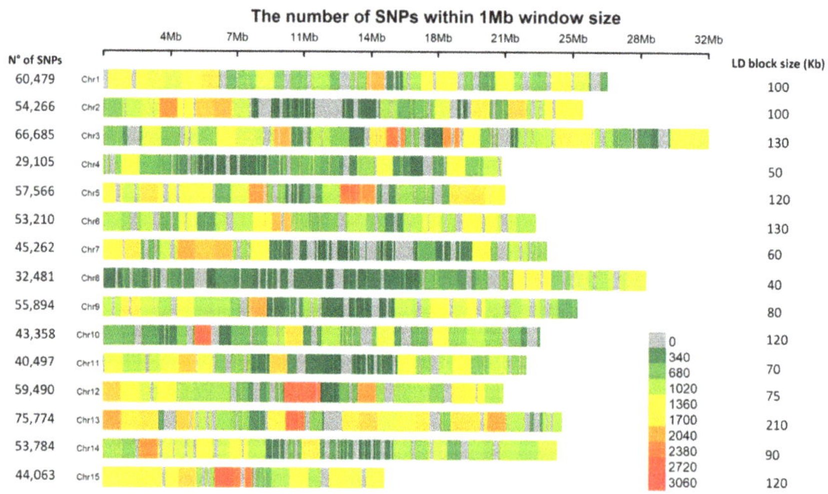

Figure 3. Single nucleotide polymorphism (SNP) density plot across the *L. usitatissimum* genome. Numbers of SNPs and LD blocks are also indicated for each of the 15 chromosomes.

2.3. Genome-Wide Association Analysis

Three GWA models were tested, including GLM-Q, GLM-PCA, and MLM-K. According to the quantile-qualtile (Q–Q) plot results, the GLM-Q model showed a strong skew toward significance for every trait (Figure S4a), indicating that the Q matrix was insufficient to account for population structure and cryptic relatedness. Conversely, the MLM-K, which only used the kinship matrix, led to an overcorrection of these confounding factors, particularly for HC (Figure S4b). The GLM-PCA was tested with 5 and 10 PCA covariates for HC and MC, accounting for 30.1 and 37.1% of the variation, respectively. Both GLM-5PCA and GLM-10PCA models performed well in controlling the rate of false positives, providing suitable statistical power to identify significant marker–trait associations for MC and HC (Figure S5). Therefore, the GLM-PCA model was applied for GWA in this study.

GWA analysis identified 12 and 17 significant associations for MC in CAR2014 and HU2015, respectively ($p < -\log_{10}(P) = 6.88$), and markers Lu5-3808878, Lu7-13225294, and Lu11-2498303 were significant in both environments (Table 1, Figure S5). Various significant SNP markers fell into the same LD blocks. For example, five other significant markers surrounded the peak SNP Lu5-3808878 (Figure S5), thus, they were considered the same QTL. Following this criterion, seven QTL were delineated on chromosomes 2, 3, 5, 7, and 11. The peak SNPs of these QTL accounted for 11.8 to 17.3% of phenotypic variation, and the combined three consistent QTL accounted for 43.6% of the MC variation (Table 1).

Table 1. Genome-wide significant peak SNPs for mucilage content (MC) and hull content (HC).

Trait	Marker	Chromosome	Allele	MAF [1]	$-\log_{10}(P)$		R^2 (%)	
					CAR2014	HU2015	CAR2014	HU2015
MC	Lu2-22298066	2	T/C	0.07	8.69	3.41ns [2]	17.32	ns [2]
	Lu3-25559600	3	G/T	0.06	7.45	4.13ns [2]	13.42	ns [2]
	Lu3-26033342	3	C/G	0.07	7.68	4.23ns [2]	13.25	ns [2]
	Lu3-7398487	3	C/T	0.41	4.96ns [2]	7.02	ns [2]	11.82
	Lu5-3808878	5	G/A	0.10	8.03	10.21	14.97	16.52
	Lu7-13225294	7	G/A	0.34	8.10	6.91	16.46	12.05
	Lu11-2498303	11	C/G	0.16	7.05	7.47	14.25	13.18
HC	Lu7-6577527	7	A/C	0.13	6.90	7.36	14.66	15.79
	Lu10-21552161	10	G/A	0.09	6.90	6.16ns [2]	16.32	ns [2]
	Lu12-5267706	12	C/T	0.06	5.91ns [2]	6.92	ns [2]	13.83
	Lu13-2803224	13	T/C	0.06	7.83	8.45	17.43	18.20

[1] MAF: minor allele frequency; [2] ns: not significant at the threshold value $-\log_{10}(P) = 6.88$.

A total of three and four significant associations were detected for HC in CAR2014 and HU2015, respectively ($p < -\log_{10}(P) = 6.88$). Markers Lu7-6577527 and Lu13-2803224 were significant in both environments (Table 1). The four QTL identified on chromosomes 7, 10, 12, and 13 explained between 13.8% and 17.8% of the HC variation. The two consistent QTL Lu7-6577527 and Lu13-2803224, accounted for a combined 33% of the HC variation (Table 1).

The peak SNPs effect for MC and HC were all significant according to the non-parametric Kruskal–Wallis test ($p < 0.05$), except for Lu3-26033342 associated with MC (Figure 4a,b and Figure S6). Accessions with a thymine (T) allele at Lu2-22298066 displayed, on average, an increase of 15.3 and 9.4 mg g^{-1} in MC, compared to accessions with a cytosine (C) allele in CAR2014 and HU2015, respectively (Figure 4a). Similarly, accessions with a "T" allele at Lu3-7398487 had, on average, 6.64 and 8.4 mg g^{-1} higher MC compared to accessions with a "C" allele in CAR2014 and HU2015, correspondingly.

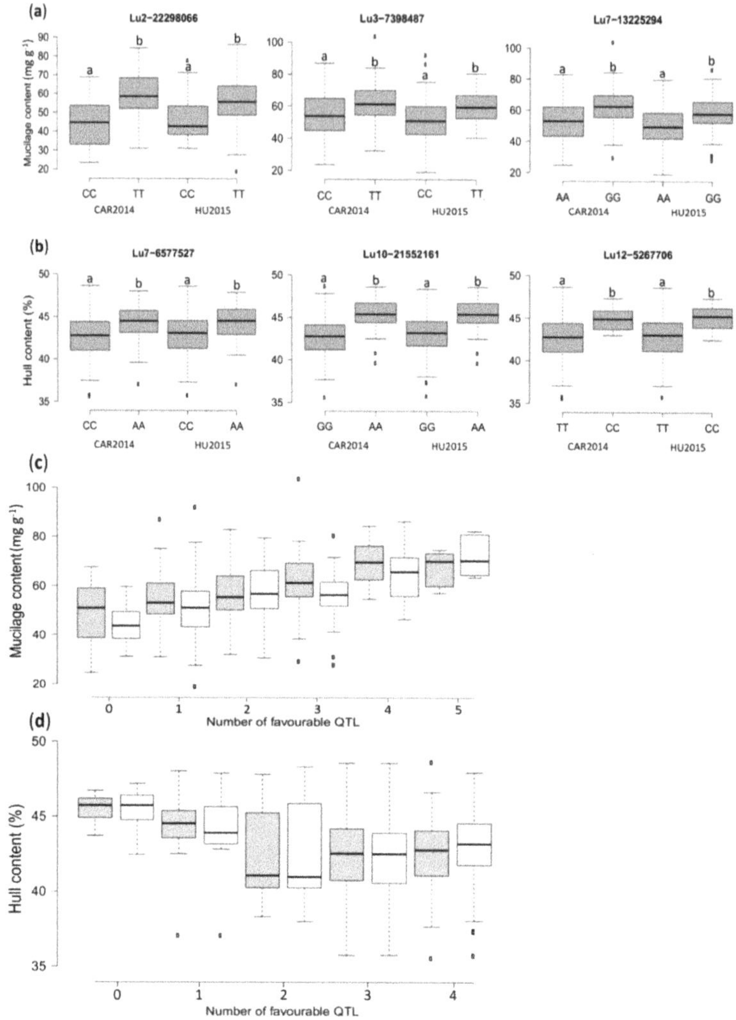

Figure 4. Box plots illustrating the phenotypic differences between flaxseed accessions carrying different alleles of the significant SNPs, and combined phenotypic effects of favorable QTL in the association panel. (**a**) Mucilage content (MC); (**b**) Hull content (HC). CAR2014 = Vilcún location 2014, HU2015 = Huichahue location 2015. Different letters indicate significant statistical differences according to the Kruskal-Wallis non-parametric test ($p < 0.05$); (**c**) QTL effect for MC; (**d**) QTL effect for HC. Grey and white boxplots represent the CAR2014 and HU2015 locations, respectively.

Accessions with a guanine (G) allele at Lu7-13225294 had, on average, 8.56 and 7.71 mg g^{-1} more mucilage compared to accessions carrying an adenine (A) allele in CAR2014 and HU2015, respectively (Figure 4a). The allelic effect for the other four peak SNPs is illustrated in Figure S6. The allelic effect of peak SNPs for HC revealed that accessions harboring a "C" allele at Lu7-6577527 had, on average, 1.4 and 1.3% less HC compared with "A" allele genotypes in CAR2014 and HU2015, correspondingly (Figure 4b). On average, HC was reduced by 1.4 and 1.3% (Lu7-6577527) to 2.6 and 2.7% (Lu13-2803224) in CAR2014 and HU2015, respectively (Figure 4b and Figure S6).

The combined QTL effect revealed that the MC of accessions harboring none of the favorable QTL alleles averaged 44.6 and 48.9 mg g^{-1}, compared to 72.1 and 67.6 mg g^{-1}, for those with five favorable alleles in CAR2014 and HU2015, respectively (Figure 4c). No accession had all seven favorable QTL alleles. The combined QTL effect for HC indicated that genotypes with none of the favorable QTL alleles averaged 45.5% and 45.3% HC compared to genotypes with four favorable QTL alleles, in which HC averaged 42.7% and 42.9% in CAR2014 and HU2015, respectively (Figure 4d).

2.4. Identification of Candidate Genes

The LD blocks harboring the peak SNPs were mined for genes relevant to MC and HC using the *L. usitatissimum* v.1.0 reference genome. A total of 204 and 118 candidate genes were identified for MC and HC, respectively (Table 2 and Table S2). Several genes ascribed to carbohydrate metabolism, seed mucilage synthesis, modification, and release, and cell wall synthesis and modification were identified at the MC QTL loci (Table 2). Five particularly promising candidate genes were identified. The SNP marker Lu3-26033342 was located 58.92 and 49.60 kb from Lus1007101 and Lus10007083 that encode the ortholog of *A. thaliana*'s *TRANSPARENT TESTA 8* (*TT8*) and *SUBTILISIN-LIKE SERINE PROTEASE* (*SBT1.7*) (Figure S5a, Table 2). In another independent QTL on chromosome 3, the SNP marker Lu3-25559600 was located 64.41 and 67.02 kb from Lus10009311 and Lus10009288 that encode the ortholog of *A. thaliana*'s *GALACTUROSYL TRANSFERASE-LIKE 5* (*GATL5*) and *MUCILAGE-MODIFIED 4* (*MUM4*). On chromosome 5, Lu5-3508878 is located 100.78 kb from Lus10008285, an ortholog of another *A. thaliana* gene implicated in mucilage transcriptional regulation, *NAC-REGULATED SEED MORPHOLOGY 1* (*NARS1*) (Figure S5a, Table 2).

Table 2. Candidate genes within LD blocks harboring peak SNPs associated with MC and HC.

Trait	Marker	Gene ID	Scaffold	*A. Thaliana* Ortholog	Gene Bank	Identity (%)	E-Value	Distance from Peak SNP (kb)
MC	Lu3-25559600	Lus10009311	318	*GATL5*	at1g02720	27	7×10^{-27}	64.41
		Lus10009288	318	*MUM4*	at1g53500	26	4×10^{-23}	67.02
		Lus10009287	318	*PME36*	at3g60730	61	2×10^{-110}	70.33
		Lus10009313	318	*SBT1.7*	at5g67360	45	0.0	75.66
	Lu3-26033342	Lus10007101	772	*TT8*	at4g09820	38	5×10^{-15}	58.92
		Lus10007083	772	*SBT1.7*	at5g67360	39	1×10^{-152}	49.60
	Lu5-3808878	Lus10008285	489	*NARS1*	at3g15510	52	9×10^{-45}	100.78
HC	Lu7-6577527	Lus10035456	151	*AGL62*	at5g60440	43	6×10^{-39}	11.40
	Lu12-5267706	Lus10018306	163	*GH17*	at2g39640	34	9×10^{-86}	39.93
	Lu13-2803224	Lus10026902	651	*DBR1*	at4g31770	68	0.0	96.87
		Lus10026926	651	*UGT79B1*	at5g54060	25	2×10^{-32}	238.19

Genes related to embryo, endosperm, and seed coat development, cell wall biogenesis/degradation, anthocyanin biosynthesis, and seed dormancy, were found at QTL loci associated with HC (Table 2 and Table S2). Among the relevant candidate genes, Lus10035456 encodes the ortholog of *A. thaliana*'s *AGAMOUS-LIKE MADS-BOX PROTEIN AGL62* (*AGL62*) and is located 11.40 kb from the SNP marker Lu7-6577527 (Figure S5b, Table 2). On chromosome 12, Lu12-5267706 was situated 39.93 kb from Lus10018306 that encodes the ortholog of *A. thaliana*'s *GLYCOSYL HYDROLASE FAMILY 17* (*GH17*). Two other interesting candidate genes, Lus10026902 and Lus10026926, were situated 96.87 and 238.19 k, respectively, from the SNP marker Lu13-2803224. Lus10026902 and Lus10026926 encode the ortholog of *A. thaliana*'s *LARIAT DEBRANCHING ENZYME* (*DBR1*) and *UDP-GLUCOSE FLAVONOL 3-O-GLUCOSYLTRANSFERASE* (*UGT79B1*), respectively.

3. Discussion

3.1. Phenotypic Variation of Mucilage and Hull Contents

Flaxseed mucilage and seed hull possess valuable nutritional and rheological attributes [33,34] but are also known to affect animal performance [7]. The presence of mucilage and fiber components (i.e., acid detergent lignin) in flaxseed meal reduces the energy uptake in both monogastric and ruminant animals [35]. Therefore, knowledge about the phenotypic variation and genetic control of seed mucilage content (MC) and hull content (HC) is pivotal to better design breeding strategies aiming to improve the overall food and feed value of flaxseed. The broad phenotypic variation of MC and HC in the association panel and the degree of additivity of the genetic components hint at the potential for improving flaxseed for either high or low MC and reduced HC through marker-assisted selection.

Kaewmanne et al. [4] reported MC ranging from 1.8 to 2.9% in seven Italian flaxseed cultivars, while Oomah et al. [16] found that MC ranged from 3.6 to 8.0% in 109 flaxseed accessions. We found a slightly wider range from 2 to 10% in our diversity panel. Little information exists for HC variation in large collections of flaxseed. In general, HC ranges from 22–27% to 36–48% were reported in mechanically treated and hand-dissected seeds, respectively [7,36], which is much higher than canola at 18.6% and soybean at 16.1% [6]. Reduction of HC can be achieved through the use of yellow-seeded cultivars, known to contain higher oil content and less HC than their brown-seeded counterparts [7,37]. Indeed, the yellow-seeded accessions displayed a lower HC compared to the brown-seeded genotypes. Nevertheless, caution should be exercised in adopting yellow-seeded flaxseed cultivars for reduced HC flaxseed because their susceptibility to natural splitting and mechanical cracking of the seed coat can negatively affect seed quality [38]. Consequently, breeding and seed tests to mechanical damage during harvesting should be conducted together in order to identify the ideal HC that would ensure seed mechanical resistance. All considered, our association panel harbored abundant phenotypic variation for dissecting the genetic landscape of MC and HC.

3.2. Population Structure and Linkage Disequilibrium

When the main factors accounting for population subdivision correlate with a trait under study (i.e., geographic distribution and flowering time), then marker–trait associations will undergo a more accentuated inflation of observed p-values as effect of the structure confounding factor [39]. In flaxseed, population structure has been assessed in varying numbers of accessions, where geographic origin and flax morphotype seemed to have been the main factors underlying population subdivisions [40–42]. In our association panel, the "red" and "blue" clusters were slightly differentiated ($F_{ST} = 0.08$), with a weak morphotypic effect on dendrogram topology, possibly due to the small number of fiber types ($n = 33$) compared to the larger number of oilseed type accessions ($n = 153$).

Linkage disequilibrium (LD) is the main factor influencing marker density requirement and mapping resolution in GWAS. Mating system and genetic diversity influence LD decay. LD decays more rapidly in outcrossing plant species than in self-pollinated plants [43] and, similarly, in wild relatives and landraces compared to modern cultivars [44]. Here, we observed a rapid LD decay for most of the chromosomes, comparable to some maize commercial elite inbred lines [45] and faster than winter-type *Brassica napus* (480 to 1283 kb, $r^2 = 0.1$) [46]. Therefore, the 200 flaxseed accessions of our diversity panel are expected to contain plentiful allelic diversity, as suggested by the generally short LD blocks for the 15 chromosomes, thereby assisting the search for candidate genes through efficient narrowing of the putative QTL regions.

3.3. Genome-Wide Association Analysis

Several general (GLM) and mixed (MLM) linear models have been proposed to control both population structure and cryptic relatedness [47–49]. In flax, MLM has been the preferred association model for multiple traits [24–26,42]. The "red" and "blue" clusters were weakly differentiated, and MC and HC between flax morphotypes was not statistically significant (Figure S2a,b), in contrast to a report comparing *indica* and *japonica* rice types assessed for 34 traits [39]. Hence, the genetic architecture of MC and HC seem to be only weakly correlated with population and family structures, and GLM-PCA was sufficient to control the rate of false positive associations.

The discovery of QTL for agronomic and economically important traits in crops is of great importance for marker-assisted breeding. This is the first report of QTL for MC in flax, likely because this trait has not been a breeding priority in the most important breeding programs of the world [18].

In the present study, GWAS identified seven QTL for MC, and their effects clearly suggest the promise of marker-assisted selection for modifying MC.

Chromosome 3's multiple MC QTL harbored candidate genes orthologous to Arabidopsis *TT8* gene, which is part of a transcription factor complex that, along with *GLABRA2* (*GL2*), regulates *MUM4* gene expression [50]. *MUM4* is required to produce rhamnose, a key substrate for mucilage biosynthesis [50], and chromosome 3 Lus10009311 is its flax ortholog. In Arabidopsis, *GATL5* encodes a glycosyltransferase involved in rhamnogalacturonan I (RG I) backbone synthesis [51]. The presence of a *L. usitatissimum* ortholog Lus10009311 in a LD block, with a peak SNP for MC, corresponds to the expected role of RG I synthesis. Arabidopsis gene *SBT1.7* triggers the activation of cell wall-modifying enzymes necessary for mucilage release upon imbibition [52]. In line with the expected seed coat mucilage dynamics, we identified two orthologous copies of this gene in two independent QTL (Table 2). Arabidopsis *PECTIN METHYLESTERASE INHIBITOR 6* (*PMEI6*) mutants were defective in seed coat mucilage release [53]. An ortholog of the Arabidopsis gene, *PECTIN METHYLESTERASE 36* (*PME36*), another family member, was located at one of the MC QTL loci identified herein. While *PME36* has not been shown to be involved in mucilage release, it might participate indirectly because it exerts a similar role to that of *PMEI6* in pectin synthesis and cell wall modification [54].

Oil content is an economically important but genetically complex trait. MC is negatively correlated with oil content, therefore, reducing MC should facilitate increasing oil content. Indeed, reduced accumulation of mucilage accompanied by increased oil content was observed in Arabidopsis *MUM4* or *GL2* mutants [55]. We observed a significant negative correlation ($r = -0.15$, $p = 0.03$) between MC and oil content in the association panel (data not shown). This is perhaps due to increased carbon allocation to the embryo in reduced or no seed coat mucilage synthesis in low MC accessions as proposed in Arabidopsis [55].

Increasing seed oil content and reducing the fiber fraction of the meal have been important goals in oil crop breeding. In *B. napus* and *L. usitatissium*, seed coat thickness or HC are negatively correlated with seed oil and protein content, as well as seed color [56–58]. QTL for seed coat color to indirectly increase oil content and minimize HC have been identified in *B. napus* and soybean [37,59,60]. In flax, a pleiotropic QTL controlling yellow seed and white flower color was recently dissected at the molecular level, but its effect on HC has not been addressed [23]. Here, we identified four QTL whose effects reduced HC by 2.6%, on average. Chromosome 7 harbored Lus10035456, which resembles the *A. thaliana* transcription factor *AGL62*. *AGL62* mutants initiated embryo and endosperm formation, but failed to form a seed coat [61]. Light seed color and low HC are thought to coincide because the biochemical pathways leading to lignin and pigment synthesis share common precursors [59]. In Arabidopsis, the core components of seed coat pigments are proanthocyanidins (PAs) [62]. Chromosome 12 encompassed three candidate genes including the ortholog of Arabidopsis *O-GLYCOSYL HYDROLASES FAMILY 17* gene. *GH17* is coexpressed with *TT12*, *AHA10*, and *BAN*, that might process glycosylated flavan-3-ol monomers, leading to accumulation of PAs in the seed coat [63]. In black seed soybean, a *UDP-GLUCOSE:FLAVONOID 3-O-GLUCOSYLTRANSFERASE* (*UGT78K1*), was isolated from the seed coat, a key enzyme that catalyzes the final step in anthocyanin biosynthesis [64]. Chromosome 13 contained Lus10026926, an ortholog of the *A. thaliana UGT79B1*, a gene also involved in anthocyanin biosynthesis. Yellow seed color stems from the blocked biosynthesis of PAs that impart the brown color to the seed coat [65]. The flaxseed meal derived from brown-seeded cultivars contains PAs that negatively affect protein digestion [66], hence low PA meal is preferred in animal ration. Additional advantages of modifying the seed color and reducing MC and HC include higher limpidity of the crude oil from the removal of gum-like residues and dark pigments, higher protein content and better feeding value of flaxseed meal for livestock and poultry [7].

Few accessions combined favorable alleles for reduced MC and HC. It should be possible to combine these attributes in a single genotype through the pyramiding of the respective favorable alleles owing to the fact that the significant QTL for both traits did not co-locate in the flax genome.

The development of yellow-seeded cultivars with low HC and either low or high MC for different industrial uses is an opportunity to increase market share and value.

4. Materials and Methods

4.1. Plant Material, Field Trials, Phenotyping, and Statistical Analyses

A total of 200 *L. usitatissimum* accessions from the Canadian flax core collection [67] were selected for this study based on their geographic distribution and genetic diversity (Table S3). The 200 genotypes were planted in 2014 and 2015 at the Agriaquaculture Nutritional Genomic Center (CGNA) experimental stations located in Vilcún (CAR2014) and Huichahue (HU2015), La Araucania region, Chile, using a completely randomized design (CRD) with three biological replicates. Genotypes were arranged in rows and columns in order to take into account spatial heterogeneity.

The seed mucilage content (MC) was determined in three biological replicates following the procedure described by Kaewmanee et al. [4] with minor modifications. A total of 2 g of seeds were incubated in 20 mL of water at 100 °C for 15 min in 50 mL Falcon tubes. Next, the tubes were shaken for 30 min at 250 rpm. The soluble extract was recovered by centrifugation at 6132 relative centrifugal force (RCF) for 30 min, and the mucilage fraction was precipitated by incubating in 30 mL of ethanol (95%) overnight at 4 °C. The seeds were recovered, and the extraction procedure was carried out twice more to maximize mucilage recovery. The mucilage pellet was weighed and expressed as milligrams of mucilage per gram of seed (mg g^{-1}).

HC was determined in three biological replicates by separating the hull from the embryos using a dissecting needle and tweezers from 50 seeds after imbibition in water for 24 h. Both fractions were dried at 90 °C for 4 h before their dry weights were measured. HC was expressed as (hull dry weight/(hull dry weight + embryo dry weight)) × 100, averaged from 50 seeds.

Variation of phenotypic data was analyzed individually for each environment using a restricted maximum likelihood (REML) analysis. Spatial correction in row and column directions was used with different variance–covariance structures. Spatial models were compared with Akaike information criterion (AIC) and Bayesian information criterion (BIC), and the most appropriate model in each environment was used to obtain a best linear unbiased estimate (BLUEs) for mucilage and hull contents in GenStat v.16 [68]. Descriptive statistics and Shapiro–Wilk normality test were conducted in the R package MVN [69]. Narrow sense heritability (h^2) was estimated using variance components from TASSEL v.5.2.31 [70]. Trait h^2 estimates were computed using the equation: $h^2 = \sigma^2_a/\sigma^2_a + \sigma^2_e$, where σ^2_a is the additive genetic variance and σ^2_e is the residual error variance [70].

4.2. Whole Genome Resequencing and SNP Calling

Genotyping by sequencing (GBS) methodology was adopted to genotype the 407 accessions from the Canadian flax core collection. The 407 individuals were grown in pots in a greenhouse with a 16 h light and 8 h dark cycle. Young leaf tissues from single plants were collected for DNA extraction using the DNeasy 96 Plant kit (Qiagen, Mississauga, ON, Canada) according to manufacturer's instructions. Genomic DNA was quantified, sheared, size-selected, and libraries were constructed for each genotype by the Michael Smith Genome Sciences Centre of the BC Cancer Agency, Genome British Columbia (Vancouver, BC, Canada) which also sequenced the libraries generating 100 bp paired-end reads on the Illumina HiSeq 2000 platform (Illumina Inc., San Diego, CA, USA). A total of 26.875 billion 100 bp pair-end reads were generated, corresponding to 6587 MB sequences and 15.5× genome equivalents of the reference genome (~370 MB) [32,71], on average, per individual.

All reads from each individual of the population was aligned to the flax reference sequence (the flax pseudomolecules v2.0) [72] using BWA (v0.6.1) [73] with default parameters. The alignment file for each individual was used as input for SNP discovery using the software package SAMtools [74]. SAMtools converts the sequence alignment files from sequence alignment/map (SAM) to sorted binary alignment/map (BAM) format and call all potential variants (SNP and indels) into the pileup files. Then, all variants were further filtered to get a set of high-quality SNPs. SNPs were filtered by the following criteria: (1) candidate SNP loci must be more than 10 bp away from each other; (2) each candidate SNP must have three mapped reads on the region; and (3) all the singleton SNPs were excluded. All steps of this procedure were implemented in an annotation-based genome-wide SNP discovery (AGSNP) pipeline [75,76]. The coordinates of all SNPs were extracted from the chromosome-based flax pseudomolecules v2.0 [72] totaling 1.7 million SNPs. Two hundred accessions, as previously mentioned, were selected for this study.

4.3. Population Structure, LD, Genome-Wide Association Study, and Candidate Genes

Population structure was estimated with 259 neutral SSR loci [41] distributed across flax's 15 chromosomes. The software STRUCTURE v.2.3.4 [47] was employed with predefined numbers of genetic clustering (K) from 1–5, using 50,000 burn-in iterations, followed by 100,000 MCMC across five independent runs for each K values. The number of clusters (K) was calculated with the Evanno method [77] implemented in the R package POPHELPER v.1.1.10 [78]. A total of 771,914 SNPs, filtered from the 1.7 million SNPs by removing those with a minor allele frequency <0.05 and >10% missing data, were used to produce a dendrogram using the neighbor-joining (NJ) algorithm implemented in TASSEL v.5.2.31 [70]. Genome-wide linkage disequilibrium (LD) and intrachromosomal LD between pairs of SNPs using the 771,914 filtered SNPs were estimated using squared allele frequency correlations (r^2) in TASSEL v.5.2.31 [70]. LD values were plotted against physical distance to determine the LD decay using the Hill and Weir [79] function. A cut-off value of $r^2 = 0.1$ was set to estimate the average LD blocks [41].

GWAS was performed in TASSEL v.5.2.31 [70] using the 771,914 filtered SNPs. Three models were evaluated, including GLM-Q, GLM-PCA, and MLM-K. The Q matrix generated from STRUCTURE was used as a cofactor to adjust for population stratification (GLM-Q). A GLM-PCA was assessed, including up to ten principal component covariates. The ten PCAs were generated in TASSEL v.5.2.31 [70] with 105,038 SNPs (MAF > 0.05 and at least 95% present among the 200 genotypes). For the MLM-K, a kinship matrix was created in TASSEL v.5.2.31 [70] with the set of 105,038 SNPs, and used as covariate to account for cryptic relatedness. A quantile–quantile (Q–Q) plot was displayed using the R package qqman [80] to evaluate the fitness and efficiency of the different models. The final Manhattan plots were also displayed using the qqman package [80]. The Bonferroni correction ($0.1/771{,}914 = -\log(P) = 6.88$) was used as threshold for the significance of marker–trait associations.

To identify candidate genes associated with significant SNPs, the Jbrowse feature of Phytozome v.12.1 (http://phytozome.jgi.doe.gov/pz/portal.html) was used to examine the *L. usitatissimum* v.1.0 genome [71] for genes relevant to MC and HC in flaxseed. As mentioned above, a cut-off value of $r^2 = 0.1$ was set to estimate the average LD block for each chromosome. The defined physical distance was used to pinpoint candidate genes on either side of the most significant SNPs. A plausible candidate gene was defined by the following criteria: (a) the gene had a function known to be related to the trait evaluated based on gene ontology term descriptions in Phytozome; (b) BLASTX searches from the Arabidopsis genome returned orthologous protein sequences with functions associated to the phenotypes of interest.

5. Conclusions

We performed GWAS using a set of 771,914 SNPs, identifying seven and four QTL for MC and HC, respectively. Above all, chromosome 3 encompassed three QTL harboring promising candidate genes for MC. Three of the QTL associated with HC contained plausible candidate genes related to seed coat and anthocyanin biosynthesis. These favorable QTL alleles will assist the design of market specific flaxseed cultivars with reduced HC while maintaining high MC for food and low MC for feed. The application of the identified SNP markers in molecular-assisted breeding for MC and HC, two complex traits whose phenotyping is labor-intensive and time-consuming, might enable a rapid transfer of favorable alleles into well adapted elite flaxseed cultivars, thus shortening the breeding cycle. Based on our results and previous gene expression studies, we hypothesize that the genetic control of mucilage and hull content in flax might share conserved pathways with Arabidopsis. Further validation of candidate genes, like *LuTT8*, *LuSBT1.7*, *LuMUM4*, and *LuAGL62*, through gene expression analysis or gene editing, will be necessary to validate the hypothesis mentioned above. Characterization of genes underlying the QTL will expand knowledge of the high complexity of cell wall dynamics involved in seed mucilage and seed coat biosynthesis in flaxseed.

Supplementary Materials: Supplementary materials can be found at http://www.mdpi.com/1422-0067/19/10/2870/s1.

Author Contributions: B.J.S.-C. designed the research experiments, performed the GWAS, interpreted the results and wrote the manuscript. S.C. performed the resequencing of the association panel, co-wrote and edited the manuscript. R.Q. planted the association panel and performed the phenotyping. H.A.G. planted the association panel and performed statistical analysis of the phenotypic data. M.O. wrote scripts and generated the figures. F.M.Y. generated the genome-wide SNP data.

Acknowledgments: This work was supported by the Agriaquaculture Nutritional Genomic Center (CGNA), the Programa Regional de Investigación Científica y Tecnológica and the Gobierno Regional de La Araucania, Chile. CGNA acknowledges the collaboration of Agriculture and Agri-Food Canada (AAFC) and the Total Utilization Flax GENomics (TUFGEN) project formerly funded by Genome Canada and other stakeholders of the Canadian flax industry.

Conflicts of Interest: The authors declare no conflict of interests.

Abbreviations

MC	mucilage content
HC	gull content
GWAS	genome-wide association study
LD	linkage disequilibrium
kb	kilobase
SNP	single nucleotide polymorphism
SSR	simple sequence repeat
CAR2014	Vilcún 2014
HU2015	Huichahue 2015
REML	restricted maximum likelihood
AIC	Akaike information criterion
BIC	Bayesian information criterion
BLUE	best linear unbiased estimation
GLM	general linear model
MLM	mixed linear model
PCA	principal component analysis
Q–Q	quantile–quantile

References

1. Rabetafika, H.N.; van Remoortel, V.; Danthine, S.; Paquot, M.; Beckler, C. Flaxseed proteins: Food uses and health benefits. *Int. J. Food Sci. Technol.* **2011**, *46*, 221–228. [CrossRef]
2. Kristensen, M.; Jensen, M.G.; Aarestrup, J.; Petersen, K.; Søndergaard, L.; Mikkelsen, M.S.; Astrup, A. Flaxseed dietary fibers lower cholesterol and increase fecal fat excretion, but magnitude of effect depend on food type. *Nutr. Metab.* **2012**, *9*, 1–8. [CrossRef] [PubMed]
3. Hunt, K.; Jones, J.K.N. The structure of linseed mucilage: Part II. *Can. J. Chem.* **1962**, *40*, 1266–1279. [CrossRef]
4. Kaewmanee, T.; Bagnasco, L.; Benjakul, S.; Lanteri, S.; Morelli, C.F.; Speranza, G.; Cosulich, M.E. Characterisation of mucilages extracted from seven Italian cultivars of flax. *Food Chem.* **2014**, *148*, 60–69. [CrossRef] [PubMed]
5. Haughn, G.; Chaudhury, A. Genetic analysis of seed coat development in Arabidopsis. *Trends Plant Sci.* **2005**, *10*, 472–477. [CrossRef] [PubMed]
6. Bhatty, R.S.; Cherdkiatgumchai, P. Compositional analysis of laboratory-prepared and commercial samples of linseed meal and of hull isolated from flax. *J. Am. Oil Chem. Soc.* **1990**, *67*, 79–84. [CrossRef]

7. Gajardo, H.A.; Quian, R.; Soto-Cerda, B. Agronomic and quality assessment of linseed advanced breeding lines varying in seed mucilage content and their use for food and feed. *Crop Sci.* **2017**, *57*, 2979–2990. [CrossRef]
8. Cherian, G.; Quezada, N. Egg quality, fatty acid composition and immunoglobulin Y content in eggs from laying hens fed full fat camelina or flax seed. *J. Anim. Sci. Biotechnol.* **2016**, *7*, 15. [CrossRef] [PubMed]
9. Oomah, B.D.; Mazza, G. Processing of flaxseed meal: Effect of solvent extraction on physicochemical characteristics. *LWT-Food Sci. Technol.* **1993**, *26*, 312–317. [CrossRef]
10. Sosulski, F.W.; Bakal, A. Isolated proteins from rapeseed, flax and sunflower meals. *Can. Inst. Food Sci. Technol. J.* **1969**, *2*, 28–32. [CrossRef]
11. Kessler, R.W.; Kohler, R. New strategies for exploiting flax and hemp. *Chemtech* **1996**, *26*, 34–42.
12. Sosulski, F.; Zadernowski, R. Fractionation of rapeseed meal into hour and hull components. *J. Am. Oil Chem. Soc.* **1981**, *58*, 96–98. [CrossRef]
13. Daun, J.K.; DeClercq, D.R. Quality of yellow and dark seeds in *Brassica campentris* canola varieties Candle and Tobin. *J. Am. Oil Chem. Soc.* **1988**, *65*, 122–126. [CrossRef]
14. Francoz, E.; Ranocha, P.; Burlat, V.; Dunand, C. Arabidopsis seed mucilage secretory cells: Regulation and dynamics. *Trends Plant Sci.* **2015**, *8*, 515–524. [CrossRef] [PubMed]
15. Venglat, P.; Xiang, D.; Qiu, S.; Stone, S.L.; Tibiche, C.; Cram, D.; Alting-Mees, M.; Nowak, J.; Cloutier, S.; Deyholos, M.; et al. Gene expression analysis of flax seed development. *BMC Plant Biol.* **2011**, *11*, 74. [CrossRef] [PubMed]
16. Oomah, B.D.; Kenaschuk, E.O.; Cui, W.; Mazza, G. Variation in the composition of water-soluble polysaccharides in flaxseed. *J. Agric. Food Chem.* **1995**, *43*, 1484–1488. [CrossRef]
17. Oomah, B.D.; Mazza, G. Effect of dehulling on chemical composition and physical properties of flaxseed. *LWT-Food Sci. Technol.* **1997**, *30*, 135–140. [CrossRef]
18. Diederichsen, A.; Raney, J.P.; Duguid, S.D. Variation of mucilage in flax seed and its relationship with other seed characters. *Crop Sci.* **2006**, *46*, 365–371. [CrossRef]
19. Spielmeyer, W.; Green, A.G.; Bittisnish, D.; Mendham, N.; Lagudah, E.S. Identification of quantitative trait loci contributing to Fusarium wilt resistance on an AFLP linkage map of flax (*Linum usitatissimum*). *Theor. Appl. Genet.* **1998**, *97*, 633–641. [CrossRef]
20. Asgarinia, P.; Cloutier, S.; Duguid, S.; Rashid, K.; Mirlohi, A.F.; Banik, M.; Saeidi, G. Mapping quantitative trait loci for powdery mildew resistance in flax (*Linum usitatissimum* L.). *Crop Sci.* **2013**, *53*, 2462–2472. [CrossRef]
21. Cloutier, S.; Ragupathy, R.; Niu, Z.; Duguid, S. SSR-based linkage map of flax (*Linum usitatissimum* L.) and mapping of QTL underlying fatty acid composition traits. *Mol. Breed.* **2011**, *28*, 437–451. [CrossRef]
22. Kumar, S.; You, F.M.; Duguid, S.; Booker, H.; Rowland, G.; Cloutier, S. QTL for fatty acid composition and yield in linseed (*Linum usitatissimum* L.). *Theor. Appl. Genet.* **2015**, *128*, 965–984. [CrossRef] [PubMed]
23. Sudarshan, G.P.; Kulkarni, M.; Akhov, L.; Ashe, P.; Shaterian, H.; Cloutier, S.; Rowland, G.; Wei, Y.; Selvaraj, G. QTL mapping and molecular characterization of the classical *D* locus controlling seed and flower color in *Linum usitatissimum* (flax). *Sci. Rep.* **2017**, *7*, 15751. [CrossRef] [PubMed]
24. Soto-Cerda, B.J.; Duguid, S.; Booker, H.; Rowland, G.; Diederichsen, A.; Cloutier, S. Genomic regions underlying agronomic traits in linseed (*Linum usitatissimum* L.) as revealed by association mapping. *J. Integr. Plant Biol.* **2014**, *56*, 75–87. [CrossRef] [PubMed]
25. Soto-Cerda, B.J.; Duguid, S.; Booker, H.; Rowland, G.; Diederichsen, A.; Cloutier, S. Association mapping of seed quality traits using the Canadian flax (*Linum usitatissimum* L.) core collection. *Theor. Appl. Genet.* **2014**, *127*, 881–896. [CrossRef] [PubMed]
26. Xie, D.; Dai, Z.; Yang, Z.; Sun, J.; Zhao, D.; Yang, X.; Zhang, L.; Tang, Q.; Su, J. Genome-wide association study identifying candidate genes influencing important agronomic traits of flax (*Linum usitatissimum* L.) using SLAF-seq. *Front. Plant Sci.* **2018**, *8*, 2232. [CrossRef] [PubMed]
27. You, F.M.; Xiao, J.; Li, P.; Yao, Z.; Jia, G.; He, L.; Kumar, S.; Soto-Cerda, B.; Duguid, S.D.; Booker, H.M.; et al. Genome-Wide Association Study and Selection Signatures Detect Genomic Regions Associated with Seed Yield and Oil Quality in Flax. *Int. J. Mol. Sci.* **2018**, *19*, 2303. [CrossRef] [PubMed]
28. Ersoz, E.S.; Yu, J.; Buckler, E.S. Applications of linkage disequilibrium and association mapping in maize. In *Molecular Genetic Approaches to Maize Improvement*; Kriz, A., Larkins, B., Eds.; Springer: Berlin, Germany, 2009; pp. 173–195.

29. Cloutier, S.; Niu, Z.; Datla, R.; Duguid, S. Development and analysis of EST-SSRs for flax (*Linum usitatissimum* L.). *Theor. Appl. Genet.* **2009**, *119*, 53–63. [CrossRef] [PubMed]
30. Cloutier, S.; Miranda, E.; Ward, K.; Radovanovic, N.; Reimer, E.; Walichnowski, A.; Datla, R.; Rowland, G.; Duguid, S.; Ragupathy, R. Simple sequence repeat marker development from bacterial artificial chromosome end sequences and expressed sequence tags of flax (*Linum usitatissimum* L.). *Theor. Appl. Genet.* **2012**, *125*, 685–694. [CrossRef] [PubMed]
31. Cloutier, S.; Ragupathy, R.; Miranda, E.; Radovanovic, N.; Reimer, E.; Walichnowski, A.; Ward, K.; Rowland, G.; Duguid, S.; Banik, M. Integrated consensus genetic and physical maps of flax (*Linum usitatissimum* L.). *Theor. Appl. Genet.* **2012**, *125*, 1783–1795. [CrossRef] [PubMed]
32. Ragupathy, R.; Rathinavelu, R.; Cloutier, S. Physical mapping and BAC-end sequence analysis provide initial insights into the flax (*Linum usitatissimum* L.) genome. *BMC Genom.* **2011**, *12*, 217. [CrossRef] [PubMed]
33. Altunkaya, A. Dermal Lubricant and Moisturizer. WO 2006/075236 A1, 20 July 2006.
34. Anttila, M.; Kankaanpää-Anttila, B.; Sepponen, M.; Timonen, H.; Autio, K. Improving of Texture of Dairy Products. WO 2008/000913 A1, 3 January 2008.
35. Kracht, W.; Dänicke, S.; Kluge, H.; Keller, K.; Matzke, W.; Henning, U.; Schumann, W. Effect of dehulling of rapeseed on feed value and nutrient digestibility of rape products in pigs. *Arch. Anim. Nutr.* **2004**, *58*, 389–404. [CrossRef] [PubMed]
36. Oomah, B.D.; Mazza, G. Fractionation of flaxseed with a batch dehuller. *Ind. Crop Prod.* **1998**, *9*, 19–27. [CrossRef]
37. Yan, X.Y.; Li, J.N.; Fu, F.Y.; Jin, M.Y.; Chen, L.; Liu, L.Z. Co-location of seed oil content, seed hull content and seed coat color QTL in three different environments in *Brassica napus* L. *Euphytica* **2009**, *170*, 355–364. [CrossRef]
38. Saedi, G.; Rowland, G.G. Seed color and linolenic acid effects on agronomic traits in flax. *Can. J. Plant Sci.* **1999**, *79*, 521–526. [CrossRef]
39. Zhao, K.; Tung, C.W.; Eizenga, G.C.; Wright, M.H.; Ali, M.L.; Price, A.H.; Norton, G.J.; Islam, M.R.; Reynolds, A.; Mezey, J.; et al. Genome-wide association mapping reveals a rich genetic architecture of complex traits in *Oryza sativa*. *Nat. Commun.* **2011**, *2*, 467. [CrossRef] [PubMed]
40. Smýkal, P.; Bačová-Kerteszová, N.; Kalendar, R.; Corander, J.; Schulman, A.H.; Pavelek, M. Genetic diversity of cultivated flax (*Linum usitatissimum* L.) germplasm assessed by retrotransposon-based markers. *Theor. Appl. Genet.* **2011**, *122*, 1385–1397. [CrossRef] [PubMed]
41. Soto-Cerda, B.J.; Diederichsen, A.; Ragupathy, R.; Cloutier, S. Genetic characterization of a core collection of flax (*Linum usitatissimum* L.) suitable for association mapping studies and evidence of divergent selection between fiber and linseed types. *BMC Plant Biol.* **2013**, *13*, 78. [CrossRef] [PubMed]
42. Chandrawati, N.S.; Kumar, R.; Kumar, S.; Singh, P.K.; Yadav, V.K.; Ranade, S.A.; Yadav, H.K. Genetic diversity, population structure and association analysis in linseed (*Linum usitatissimum* L.). *Physiol. Mol. Biol. Plants* **2017**, *23*, 207–219. [CrossRef] [PubMed]
43. Abdurakhmonov, I.; Abdukarimov, A. Application of association mapping to understanding the genetic diversity of plant germplasm resources. *Int. J. Plant Genom.* **2008**, *2008*, 574297. [CrossRef] [PubMed]
44. Xu, J.; Ranc, N.; Munos, S.; Rolland, S.; Bouchet, J.P.; Desplant, N.; Le Paslier, M.C.; Liang, Y.; Brunel, D.; Causse, M. Phenotypic diversity and association mapping for fruit quality traits in cultivated tomato and related species. *Theor. Appl. Genet.* **2013**, *126*, 567–581. [CrossRef] [PubMed]
45. Jung, M.; Ching, A.; Bhattramakki, D.; Dolan, M.; Tingey, S.; Morgante, M.; Rafalski, A. Linkage disequilibrium and sequence diversity in a 500-kbp region around the adh1 locus in elite maize germplasm. *Theor. Appl. Genet.* **2004**, *109*, 681–689. [CrossRef] [PubMed]
46. Hatzig, S.V.; Frisch, M.; Breuer, F.; Nesi, N.; Ducournau, S.; Wagner, M.H.; Leckband, G.; Abbadi, A.; Snowdon, R.J. Genome-wide association mapping unravels the genetic control of seed germination and vigor in *Brassica napus*. *Front. Plant Sci.* **2015**, *6*, 221. [CrossRef] [PubMed]
47. Pritchard, J.K.; Stephens, M.; Rosenberg, N.A.; Donnelly, P. Association mapping in structured populations. *Am. J. Hum. Genet.* **2000**, *67*, 170–181. [CrossRef] [PubMed]
48. Price, A.L.; Patterson, N.J.; Plenge, R.M.; Weinblatt, M.E.; Shadick, N.A.; Reich, D. Principal components analysis corrects for stratification in genome-wide association studies. *Nat. Genet.* **2006**, *38*, 904–909. [CrossRef] [PubMed]

49. Yu, J.; Pressoir, G.; Briggs, W.; Vroh, B.; Yamasaki, M.; Doebley, J.; McMullen, M.D.; Gaut, B.S.; Nielsen, D.M.; Holland, J.B.; et al. A unified mixed-model method for association mapping that accounts for multiple levels of relatedness. *Nat. Genet.* **2006**, *38*, 203–208. [CrossRef] [PubMed]
50. Western, T.L.; Young, D.S.; Dean, G.H.; Tan, W.L.; Samuels, A.L.; Haughn, G.W. MUCILAGE-MODIFIED4 encondes a putative pectin biosynthetic enzyme developmentally regulated by APETALA2, TRANSPARENT TESTA GLABRA1, and GLABRA2 in the *Arabidopsis* seed coat. *Plant Physiol.* **2004**, *134*, 296–306. [CrossRef] [PubMed]
51. Kong, Y.; Zhou, G.; Abdeen, A.A.; Schafhauser, J.; Richardson, B.; Atmodjo, M.A.; Jung, J.; Wicker, L.; Mohnen, D.; Western, T.; Hahn, M.G. GALACTURONOSYLTRANSFERASE-LIKE5 is involved in the production of *Arabidopsis* seed coat mucilage. *Plant Physiol.* **2013**, *163*, 1203–1217. [CrossRef] [PubMed]
52. Rautengarten, C.; Usadel, B.; Neumetzler, L.; Hartmann, J.; Büssis, D.; Altmann, T. A subtilisin-like serine protease essential for mucilage release from Arabidopsis seed coats. *Plant J.* **2008**, *54*, 466–480. [CrossRef] [PubMed]
53. Saez-Aguayo, S.; Ralet, M.C.; Berger, A.; Botran, L.; Ropartz, D.; Marion-Poll, A.; North, H.M. PECTIN METHYLESTERASE INHIBITOR6 promotes Arabidopsis mucilage release by limiting methylesterification of homogalacturonan in seed coat epidermal cells. *Plant Cell* **2013**, *25*, 308–323. [CrossRef] [PubMed]
54. Louvet, R.; Cavel, E.; Gutierrez, L.; Guénin, S.; Roger, D.; Gillet, F.; Guerineau, F.; Pelloux, J. Comprehensive expression profiling of the pectin methylesterase gene family during silique development in *Arabidopsis thaliana*. *Planta* **2006**, *224*, 782–791. [CrossRef] [PubMed]
55. Shi, L.; Katavic, V.; Yu, Y.; Kunst, L.; Haughn, G. Arabidopsis glabra2 mutant seeds deficient in mucilage biosynthesis produce more oil. *Plant J.* **2012**, *69*, 37–46. [CrossRef] [PubMed]
56. Wang, R.; Li, J.N.; Chen, L.; Tang, Z.L.; Zhang, X.K. Genetic correlation analysis for main characters in yellow-seeded rapeseed lines (*Brassica napus* L.). *Chin. J. Oil Crop Sci.* **2003**, *25*, 8–11.
57. Khan, N.A.; Booker, H.; Yu, P. Molecular structures and metabolic characteristics of protein in brown and yellow flaxseed with altered nutrient traits. *J. Agri. Food Chem.* **2014**, *62*, 6556–6564. [CrossRef] [PubMed]
58. Qu, C.; Hasan, M.; Lu, K.; Liu, L.; Zhang, K.; Fu, F.; Wang, M.; Liu, S.; Bu, H.; Wang, R.; et al. Identification of QTL for seed coat colour and oil content in *Brassica napus* by association mapping using SSR markers. *Can. J. Plant Sci.* **2015**, *95*, 387–395. [CrossRef]
59. Badani, A.G.; Snowdon, R.J.; Wittkop, B.; Lipsa, F.D.; Baetzel, R.; Horn, R.; De Haro, A.; Font, R.; Lühs, W.; Friedt, W. Colocalization of a partially dominant gene for yellow seed colour with a major QTL influencing acid detergent fibre (ADF) content in different crosses of oilseed rape (*Brassica napus*). *Genome* **2006**, *49*, 1499–1509. [CrossRef] [PubMed]
60. Zhang, D.; Sun, L.; Li, S.; Wang, W.; Ding, Y.; Swarm, S.A.; Li, L.; Wang, X.; Tang, X.; Zhang, Z.; et al. Elevation of soybean seed oil content through selection for seed coat shininess. *Nat. Plants* **2018**, *4*, 30–35. [CrossRef] [PubMed]
61. Roszak, P.; Köhler, C. Polycomb group proteins are required to couple seed coat initiation to fertilization. *Proc. Natl. Acad. Sci. USA* **2011**, *108*, 20826–20831. [CrossRef] [PubMed]
62. Mizzotti, C.; Ezquer, I.; Paolo, D.; Rueda-Romero, P.; Guerra, R.F.; Battaglia, R.; Rogachev, I.; Aharoni, A.; Kater, M.M.; Caporali, E.; et al. SEEDSTICK is a master regulator of development and metabolism in the Arabidopsis seed coat. *PLoS Genet.* **2014**, *10*, e1004856. [CrossRef] [PubMed]
63. Marinova, K.; Pourcel, L.; Weder, B.; Schwarz, M.; Barron, D.; Routaboul, J.M.; Debeaujon, I.; Klein, M. The Arabidopsis MATE transporter TT12 acts as a vacuolar flavonoid/H^+-antiporter active in proanthocyanidin-accumulating cells of the seed coat. *Plant Cell* **2007**, *19*, 2023–2038. [CrossRef] [PubMed]
64. Kovinich, N.; Saleem, A.; Arnason, J.T.; Miki, B. Functional characterization of a UDP-glucose:flavonoid 3-O-glucosyltransferase from the seed coat of black soybean (*Glycine max* (L.) Merr.). *Phytochemistry* **2010**, *71*, 1253–1263. [CrossRef] [PubMed]
65. Xu, W.; Dubos, C.; Lepiniec, L. Transcriptional control of flavonoid biosynthesis by MYB–bHLH–WDR complexes. *Trends Plant Sci.* **2015**, *20*, 176–185. [CrossRef] [PubMed]
66. Mole, S.; Waterman, P.G. Tannic acid and proteolytic enzymes: Enzyme inhibition or substrate deprivation? *Phytochemistry* **1986**, *26*, 99–102. [CrossRef]
67. Diederichsen, A.; Kusters, P.M.; Kessler, D.; Bainas, Z.; Gugel, R.K. Assembling a core collection from the flax world collection maintained by Plant Gene Resources of Canada. *Genet. Resour. Crop Evol.* **2013**, *60*, 1479–1485. [CrossRef]

68. VSN International. *Genstat for Windows*, 18th ed.; VSN International: Hemel Hempstead, UK, 2015; Available online: http://www.Genstat.co.uk (accessed on 10 May 2015).
69. Korkmaz, S.; Goksuluk, D.; Zararsiz, G. MVN: An R Package for Assessing Multivariate Normality. *R J.* **2014**, *6*, 151–162.
70. Bradbury, P.J.; Zhang, Z.; Kroon, D.E.; Casstevens, T.M.; Ramdoss, Y.; Buckler, E.S. TASSEL: Software for association mapping of complex traits in diverse samples. *Bioinformatics* **2007**, *23*, 2633–2635. [CrossRef] [PubMed]
71. Wang, Z.; Hobson, N.; Galindo, L.; Zhu, S.; Shi, D.; McDill, J.; Yang, L.; Hawkins, S.; Neutelings, G.; Datla, R.; et al. The genome of flax (*Linum usitatissimum*) assembled de novo from short shotgun sequence reads. *Plant J.* **2012**, *72*, 461–473. [CrossRef] [PubMed]
72. You, F.M.; Xiao, J.; Li, P.; Yao, Z.; Jia, G.; He, L.; Zhu, T.; Luo, M.C.; Wang, X.; Deyholos, M.K.; et al. Chromosome-scale pseudomolecules refined by optical, physical and genetic maps in flax. *Plant J.* **2018**, *95*, 371–384. [CrossRef] [PubMed]
73. Li, H.; Durbin, R. Fast and accurate short read alignment with Burrows-Wheeler Transform. *Bioinformatics* **2009**, *25*, 1754–1760. [CrossRef] [PubMed]
74. Li, H.; Handsaker, B.; Wysoker, A.; Fennell, T.; Ruan, J.; Homer, N.; Marth, G.; Abecasis, G.; Durbin, R. The sequence alignment/map format and SAMtools. *Bioinformatics* **2009**, *25*, 2078–2079. [CrossRef] [PubMed]
75. Kumar, S.; You, F.M.; Cloutier, S. Genome wide SNP discovery in flax through next generation sequencing of reduced representation libraries. *BMC Genom.* **2012**, *13*, 684. [CrossRef] [PubMed]
76. You, F.M.; Deal, K.R.; Wang, J.; Britton, M.T.; Fass, J.N.; Lin, D.; Dandekar, A.M.; Leslie, C.A.; Aradhya, M.; Luo, M.C.; et al. Genome-wide SNP discovery in walnut with an AGSNP pipeline updated for SNP discovery in allogamous organisms. *BMC Genom.* **2012**, *13*, 354. [CrossRef] [PubMed]
77. Evanno, G.; Regnaut, S.; Goudet, J. Detecting the number of clusters of individuals using the software STRUCTURE: A simulation study. *Mol. Ecol.* **2005**, *14*, 2611–2620. [CrossRef] [PubMed]
78. Francis, M.R. POPHELPER: An R package and web app to analyse and visualise population structure. *Mol. Ecol. Resour.* **2017**, *1*, 27–32. [CrossRef] [PubMed]
79. Hill, W.G.; Weir, B.S. Variances and covariances of squared linkage disequilibria in finite populations. *Theor. Popul. Biol.* **1998**, *33*, 54–78. [CrossRef]
80. Turner, S.D. QQMAN: An R package for visualizing GWAS results using Q-Q and manhattan plots. *biorXiv* **2014**. [CrossRef]

© 2018 by the authors. Licensee MDPI, Basel, Switzerland. This article is an open access article distributed under the terms and conditions of the Creative Commons Attribution (CC BY) license (http://creativecommons.org/licenses/by/4.0/).

International Journal of
Molecular Sciences

MDPI

Review

Computational and Experimental Tools to Monitor the Changes in Translation Efficiency of Plant mRNA on a Genome-Wide Scale: Advantages, Limitations, and Solutions

Irina V. Goldenkova-Pavlova [1,*], Olga S. Pavlenko [1], Orkhan N. Mustafaev [2], Igor V. Deyneko [1], Ksenya V. Kabardaeva [1] and Alexander A. Tyurin [1]

1. Group of Functional Genomics, Institute of Plant Physiology, Russian Academy of Sciences, Botanicheskaya str. 35, Moscow 127276, Russia; helliga.p@gmail.com (O.S.P.); igor.deyneko@inbox.ru (I.V.D.); kabardaewa@yandex.ru (K.V.K.); alexjofar@gmail.com (A.A.T.)
2. Department of Biophysics and Molecular Biology, Baku State University, Zahid Khalilov Str. 23, Baku AZ 1148, Azerbaijan; orkhan@bioset.org

* Correspondence: irengold58@gmail.com; Tel.: +7-499-678-5356

Received: 27 November 2018; Accepted: 19 December 2018; Published: 21 December 2018

Abstract: The control of translation in the course of gene expression regulation plays a crucial role in plants' cellular events and, particularly, in responses to environmental factors. The paradox of the great variance between levels of mRNAs and their protein products in eukaryotic cells, including plants, requires thorough investigation of the regulatory mechanisms of translation. A wide and amazingly complex network of mechanisms decoding the plant genome into proteome challenges researchers to design new methods for genome-wide analysis of translational control, develop computational algorithms detecting regulatory mRNA contexts, and to establish rules underlying differential translation. The aims of this review are to (i) describe the experimental approaches for investigation of differential translation in plants on a genome-wide scale; (ii) summarize the current data on computational algorithms for detection of specific structure–function features and key determinants in plant mRNAs and their correlation with translation efficiency; (iii) highlight the methods for experimental verification of existed and theoretically predicted features within plant mRNAs important for their differential translation; and finally (iv) to discuss the perspectives of discovering the specific structural features of plant mRNA that mediate differential translation control by the combination of computational and experimental approaches.

Keywords: regulation and efficiency of translation; genome-wide scale; experimental approaches; computational algorithms; features of plant mRNAs

1. Introduction

The genomic information in plants, similar to other eukaryotes, is implemented via a successive series of biological processes, including transcription and translation as the key events. The current experimental omics tools for genomic monitoring of plant gene expression allow tracking the flow of genetic information from genome to proteome and to metabolome. New experimental approaches, for example, RNA-Seq and DNA microarrays, have given insight into many key mechanisms involved in transcription regulation in plants: the first stage of gene expression and the easiest to study in terms of experimental methodology. The studies of transcriptomes, i.e., the qualitative and quantitative estimation of expression of the entire gene pool on a genome-wide scale, have given convincing evidence of dynamic changes in the transcriptomes of various plant species in both growth and development processes and the impact of environmental factors. Comparative omics studies in plants clearly demonstrate a very modest correlation between the levels of transcription (abundance of individual mRNAs) and translation (the levels of the corresponding proteins in the proteome). Of note, the observed fluctuations in the levels of a transcript do not always lead to changes in the levels of the corresponding protein [1]. This suggests an intricate nature of the mechanisms providing the decoding of a genome, which involve not only differential transcription, but also differential translation.

Translation is a complex biological process with numerous players, including mRNAs, tRNAs, ribosomes, and manifold protein factors. Undoubtedly, each is important for efficient translation. The mRNAs themselves comprise different regions, namely, the 5′ untranslated region (5′UTR) and coding region (CDS) and 3′ untranslated region (3′UTR), which modulate translation at a number of "checkpoints": translation initiation, elongation, and termination. In the current view, numerous regulatory elements may be concealed in the nucleotide contexts of these mRNA regions and each of them individually or in combination can determine the development of any transcript in translational process [2].

The paradox of misfit between the levels of mRNAs and their protein products observable in different plant species at all stages of their growth and development as well as upon the impact of various environmental factors focuses the attention of researchers on two key problems, namely (i) detection of the specific sets of differentially-translated transcripts, i.e., the sets of transcripts that are efficiently translated under certain conditions, and the sets of transcripts with repressed or unchanged translation under the same conditions and (ii) clarification of the particular regions or specific structural features of the mRNA nucleotide composition that mediate this differential translational control.

This review focuses on the experimental methods for genome-wide analysis of translational control, computational algorithms to search and analyze various regulatory contexts within mRNAs, and approaches for subsequent experimental verification of their correlation with mRNA translation in plants. Currently, we cannot refer to deficiency in publications comprehensively reporting the basic protocols of various methods for genome-wide analyses of translational control in general, including the methods applicable to plant objects. However, reviews that consider and discuss the three key components of the general strategy for identification of regulatory contexts in mRNA that may play a key role in differential translation are still absent in the scientific literature. Our goals here are (i) to consider the experimental approaches aiming to clarify differential translation on a plant genome-wide scale; (ii) to summarize the current data on the computational algorithms used for detection of the specific structural and functional features of key determinants within plant mRNAs and their interrelation with the translation efficiency; (iii) to highlight the methods for experimental verification of existed data and theoretical predictions of the intrinsic features of plant mRNAs important for their differential translation; and (iv) to discuss the ways of decoding the specific structural features of plant mRNA that mediate differential translational control by combining computational and experimental approaches. In general, this review discusses the main and critical steps for each method in this general strategy, areas of their application, and the main results obtained using plant objects and their contribution to our knowledge about the fine mechanisms of translation in plants.

2. Experimental Approaches to Determine Differentially-Translated mRNAs in Plants

Initially, proteomics methods were used to identify the correlation between the observed fluctuations in the expression of a transcript and the actual level of peptides in plants [3]. However, the proteomics approaches have certain limitations in the case of a spatiotemporal study of a large pool of translated mRNAs and are mainly applied for assessing translation of the known peptides and proteins. Moreover, the methods of proteomics are laborious and expensive, while preparation of the specimens, quantification of proteome, and subsequent peptide sequencing require specialized technical experience [4]. Advances in high-throughput technologies, such as microarrays and deep sequencing, have made it possible to develop the new experimental approaches to studying mRNA translation efficiency on a global scale. Three basic experimental approaches are currently used for these purposes: (a) polysome profiling; (b) translating ribosome affinity purification (TRAP); and (c) ribosome profiling or Ribo-Seq. These approaches are based on (i) the production of the mRNA pool with the ribosomes arrested on them; (ii) separation of actively-translated mRNAs (polysomal mRNAs and mRNAs bound to several ribosomes), moderately translated mRNAs (monosomal mRNAs and mRNAs bound to one ribosome), and untranslated mRNAs (steady-state mRNAs that are not bound to ribosomes); and (iii) subsequent quantitative assessment of an individual transcript or an mRNA population represented in polysomal complexes relative to the total amount of the transcript in the assayed plant specimens. Note that polysomes are several ribosomes performing translation from one mRNA and this process is regulated for individual mRNAs.

2.1. Profiling Polysomes

The translational status of the mRNA pool on a genome-wide scale can be estimated using polysome profiling. The basic protocols for the polysome profiling in plants are described in several publications [5]. Simple protocols have been additionally designed and verified for individual plant species, including *Arabidopsis thaliana*, *Nicotiana benthamiana*, *Solanum lycopersicum*, and *Oryza sativa*, as well as for individual plant tissues [5]. This method is based on the separation of the polysomal mRNAs, monosomal mRNAs, and steady-state mRNAs using sucrose density gradient centrifugation, referred to as polysome fractionation assays (Figure 1). Then, the transcripts (mRNAs) associated with each mRNA pool are analyzed by hybridization on microarrays or undergo RNA sequencing. Assembly, mapping, and in silico analysis of the sequencing data for different pools (polysomes, monosomes, and steady-state mRNAs) provide the researcher with the initial lists of the transcripts with different translation activities [6,7].

According to the experimental data, the results of polysome profiling can be used for a quantitative estimation of mRNA translation efficiency both at different plant growth and developmental stages and under the effect of adverse environmental factors [6,8] or for assessment of quantitative changes in the translational status of individual mRNAs [9]. As a rule, the polysome score (PS) or polysome ratio (PR) are used for this purpose; they are computed as the relative abundances of RNAs in polysomes versus RNAs in nonpolysomes or versus total mRNA. Where total mRNA is the total mRNA level in polysomal and nonpolysomal fractions, respectively [6,8,9].

The polysome profiling appeared rather efficient in the studies on differential translation regulation of specific plant mRNAs under the influence of several abiotic environmental factors [2,6,10]. For example, it has been convincingly demonstrated that the main part of the transcripts under stress displays different degrees of translation repression; moreover, a specific set of transcripts that avoids such repression and retain their transcriptional activity was detected. Below are several examples that in our view, illustrate the abilities of this method in clarifying the mechanisms of translation control in plants. In particular, it is shown that the shares of individual mRNA species in *A. thaliana* polysomal fractions under controlled growth upon a moderate dehydration stress vary from 5% to 95% and that this stress causes a decrease in the ribosome load for over 60% of all mRNAs [2]. The results of genome-wide assay of the relative amounts of individual mRNAs in polysomal versus nonpolysomal fractions under heat shock in the *A. thaliana* cell culture gave the set of genes with different translational responses, i.e., the genes that either considerably increased or considerably decreased the amounts of their mRNAs in polysomal fractions [10]. These results formed the background for further identification of the new cis-regulatory elements in 5′UTRs that influenced differential translation in response to heat shock in *A. thaliana* [8].

In another study, polysome profiling was used for a global assessment of the translation efficiency of mRNA pools during the growth and development of *A. thaliana* leaves. It was demonstrated that the degree of association of each mRNA with the polysomal fraction was different and considerably (from a strong repression to activation at a constant level) changed throughout these processes. Analysis of the functional categories of the mRNAs associated with polysomal fraction showed that the translation control, being of physiological significance during plant growth and development, was especially pronounced in the mRNAs associated with signaling and protein synthesis. In general, these results emphasize the importance of the dynamic changes in mRNA translation during plant growth and development and suggest that mRNA translation may be controlled via complex mechanisms underlying the response to each factor [6].

Although polysome profiling has been successfully used for a global study of plant mRNA translation efficiency, this method still has some limitations [11]. One of these, it cannot precisely determine the ribosome density, i.e., the number of ribosomes per mRNA, because the mRNA–ribosome complexes from the same differential centrifugation fractions may contain a different number of ribosomes. Moreover, polysome profiling fails to determine the actual ribosome distribution along the transcript, i.e., it is impossible to determine a mRNA region (5′UTR, CDS, or 3′UTR) in which reside the arrested ribosomes. This is very important since it allows for assessing of the translation stage (initiation, elongation, or termination) associated with differential translation of an individual transcript. As a consequence, this makes it not possible to specifically search for the regulatory determinants in particular mRNA regions important for an efficient translation.

Nonetheless, these limitations of the polysome profiling technique do not diminish its tremendous potential for the study of the fine mechanisms of translation in plants on a global scale. This method not only makes it possible to determine the correlations between the observed translational and transcriptional fluctuations under normal conditions and under stress factors, but also provides researchers with general information useful for further insights into the rules of mRNA decoding, i.e., allows defining the pools of transcripts with different translation efficiency and to find regulatory contexts of mRNAs or their combinations important for translation efficiency using computational analysis (this will be considered below in more detail). According to the available experimental data, polysome profiling is, as a rule, applicable to the search for actively-translated mRNAs and the subsequent analysis, although the understanding of the mechanisms associated with the repression of translation in a certain pool of transcripts is of the same importance; perhaps, researchers will focus on this area in future. It should be also emphasized that most studies utilizing polysome profiling performed so far, involve the plants with annotated genomes. However, the use of this method is not limited to the plant species with annotated genomes and can be extended to other plant species, including those genomes that have not been yet determined or those already sequenced but poorly annotated.

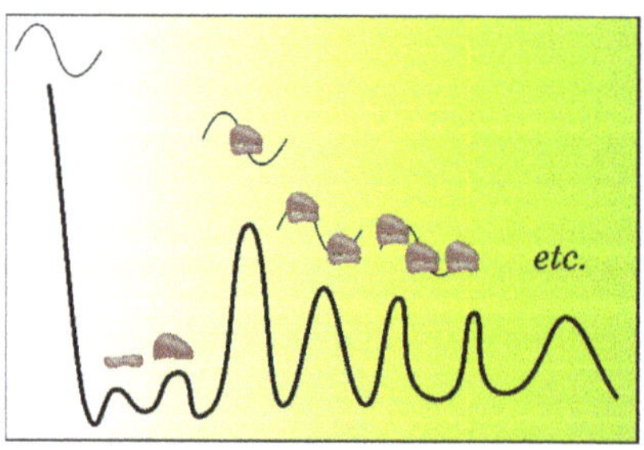

Figure 1. Polysome profiling in a sucrose density gradient. Separation of the transcripts depending on the ribosome loading: the first peak corresponds to the mRNAs unbound to ribosomes; second and third peaks, to the ribosome small and large subunits, respectively; and the fourth and subsequent peaks, to the mRNAs with different ribosome loadings.

2.2. Translating Ribosome Affinity Purification (TRAP) and TRAP-Seq

The experimental approach referred to as translating ribosome affinity purification (TRAP) is a modification of the traditional polysome profiling procedure and was for the first time described for *A. thaliana* [12]. This method utilizes the plant transgenic lines that express an epitope-tagged variant of ribosomal protein L18 (usually referred to as RPL18). As a rule, these plant transgenic lines express FLAG epitope-tagged RPL18 in the N-terminal region [12,13]. The cell lysates of transgenic plants are produced under the conditions that stabilize the ribosomes on RNA and block translation. The transcripts bound to the ribosomes that carry the labeled RPL18 are selectively separated using the absorption on anti-FLAG-M2 agarose. This enables ribosome capture from crude cell extracts by a single-stage immune precipitation (Figure 2) and, as a rule, allows the pool of RNAs (designated as TRAP RNA) that are actively translated to be obtained.

This method is described and discussed in detail in several papers [3,13–15]. Note that both the traditional polysome profiling approach and TRAP give analogous proportions of the small and large polysomes (i.e., ribosome profiles) [13]. Both approaches also have similar limitations on their application, namely, in the assessment of the number of ribosomes per mRNA length and the distribution of ribosomes along the transcript (see above). However, note that a wide use of the experimental TRAP approach is also limited by the available plant transgenic lines but, nonetheless, the use of transgenic lines gives certain advantages as compared with the traditional polysome profiling. This advantage consists in the possibility of not only constitutive, but also tissue-specific RPL18 expression by using different tissue-specific promoters [14]. Thanks to the tissue-specific RPL18 expression, TRAP is applicable to profiling of actively-translated RNAs in different populations of plant cells, namely, in (i) different root cells (epidermis, cortex, or endodermis); (ii) companion phloem cells, meristem cells, and leaf mesophyll cells; and (iii) microspores, pollen, and other plant tissues and cell types [14]. For example, the use of APETALA1, APETALA3, and AGAMOUS for expression of FLAG-RPL18 in early flower development allowed for the discovery of new levels of the expression control in developing flowers associated with differential translation [16]. A systemic analysis of the mRNAs in different specimens relative to the pollen grains within buds and in vitro-germinated pollen tubes has been performed with the help of the *A. thaliana* transgenic lines expressing epitope-tagged RPL18 under the control of ProLAT52 promoter, which allowed for the identification of a cohort of the transcripts that regulate late stages of pollination in flowering plants; this paves the way for better understanding of the pollen-based mechanisms that promote fertilization [15]. It should be emphasized that the in vivo proteomic studies of pollen tubes are extremely complicated because of the difficulties with pollen collection; the selective immune purification of the transcripts associated with the polysomes in pollen tubes in this case assisted in identification of the genes important for the in vivo pollen biology. Thus, the TRAP approach has an important advantage for efficient isolation of the population of mRNA complexes from particular cell types.

The sensitive moment when using TRAP approach is during the selection of the transgenic line that expresses FLAG-RPL18, which is extremely important for a successful analysis of the tissue-specific responses. A position effect associated with the T-DNA integration site in the genome of transgenic plants is known. In this regard, the new transgenic lines intended for this research should be selected bearing in mind the presence of known tissue-specific genes in the corresponding tissues or cell types. This will ensure selection of the most appropriate line for further analysis.

According to the current opinion, not only stable plant transformants, but also a transient expression of FLAG epitope-tagged RPL18 can be used for identification of the differentially-translated mRNA pools in plant genomes, for example, utilizing the agroinfiltration of *Medicago truncatula* hairy root cultures or of *N. benthamiana* leaves by *Agrobacterium rhizogenes*.

The FLAG tag may be also added to other proteins in order to determine their role in translation. For example, the expression of tagged oligouridylate binding protein 1 (UBP1) with subsequent immune purification of the mRNA–protein complexes (mRNPs) clarified the role of this protein in the dynamic and reversible aggregation of translationally repressed mRNAs in hypoxia [17]. In particular, UBP1 constitutively binds a subpopulation of the mRNAs with the 3′UTRs enriched for uracil under normoxic conditions. In hypoxia, UBP1 is associated with non-uracil-rich mRNAs, which increases its aggregation in microscopically-visible cytoplasmic foci, referred to as UBP1 stress granules (SGs). This UBP1–mRNA association leads to a global decrease in the protein synthesis. The translation limitation for the transcripts associated into SGs reduces the energy spending, thereby determining the priority in synthesis of the proteins that enhance plant survival in stress. The UBP1 SGs rapidly disaggregate during reoxygenation, which coincides with the mRNA return to polysomes. In this process, the mRNAs that are significantly induced and translated in hypoxia to a considerable degree manage to avoid UBP1 sequestration. Thus, it has been shown that the SG-nucleating RNA-binding UBP1 is a component of the mechanism that post-translationally reprograms plant gene expression, thereby enhancing plant survival in hypoxia [17].

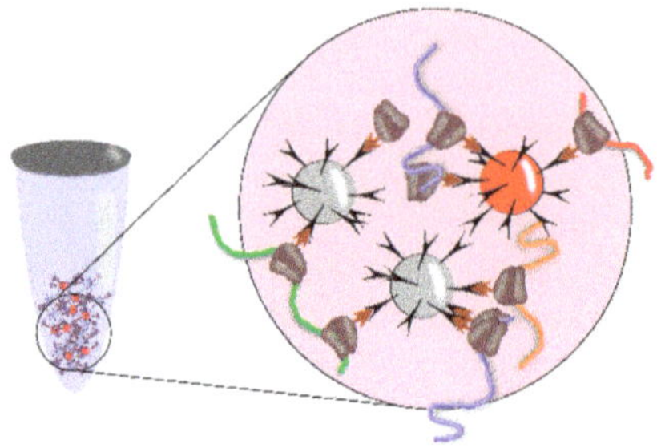

Figure 2. Polysome profiling using translating ribosome affinity purification (TRAP). General principle of selective separation on anti-FLAG-M2 agarose of the transcripts bound to ribosomes carrying the epitope-tagged variant of ribosomal protein. Brown arrows denote the FLAG epitope in ribosomal protein and black icons denote the anti-FLAG on agarose beads.

2.3. Ribosome Profiling, or Ribo-Seq

Ribosome Profiling (RP), or Ribo-Seq, elaborated by Ingolia, Newman, and Weissman in 2009 [18], is based on the isolation and sequencing of the mRNA fragments protected by ribosome. This gives a "snapshot" of the ribosome positions along mRNA on a genome-wide scale, i.e., gives the possibility to determine both the number and positions of the ribosomes in the mRNA coding region in vivo (Figure 3).

As a rule, many studies use the RP experimental protocol, which comprises five interrelated stages: (i) preparation of RNA specimens with the arrested ribosomes; (ii) controlled hydrolysis of these specimens by RNase to generate small RNA fragments associated with a ribosome (referred to as footprints); (iii) their subsequent isolation; (iv) preparation of purified footprints with a size of 28–30 nucleotides; and (v) construction of the library and its high-throughput sequencing, as a rule, with the help of short-read sequencers. The deep sequencing reads of the footprints are analyzed using bioinformatics methods and the translation efficiency is derived by normalizing the number of reads of the footprints to the number of reads of the total transcriptome by RNA-Seq.

As is mentioned above, the first experimental protocol for ribosome profiling was described in 2009 [18] and has been constantly developed, in particular, for its application to different organisms [18,19], including plants [20,21] and plant organelles—chloroplasts [20] and mitochondria [22]. The individual protocols differ in the particular details providing optimization of each of the five interrelated stages, including the differences in tissue and cell processing; pH and composition of the buffer for cell lysis; prepurification of polysomes before RNase hydrolysis (done or omitted); type of RNase used for generating monosomes [23]; and the methods used to purify the monosome fractions and construct sequencing libraries. Ribosome affinity purification (TRAP method), including the tissue-specific purification, can be also used as the starting point for ribosome profiling [20].

In general, the RP results allow for determination of the precise positions of the translating ribosomes on mRNA with an unprecedented resolution, to a single nucleotide. The specialized software for analysis, interpretation, and visualization of RP data is currently available (for detailed review, see [24]). By assessing the relative number and location of ribosomes on mRNA, the researcher can estimate the general translation pattern i.e., to assess the translation efficiency, which is calculated as the ratio of translation (the data on the number of footprint reads in individual mRNA) to transcription (RNA-Seq data at the level of individual mRNA) (Figure 4). Note that it is possible not only to directly quantify the mRNAs that will be translated into proteins, but also to detect the new types of contexts in the plant mRNAs associated with translation, for example, uORFs (upstream ORFs) and frameshifts; to precisely determine the translation initiation site (TIS) of the main ORF; and to find new translated ORFs, including those residing in intergenic RNAs or putative noncoding short RNAs (ncRNAs) (Figure 4) [3,24,25]. The researcher gets these additional options thanks to the fact that the 80S ribosomes associate only with the portion of the transcript that will be most likely decoded into the protein product. The 80S ribosome and transcript will associate not only in CDS, but also in 5′UTRs if they contain an uORF, i.e., short translated reading frame located upstream of the main ORF (CDS), which may have an important role in translation regulation. Another most important aspect that can be studied in terms of the RP experimental data is assessment of the dynamics of ribosome movement along individual mRNAs and the rate at which certain codons are translated. This is possible because three nucleotide bases in the sequenced footprints are reflected in a periodic mode as a consequence of the ribosome movement along the mRNA coding region, since the ribosome moves along the overall coding sequence in a codon-wise manner, the 5′ region of ribosome footprints tend to be mapped at the same position of each codon.

Find below several examples which in our view illustrate the distinctive capabilities of RP in clarification of the fine mechanisms underlying the translational control in plants, such as the detection of new ORFs, including those annotated as noncoding RNAs and pseudogenes. In particular, the study of translation regulation under normoxic and sublethal hypoxic conditions (hypoxia) in *A. thaliana* shoots with the help of RP not only detected an inhibitory effect of the uORF on the translation of downstream protein coding regions in normoxia, which was further modulated by hypoxia, but also determined the alternatively spliced mRNAs as well as the fact that ribosomes were associated with certain noncoding RNAs [21]. An RP study of the maize shoots under drought showed a statistically significant change in the translation efficiency of 931 genes, which according to further analysis of the transcripts was associated with the nucleotide composition of the sequence, including GC content, length of coding sequences, and normalized minimum free energy. In addition, potential translation of 3036 open reading frames (uORFs) in 2558 genes was detected; the authors believe that these uORFs are able to influence the translation efficiency of the downstream main open reading frames (ORFs) [26]. In another study, the Ribo-Seq data detected 27 and 37 translated sORFs (short ORFs) among the annotated noncoding ncRNAs and pseudogenes of *A. thaliana*, respectively [27]. Moreover, 187 translated uORFs were identified with a high degree of reliability. In addition, the events of translation from the start codons other than AUG were identified in the dataset among both annotated genes and uORFs. They also demonstrated that 15 of the 19 detected single-exon sORFs had homologs in various flowering plants, which suggests their functional significance [27].

Lukoszek et al. [28] used RNA-Seq and Ribo-Seq to assess reprogramming of the *A. thaliana* global gene expression during a long-term heat shock (3 h at 37 °C) at both the transcriptional and translational levels. They have shown that translation is globally impaired in the early period of the heat impact (15 to 45 min), while the stress response appears mainly at the expense of transcriptional programs. In this process, a long-term stress impact (3 h) activated translational programs, which eventually form the adaptive response. The transcripts regulated via translation display a number of common characteristics, namely, the presence of relatively conserved A/G-rich motifs in their 5′UTRs or 3′UTRs that are similar to the sequences identified as protein-binding nucleotide motifs. Another specific feature widespread among the genes upregulated in heat stress is that they are less inclined to form secondary structures, which is likely to ensure their binding with ribosomes and to enhance translation. In addition, several transcripts prevalently induced by heat contain a putative G2 quadruplex in their 5′UTRs. Note that an increased number of reads for RP footprints in quadruplex structures correlates with an expanded expression of the downstream CDSs. This suggests an important role of these structures in translation activation of the downstream ORF according to yet unknown mechanism [28]. Ribosome profiling has been used to analyze translation of the chloroplast transcripts in maize shoots in response to changes in light conditions. According to the experimental data, all chloroplast mRNAs except for psbA maintain similar numbers of ribosomes after short-term changes in light conditions but nonetheless are more efficiently translated in the light. On the other hand, the psbA mRNA displays a sharp increase in the ribosomes over several minutes after the plants are transferred to light and restores a low ribosome loading during 1 h in the dark, which correlates with the need to replace the damaged psbA in photosystem II. These results emphasize the unique translational response of psbA in mature chloroplasts, indicate the particular light-regulated steps in the context of photosystem II activity maintenance, and provide the background for the study into the mechanisms underlying both the psbA-specific and genome-wide effects of the light on the translation in chloroplasts [29].

The RP technology was also used to study several aspects in the translation of *A. thaliana* mitochondrial genome in a dynamic mode. As has been shown, the mitochondrial mRNAs are differentially-translated; in this process, the translational levels of the transcripts encoding the subunits of mitochondrial protein complexes, in particular, complex V, proportionally correlate with the stoichiometry of respiratory chain subunits. In general, the mitochondrial translation is shown to be controlled at the level of individual mRNAs and is directly involved in the activity regulation of plant mitochondria [22].

Note that Ribo-Seq technology is currently at a relatively early stage of its development, which leads to some experimental difficulties and technical artifacts influencing the Ribo-Seq data interpretation [18]. In particular, the RP results may display statistically significant differences associated with the modifications of one of the five stages in the basic protocol, such as the conditions of cell lysis, composition of buffer solutions, selection of nucleases and the absence of pronounced specificity to the sequences to be cleaved, and construction of the library; even more so as these details in many cases are not analyzed in a systematic manner [19,23]. This suggests the need to systematically study the effects of the corresponding experimental parameters of the used RP protocols [19].

The RP technique also has its limitations. According to the current scientific consensus, the basic limitation of RP approach is a static position of ribosomes along the mRNA. This prevents distinguishing between the ribosomes involved in translation from the ribosome in a steady-state [19]. Thus, the methods used in the majority of studies involving RP can overestimate the translation efficiency because of the data related to monosomes, in which mRNA is also protected by a ribosome (Figure 5A) [18,23,26]. Underrepresentation of the transcript regions with ribosome stacking is also possible; this is associated with the stacked polysomes and may prevent hydrolysis in monosomes, because of inaccessibility to RNases (Figure 5B) [4,30]. Correspondingly, the recommendation for an additional direct measurement of the polysome-protected mRNA looks most reasonable to overcome this limitation of the RP approach [26].

Note that the Ribo-Seq technique now is mainly applicable to study translation of the organisms with annotated genomes since the deep sequencing data for footprints are represented by very short reads (28–30 nucleotides), which, as a rule, are analyzed by mapping onto the genome data (Figure 4).

However, the current limitations of this method can be bypassed and this experimental technology will remain a useful tool in the omics [30]. RP data with high resolution is a priceless resource for studying noncanonical start codons and alternative start sites and can be useful for characterizing translation of different isoforms of transcripts, identifying new translated ORFs and their quantifying, and, in general, for improving the genome annotation of poorer characterized organisms. Additional ribosome profiling can also be a proxy for the proteome or assist in proteomics studies [27,30].

Completing this section, note that the genome-wide profiling of the transcripts associated with ribosomes utilizing one of the above experimental approaches may highlight the new aspects in gene expression unvisualizable by an ordinary profiling of the total cellular mRNA. In Table 1, we attempted to consolidate the advantages, limitations, and areas of applicability of the discussed experimental approaches to the study of differential translation on a genome-wide scale. Undoubtedly, selection of the appropriate experimental approach depends on the particular aims set by a researcher.

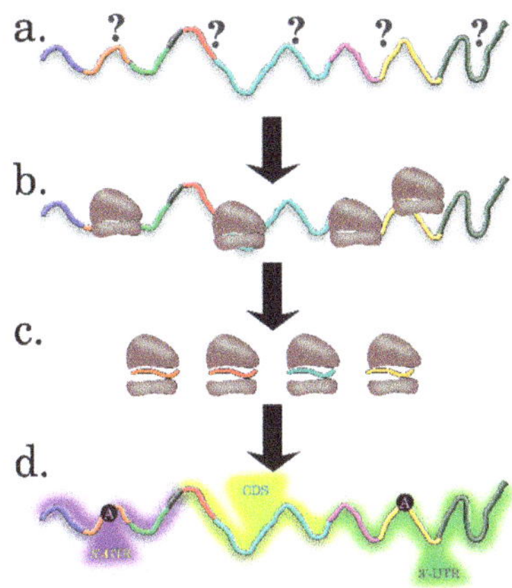

Figure 3. Scheme of application of ribosome profiling to functional characterization of mRNA regions. (**a**) Scheme of an mRNA with unknown ribosome positions. (**b**) The mRNA with arrested ribosomes in the transcript regions potentially important for efficient translation. (**c**) Formation of the ribosome footprints by RNase hydrolysis. The resulting footprints characterize the translational functionality of a certain mRNA region. The footprints shown with different colors correspond to different mRNA regions. (**d**) Result of analysis of the precise positions of translating ribosomes along mRNA, where A is the identified alternative open reading frames in 5′UTRs or 3′UTRs and CDS (coding sequence) is the main reading frame of the transcript.

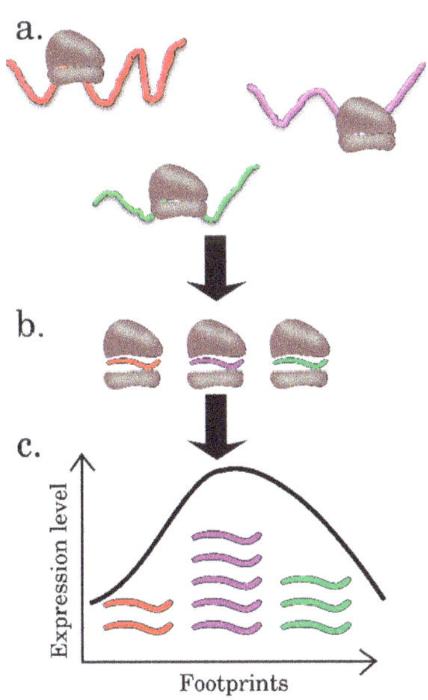

Figure 4. Principle of analysis, interpretation, and visualization of the ribosome profiling data. (**a**) The ribosomes arrested on transcripts (**b**) form ribosome footprints after RNase hydrolysis. (**c**) The footprints mapped onto genome can be associated with particular sequences to assess the relative amount and positions of ribosomes on the transcripts on a genome-wide scale.

Table 1. Comparative characterization of the experimental approaches producing pools of differentially-translated mRNAs.

Experimental Approach	Basic Protocol	Advantages	Limitations	References
Polysome Profiling	Separation of transcripts with different ribosome loading by ultracentrifugation; supplemented by sequencing of different mRNA fractions, including transcriptome-wide analysis	Simplicity and possibility to analyze the plant species with annotated and unannotated genomes	Does not assess the number and location of ribosomes on each transcript	[4,5,21]
Translating Ribosome Affinity Purification	Separation of the transcripts with different ribosome loadings by absorption on anti-FLAG-M2 agarose; supplemented by mRNA sequencing, including transcriptome-wide analysis	Profiling of actively-translated RNAs from different plant tissues and particular cell types; Identification of only the mRNAs bound to ribosomes, which makes it possible to avoid the potential confusion with the transcripts associated with the other RNA-binding proteins	Cannot estimate the number and location of ribosomes on each transcript; Requires production and accurate selection of the plant transgenic lines that express an epitope-tagged variant of ribosomal protein L18	[3,12,14,15]
Ribosome Profiling	Isolation and sequencing of the ribosome-protected mRNA fragments; modification of the protocol is necessary for individual species	Identification of the ribosome number and location on each transcript; Detection of new translated ORFs and noncanonical translation start sites	Requires significant material, time and labor investments; A considerable amount of biological material is necessary; The transcript regions with stacked ribosomes may be underrepresented; Incorrect trace identification may result from the RNA interaction with RNA-binding proteins of an analogous size; Applicable only to the plants with a well-annotated genome	[4,18–20,24]

Note: The key advantages and limitations are shown for each approach; see the text for a comprehensive description.

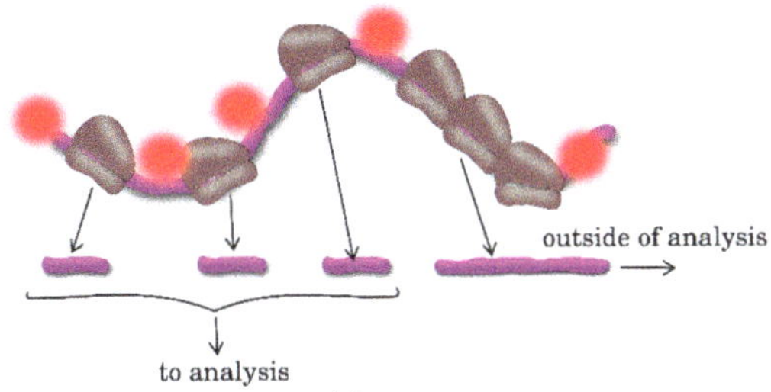

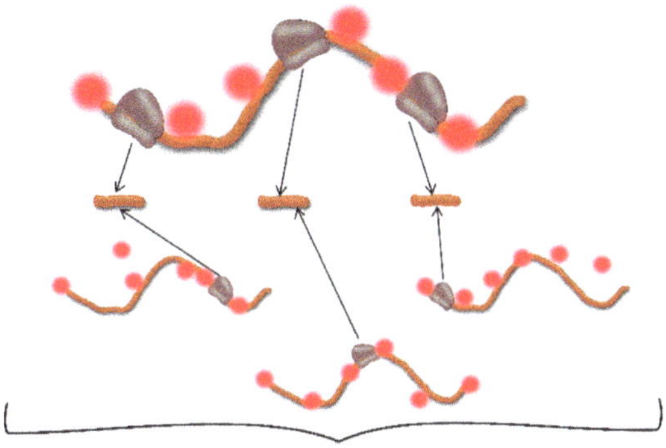

Figure 5. Limitations in the use of ribosome profiling: (**a**) overestimation of translation efficiency because of the footprints of monosomes, where mRNA is also protected by ribosome and (**b**) underrepresentation of the transcript region with stacked ribosomes; carefully stacked polysomes are inaccessible to RNases, thus cannot be digested into ribosome footprints of the tested size (28–30 nucleotides). Red spots denote the region attacked by RNase.

3. Computational Algorithms for Predicting the Features of Plant mRNAs Important for Differential Translation

The above described experimental approaches make it possible to detect the specific pools of transcripts with characteristic differential translation. Several computational resources are useful for identification of regions of specific structural features in mRNA nucleotide composition that can mediate differential translational control.

In this section, we summarized the resources and some computational algorithms that have been used to form the samples of target plant transcript sequences and to predict their peculiar characteristics, as well as their main functions and domains of application. Note that the resources and the corresponding software are rather numerous and, in fact, require a separate review. Here, we consider only those that have given the data on and predictions of regulatory contexts in transcripts with further experimental confirmation.

3.1. Preparing Datasets for Analysis

The key preparatory stage in the in silico predictions is a construction of the most representative sets of sequences for the transcript pools differing in their translation efficiency. Note that the researcher needs not only full-sized transcript sequences (cDNA), but also the sequences of individual regions of these transcripts, namely, coding (CDS) and untranslated (5′UTR and 3′UTR) regions, which, as mentioned above, can also contribute to translation efficiency. Currently, many internet resources have been elaborated that allow sets of such sequences to be downloaded, including the sequences for plants. In particular, TAIR is the information source for the model plant *A. thaliana* (https://www.arabidopsis.org/download/index-auto.jsp?dir=%2Fdownload_files%2FSequences%2FTAIR10_blastsets) [31], which is widely used for loading 5′UTR, CDS, 3′UTR, and cDNA sequences using the tools "Download", "Sequences", and TAIR10 blastsets [32–34]. Another information resource containing CDS, cDNA, 5′UTR, and 3′UTR sequences of the representatives of six key kingdoms of the living organisms, including plants, is JetGene. JetGene is publicly available at http://jetgene.bioset.org/; its data are stored and updated at the Ensembl server [35]. The intuitively clear and friendly JetGene interface allows the cDNA, 5′UTR, CDS, and 3′UTR sequences to be extracted in a FASTA format, including the specific samples on user request. Note that only the sequences with complete information about the full-sized transcripts are in most cases selected for further analysis.

Once the sets of sequences (5′UTR, CDS, and 3′UTR) for the pools of differentially-translated transcripts are obtained, the researcher has to select for analysis the regions of transcripts and regulatory sequences that may be potentially involved in translation modulation. According to the current opinion, the complex multilevel information is encoded in the full-sized mRNA sequence (transcript) in general and in its individual parts—5′UTR, CDS, and 3′UTR (Figure 6). This gives researchers the grounds to include all these regions into in silico analysis to characterize the differentially-translated plant transcripts. Note that translation initiation is, as a rule, the stage limiting the translation rate and 5′UTR plays here the decisive role. The length, nucleotide composition, secondary structures, and regulatory elements of a smaller size, such as upstream start codons (uAUGs), uORFs, nucleotide motifs, and several other features in the 5′UTRs of transcripts, are closely examined in terms of their contributions to the translation efficiency. In this process, the probability to find the potential regulatory regions and contexts and to clarify how their properties influence the translation efficiency will be higher if more traits of this kind are involved in the initial in silico analysis.

The further aims of the researcher could be (i) to assess the variations in distribution of individual traits in the sequences from the examined transcript pools and to figure out the statistically significant differences that are positively correlated with the translation efficiency; (ii) to find and determine the statistically significant representation of the potential regulatory contexts in the transcripts with different translation efficiencies; and (iii) to identify the specific regulatory sequences if they are present in the examined pools.

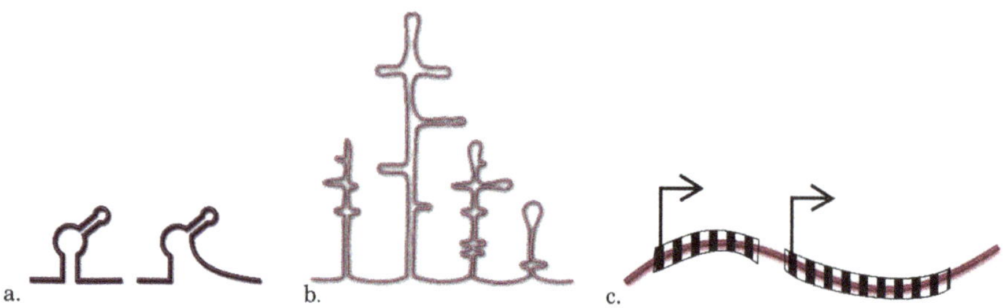

Figure 6. Examples of some mRNA cis-regulatory elements: (**a**) riboswitches; (**b**) internal ribosome entry sites (IRESs); and (**c**) alternative open reading frames.

3.2. Statistical Methods

The methods of mathematical statistics have been rather efficiently used for solution of the first task. As a rule, basic and extended statistical analyses are performable with the help of the available standard programs, such as Excel, STATA, and IBM SPSS Statistics 20 [26,32–34]. For example, the genome-wide monitoring of the changes in the translation efficiency of individual mRNAs in *A. thaliana* shoots after heat shock have demonstrated translation repression for the majority of mRNAs; however, some mRNAs still followed the differential translation pattern. Analysis of the differentially-translated mRNA sequences demonstrated that only some characteristics, such as the

G + C content in 5′UTR and cDNA length, are putatively involved in the mechanisms providing discrimination of the mRNA loading with ribosomes and are associated with differential translation of a certain transcript cohort in response to a high temperature. In particular, the translationally active mRNAs have a low G + C content (on the average, 36%) versus the transcripts with repressed translation (42%). This selection mechanism also influences the differential polysomal loading of the transcripts associated with stress and, as a consequence, the efficiency of their translation [32].

In general, the methods of mathematical statistics have made it possible to (i) find the characteristics that are representative for the analyzed mRNA, (ii) discard the characteristics the effect of which can result from a bias to the group of particular genes, and (iii) determine the statistically significant differences displaying a positive correlation with the relative translation efficiency.

3.3. Methods for Identification of RNA Motifs

The methods that are used to identify potential regulatory motifs in mRNA sequences assess the statistical significance of represented potential regulatory motifs in the mRNAs with different translation efficiencies. These methods are in general the same or very similar to those for DNA motifs except for the methods addressing the DNA conformational properties. The latter utilize physical parameters of DNA double helix and can be plied both to prokaryotic [36] and eukaryotic [37] genomes. Having appropriate parameters to convert letter representation of RNA into numerical representation, the same methodology could be applied to analysis of mRNA. Conventional approaches are based on accounting for conserved nucleotides within a short motif. One of the most frequently used programs for the detection of motifs in the transcript pools with different translation efficiencies is MEME, which is based on the maximal likelihood optimization [38]. Ease of use and a wide set of the accompanying programs for visualization and further search are advantages of this program. The MEME suite comprises four main sections, namely, motif discovery, motif enrichment analysis, motif search, and motif comparison, altogether 14 different tools. This toolkit allows the researcher to both determine motifs de novo and to scan a dataset of sequences for the matches of the already known motifs. MEME shows a schematic arrangement of the found motifs on the initial sequence, constructs a graphical representation for them, and computes statistical significance (p-value) for these motifs.

In particular, MEME suite has allowed identification of a nine-nucleotide-long element present in both the 5′UTRs and 3′UTRs of numerous *A. thaliana* and *Gynandropsis gynandra* transcripts; the authors named it MEM2. Later, it was experimentally confirmed that the MEM2 motif residing in the 5′UTR was necessary for preferential protein accumulation in the mesophyll cells. It is assumed that this motif can be involved in the mechanism guiding preferential cellular accumulation of several enzymes necessary for C4 photosynthesis, which provides a more efficient carbon capture as compared with the ancestral C3 pathway [39]. The MEME suite has been used in a comparative analysis of the 5′UTR sequences for steady-state and polysomal *A. thaliana* mRNAs and allowed for discovery of two motifs (TAGGGTTT and AAAACCCT) present in many genes, which potentially suggests their contribution to the translation efficiency. Furthermore, it has been experimentally shown that only one of these motifs, TAGGGTTT, regulates gene expression at the level of translation [33].

However, the search for the motifs using this tool also has some limitations. Among the serious disadvantages of this program is the trend to find very long motifs (over 20 nucleotides), these motifs are present only on a small subgroup of sequences and/or frequently repeated motifs in one or just a few sequences. Although statistical significance (p-value) of such motifs is very high, the motifs themselves, as a rule, are rarely of any biological/practical interest and represent statistical artifacts.

Most likely, these limitations are the main reason why several studies of motifs failed to bring any positive results [8,32,33]. Correspondingly, other computational approaches were used for this search and their statistically significant representation in the transcripts with different translation efficiency, for example, by comparing the frequencies of mono-, di-, and trinucleotide sequences. Statistical tests, for example, the Kolmogorov–Smirnov or Fisher test, allow the detection of statistically significant differences in such nucleotide distributions. Moreover, the use of linear or logistic regression makes it possible to detect not only the individual contribution of each sequence, but also the effect of their combinations. In particular, partial least regression analysis has been applied to the detection of the short regions residing in the neighborhood of the 5′-proximal region of 5′UTR that can play an important role in differential translation in response to heat shock [8]. However, the linear or logistic regression methods are also not free from limitations. For example, it is not practically feasible to analyze motifs with a length of four nucleotides or longer, because their frequencies sharply decrease and, as a consequence, the computation of statistical characteristics becomes too complicated. In addition, these methods do not take into account the locations of motifs on sequences, which in terms of biology mean the equal contributions of the codons residing far from the translation start codon and in the immediate proximity.

3.4. Detection of Other Context Features of RNA

Statistical approaches are rather efficient when a set of potential characteristic features is determined for a pool of sequences and is used as a reference in the analysis. However, the specific cis-regulatory sequences in mRNAs that can modulate translation are identified using specialized approaches and/or resources for their prediction. The examples below illustrate the approaches to predict cis-regulatory sequences in the case studies of internal ribosome entry sites (IRESs) and upstream ORFs (uORFs), first and foremost, conserved peptide uORFs (CPuORFs).

IRES are the nucleotide sequences that mediate translation initiation of alternative reading frames (aORFs) under stress conditions, when the trivial cap-dependent translation mechanism is inhibited without the corresponding changes in gene transcription [40]. In general, the IRESs of plant mRNAs, unlike the IRESs of viruses, display considerable diversity in both nucleotide composition and structure [41]. Despite this diversity, characteristic functional modules are distinguishable in the IRESs, namely, (i) the presence of several start codons and their localization and (ii) the fact that some IRESs carry short conserved modules, which are recognized by the plant translational machinery and are directly involved in the immobilization of ribosome small subunit [42]. Polypurine blocks residing close to the start codon, which may be directly involved in the immobilization of ribosome small subunit, are an example of such conserved motifs [43].

The mRNAs potentially carrying IRESs can be selected by analyzing the experimental data obtained by polysome or ribosome profiling followed by deep sequencing and/or by mass spectrometry analysis. First and foremost, such mRNAs must retain a high level of their translational activity under the impact of adverse environmental factors and carry additional alternative start codons. The following strategy is appropriate for further selection and analysis of the mRNAs carrying IRESs. (i) Interspecific comparison of the transcript sequences of homologous genes, which allows for identification of the conserved region in the vicinity (30–50 nucleotides) of the alternative start codon followed by (ii) assessment of the context of the alternative start codon, the optimal neighborhood of which may suggest that translation can be potentially started from it. This strategy has been successfully implemented for predicting translation initiation of a short aORF with involvement of a polypurine block via internal ribosome entry [43]. Note that a commonly accepted confirmation for an IRES activity is still its ability to provide a coordinated translation of reporter genes within a bicistronic transcript (see below).

The advent of RP and high-throughput sequencing technologies made it possible to determine the translation start sites and to discover numerous mRNAs with aORFs that (i) may have a putatively inert sequence that acts as a mere translation barrier upstream of the main ORF or (ii) may encode short peptides referred to as CPuORFs [44]. The main difference between CPuORFs and the other ORFs is their length and, although there are no strictly defined frames, the ORFs shorter than 200–250 codons are regarded as short. In general, the search for CPuORFs is analogous to the approach for prediction of main ORFs and the strategy utilizing interspecific comparison of CPuORF sequences to identify the conserved regions is in most cases used for this search and estimation of the coding potential. This strategy is based on revealing the homology between such short peptide sequences and to a considerable degree depends on the range of the species selected for comparison. For example, it is quite possible that the fact of preservation of CPuORFs within the plant species selected for comparison is insufficient to reveal the conserved regions. In this case, the analyzed CPuORFs will not be identified although their sequence is sufficiently conserved among the other species. When comparing the CPuORF sequences among closely related species, the observed similarity between short uORF peptide sequences may result from nucleotide sequence retention rather than conservation of these peptides. In order to overcome the problems associated with selection of the species for comparative analysis, a new method for CPuORF identification, BAIUCAS, was developed and tested [45]. BAIUCAS utilizes sets of EST (expressed sequence tag) data for thousands of plant species to search for homologs. The BAIUCAS algorithm consists of six successive stages: (i) exhaustive search for uORFs; (ii) search for homologs of CPuORF amino acid sequences over EST databases using tBLASTn; (iii) selection of CPuORFs based on the conservation of the stop codon position; (iv) selection of the CPuORFs conserved in a wide range of species, i.e., the CPuORFs with the conserved stop codons detected in each of several taxonomic categories; and (v) and (vi) are filtration stages, which excludes the "false" conserved CPuORFs [45]. Using this approach to the search for CPuORFs, 16 *A. thaliana* CPuORFs were identified; five of them are the new CPuORFs involved in the translation regulation of the main ORF, which has been experimentally confirmed [46].

The list of the computational approaches that have been so far successfully used in decoding the specific structural features in nucleotide composition of the plant mRNAs that mediate differential translation control is rather short. One of the possible efficient computational tools for analyzing the translatomes and predicting numerous regulatory codes in transcript sequences could be artificial neural networks (ANNs). This assumption relies on the facts that (i) most neural network architectures is theoretically able to approximate any function, i.e., it is potentially possible using ANNs to construct a model for almost any biological pattern, and (ii) the capabilities of the supercomputers have reached the appropriate level to model biological processes using neural networks. However, a positive result of analysis depends first on an adequately selected architecture of the network and second on the amount and composition of the training sample and a training strategy. Several recent reports confirm the utility of ANNs in deciphering the molecular mechanisms involved in decoding the eukaryotic genome. For example, the ANNs constructed based on RP data have been used to predict the yield of protein products [47]; to search for the motifs potentially able to influence translation [48]; to extract the biologically important information from omics data [49]; and to simulate the interaction between nucleic acids and different types of ligands (protein and peptides) [50]. The ANN potential is broad enough and it can be expected its broad application to the research in diverse and multilevel mechanism underlying translation in plants.

4. Approaches for Experimental Verification of the Systemic Experimental Data and Theoretical Predictions of Intrinsic Features of Plant mRNAs Important for Differential Translation

As a rule, the experimental results on determination of the pools of differentially-translated plant mRNAs as well as the predictions of the regions and nucleotide contexts in differentially-translated transcripts require experimental confirmation of their functionality and contribution to the translation efficiency. In this section, we will consider the main experimental approaches that allow for convincing confirmation of which particular regions in individual mRNAs and/or features within plant transcripts are important for modulation of translation. It should be emphasized that methods for high-throughput experimental verification, i.e., concurrent analysis of a large pool of regulatory regions, are yet not available; thus, potential regulatory sequences are examined for each individual transcript [4].

4.1. Direct Measurments of Individual Transcripts

Several tools for determining the changes of individual transcripts at the level of translation are available, including (i) the systems for in vitro translation based on measuring the incorporation of labeled amino acids, such as FUNCAT (fluorescent noncanonical amino acid tagging), SILAC (stable isotope labeling by amino acids in cell culture), BONCAT (bioorthogonal noncanonical amino acid tagging), QuaNCAT (quantitative noncanonical amino acid tagging), and PUNCH-P (puromycin-associated nascent chain proteomics), or cell-free protein expression systems, such as the wheat-germ extract, which contain all factors necessary for translation of the target transcript; (ii) toeprinting, or the primer extension inhibition assay, utilizing reverse transcription to study the interaction of ribosomes with the target transcript; (iii) enzyme immunoassays, in particular, Western blot hybridization; and (iv) mass spectrometry-based methods for identifying the changes in the proteome or their combinations with in vitro translation methods, for example, PUNCH-P. The review by Mazzoni-Putman et al. [4] describes the principles, advantages, and limitations of these methods in detail.

However, these research methods are in most cases suitable for assessing the general changes in translation, require considerable time, amount of reagents, and specialized equipment; correspondingly, most research works now use the strategy of reporter systems for studying the structure–function characteristics of the target sequences. The strategy of reporter genes considerably enhances such research since it is much easier to record the protein product of a reporter gene as compared with a studied gene. It should be also emphasized that the reporter genes code for the proteins that display either unique specific features or unique enzyme activities, allowing their products to be easily isolated from the totality of intracellular and extracellular proteins. Thanks to these advantages of reporter systems over the other methods for studying the regulation of gene expression, they have been widely used for experimental verification of the regions and nucleotide contexts in differentially-translated transcripts. For studies of this type, expression cassettes are constructed that carry the reporter gene sequence with the expression controlled by a particular regulatory region or sequence selected by researcher (Figure 7). Researchers have at their disposal several reporter systems that have proved their efficiency in the studies of potential regulatory sequences or the nucleotide contexts that modulate translation in plant systems, in particular, β-glucuronidase (GUS); different variants of fluorescent proteins (for example, GFP and RFP); luciferases (Renilla luciferase, RLuc, and firefly luciferase, FLuc); and thermostable lichenase (LicBM) [51,52]. Commercial substrates and kits as well as the quantification methods for assessing the corresponding protein products are available for these reporter systems. The main approaches that have been applied in the studies of various regulatory regions or nucleotide contexts within transcripts with the use of reporter systems are illustrated below.

When experimentally confirming the role of the full-sized 5′UTRs in transcripts, these sequences are cloned upstream of the 5′ region of a reporter gene (Figure 7). Quantitative estimate of the reporter gene protein product when using different target 5′UTRs versus the known translational enhancers of various origins makes it possible to assess their contribution to the translation efficiency [51]. For example, *A. thaliana* mRNAs that are stably translated under any growth and environmental conditions have been found by polysome profiling. Testing of the translation capability of mRNA 5′UTRs of candidate genes using the reporter gene strategy has convincingly demonstrated that the 5′UTR of 47 cold-regulated genes are an efficient translational enhancers, which enables a stable high level translation under any conditions of plant growth. This suggests the utility of this method for plant biotechnology, for example, when engineering plants producing biologically active substances or the plants resistant to some stress factors, including the schemes that involve genome editing technologies [9].

Recombinant 5′UTR sequences are designed for identification of the cis-regulatory elements in these mRNA regions, first and foremost, the motifs or specific nucleotide contexts potentially able to influence differential translation, using for this purpose a (i) combinatorial approach, (ii) site-specific mutagenesis, (iii) translation assessment of the second (3′-terminal) ORF of a bicistronic transcript, (iv) deletion analysis, (v) frameshift mutations, or a combination of these methods (Figure 7).

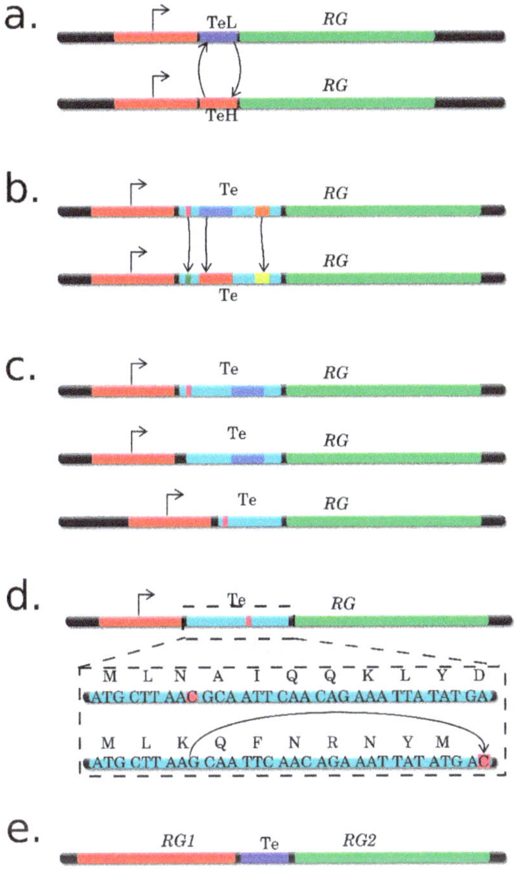

Figure 7. Approaches to experimental verification of the tested regulatory elements using the strategy of reporter systems. (**a**) Combinatorial approach. Mutual substitutions of the predicted regulatory motifs TeL and TeH (denoted with arrows) are introduced into the pairs of 5′UTRs of the same length but considerably differing in the experimentally confirmed translation efficiencies. TeL and TeH are the tested elements characteristic of the transcripts with low and high translation efficiencies, respectively. (**b**) Site-specific mutagenesis. The native regulatory sequence is above and the mutant regulatory sequence is below; different colors denote the region used for mutagenesis; direct arrows indicate the substituted regions in two sequences. (**c**) Deletion analysis. The native regulatory sequence is above; different colors denote the regions used for deletions: the regulatory region with deletion in the 5′ region is in the middle (deletion of the pink region of the native sequence) and the regulatory region with deletion in the 3′ region is below (deletion of the blue region of the native sequence). (**d**) Frameshift analysis. The region for introducing frameshift is dashed; the native nucleotide and amino acid sequences are above and the mutant, below. The nucleotides colored red were frameshifted. Simultaneous introduction of deletions and insertions to positions −1 and +1 changes the amino acid composition of peptide sequence encoded by the alternative open reading frame preserving the presence of the overlapping peptide and its length. (**e**) Bicistronic construct for studying the functionality of IRESs. Two reporter genes are translated from the bicistronic construct in a coordinated manner; translation of one of them (RG2) is controlled by the tested element (Te) and the other (RG1) is translated according to the classical cap-dependent mechanism. In all panels, bent arrows denote the transcription start point; Te, tested element; and RG, reporter gene.

4.2. Site-Specific Substitutions

The combinatorial approach utilizes the fact that mutual substitutions of the predicted regulatory motifs are introduced into the pairs of 5′UTRs with the same lengths but with experimentally confirmed significant difference in translation efficiency (Figure 7a). Matsuura et al. [8] successfully used this approach to identify new cis-regulatory elements in 5′UTRs that influence the differential translation of *A. thaliana* transcripts in response to heat shock (HS). The genome-wide analysis of the changes in polysome loading of the transcripts in Arabidopsis cell culture allowed for selection of a set of genes with different translational responses to HS. The 5′UTR nucleotide sequences of the transcripts that change the level of reporter protein in the protoplasts affected by HS were used to predict the regulatory elements in 5′UTRs with the help of partial least square (PLS) method. These computational predictions suggested that two short regions residing in the vicinity of 5′-proximal region of the 5′UTR can play an important role in the relative activity of reporter protein and, thus, may be regarded as cis-regulatory region candidates. In order to experimentally confirm the predictions on the importance of these 5′UTR regions in differential translation control, a series of mutual substitutions of these regions in the pairs of 5′UTRs with equal length but different translation efficiencies were analyzed. Analysis of the reporter gives convincing evidence that the 5′-proximal region of the 5′UTR plays a key role and that certain specific determinants in 5′UTR mediate the differential translation in response to HS [8].

Site-specific mutagenesis makes it possible to introduce substitutions of one or a group of nucleotides within strictly defined regions of nucleic acid sequences. Comparison of the levels of a reporter protein translated when controlled by a native regulatory region (for example, 5′UTR) and the same regulatory regions but with the introduced mutations (Figure 7b) demonstrates how the modification of the primary sequences of nucleotide contexts and/or their secondary structures modulates the translation of specific mRNAs. This approach, along with other molecular methods (Western blotting, qPCR, polysome fractionation, and so on), has emerged to be most efficient when studying both the mechanism underlying formation of RNA G-quadruplex in the 5′UTR of the SUPPRESSOR OF MAX2 1-LIKE4/5 (SMXL4/5) mRNAs and the clarification of the role of a specialized structure, the regulator of phloem formation. In particular, a novel zinc finger protein, JUL, was identified; it specifically bound to consecutive guanine repeats in the 5′UTR of SMXL4/5 and induced RNA G-quadruplex. Moreover, convincing experimental data that both JUL1 and G-quadruplex are necessary for strong translation suppression rather than a single-stranded G-rich element have been obtained using the strategy of reporter systems. This suggests that the suppression of translation is caused by either JUL1-mediated formation of G-quadruplex or the G-quadruplex/JUL1 complex recruits an unknown translational suppressor [53].

Site-specific mutagenesis has emerged to be efficient for clarifying the role of the TAGGGTTT motif, overrepresented in the 5′UTRs of the transcripts regulated at the level of translation. A comparative study of two constructs, one with a native 5′UTR carrying this motif and the other with the 5′UTRs carrying mutations in this motif, has shown that the transcripts with the native 5′UTR are more efficiently translated provided that the number of transcript are equal. Thus, it is experimentally proved using reporter genes that the TAGGGTTT cis-element regulates expression of the gene particularly at the level of translation [33].

4.3. Analysis Using Deletions

Deletion analysis of 5′UTRs, implying construction of truncated variants of sequences and their subsequent fusion with a reporter gene, makes it possible to identify nucleotide contexts decisive for maintaining the structure of RNA that can be in two particular conformations, as for example, in riboswitches (Figure 7c). Note that an important specific feature of the riboswitches is their ability to both activate and inhibit translation from the controlled ORF, due to the presence of a specific regulatory region, the aptamer. This region with a particular secondary and sometimes tertiary structure, has the properties of a receptor for small molecules (ligands). In the overwhelming majority of cases, riboswitches reside in 5′UTRs. A deletion analysis of the *A. thaliana* phytoene synthase (PSY) 5′UTR has shown that the long 5′UTR variants (403 and 330 nucleotides) with two predicted sequences of convertible RNA conformations, similar to riboswitches, inhibit translation of the reporter gene, in contrast to the short variant (252 nucleotides) of the PSY 5′UTR, lacking such hairpin structure. This allows the short 5′UTR variant to pass the translational control and rapidly elevate the protein levels [54].

4.4. Translation from Alternative ORFs

Translation of the alternative ORFs may be provided by a specific mRNA region, referred to as IRES (internal ribosome entry site). The functionality of an IRES can be experimentally confirmed by constructing bicistronic transcripts (Figure 7e) [40,41]. Note that the use of bicistronic constructs when studying the IRES functionality is determined, in particular, by the need to normalize a relative efficiency of the IRES-guided translation as compared with the cap-dependent translation. For example, functionality of the initially predicted IRES in the mRNA for a tobacco heat shock protein [55] was experimentally confirmed utilizing the ability of coordinated expression of reporter genes within a bicistronic transcript [56]. A structural analysis of the 5′UTR of a maize heat shock protein (Hsp101) mRNA has shown the presence of three stem loops towards the 5′ end, which suggested the functioning of the 5′UTR structure as an IRES. This assumption was experimentally confirmed by comparing bicistronic reporter constructs; in particular, it has been shown that the overall hsp101 5′UTR sequence (150 nucleotides) acts as an IRES, since the deletion of 17 nucleotides from the 5′ end decreases the translation efficiency of the reporter gene transcript by 87% as compared with the control sequence [57]. A functional analysis of the bicistronic constructs has revealed IRESs in the *A. thaliana* WUS mRNA, coding for the homeodomain transcription factor WUSCHEL (WUS), as well as the role of an additional protein, AtLa1 (an RNA-binding factor). As demonstrated, AtLa1 initiates an IRES-dependent WUS mRNA translation by binding 5′UTR (WUS is responsible for supporting the *A. thaliana* apical meristems under stress conditions) [58].

Alternative ORFs are among the most abundant regulatory elements in mRNAs; they are frequently present in the 5′ leader regions of eukaryotic mRNAs (designated uORFs). Such uORFs may negatively modulate the translation efficiency of the downstream main ORF. According to the current estimate, approximately 20% of the *A. thaliana* protein-coding genes contain uORFs in their mRNA 5′UTRs [59]. Initially, such regulatory sequences are either predicted via searching for the conserved peptide uORFs (CPuORFs) [45] or experimentally detected, in particular, by ribosome profiling during, for example, plant cell response to a stress [43,60,61]. The highest interest of researchers is associated with the function of the regulatory sequences, such as CPuORFs, which are able to act in a sequence-dependent manner or cause ribosomal arrest, thereby modulating translation of the main ORF.

4.5. Frameshift Mutations

The approach of frameshift mutations utilizes concurrent introduction of deletions and insertions at −1 and +1 positions; this procedure changes only the amino acid composition of a peptide sequence coded for in CPuORFs but retains the presence and unchanged length of the overlapping CPuORFs (Figure 7d). This method has been used to analyze 16 recently predicted conserved CPuORFs of *A. thaliana* for assessing a sequence-dependent effect of each CPuORF on expression of the main ORF. A comparative analysis of the reporter protein activity of the variants when the translation is controlled by either native CPuORFs or the CPuORFs with introduced frameshift mutations has identified five novel CPuORFs that repress the expression of the main ORF in a sequence-dependent manner. Moreover, it has been convincingly demonstrated that the C-terminal regions of four of these CPuORF-encoded peptides play a crucial role in repressing the translation of the main ORF [46]. The functionality of three *A. thaliana* CPuORFs in arresting ribosomes during translation was tested in another study. This mechanism of CPuORF action was clarified using toeprinting analysis and the additional experimental evidence was obtained by constructing the following three types of reporter constructs. (i) With the CPuORF initiation codon removed from each reporter construct of the native CPuORF by replacing AUG with AAG; (ii) with frameshift only mutations, introduced to the CPuORF sequences; and (iii) with both removed initiation codon and frameshift mutations in CPuORF sequences. A comparative testing of all types of reporter constructs has shown that removal of the initiation codon from CPuORFs considerably increases the reporter gene expression; the frameshift mutations in CPuORFs also efficiently increase the reporter gene expression, although to a lower degree as compared with the removal of initiation codon; while the simultaneous presence of frameshift mutations and absence of the initiation codon have almost no effect on the reporter gene expression. These results clearly demonstrate that (i) the peptide sequences are partially responsible for strong repressive effects of these CPuORFs on the main ORF expression; (ii) repression of the main ORF expression (in this case, the ORF of reporter transcript) depends on CPuORF translation; and (iii) these CPuORFs induce ribosomal arrest and thereby considerably inhibit expression of the main ORF [62]. Thus, it is possible not only to insert regulatory sequences that control the reporter gene translation into the constructs carrying this reporter, but also to introduce manifold modifications, which allows their functional role in a key biological process, translation, to be assessed.

4.6. Characteristics and Reasoning for Selection of a Verification Method(s)

It should be emphasized that the selection of a method for assembling reporter constructs is of highest importance, since it is necessary to clone the target regulatory sequences with a reporter gene without the introduction of additional nucleotides, which may influence the translation modulation. The classical restriction–ligation cloning method does not allow generation of such constructs and requires several cloning stages. Several more economical and efficient technologies facilitating and accelerating the design of such constructs have been recently proposed. These technologies make it possible to produce seamless fusions of a "regulatory sequence–reporter gene". Most of them utilize the recombination between homologous sequences residing at the ends of the DNA fragments to be assembled [63]. For example, the Gateway®cloning system is based on the well-characterized site-specific recombinase system of the lambda phage for recombination of DNA segments. The DNA segments are flanked by the ATT recombination sites, which provide seamless assembly of almost any sequence [64]. However, this system also has certain limitations, namely, the need that the sequences overlap for at least 15 nucleotides at their ends. Correspondingly, the assembly of nonoverlapping DNA fragments requires additional terminal extensions or the use of bridge oligonucleotides [65]. Moreover, in our view, the approaches, such as Golden Gate system [66,67] or CPEC (circular polymerase extension cloning) strategy are more purposeful for designing the regulatory sequence–reporter gene constructs [68]. The Golden Gate method utilizes the IIS type restriction fragments to generate 4-nucleotide sticky ends, flanking each DNA portion, which then can be effectively ligated using T4 ligase. The assembly is performed in a single reaction and gives a seamless or nearly seamless target construct. This is determined by that the IIS type recognition sites are removed during ligation leaving only four nucleotide which positions may be determined by researcher [69,70].

The principle underlying the CPEC method utilizes the polymerase extension mechanism and one DNA polymerase for the in vitro assembly and cloning of sequences in any vector in a single-stage reaction. CPEC allows for an integrated, combinatorial, or multifragment assembly of sequences as well as for routine cloning procedures [68]. Thus, the Golden Gate and CPEC technologies have important advantages suggesting their utility in designing of the reporter constructs intended for studying and experimentally verifying the role of the regulatory regions in transcripts during translation.

As for the functional assessment of the constructs carrying a target regulatory sequence fused with a reporter gene, two basic experimental approaches are used: they utilized (i) a transient (temporary) and/or (ii) stable (constant) expression of reporter constructs in plants [4,71]. In the case of the transient expression of reporter constructs, transfection of the protoplasts derived from leaves of *N. tabacum* L. cv. BY2, lettuce, or *A. thaliana* are used as well as agroinfiltration of leaves (*N. benthamiana*,

N. excelsior, *N. tabacum* var. xanth, or *A. thaliana*) or plant cell suspension culture (*N. tabacum* BY2 or *A. thaliana* T87) [4,71,72] (Supplement Table S1). In the current view, the transient expression of reporter constructs is regarded as a more efficient approach because of lower material and time expenditures. The protocols for generation of protoplasts and agroinfiltration have been elaborated for many model and non-model plants, widening the range of the used plant objects. However, there are some limitations in the application of transient expression associated with the variation of protoplasts in the transformation efficiency and with the delivery of constructs to plant cells in agroinfiltration [4,73].

A stable expression of reporter constructs in plants requires production of transgenic plants or transgenic plant cell suspension cultures. This makes it possible to solve increasingly more complex problems of translational control, such as translation regulation under different growth and stress conditions or long-term physiological effects of certain changes in a sequence that modulate translation. In particular, polysome fractionation of control and transgenic plants makes it possible to confirm that the transcript of a reporter gene controlled by a tested regulatory sequence is actually associated with the polysome fraction. Thus, it is possible to assess the translational status of the mRNA of a reporter gene fused with the tested regulatory sequence, including under the effect of stress factors [8].

When using the methods involving both stable and transient expressions, the choice of an adequate control is of a paramount importance to ascertain that the change in expression of the reporter protein is actually associated with the change in translation (rather than with transcription, protein stability, and so on) [4,51].

According to the available experimental data, the strategy of reporter systems is in demand for a wide range of studies into individual regulatory sequences or their nucleotide contexts. The use of this strategy gives the unique data on the functional role of target sequences in translation efficiency; however, this strategy is, as a rule, supplemented with other methods. The choice of a particular method depends on the regulatory sequence or its context to be studied be it a full-sized 5′UTR or its regulatory elements, such as RNA G-quadruplex, IRES, uORFs, and so on, which is in part summarized above and is comprehensively described in the corresponding publications.

5. Conclusions

Translation plays a key role in the overall implementation of genetic information and the new knowledge about the intricate and multilevel information encoded in the mRNA sequence are of a paramount importance. The research into translation has revealed many new and interesting facts about the structural and functional role of the mRNA regulatory sequences that mediate differential translation. In particular, this success has been determined by the use of state-of-the-art technologies for assessing the translational statuses of individual mRNA species on a genome-wide scale in combination with computational algorithms and the methods for experimental verification, summarized here (Figure 8).

The knowledge on roles of regulatory contexts in mRNA for translation efficiency as well as the combinations of these contexts will require improvement of both experimental approaches and theoretical algorithms. The researchers must have the opportunity not only to precisely determine the correlation between the observed fluctuations in expression of a transcript and the actual content of the corresponding protein in plants, but also to precisely define and estimate the contributions of individual regulatory contexts and their combinations within mRNAs. Correspondingly, the need for development of an integrated information resource for this purpose looks very reasonable. This resource would comprise the information about (i) the experimental methods for assessing the changes in translation on a genome-wide scale, including their modifications and applicability to different plant species; (ii) the resources for analyzing, interpreting, and visualizing the polysome and ribosome profiling data; (iii) the resources for constructing the target sequences of plant transcripts and predicting their characteristics; (iv) the methods for experimental verification of the regulatory codes of the plant transcripts involved in translation modulation; and so on. This will form the background for coordination of the numerous studies and the insight into the fine mechanisms underlying the control of biological processes at the point of translation in plants. Also it will expand the capabilities for future studies and the potential of applications of the mRNA regulatory contexts, including their use in engineering novel plant genotypes carrying the best combinations of the corresponding alleles and the generation of new of transgenes, including the use of genome editing technologies.

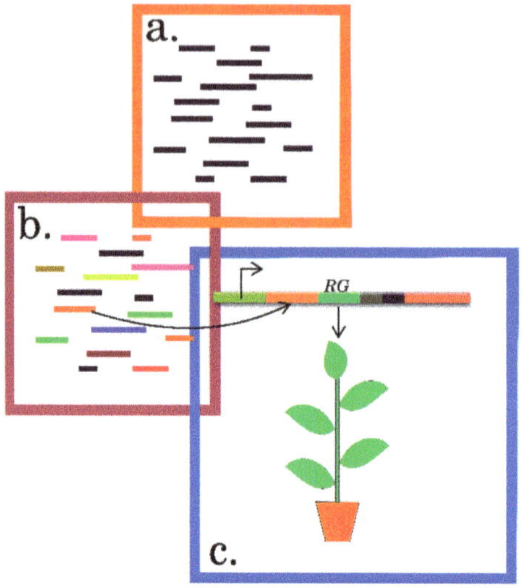

Figure 8. General strategy for identification, prediction, and experimental verification of the functional elements within transcripts that mediate their efficient translation. (**a**) Primary data on the transcripts differing in their translation efficiencies are obtained by sequencing the pools of the transcripts generated by polysome profiling, TRAP, or ribosome profiling. (**b**) In silico analysis of the transcript sequences identified the functional elements of transcripts. (**c**) Experimental verification of the predicted functional elements within transcripts (case study of reporter systems).

Supplementary Materials: Supplementary materials can be found at http://www.mdpi.com/1422-0067/20/1/33/s1.

Author Contributions: I.V.G.-P., O.S.P., and A.A.T. literature review and preparation of material for the second section of the review; I.V.G.-P., O.N.M., I.V.D., and A.A.T. literature review and preparation of material for the third section of the review; I.V.G.-P., O.S.P., K.V.K., and A.A.T. literature review and preparation of material for the fourth section of the review; A.A.T. preparation of original figures; and I.V.G.-P. wrote the manuscript. All authors were fully involved in preparing and revising the manuscript critically at its current state. All authors approved the final version for publishing.

Funding: The authors thank the Russian Science Foundation (grant no. 18-14-00026) for supporting this project.

Conflicts of Interest: The authors declare no conflicts of interest.

References

1. Baerenfaller, K.; Grossmann, J.; Grobei, M.A.; Hull, R.; Hirsch-Hoffmann, M.; Yalovsky, S.; Zimmermann, P.; Grossniklaus, U.; Gruissem, W.; Baginsky, S. Genome-scale proteomics reveals *Arabidopsis thaliana* gene models and proteome dynamics. *Science* **2008**, *320*, 938–941. [CrossRef] [PubMed]
2. Kawaguchi, R.; Bailey-Serres, J. mRNA sequence features that contribute to translational regulation in Arabidopsis. *Nucleic Acids Res.* **2005**, *33*, 955–965. [CrossRef] [PubMed]
3. Sablok, G.; Powell, J.J.; Kazan, K. Emerging Roles and Landscape of Translating mRNAs in Plants. *Front. Plant Sci.* **2017**, *8*, 1443. [CrossRef] [PubMed]
4. Mazzoni-Putman, S.M.; Stepanova, A.N. A Plant Biologist's Toolbox to Study Translation. *Front. Plant Sci.* **2018**, *9*, 873. [CrossRef] [PubMed]

5. Lecampion, C.; Floris, M.; Fantino, J.R.; Robaglia, C.; Laloi, C. An Easy Method for Plant Polysome Profiling. *J. Vis. Exp.* **2016**. [CrossRef] [PubMed]
6. Sharma, V.; Salwan, R.; Sharma, P.N.; Gulati, A. Integrated Translatome and Proteome: Approach for Accurate Portraying of Widespread Multifunctional Aspects of Trichoderma. *Front. Microbiol.* **2017**, *8*, 1602. [CrossRef] [PubMed]
7. Yamasaki, S.; Matsuura, H.; Demura, T.; Kato, K. Changes in Polysome Association of mRNA Throughout Growth and Development in *Arabidopsis thaliana*. *Plant Cell Physiol.* **2015**, *56*, 2169–2180. [CrossRef]
8. Matsuura, H.; Takenami, S.; Kubo, Y.; Ueda, K.; Ueda, A.; Yamaguchi, M.; Hirata, K.; Demura, T.; Kanaya, S.; Kato, K. A computational and experimental approach reveals that the 5'-proximal region of the 5'-UTR has a Cis-regulatory signature responsible for heat stress-regulated mRNA translation in Arabidopsis. *Plant Cell Physiol.* **2013**, *54*, 474–483. [CrossRef]
9. Yamasaki, S.; Sanada, Y.; Imase, R.; Matsuura, H.; Ueno, D.; Demura, T.; Kato, K. *Arabidopsis thaliana* cold-regulated 47 gene 5′-untranslated region enables stable high-level expression of transgenes. *J. Biosci. Bioeng.* **2018**, *125*, 124–130. [CrossRef]
10. Matsuura, H.; Ishibashi, Y.; Shinmyo, A.; Kanaya, S.; Kato, K. Genome-wide analyses of early translational responses to elevated temperature and high salinity in *Arabidopsis thaliana*. *Plant Cell Physiol.* **2010**, *51*, 448–462. [CrossRef]
11. Juntawong, P.; Hummel, M.; Bazin, J.; Bailey-Serres, J. Ribosome profiling: A tool for quantitative evaluation of dynamics in mRNA translation. *Methods Mol. Biol.* **2015**, *1284*, 139–173. [CrossRef] [PubMed]
12. Zanetti, M.E.; Chang, I.F.; Gong, F.; Galbraith, D.W.; Bailey-Serres, J. Immunopurification of polyribosomal complexes of Arabidopsis for global analysis of gene expression. *Plant Physiol.* **2005**, *138*, 624–635. [CrossRef] [PubMed]
13. Mustroph, A.; Juntawong, P.; Bailey-Serres, J. Isolation of plant polysomal mRNA by differential centrifugation and ribosome immunopurification methods. *Methods Mol. Biol.* **2009**, *553*, 109–126. [CrossRef]
14. Mustroph, A.; Zanetti, M.E.; Girke, T.; Bailey-Serres, J. Isolation and analysis of mRNAs from specific cell types of plants by ribosome immunopurification. *Methods Mol. Biol.* **2013**, *959*, 277–302. [CrossRef] [PubMed]
15. Lin, S.Y.; Chen, P.W.; Chuang, M.H.; Juntawong, P.; Bailey-Serres, J.; Jauh, G.Y. Profiling of translatomes of in vivo-grown pollen tubes reveals genes with roles in micropylar guidance during pollination in Arabidopsis. *Plant Cell* **2014**, *26*, 602–618. [CrossRef] [PubMed]
16. Jiao, Y.; Meyerowitz, E.M. Cell-type specific analysis of translating RNAs in developing flowers reveals new levels of control. *Mol. Syst. Biol.* **2010**, *6*, 419. [CrossRef] [PubMed]
17. Sorenson, R.; Bailey-Serres, J. Selective mRNA sequestration by OLIGOURIDYLATE-BINDING PROTEIN 1 contributes to translational control during hypoxia in Arabidopsis. *Proc. Natl. Acad. Sci. USA* **2014**, *111*, 2373–2378. [CrossRef] [PubMed]
18. Ingolia, N.T.; Ghaemmaghami, S.; Newman, J.R.; Weissman, J.S. Genome-wide analysis in vivo of translation with nucleotide resolution using ribosome profiling. *Science* **2009**, *324*, 218–223. [CrossRef]
19. McGlincy, N.J.; Ingolia, N.T. Transcriptome-wide measurement of translation by ribosome profiling. *Methods* **2017**, *126*, 112–129. [CrossRef]
20. Chotewutmontri, P.; Stiffler, N.; Watkins, K.P.; Barkan, A. Ribosome Profiling in Maize. *Methods Mol. Biol.* **2018**, *1676*, 165–183. [CrossRef]
21. Juntawong, P.; Girke, T.; Bazin, J.; Bailey-Serres, J. Translational dynamics revealed by genome-wide profiling of ribosome footprints in Arabidopsis. *Proc. Natl. Acad. Sci. USA* **2014**, *111*, E203–212. [CrossRef] [PubMed]
22. Planchard, N.; Bertin, P.; Quadrado, M.; Dargel-Graffin, C.; Hatin, I.; Namy, O.; Mireau, H. The translational landscape of Arabidopsis mitochondria. *Nucleic Acids Res.* **2018**, *46*, 6218–6228. [CrossRef] [PubMed]
23. Gerashchenko, M.V.; Gladyshev, V.N. Ribonuclease selection for ribosome profiling. *Nucleic Acids Res.* **2017**, *45*, e6. [CrossRef] [PubMed]
24. Eastman, G.; Smircich, P.; Sotelo-Silveira, J.R. Following Ribosome Footprints to Understand Translation at a Genome Wide Level. *Comput. Struct. Biotechnol. J.* **2018**, *16*, 167–176. [CrossRef] [PubMed]
25. Hsu, P.Y.; Benfey, P.N. Small but Mighty: Functional Peptides Encoded by Small ORFs in Plants. *Proteomics* **2018**, *18*, e1700038. [CrossRef] [PubMed]
26. Lei, L.; Shi, J.; Chen, J.; Zhang, M.; Sun, S.; Xie, S.; Li, X.; Zeng, B.; Peng, L.; Hauck, A.; et al. Ribosome profiling reveals dynamic translational landscape in maize seedlings under drought stress. *Plant J.* **2015**, *84*, 1206–1218. [CrossRef]

27. Hsu, P.Y.; Calviello, L.; Wu, H.L.; Li, F.W.; Rothfels, C.J.; Ohler, U.; Benfey, P.N. Super-resolution ribosome profiling reveals unannotated translation events in Arabidopsis. *Proc. Natl. Acad. Sci. USA* **2016**, *113*, E7126–E7135. [CrossRef]
28. Lukoszek, R.; Feist, P.; Ignatova, Z. Insights into the adaptive response of *Arabidopsis thaliana* to prolonged thermal stress by ribosomal profiling and RNA-Seq. *BMC Plant Biol.* **2016**, *16*, 221. [CrossRef]
29. Chotewutmontri, P.; Barkan, A. Multilevel effects of light on ribosome dynamics in chloroplasts program genome-wide and psbA-specific changes in translation. *PLoS Genet.* **2018**, *14*, e1007555. [CrossRef]
30. Andreev, D.E.; O'Connor, P.B.; Loughran, G.; Dmitriev, S.E.; Baranov, P.V.; Shatsky, I.N. Insights into the mechanisms of eukaryotic translation gained with ribosome profiling. *Nucleic Acids Res.* **2017**, *45*, 513–526. [CrossRef]
31. Lamesch, P.; Berardini, T.Z.; Li, D.; Swarbreck, D.; Wilks, C.; Sasidharan, R.; Muller, R.; Dreher, K.; Alexander, D.L.; Garcia-Hernandez, M.; et al. The Arabidopsis Information Resource (TAIR): Improved gene annotation and new tools. *Nucleic Acids Res.* **2012**, *40*, D1202–1210. [CrossRef]
32. Yanguez, E.; Castro-Sanz, A.B.; Fernandez-Bautista, N.; Oliveros, J.C.; Castellano, M.M. Analysis of genome-wide changes in the translatome of Arabidopsis seedlings subjected to heat stress. *PLoS ONE* **2013**, *8*, e71425. [CrossRef] [PubMed]
33. Liu, M.J.; Wu, S.H.; Chen, H.M.; Wu, S.H. Widespread translational control contributes to the regulation of Arabidopsis photomorphogenesis. *Mol. Syst. Biol.* **2012**, *8*, 566. [CrossRef] [PubMed]
34. Basbouss-Serhal, I.; Pateyron, S.; Cochet, F.; Leymarie, J.; Bailly, C. 5′ to 3′ mRNA Decay Contributes to the Regulation of Arabidopsis Seed Germination by Dormancy. *Plant Physiol.* **2017**, *173*, 1709–1723. [CrossRef] [PubMed]
35. Mustafaev, O.; Sadovskaya, N.S.; Tyurin, A.A.; Goldenkova-Pavlova, I.V. JetGene: An integrated database for analysis of regulatory regions or nucleotide contexts in plant differentially translated transcripts. manuscript in preparation.
36. Deyneko, I.V.; Kel, A.E.; Bloecker, H.; Kauer, G. Signal-theoretical DNA similarity measure revealing unexpected similarities of *E. coli* promoters. *In Silico Biol.* **2005**, *5*, 547–555. [PubMed]
37. Deyneko, I.V.; Kalybaeva, Y.M.; Kel, A.E.; Blöcker, H. Human-chimpanzee promoter comparisons: Property-conserved evolution? *Genomics* **2010**, *96*, 129–133. [CrossRef] [PubMed]
38. Bailey, T.L.; Johnson, J.; Grant, C.E.; Noble, W.S. The MEME Suite. *Nucleic Acids Res.* **2015**, *43*, W39–49. [CrossRef]
39. Williams, B.P.; Burgess, S.J.; Reyna-Llorens, I.; Knerova, J.; Aubry, S.; Stanley, S.; Hibberd, J.M. An Untranslated cis-Element Regulates the Accumulation of Multiple C4 Enzymes in *Gynandropsis gynandra* Mesophyll Cells. *Plant Cell* **2016**, *28*, 454–465. [CrossRef]
40. Merchante, C.; Stepanova, A.N.; Alonso, J.M. Translation regulation in plants: An interesting past, an exciting present and a promising future. *Plant J.* **2017**, *90*, 628–653. [CrossRef]
41. Thompson, S.R. So you want to know if your message has an IRES? *Wiley Interdiscip. Rev. RNA* **2012**, *3*, 697–705. [CrossRef]
42. Cobbold, L.C.; Spriggs, K.A.; Haines, S.J.; Dobbyn, H.C.; Hayes, C.; de Moor, C.H.; Lilley, K.S.; Bushell, M.; Willis, A.E. Identification of internal ribosome entry segment (IRES)-trans-acting factors for the Myc family of IRESs. *Mol. Cell. Biol.* **2008**, *28*, 40–49. [CrossRef] [PubMed]
43. Sheshukova, E.V.; Komarova, T.V.; Ershova, N.M.; Shindyapina, A.V.; Dorokhov, Y.L. An Alternative Nested Reading Frame May Participate in the Stress-Dependent Expression of a Plant Gene. *Front. Plant Sci.* **2017**, *8*, 2137. [CrossRef] [PubMed]
44. Andrews, S.J.; Rothnagel, J.A. Emerging evidence for functional peptides encoded by short open reading frames. *Nat. Rev. Genet.* **2014**, *15*, 193–204. [CrossRef] [PubMed]
45. Takahashi, H.; Takahashi, A.; Naito, S.; Onouchi, H. BAIUCAS: A novel BLAST-based algorithm for the identification of upstream open reading frames with conserved amino acid sequences and its application to the Arabidopsis thaliana genome. *Bioinformatics* **2012**, *28*, 2231–2241. [CrossRef] [PubMed]

46. Ebina, I.; Takemoto-Tsutsumi, M.; Watanabe, S.; Koyama, H.; Endo, Y.; Kimata, K.; Igarashi, T.; Murakami, K.; Kudo, R.; Ohsumi, A.; et al. Identification of novel *Arabidopsis thaliana* upstream open reading frames that control expression of the main coding sequences in a peptide sequence-dependent manner. *Nucleic Acids Res.* **2015**, *43*, 1562–1576. [CrossRef] [PubMed]
47. Tunney, R.; McGlincy, N.J.; Graham, M.E.; Naddaf, N.; Pachter, L.; Lareau, L.F. Accurate design of translational output by a neural network model of ribosome distribution. *Nat. Struct. Mol. Biol.* **2018**, *25*, 577–582. [CrossRef] [PubMed]
48. Hill, S.T.; Kuintzle, R.; Teegarden, A.; Merrill, E., 3rd; Danaee, P.; Hendrix, D.A. A deep recurrent neural network discovers complex biological rules to decipher RNA protein-coding potential. *Nucleic Acids Res.* **2018**, *46*, 8105–8113. [CrossRef] [PubMed]
49. Ching, T.; Zhu, X.; Garmire, L.X. Cox-nnet: An artificial neural network method for prognosis prediction of high-throughput omics data. *PLoS Comput. Biol.* **2018**, *14*, e1006076. [CrossRef]
50. Cao, Z.; Zhang, S. Simple tricks of convolutional neural network architectures improve DNA-protein binding prediction. *Bioinformatics* **2018**. [CrossRef]
51. Tyurin, A.A.; Kabardaeva, K.V.; Gra, O.A.; Mustafaev, O.M.; Sadovskaya, N.S.; Pavlenko, O.S.; Goldenkova-Pavlov, I.V. Efficient expression of a heterologous gene in plants depends on the nucleotide composition of mRNA's 5′-region. *Russ. J. Plant. Physiol.* **2016**, *4*, 511–522. [CrossRef]
52. Anami, S.; Njuguna, E.; Coussens, G.; Aesaert, S.; Van Lijsebettens, M. Higher plant transformation: Principles and molecular tools. *Int. J. Dev. Biol.* **2013**, *57*, 483–494. [CrossRef] [PubMed]
53. Cho, H.; Cho, H.S.; Nam, H.; Jo, H.; Yoon, J.; Park, C.; Dang, T.V.T.; Kim, E.; Jeong, J.; Park, S.; et al. Translational control of phloem development by RNA G-quadruplex-JULGI determines plant sink strength. *Nat. Plants* **2018**, *4*, 376–390. [CrossRef] [PubMed]
54. Alvarez, D.; Voss, B.; Maass, D.; Wust, F.; Schaub, P.; Beyer, P.; Welsch, R. Carotenogenesis Is Regulated by 5′UTR-Mediated Translation of Phytoene Synthase Splice Variants. *Plant Physiol.* **2016**, *172*, 2314–2326. [CrossRef]
55. Dorokhov, Y.L.; Skulachev, M.V.; Ivanov, P.A.; Zvereva, S.D.; Tjulkina, L.G.; Merits, A.; Gleba, Y.Y.; Hohn, T.; Atabekov, J.G. Polypurine (A)-rich sequences promote cross-kingdom conservation of internal ribosome entry. *Proc. Natl. Acad. Sci. USA* **2002**, *99*, 5301–5306. [CrossRef] [PubMed]
56. Ali, Z.; Schumacher, H.M.; Heine-Dobbernack, E.; El-Banna, A.; Hafeez, F.Y.; Jacobsen, H.J.; Kiesecker, H. Dicistronic binary vector system-A versatile tool for gene expression studies in cell cultures and plants. *J. Biotechnol.* **2010**, *145*, 9–16. [CrossRef] [PubMed]
57. Jimenez-Gonzalez, A.S.; Fernandez, N.; Martinez-Salas, E.; Sanchez de Jimenez, E. Functional and structural analysis of maize hsp101 IRES. *PLoS ONE* **2014**, *9*, e107459. [CrossRef]
58. Cui, Y.; Rao, S.; Chang, B.; Wang, X.; Zhang, K.; Hou, X.; Zhu, X.; Wu, H.; Tian, Z.; Zhao, Z.; et al. AtLa1 protein initiates IRES-dependent translation of WUSCHEL mRNA and regulates the stem cell homeostasis of Arabidopsis in response to environmental hazards. *Plant Cell Environ.* **2015**, *38*, 2098–2114. [CrossRef]
59. Jorgensen, R.A.; Dorantes-Acosta, A.E. Conserved Peptide Upstream Open Reading Frames are Associated with Regulatory Genes in Angiosperms. *Front. Plant Sci.* **2012**, *3*, 191. [CrossRef]
60. Tanaka, M.; Sotta, N.; Yamazumi, Y.; Yamashita, Y.; Miwa, K.; Murota, K.; Chiba, Y.; Hirai, M.Y.; Akiyama, T., Onouchi, H.; et al. The Minimum Open Reading Frame, AUG-Stop, Induces Boron-Dependent Ribosome Stalling and mRNA Degradation. *Plant Cell* **2016**, *28*, 2830–2849. [CrossRef]
61. Xu, G.; Yuan, M.; Ai, C.; Liu, L.; Zhuang, E.; Karapetyan, S.; Wang, S.; Dong, X. uORF-mediated translation allows engineered plant disease resistance without fitness costs. *Nature* **2017**, *545*, 491–494. [CrossRef]
62. Hayashi, N.; Sasaki, S.; Takahashi, H.; Yamashita, Y.; Naito, S.; Onouchi, H. Identification of *Arabidopsis thaliana* upstream open reading frames encoding peptide sequences that cause ribosomal arrest. *Nucleic Acids Res.* **2017**, *45*, 8844–8858. [CrossRef] [PubMed]
63. Li, M.Z.; Elledge, S.J. Harnessing homologous recombination in vitro to generate recombinant DNA via SLIC. *Nat. Methods* **2007**, *4*, 251–256. [CrossRef] [PubMed]
64. Smedley, M.A.; Harwood, W.A. Gateway(R)-compatible plant transformation vectors. *Methods Mol. Biol.* **2015**, *1223*, 3–16. [CrossRef] [PubMed]
65. Tsvetanova, B.; Peng, L.; Liang, X.; Li, K.; Yang, J.P.; Ho, T.; Shirley, J.; Xu, L.; Potter, J.; Kudlicki, W.; et al. Genetic assembly tools for synthetic biology. *Methods Enzymol.* **2011**, *498*, 327–348. [CrossRef] [PubMed]

66. Engler, C.; Gruetzner, R.; Kandzia, R.; Marillonnet, S. Golden gate shuffling: A one-pot DNA shuffling method based on type IIs restriction enzymes. *PLoS ONE* **2009**, *4*, e5553. [CrossRef] [PubMed]
67. Engler, C.; Kandzia, R.; Marillonnet, S. A one pot, one step, precision cloning method with high throughput capability. *PLoS ONE* **2008**, *3*, e3647. [CrossRef] [PubMed]
68. Quan, J.; Tian, J. Circular polymerase extension cloning. *Methods Mol. Biol.* **2014**, *1116*, 103–117. [CrossRef] [PubMed]
69. Engler, C.; Marillonnet, S. Golden Gate cloning. *Methods Mol. Biol.* **2014**, *1116*, 119–131. [CrossRef]
70. Sarrion-Perdigones, A.; Vazquez-Vilar, M.; Palaci, J.; Castelijns, B.; Forment, J.; Ziarsolo, P.; Blanca, J.; Granell, A.; Orzaez, D. GoldenBraid 2.0: A comprehensive DNA assembly framework for plant synthetic biology. *Plant Physiol.* **2013**, *162*, 1618–1631. [CrossRef]
71. Vyacheslavova, A.O.; Berdichevets, I.N.; Tyurin, A.A.; Shimshilashvili, K.R.; Mustafaev, O.N.; Goldenkova-Pavlova, I.V. Expression of heterologous genes in plant systems: New possibilities. *Russ. J. Genet.* **2012**, *48*, 1067–1079. [CrossRef]
72. Agarwal, P.; Garg, V.; Gautam, T.; Pillai, B.; Kanoria, S.; Burma, P.K. A study on the influence of different promoter and 5′UTR (URM) cassettes from *Arabidopsis thaliana* on the expression level of the reporter gene beta glucuronidase in tobacco and cotton. *Transgenic Res.* **2014**, *23*, 351–363. [CrossRef] [PubMed]
73. Tyurin, A.A.; Kabardaeva, K.V.; Berestovoy, M.A.; Sidorchuk, Y.V.; Fomenkov, A.A.; Nosov, A.V.; Goldenkova-Pavlova, I.V. Simple and reliable system for transient gene expression for the characteristic signal sequences and the estimation of the localization of target protein in plant cell. *Russ. J. Plant. Physiol.* **2017**, *64*, 672–679. [CrossRef]

© 2018 by the authors. Licensee MDPI, Basel, Switzerland. This article is an open access article distributed under the terms and conditions of the Creative Commons Attribution (CC BY) license (http://creativecommons.org/licenses/by/4.0/).

International Journal of
Molecular Sciences

MDPI

Brief Report

One Major Challenge of Sequencing Large Plant Genomes Is to Know How Big They Really Are

Jaroslav Doležel *, Jana Čížková, Hana Šimková and Jan Bartoš

Institute of Experimental Botany, Centre of the Region Haná for Biotechnological and Agricultural Research, Šlechtitelů 31, CZ-78371 Olomouc, Czech Republic; cizkova@ueb.cas.cz (J.Č.); simkovah@ueb.cas.cz (H.Š); bartos@ueb.cas.cz (J.B.)

* Correspondence: dolezel@ueb.cas.cz; Tel.: +420-585-238-703; Fax: +420-585-238-704

Received: 26 October 2018; Accepted: 6 November 2018; Published: 11 November 2018

Abstract: Any project seeking to deliver a plant or animal reference genome sequence must address the question as to the completeness of the assembly. Given the complexity introduced particularly by the presence of sequence redundancy, a problem which is especially acute in polyploid genomes, this question is not an easy one to answer. One approach is to use the sequence data, along with the appropriate computational tools, the other is to compare the estimate of genome size with an experimentally measured mass of nuclear DNA. The latter requires a reference standard in order to provide a robust relationship between the two independent measurements of genome size. Here, the proposal is to choose the human male leucocyte genome for this standard: its 1C DNA amount (the amount of DNA contained within unreplicated haploid chromosome set) of 3.50 pg is equivalent to a genome length of 3.423 Gbp, a size which is just 5% longer than predicted by the most current human genome assembly. Adopting this standard, this paper assesses the completeness of the reference genome assemblies of the leading cereal crops species wheat, barley and rye.

Keywords: flow cytometry; genome size; nuclear DNA content; reference genome assembly; standardization

1. Introduction

The more that is known regarding the organization and function of plant and animal genomes, the more it becomes clear that a full understanding of genome function will require the acquisition of a complete sequence. The enormous throughput offered by current short read DNA sequencing technologies allows for the sequencing of genomes of any size and at a high sequencing depth. While this enables the ready assembly of single and low-copy sequences, the inclusion within the assembly of repetitive sequence is a non-trivial challenge, and, together with sequence redundancy due to polyploidy, represent a major obstacle to the acquisition of gap-free long-range genome sequences.

A reference genome assembly aims to faithfully represent a complete genome sequence, ideally with each chromosome being represented by a single, gap-less pseudomolecule. The level of completeness of an assembly remains difficult to ascertain, however, especially in the case of complex genomes, in which tracts of repetitive DNA, segmental duplications and, in the case of polyploid genomes, the presence of homoeologs, are all inimical to the elaboration of a "correct" assembly: the result is that gaps, mis-assemblies and collapsed tandem repeats feature in most published genome sequences. A much-used computational method to size a nuclear genome relies on the concept of k-mer frequencies [1,2]. An alternative may be to determine the number of full-length LTR-retrotransposons. As their number increases linearly with genome size, at least in grass species, it may serve as a measure of assembly quality [3]. Genome size of unknown species might then be obtained by extrapolation, using data from species whose genome size is known. However, as both approaches rely on sequence data, the only truly independent way to determine genome size is to experimentally determine the quantity of DNA present in the nuclei.

2. Estimation of Genome Size

Two experimental approaches have been developed to estimate nuclear DNA amounts: biochemical and cytometric. The former seeks to quantify the DNA harbored within a known mass of plant tissue [4]; its weakness lies in the errors inherent in the estimation of the number of nuclei present in the sample, in the unknown proportion of nuclei present at each of the various different cell cycle stages and the non-estimable proportion of endo-reduplicated nuclei present. As a result, cytometry-based estimations tend to be preferred, since these are designed to quantify the DNA present in a population of nuclei at a known cell cycle stage [5]. The attempt by [6] to derive relative nuclear DNA amounts present in several plant species using Feulgen micro-densitometry led to the development of the now universally understood C-value terminology, where un-replicated haploid nuclei contain a 1C DNA amount; the terminology has been refined in recent years [7]. Feulgen microdensitometry was phased out during the 1980s as a result of the throughput benefits offered by flow cytometry, which offers the possibility of analyzing large numbers of isolated nuclei in a short time [5].

It is important to note that flow cytometry does not quantify nuclear DNA directly, but rather achieves this by capturing the signal emitted from fluorochrome-stained nuclei. In order to determine a nuclear DNA amount in absolute units, the fluorescence of an unknown sample has to be compared with that of a reference standard of known genome size [8]. To avoid errors due to non-linearity, an ideal reference standard should not differ in size by more than two or three-fold from the test sample, implying that a set of reference standards is needed in order to cope with the large range of genome size encountered among higher organisms. The question then becomes how to calibrate these reference standards if none of the candidate species has itself been completely sequenced.

3. Standardization

Not unexpectedly, the issue applies as much to animal to plant or fungal genomes. To enable a comparison of data obtained by different laboratories, Tiersch et al. [9] calibrated a set of animal reference standards, choosing human male leukocytes (7 pg DNA/2C) as the primary reference; the 7 pg figure was based on estimates derived from Feulgen micro-densitometry [10]. The experiments derived a 2C value of 2.5 pg DNA for domestic chicken, which was close to the value given by [11]. The domestic chicken genome has been adopted since this time as the most widely used reference standard for the sizing of animal genomes [12]. In an effort to enable comparisons between animal and plant genomes, Doležel et al. [8] recommended a set of plant reference standards (Table 1), also calibrated with respect to the human male leukocyte genome, assuming the 7 pg value assigned by Tiersch et al. [9]. Over the past three decades, hundreds of genome size estimates have been published, based mainly on the 7 pg value. The question is how close to reality these estimates really are, which relates in the main to how accurate the 7 pg figure is. According to the arguments made by Doležel and Greilhuber [13], the value most probably over-estimates the true value by 5–10%.

Table 1. Plant DNA reference standards calibrated for the estimation of nuclear DNA amounts in absolute units [8].

Plant Species and Cultivar *	2C DNA Content (pg DNA) **
Raphanus sativus L. 'Saxa'	1.11
Solanum lycopersicum L. 'Stupické polní rané'	1.96
Glycine max Merr. 'Polanka'	2.50
Zea mays L. 'CE-777'	5.43
Pisum sativum L. 'Ctirad'	9.09
Secale cereale L. 'Daňkovské'	16.19
Vicia faba L. 'Inovec'	26.90
Allium cepa L. 'Alice'	34.89

* Seeds of the reference standards can be obtained from the corresponding author free of charge at dolezel@ueb.cas.cz. Since the year 2000, seed samples were provided to 615 research projects worldwide. ** Estimated after considering 7 pg DNA/2C for human [9].

4. The Human Genome as a Universal Reference Standard

Seventeen years have passed since the joint announcement of the human genome sequence [14,15]. This period has seen a number of attempts to complete the assembly, applying a variety of technologies [16,17]. All of these have reported a smaller genome size than what has, as of the end of 2017, been suggested in GRCh38.p12, the most recently released Genome Reference Consortium version, which comprises 3,257,319,537 bp. Assuming the Doležel et al. [18] conversion of 1 pg = 0.978 Gbp, 3.5 pg 1C DNA is equivalent to 3,423,000,000 bases. Thus, the 7 pg value represents an ~5.1% over-estimate of the GRCh38.p12 assembly prediction. This difference lies at the lower end of the error range predicted by Doležel and Greilhuber [13]. Given that the human reference genome is still incomplete, the expectation is that the gap between the 7 pg figure and the "real" human genome size will continue to diminish. Nevertheless, a 5% error is not dissimilar to the variation observed between estimates of nuclear DNA amounts of a given species produced by different laboratories [19,20]. Thus, the recommendation remains that the 7 pg figure continue to be used as the reference for measuring 2C values of both animal and plant genomes.

5. Sizing the Large Triticeae Genomes

Three species belonging to the tribe Triticeae—namely, bread wheat (*Triticum aestivum*), barley (*Hordeum vulgare*) and to a lesser extent, cereal rye (*Secale cereale*)—provide a major proportion of the calories used by humans and their livestock across the temperate world. The acquisition of their genome sequences will facilitate marker- and genomics-assisted breeding, gene editing and other novel breeding technologies currently under development. Reference genome sequences have been published for barley [21], wild emmer wheat (*T. dicoccoides*) [22] and hexaploid bread wheat (*T. aestivum*) [3], and one for cereal rye is currently being finalized (Nils Stein, pers. comm.). Here, flow cytometry was utilized to assess the nuclear DNA content of wild emmer, bread wheat, barley and cereal rye. To minimize errors due to copy number variants and intraspecific differences in genome size, the accessions of each species were those used for the acquisition of their genome sequences. The cereal rye cultivar Daňkovské (16.19 pg/2C) and garden pea (*Pisum sativum*) cultivar Ctirad (9.09 pg/2C) were used as reference standards (Table 1). Rye was selected out of the calibrated reference standards (Table 1) as its 2C value was close to 2 C DNA amounts of tetraploid and hexaploid wheat and barley. However, this standard could not be used for another accession of rye and thus pea was employed as the second standard. The outcomes are summarized in Table 2.

Table 2. Estimation of nuclear DNA amounts in the four Triticeae species.

Species and Genotype	2C Nuclear DNA Content (pg) *		Reference Standard
	Mean	± SD	
Triticum aestivum cv. Chinese Spring	33.91	0.27	*Secale cereale* cv. Daňkovské
Triticum dicoccoides cv. Zavitan	25.11	0.16	*Secale cereale* cv. Daňkovské
Hordeum vulgare cv. Morex	10.31	0.09	*Secale cereale* cv. Daňkovské
Secale cereale inbred line Lo7	15.95	0.11	*Pisum sativum* cv. Ctirad

* Considering 7 pg DNA/2C for human [9].

6. Completeness of the Current Triticeae Reference Genome Assemblies

To estimate the completeness of reference genome assemblies of the four test-species, the sizes predicted by each of their assemblies were compared with their estimated genome sizes as derived by flow cytometry. Taking the [9] figure of 7 pg DNA/2C, the conclusion was that the Triticeae assemblies represent at least 85% of their full genome (Table 3). However, adopting the GRCh38.p12 with 1C genome size of 3,257,319,537 bases as the reference, increased the coverage to at least 88%. It should be noted, however, that the data on the size of genome assembly do not inform about its quality, i.e., the correct ordering and orientation DNA contigs. This parameter needs to be assessed using other methods than flow cytometry.

Table 3. The estimated level of completeness of the four Triticeae reference genome assemblies.

Species	Reference Genome Assembly (Gbp) *	Flow Cytometric Estimation of 1C Genome Size **			
		GRCh38.12		[9]	
		Genome Size (Gbp)	Assembly Coverage (%)	Genome size (Gbp)	Assembly Coverage (%)
H. vulgare	4.79	4.88	98	5.04	95
S. cereale	6.67	7.42	90	7.80	86
T. dicoccoides	10.50	11.87	88	12.28	85
T. aestivum	14.50	16.03	90 ***	16.58	87

* Reference genome assemblies: *H. vulgare* [21], *T. dicoccoides* [22], *T. aestivum* [3], *S. cereale* (Nils Stein, pers. comm.). ** Two different values were used for human 1C genome size as a primary reference standard: 3,257,319,537 bp (GRCh38.p12) and 3,423,000,000 bp [9]. *** Slightly higher value (92%) was estimated by the International Wheat Genome Sequencing Consortium [3] when considering human genome size of 3,253,848,404 bases (Human Genome Assembly GRCh38.p11).

7. Concluding Remarks and Recommendations

Cytometric methods suitable for the estimation of nuclear genome size independent of DNA sequence data require a reference standard of known genome size. The most widely used animal and plant DNA reference standards have been calibrated from the human male leucocyte genome, assuming its length to be 3.42 Gbp/1C (and its 2C content to be 7 pg DNA), even though the length estimate is 5.1% greater than what the most current assembly predicts; however, given that the GRCh38.p12 assembly is most probably still incomplete, the real difference may be smaller than this. Thus, for the moment, it would seem reasonable to continue with this figure. The use of an agreed standard will facilitate comparisons between results obtained in different laboratories. Once the human genome size is known to a yet higher level of precision, it will be straightforward to recalculate the size of genomes estimated to date.

8. Materials and Methods

Grain of hexaploid bread wheat cultivar (cv.) Chinese Spring were obtained from *P. Sourdille* (INRA Clermont-Ferrand, Clermont-Ferrand, France), those of *T. dicoccoides* (accession Zavitan) from A. Distelfeld (Tel Aviv University, Tel Aviv, Israel), those of barley cv. Morex from Nils Stein (IPK, Gatersleben, Germany) and those of cereal rye inbred line Lo7 from Eva Bauer (Technische Universität Munich, Munich, Germany). Grains of cereal rye cv. Daňkovské and seed of pea cv. Ctirad were obtained from, respectively, the Oseva Agro (Brno, Czech Republic) and Semo (Smržice, Czech Republic) breeding stations. Plants were raised in garden compost in pots and maintained in a greenhouse until they reached a height of 10–15 cm. Nuclei were extracted from leaves and suspended in preparation for flow cytometry following the methods given by [23]. Briefly, 10 mg of leaf tissue of each of the sample species and one of the two reference standards were chopped together in a 1 mL volume of LB01 solution [23] using a razor blade. The resulting homogenate was filtered through a 50-μm nylon mesh. The filtrate was made up to 50 μg/mL RNase and 50 μg/mL propidium iodide, and subjected to flow cytometry using a CyFlow Space flow cytometer (Sysmex Partec GmbH, Görlitz, Germany) equipped with a 532 nm green laser. The gain of the instrument was adjusted so that the peak representing G1 nuclei of the standard was positioned approximately on channel 100 on a histogram of relative fluorescence intensity when using a 512-channel scale. Five individual plants per each test species were sampled, and each sample was analyzed three times, each time on a different day. A minimum of 5000 nuclei per sample was analyzed and 2C DNA contents (in pg) were calculated from the means of the G1 peak positions by applying the formula (sample G1 peak mean) × (standard 2C DNA content)/(standard G1 peak mean). DNA contents in pg were converted to genome lengths in bp using the factor suggested by Doležel et al. [18], i.e., 1 pg DNA = 0.978 Gbp.

Author Contributions: J.D. conceived the project and drafted the manuscript, J.Č. performed the flow cytometry, H.Š. and J.B. contributed to discussions.

Funding: This research was financially supported by the Czech Republic Ministry of Education, Youth and Sports (award LO1204 from the National Program of Sustainability I).

Acknowledgments: We thank Assaf Distelfeld, Pierre Sourdille, Nils Stein and Eva Bauer for the gift of grain of the various Triticeae species, and T. Ryan Gregory for input regarding animal DNA content reference standards. The authors thank the International Wheat Genome Sequencing Consortium (IWGSC) for pre-publication access to IWGSC RefSeq v1.0 and to Nils Stein and the rye genome sequencing team for pre-publication access to the rye Lo7 genome assembly. This paper is dedicated to our dear colleague and friend, the late Jan Suda, who pioneered the use of DNA flow cytometry in plant taxonomy, biosystematics and population biology and contributed significantly to the advancement of botany.

Conflicts of Interest: The authors declare no conflicts of interest.

References

1. He, K.; Lin, K.; Wang, G.; Li, F. Genome sizes of nine insect species determined by flow cytometry and k-mer analysis. *Front Physiol.* **2016**, *7*, 569. [CrossRef] [PubMed]
2. Sun, H.; Ding, J.; Piednoël, M.; Schneeberger, K. findGSE: Estimating genome size variation within human and *Arabidopsis* using k-mer frequencies. *Bioinformatics* **2018**, *34*, 550–557. [CrossRef] [PubMed]
3. International Wheat Genome Sequencing Consortium. Shifting the limits in wheat research and breeding using a fully annotated reference genome. *Science* **2018**, *361*, eaar7191. [CrossRef] [PubMed]
4. Van't Hof, J. Cell population kinetics of excised roots of *Pisum sativum*. *J. Cell Biol.* **1965**, *27*, 179–189. [CrossRef] [PubMed]
5. Doležel, J.; Bartoš, J. Plant DNA flow cytometry and estimation of nuclear genome size. *Ann. Bot.* **2005**, *95*, 99–110. [CrossRef] [PubMed]
6. Swift, H. The constancy of desoxyribose nucleic acid in plant nuclei. *Proc. Natl. Acad. Sci. USA* **1950**, *36*, 643–654. [CrossRef] [PubMed]
7. Greilhuber, J.; Doležel, J.; Lysák, M.A.; Bennett, M.D. The origin, evolution and proposed stabilization of the terms 'genome size', and 'C-value' to describe nuclear DNA contents. *Ann. Bot.* **2005**, *95*, 255–260. [CrossRef] [PubMed]
8. Doležel, J.; Greilhuber, J.; Suda, J. Estimation of nuclear DNA content in plants using flow cytometry. *Nat. Protoc.* **2007**, *2*, 2233–2244. [CrossRef] [PubMed]
9. Tiersch, T.R.; Chandler, R.W.; Wachtel, S.S.; Elias, S. Reference standards for flow cytometry and application in comparative studies of nuclear DNA content. *Cytometry* **1989**, *10*, 706–710. [CrossRef] [PubMed]
10. Shapiro, H.S. Deoxyribonucleic acid content per cell of various organisms. In *Handbook of Biochemistry and Molecular Biology*; Fasman, G.D., Ed.; CRC Press: Cleveland, OH, USA, 1976; Volume 2, pp. 284–306.

11. Rasch, E.M.; Barr, H.J.; Rasch, R.W. The DNA content of sperm of *Drosophila melanogaster*. *Chromosoma* **1971**, *33*, 1–18. [CrossRef] [PubMed]
12. Gregory, T.R. Animal Genome Size Database. 2005. Available online: http://www.genomesize.com (accessed on 25 October 2018).
13. Doležel, J.; Greilhuber, J. Nuclear genome size: Are we getting closer? *Cytometry* **2010**, *77*, 635–642. [CrossRef] [PubMed]
14. International Human Genome Sequencing Consortium. Initial sequencing and analysis of the human genome. *Nature* **2001**, *409*, 860–921. [CrossRef] [PubMed]
15. Venter, J.C.; Adams, M.D.; Myers, W.W.; Li, P.W.; Mural, R.J.; Sutton, G.G. The sequence of the human genome. *Science* **2001**, *291*, 1304–1351. [CrossRef] [PubMed]
16. Seo, J.S.; Rhie, A.; Kim, J.; Lee, S.; Sohn, M.H.; Kim, C.U.; Hastie, A.; Cao, H.; Yun, J.Y.; Kim, J.; et al. De novo assembly and phasing of a Korean human genome. *Nature* **2016**, *538*, 243–247. [CrossRef] [PubMed]
17. Jain, M.; Koren, S.; Miga, K.H.; Quick, J.; Rand, A.C.; Sasani, T.A.; Tyson, J.R.; Beggs, A.D.; Dilthey, A.T.; Fiddes, I.T.; et al. Nanopore sequencing and assembly of a human genome with ultra-long reads. *Nat. Biotechnol.* **2018**, *36*, 338–345. [CrossRef] [PubMed]
18. Doležel, J.; Bartoš, J.; Voglmayr, H.; Greilhuber, J. Nuclear DNA content and genome size of trout and human. *Cytometry* **2003**, *51*, 127–128. [CrossRef] [PubMed]
19. Doležel, J.; Greilhuber, J.; Lucretti, S.; Meister, A.; Lysák, M.A.; Nardi, L.; Obermayer, R. Plant genome size estimation by flow cytometry: Inter-laboratory comparison. *Ann. Bot.* **1998**, *82*, 17–26. [CrossRef]
20. Praca-Fontes, M.M.; Carvalho, C.R.; Clarindo, W.R.; Cruz, C.D. Revisiting the DNA C-values of the genome size-standards used in plant flow cytometry to choose the "best primary standards". *Plant Cell Rep.* **2011**, *30*, 1183–1191. [CrossRef] [PubMed]
21. Mascher, M.; Gundlach, H.; Himmelbach, A.; Beier, S.; Twardziok, S.O.; Wicker, T.; Radchuk, V.; Dockter, C.; Hedley, P.E.; Russell, J.; et al. A chromosome conformation capture ordered sequence of the barley genome. *Nature* **2017**, *544*, 427–433. [CrossRef] [PubMed]
22. Avni, R.; Nave, M.; Barad, O.; Baruch, K.; Twardziok, S.O.; Gundlach, H.; Hale, I.; Mascher, M.; Spannagl, M.; Wiebe, K.; et al. Wild emmer genome architecture and diversity elucidate wheat evolution and domestication. *Science* **2017**, *357*, 93–97. [CrossRef] [PubMed]
23. Doležel, J.; Binarová, P.; Lucretti, S. Analysis of nuclear DNA content in plant cells by flow cytometry. *Biol. Plant.* **1989**, *31*, 113–120. [CrossRef]

© 2018 by the authors. Licensee MDPI, Basel, Switzerland. This article is an open access article distributed under the terms and conditions of the Creative Commons Attribution (CC BY) license (http://creativecommons.org/licenses/by/4.0/).

International Journal of
Molecular Sciences

MDPI

Article

Structural and Comparative Analysis of the Complete Chloroplast Genome of *Pyrus hopeiensis*—"Wild Plants with a Tiny Population"—and Three Other *Pyrus* Species

Yongtan Li [1,2,†], Jun Zhang [1,2,†], Longfei Li [3], Lijuan Gao [3], Jintao Xu [3] and Minsheng Yang [1,2,*]

1 Institute of Forest Biotechnology, Forestry College, Agricultural University of Hebei, Baoding 071000, China; liyongt37@126.com (Y.L.); zhangjunem@126.com (J.Z.)
2 Hebei Key Laboratory for Tree Genetic Resources and Forest Protection, Baoding 071000, China
3 Changli Institute for Pomology, Hebei Academy of Agricultural and Forestry Science, Changli 066600, China; shv266@163.com (L.L.); gaolijuan0306@163.com (L.G.); swxujintao@aliyun.com (J.X.)
* Correspondence: yangms100@126.com; Tel.: +86-0312-752-8715
† These authors contributed to the work equally and should be regarded as co-first authors.

Received: 2 October 2018; Accepted: 16 October 2018; Published: 20 October 2018

Abstract: *Pyrus hopeiensis* is a valuable wild resource of *Pyrus* in the Rosaceae. Due to its limited distribution and population decline, it has been listed as one of the "wild plants with a tiny population" in China. To date, few studies have been conducted on *P. hopeiensis*. This paper offers a systematic review of *P. hopeiensis*, providing a basis for the conservation and restoration of *P. hopeiensis* resources. In this study, the chloroplast genomes of two different genotypes of *P. hopeiensis*, *P. ussuriensis* Maxin. cv. Jingbaili, *P. communis* L. cv. Early Red Comice, and *P. betulifolia* were sequenced, compared and analyzed. The two *P. hopeiensis* genotypes showed a typical tetrad chloroplast genome, including a pair of inverted repeats encoding the same but opposite direction sequences, a large single copy (LSC) region, and a small single copy (SSC) region. The length of the chloroplast genome of *P. hopeiensis* HB-1 was 159,935 bp, 46 bp longer than that of the chloroplast genome of *P. hopeiensis* HB-2. The lengths of the SSC and IR regions of the two *Pyrus* genotypes were identical, with the only difference present in the LSC region. The GC content was only 0.02% higher in *P. hopeiensis* HB-1. The structure and size of the chloroplast genome, the gene species, gene number, and GC content of *P. hopeiensis* were similar to those of the other three *Pyrus* species. The IR boundary of the two genotypes of *P. hopeiensis* showed a similar degree of expansion. To determine the evolutionary history of *P. hopeiensis* within the genus *Pyrus* and the Rosaceae, 57 common protein-coding genes from 36 Rosaceae species were analyzed. The phylogenetic tree showed a close relationship between the genera *Pyrus* and *Malus*, and the relationship between *P. hopeiensis* HB-1 and *P. hopeiensis* HB-2 was the closest.

Keywords: *Pyrus hopeiensis*; cp genome; IR boundary; phylogeny

1. Introduction

Pyrus belongs to the *Pyrus* ssp. of the Maloideae subfamily (Rosaceae), which mainly includes temperate fruit trees. There are more than 30 species in this genus, and 13 species are present in China [1,2]. The pear has been cultivated for over 3000 years in China. It is the third most commonly cultivated fruit tree after apple and citrus. China is the world's largest pear producer, accounting for 71.2% of the world's total pear area. Wild resources of *Pyrus* are extremely precious. They are mostly distributed in the valleys, hillsides, and forest margins, and are characterized by cold resistance, drought resistance, disease resistance, barren tolerance, saline–alkali tolerance, and strong adaptability.

They provide material for screening quality rootstocks and for molecular breeding. The flowers, leaves, and fruit also have high ornamental value. The fruit of the pear is rich in fruit acids, vitamins, sugars, and many mineral elements essential for human life. It is sweet and refreshing and can be used to make dried and preserved pears, wine, and other products. Furthermore, it is regarded as having a high medicinal value and is used to reduce fevers, moisten the lungs, provide cough relief, and eliminate phlegm. Therefore, *Pyrus* is a valuable wild resource with a high exploitation value.

P. hopeiensis is a rare wild resource of the genus *Pyrus* in the subfamily Rosaceae [3,4], which has been listed as one of the "wild plants with a tiny population" in China. It can be found at the edges of hillside jungles at 100–800 m above sea level. At present, only a few genotypes have been found in Changli, Hebei Province. So far, there have been few studies about *P. hopeiensis*. Successful sequencing of the chloroplast genome of *P. hopeiensis* in our study provides a foundation for further study of its chloroplast molecular biology and can effectively promote genetic breeding and help clarify the molecular evolution of *P. hopeiensis*. It also provides some basis for the evolutionary analysis and classification of the genus *Pyrus*. What is more, this study gives a systematic review of *P. hopeiensis* that is useful for the conservation and restoration of wild *P. hopeiensis* resources.

Chloroplasts are the main site of photosynthesis, where fatty acids, starch, pigments, and other materials are synthesized [5]. They are independent of the plant nuclei and have a highly conserved genomic structure. Chloroplast DNA (cpDNA) has the beneficial characteristics of multiple copies, low molecular weight, and simple structure. Unlike the nuclear genome, which contains more repetitive sequences, and the mitochondrial genome, which is frequently rearranged, the chloroplast genome is rather conservative. The main mutation types are substitution and base insertion or deletion, and the mutation rate is low. Additionally, the chloroplast genome is of moderate size, making it easier to sequence than complex nuclear genomes. The chloroplast genome is maternally inherited in angiosperms with an independent evolutionary route [6]. Phylogenetic trees can be constructed using cpDNA data only, and the chloroplast genome shows good collinearity among plant groups. Sequencing data are relatively easy to analyze and the chloroplast genome structure sequence information can be used to study the species origin, evolution, and relationships between different species. In recent years, with the development of high-throughput sequencing technology, more chloroplast genomes of Rosaceae have been sequenced. In this study, the chloroplast genomes of two genotypes of *P. hopeiensis* and three other *Pyrus* (*P. ussuriensis* Maxin. cv. Jingbaili, *P. communis* L. cv. Early Red Comice, and *P. betulifolia*) were sequenced and compared with other Rosaceae plants. The genome structure and phylogeny of *Pyrus* were elucidated.

2. Results and Analysis

2.1. Basic Characteristics of Chloroplast Genome of P. hopeiensis

The chloroplast genome of *P. hopeiensis* has a typical tetrad structure, including paired IRa and IRb sequences, encoding in opposite directions, and large and small single copy regions (Figure 1). The total chloroplast genome of *P. hopeiensis* HB-1 was 159,935 bp in length, 46 bp smaller than that of *P. hopeiensis* HB-2. The large single copy (LSC) region was 87,961 bp long, 46 bp smaller than that of *P. hopeiensis* HB-2. The length of the small single copy (SSC) region was 19,200 bp and the IR region was 26,387 bp, the same as those of *P. hopeiensis* HB-2. There was little difference in length between the two *P. hopeiensis* genotypes, and what difference existed was in the LSC region.

A total of 117 genes (Table 1) from the chloroplast genome of *P. hopeiensis* HB-1 were annotated, including 77 protein coding genes, 31 tRNAs, eight rRNAs, and two pseudogenes (*clpP* and *atpF*). *P. hopeiensis* HB-1 lacked only the *MATK* protein-coding gene that was associated with biosynthesis in *P. hopeiensis* HB-2. These 77 protein-coding genes can be divided into four categories. The first contains 28 self-replicating genes, including three subunits encoding the synthesis of chloroplast RNA polymerase. The second category contains 40 genes related to photosynthesis, including light systems I and II, a cytochrome b 6/f protein complex, and ATP synthase and other biosynthesis genes, including cytochrome-related genes (*P. hopeiensis* HB-1 contains three genes and *P. hopeiensis* HB-2 contains four genes). The fourth category contains five unknown genes, such as the *ycf* gene. The IR region of *P. hopeiensis* contains 32 genes, of which the *ndhB* gene is present only in the IRb region and is absent in the IRa region. In addition, *rps12* is a mitotic gene with its 5′ terminal located at the LSC region and 3′ end with a copy is located in each of the two IR regions. This phenomenon is common in higher plants.

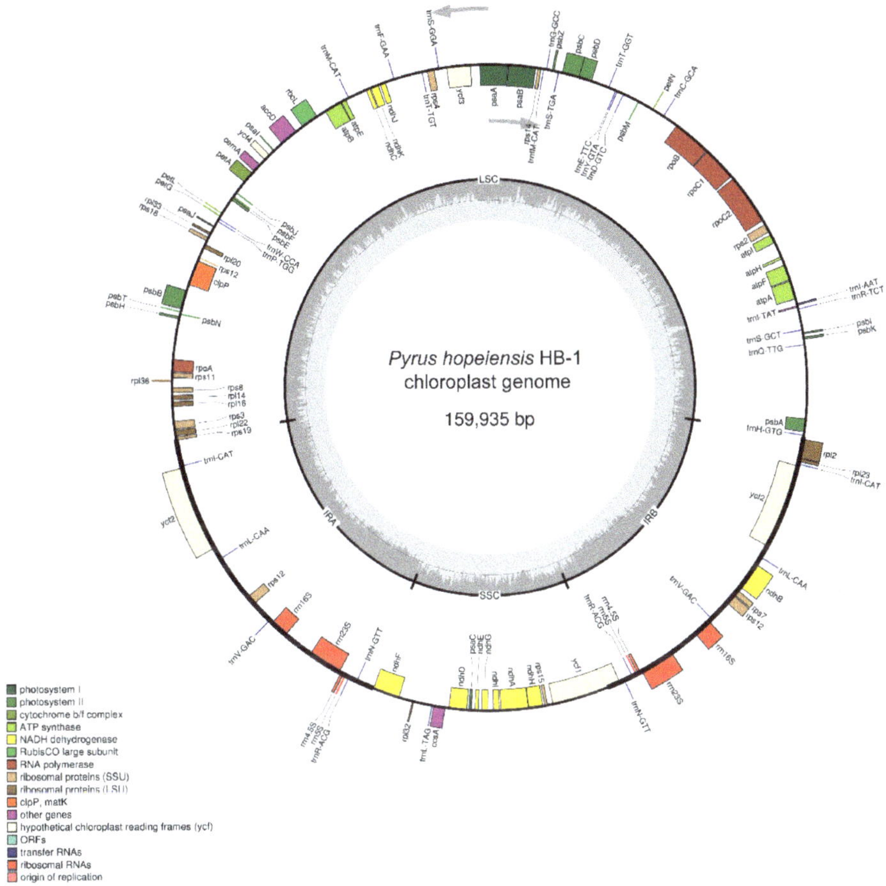

Figure 1. Gene maps of *Pyrus hopeiensis* HB-1chloroplast genomes.

The chloroplast genome of *P. hopeiensis* HB-1 contains 11 genes that harbor introns and one more *trnI-TAT* gene than *P. hopeiensis* HB-2. Of these 11 genes, two are tRNA genes (*trnI-TAT* and *trnI-AAT*) and nine are protein-coding genes (*rpoC1*, *rpl22*, *rpl22*, *ndhA*, *rpl2*, *rpl2*, *rps12*, *rps12*, *rps12*, *ycf3*), of which *ycf3* contains two introns. Excepting the intron length of the *trnI-AAT* gene, *P. hopeiensis* HB-2 is larger than *P. hopeiensis* HB-1. The exon and intron lengths of the two *P. hopeiensis* genotypes are identical. Among the coding genes, *ndhA* was the longest at 1125 bp, and rpl22 was the shortest at 63 bp.

Table 1. Genes of the cp genome of *P. hopeiensis* HB-1.

Functions	Family Name	Code	List of Genes
Self-replication	Small subunit of ribosome	*rps*	*rps2, rps3, rps4, rps7* [a], *rps8, rps11, rps12* [a b e], *rps14, rps15, rps18, rps19*
	rRNA Genes	*rrn*	*rrn4.5S* [a], *rrn5S* [a], *rrn16S* [a], *rrn23S* [a]
	Large subunit of ribosome	*rpl*	*rpl2* [a b], *rpl14, rpl16, rpl20, rpl22* [b], *rpl23* [a], *rpl32, rpl33, rpl36*
	DNA dependent RNA polymerase	*rpo*	*rpoA, rpoB, rpoC1* [b], *rpoC2*
	tRNA Genes	*trn*	*trnC-GCA, trnD-GTC, trnE-TTC, trnF-GAA, trnfM-CAT, trnG-GCC, trnH-GTG, trnI-TAT* [b], *trnI-CAT* [a], *trnI-AAT* [b], *trnL-CAA* [a], *trnL-TAG, trnM-CAT, trnN-GTT* [a], *trnQ-TTG, trnP-TGG, trnR-TCT, trnR-ACG* [a], *trnS-GCT, trnS-TGA, trnS-GGA, trnT-GGT, trnT-TGT, trnV-GAC* [a], *trnW-CCA, trnY-GTA*
Genes for photosynthesis	Subunits of ATP synthase	*atp*	*atpA, atpB, atpE, atpF* [d], *atpH, atpI*
	Subunits of protochlorophyllide reductase	*chl*	
	Subunits of NADH-dehydrogenase	*ndh*	*ndhA* [b], *ndhB* [b], *ndhC, ndhD, ndhE, ndhF, ndhG, ndhH, ndhI, ndhJ, ndhK*
	Subunits of cytochrome b/f complex	*pet*	*petA, petG, petL, petN*
	Subunits of photosystem I	*psa*	*psaA, psaB, psaC, psaJ, psaI*
	Subunits of photosystem II	*psb*	*psbA, psbB, psbC, psbD, psbE, psbF, psbH, psbI, psbJ, psbK, psbM, psbN, psbT, psbZ*
	Subunit of rubisco	*rbc*	*rbcL*
Other genes	Subunit of Acetyl-CoA-carboxylase	*acc*	*accD*
	Envelop membrane protein	*cem*	*cemA*
	c-type cytochrome synthesis gene	*ccs*	*ccsA*
	Protease	*clp*	*clpP* [d]
	Translational initiation factor	*inf*	
	Maturase	*mat*	*matK*
	Elongation factor	*tuf*	
Genes of unknown function	Conserved open reading frames	*ycf*	*ycf1, ycf2* [a], *ycf3* [c], *ycf4*

[a]—Two gene copies in IRs; [b]—Gene containing a single intron; [c]—Gene containing two introns; [d]—Pseudogene; [e]—Gene divided into two independent transcription units.

2.2. Comparison of the Basic Characteristics of the Chloroplast Genome in Five Pyrus Species

The chloroplast genome of *Pyrus* is a typical ring structure 159,834–160,059 bp in length (Table 2). The chloroplast genome of *P. communis* L. cv. Early Red Comice is the shortest, and the longest is *P. ussuriensis* Maxin. cv. Jingbaili. The LSC length of *Pyrus* is 87,793–88,074 bp, the longest in *P. ussuriensis* Maxin. cv. Jingbaili and the shortest in *P. communis* L. cv. Early Red Comice. The SSC length is 19,201–19,261 bp, with the longest in *P. betulifolia* and the shortest in both genotypes of *P. hopeiensis*.

Table 2. Comparison of the basic characteristics of chloroplast Genome in five *Pyrus* species.

	Pyrus hopeiensis **HB-1**	*Pyrus hopeiensis* **HB-2**	*Pyrus ussuriensis* **Maxin. cv. Jingbaili**	*Pyrus communis* **L. cv. Early Red Comice**	*Pyrus betulifolia*
Length (bp)	**159,935**	**159,981**	**160,059**	**159,834**	**160,058**
GC content (%)	36.59	36.57	36.57	36.58	36.57
AT content (%)	63.41	63.43	63.43	63.42	63.43
LSC length (bp)	87,962	88,008	88,075	87,794	88,025
SSC length (bp)	19,201	19,201	19,212	19,260	19,261
IR length (bp)	26,386	26,386	26,386	26,390	26,386
Gene number	118	119	117	114	120
Pseudogene number	2	2	2	2	2
Gene number in IR regions	32	32	31	31	32
Protein-coding gene number	77	78	75	74	77
Protein-coding gene (%)	64.25	65.55	64.10	64.91	64.17
rRNA gene number	8	8	8	8	8
rRNA (%)	6.78	6.72	6.84	7.02	6.67
tRNA gene number	31	31	32	30	33
tRNA (%)	26.27	26.05	27.35	26.32	26.50

The IR regions were of similar length in four of the five *Pyrus* species, and only differed by 4 bp, except in *P. communis* L. cv. Early Red Comice. The GC content was similar, 36.57–36.59%. The number of protein-coding genes ranged from 74 to 78. The chloroplast genome of the five *Pyrus* species contained 15 genes, including introns (nine protein-coding genes and six tRNAs, see Table 3), and the *ycf3* gene contained two introns. The intron in *P. betulifolia* contained 13 genes, followed by 12 in *P. ussuriensis* Maxin. cv. Jingbaili, 11 in *P. hopeiensis* HB-1, 10 in *P. hopeiensis* HB-2, and 10 in *P. communis* L. cv. Early Red Comice. In the five *Pyrus* species, nine genes contained introns (*rpoC1*, *ycf3*, *rpl22*, *rpl2*, *ndhA*, *ndhB*, *rpl2*, *rps12*, and *rps12*), all of which were protein-coding genes. The intron in the *ndhA* gene was the longest, at 1125–1169 bp. Other than the short intron length of the *trnI-TAT* gene in *P. ussuriensis* Maxin. cv. Jingbaili, *rpl22* harbored the smallest intron of the other four *Pyrus* species, at 63–83 bp. The results showed that the size, structure, sequence, and GC content of the chloroplast genome of *P. hopeiensis* were similar to those of the other three *Pyrus* species, which was characteristic of the slow evolution of the genus *Pyrus* [7] in comparison to the chloroplast genomes of *P. hopeiensis*, *P. communis* L. cv. Early Red Comice, *P. ussuriensis* Maxin. cv. Jingbaili, and *P. betulifolia*.

Table 3. Statistics of gene introns in the chloroplast genome of five *Pyrus* species.

Gene	Strand	*Pyrus hopeiensis* **HB-1**	*Pyrus hopeiensis* **HB-2**	*Pyrus ussuriensis* **Maxin. cv. Jingbaili**	*Pyrus communis* **L. cv. Early Red Comice**	*Pyrus betulifolia*
trnI-TAT	−	√	×	√	×	√
trnI-TAT	+	×	×	√	×	√
trnN-ATT	+	×	×	√	√	×
rpoC1	−	√	√	√	√	√
ycf3	−	√	√	√	√	√
rpl22	−	√	√	√	√	√
rpl2	−	√	√	√	√	√
ndhA	−	√	√	√	√	√

Table 3. *Cont.*

Gene	Strand	*Pyrus hopeiensis* HB-1	*Pyrus hopeiensis* HB-2	*Pyrus ussuriensis* Maxin. cv. Jingbaili	*Pyrus communis* L. cv. Early Red Comice	*Pyrus betulifolia*
ndhB	+	√	√	√	√	√
rpl2	+	√	√	√	√	√
rps12	−	√	√	√	√	√
rps12	+	√	√	√	√	√
trnI-AAT	+	√	√	×	×	×
trnL-TAG	−	×	×	×	×	√
trnY-ATA	+	×	×	×	×	√
Total	15	11	10	12	10	13

2.3. Chloroplast Gene Gain–Loss Events

The chloroplast genome structure of most higher plants is stable, and the number, sequence, and composition of genes are conserved. However, the loss of chloroplast genome genes is common. For example, the chloroplast genome of the sweet orange has lost the *infA* gene [8]; the *ycf1*, *ycf2*, and *accD* genes have been lost in Gramineae [9], and the chloroplast genome of some legumes has been recombined several times, resulting in the deletion of a copy of the IR region [10]. The *rpl22* gene was detected in the chestnut nuclear genome [11], presumably having derived from the chloroplast genome. In addition, the *rpl32* gene was transferred to the nuclear genome in the poplar [12], and 17 similar chloroplast regions were found in the mitochondrial genome of papaya, suggesting that the chloroplast gene may have transferred to the mitochondria [13]. However, there have been few studies on how these genes are lost, transferred, and integrated into the nuclear and mitochondrial genomes.

In this study, we compared the gain–loss events of eight *Pyrus* species (five *Pyrus* species that were sequenced in this study and three other *Pyrus* species, *P. pyrifolia*, *P. spinose*, and *P. pashia*, which were downloaded from NCBI) (Tables 4 and 5). The *psbL*, *psbl*, *trnI-GAU*, *trnA*, *trnL*, and *trnY-AUA* genes were most readily lost through evolution, followed by *trnA-UGC*, *trnG-UCC*, *trnI-AAU*, *trnI-GAU*, *trnK-UUU*, *trnN-AUU*, and *trnV-UAC*. *P. hopeiensis* HB-1 contained one less *MATK* gene than *P. hopeiensis* HB-2. Compared with the other three sequenced *Pyrus* species, *trnI-AAU* was only present in *P. hopeiensis* HB-1 and *P. hopeiensis* HB-2. The *atpB* gene was only lost in *P. ussuriensis* Maxin. cv. Jingbaili and *P. communis* L. cv. Early Red Comice, and *petB*, *petD*, *rps16* and *trnL-UAA* were present in the five *Pyrus* species sequenced here, but missing in *P. pyrifolia*, *P. spinosa*, and *P. pashia*.

2.4. Synonymous (KS) and Nonsynonymous (KA) Substitution Rate Analysis

Nucleotide mutations that do not cause amino acid changes are known as synonymous mutations, whereas nonsynonymous mutations do cause changes to the amino acid sequence. The Ka/Ks ratio (or dN/ds) of nonsynonymous substitution (Ka) and synonymous substitution (Ks) is the selection pressure of an encoded protein, which can be used to determine whether the gene encoded by the protein is under selection pressure. If $Ka/Ks > 1$, the protein is considered to be positively selected; if $Ka/Ks = 1$, the protein is neutral; and if $Ka/Ks < 1$, the protein is considered to have undergone purifying selection. It is generally believed that synonymous mutations are not subject to natural selection, whereas nonsynonymous mutations are.

Table 4. Genes from the chloroplast genomes of *Pyrus*.

Specie	*atpB*	*matK*	*petB*	*petD*	*psaC*	*psbI*	*psbL*	*psbl*	*Rpl20*	*Rpl36*	*rps16*	*ycf1*	*trnI-GAU*	*trnA*
Pyrus ussuriensis Maxin. cv. Jingbaili	0	1	0	0	1	1	0	0	1	1	0	0	0	0
Pyrus communis L. cv. Early Red Comice	0	1	0	0	1	1	0	0	0	1	0	0	0	0
Pyrus hopeiensis HB-1	1	0	0	0	1	1	0	0	1	1	0	1	0	0
Pyrus hopeiensis HB-2	1	1	0	0	1	1	0	0	1	1	0	1	0	0
Pyrus betulifolia	1	1	0	0	1	1	0	0	1	1	0	1	0	0
Pyrus pyrifolia	1	1	1	1	0	0	0	1	1	0	1	1	1	1
Pyrus spinosa	1	1	1	1	1	1	0	0	1	1	1	1	0	0
Pyrus pashia	1	1	1	1	1	1	1	0	1	0	1	2	0	0
Total number of missing genes	2	1	5	5	1	1	7	7	1	2	5	2	7	7

Table 5. Genes from the chloroplast genomes of *Pyrus*.

Specie	*trnA-UGC*	*trnG-GCC*	*trnG-UCC*	*trnI-AAU*	*trnI-GAU*	*trnI-UAU*	*trnK-UUU*	*trnL*	*trnL-UAA*	*trnN-AUU*	*trnV-UAC*	*trnY-AUA*
Pyrus ussuriensis Maxin. cv. Jingbaili	0	1	0	0	0	2	0	0	0	1	0	0
Pyrus communis L. cv. Early Red Comice	0	1	0	0	0	0	0	0	0	1	0	0
Pyrus hopeiensis HB-1	0	1	0	1	0	1	0	0	0	0	0	0
Pyrus hopeiensis HB-2	0	1	0	1	0	1	0	0	0	0	0	0
Pyrus betulifolia	0	1	0	0	0	2	0	0	0	0	0	1
Pyrus pyrifolia	1	1	0	0	1	0	1	1	1	0	1	0
Pyrus spinosa	0	0	1	0	0	0	0	0	1	0	0	0
Pyrus pashia	2	1	1	0	2	0	1	0	1	0	1	0
Total number of missing genes	6	1	6	6	6	4	6	7	5	6	6	7

Compared to *P. hopeiensis* HB-1, *psaJ*, *rpl20*, *rps18*, and *ycf1* in *P. hopeiensis* HB-2 were subject to negative selection, and no positive selection gene (Figure 2 and Table S1–S4) was found. In *P. betulifolia*, *atpE*, *ndhF*, *ndhI*, *rps18*, and *ycf2* were subject to positive selective pressure, whereas *ndhD*, *ndhH*, *ndhK*, *rpl20*, *rpl22*, *rpoC2*, *rps11*, and *ycf1* were subject to negative selection. The *psbC*, *psbK*, *rpoA*, *rps14*, *rps18*, and *ycf2* genes were subject to positive selective pressure in *P. communis* L. cv. Early Red Comice. Moreover, *accD*, *atpA*, *atpE*, *cemA*, *matK*, *ndhA*, *ndhD*, *ndhF*, *ndhH*, *petA*, *psaA*, *psaB*, *rbcL*, *rpl22*, *rpoB*, *rpoC2*, *rps11*, *rps2*, *rps3*, and *ycf4* were subject to negative selection. In *P. ussuriensis* Maxin. cv. Jingbaili, *atpE*, *atpI*, *cemA*, *ndhF*, *rps18*, and *ycf2* were subject to positive selective pressure, and *ndhD*, *ndhH*, *psaA*, *psbC*, *rpl20*, *rpl22*, *rpoC2*, *rps11*, and *ycf4* were subject to negative selection. Compared with *P. betulifolia*, *atpE* was subject to positive selective pressure in *P. betulifolia* and *P. ussuriensis* Maxin. cv. Jingbaili, whereas *atpE* was subject to negative selection in *P. communis* L. cv. Early Red Comice. This shows that the chloroplast genome of *Pyrus* has been affected by different environmental pressures during evolution, which may account for the different gene numbers among the five *Pyrus* species.

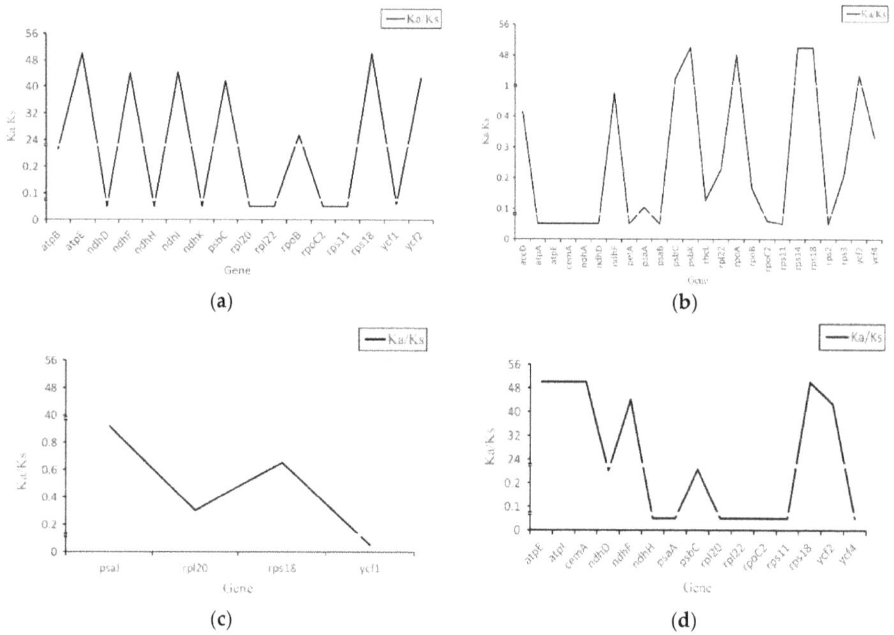

Figure 2. Ka/Ks value of five *Pyrus* species. (**a**)–(**d**) represent the Ka/Ks values of *Pyrus betulifolia*, *Pyrus communis* L. cv. Early Red Comice, *Pyrus ussuriensis* Maxin. cv. Jingbaili, and *Pyrus hopeiensis* HB-2, respectively, with respect to *Pyrus hopeiensis* HB-1.

2.5. Indel Identification and Relationship of the Five Pyrus cp Genomes

The nucleotide bases in coding and non-coding regions have different evolutionary mutation rates. DNA variations located in coding regions can lead to large phenotypic and functional variations; moreover, these often have a slower mutation rate, making them suitable for phylogenetic studies of higher order elements (families, orders, and higher). Mutations in non-coding regions have little effect on phenotype and fewer functional restrictions, and as they take no part in the transcription/translation process, they have a relatively high nucleotide replacement rate and hence rapid evolution, making them suitable for the phylogenetic study of lower order elements (species, genus) [14].

The chloroplast genome data of five *Pyrus* species were compared with those of *P. hopeiensis* HB-1 by multiple sequence alignment using MAFFT. All differentially expressed sites were extracted using a script from the comparison results, and differences in sites of indels $\geq$ 5 bp were screened out. The location of different chloroplast genome sites was determined and ggplot in R was used to create graphic plots that were then optimized using AI. The results indicated 15 mutation sites in *P. hopeiensis* (Figure 3), which included 11 insertion and four deletion sites. All of these mutation sites were located in the LSC region; three were located in gene regions and 12 in intergenic regions. Among these, the longest was located in the *ndh-trnM-CAT* region, and as many as six mutations were located in the intergenic region *rpl18–rps20*. A total of 96 mutation sites were detected in the other four *Pyrus* species, 81 of which were located in the LSC region of the chloroplast genome and 11 in the SSC region, whereas only two mutation sites were found in the IRa and IRb regions in *P. communis* L. cv. Early Red Comice. There were more mutation sites in the SC region, and the IR region was more conserved. Indels were mainly located in the intergenic regions, and three indel loci were detected in the intron region (*rpl22*, *trnN-ATT*, *ndhA*). Because the protein-coding region is arranged by triplet codons, the tolerance of indels is poor. Therefore, only five indel loci were detected in the protein-coding region (*trnL-TAT*, *trnN-ATT*, *rps18*, *rps19* and *ycf1*), but no indel loci were detected in the rRNA region. A comparison of the occurrence of these indel loci among the four *Pyrus* species revealed 15 indel loci in the chloroplast genome of *P. hopeiensis*, 32 in *P. ussuriensis* Maxin. cv. Jingbaili, 57 in *P. communis* L. cv. Early Red Comice, and 31 in *P. betulifolia*. The insertion or deletion frequency in the chloroplast genome of *P. hopeiensis* HB-2 was less than that in *P. hopeiensis* HB-1. The *psbA-trnQ_TTG* and *rpl18-rps20* intergenic regions were the most variable regions with seven loci, followed by *trnT-TGT_trnF_GAA* (six) and the *trnI-TAT* gene-coding region (six). The largest indels were located in *psbA_trnQ-TTG* in the chloroplast genome of *P. communis* L. cv. Early Red Comice.

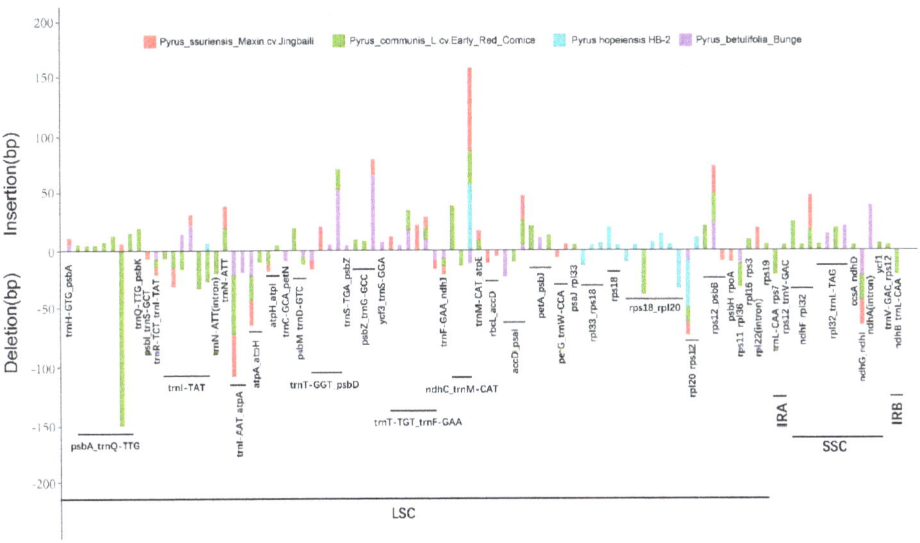

Figure 3. Indels (≥5 bp) identified based on multiple sequence alignment of five *Pyrus* cp genomes. Insertions are shown above and deletions below the horizontal axis. Indel distribution was positioned using *Pyrus hopeiensis* HB-1 as a reference.

2.6. Codon Preference Analysis

Codons have an important role in the transmission of genetic information. Codon use is not equal in many species, and the phenomenon of a specific codon use frequency being higher than that of its synonymous codon is known as codon preference [15]. Codon preference is formed during the long-term evolution of organisms, with different species having different codon preferences. Codon use is affected by natural selection, mutagenesis, tRNA abundance, the composition of base groups, hydrophilicity of codons, gene length, and expression levels [16]. Analysis of the codon use preferences of a species improves our understanding of the transmission of genetic information and the development of evolutionary and phylogenetic models.

The annotated files of plant genomes, including *P. pashia*, *P. pyrifolia*, *Malus prunifolia*, *Prunus mume*, and *Chaenomeles japonica*, were selected from the NCBI database, including the sequence files encoding CDS and proteins. According to the full-length CDS criterion, sequences with lengths <300 nt were deleted. The codon-use frequency of each genome was extracted from the annotated files of each genome and the corresponding frequency ratio was calculated. The final statistical results were clustered and mapped using the pheatmap package in R. The results showed obvious codon use preferences for both types of *P. hopeiensis*, among which ATT, AAA, GAA and AAT, and TTT were used most frequently (Figure 4). Statistical analysis of all the codons of *P. hopeiensis*, the three other *Pyrus* species, and the other Rosaceae showed a high A/T preference in the third chloroplast codon. This is common in the chloroplast genomes of higher plants [17–21].

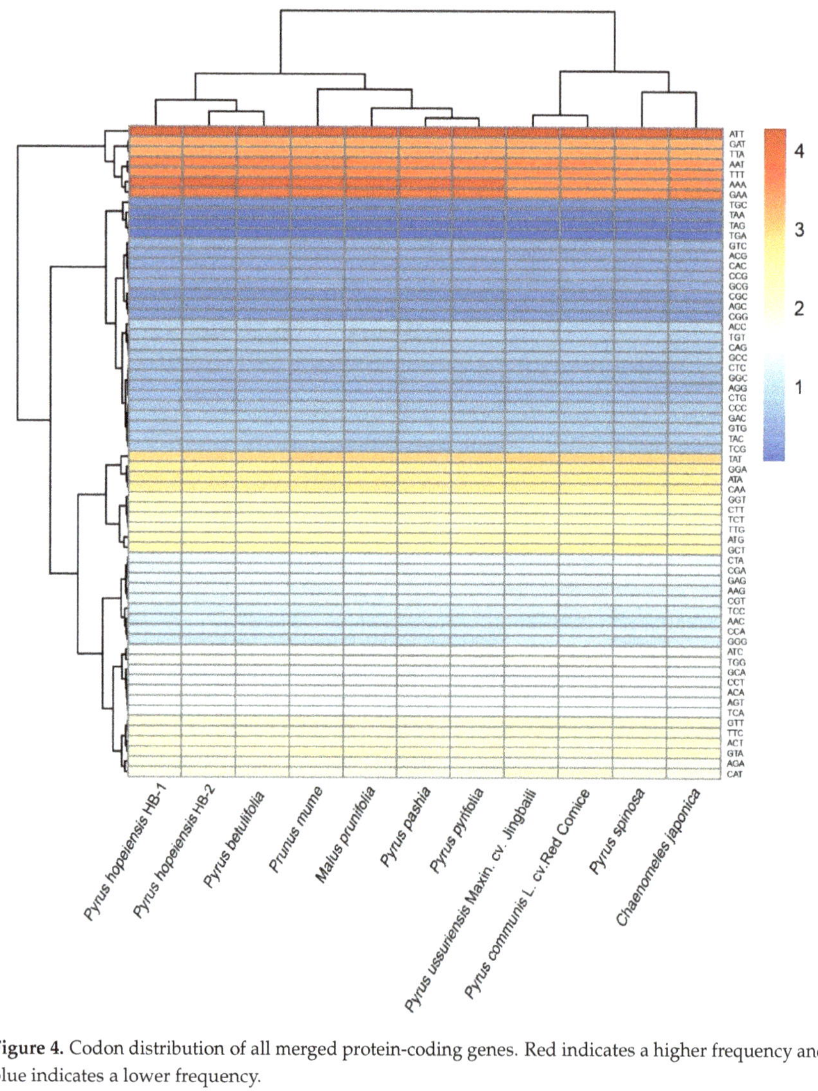

Figure 4. Codon distribution of all merged protein-coding genes. Red indicates a higher frequency and blue indicates a lower frequency.

2.7. Comparison of the Genome Structure in Rosaceae cp Genomes

The chloroplast genome structure of most higher plants is relatively stable and the number, sequence, and composition of their genes are conserved. However, because different plant groups have different evolutionary histories and genetic backgrounds, the chloroplast genome size, genome structure, and gene numbers vary. Insertion/deletion is the most frequent type of microstructural variation in the chloroplast genome, and it occurs frequently in some segments where the variation is high, such as *trnH-psbA* and *trnS-G*. In Rosaceae, an insertion/deletion of 277 bp in the intergenic region of the *trnS-G* gene was reported in peach plants [22], and an insertion/deletion of 198 bp in the intergenic region of *trnL-F* was identified in *P. mume* [23].

The collinear method was used to analyze and compare the chloroplast genomes of the two genotypes of *P. hopeiensis*, the other three sequenced *Pyrus*, and other related Rosaceae (*P. pashia*, *P. pyrifolia*, *P. spinosa*, *M. prunifolia*, *P. mume*, and *C. japonica*). The results showed optimal collinearity between *P. hopeiensis* HB-1 and *P. hopeiensis* HB-2, and only a few sites contained insertions and deletions (Figure 5). Compared with the other Rosaceae, the genome structure and gene sequences were highly conserved, with more linear relationships indicating high chloroplast genome homology among the different plants.

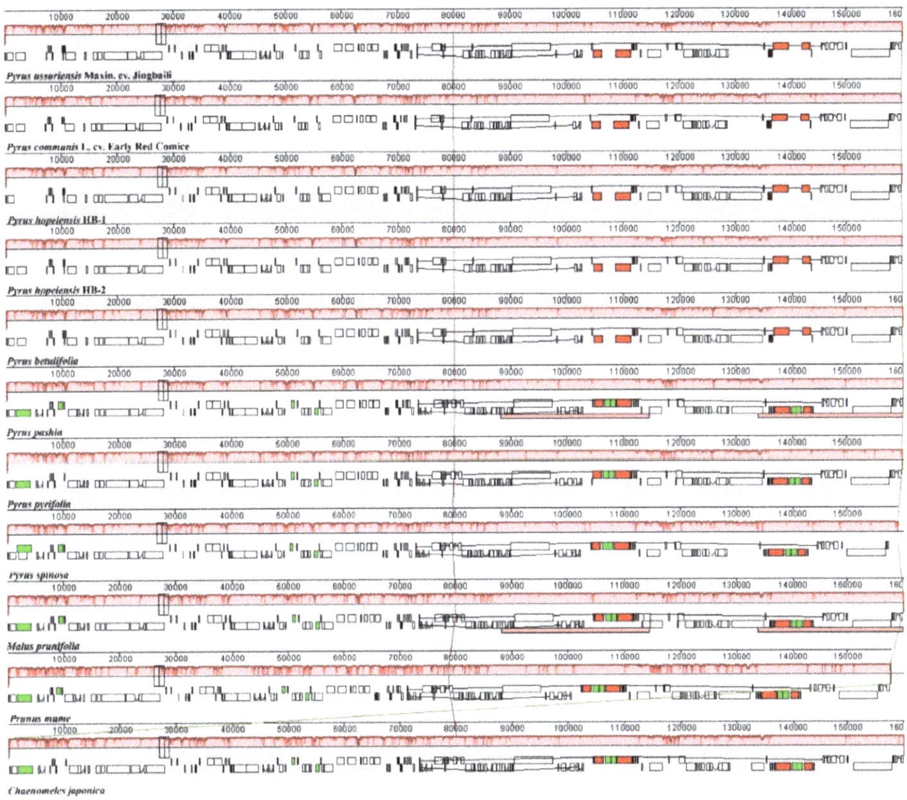

Figure 5. Co-linear analysis of various plant chloroplast genomes.

2.8. IR Contraction Analysis

The IR region is considered to be consistent and stable in the chloroplast genome. However, in the evolution of species, border region contraction and expansion are common. In this study, the IR boundaries of both genotypes of *P. hopeiensis* were compared. The IRa/LSC boundary extended into the *rps19* gene, and 120 bp of *rps19* extended into the IRa region. The IRa/SSC boundary extended into the *ndhF* gene, and 12 bp of *ndhF* extended into the IRa region. The IRb/SSC boundary extended into 1074 bp of *ycf1* and the IRb/LSC border extended into the *rpl2* gene, with the *trnH-GTG* gene located downstream.

The IR boundaries were compared among the Rosaceae, including the five *Pyrus* species sequenced, and *P. pashia*, *P. pyrifolia*, *P. spinosa*, *M. prunifolia*, *P. mume*, and *C. japonica* (Figure 6). The IRa/LSC boundary of these plants extended to the *rps19* gene. The IRa/SSC boundaries were located upstream of the *ndhF* gene, except in *M. prunifolia*, whose IRa/SSC junction lost *ndhF* but extended to *ycf1*. The *P. spinosa* IRb/LSC boundary had no *rpl2*. All IRb/SSC boundaries expect those of *P. ussuriensis* Maxin. cv. Jingbaili, *P. communis* L. cv. Early Red Comice, and *P. Spinosa* extended to *ycf1*. The IRb/SSC boundary lost *ycf1*. These findings were similar to those in the Actinidiaceae, Theaceae, and Primulaceae, but differed markedly from those in Ericaceae. For the IRb/LSC boundary, all but that of *P. spinosa* extended into the *rpl2* gene, and the IRb/LSC boundary of the *trnH-GTG* gene located downstream extended into the *rpl23* gene in *P. spinosa*.

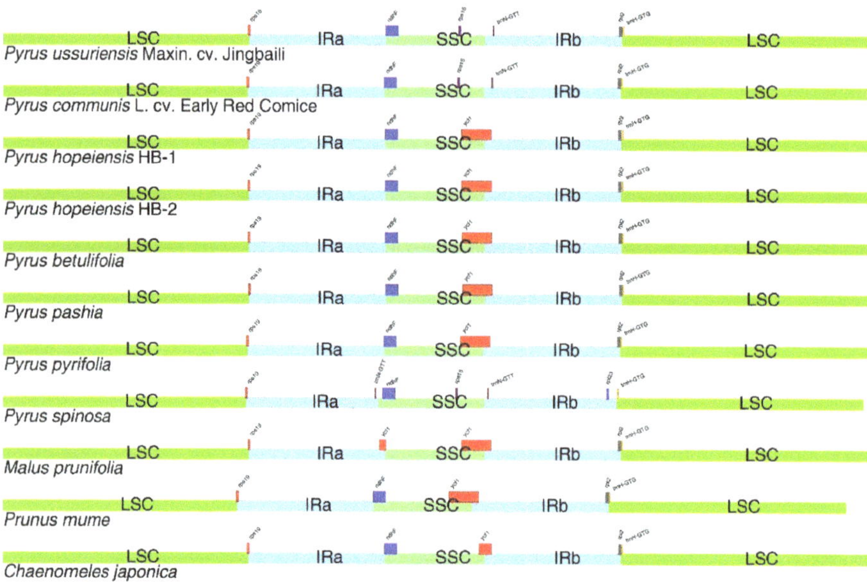

Figure 6. IR contraction analysis of Rosaceae.

2.9. Phylogenetic Analysis

To gain an insight into the position of *Pyrus* within the Rosaceae, a molecular phylogenetic tree was constructed using 57 protein-coding genes (*accD*, *atpA*, *atpE*, *atpH*, *atpI*, *csA*, *cemA*, *ndhA*, *ndhB*, *ndhC*, *ndhD*, *ndhE*, *ndhG*, *ndhH*, *ndhJ*, *ndhK*, *petA*, *petG*, *petL*, *petN*, *psaA*, *psaB*, *psaI*, *psbA*, *psbB*, *psbC*, *psbD*, *psbE*, *psbF*, *psbH*, *psbJ*, *psbM*, *psbN*, *psbT*, *rbcL*, *rpl14*, *rpl16*, *rpl2*, *rpl22*, *rpl23*, *rpl33*, *rpoA*, *rpoC1*, *rps11*, *rps12*, *rps14*, *rps15*, *rps18*, *rps19*, *rps2*, *rps3*, *rps4*, *rps7*, *rps8*, *ycf2*, *ycf3*, *ycf4*)from the chloroplast genomes of 36 Rosaceae, which were downloaded from GenBank, and using *Arabidopsis thaliana* as the outgroup. The resulting phylogenetic tree was consistent with the traditional plant morphological taxonomy (Figure 7), which can be divided into three sections: Maloideae, Prunoideae, and Rosoideae. The Maloideae include *Pyrus*, *Malus*, *Sorbus* L., and *Eriobotrya*. *Prunus* lies within the Prunoideae, and *Fragaria* is included in the Rosoideae. The phylogenetic relationship of the Prunoideae was closer than that of the Rosoideae to the Maloideae, and the relationship between *Malus* and *Pyrus* was the closest.

Within *Pyrus*, the relationship between *P. hopeiensis* HB-1 and *P. hopeiensis* HB-2 was the closest, and the relationship between *P. betulifolia* and *P. ussuriensis* Maxin. cv. Jingbaili was closer than that between other *Pyrus* and *P. hopeiensis*. In addition, it can be seen from the evolutionary tree that Rosoideae is a subfamily that split off from the evolutionary tree.

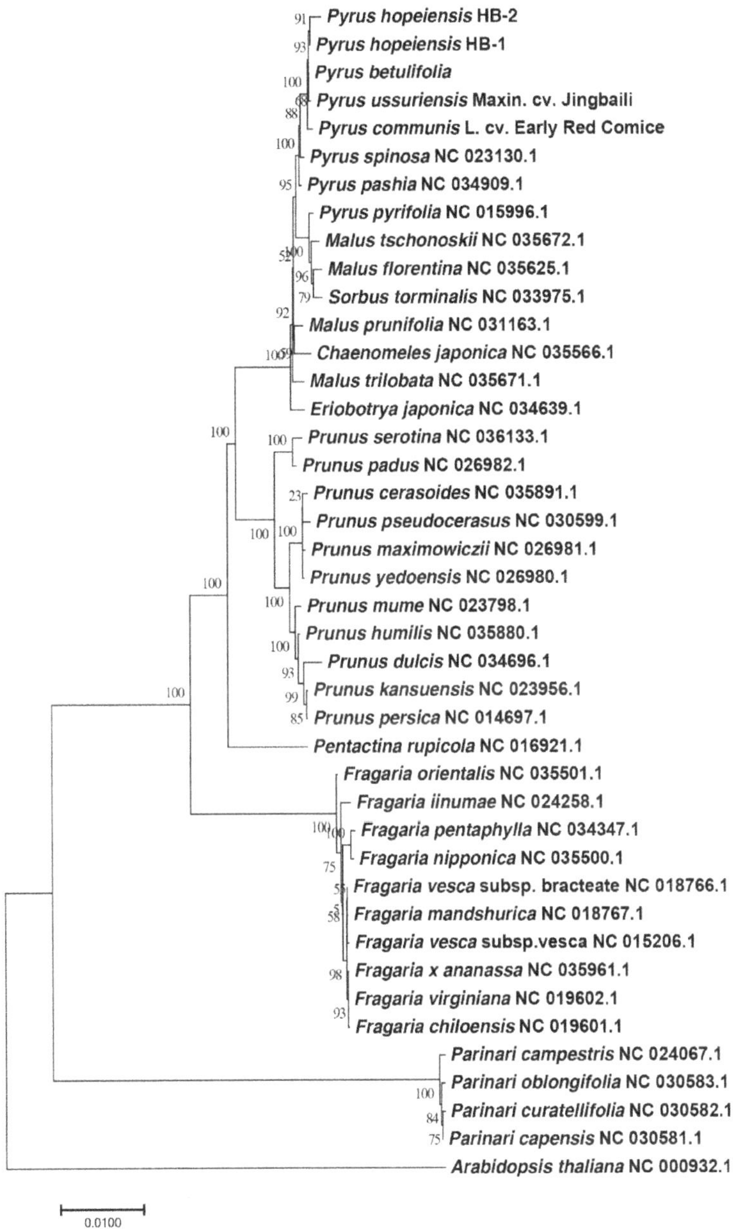

Figure 7. The ML phylogenetic tree of the Rosaceae clade based on same protein-coding genes. Numbers above or below the nodes are bootstrap support values.

3. Discussion

Pyrus hopeiensis is a valuable wild resource of *Pyrus*, which belongs to the family Rosaceae. Because of its limited distribution and population decline, *P. hopeiensis* is listed among "the wild plants with tiny population" in China. It belongs to one of 13 species of *Pyrus* present in China. In this study, the chloroplast genomes of the two genotypes *P. hopeiensis* HB-1 and *P. hopeiensis* HB-2 and those of three other major pear plants, *P. ussuriensis* Maxin. cv. Jingbaili, *P. communis* L. cv. Early Red Comice, and *P. betulifolia*, were analyzed using high-throughput sequencing for comparative analysis. The chloroplast genome of *Pyrus*, like those of most higher plants, is a typical tetrad consisting of two reverse repeat IR regions and small and large single copy fragments [24]. There was a 46 bp difference in the chloroplast genome size between the two *P. hopeiensis*, which was located in the LSC region. Compared with the other three *Pyrus* species, the total genome length was <225 bp, and the gene number, gene type, gene sequence, LSC, IR, and SSC lengths and GC content were similar. This strongly suggests that chloroplast genomes are highly conserved [25]. The encoded genes of the chloroplast genome are divided into three categories based on their functions. The first is related to chloroplast gene expression, such as tRNAs, rRNAs, and three subunits encoding chloroplast RNA polymerase synthesis. The second is related to photosynthesis, and the third consists of other biosynthesis genes and some genes of unknown function, such as mat and *ycf* [26]. The chloroplast genes of *Pyrus* are similar in composition.

The genomic sequence of the *P. hopeiensis* HB-1 chloroplast was used as a reference sequence to detect single-nucleotide polymorphisms (SNPs) and indels in the other four *Pyrus* species. The results showed a significantly higher variation in the non-coding region than in the coding region and more mutation sites in the intergenic region of the *psbA-trnQ_TTG*, *rpl18-rps20*, and *trnT-TGT_trnF_GAA* genes, which could be used for evolutionary analysis of *Pyrus*. The chloroplast genomes of *Pyrus* show obvious codon preference and similar codon use frequency. Furthermore, the third chloroplast codon has a higher A/T preference. This phenomenon is common in the chloroplast genomes of other higher plants [27]. Although the IR region is highly conserved, the expansion and contraction of the IR region is a common characteristic of the chloroplast genome. The degree of expansion of the IR/SC boundary is similar among the five *Pyrus* species, the two *P. hopeiensis* genotypes contain few genes with different extension positions, and any differences are very small, which is useful in the classification of *Pyrus*, as it can be used as a basis to identify the evolution of the chloroplast genome. The classification and identification of pear species in this study can be utilized for the preservation of pear germplasm resources. The initial identification of species and varieties of *Pyrus* was mainly based on morphological features (leaves, petioles, floral organs, sepals, hairs, fruits, and ventricles) and geographical distribution. For example, based on an investigation of morphological characteristics and natural distribution, Chinese taxonomists believe that *P. hopeiensis*, *P. phaeocarpa*, *P. sinkiangensis*, and *P. serrulata* were all formed by natural crosses [28]. Yu [29] divided *Pyrus* from China into 13 species based on their serrated leaf margins, and these included *P. hopeiensis*, *P. betulifolia*, *P. ussuriensis*, *P. phaeocarpa*, *P. bretschneideri*, *P. pyrifolia*, *P. pashia*, *P. armeniacaefolia*, *P. calleryana*, *P. pseudopashia*, *P. serrulata*, *P. sinkiangensis*, and *P. xerophila*. Anatomical studies in Wang Yingzhong [30] revealed that the anatomical structures of *P. betulifolia* and *P. ussuriensis*, and *P. bretschneideri*, *P. pyrifalia*, and *P. communis* were similar. The results showed that the relationship between *P. ussuriensis* and *P. betulifolia*, *P. bretschneideri*, and *P. pyrifalia* was close. However, it is easy to cross *Pyrus* species and there are no obvious differences in the biological and morphological characteristics among species and varieties, which greatly increases the difficulty of establishing its phylogenetic evolution and classification.

Pollen morphological identification, cytological markers, isozymes, and other methods have also been studied with a view to classifying *Pyrus*. The pollen morphology of *P. sinkiangensis* is similar to that of the Western pear, indicating a close relationship [31]. The pollen morphology of the Western pear is obviously different from that of the Oriental pear. The pollen morphology of *P. calleryana* has many primitive characteristics, and it is a primitive species of *Pyrus* in China. The pollen morphology of *P. bretschneideri*, also present in China, has the characteristics of both *P. pyrifalia* and *P. ussuriensis*, and may be a natural hybrid of *P. pyrifalia* and *P. ussuriensis*. Cytological markers enabled the analysis of the number, banding, karyotype, and meiosis behavior of the chromosomes. *P. phaeocarpa* has a similar karyotype to that of *P. betulifolia*, and those of *P. sinkiangensis*, *P. hopeiensis*, and *P. serrulata* were also similar [32,33]. In 1983, Lin Bonian and Shen Dexu [34] proved, through the use of the peroxidase isozyme, that *P. bretschneideri* and *P. pyrifolia* were closely related. However, these methods have few characteristic sites, poor polymorphism, and low accuracy, and provide a limited amount of information. To date, the relationships among *Pyrus* species, their origin, evolution of cultivation systems, and the origin of some suspicious species and hybrids remain unclear.

In recent years, molecular markers based on DNA, such as restriction fragment length polymorphisms (RFLPs) and simple sequence repeats (SSRs) have been used to investigate the genetic relationships, genetic diversity, and germplasm of *Pyrus*. However, there remain some deficiencies in the study of the interspecific relationships and origins of hybrids. Results based on random amplification of polymorphic DNA (RAPD) showed that the origin of *P. sinkiangensis* involved the crossing of many Eastern and Western pear species and that the genetic relationship between *P. bretschneideri* and *P. pyrifolia* is very close [35]. RAPD, inter sequence simple repeats (ISSR), and other DNA markers showed that *P. hopeiensis*, *P. betulifolia* and *P. phaeocarpa* were closely related to each other. In the same way, *P. phaeocarpa* is considered to be a hybrid of *P. betulifolia* and *P. ussuriensis*, whereas *P. hopeiensis* is a hybrid of *P. phaeocarpa* and *P. ussuriensis*. In a study using RAPD, *P. hopeiensis* and *P. phaeocarpa* shared some spectral bands with *P. betulifolia* and *P. ussuriensis* [36–38]. Zheng et al. identified a close relationship between *P. ussuriensis* and *P. hopeiensis* using internal transcribed spacer (ITS) sequences, which is consistent with the results in our study [39].

Because the chloroplast genome is the second-largest genome after the nuclear genome, it is maternally inherited in most angiosperms; thus it reflects the maternal evolutionary history, and this helps us to understand the maternal ancestors of suspected hybrids. The coding and non-coding regions of the chloroplast genome evolve at different rates, making them suitable for systematic research at different levels. The coding region is highly conserved and is only suitable for phylogenetic studies of families, orders, and higher taxonomic levels, whereas the non-coding regions are less constrained by function and the rapid evolutionary rate is suitable for plant phylogenetic studies at interspecific and subspecies levels. At present, the successful design of a set of universal primers for the chloroplast gene non-coding regions (such as *trnS-psbcc*, *trnL-trnF* and *accD-pasI*) [40] has made the study of chloroplast non-coding regions a hot topic in studies of the systematic relationship of *Pyrus*. Phylogenetic trees based on combinations of the sequences of *trnL-trnF* and *accD-psaI* in the chloroplast non-coding regions have further confirmed the theory of an independent evolution of the Oriental pear and the Western pear from the background of matrilineal evolution, and have shown the close relationship between *P. bretschneideri* and *P. pyrifolia* [41]. A study of the *trnL-trnF* region of cpDNA showed that *P. sinkiangensis* is closely related to the Western pear and the Oriental pear; the relationship between *P. betulifolia* and *P. ussuriensis* is close; *P. bretschneideri* is a hybrid of *P. ussuriensist*, *P. phaeocarpa*, and *P. pyrifolia*; and that the Western pear and Oriental pear are related to each other [42]. The sequences of these regions are highly conserved, with only limited sites available to provide information to unravel the phylogeny of *Pyrus*. However, no comprehensive and systematic cpDNA sequence analysis of *Pyrus* exists in China. To further our understanding of the inter-species relationships of *Pyrus* and to reveal the origin of hybrids and explore the evolutionary model of Eastern and Western pears, a wider range of representative species and varieties of Eastern and Western pears must be selected, and the nuclear gene fragments inherited by their parents should be combined, especially the low copy nuclear gene introns.

4. Materials and Methods

4.1. Plant Materials

In early May 2017, fresh leaves were collected from *P. hopeiensis* HB-1, *P. hopeiensis* HB-2, and three local *Pyrus* species, *P. ussuriensis* Maxin. cv. Jingbaili (which belongs to the *P. ussuriensis* family), *P. communis* L. cv. Early Red Comice (a high-quality variety of Western pear native to the United Kingdom) and *P. betulifolia* (most widely used in northern China as pear rootstocks) in Changli, Hebei Province and were stored before being transported to ORI-GENE Ltd., a science and technology company based in Beijing, China, for chloroplast genome sequencing.

4.2. DNA Sequencing, Genome Assembly, and Validation

The total DNA of fresh young leaves was extracted using a plant DNA extraction kit (Tiangen, Beijing, China). Agarose gel electrophoresis was used to detect DNA integrity, and purity and concentration were ascertained. The Illumina HiSeq platform was used to sequence the total DNA. After sequencing, the raw data was initially screened to remove low quality regions affecting the data quality and subsequent analysis needed to obtain the expected clean data. The SOAPdenovo2.01 [43] oligonucleotide analysis package was used to assemble the contig sequence. BLAT36 [44] was used to locate the assembled long sequence on the chloroplast reference genome of the relative species and to obtain the relative position of the contig sequence to enable splicing of the contig according to its relative position, and to correct assembly errors. A full-length frame map of the chloroplast genome was obtained. GapCloser software was used to fill gaps on the frame map sequence with high-quality short sequences. Any remaining gaps and suspected regions were supplemented and confirmed by generation sequencing, and the small single copy (SSC) and inverted repeat (IR) region junctions were verified. Finally, a complete ring chloroplast genome sequence was obtained.

4.3. Gene Annotation

CpGAVAS [45] was used to annotate the gene and the final annotation results were obtained by artificial correction. First, the results of Blastx, BLASTn, protein2genome, and est2genome [46] were integrated to predict the protein coding gene and the rRNA gene. Then, tRNA was identified using tRNAscan [47] and ARAGORN [48]. Finally, the reverse repeat region IR was identified using Vmatch [49]. Chloroplast genome mapping was performed using OrganellarGenomeDRAW [50] (http://ogdraw.mpimp-golm.mpg.de/index.shtml) based on the annotated results.

The protein and coding sequences (CDS) of each sample were extracted from the annotated files of each sample and the pairwise protein sequences aligned using MUSCLE software. The aligned protein sequences were converted to DNA sequences using PAL2NAL. KaKs_Calculator2.0 [51] software (https://sourceforge.net/projects/kakscalculator2) was then used to calculate Ka/Ks, which was used to analyze the selection pressure on different *Pyrus* species during the evolutionary process. Chloroplast genome sequences of 36 Rosaceae species were selected from NCBI and 57 common protein coding genes were used to explore the evolution of the *Pyrus* chloroplast genome, using *Arabidopsis thaliana* as the outgroup. The taxonomic status was confirmed. The annotated files of all of the genomes were downloaded from NCBI and the protein sequences of any genes shared among the chloroplast genomes of all of the species were extracted. Each gene was placed in a file in which each genome contained only one protein sequence. MUSCLE was used to make multiple sequence alignments for each file. The first and last sequences were aligned according to the genome source to obtain a growing alignment sequence: final.fa. MEGA7.0 software was then used to construct a neighbor-joining tree and the CGView Server was used to analyze the genetic variation among the chloroplast genomes of the five *Pyrus* species.

5. Conclusions

In this study, we reported the de novo sequencing results of *P. hopeiensis* chloroplast genomes. The length of the chloroplast genome of *P. hopeiensis* HB-1 is 159,935 bp, which is 46 bp longer than that of *P. hopeiensis* HB-2. The SSC and IR regions of the two Pyrus genotypes were the same length, with the only difference present in the LSC region. A total of 118 genes were identified in *P. hopeiensis* HB-1, and it only lacked the *MATK* protein-coding gene that was associated with biosynthesis in *P. hopeiensis* HB-2. The GC content of *P. hopeiensis* HB-1 was only 0.02% higher than that of *P. hopeiensis* HB-2. A total of 11 genes in the chloroplast genome of *P. hopeiensis* HB-1 contained introns, and an additional *trnI-TAT* gene not present in *P. hopeiensis* HB-2. *ycf3* is the only gene that contained two introns. The chloroplast genome structure and size, gene species, gene number, and GC content of *P. hopeiensis* were similar to those of the other three *Pyrus* species investigated. Almost all of the protein coding sequences and amino acid codons showed an obvious codon preference. Selection pressure analysis revealed that the chloroplast genomes of different pears were affected by different environmental pressures during the evolutionary process, which may account for the differences in gene numbers among the five *Pyrus* species. Phylogenetic analysis strongly supports the status of *Pyrus* in the Rosaceae. This study adds to our knowledge of the molecular evolution of *Pyrus*, and will be of use for the genetic breeding and chloroplast engineering of *Pyrus*.

Supplementary Materials: Supplementary materials can be found at http://www.mdpi.com/1422-0067/19/10/3262/s1.

Author Contributions: M.Y. designed the research; Y.L., J.Z. and J.X. collected the samples; Y.L., L.L. and L.G. performed the experiments and analysis; Y.L. and J.Z. wrote the manuscript; all authors revised the manuscript.

Funding: This research received no external funding.

Acknowledgments: This research was supported by the National Key Research and Development Plan "Research on protection and restoration of typical small populations of wild plants" (Grant No. 2016YFC0503106).

Conflicts of Interest: The authors declare no conflict of interest.

References

1. Chai, M.L.; Shen, D.X. Status and Prospects of Pear Breeding in China. *J. Fruit Sci.* **2003**, *5*, 379–383.
2. Yu, D.J. *Taxonomy of the Fruit Tree in China*; China Agriculture Press: Beijing, China, 1979; p. 122.
3. Jiang, X.F.; Chu, Q.G.; Zhang, C.S. Studies on the classification and evolution of the genus pyrus in China. *J. Laiyang Agric. Coll.* **1992**, *9*, 18–21.
4. Pu, F.S.; Huang, L.S.; Sun, B.J.; Li, S.L. Study on the chromosome number of wild and cultivated pears (Pyrus sp.) in China. *Acta Hortic. Sin.* **1985**, *12*, 155–158.
5. Neuhaus, H.E.; Ernes, M.J. Nonphotosyntiietic Metabolism in Plastids. *Annu. Rev. Plant Physiol. Plant Mol. Biol.* **2000**, *51*, 111–140. [CrossRef] [PubMed]
6. Dumolin, S.; Demesure, B.; Etit, R.J. Inheritance of chloroplast and mitochondrial genomes in pedunculate oak investigated with an efficient PCR method. *Theor. Appl. Genet.* **1995**, *91*, 1253–1256. [CrossRef] [PubMed]
7. Xing, S.C. Progress in Chloroplast Genome Analysis. *Prog. Biochem. Biophys.* **2008**, *35*, 21–28.
8. Bausher, M.G.; Singh, N.D.; Lee, S.B.; Jansen, R.K.; Daniell, H. The complete chloroplast genome sequence of *Citrus sinensis* (L.) Osbeck var 'Ridge Pineapple': Organization and phylogenetic relationships to other angiosperms. *BMC Plant Biol.* **2006**, *6*, 21. [CrossRef] [PubMed]
9. Guisinger, M.M.; Chumley, T.W.; Kuehl, J.V.; Boore, J.L.; Jansen, R.K. Implications of the plastid genome sequence of *Typha* (Typhaceae, Poales) for understanding genome evolution in Poaceae. *J. Mol. Evol.* **2010**, *70*, 149–166. [CrossRef] [PubMed]
10. Tao, X.L.; Ma, L.C.; Nie, B.; Wang, Y.R.; Liu, Z.P. The draft and characterization of the complete chloroplast genome of *Vicia sativa* cv. Lanjian No. 3. *Pratacult. Sci.* **2017**, *34*, 321–330.
11. Jansen, R.K.; Saski, C.; Lee, S.B.; Hansen, A.K.; Daniell, H. Complete plastid genome sequences of three Rosids (*Castanea, Prunus, Theobroma*): Evidence for at least two independent transfers of rpl22 to the nucleus. *Mol. Biol. Evol.* **2011**, *28*, 835–847. [CrossRef] [PubMed]

12. Ueda, M.; Fujimoto, M.; Arimura, S.I.; Murata, J.; Tsutsumi, N.; Kadowaki, K.I. Loss of the rp132 gene from the chloroplast genome and subsequent acquisition of a preexisting transit peptide within the nuclear gene in populus. *Gene* **2007**, *402*, 51–56. [CrossRef] [PubMed]
13. Cheng, H.; Ge, C.F.; Zhang, H.; Qiao, Y. Advances on Chloroplast Genome Sequencing and Phylogenetic Analysis in Fruit Trees. *J. Nucl. Agric. Sci.* **2018**, *32*, 58–69.
14. Clegg, M.T.; Gaut, B.S.; Learn, G.H.; Morton, B.R. Rates and patterns of chloroplast DNA evolution. *Proc. Natl. Acad. Sci. USA* **1994**, *91*, 6795–6801. [CrossRef] [PubMed]
15. Jiang, W.; Lyu, B.B.; He, J.H.; Wang, J.B.; Wu, X.; Wu, G.G.; Bao, D.P.; Chen, M.J.; Zhang, J.S.; Tan, Q.; et al. Codon usage bias in the straw mushroom Volvariella volvacea. *Chin. J. Biotechnol.* **2014**, *30*, 1424–1435. (In Chinese)
16. Li, Y.; Kuang, X.J.; Zhu, X.X.; Zhu, Y.J.; Sun, C. Codon usage bias of Catharanthus roseus. *China J. Chin. Mater. Med.* **2016**, *41*, 4165–4168.
17. Zuo, L.H.; Shang, A.Q.; Zhang, S.; Yu, X.Y.; Ren, Y.C.; Yang, M.S.; Wang, J.M. The first complete chloroplast genome sequences of Ulmus species by de novo sequencing: Genome comparative and taxonomic position analysis. *PLoS ONE* **2017**, *12*, e0171264. [CrossRef] [PubMed]
18. Yang, M.; Zhang, X.; Liu, G.; Yin, Y.; Chen, K.; Yun, Q.; Zhao, D.; Al-Mssallem, I.S.; Yu, J. The complete chloroplast genome sequence of date palm (*Phoenix dactylifera* L.). *PLoS ONE* **2010**, *5*, el2762. [CrossRef] [PubMed]
19. Nie, X.; Lv, S.; Zhang, Y.; Du, X.; Wang, L.; Biradar, S.S.; Tan, X.; Wan, F.; Weining, S. Complete chloroplast genome sequence of a major invasive species, crofton weed (*Ageratina adenophord*). *PLoS ONE* **2012**, *7*, e36869. [CrossRef] [PubMed]
20. Tangphatsornruang, S.; Sangsrakru, D.; Chanprasert, J.; Uthaipaisanwong, P.; Yoocha, T.; Jomchai, N.; Tragoonrung, S. The chloroplast genome sequence of mung bean (*Vigna radiatd*) determined by high-throughput pyrosequencing: Structural organization and phylogenetic relationships. *DNA Res.* **2010**, *17*, 11–22. [CrossRef] [PubMed]
21. Yi, D.K.; Kim, K.J. Complete chloroplast genome sequences of important oil seed crop *Sesamum indicum* L. *PLoS ONE* **2012**, *7*, e35872. [CrossRef] [PubMed]
22. Quan, X.; Zhou, S.L. Molecular identification of species in Prunus sect. Persica (Rosaceae), with emphasis barcodes for plants. *J. Syst. Evol.* **2011**, *49*, 138–145. [CrossRef]
23. Wand, L.; Dong, W.P.; Zhou, S.L. Structural Mutations and Reorganizations in Chloroplast Genomes of Flowering Plants. *Acta Bot. Boreal. Occident. Sin.* **2012**, *32*, 1282–1288.
24. Yu, D.J.; Gu, C.Z. *Flora Reipublicae Popularis Sinicae*; Bejing Science Press: Beijing, China, 1974; Volume 36. (In Chinese)
25. Jansen, R.K.; Raubeson, L.A.; Boore, J.L.; Pamphilis, C.W.; Chumley, T.W.; Haberle, R.C.; Wyman, S.K.; Alverson, A.J.; Peery, R.; Herman, S.J.; et al. Methods for obtaining and analyzing whole chloroplast genome sequences. *Methods Enzymol.* **2005**, *395*, 348–384. [PubMed]
26. Wu, Y.; Zhou, H. Research progress of sugarcane chloroplast genome. *J. South. Agric.* **2013**, *44*, 17–22.
27. Luo, R.; Liu, B.; Xie, Y.; Li, Z.; Huang, W.; Yuan, J.; He, G.; Chen, Y.; Pan, Q.; Liu, Y.; et al. SOAPdenovo2: An empirically improved memory-ecient short-read de novo assembler. *Gigascience* **2012**, *1*, 18. [CrossRef] [PubMed]
28. Mardis, E.R. The impact of next-generation sequencing technology on genetics. *Trends Genet.* **2008**, *24*, 133–141. [CrossRef] [PubMed]
29. Szmidt, A.E.; Aldén, T.; Hällgren, J.E. Paternal inheritance of chloroplast DNA in larix. *Plant Mol. Biol.* **1987**, *9*, 59–64. [CrossRef] [PubMed]
30. Palmer, J.D. Comparative organization of chloroplast genomes. *Annu. Rev. Genet.* **1985**, *19*, 325–354. [CrossRef] [PubMed]
31. Nock, C.J.; Waters, D.L.; Edwards, M.A.; Bowen, S.G.; Rice, N.; Cordeiro, G.M.; Henry, R.J. Chloroplast genome sequences from total DNA for plant identification. *Plant Biotechnol. J.* **2010**, *9*, 328–333. [CrossRef] [PubMed]
32. Pu, F.S.; Lin, S.H.; Chen, R.Y.; Song, W.Q.; Li, X.L. Studies on karyotype of pyrus in China(II). *Acta Horticult. Sin.* **1986**, *13*, 87–90.
33. Pu, F.S.; Lin, S.H.; Song, W.Q.; Chen, R.Y.; Li, X.L. Studies on karyotype of pyrus in China(I). *J. Wuhan Bot. Res.* **1985**, *3*, 381–387.

34. Lin, B.N.; Shen, D.X. Studies on the germplasmic characteristics of Pyrus by use of isozymic patterns. *Acta Agric. Univ. Zhejiang Gensis* **1983**, *9*, 235–242.
35. Teng, Y.W.; Tanabe, K.; Tamura, F.; Itai, A. Genetic relationships of pear cultivars in Xin Jiang, China, as measured by RAPD markers. *J. Horticult. Sci. Biotechnol.* **2001**, *76*, 771–779. [CrossRef]
36. Teng, Y.; Tanabe, K.; Tamura, F.; Itai, A. Genetic relationships of Pyrus species and cultivars native to East Asia revealed by randomly amplified polymorphic DNA markers. *J. Am. Soc. Hortic. Sci. Biotech.* **2002**, *127*, 262–270.
37. Teng, Y.W.; Tanabe, K. Reconsideration on the origin of cultivated pears native to East Asia. 4thInternational Symposium of Taxonomy and Nomenclature of Cultivated Plants. *Acta Hortic.* **2004**, *634*, 175–182. [CrossRef]
38. Teng, Y.W. Advances in the research on phylogeny of the genus Pyrus and the origin of pear cultivars native to East Asia. *J. Fruit Sci.* **2017**, *34*, 370–378.
39. Zheng, X.Y.; Cai, D.Y.; Yao, L.H.; Teng, Y.W. Non-concerted ITS evolution, early origin and phylogenetic utility of ITS pseudogenes in Pyrus. *Mol. Phylogenet. Evol.* **2008**, *48*, 892–903. [CrossRef] [PubMed]
40. Taberlet, P.; Gielly, L.; Pautou, G.; Bouvet, J. Universal primers for amplification of three non-coding regions of chloroplast DNA. *Plant Mol. Biol.* **1991**, *17*, 1105–1109. [CrossRef] [PubMed]
41. Hu, C.Y.; Zheng, X.Y.; Teng, Y.W. Characterization and Phylogenetic Utility of Non-coding Chloroplast Regions trnL-trnF and accD-psaI in Pyrus. *Acta Hortic. Sin.* **2011**, *38*, 2261–2272.
42. Liu, Y. *Studies on Chloroplast DNA Diversity of Chinese Pear*; Capital Normal University: Beijing, China, 2006.
43. Raghvendra, A.S. *Photosynhcsis: A Comprehensive Treatise*; Cambridge University Press: Cambridge, UK, 1998; pp. 72–86.
44. Kent, W.J. BLAT-the BLAST-like alignment tool. *Genome Res.* **2002**, *12*, 656–664. [CrossRef] [PubMed]
45. Liu, C.; Shi, L.; Zhu, Y.; Chen, H.; Zhang, J.; Lin, X.; Guan, X. CpGAVAS, an integrated web server for the annotation, visualization, analysis, and GenBank submission of completely sequenced chloroplast genome sequences. *BMC Genom.* **2012**, *13*, 715. [CrossRef] [PubMed]
46. Mott, R. EST_GENOME: A program to align spliced DNA sequences to unspliced genomic DNA. *Comput. Appl. Biosci.* **1997**, *13*, 477–478. [CrossRef] [PubMed]
47. Lowe, T.M.; Eddy, S.R. tRNAscan-SE: A program for improved detection of transfer RNA genes in genomic sequence. *Nucleic Acids Res.* **1997**, *25*, 955–964. [CrossRef] [PubMed]
48. Laslett, D.; Canback, B. ARAGORN, a program to detect tRNA genes and tmRNA genes in nucleotide sequences. *Nucleic Acids Res.* **2004**, *32*, 11–16. [CrossRef] [PubMed]
49. Abouelhoda, M.I.; Kurtz, S.; Ohlebusch, E. Replacing suffix trees with enhanced suffix arrays. *J. Disc. Algo.* **2004**, *2*, 53–86. [CrossRef]
50. Lohse, M.; Drechsel, O.; Kahlau, S.; Bock, R. OrganellarGenomeDRAW—A suite of tools for generating physical maps of plastid and mitochondrial genomes visualizing expression data sets. *Nucleic Acids Res.* **2013**, *41*, W575–W581. [CrossRef] [PubMed]
51. Wang, D.P.; Zhang, Y.B.; Zhang, Z.; Zhu, J.; Yu, J. KaKs_Calculator 2.0: A toolkit incorporating gamma-series methods and sliding window strategies. *Genom. Proteom. Bioinform.* **2010**, *8*, 77–80. [CrossRef]

© 2018 by the authors. Licensee MDPI, Basel, Switzerland. This article is an open access article distributed under the terms and conditions of the Creative Commons Attribution (CC BY) license (http://creativecommons.org/licenses/by/4.0/).

International Journal of
Molecular Sciences

MDPI

Article

Natural History of a Satellite DNA Family: From the Ancestral Genome Component to Species-Specific Sequences, Concerted and Non-Concerted Evolution

Alexander Belyayev [1,*], Jiřina Josefiová [1], Michaela Jandová [1], Ruslan Kalendar [2,3], Karol Krak [1,4] and Bohumil Mandák [1,4]

1 The Czech Academy of Sciences, Institute of Botany, Zámek 1, 252 43 Průhonice, Czech Republic; jirina.josefiova@ibot.cas.cz (J.J.); michaela.jandova@ibot.cas.cz (M.J.); karol.krak@ibot.cas.cz (K.K.); bohumil.mandak@ibot.cas.cz (B.M.)
2 Department of Agricultural Sciences, University of Helsinki, P.O. Box 27 (Latokartanonkaari 5), 00014 Helsinki, Finland; ruslan.kalendar@helsinki.fi
3 RSE "National Center for Biotechnology", 13/5, Kurgalzhynskoye road, Astana 010000, Kazakhstan
4 Faculty of Environmental Sciences, Czech University of Life Sciences Prague, Kamýcká 129, 165 00 Praha-Suchdol, Czech Republic
* Correspondence: alexander.belyayev@ibot.cas.cz; Tel.: +420-271-015-461; Fax: +420-267-750-031

Received: 25 January 2019; Accepted: 6 March 2019; Published: 9 March 2019

Abstract: Satellite DNA (satDNA) is the most variable fraction of the eukaryotic genome. Related species share a common ancestral satDNA library and changing of any library component in a particular lineage results in interspecific differences. Although the general developmental trend is clear, our knowledge of the origin and dynamics of satDNAs is still fragmentary. Here, we explore whole genome shotgun Illumina reads using the RepeatExplorer (RE) pipeline to infer satDNA family life stories in the genomes of *Chenopodium* species. The seven diploids studied represent separate lineages and provide an example of a species complex typical for angiosperms. Application of the RE pipeline allowed by similarity searches a determination of the satDNA family with a basic monomer of ~40 bp and to trace its transformation from the reconstructed ancestral to the species-specific sequences. As a result, three types of satDNA family evolutionary development were distinguished: (i) concerted evolution with mutation and recombination events; (ii) concerted evolution with a trend toward increased complexity and length of the satellite monomer; and (iii) non-concerted evolution, with low levels of homogenization and multidirectional trends. The third type is an example of entire repeatome transformation, thus producing a novel set of satDNA families, and genomes showing non-concerted evolution are proposed as a significant source for genomic diversity.

Keywords: satellite DNA; genome evolution; plants; next-generation sequencing; high order repeats

1. Introduction

Genome evolution can be defined as the multifactorial process of variation of nuclear genome components over time. The process is heterogeneous, and different genomic fractions evolve at different rates. The most rapid changes were recorded for repeatomes, which form the basis of most eukaryotic genomes and consist of repeated and repeat-derived sequences [1,2]. As a subject of concerted evolution, the repeatomes of diverging species mostly change non-independently in a concerted way those results in a sequence similarity of repeating units greater within than among species [3]. Repetitive DNA complexes play an important role in evolutionary genome transformation, and determination of their origin, composition and dynamics is crucial for understanding genomic diversity [4].

The repeatome consists of several large classes, among which transposable elements (TEs) and satellite DNA (satDNA) predominate [5,6]. The latter consists of long, late-replicating, non-coding arrays of tandemly arranged monomers [5,7]. These sequences are often species or genus specific and are considered the most variable fraction of the eukaryotic genome, thus reflecting trajectories of short-term evolutionary change [8–11]. Recent studies suggest that satDNA, which is predominantly concentrated in the heterochromatic regions of chromosomes, is involved in various functions ranging from chromosome organization and pairing to cell metabolism and adjustment of gene functions [12–16]. Despite their particular importance for understanding genome functioning and restructuring during micro- and macroevolutionary processes and the growing awareness of their structure and functional significance, knowledge on the origin and dynamics of satDNA is fragmentary, especially in non-model species.

It is generally accepted that an intraspecific monomer change in various satDNA families is permanent [17]. Related species share a common satDNA library that was present in the common ancestor. Differential amplification of satellites from this library and acquisition of mutations in diverse lineages results in interspecific differences in that fraction [18]. Spreading of a new variant processed by non-Mendelian molecular mechanisms is followed by the fixation of the new variant within a population by sexual reproduction [19–21]. Thus, intraspecific homogenization of the satDNA family and fixation of species-specific polymorphisms occur simultaneously [22], and the main trend of satDNA conversion can be considered as a transformation from the common ancestral to the species-specific tandem repeats. The process appears to be a significant part of speciation at the molecular level [4]. Recently, the possibility of unraveling details of this ubiquitous phenomenon by next-generation sequencing (NGS) technology appeared through comparative analysis of the entire species repeatome. Importantly, this method is applicable not only for model organisms but also for a wide range of wild species, which allows the construction of a generalized model.

In the present study, we sought to explore NGS data using the RepeatExplorer (RE) pipeline [23] to infer satDNA evolutionary dynamics in the genomes of *Chenopodium* s. str. (also referred to as the *Chenopodium album* aggregate). Species of the *C. album* aggregate are distributed worldwide, with the highest species diversity in temperate areas [24]. The majority of these diploid-polyploid species are phenotypically exceptionally plastic [25], in some cases widely distributed and able to grow under a wide range of conditions [26]. We focused on diploid species ($2n = 2x = 18$) of the aggregate that represent separate lineages. Specifically,: (i) "clade A" are the species native to America and East Asia (the latter area being represented by *C. bryoniifolium* Bunge); (ii) "clade B" of the Eurasian temperate species *C. ficifolium* Sm. and the boreal species *C. suecicum* Murr.; (iii) "clade D" comprising the only East and Central Asian species, *C. acuminatum* Willd; (iv) "clade E" represented by the Central Asian *C. pamiricum* Iljin and *C. iljinii* Golosk.; (v) "clade H" comprising presumably European and southwest Asian species *C. vulvaria* L; and clades C, F and G consist of polyploid species. By the existence of basic diploid lineages, the origin of the majority of Eurasian polyploid species can be explained as hybridization among the diploid lineages that created subgenomic combinations of individual polyploid taxa (see [27] for details) (Figure 1). This group was selected based on the following two criteria: (i) analyzed species of the genus *Chenopodium* provide an example of a diploid/polyploid complex [26,27] that is very typical for angiosperms and, to a certain extent, can be regarded as a standard model for the divergent evolution of higher plants; and (ii) a basic repeat unit with pan-chromosomal distribution and also related to the satellite monomer of *Beta corolliflora* was previously found in the genome of a *Chenopodium* species [28,29]. This combination of favorable factors makes the study promising for describing satDNA family evolution in a typical group of flowering plants. Given the worldwide distribution of the *C. album* aggregate and its tens of millions of years of evolution [27], we hypothesize the presence of different types of satDNA family transformations in diverged lineages.

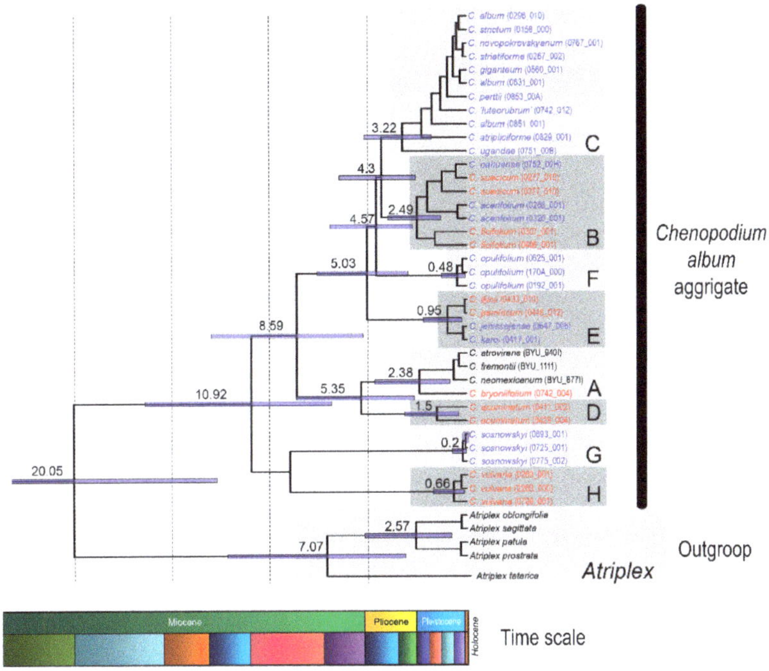

Figure 1. Phylogenetic tree calculated using Bayesian inference within the *C. album* aggregate estimated based on the concatenated dataset of three chloroplast DNA spacers (adapted from [27]). Major evolutionary lineages (A–H) are marked by grey rectangles. The numbers above branches correspond to the ages of the particular clades (in millions of years) as inferred by the analysis in BEAST2. Positions of explored diploid species are shown in red. Polyploid species are shown in blue. The schematic stratigraphic time scale (Miocene–Holocene) is shown at the bottom of the figure.

2. Results

2.1. Clustering Results and Identification of satDNA Clusters

Application of the RE pipeline clustering tool for Illumina reads of seven diploid *Chenopodium* species (Table 1) (genome coverage 41.3–58.2%) resulted in the identification of clusters that represent different families of TEs, their derivatives and satDNAs. The latter was the main aim of our research. Several valuable outcomes from the present study are shown in Table 2. *C. vulvaria* and *C. acuminatum* possess the smallest genomes in the group, while those of *C. ficifolium* and *C. suecicum* are the largest. *C. vulvaria* exceeds all investigated species in the number of RE clusters and RE singlets, which emphasizes for its genome diversity.

Table 1. The accessions and geographic origin of *Chenopodium* diploid species used for satDNA cluster analysis (NGS), probe preparation (cloning) and fluorescent in situ hybridization (FISH).

Species	Clade	ID Number	Locality
C. acuminatum	D	429/3	China, Burquin
C. bryoniifolium	A	742/4	Russian Federation, Nakhodka
C. ficifolium	B	330/2	Czech Republic, Slatina
C. iljinii	E	433/9	China, Hoboksar
C. pamiricum	E	830/3C	Tajikistan, Gorno-Badakhshan
C. suecicum	B	328/10	Czech republic, Švermov
C. vulvaria	H	771/1	Iran, Shahr

Table 2. Summary of chromosome parameters, genome size, RE clusters and percentage of CficCl-61-40 satDNA family in the genomes of *C. album* aggregate diploid species.

Species	Chr. Numb. 2*n*	Chr. Size μm	Genome Size 2C Values Mb [26]	RE Clusters #	RE Singlets #	CficCl-61-40 % in Genome
C. acuminatum	18	0.8–1.5	960	393251	34269	3.80
C. bryoniifolium	18	0.7–0.9	1200	307778	38905	2.25
C. ficifolium	18	1.5–4.5	1785	369861	20661	0.31
C. iljinii	18	1.7–3.3	1144	327760	82679	0.42
C. pamiricum	18	1.2–2.5	1154	249599	42427	0.25
C. suecicum	18	2.5–5.0	1775	369583	72167	0.27
C. vulvaria	18	1.5–2.0	924	542674	93278	0.79

The satellite monomer of ~40 bp was found during the RE analysis of satDNA in each genome of analyzed species. According to BLAST results, these monomers were related to each other and to the tribe-specific repetitive sequence (GenBank ID HM641822.1), found in *Chenopodium quinoa* by Kolano et al. [29], and to the satellite sequence with the GenBank ID AJ288880.1, which was found in *Beta corolliflora* by Gao et al. [28] (S1). It was thus assumed that in the genomes of the several *Chenopodium* diploids under study, the most abundant and the evolutionarily oldest component (*Chenopodium* and *Beta* diverged approximately in the Paleogene) is present. In the remainder of this paper, tandem arrays from the genomes of *Chenopodium* diploid species related to GenBank accession HM641822.1 will be termed the "CficCl-61-40 satDNA family". This refers to the analysis of NGS data from the *C. ficifolium* genome, RE Cluster #61 (the single cluster in genome of *C. ficifolium* that contains the basic repeat unit), with a length of 40 bp. A further thorough analysis of the interspecies divergence of the sequences of this family was also conducted to identify the main phases of transformation over time.

2.2. Sequence Analysis in the CficCl-61-40 satDNA Family

Among the multitude of clusters produced by RE pipeline, a BLAST search determined a single cluster that belongs to the CficCl-61-40 satDNA family in the genomes of *C. acuminatum, C. bryoniifolium, C. ficifolium, C. iljinii, C. pamiricum*, and *C. suecicum* and seven clusters in genome of *C. vulvaria* (supplementary data 1, Figure 2). The highest percentages of the CficCl-61-40 satDNA family were observed in the *C. acuminatum* and *C. bryoniifolium* genomes (Table 2). Subsequent tandem repeat finder (TRF) analysis allows determination of consensus monomer(s) (supplementary data 1). The algorithm of TRF looks for tandem repeats that are often hidden in larger homologous regions or which may fall well below the level of significance required for other programs to report a match. The detection criteria are based on a stochastic model of tandem repeats specified by percent identity and frequency of insertions and deletions rather than some minimal alignment score and align repeat copies against a consensus sequence, revealing patterns of common mutations [30]. Nucleotide sequence divergence among monomers within satDNA arrays is usually quite low, generally, not exceeding a few percent, and for the purpose of sequence analysis, it is acceptable to manipulate with the satDNA consensus sequence [17]. For *C. ficifolium, C. pamiricum* and *C. suecicum* a single monomer of ~40 bp was detected. However, for *C. acuminatum, C. bryoniifolium, C. iljinii*, and *C. vulvaria*, several derivatives from CficCl-61-40 satDNA family monomers were found inside the single cluster. The following two levels of CficCl-61-40 satDNA family variability in the genomes of *C. album* aggregate diploid species were thus observed: (i) at the inter-cluster level, namely single or multiple RE clusters, and (ii) at the intra-cluster level, namely single monomer or a set of related monomers of different lengths detected by TRF.

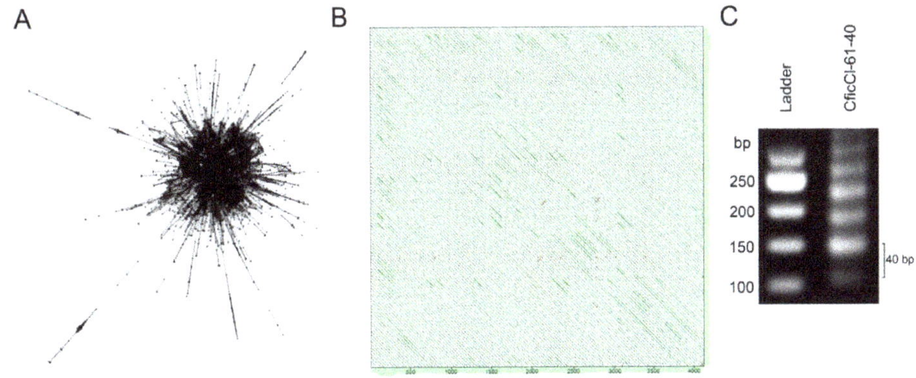

Figure 2. RepeatExplorer (RE) analysis of next-generation sequencing (NGS) data in *Chenopodium* diploids. (**A**) Cluster 61 of *C. ficifolium* demonstrate layouts that are typical for tandem repeats where nodes represent the sequence reads and edges between the nodes correspond to similarity hits; (**B**) Self-to-self comparisons of the contig 25 cluster 61 displayed as dot plots (genomic similarity search tool YASS program output) where parallel lines indicate tandem repeats (the distance between the diagonals equals the lengths of the motifs ~40 bp); (**C**) Agarose gel electrophoresis of PCR products obtained with primers designed from consensus monomer sequence of *C. ficifolium* (Cluster 61) showing typical ladder structure of tandem array.

2.3. Reconstruction of the Ancestral Monomer

BLAST-detected relatedness between satellite monomers of the CficCl-61-40 satDNA family allowed determination of the major part of the ancestral monomer. For this reconstruction, satellites of *C. bryoniifolium* (consensus monomer from one RE cluster) and *C. vulvaria* (consensus monomers from seven RE clusters) that show relatedness to both *C. quinoa* and *B. corolliflora* satellites were aligned. DNA fragments with 100% BLAST matches in combination formed the most conservative fragment of the basic monomer (supplementary data 2). This approach is quite similar to the method of ancient paralogs (LUCA) [31,32]. The sequence of 37 bp was as follows: TCAAACAAAGCTAATTGAATCAAATGAAAGTCAAATG. This sequence was used as a basis for the subsequent comparison of the monomer divergence in *Chenopodium* lineages. Analysis of basic satellite alterations revealed point mutations, indels, and shifts that were present with different frequencies in the genomes of the studied diploid *Chenopodium* species (supplementary data 1). K-mer based distance estimation revealed a phylogenetically reliable tree with the ancestral monomer as a base, *B. corolliflora* is located separately and rather close to the root, the analyzed diploids that form fairly natural groups with species of clades B and E located nearby, *C. bryoniifolium*, *C. acuminatum* aside, and polyploid *C. quinoa* at the maximum distance from the ancestral monomer (Figure 3).

Clade H (*C. vulvaria*) deserves separate attention. The RE pipeline divided the variety of CficCl-61-40 satDNA family sequences in the genome of *C. vulvaria* into seven clusters (supplementary data 1), indicating valuable heterogeneity. On one hand, all the basic monomers of the clusters contain BLAST-recognizable fragments of the ancestral monomer. On the other hand, the observed variability exceeds that for all clades taken together (Figure 3). An important question is whether all these clusters from the *C. vulvaria* genome belong to the same CficCl-61-40 satDNA family. RE output includes not only the row of clusters but also detailed cluster characteristics, including the cluster neighborhoods of connected components. The analysis showed that all clusters that we classified as belonging to the CficCl-61-40 satDNA family are related to each other and to the repetitive sequences with the GenBank IDs HM641822.1 and AJ288880.1. Additionally, these satDNA clusters possess a limited number of similarity hits with TEs clusters (mainly with Ty3-*gypsy* retrotransposons) which may indicate for splitting of satDNA arrays by the insertion of TEs.

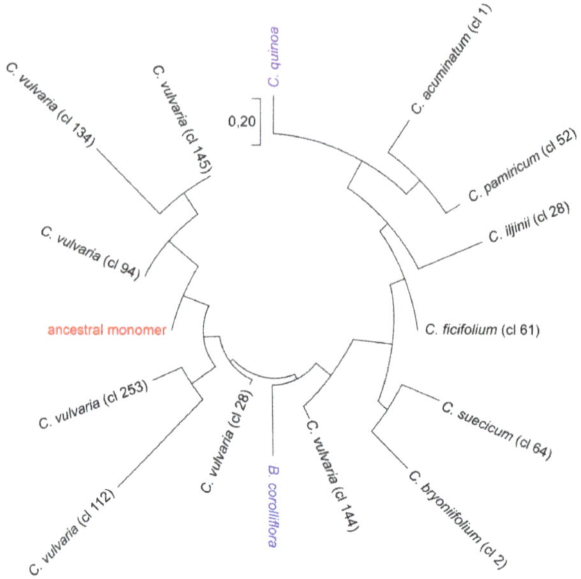

Figure 3. Phylogenetic relationships of the CficCl-61-40 satDNA family sequences. Phylogenetic tree based on the k-mer analysis.

2.4. High Order Repeat (HOR) Detection in the CficCl-61-40 satDNA Family and Determination of Its Physical Counterpart

TRF analysis of the CficCl-61-40 satDNA family in seven diploid species of *Chenopodium* revealed different structures of the arrays. In *C. ficifolium*, *C. pamiricum*, and *C. suecicum*, uniform tandem arrays with basic satellite motifs of ~40 bp (87–96% matches between monomers and copy numbers of 79.2–153.4) were identified by TRF. In *C. acuminatum*, *C. bryoniifolium*, *C. iljinii* and *C. vulvaria*, derivatives from CficCl-61-40 satDNA family repeats ranging up to 332 bp and of different repeatability were found (supplementary data 1). It was proposed that in the latter species, HORs could be formed by concurrent amplification and homogenization of modified monomers.

Here, it is necessary to elucidate the TRF algorithm using an example of the detection of a 117 bp monomer in the genome of *C. acuminatum* (later used as a probe in fluorescent in situ hybridization (FISH) experiments). Analysis of the RE Cluster-1 sequence by TRF produced a table of monomers with the most frequent of 117 bp (consensus size) (supplementary data 1). However, when the consensus sequence was manually analyzed, it decomposed into three 39 bp long subrepeats. Nevertheless, it can be argued that the 117 bp fragment is the basic monomer and that the formation of a HOR unit is based on an ~40 bp monomer. The program finds likely patterns (monomers) and then refines them into a consensus sequence. Patterns are detected by a high percentage of matches at the candidate pattern length. For 39 bp not enough matches were found, but a very high number for 117 bp. This indicates that the unit of duplication was 117 bp and not 39 bp. Furthermore, the mismatches and indels are more consistent with a 117 bp monomer than with a 39 bp monomer (Gary Benson, personal communication). Following sequencing of physical counterparts of CacuCl-1-117 consensus sequence (see below) revealed that the physical components of the CacuCl-1-117 HOR unit did not coincide completely (as in consensus) but varied within the interval of 82% to 86% similarity, which confirmed the accuracy of the TRF algorithm. Additionally, it can be considered that the TRF analysis of all RE clusters belonging to the CficCl-61-40 satDNA family was performed with the same parameters, and in genomes of tree species, only homogeneous arrays were identified while the four other arrays were heterogeneous, which reflects the real structure of satDNA.

A total of three to four different proposed HOR units were detected in the genomes of *C. acuminatum*, *C. bryoniifolium* and *C. iljinii*. However, approximately 23 such units were found in genome of *C. vulvaria* (supplementary data 1). The genome of *C. vulvaria* is thus again the most variable according to this parameter.

While there are multiple studies that demonstrate that RE is efficient in repeat identification using NGS, there are some limitations regarding sequence analysis of satellite repeats. The most important one was that generation of consensus sequences by assembling reads to contigs. While this works well for most dispersed repeats like TEs, this is problematic for satellites due to their tandem structure. Consequently, contigs vary in their coverage by reads and their sequences could be partially chimeric (producing sequence variant combinations that in fact do not exist in the genome). To confirm the existence in genomes the physical counterparts of computer-generated consensus monomers we analyzed the sequence variation of CficCl-61-40 and proposed HOR units CacuCl-1-117, CvulCl-28-118, CvulCl-28-397, CvulCl-112-117, CvulCl-134-117 and Cvul-145-129 by cloning. We then compared the obtained sequences with the consensus sequence from the TRF output (supplementary data 3, Figure 4). For all monomers, we obtained several clones that differed from each other as well as from the consensus sequence (supplementary data 4). The CficCl-61-40 monomer is rather uniform with a few point mutations and sequence similarity between clones. The consensus sequence ranged from 90.2% to 95%. For the four obtained clones of the CacuCl-1-117 monomer, two sequence types were found with generally higher similarity to the consensus sequences as well as to each other (similarity value ranges 89.8–91.5 and 90.7–99.2, respectively). This once again confirmed the correctness of the TRF algorithm.

More variability was detected for the proposed HOR units in the *C. vulvaria* genome, which once again highlights the complexity of the satDNA fraction in this species. Thus, among tree clones obtained for the CvulCl-28-118-proposed HOR unit, two sequence types were found with generally high similarity to each other than to the consensus monomer (similarity value ranges 88.2–98.3 and 76.4–79.1, respectively). For CvulCl-28-397-proposed HOR unit sequences amplified by primers (supplementary data 3) also shows more relatedness to each other than to consensus sequence (supplementary data 4). For the CvulCl-112-117- and CvulCl-134-117-proposed HOR units, two types were found among cloned sequences. One showed high relatedness to the consensus monomer (83.3%–90.7%) and the other clones were 100% related to each other and less to the consensus monomer and to the first variant (supplementary data 4). This most likely suggests that several related HOR units could be formed simultaneously. For Cvul-145-129-proposed HOR unit clones possess high similarity to the consensus sequences as well as to each other (82.9%–100.0%). Part of the cloned sequences was submitted to GenBank (accession numbers MH257681–MH257687). However, it should also be noted that we were not able to amplify part of the proposed HOR units generated by TRF analysis (for example CvulCl-28-355 and Cvul-134-148) (supplementary data 1, far right line on Figure 4). These sequences could be attributed most likely to computer-generated chimeric sequences (i.e., method error). However, for the majority of the proposed HOR units its physical counterparts were discovered in the genomes.

2.5. Comparison of CficCl-61-40 and Proposed HOR Unit CacuCl-1-117 Chromosomal Distribution

For further confirmation of the existence of HOR units' in the genomes of *Chenopodium* species an alternative method of FISH was used. Two distant clusters according to phylogenetic analysis, cluster 61 of *C. ficifolium* (CficCl-61-40) and cluster 1 of *C. acuminatum* (CacuCl-1-117) from the RE output, were selected as sources for in situ probes for comparative molecular cytogenetic analysis (Figure 5, supplementary data 1 and 5). FISH experiments were performed to verify if (i) *C. acuminatum*-specific tandem repeats that were proposed to be HOR units (CacuCl-1-117) are species-specific and do not hybridize to chromosomes of the other six species and (ii) if the chromosomal positions of the *C. acuminatum*-specific tandem repeat (CacuCl-1-117) are similar to or different from the positions of the tribe-specific repeat (CficCl-61-40) on the chromosomes of

C. acuminatum. It should be noted, however, that accurate FISH-based karyotyping and chromosome mapping of CficCl-61-40 satDNA family tandem repeats is challenging in *Chenopodium* due to the small chromosome sizes and to the large number of clusters (Table 2, Figure 5).

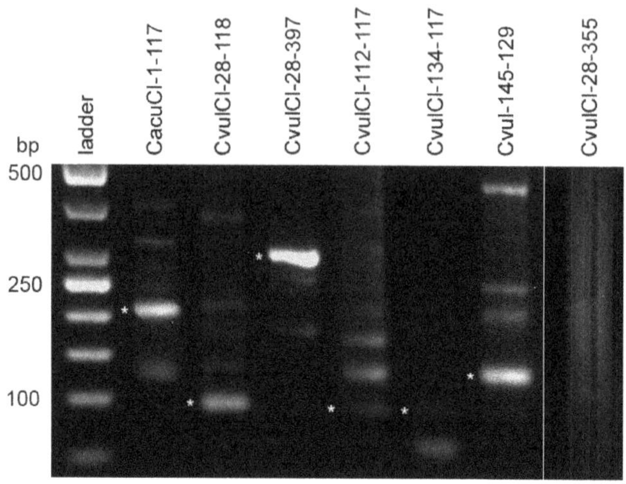

Figure 4. Agarose gel electrophoresis of PCR products obtained with primers designed from consensus monomer sequence of proposed high order repeat (HOR) units for determination of their physical counterparts. Cloned DNA fragments are shown by asterisks. The far-right line is an example of negative amplification of a computer-generated proposed HOR unit.

FISH experiments confirmed species specificity and sometimes separate chromosomal positions of newly formed HOR units. Probe CficCl-61-40 hybridized to the chromosomes of all analyzed species except *C. iljinii* (similarity, copy number, or both of the particular FISH probe in *C. iljinii* genome is likely much less in comparison with other species), which demonstrates the presence of a tribe-specific satellite, while CacuCl-1-117 hybridized only to chromosomes of *C. acuminatum* with no signal on the chromosomes of the other six species (Figure 5, supplementary data 5, the minor green signal in *C. pamiricum*, *C. suecicum* and *C. vulvaria* in supplementary data 5 is epifluorescence). In addition, the simultaneous hybridization of CacuCl-1-117 and CficCl-61-40 on the chromosomes of *C. acuminatum* shows that in many cases these tandem arrays form separate clusters that create a species-specific chromosomal pattern (Figure 5).

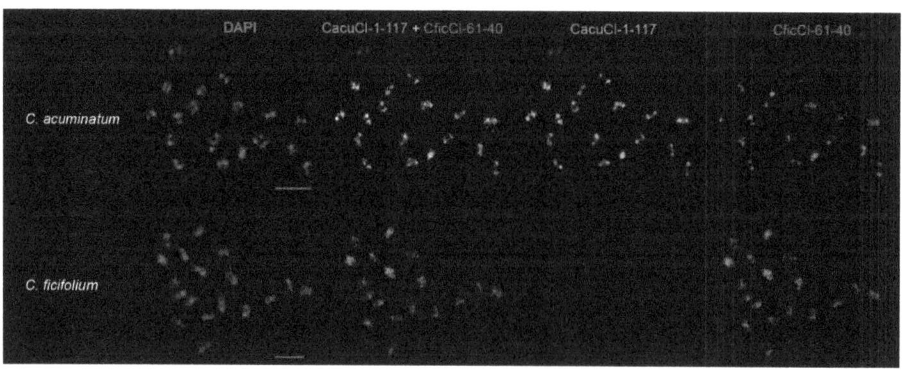

Figure 5. Chromosomal distribution CficCl-61-40 satDNA family sequences. CficCl-61-40 is labelled red; *C. acuminatum*-specific HOR unit CacuCl-1-117 is labelled green. Bar represent 5 μm.

3. Discussion

Application of the RE pipeline for analysis of whole genome shotgun Illumina reads from the genomes of seven diploid *Chenopodium* species from divergent lineages revealed that the investigated CficCl-61-40 satDNA family is the most abundant and oldest component of the *Chenopodium* genome, given that related sequences were found in both *Chenopodium* and *Beta* species. Regarding these two genera, it is essential to note that the genome of *Beta* should be recognized as more static, at least because it contains many fewer species (approximately 7–8 species in total, [33]) in comparison with *Chenopodium* (approximately 150 species [24]). Alignment of the satellite monomers allowed identification of the ancestral DNA fragment of 37 bp that showed 100% identity between *B. corolliflora* from one side and *C. bryoniifolium* and *C. vulvaria* from the other (supplementary data 2). The latter two are species that split off early and possess a modified sequence that is still recognizable by BLAST as a ~40 bp variant of the ancestral monomer. The identified DNA fragment served as a benchmark for our subsequent analyses, in which we intended to characterize intra-unit evolutionary transformations in the diverse *Chenopodium* lineages.

Remarkably, the evolutionary history of the *C. album* aggregate revealed by cpDNA spacers and two low-copy genes [27] correlates fairly well with significant paleoclimatic events. Thus, the early differentiation coincides with the beginning of the Miocene Climatic Optimum in the Burdigalian Age (approximately 20 Mya) (Figure 1). Clade H (*C. vulvaria*) separated upon transition between the Serravallian and Tortonian Ages, ~11 Mya. However, the main lineages were formed in the Pliocene, when due to a cooler and dry, seasonal climate, grasslands spread on all continents, and savannahs and deserts appeared in Asia and Africa. Subsequent speciation within the lineages and the appearance of the majority of polyploids occurred in the Quaternary Period, when the glacial and interglacial epochs succeeded each other. During this time, since there were no places on Earth with identical climate history and since the species of aggregate were spread widely, the CficCl-61-40 satDNA arrays evolved divergently. Excluding clade H, which split off early and is now very different, k-mer-based distance estimation of basic monomer show the most significant differences in genomes of species from clades A and D. It is most likely that both lineages separated early from the ancestral group and evolved independently. This is consistent with the present species distribution ranges and with molecular phylogenetic data [26,27]. However, the pace of evolution of these clades was probably different and is most likely connected with the climatic history of the species distribution areas. In clades B and E, the species are much more similar in the CficCl-61-40 satDNA family structure (Figure 2).

The concept of "molecular drive" [19] postulates that mutations can gradually spread throughout a satDNA family by several of ubiquitous mechanisms of DNA turnover (homogenization) and become fixed in a population. SatDNA families can show a rapid rate of inter-specific evolutionary changes concerning DNA sequence and high levels of conservation between species separated for long evolutionary times [22,34,35]. Although these trends are also true for the CficCl-61-40 satDNA family when monomers are homogenized on the species level in the genomes of different *Chenopodium* lineages, each of them has its own mode and tempo. Although the genome of *C. vulvaria* presents an exception, it seems that concerted evolution does not operate there. This example of non-concerted evolution will be discussed below.

In addition to mutations in basic satellite monomers, a distinct trend toward increased complexity and length of the monomer (HOR unit formation) was recorded in the species of Clades A, D, E and H of the *C. album* aggregate. HORs occur by concurrent amplification and homogenization of different monomers in the original satDNA when a complex monomer is first formed, after which it merges into a more complex HOR unit [17]. The origin of such structures has been described for the alpha satellite of primates [36], for the satellite families in bovids [37,38] and for the plant species *Vicia grandiflora* [39]. A detailed analysis of the CficCl-61-40 satDNA family tandem arrays in the genomes of *C. acuminatum*, *C. bryoniifolium*, *C. iljinii* and *C. vulvaria* along with the basic ~40 bp monomer revealed related but longer monomers of up to 332 bp, suggesting the generation of new species-specific HOR units. Cloning of PCR-amplified DNA fragments in most cases confirmed the accuracy of the monomer/array compilation produced by the RE pipeline, and the physical counterparts were mostly in agreement with the consensus sequences. However, the exact satDNA array structure of the species could be determined by complete genome sequencing, assembly and annotation [40].

FISH experiments further prove the genesis of species-specific HOR units and their separate locations on the chromosomes. CficCl-61-40 arrays were thus found in all species. On the other hand, related CacuCl-1-117 arrays were found exclusively in *C. acuminatum*, where they form multiple, sometimes separate chromosome clusters, thus creating a species-specific chromosomal pattern (Figure 5). Formation of HOR units based on two or more monomers has been reported in primates and bovids (for a review, see [17]). We observed a similar process but based on the single tribe-specific monomer when unequal changes in the initial sequence in diverging satDNA sets led to monomer alterations with the subsequent merging of the modified monomers in a complex HOR unit. A similar process (i.e., HOR formation based on one initial repeated unit in *Vicia* sp.) was reported by Macas et al. [39]. Presumably, the process of HOR formation on the basis of a single monomer can take more time (in our research, it appears predominantly in ancient species) than that involving two or several monomers, although it apparently contributes to satDNA divergence.

We might next ask whether the formation of HORs is common for plant satDNA evolution. As another example of supposed HOR formation in plants, we can provide a complex structure of the *Hieracium* species centromeric tandem array [41]. Analysis of both RE clusters and the sequenced physical counterparts revealed a complex structure with 21 repetitive elements identified by TRF (ranging from 21 bp to 348 bp) and with two abandoned motifs of 21 and 23 bp. Eventually, we can also observe the stages of HOR formation based on the two short monomers in centromeric regions. It is essential to note that although chromosome segregation machinery is highly conserved across all eukaryotes, centromeric DNA evolves rapidly, and discovered tandem repeats are absent in related *Pilosella* species. Incompatibilities between rapidly evolving centromeric components may be responsible for both the organization of centromeric regions and the reproductive isolation of emerging species [42].

The above examples and the fact that the presented species refer to different large clades of flowering plants suggests that the HOR formation process may not only occur in the *Chenopodium*, *Hieracium*, and *Vicia* genomes but that this mechanism is also ubiquitous for at least angiosperms and could underlie satDNA divergence in related plant species, as it does in animal genomes. It should also be noted that HOR formation is presumably a species-specific event; in clade B (*C. ficifolium* and *C. suecicum*), neither species showed any sign of HORs. In contrast, in clade E, CficCl-61-40 satDNA family arrays of *C. pamiricum* are uniform, while HORs were detected in the *C. iljinii* genome. However, it is still not clear what triggers the HOR formation in a particular genome [17].

In generalizing the life history of the CficCl-61-40 satDNA family stretching from the ancestral basic repeat unit to species-specific sequences, it is worth noting that the family consists of an extensive group of related, divergent repeats. It is a dominant and old component of *Chenopodium* species genomes and can be characterized by a high complexity of evolution. Independently amplified in each genome, it ultimately acquires lineage-specific profiles due to differential stochastic amplifications, contractions or both. Additionally, in several lineages, a clear trend toward increased complexity and satellite monomer length was observed. Long tandem arrays are characterized by HOR units whose organization and nucleotide sequence are specific for a particular species. Analysis of the sequence organization of these diverged subsets provides a framework for considering mechanisms of sequence diversity generation and for understanding the evolutionary processes of satDNA family homogenization and polymorphism [37]. Homogenization of satellite repeats driven by molecular mechanisms of nonreciprocal sequence transfer occurs simultaneously, which makes satDNA evolve mostly in a concerted manner [3]. Nevertheless, as mentioned above, the small genome of *C. vulvaria* (2C value 0.945 pg) is an exception to this rule. The observed variability indicates a low level of CficCl-61-40 satDNA family homogenization, with multidirectional trends in the *C. vulvaria* genome (non-concerted evolution). Although the data are unusual, our unpublished results on the NGS-based qualitative analysis of TEs in genomes of the same *Chenopodium* diploid species (where we observed that *C. vulvaria* possesses a unique pool of different and diverse retrotransposons [43]) make it possible to hypothesize a link between the TE dynamics and abnormalities in the homogenization of satDNA families, given that satDNA could be a target for TE insertions [44] and evolve further to species-specific tandem repeats [45]. Suppression of concerted evolution resembles those described for termites by Luchetti et al. [46]. This was proposed to be evoked by the limited number of reproducers, especially considering that *C. vulvaria* is an ancient species, restricted to nutrient-rich bare soil largely of anthropogenic impact and not tolerant of competition [47]. Specific habitats may presumably cause abnormal repeatome composition that, in turn, may support the models assuming that genotypes from marginal populations are evolutionarily significant [48–51]. Despite the causes, discovered suppression of homogenization itself may result in alteration of satDNA libraries, ultimately leading to spontaneous transformation of the entire repeatome, thus producing a novel set of satDNA families for the next round of the conversion cycle, and genomes undergoing non-concerted evolution can be proposed as a significant source of genomic diversity.

4. Material and Methods

4.1. Plant Material, DNA Extraction, Library Preparation and Illumina Sequencing

For both preparation of the DNA libraries and cytogenetic experiments, plants of the diploid species *C. acuminatum*, *C. bryoniifolium*, *C. ficifolium*, *C. iljinii*, *C. pamiricum*, *C. suecicum* and *C. vulvaria*, which represent the main lineages of the *C. album* aggregate as described in Mandák et al. [27], were used (Table 1). For our research, we sampled genotypes that, according to our previous data, have average parameters for the lineage [27]. All plants were cultivated at the experimental garden of the Institute of Botany, Czech Academy of Sciences, Průhonice, Czech Republic (49.9917° N, 14.5667° E, ca. 320 m above sea level). Leaves were collected, and DNA was extracted using the DNeasy Plant Mini Kit (Qiagen, Venlo, The Netherlands) according to the manufacturer's instructions. For in situ hybridization experiments, root tips of young, fine roots were collected and fixed as described in Mandák et al. [26] and stored until use. For all analyzed accessions, the ploidy level was verified by flow cytometry as described in Vít et al. [52].

One individual per species was used for library preparation and NGS. One microgram of extracted DNA was sheared to fragments of approximately 500 to 600 bp using a Bioruptor Pico sonication device (Diagenode, Liège, Belgium). The NEBNext adaptors for Illumina were ligated to the resulting fragments using the NEBNext Ultra DNA Library Prep Kit for Illumina (New England BioLabs, Ipswich, MA, USA), following the manufacturer's instructions. The QIAquick PCR Purification Kit (Qiagen) was used to clean the samples from unbound adapters and to concentrate the samples to a total volume of 30 μL. Afterwards, the samples were loaded onto a 1% agarose gel in low EDTA/TAE buffer. Fragments with sizes ranging from 500 to 750 bp were excised and purified using the Zymoclean Gel DNA Recovery Kit (Zymo Research, Irvine, CA, USA) and eluted into 20 μL of ddH2O Concentration was estimated with a Qubit fluorometer using the Qubit HS Assay kit (Thermo Scientific, Waltham, MA, USA). The individual libraries (corresponding to individual species) were enriched and indexed by unique barcodes using PCR with NEBNext Q5 HotStart HiFi PCR Master Mix and NEBNext Multiplex Oligos for Illumina (New England BioLabs) according to the manufacturer's instructions. The enriched libraries were purified twice using AMPure magnetic beads (Beckman Coulter, Pasadena, CA, USA). The bead:library ratio was 0.7:1 in the first purification and 1:1 in the second purification. The libraries were verified on 1% agarose gels after each purification step. Concentration was measured using the Qubit HS Assay kit (Thermo Scientific) after the final purification step. Libraries of all seven species were pooled and sequenced on an Illumina MiSeq system at Macrogen Inc., to obtain 2*x* 300 bp paired-end reads.

4.2. Clustering of Repeatome Elements

To process Illumina NGS data and to compare the repetitive DNA fraction of the studied species, a public web server running RE version 1 (http://www.repeatexplorer.org) (České Budějovice, Czech Republic) was used [53]. The discovery and characterization of repetitive elements in the genome was performed using "clustering" tools. An all-to-all sequence comparison of sequencing reads was performed using the mgblast tool. All hits with similarities above 90% over at least 55% of the sequence length were recorded, thus identifying a set of related DNA fragments. The information on similarity hits was used for construction of a graph in which nodes represent sequence reads and the edges between nodes correspond to similarity hits (Figure 2A). This algorithm was first applied to each species separately and subsequently for the seven species in conjunction for the comparative analysis of repeatome quantitative values. For comparative analysis, sampling was performed proportionally to the genome size of the species (Table 2) [26].

4.3. Satellite DNA Clusters Screening for Tandem Repeats

All genomically abundant clusters containing at least 0.01% of the input reads were examined manually to select those that potentially possess tandemly organized DNA. Primary selection of the clusters was performed based on their form (Figure 2A). Contigs of each selected cluster were analyzed using the following publicly available online tools: (i) the YASS genomic similarity tool, which enables searches of more fuzzy repeats for potential tandem organization (http://bioinfo.lifl.fr/yass/yass.php) [54], with each contig compared against itself and visualized by dot plots (Figure 2B); (ii) BLAST was used to confirm that the cluster belongs to the CficCl-61-40 satDNA family (supplementary data 1); and (iii) primers were designed from the consensus sequence for PCR conformation of the typical tandem array structure (Figure 2C). Each search was performed for each of the analyzed species.

4.4. Sequence Analysis

NGS clusters of *C. acuminatum*, *C. bryoniifolium*, *C. ficifolium*, *C. iljinii*, *C. pamiricum*, *C. suecicum* and *C. vulvaria* falling into the CficCl-61-40 satDNA family were investigated on the intra-unit (analysis of changes in single monomer) and inter-unit (analysis of changes in array components) levels with TRF software (https://tandem.bu.edu/trf/trf.html). As a result, performance tables with data on monomer sizes, copy numbers, percent matches, percent indels and consensus patterns were obtained (supplementary data 1).

For the reconstruction of phylogenetic relationships among the analyzed monomers k-mer based distance estimation was performed [55]. We have chosen the k-mer value equal to 9, as the most optimal for the analyzed sequences. For calculation of distances method based on fractional common k-mer count was used [56]. The phylogenetic relationships among the sequences are then reconstructed from the pairwise distance matrix [57]. The distance matrix thus obtained can be used to construct a phylogenetic tree using the Minimum Evolution method. The construction of the phylogenetic tree was performed in the MEGA program (Figure 3) [58]. The ancestral monomer (root) was reconstructed as follows: nucleotide-BLAST was used to align contigs of each cluster that, according to BLAST searches, show relatedness between satellite monomers of *Chenopodium* and *Beta* species. DNA fragments with 100% similarity were selected and aligned with each other (supplementary data 2). As a result, a fragment of the ancestral monomer was reconstructed.

4.5. Detection Physical Counterparts of Basic Monomer and Proposed HOR Units

RE identifies consensus sequences of the most abundant repetitive elements in the genome. However, these consensuses are only virtual assemblies of short reads originating from many different interspersed loci. To reveal the sequences' physical counterparts and sequence variation within the selected repetitive elements that are proposed to be HOR units, primers were designed based on the consensus sequences (supplementary data 3). PCRs were performed in 25 μL reactions and contained

1× TopBio Plain PP Master Mix (TopBio, Vestec, Czech Republic), each primer at 0.2 mM and 10 to 50 ng of genomic DNA. The cycling conditions were as follows: 4 min at 95 °C followed by 35 cycles of 95 °C for 30 s, sequence-specific annealing temperature for 30 s and 72 °C for 2.5 min, and a final extension at 72 °C for 10 min. The PCR results were verified on a 1% agarose gel (Figure 4). The PCR products of clusters were excised from the gels, cloned and sequenced at GATC Biotech (Konstanz, Germany) according to standard protocols.

4.6. FISH Procedure

FISH analysis was performed to further confirm the physical existence of the HOR units in the genome. Root tips were pre-treated in 0,002 M 8-hydroxyquinolin for 3 h in dark and fixed in 3:1 (v/v) 100% ethanol:acetic acid. The fixed root meristems were thoroughly washed in water and enzyme buffer (10 mM citrate buffer at pH 4.6) and partially digested in 0,3% (w/v) cytohelicase, pectolyase and cellulase (Sigma) at 37 °C for 3 h followed by washes in water [27]. The material, in a water drop, was carefully transferred onto a grease-free microscope slide and the cells were spread according to the technique of Pijnacker and Ferwerda [59] with modifications as previously described [60].

FISH experiments were performed with clones CficCl-61-40 X-1 and CacuCl-1-117 C-2 as probes labelled with Cy3 (Amersham, Amersham, Buckinghamshire, UK) and biotin (Roche, Basel, Switzerland) according to a standard oligolabeling protocol [61]. For evaluation of probe-specific chromosomal pattern probes were hybridized simultaneously to chromosomes of *C. acuminatum*, *C. bryoniifolium*, *C. ficifolium*, *C. iljinii*, *C. pamiricum*, *C. suecicum* and *C. vulvaria* (Figure 5, supplementary data 5). FISH was performed on ThermoBrite programmable temperature-controlled slide processing system at 63 °C for 3 h. Slides were stained with DAPI and mounted in antifade mountant (Vector Laboratories, Burlingame, CA, USA) and were examined and photographed on Zeiss Axio Imager.Z2 microscope system. Chromosome measurements were obtained by the analysis of metaphase plates using the computer application MicroMeasure version 3.3 [62].

5. Conclusions

Application of the RE pipeline for analysis of whole genome shotgun Illumina reads from the genomes of seven diploid plant species from divergent lineages allowed us to distinguish three types of satDNA family evolutionary development: (i) concerted evolution with mutation and recombination events (most conserved); (ii) concerted evolution with a trend toward increased complexity and length of the satellite monomer (HOR formation); and (iii) non-concerted evolution, with low levels of homogenization and multidirectional trends.

Supplementary Materials: Supplementary materials can be found at http://www.mdpi.com/1422-0067/20/5/1201/s1. **Supplementary data 1.** Occurrence of CficCl-61-40 satDNA family in genomes of *Chenopodium* diploid species revealed by RepeatExplorer pipeline and formations of high order repeat (HOR) units. **Supplementary data 2.** Reconstruction of the major part of the ancestral monomer. **Supplementary data 3.** Repetitive elements selected for sequence characterization and in situ hybridization and primers used for amplification. **Supplementary data 4.** Pairwise comparison of sequence variation within the CficCl-61-40 (**A**,**B**), and proposed HOR units CacuCl-1-117 (**C**,**D**), CvulCl-28-118 (**E**,**F**), CvulCl-28-397 (**G**,**H**), CvulCl-112-117 (**I**,**J**), CvulCl-134-117 (**K**,**L**) and Cvul-145-129 (**M**,**N**). **Supplementary data 5.** Chromosomal distribution CficCl-61-40 satDNA family sequences. CficCl-61-40 is labelled red; *C. acuminatum*-specific HOR unit CacuCl-1-117 of 117 bp is labelled green.

Author Contributions: A.B. conceived the idea for the study. B.M., K.K. collected plant material. A.B., K.K., R.K, M.J., J.J. performed or supervised the wet lab work. A.B., K.K., R.K. analyzed the data. A.B., B.M., K.K. wrote the manuscript and supplements.

Funding: This work was supported by the Czech Science Foundation (grant no. 13-02290S) and as part of a long-term research development project RVO 67985939. This work was also supported for R.K. by the Science Committee of the Ministry of Education and Science of the Republic of Kazakhstan in the framework of program funding for research (AP05130266, BR05236574 and BR06349586).

Acknowledgments: We thank Gary Benson for helpful comments. English language in original version was edited by Springer Nature Language Services (certificate number 2ZF7JW42), and in reversed version by "Kielentarkistuksen tilauslomake/Work Order for Revisions" University of Helsinki.

Conflicts of Interest: The authors declare no conflicts of interest.

References

1. Bennetzen, J.L. The structure and evolution of angiosperm nuclear genomes. *Curr. Opin. Plant Biol.* **1998**, *1*, 103–108. [CrossRef]
2. Maumus, F.; Quesneville, H. Ancestral repeats have shaped epigenome and genome composition for millions of years in *Arabidopsis thaliana*. *Nat. Commun.* **2014**, *5*, 4104. [CrossRef] [PubMed]
3. Elder, J.F.; Turner, B.J. Concerted evolution of repetitive DNA sequences in eukaryotes. *Q. Rev. Biol.* **1995**, *70*, 297–320. [CrossRef] [PubMed]
4. Garrido-Ramos, M.A. Satellite DNA: An Evolving Topic. *Genes* **2017**, *8*, 230. [CrossRef] [PubMed]
5. Biscotti, M.A.; Olmo, E.; Heslop-Harrison, J.S. Repetitive DNA in eukaryotic genomes. *Chromosome Res.* **2015**, *23*, 415–420. [CrossRef] [PubMed]
6. Wei, K.H.-C.; Lower, S.E.; Caldas, I.V.; Sless, T.J.; Barbash, D.A.; Clark, A.G. Variable rates of simple satellite gains across the *Drosophila* phylogeny. *Mol. Biol. Evol.* **2018**, *35*, 925–941. [CrossRef]
7. Satović, E.; Vojvoda Zeljko, T.; Luchetti, A.; Mantovani, B.; Plohl, M. Adjacent sequences disclose potential for intra-genomic dispersal of satellite DNA repeats and suggest a complex network with transposable elements. *BMC Genom.* **2016**, *17*, 997. [CrossRef]
8. Charlesworth, B.; Sniegowski, P.; Stephan, W. The evolutionary dynamics of repetitive DNA in eukaryotes. *Nature* **1994**, *371*, 215–220. [CrossRef]
9. Raskina, O.; Barber, J.C.; Nevo, E.; Belyayev, A. Repetitive DNA and chromosomal rearrangements: Speciation-related events in plant genomes. *Cytogenet. Gen. Res.* **2008**, *120*, 351–357. [CrossRef]
10. Emadzade, K.; Jang, T.S.; Macas, J.; Kovařík, A.; Novák, P.; Parker, J.; Weiss-Schneeweiss, H. Differential amplification of satellite PaB6 in chromosomally hypervariable *Prospero autumnale* complex (Hyacinthaceae). *Ann. Bot.* **2014**, *114*, 1597–1608. [CrossRef]
11. Dodsworth, S.; Chase, M.W.; Kelly, L.J.; Leitch, I.J.; Macas, J.; Novák, P.; Piednoël, M.; Weiss-Schneeweiss, H.; Leitch, A.R. Genomic repeat abundances contain phylogenetic signal. *Syst. Biol.* **2015**, *64*, 112–126. [CrossRef]
12. Martienssen, R.A. Maintenance of heterochromatin by RNA interference of tandem repeats. *Nat. Genet.* **2003**, *35*, 213–214. [CrossRef] [PubMed]
13. Kloc, A.; Martienssen, R. RNAi, heterochromatin and the cell cycle. *Trends Genet.* **2008**, *24*, 511–517. [CrossRef] [PubMed]
14. Mehrotra, S.; Goyal, V. Repetitive sequences in plant nuclear DNA: Types, distribution, evolution and function. *Genom. Proteom. Bioinform.* **2014**, *12*, 164–171. [CrossRef] [PubMed]
15. Garrido-Ramos, M.A. SatDNA in plants: More than just rubbish. *Cytogenet. Genome Res.* **2015**, *146*, 153–170. [CrossRef] [PubMed]
16. Meštrović, N.; Mravinac, B.; Pavlek, M.; Vojvoda-Zeljko, T.; Šatović, E.; Plohl, M. Structural and functional liaisons between transposable elements and satellite DNAs. *Chromosome Res.* **2015**, *23*, 583–596. [CrossRef] [PubMed]
17. Plohl, M.; Meštrović, N.; Mravinac, B. Satellite DNA evolution. *Genome Dyn.* **2012**, *7*, 126–152.
18. Salser, W.; Bowen, S.; Browne, D.; el-Adli, F.; Fedoroff, N.; Fry, K.; Heindell, H.; Paddock, G.; Poon, R.; Wallace, B.; et al. Investigation of the organization of mammalian chromosomes at the DNA sequence level. *Fed. Proc.* **1976**, *35*, 23–35.
19. Dover, G. Molecular drive. *Trends Genet.* **2002**, *18*, 587–589. [CrossRef]
20. Plohl, M.; Luchetti, A.; Mestrovic, N.; Mantovani, B. Satellite DNAs between selfishness and functionality: Structure, genomics and evolution of tandem repeats in centromeric (hetero) chromatin. *Gene* **2008**, *409*, 72–82. [CrossRef]
21. Samoluk, S.S.; Robledo, G.; Bertioli, D.; Seijo, J.G. Evolutionary dynamics of an at-rich satellite DNA and its contribution to karyotype differentiation in wild diploid *Arachis* species. *Mol. Genet. Genom.* **2017**, *292*, 283–296. [CrossRef] [PubMed]
22. Ugarkovic, D.; Plohl, M. Variation in satellite DNA profiles-causes and effects. *EMBO J.* **2002**, *2*, 5955–5959. [CrossRef]
23. Novák, P.; Neumann, P.; Macas, J. Graph-based clustering and characterization of repetitive sequences in next-generation sequencing data. *BMC Bioinform.* **2010**, *11*, 378. [CrossRef] [PubMed]

24. Chu, G.-L.; Mosyakin, S.L.; Clemants, S.E. Chenopodiaceae. In *Flora of China. Volume 5: Ulmaceae through Basellaceae*; Wu, Z., Raven, P.H., Hong, D., Eds.; Missouri Botanical Garden Press: St. Louis, MI, USA, 2003; pp. 351–414.
25. Habibi, F.; Vít, P.; Rahiminejad, M.; Mandák, B. Towards a better understanding of the *C. album* aggregate in the Middle East: A karyological, cytometric and morphometric investigation. *J. Syst. Evol.* **2018**, *56*, 231–242. [CrossRef]
26. Mandák, B.; Krak, K.; Vít, P.; Pavlíková, Z.; Lomonosova, M.N.; Habibi, F.; Lei, W.; Jellen, E.N.; Douda, J. How genome size variation is linked with evolution within *Chenopodium* sensu lato. *Perspect. Plant Ecol. Evol. System.* **2016**, *23*, 18–32. [CrossRef]
27. Mandák, B.; Krak, K.; Vít, P.; Lomonosova, M.N.; Belyayev, A.; Habibi, F.; Wang, L.; Douda, J.; Storchova, H. Hybridization and polyploidization within the *Chenopodium album* aggregate analyzed by means of cytological and molecular markers. *Mol. Phylogenet. Evol.* **2018**, *129*, 189–201. [CrossRef]
28. Gao, D.; Schmidt, T.; Jung, C. Molecular characterization and chromosomal distribution of species-specific repetitive DNA sequences from *Beta corolliflora*, a wild relative of sugar beet. *Genome* **2000**, *43*, 1073–1080. [CrossRef]
29. Kolano, B.; Gardunia, B.W.; Michalska, M.; Bonifacio, A.; Fairbanks, D.; Maughan, P.J.; Coleman, C.E.; Stevens, M.R.; Jellen, E.N.; Maluszynska, J. Chromosomal localization of two novel repetitive sequences isolated from the *Chenopodium quinoa* Willd. *Genome* **2011**, *54*, 710–717. [CrossRef]
30. Benson, G. Tandem repeats finder: A program to analyze DNA sequences. *Nucleic Acids Res.* **1999**, *27*, 573–580. [CrossRef]
31. Gogarten, J.P.; Kibak, H.; Dittrich, P.; Taiz, L.; Bowman, E.J.; Bowman, B.J.; Manolson, M.F.; Poole, R.J.; Date, T.; Oshima, T.; et al. Evolution of the Vacuolar H+-Atpase: Implications for the Origin of Eukaryotes. *Proc. Natl. Acad. Sci. USA* **1989**, *86*, 6661–6665. [CrossRef]
32. Iwabe, N.; Kuma, K.; Hasegawa, M.; Osawa, S.; Miyata, T. Evolutionary Relationship of Archaebacteria, Eubacteria, and Eukaryotes Inferred from Phylogenetic Trees of Duplicated Genes. *Proc. Natl. Acad. Sci. USA* **1989**, *86*, 9355–9359. [CrossRef] [PubMed]
33. Kadereit, G.; Hohmann, S.; Kadereit, J.W. A synopsis of *Chenopodiaceae* subfam. *Betoideae* and notes on the taxonomy of *Beta*. *Willdenowia* **2006**, *36*, 9–19. [CrossRef]
34. Koukalova, B.; Moraes, A.P.; Renny-Byfield, S.; Matyasek, R.; Leitch, A.; Kovarik, A. Fall and rise of satellite repeats in allopolyploids of *Nicotiana* over c. 5 million years. *New Phytol.* **2009**, *186*, 148–160. [CrossRef] [PubMed]
35. Plohl, M.; Petrović, V.; Luchetti, A.; Ricci, A.; Satović, E.; Passamonti, M.; Mantovani, B. Long-term conservation vs. high sequence divergence: The case of an extraordinarily old satellite DNA in bivalve mollusks. *Heredity* **2009**, *104*, 543–551. [CrossRef] [PubMed]
36. Willard, H.F.; Waye, J.S. Hierarchical order in chromosome-specific human alpha satellite DNA. *Trends Genet.* **1987**, *3*, 192–198. [CrossRef]
37. Gallagher, D.S.; Modi, W.S.; Ivanov, S. Concerted Evolution and Higher-Order Repeat Structure of the 1.709 (Satellite IV) Family in Bovids. *J. Mol. Evol.* **2004**, *58*, 460–465. [CrossRef] [PubMed]
38. Adega, F.; Chaves, R.; Guedes-Pinto, H.; Heslop-Harrison, J.S. Physical organization of the 1.709 satellite IV DNA family in Bovini and Tragelaphini tribes of the Bovidae: Sequence and chromosomal evolution. *Cytogenet. Genome Res.* **2006**, *114*, 140–146. [CrossRef]
39. Macas, J.; Navrátilová, A.; Koblížková, A. Sequence homogenization and chromosomal localization of VicTR-B satellites differ between closely related *Vicia* species. *Chromosoma* **2006**, *115*, 437–447. [CrossRef]
40. Jarvis, D.E.; Ho, Y.S.; Lightfoot, D.J.; Schmöckel, S.M.; Li, B.; Borm, T.J.; Ohyanagi, H.; Mineta, K.; Michell, C.T.; Saber, N.; et al. The genome of Chenopodium quinoa. *Nature* **2017**, *542*, 307–312. [CrossRef]
41. Belyayev, A.; Paštová, L.; Fehrer, J.; Josefiová, J.; Chrtek, J.; Mráz, P. Mapping of *Hieracium* (*Asteraceae*) chromosomes with genus-specific satDNA elements derived from next-generation sequencing data. *Plant Syst. Evol.* **2018**, *304*, 387–396. [CrossRef]
42. Henikoff, S.; Ahmad, K.; Malik, H.S. The centromere paradox: Stable inheritance with rapidly evolving DNA. *Science* **2001**, *293*, 1098–1102. [CrossRef] [PubMed]
43. Belyayev, A.; Josefiová, J.; Jandová, M.; Krak, K.; Mandák, B. Transposable elements dynamics in the evolution of *Chenopodium album* aggregate. in preparation.

44. Kejnovský, E.; Michalovova, M.; Steflova, P.; Kejnovska, I.; Manzano, S.; Hobza, R.; Kubat, Z.; Kovarik, J.; Jamilena, M.; Vyskot, B. Expansion of microsatellites on evolutionary young Y chromosome. *PLoS ONE* **2013**, *8*, e45519. [CrossRef] [PubMed]
45. Li, X.-M.; Lee, B.S.; Mammadov, A.C.; Koo, B.C.; Mott, I.W.; Wang, R.R.-C. CAPS markers specific to Eb, Ee, and R genomes in the tribe Triticeae. *Genome* **2007**, *50*, 400–411. [CrossRef] [PubMed]
46. Luchetti, A.; Marini, M.; Mantovani, B. Non-concerted evolution of the RET76 satellite DNA family in *Reticulitermes* taxa (Insecta, Isoptera). *Genetica* **2006**, *128*, 123–132.
47. Groom, Q.J. Piecing together the biogeographic history of *Chenopodium vulvaria* L. using botanical literature and collections. *Peer J.* **2015**, *3*, e723. [CrossRef] [PubMed]
48. Mayr, E. *Populations Species and Evolution: An Abridgment of Animal Species and Evolution*; Belknap Press: Cambridge, UK, 1970.
49. Grant, V. *Plant Speciation*, 2nd ed.; Columbia University Press: New York, NY, USA, 1981.
50. Husband, B.C. Chromosomal variation in plant evolution. *Am. J. Bot.* **2004**, *91*, 621–625. [CrossRef]
51. Belyayev, A. Bursts of transposable elements as an evolutionary driving force. *J. Evol. Biol.* **2014**, *27*, 2573–2584. [CrossRef]
52. Vít, P.; Krak, K.; Trávníček, P.; Douda, J.; Lomonosova, M.N.; Mandák, B. Genome size stability across Eurasian *Chenopodium* species (*Amaranthaceae*). *Bot. J. Linn. Soc.* **2016**, *182*, 637–649. [CrossRef]
53. Novák, P.; Neumann, P.; Pech, J.; Steinhaisl, J.; Macas, J. RepeatExplorer: A Galaxy-based web server for genome-wide characterization of eukaryotic repetitive elements from next-generation sequence reads. *Bioinformatics* **2013**, *29*, 792–793. [CrossRef]
54. Noe, L.; Kucherov, G. YASS: Enhancing the sensitivity of DNA similarity search. *Nucleic Acids Res.* **2005**, *33*, W540–W543. [CrossRef]
55. Vinga, S.; Almeida, J. Alignment-free sequence comparison-a review. *Bioinformatics* **2003**, *19*, 513–523. [CrossRef] [PubMed]
56. Edgar, R.C. MUSCLE: A multiple sequence alignment method with reduced time and space complexity. *BMC Bioinform.* **2003**, *5*, 113.
57. Kalendar, R.; Tselykh, T.; Khassenov, B.; Ramanculov, E.M. Introduction on using the FastPCR software and the related Java web tools for PCR, in silico PCR, and oligonucleotide assembly and analysis. *Met. Mol. Biol.* **2017**, *1620*, 33–64. [CrossRef]
58. Kumar, S.; Stecher, G.; Li, M.; Knyaz, C.; Tamura, K. MEGA X: Molecular Evolutionary Genetics Analysis across Computing Platforms. *Mol. Biol. Evol.* **2018**, *35*, 1547–1549. [CrossRef] [PubMed]
59. Pijnacker, L.P.; Ferwerda, M.A. Giemsa C-banding of potato chromosomes. *Can. J. Genet. Cytol.* **1984**, *26*, 415–419. [CrossRef]
60. Belyayev, A.; Raskina, O.; Nevo, E. Chromosomal distribution of reverse transcriptase containing retroelements in two *Triticeae* species. *Chromosome Res.* **2001**, *9*, 129–136. [CrossRef] [PubMed]
61. Feinberg, A.P.; Vogelstein, B. A technique for radiolabeling DNA restriction endonuclease fragments to high specific activity. *Anal. Biochem.* **1983**, *132*, 6–13. [CrossRef]
62. Reeves, A. MicroMeasure: A new computer program for the collection and analysis of cytogenetic data. *Genome* **2001**, *44*, 439–443. [CrossRef]

© 2019 by the authors. Licensee MDPI, Basel, Switzerland. This article is an open access article distributed under the terms and conditions of the Creative Commons Attribution (CC BY) license (http://creativecommons.org/licenses/by/4.0/).

International Journal of
Molecular Sciences

MDPI

Article

Multifaceted Role of *PheDof12-1* in the Regulation of Flowering Time and Abiotic Stress Responses in Moso Bamboo (*Phyllostachys edulis*)

Jun Liu, Zhanchao Cheng, Lihua Xie, Xiangyu Li and Jian Gao *

International Center for Bamboo and Rattan, Key Laboratory of Bamboo and Rattan Science and Technology, State Forestry Administration, Beijing 100102, China; liujun_0325@163.com (J.L.); chengzhan_chao@126.com (Z.C.); xielihua0227@163.com (L.X.); leerduo727@163.com (X.L.)

* Correspondence: gaojianicbr@163.com or gaojian@icbr.ac.cn; Tel.: +86-010-8478-9801

Received: 23 December 2018; Accepted: 17 January 2019; Published: 19 January 2019

Abstract: DNA binding with one finger (Dof) proteins, forming an important transcriptional factor family, are involved in gene transcriptional regulation, development, stress responses, and flowering responses in annual plants. However, knowledge of Dofs in perennial and erratically flowering moso bamboo is limited. In view of this, a Dof gene, *PheDof12-1*, was isolated from moso bamboo. *PheDof12-1* is located in the nucleus and has the highest expression in palea and the lowest in bract. Moreover, *PheDof12-1* expression is high in flowering leaves, then declines during flower development. The transcription level of *PheDof12-1* is highly induced by cold, drought, salt, and gibberellin A3 (GA_3) stresses. The functional characteristics of *PheDof* are researched for the first time in Arabidopsis, and the results show that transgenic Arabidopsis overexpressing *PheDof12-1* shows early flowering under long-day (LD) conditions but there is no effect on flowering time under short-day (SD) conditions; the transcription levels of *FT*, *SOC1*, and *AGL24* are upregulated; and *FLC* and *SVP* are downregulated. *PheDof12-1* exhibits a strong diurnal rhythm, inhibited by light treatment and induced in dark. Yeast one-hybrid (Y1H) assay shows that *PheDof12-1* can bind to the promoter sequence of *PheCOL4*. Taken together, these results indicate that *PheDof12-1* might be involved in abiotic stress and flowering time, which makes it an important candidate gene for studying the molecular regulation mechanisms of moso bamboo flowering.

Keywords: *Phyllostachys edulis*; Dof transcription factor; flowering time; abiotic stress; gene expression

1. Introduction

DNA binding with one finger (Dof) transcription factors (TFs) are a family of plant-specific transcription factors. The proteins generally contain 50–52 highly conserved amino acids, including a C_2C_2-type zinc-finger motif at the N-terminal end [1]. Dof transcription factors have been shown to be widely distributed in the plant kingdom. The cDNA sequence of Dof was first obtained from *Zea mays* [2]. Since then, many Dofs have been cloned from various plant species [3–5]. In previous studies, it is suggested that Dof proteins are involved in the regulation of a variety of biological processes, including seed germination, floral organ abscission, hormone signaling, and cell cycles. In *Arabidopsis*, *DAG1* and *DAG2* can promote seed germination [6,7], *DOF6* acts as a negative regulator of seed germination and interacts with TCP14 [8], and *AtDOF4.7* participates in the transcriptional regulation of floral organ abscission via an effect on cell wall hydrolase gene expression [9]. In addition, some Dof genes (*AtDof2.4*, *AtDof5.8*, and *AtDof5.6/HCA2*) are expressed in the early development of vascular cells [10]. In rice, *OsDof3* is involved in gibberellin-regulated expression [11]. Moreover, Dof TFs such as maize *Dof1* and *Dof2* are also involved in the control of carbon and nitrogen metabolism through the regulation of phosphoenolpyruvate carboxykinase (PECPK), glutamine synthase (GS), and glutamate synthase (GLU) [7,12–16].

Genetic and molecular studies have suggested that Dof transcription factors participate in different stresses, light responsiveness, and flowering regulation. In *Brachypodium distachyon*, *BdCBF1*, *BdCBF2*, and *BdCBF3* contribute to cold, drought, and salt stresses by regulating downstream targets such as DEHYDRIN5.1 (*Dhn5.1*) and *COR* genes [17]. Overexpressing *SlCDF3* shows increased transgenic Arabidopsis drought and salt tolerance [18]. In Chinese cabbage, most *BraDof* genes are induced by cold, heat, high salinity, and drought stresses [19]. Moreover, Dof proteins are involved in photoperiod flowering. In Arabidopsis, cycling Dof factor-1 (CDF1) binds to the *COSTANS* (*CO*) and *FLOWERING LOCUS T* (*FT*) promoter regions to block transactivation of these two flowering genes, whereas this inhibition could be released based on the GIGANTEA-FLAVIN-BINDING, KELCH REPEAT, F-BOX1(GI-FKF1) complex-mediated degradation of CDF1 under long-day (LD) conditions [20]. In addition, *CDF2*, *CDF3*, and *CDF5* repress flowering of Arabidopsis by decreasing the mRNA level of *CO* [21]. In rice, overexpressing *OsDof12* promotes early flowering under LD conditions by upregulating the expression of *Hd3a* and *OsMADS14* [22]. Although a large number of Dofs have been extensively studied in annual plants [23,24], the knowledge of Dofs in moso bamboo is limited.

Moso bamboo (*Phyllostachys edulis*) is a perennial plant characterized by a long vegetative stage that flowers synchronously followed by widespread death [25]. In this case, studying the mechanism of moso bamboo flowering time is very important and challenging, and it is quite difficult to determine the key regulatory gene. Moreover, the growth of moso bamboo in the wild is severely threatened by various environmental conditions such as drought, salinity and cold, which severely limit the growth and distribution of moso bamboo and affect the yield and quality of winter shoots, as well as new bamboo yield in the following year and the yield of wood harvesting of the subsequent years [26–28]. In addition, recent research on Dofs is mainly in annual plants, and is limited in perennials. Therefore, researching the role of Dofs in moso bamboo is necessary, especially in terms of abiotic stress and flowering time. In this study, a Dof gene (*PheDof12-1*) is isolated from moso bamboo, induced by cold, drought, salt, and gibberellin (GA_3) stresses. The functional characteristics of *PheDof12-1* are researched for the first time by ectopic expression in Arabidopsis, and transgenic Arabidopsis overexpressing homozygous *PheDof12-1* show early flowering under long-day (LD) conditions, binding to the promoter sequence of PheCOL4 with a strongly diurnal pattern. These results provide new insights into the functions of the Dof transcription factor in the regulation of photoperiod flowering time and abiotic stress in moso bamboo.

2. Results

2.1. Isolation and Analysis of *PheDof12-1*

Based on the moso bamboo genome database, *PheDof12-1* was isolated from moso bamboo. The full-length CDS of *PheDof12-1* is 1299 bp, encoding 432-amino acids, with predicted molecular weight (MW) and isoelectric point (pI) of 46.37 kDa and 8.32, respectively. Structure analysis showed that PheDof12-1 contains one intron and two exons (Figure 1A). The deduced proteins contain the conserved zf-Dof domain. Furthermore, phylogenetic analysis of PheDof12-1 and homologous proteins from other plants shows that PheDof12-1 and other Dofs from monocotyledons belong to the same clade (Figure 1B). The amino acid sequence of PheDof12-1 shows 83% and 84% identity with *Brachypodium* (XP_003558722) and rice (XP_015690912), respectively. This result was consistent with the findings in the stated phylogeny and classification of plants. All these proteins contain the conserved zf-Dof domain (Supplementary Figure S1).

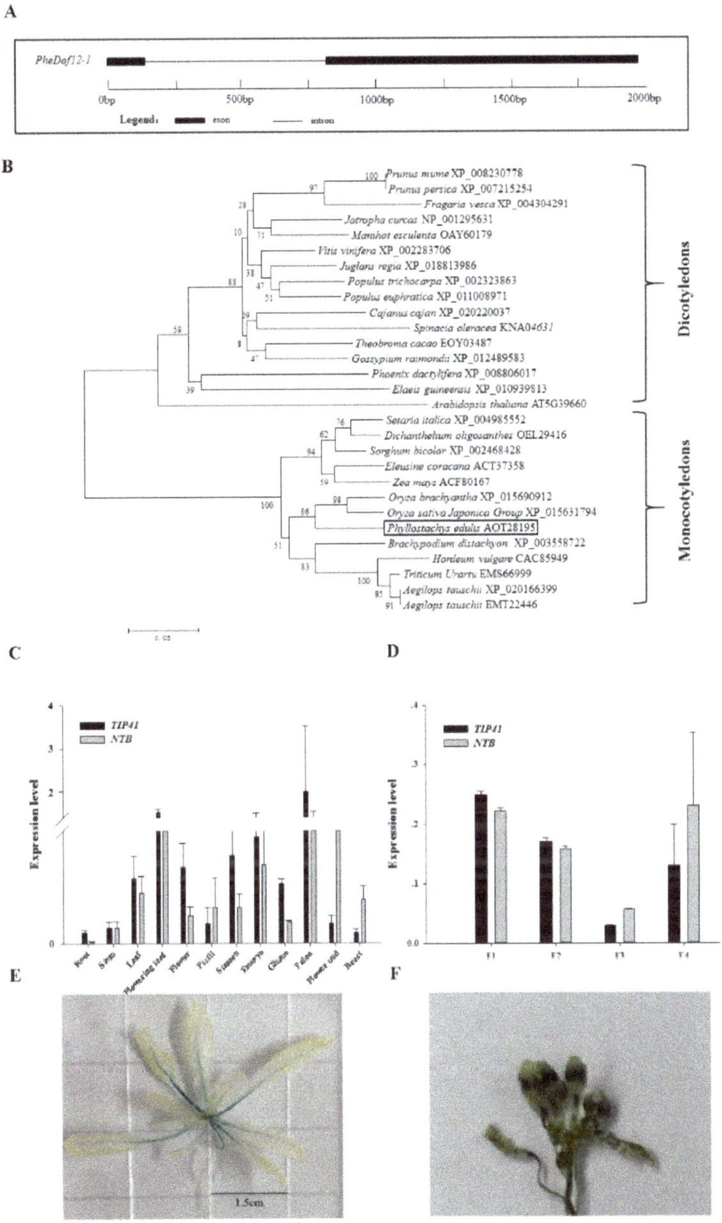

Figure 1. Characterization and preliminary expression analysis of *PheDof12-1*. (**A**) Gene structure of *PheDof12-1*. (**B**) Phylogenetic analysis of PheDof12-1 with other DNA binding with one finger (Dof) proteins. (**C**) qRT-PCR analysis of *PheDof12-1* in different tissues of moso bamboo. (**D**) Expression profile of *PheDof12-1* in different flower developmental stages: F1: floral bud formation stage; F2: inflorescence growing stage; F3: blooming stage; F4: flowers are withered. (**E**) Glucuronidase (GUS) staining of $Pro_{PheDof12\text{-}1}$-*GUS* in transgenic Arabidopsis seedling. (**F**) GUS staining of $Pro_{PheDof12\text{-}1}$-*GUS* plants showing PheDof12-1 localization in flower and pollen.

2.2. Tissue-Specific Gene Expression

In order to analyze the expression of *PheDof12-1* in different tissues (root, stem, leaf, flowering leaf, flower) and floral organs (pistil, stamen, embryo, glume, palea, flower bud, bract), RNA was isolated to perform qRT-PCR. The results show that the transcription level of *PheDof12-1* in flowering leaf is significantly higher than in other tissues. In different flower organs, the expression of *PheDof12-1* was highest in palea, and lowest in bract (Figure 1C). In developing flowers, *PheDof12-1* had higher transcript accumulation at the floral bud formation stage (F1) (Figure 1D), and decreased gradually at flower development, which was consistent with the previously reported detection of *PheDof1* at early stages of flower formation and development [29]. We further generated $Pro_{PheDof12\text{-}1}$-*GUS* transgenic lines, and glucuronidase (GUS) staining was detected in the vasculature of cotyledons and hypocotyls, true leaves, roots, flower, and pollen (Figure 1E,F). The results demonstrate that *PheDof12-1* is expressed in different tissues and at different flower development stages, suggesting that it is dynamic during plant development and may play an important role in moso bamboo growth and development.

2.3. Expression Patterns of PheDof12-1 under Stress Treatments

Previous reports have shown that Dof TFs are involved in abiotic stress [30]. To determine the expression pattern of *PheDof12-1* in moso bamboo under different stresses, we performed detailed qRT-PCR with *TIP41* and *NTB* as internal reference genes. The results show that *PheDof12-1* was responded to cold, drought, and salt stresses. In drought stress, *PheDof12-1* was induced and upregulated at each time point, and levels of transcripts in leaves and stems were slightly elevated, but a sharp increase occurred after 1 h in roots, peaking at 70.9-fold. This implies that *PheDof12-1* is induced and has a positive function in response to drought stress (Figure 2A–C). In cold treatment using *NTB* as a reference gene, the expression of *PheDof12-1* rapidly increased in leaves, reaching 86.1-fold at 24 h (Figure 2I). Regarding salt treatment, the maximum increase was observed at 12 h, reaching 12.5-fold in leaves when *TIP41* was used as the reference gene (Figure 2F), but the expression level was first induced and then decreased in roots. To further investigate the functions of *PheDof12-1*, we initially analyzed the effects of gibberellin A3 (GA_3) and abscisic acid (ABA) on its expression (Figure 2J,K). In GA_3 stress, the transcription level of *PheDof12-1* was induced and upregulated at almost every time point, peaking at 15.0-fold at 24 h. Under ABA treatment, the translation level of *PheDof12-1* initially decreased and then increased, was lowest at 6 h, dropping to undetectable levels, and reached a peak at 48 h. All of these data indicate that *PheDof12-1* takes part in the hormones and different abiotic stresses of moso bamboo.

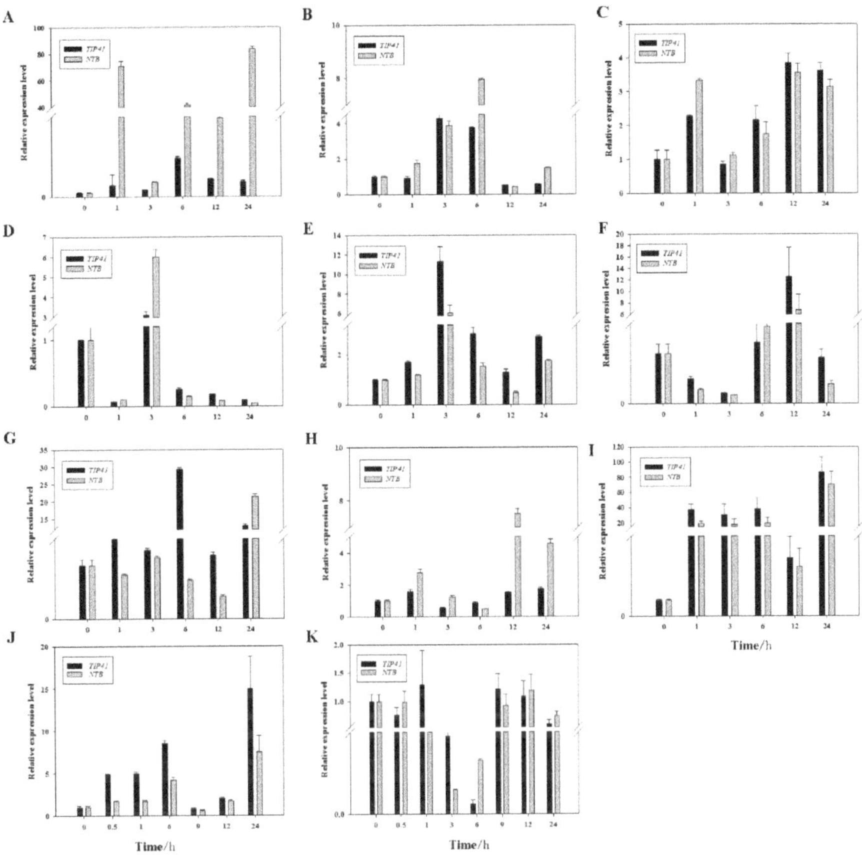

Figure 2. Relative expression of *PheDof12-1* in different tissues of moso bamboo under drought: (**A**) root, (**B**) stem, (**C**) leaf; salt: (**D**) root, (**E**) stem, (**F**) leaf; under cold: (**G**) root, (**H**) stem, (**I**) leaf; and under (**J**) gibberellin A3 (GA_3) and (**K**) abscisic acid (ABA) treatments.

2.4. Overexpression of PheDof12-1 Promotes Early Flowering in Arabidopsis

In order to verify the subcellular localization of PheDof12-1, we further amplified its coding region and fused it to the N-terminal of the eGFP vector. The subcellular localization assay indicated that PheDof12-1 was localized in the nucleus, in accordance with its function as a transcription factor (Figure 3B). To study the genetic functions of *PheDof12-1*, we transformed it in Arabidopsis. The overexpressed plants showed an early flowering phenotype under LD conditions (Figure 3A), whereas *PheDof12-1* overexpression had no effect on flowering time under SD conditions (not shown). The flowering time was about 10 days earlier than wild-type, and the number of rosette leaves of overexpressed lines was smaller than that of wild Arabidopsis (Figure 3C). We further investigated the transcription levels of *FT*, *SUPPRESSOR OF OVEREXPRESSION OF CONSTANS1* (*SOC1*), *AGAMOUS-LIKE 24* (*AGL24*), *FLOWERING LOCUS C* (*FLC*), and *SHORT VEGETATIVE PHASE* (*SVP*) in the T3 generation to ascertain the downstream effects of this construct. *FT*, *SOC1*, and *AGL24* were upregulated, while *FLC* and *SVP* expression were rather low compared with wild-type (Figure 3D). These data suggest that *PheDof12-1* might regulate flowering by controlling the expression of *FT*, *SOC1*, *AGL24*, *FLC*, and *SVP*.

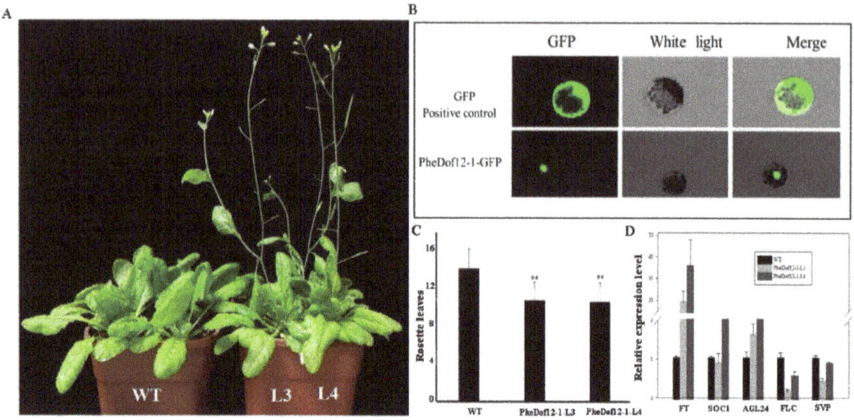

Figure 3. Analysis of an early flowering phenotype by overexpression of PheDof12-1 in Arabidopsis. (**A**) Phenotypes of overexpressing *PheDof12-1* transgenic lines (L3, L4) and wild-type (WT) plants as control under long-day (LD) conditions. (**B**) Subcellular localization of PheDof12-1. (**C**) Flowering time scored as number of rosette leaves at flowering of wild-type and transgenic plants under LD conditions. (**D**) Transcription levels of *FT*, *SOC1*, *AGL24*, *FLC*, and *SVP* in wild-type and transgenic plants. Arabidopsis *Actin* was used as the internal reference gene. Error bars indicate standard deviations. Asterisks indicate statistically significant difference between wild-type and transgenic plants ($p < 0.01$ by Student's *t*-test).

2.5. PheDof12-1 Interacts with Photoperiod-Related Regulators

In Arabidopsis, CDFs are transcriptional repressors that bind to *CO* and *FT* promoters to repress their transcription [20]. To explore whether PheDof12-1 can form heterodimers with other proteins, an interaction prediction was performed using STRING (https://stringdb.org/) based on the interaction network of rice orthologous genes. As shown in Figure 4E, PheDof12-1 interacted with 10 identified proteins. Among them, the B-box protein (PH01004196G0130), Dof transcription factor (PH01001184G0160), grain size gene (*PheSLR1*) [31], photoperiodic flowering response gene (*PH01002431G00090*) [32], and drought-induced protein (PH01000199G0750) were identified, suggesting that PheDof12-1 may be involved in growth and development, photoperiodic response, and abiotic stress.

CDF1, *CDF2*, *CDF3*, and *CDF5* had high mRNA levels at the beginning of the light period in Arabidopsis [21], and CDFs displayed a similar expression pattern in *Populus* [33]. So, we detected the expression patterns under photoperiod treatments. The results show that *PheDof12-1* was similarly expressed under both LD and SD conditions. The transcription level of *PheDof12-1* decreased with the increased light time, reaching the minimum value before dark (Figure 4A,B), with high mRNA levels at the beginning of the light period, which was consistent with the expression patterns of CDFs in Arabidopsis and *Populus*. The highly similar expression pattern of CDFs in *Populus*, Arabidopsis, and moso bamboo suggests a functional conservation.

CO and CO-like (COL) proteins are members of the B-box family, playing a central role in the photoperiod response pathway by mediating between the circadian clock and the floral integrators [34]. CDFs are transcriptional repressors that bind to the *CO* promoter to repress its transcription [20]. PheDof12-1 interacted with B-box proteins by interaction prediction; moreover, *PheDof12-1* and *PheCOL4* had similar expression patterns under photoperiod treatments (Figure 4C,D), suggesting that PheDof12-1 may interact with PheCOL4 in moso bamboo. To examine whether the PheDof12-1 protein regulated *PheCOL4* expression by directly binding to the promoter region, the *PheCOL4* promoter sequence was investigated. We performed a targeted yeast one-hybrid (Y1H) assay using PheDof12-1, and PheCOL4 was inserted upstream of the reporter plasmid pHIS2 and cotransfected into the yeast cells with the AD-PheCOL12-1 effector plasmid. The binding of PheCOL12-1 and the promoter of *PheCOL4* was indicated by the growth of transfected yeast cells on a nutrient-deficient medium (synthetic dextrose (SD)/-Trp-Leu-His) plus 3-amino-1, 2, 4-triazole (3-AT) and 5-bromo-4-chloro-3-indoxyl-α-D-galactopyranoside (X-α-Gal). The results show that all transformants tested were found to grow well on the SD/-Leu/-Trp medium when transferred onto SD/-Trp/-Leu/-His/3-AT/X-α-Gal plates for 3 days; only the yeast cells of AD-PheDof12-1 + pHIS2-PheCOL4 vectors and the positive control grew strong and turned blue (Figure 4F). This result suggests that PheDof12-1 could bind to the promoter of PheCOL4 and regulate *PheCOL4* expression in moso bamboo.

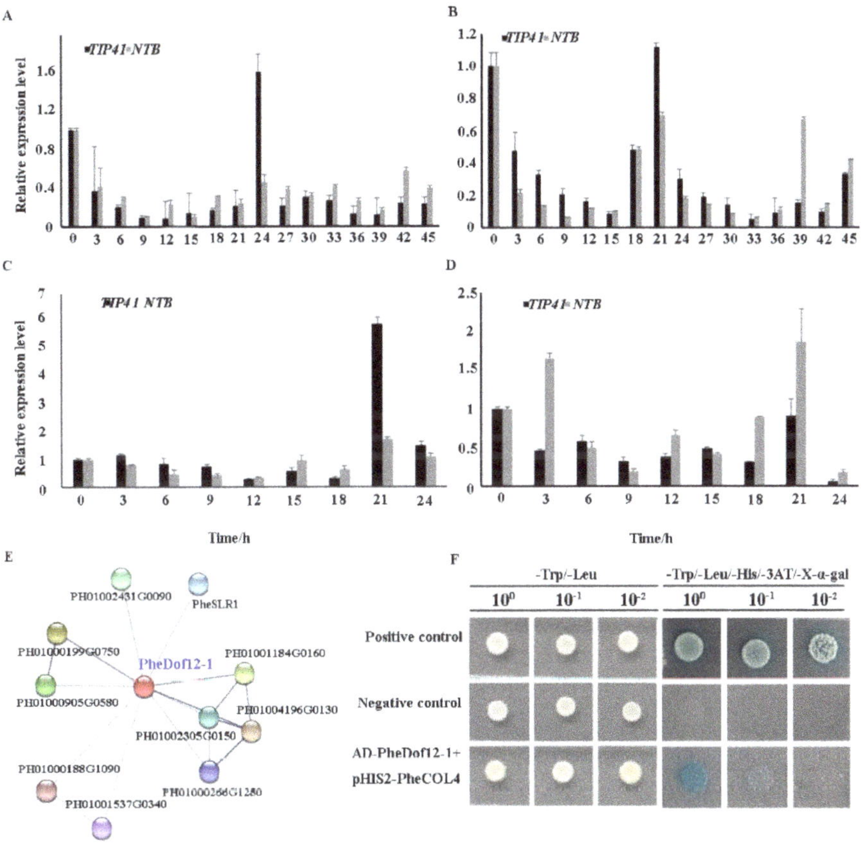

Figure 4. PheDof12-1 protein binds to the promoter region of *PheCOL4*. Relative expression of *PheDof12-1* under (**A**) LD and (**B**) SD conditions. Transcription level of *PheCOL4* under (**C**) LD and (**D**) SD conditions. (**E**) Interaction network of PheDof12-1 in moso bamboo. Colored balls (protein nodes) in the network were used as a visual aid to indicate different input proteins and predicted interactions. Enlarged protein nodes indicate the availability of 3D protein structure information. Gray lines connect proteins that are associated by recurring text mining evidence. (**F**) Yeast one-hybrid (Y1H) assay for AD-PheDof12-1 and pHIS2-PheCOL4. The reporter pHIS2 vector carrying the corresponding fragment and the effector AD-PheDof12-1 vector were cotransfected into yeast Y187 cells. Growth of the transfected yeast cells on a 3-AT and X-α-Gal medium indicates that the PheDof12-1 protein can bind to the PheCOL4 promoter.

3. Discussion

Moso bamboo is a perennial plant characterized by rapid growth and a long vegetative stage that lasts for decades or even longer before flowering [25]. Dof proteins are a group of plant-specific TFs that are involved in diverse plant-specific biological processes [16]. In addition, recent research on *Dofs* is mainly in annual plants, and is limited in perennials. Therefore, researching the roles of *Dofs* in moso bamboo is necessary. In this study, a *Dof* gene, *PheDof12-1*, is identified from moso bamboo as a nucleus-localized transcription factor that contains typical zf-dof domains.

In recent decades, reports have indicated that Dof transcription factors are involved in stress response. In Arabidopsis, the expression level of *AtCDF3* is upregulated by cold, drought, high salinity, and ABA treatment [30], and overexpression of *35S::SlCDF1* and *35S::SlCDF3* increases Arabidopsis's tolerance to salt and drought stresses [18]. In wheat, *TaDof14* and *TaDof15* are significantly induced under drought treatment [35]. Previous research has suggested that drought or other environmental stresses are functional in the flowering stage of bamboo, and the transcription levels of *Dof* genes are upregulated in drought stress [36]. In addition, studying the tolerance of *PheDof12-1* will help to characterize moso bamboo cultivars such as salt, cold, and drought tolerance. In this study, *PheDof12-1* exhibited differential expression patterns under the conditions of drought, cold, salt, and ABA and GA_3 treatments. Through the drought, cold, salt, and GA_3 stresses, the expression pattern of *PheDof12-1* is basically upregulated in roots, stems, and leaves, indicating that it might participate in abiotic stress and hormone pathways, which is consistent with previous reports [36,37]. The results provide a better understanding of the stress tolerance of *PheDof12-1* in moso bamboo.

Hd1/CO and *Hd3a/FT* are conserved genetic pathways that regulate photoperiodic flowering between rice and Arabidopsis by their genomic comparison [38]. In Arabidopsis, *CDF1–CDF3* are suggested to participate in photoperiodic flowering [39]. *JcDof3* is a circadian clock regulated gene involved in the regulation of flowering time in *Jatropha curcas* [40]. In rice, *OsDof12* and *CDF1* belong to the same group [41], and overexpression of *OsDof12* resulted in early flowering by increasing the expression of *Hd3a* and *OsMADS14* under LD conditions [22]. *PheDof12-1* is the homologous gene of *OsDof12*, and Dof-Hd3a-MADS-flowering may play an important role in moso bamboo flowering [36]. Therefore, we researched the function of *PheDof12-1* in flowering time by ectopic expression in Arabidopsis for the first time, and the transgenic lines overexpressing *PheDof12-1* show earlier flowering than the wild-type plants under LD conditions. In addition, *FT*, *SOC1*, and *AGL24* are upregulated and *FLC* and *SVP* are downregulated in the transgenic lines. *FT* promotes flowering [42], which is activated by *CO* in the phloem [43]. SOC1 is a core regulator of flowering in Arabidopsis, which can interact with SVP and AGL24 proteins, but SVP and AGL24 have opposite effects on flowering time, acting as floral repressor and inducer, respectively [44]. *FLC* encodes a MADS domain-containing transcription factor that acts as an inhibitor of flowering [45]. This leads us to suspect that *PheDof12-1* promotes flowering time by regulating *FT*, *SOC1*, *AGL24*, *FLC*, and *SVP* directly or indirectly, suggesting that it might retain some function in the control of flowering time through similar molecular mechanisms to those observed when expressed in Arabidopsis.

Diurnal oscillation of the transcription levels of *CDFs* has been reported in Arabidopsis and other species [21,23]. In Arabidopsis, *CDF1–CDF3* and *CDF5* show maximum expression at the beginning of the light period, decreasing to a minimum between 16 and 20 h, then rising again during dawn [21]. In tomato, *SlCDF1* and *SlCDF3* exhibit maximum expression at the beginning of the day, while *SlCDF2*, *SlCDF4*, and *SlCDF5* exhibit maximum levels during the night [18]. In rice, *OsDof12* is strongly inhibited by dark treatment [22]. In the study, *PheDof12-1* exhibited significantly diurnal expression patterns with high mRNA levels at the beginning of the light period under LD and SD conditions, supporting the assumption that it is a true homologue of the Arabidopsis CDFs. In Arabidopsis, CDFs can bind to the *CO* promoter to repress its transcription [20], and PttCDF3 can bind directly to the *PttCO2* promoter in *Populus* [33]. In moso bamboo, the diurnal expression pattern of *PheCOL4* is consistent with *PheDof12-1*, and Y1H analysis shows that PheDof12-1 binds directly to the promoter of *PheCOL4*. These results support the hypothesis that flowering regulator CO, a target of CDFs, is controlled precisely [21], which is similar to the situation in Arabidopsis and *Populus*.

4. Materials and Methods

4.1. Plant Materials and Treatments

Moso bamboo seeds were harvested from Guilin in the Guangxi Zhuang Autonomous Region, China. Seedlings were grown in an illumination incubator under long-day conditions (16 h light/8 h dark) at day/night temperatures of 25/18 °C, and watered with Hoagland nutrient solution. For drought and salt stress, the seedlings were watered with 50% Hoagland's solution with 20% polyethylene glycol 6000 (PEG 6000) and 250 mM NaCl. For low temperature treatment, the plants were transferred to a growth chamber at 4 °C, and plant leaf, stem, and root tissues were collected [46]. For abscisic acid (ABA) and gibberellin A3 (GA_3) treatments, the seedlings were watered with 200 μM ABA [47] and 200 μM GA_3 solution [48]. To detect the transcriptional level of *PheDof12-1* in photoperiod treatments, leaves were collected for analysis from plants exposed to LD (16 h light/8 h dark) and SD (16 h light/8 h dark) treatments [21]. All samples were immediately frozen in liquid nitrogen and stored at −80 °C until further analysis.

4.2. Bioinformatic Analysis

The sequences were downloaded from BambooGDB (http://forestry.fafu.edu.cn/db/PhePacBio/) [49]. Molecular weight (MW) and isoelectric point (pI) were analyzed using ProtParam (http://web.expasy.org/protparam/) [50]. The structure was shown using Gene Structure Display Server software (http://gsds1.cbi.pku.edu.cn/index.php) [51]. To search the database, the Basic Local Alignment Search Tool (BLAST) network service from the National Center for Biotechnology Information (NCBI) web server was applied. Homologue alignment was obtained using Clustal 1.83, and a phylogenetic tree was constructed by MEGA6.0 [52] using the following parameters: NJ method, complete deletion, and bootstrap with 1000 replicates.

4.3. Vector Construction and Plant Transformation

The subcellular localization was performed by transfecting GFP-tagged PheDof12-1 into Arabidopsis sheath protoplasts [53] (Supplementary Table S1). The full-length cDNA of *PheDof12-1* was fused in frame with the GFP cDNA and ligated between the CaMV 35 S promoter and the nopaline synthase terminator. The fluorescence signals were examined using a confocal laser scanning microscope (Leica Microsystems, Wiesler, Germany).

The full-length coding sequence of *PheDof12-1* was cloned into the pCAMBIA 2300 vector under the control of the modified CaMV 35S promoter (Supplementary Table S1). The pCAMBIA 2300-*PheDof12-1* vector was introduced into *Agrobacterium umefaciens* strain CV3101 for Arabidopsis transformation in the Col-0 background by the floral dipping method [54]. Putative transgenic plants were screened on 50% Murashige and Skoog (MS) solid medium supplemented with 50 mg/L kanamycin, and homozygous T3 or T4 seeds were used.

In order to analyze the spatial expression patterns of PheDof12-1, a 2 kb region upstream of the PheDof12-1 transcription start site was cloned and fused to the pCAMBIA2391Z vector to generate the $Pro_{PheDof12\text{-}1}$-*GUS* reporter, which was transformed into wild-type (WT) plants (Supplementary Table S1). For GUS staining, $Pro_{PheDof12\text{-}1}$-*GUS* transgenic plants were used as previously reported [55].

4.4. Gene Expression Analysis

Total RNA was extracted from the frozen samples using Trizol reagent (Invitrogen, Carlsbad, CA, USA) according to the manufacturer's instructions and treated with DNase I (TaKaRa, Tokyo, Japan) to remove genomic DNA contamination. Then, for each sample, the first-strand cDNA was synthesized using a PrimeScript™ RT Reagent Kit (TaKaRa). The expression profiles of PheDof12-1 in different tissues, abiotic stress, and photoperiod treatments were analyzed by quantitative RT-PCR (qRT-PCR). *TIP41* and *NTB* were used as internal housekeeping genes [56]. The qRT-PCR reactions were carried out using a Light Cycler 480 System (Roche, Basel, Switzerland) and a SYBR Premix EX TaqTMkit (Roche, Mannheim, Germany). All reactions were performed in triplicate, both technical and biological, and data were analyzed using the Roche manager software. The primer sequences are listed in Supplementary Table S1.

4.5. Yeast One-Hybrid Assay

To perform the Y1H assay, the full length of PheDof12-1 was cloned into the pGADT7-Rec2 bait vector, and the promoter sequence of PheCOL4 was cloned into the pHIS2 prey vector (Supplementary Table S1). The lithium acetate method was used to transform into the Y187 strain. The transformed yeast cells were selected on SD/-Trp/-Leu and SD/-Trp/-Leu/-His/-3AT/X-α-Gal plates at 30 °C for 3–5 days.

5. Conclusions

In conclusion, the present study provides new notions about the function of Dof TFs in moso bamboo and shows PheCOL12-1 as a key factor with multiple roles related to abiotic stress, and the developmental program underlying the transition from the vegetative to the reproductive phase under LD conditions. PheCOL12-1 is a nucleus-localized transcription factor that regulates photoperiodic-related regulators. These findings not only increase our understanding of the functional roles of Dof proteins in the regulation of abiotic stress and flowering time, but also provide an important candidate gene for studying molecular regulation mechanisms of moso bamboo flowering.

Supplementary Materials: Supplementary materials can be found at http://www.mdpi.com/1422-0067/20/2/424/s1.

Author Contributions: J.L. and J.G. designed the experiments; X.L., and L.X. performed the tissue and organ collection; J.L. writing—original draft preparation; Z.C. writing—review and editing; J.G. review and funding acquisition.

Funding: This work was supported by the National Natural Science Foundation of China (grant number 31570673).

Conflicts of Interest: Authors declare that there is no competing interest.

Abbreviations

LDs	Long days
SDs	Short days
DOF	DNA binding with One Finger
PCR	Polymerase chain reaction
CO	CONSTANS
FT	Flowering locus T
TFs	Transcription Factors
GI	GIGANTEA
FKF1	FLAVIN-BINDING, KELCH REPEAT, F-BOX1
CDF	Cycling Dof Factor
Hd3a	Heading date 3a
MADS	MCM1, AGAMOUS, DEFICIENS and SRF
GFP	Green Xuorescent protein
ABA	Abscisic acid
GA	Gibberellin
GUS	Glucuronidase
COL	CO-Like
FLC	FLOWERING LOCUS C
SOC1	SUPPRESSOR OF OVEREXPRESSION OF CONSTANS1

SVP	SHORT VEGETATIVE PHASE
AGL24	AGAMOUS-LIKE 24
3-AT	3-amino-1, 2, 4-triazole
X-α-Gal	5-bromo-4-chloro-3-indoxyl-α-D-galactopyranoside

References

1. Yanagisawa, S. The Dof family of plant transcription factors. *Trends Plant Sci.* **2002**, *7*, 555–560. [CrossRef]
2. Yanagisawa, S.; Izui, K. Molecular cloning of two DNA-binding proteins of maize that are structurally different but interact with the same sequence motif. *J. Biol. Chem.* **1993**, *268*, 16028–16036. [PubMed]
3. Plesch, G.; Ehrhardt, T.; Muellerroeber, B. Involvement of TAAAG elements suggests a role for Dof transcription factors in guard cell-specific gene expression. *Plant J.* **2001**, *28*, 455–464. [CrossRef] [PubMed]
4. Tanaka, M.; Takahata, Y.; Nakayama, H.; Nakatani, M.; Tahara, M. Altered carbohydrate metabolism in the storage roots of sweetpotato plants overexpressing the *SRF1* gene, which encodes a Dof zinc finger transcription factor. *Planta* **2009**, *230*, 737–746. [CrossRef] [PubMed]
5. Wen, C.L.; Cheng, Q.; Zhao, L.; Mao, A.; Yang, J.; Yu, S.; Weng, Y.; Xu, Y. Ientification and characterisation of Dof transcription factors in the cucumber genome. *Sci. Rep.* **2016**, *6*, 23072–23083. [CrossRef]
6. Papi, M.; Sabatini, S.; Bouchez, D.; Camilleri, C.; Costantino, P.; Vittorioso, P. Identification and disruption of an Arabidopsis zinc finger gene controlling seed germination. *Genes Dev.* **2000**, *14*, 28–33. [CrossRef] [PubMed]
7. Gualberti, G.; Papi, M.; Bellucci, L.; Ricci, I.; Bouchez, D.; Camilleri, C.; Costantino, P.; Vittorioso, P. Mutations in the Dof zinc finger genes *DAG2* and *DAG1* influence with opposite effects the germination of Arabidopsis seeds. *Plant Cell* **2002**, *14*, 1253–1263. [CrossRef] [PubMed]
8. Rueda-Romero, P.; Barrero-Sicilia, C.; Gomez-Cadenas, A.; Carbonero, P.; OnateSanchez, L. *Arabidopsis thaliana* DOF6 negatively affects germination in non-after-ripened seeds and interacts with TCP14. *J. Exp. Bot.* **2012**, *63*, 1937–1949. [CrossRef]
9. Wei, P.C.; Tan, F.; Gao, X.Q.; Zhang, X.Q.; Wang, G.Q.; Xu, H.; Li, L.J.; Chen, J.; Wang, X.C. Overexpression of AtDOF4. 7, an Arabidopsis DOF family transcription factor, induces floral organ abscission deficiency in Arabidopsis. *Plant Physiol.* **2010**, *153*, 1031–1045. [CrossRef]
10. Guo, Y.; Qin, G.; Gu, H.; Qu, L.J. *Dof5. 6/HCA2*, a Dof transcription factor gene, regulates interfascicular cambium formation and vascular tissue development in *Arabidopsis*. *Plant Cell* **2009**, *21*, 3518–3534. [CrossRef]
11. Washio, K. Functional dissections between GAMYB and Dof transcription factors suggest a role for protein-protein associations in -the gibberellin-mediated expression of the *RAmy1A* gene in the rice aleurone. *Plant Physiol.* **2003**, *133*, 850–863. [CrossRef]
12. Yanagisawa, S.; Sheen, J. Involvement of maize Dof zinc finger proteins in tissue-specific and light-regulated gene expression. *Plant Cell* **1998**, *10*, 75–89. [CrossRef] [PubMed]
13. Yanagisawa, S. Dof1 and Dof2 transcription factors are associated with expression of multiple genes involved in carbon metabolism in maize. *Plant J.* **2000**, *21*, 281–288. [CrossRef] [PubMed]
14. Rueda-López, M.; Crespillo, R.; Cánovas, F.M.; Avila, C. Differential regulation of two glutamine synthetase genes by a single Dof transcription factor. *Plant J.* **2008**, *56*, 73–85. [CrossRef] [PubMed]
15. Kurai, T.; Wakayama, M.; Abiko, T.; Yanagisawa, S.; Aoki, N.; Ohsugi, R. Introduction of the *ZmDof1* gene into rice enhances carbon and nitrogen assimilation under low-nitrogen conditions. *Plant Biotechnol. J.* **2011**, *9*, 826–837. [CrossRef] [PubMed]
16. Yanagisawa, S.; Akiyama, A.; Kisaka, H.; Uchimiya, H.; Miwa, T. Metabolic engineering with Dof1 transcription factor in plants: Improved nitrogen assimilation and growth under low-nitrogen conditions. *Proc. Natl. Acad. Sci. USA* **2004**, *101*, 7833–7838. [CrossRef] [PubMed]
17. Ryu, J.Y.; Hong, S.Y.; Jo, S.H.; Woo, J.C.; Lee, S.; Park, C.M. Molecular and functional characterization of cold responsive C-repeat binding factors from *Brachypodium distachyon*. *BMC Plant Biol.* **2014**, *14*, 15. [CrossRef]
18. Corrales, A.R.; Nebauer, S.G.; Carrillo, L.; Fernández-Nohales, P.; Marqués, J.; Renau-Morata, B.; Granell, A.; Pollmann, S.; Vicente-Carbajosa, J.; Molina, R.V.; et al. Characterization of tomato Cycling Dof Factors reveals conserved and new functions in the control of flowering time and abiotic stress responses. *J. Exp. Bot.* **2014**, *65*, 995–1012. [CrossRef]

19. Ma, J.; Li, M.Y.; Wang, F.; Tang, J.; Xiong, A.S. Genome-wide analysis of Dof family transcription factors and their responses to abiotic stresses in Chinese cabbage. *BMC Genom.* **2015**, *16*, 33. [CrossRef]
20. Song, Y.H.; Smith, R.W.; To, B.J.; Millar, A.J.; Imaizumi, T. *FKF1* conveys timing information for *CONSTANS* stabilization in photoperiodic flowering. *Science* **2012**, *336*, 1045–1049. [CrossRef]
21. Fornara, F.; Panigrahi, K.C.; Gissot, L.; Sauerbrunn, N.; Rühl, M.; Jarillo, J.A.; Coupland, G. Arabidopsis DOF transcription factors act redundantly to reduce *CONSTANS* expression and are essential for a photoperiodic flowering response. *Dev. Cell* **2009**, *17*, 75–86. [CrossRef] [PubMed]
22. Li, D.; Yang, C.; Li, X.; Gan, Q.; Zhao, X.; Zhu, L. Functional characterization of rice *OsDof12*. *Planta* **2009**, *229*, 1159–1169. [CrossRef] [PubMed]
23. Iwamoto, M.; Higo, K.; Takano, M. Circadian clock- and phytochrome-regulated Dof-like gene, *Rdd1*, is associated with grain size in rice. *Plant Cell Environ.* **2009**, *32*, 592–603. [CrossRef] [PubMed]
24. Ahmad, M.; Rim, Y.; Chen, H.; Kim, J.K. Functional characterization of Arabidopsis Dof transcription factor AtDof4.1. *Russ. J. Plant Physiol.* **2013**, *60*, 116–123. [CrossRef]
25. Lin, X.C.; Chow, T.Y.; Chen, H.H.; Liu, C.C.; Chou, S.J.; Huang, B.L.; Kuo, C.I.; Wen, C.K.; Huang, L.C.; Fang, W. Understanding bamboo flowering based on large-scale analysis of expressed sequence tags. *Genet. Mol. Res.* **2010**, *9*, 1085–1093. [CrossRef] [PubMed]
26. Wu, M.; Liu, H.; Han, G.; Cai, R.; Pan, F.; Xiang, Y. A moso bamboo *WRKY* gene *PeWRKY83* confers salinity tolerance in transgenic Arabidopsis plants. *Sci. Rep.* **2017**, *7*, 11721. [CrossRef] [PubMed]
27. Chen, J.; Shafi, M.; Li, S.; Wang, Y.; Wu, J.; Ye, Z.; Peng, D.; Yan, W.; Liu, D. Copper induced oxidative stresses, antioxidant responses and phytoremediation potential of moso bamboo (*Phyllostachys pubescens*). *Sci. Rep.* **2015**, *5*, 13554. [CrossRef]
28. Zhang, P.; Wang, J.; Zhang, H. Measures of water management and increasing drought resistance of moso forests in Anji County, Zhejiang Province. *World Bamboo Rattan* **2008**, *6*, 23–24. [CrossRef]
29. Ge, W.; Zhang, Y.; Cheng, Z.; Hou, D.; Li, X.; Gao, J. Main regulatory pathways, key genes and microRNAs involved in flower formation and development of moso bamboo (*Phyllostachys edulis*). *Plant Biotechnol. J.* **2017**, *15*, 82–96. [CrossRef]
30. Corrales, A.R.; Carrillo, L.; Lasierra, P.; Nebauer, S.G.; Dominguez-Figueroa, J.; Renau-Morata, B.; Pollmann, S.; Granell, A.; Molina, R.V.; Vicente-Carbajosa, J.; et al. Multifaceted role of cycling DOF factor 3 (CDF3) in the regulation of flowering time and abiotic stress responses in Arabidopsis. *Plant Cell Environ.* **2017**, *40*, 748–764. [CrossRef]
31. Sun, L.; Li, X.; Fu, Y.; Zhu, Z.; Tan, L.; Liu, F.; Sun, X.; Sun, X.; Sun, C. GS6, a member of the GRAS gene family, negatively regulates grain size in rice. *J. Integr. Plant Biol.* **2013**, *55*, 938–949. [CrossRef] [PubMed]
32. Murakami, M.; Ashikari, M.; Miura, K.; Yamashino, T.; Mizuno, T. The evolutionarily conserved OsPRR quintet: Rice pseudo-response regulators implicated in circadian rhythm. *Plant Cell Physiol.* **2003**, *44*, 1229–1236. [CrossRef] [PubMed]
33. Ding, J.; Böhlenius, H.; Rühl, M.G.; Chen, P.; Sane, S.; Zambrano, J.A.; Zheng, B.; Eriksson, M.E.; Nilsson, O. *GIGANTEA-like* genes control seasonal growth cessation in Populus. *New Phytol.* **2018**, *218*, 1491–1503. [CrossRef] [PubMed]
34. Searle, L.; Coupland, G. Induction of flowering by seasonal changes in photoperiod. *Embo J.* **2004**, *23*, 1217–1222. [CrossRef] [PubMed]
35. Shaw, L.M.; McIntyre, C.L.; Gresshoff, P.M.; Xue, G.P. Members of the Dof transcription factor family in *Triticum aestivum* are associated with light-mediated gene regulation. *Funct. Integr. Genom.* **2009**, *9*, 485–498. [CrossRef] [PubMed]
36. Gao, J.; Zhang, Y.; Zhang, C.; Qi, F.; Li, X.; Mu, S.; Peng, Z. Characterization of the floral transcriptome of moso bamboo (*Phyllostachys edulis*) at different flowering developmental stages by transcriptome sequencing and RNA-seq analysis. *PLoS ONE* **2014**, *9*, e98910. [CrossRef] [PubMed]
37. Cheng, Z.; Hou, D.; Liu, J.; Li, X.; Xie, L.; Ma, Y.; Gao, J. Characterization of moso bamboo (*Phyllostachys edulis*) Dof factors in floral development and abiotic stress responses. *Genome* **2018**, *61*, 151–156. [CrossRef] [PubMed]
38. Izawa, T.; Takahashi, Y.; Yano, M. Comparative biology comes into bloom: Genomic and genetic comparison of flowering pathways in rice and Arabidopsis. *Curr. Opin. Plant Biol.* **2003**, *6*, 113–120. [CrossRef]
39. Imaizumi, T.; Schultz, T.F.; Harmon, F.G.; Ho, L.A.; Kay, S.A. FKF1 F-box protein mediates cyclic degradation of a repressor of *CONSTANS* in *Arabidopsis*. *Science* **2005**, *309*, 293–297. [CrossRef]

40. Yang, J.; Yang, M.F.; Zhang, W.P.; Chen, F.; Shen, S.H. A putative flowering-time-related Dof transcription factor gene, *JcDof3*, is controlled by the circadian clock in *Jatropha curcas*. *Plant Sci.* **2011**, *181*, 667–674. [CrossRef]
41. Lijavetzky, D.; Carbonero, P.; Vicentecarbajosa, J. Genome-wide comparative phylogenetic analysis of the rice and Arabidopsis Dof gene families. *BMC Evol. Biol.* **2003**, *3*, 17. [CrossRef] [PubMed]
42. Kardailsky, I.; Shukla, V.K.; Ahn, J.H.; Dagenais, N.; Christensen, S.K.; Nguyen, J.T.; Chory, J.; Harrison, M.J.; Weigel, D. Activation tagging of the floral inducer *FT*. *Science* **1999**, *286*, 1962–1965. [CrossRef] [PubMed]
43. An, H.; Roussot, C.; Suárez-López, P.; Corbesier, L.; Vincent, C.; Piñeiro, M.; Hepworth, S.; Mouradov, A.; Justin, S.; Turnbull, C.; et al. *CONSTANS* acts in the phloem to regulate a systemic signal that induces photoperiodic flowering of Arabidopsis. *Development* **2004**, *131*, 3615–3626. [CrossRef] [PubMed]
44. Lee, J.; Oh, M.; Park, H.; Lee, I. SOC1 translocated to the nucleus by interaction with AGL24 directly regulates *LEAFY*. *Plant J.* **2010**, *55*, 832–843. [CrossRef] [PubMed]
45. Michaels, S.D.; Amasino, R.M. FLOWERING LOCUS C encodes a novel MADS domain protein that acts as a repressor of flowering. *Plant Cell* **1999**, *11*, 949–956. [CrossRef] [PubMed]
46. Wu, H.; Lv, H.; Li, L.; Liu, J.; Mu, S.; Li, X.; Gao, J. Genome-Wide Analysis of the AP2/ERF Transcription Factors Family and the Expression Patterns of *DREB* Genes in Moso Bamboo (*Phyllostachys edulis*). *PLoS ONE* **2015**, *10*, e0126657. [CrossRef] [PubMed]
47. Wang, H.; Zhao, S.; Gao, Y.; Yang, J. Characterization of Dof transcription factors and their responses to osmotic stress in Poplar (*Populus trichocarpa*). *PLoS ONE* **2017**, *12*, e0170210. [CrossRef]
48. Huang, R.; Li, L.; Liu, J.; Gao, H.; Li, X.P. Sequence of *PheWRKY9-1* gene in *Phyllostachys edulis* and analysis of related expression. *Mol. Plant Breed.* **2018**, *14*, 4569–4575. [CrossRef]
49. Wang, T.; Wang, H.; Cai, D.; Gao, Y.; Zhang, H.; Wang, Y.; Lin, C.; Ma, L.; Gu, L. Comprehensive profiling of rhizome-associated alternative splicing and alternative polyadenylation in moso bamboo (*Phyllostachys edulis*). *Plant J.* **2017**, *91*, 684–699. [CrossRef]
50. Gasteiger, E.; Gattiker, A.; Hoogland, C.; Ivanyi, I.; Appel, R.D.; Bairoch, A. ExPASy: The proteomics server for in-depth protein knowledge and analysis. *Nucleic Acids Res.* **2003**, *31*, 3784–3788. [CrossRef]
51. Guo, A.Y.; Zhu, Q.H.; Chen, X.; Luo, J.C. GSDS:a gene structure display server. *Yi Chuan* **2007**, *29*, 1023–1026. [CrossRef] [PubMed]
52. Tamura, K.; Stecher, G.; Peterson, D.; Filipski, A.; Kumar, S. MEGA6: Molecular evolutionary genetics analysis version 6.0. *Mol. Biol. Evol.* **2013**, *30*, 2725–2729. [CrossRef]
53. Yoo, S.D.; Cho, Y.H.; Sheen, J. Arabidopsis mesophyll protoplasts: A versatile cell system for transient gene expression analysis. *Nat. Protoc.* **2007**, *2*, 1565–1572. [CrossRef]
54. Clough, S.J.; Bent, A.F. Floral dip: A simplified method for Agrobacterium-mediated transformation of *Arabidopsis thaliana*. *Plant J.* **1998**, *16*, 735–743. [CrossRef] [PubMed]
55. Jefferson, R.A.; Kavanagh, T.A.; Bevan, M.W. GUS fusions: Betaglucuronidase as a sensitive and versatile gene fusion marker in higher plants. *Embo J.* **1987**, *6*, 3901–3907. [CrossRef]
56. Fan, C.; Ma, J.; Guo, Q.; Li, X.; Wang, H.; Lu, M. Selection of reference genes for quantitative Real-Time PCR in bamboo (*Phyllostachys edulis*). *PLoS ONE* **2013**, *8*, e56573. [CrossRef] [PubMed]

© 2019 by the authors. Licensee MDPI, Basel, Switzerland. This article is an open access article distributed under the terms and conditions of the Creative Commons Attribution (CC BY) license (http://creativecommons.org/licenses/by/4.0/).

International Journal of *Molecular Sciences*

MDPI

Article

Identification and Characterization of the *EXO70* Gene Family in Polyploid Wheat and Related Species

Jia Zhao, Xu Zhang, Wentao Wan, Heng Zhang, Jia Liu, Mengli Li, Haiyan Wang, Jin Xiao * and Xiue Wang *

State Key Laboratory of Crop Genetics and Germplasm Enhancement, Cytogenetics Institute, Nanjing Agricultural University/JCIC-MCP, Nanjing 210095, China; 2015201054@njau.edu.cn (J.Z.); 2018201061@njau.edu.cn (X.Z.); 2016201003@njau.edu.cn (W.W.); 2016201031@njau.edu.cn (H.Z.); 2018201063@njau.edu.cn (J.L.); 2016101134@njau.edu.cn (M.L.); hywang@njau.edu.cn (H.W.)

* Correspondence: xiaojin@njau.edu.cn (J.X.); xiuew@njau.edu.cn (X.W.); Tel.: +86-25-8439-9893 (J.X.); +86-25-8439-5308 (X.W.)

Received: 18 November 2018; Accepted: 21 December 2018; Published: 24 December 2018

Abstract: The *EXO70* gene family is involved in different biological processes in plants, ranging from plant polar growth to plant immunity. To date, analysis of the *EXO70* gene family has been limited in *Triticeae* species, e.g., hexaploidy *Triticum aestivum* and its ancestral/related species. By in silico analysis of multiple *Triticeae* sequence databases, a total of 200 *EXO70* members were identified. By homologue cloning approaches, 15 full-length cDNA of *EXO70s* were cloned from diploid *Haynaldia villosa*. Phylogenetic relationship analysis of 215 *EXO70* members classified them into three groups (*EXO70.1*, *EXO70.2*, and *EXO70.3*) and nine subgroups (*EXO70A* to *EXO70I*). The distribution of most *EXO70* genes among different species/sub-genomes were collinear, implying their orthologous relationship. The *EXO70A* subgroup has the most introns (at least five introns), while the remaining seven subgroups have only one intron on average. The expression profiling of *EXO70* genes from wheat revealed that 40 wheat *EXO70* genes were expressed in at least one tissue (leaf, stem, or root), of which 25 wheat *EXO70* genes were in response to at least one biotic stress (stripe rust or powdery mildew) or abiotic stress (drought or heat). Subcellular localization analysis showed that ten EXO70-V proteins had distinct plasma membrane localization, EXO70I1-V showed a distinctive spotted pattern on the membrane. The 15 *EXO70-V* genes were differentially expressed in three tissue. Apart from *EXO70D2-V*, the remaining *EXO70-V* genes were in response to at least one stress (flg22, chitin, powdery mildew, drought, NaCl, heat, or cold) or phytohormones (salicylic acid, methyl jasmonate, ethephon, or abscisic acid) and hydrogen peroxide treatments. This research provides a genome-wide glimpse of the *Triticeae EXO70* gene family and those up- or downregulated genes require further validation of their biological roles in response to biotic/abiotic stresses.

Keywords: EXO70; *Haynaldia villosa*; gene family; phylogenetic relationship; subcellular localization; expression profiling

1. Introduction

The exocyst complex is an evolutionarily conserved octameric tethering factor, which mediates the fusion of post-Golgi secretory vesicle with the plasma membrane (PM) and plays a major role in exocytosis [1,2]. EXO70 is a key member of the exocyst complex and has been found to be widely present in yeast, mammals and plants [3]. In yeast and mammals, the *EXO70* only has a single copy, while plants have multiple copies of *EXO70* genes [4] ranging from 21 to 47 *EXO70* members in potatoes (*Symphytum tuberosum*), *Arabidopsis*, *Populus trichocarpa* and rice [1,5,6]. The *EXO70s* of land plants possibly originated from three ancient *EXO70* genes and thus can be divided into three groups

EXO70.1, *EXO70.2* and *EXO70.3*. They have been further duplicated independently in the moss, lycophyte and angiosperm lineages, and in the subsequent lineage-specific multiplications which are represented by nine subgroups (*EXO70A-EXO70I*) [4,7–9].

The function of *EXO70* has been extensively studied in yeast, and mammals [3,10–15]. In plants, *EXO70s* have been proven to play diverse roles in regulating plant growth and coping with adverse biotic/abiotic stresses. In *Arabidopsis*, *EXO70A1* has been implicated in a wide range of developmental processes, including the differentiation of tracheary elements [16,17], the development of seed coat, root hair and stigmatic papillae [18], the recycling of the auxin efflux carrier proteins (PIN1 and PIN2) [19] and the formation of the Casparian strip [20]. *EXO70C1* and *EXO70C2* regulated the polarized growth and maturation of the pollen tube [21,22]. *EXO70H4* regulates trichome cell wall maturation by mediating the secretion and accumulation of callose and silica [23,24]. The rice *OsEXO70A1* is necessary for vascular bundle differentiation and assimilation of mineral nutrients [5]. The legumes EXO70J7, EXO70J8 and EXO70J9 are members of an atypical subgroup of EXO70 proteins (EXO70J) that regulate leaf senescence and nodule formation [25]. In *Nicotiana benthamiana*, silencing all the paralogue genes in subgroups EXO70A (six), C (three), D (four) and G (six) resulted in a smaller leaf phenotype [6].

Evidence has accumulated for the critical role of *EXO70s* in plant-pathogen interactions or responses to abiotic stresses. In *N. benthamiana*, the silencing of two *EXO70B* paralogues led to increased susceptibility to *Phytophthora infestans* [6]. Three of the 23 members of the *Arabidopsis EXO70* gene family (*EXO70B1*, *EXO70B2* and *EXO70H1*) have been proven to be involved in plant immunity [26]. *AtEXO70B1* and *AtEXO70B2* belong to the same subgroup and are both involved in plant immunity, of which *AtEXO70B1* a negative regulator, while *AtEXO70B2* is a positive regulator. *AtEXO70B1* underwent autophagic transport, and the loss-of-function of *exo70B1* led to reduced numbers of internalized autophagosomes, accumulation of salicylic acid (SA), and finally, ectopic hypersensitive responses and enhanced resistance to several pathogens. AtEXO70B1's regulation of disease resistance, either by interacting with TIR-NBS2, a truncated version of the classical nucleotide binding (NB) domain and a leucine-rich repeat (LRR)-containing (NLR) intracellular immune receptor-like protein [27–29], or by interacting with RIN4, a well-known regulator of pathogen-associated molecular pattern (PAMP)-triggered immunity (PTI) [29]. AtEXO70B2 regulated innate immunity via interacting with a negative PTI regulator, AtPUB22, which mediated the ubiquitination and degradation of AtEXO70B2 and contributed to PTI. The *exo70B2* mutants showed aberrant papillae with halos and were susceptible to different PAMPs and pathogens [30,31]. *AtEXO70B1* and *AtEXO70B2* also contribute to the abiotic stress response; both were positive regulators of stomatal movement. The response to mannitol (drought) treatments is in either an abscisic acid (ABA)-dependent or -independent manner [32,33]. *AtEXO70H1* is a homolog of *AtEXO70B2* and is also involved in plant immunity [31]. Three of the 47 *EXO70* members of rice (*OsEXO70E1*, *OsEXO70F2* and *OsEXO70F3*) were reported to participate in plant immunity [5]. OsEXO70E1 is attributed to planthopper resistance by interacting with a broad resistance protein, *Bph6*. Interaction of the two proteins increased exocytosis and blocked the feeding of a planthopper by cell wall thickening at the infection sites [34]. The importance of *EXO70* in plant immunity was also shown by the fact that some of the *EXO70s* were targets of the secreted effectors of the plant pathogen. Both OsEXO70F2 and OsEXO70F3 were targets of the *Magnaporthe oryzae* effector AVR-Pii, and OsEXO70F3 was proven to play an important role in *Pii*-dependent resistance by interacting with AVR-Pii [35].

Accumulated evidence has shown that a large number of *EXO70s* exist in plants; however, only a few have had their biological roles elucidated [5,35,36]. Due to the huge genome size and complexity [37], the knowledge of the *EXO70* gene family from the *Triticeae* species is rather limited. In the last five years, the genome sequences of wheat and its ancestor species have been released, which makes genome-wide identification of a gene family in the *Triticeae* species feasible [38–43].

Haynaldia villosa L. (2*n* = 2x = 14, VV) is a diploid wild relative of wheat. Previous studies showed that *H. villosa* is a valuable genetic resource harboring many elite traits, such as resistance to several wheat diseases and tolerance to abiotic stresses [44–46]. In the present study, different members of the *EXO70* gene family are identified by browsing the released genome sequences of the *Triticea* species. Specific primer pairs are designed, *EXO70s* are cloned from *H. villosa* and their potential functions are elucidated by expression profiles based on in silico analysis and quantitative RT-PCR (qRT-PCR). The obtained results would help us to understand the evolution and diversification of the *EXO70s* among *Triticeae* species and their potential roles in plant immunity and responses to abiotic stresses.

2. Results

2.1. Identification and Phylogenetic Relationship Analysis of the EXO70 Gene Family in Triticeae Species

In total, 200 *EXO70* genes were identified from the public database of five *Triticeae* species. Among them, there were 26 each from *T. urartu*, *Ae. Tauschii* and *H. vulgare*; 47 from *T. dicoccoides*; and 75 from common wheat (*T. aestivum*), respectively. Fifteen *EXO70s* from *H. villosa* were obtained by homology cloning (Figure 1a). The evolutionary relationship of the above 215 *Triticeae EXO70s*, along with 22 from *Brachypodium distachyon*, 41 from rice and 23 from *Arabidopsis*, were phylogenetically analyzed (Figure 1b, Table S1). These *EXO70s* were divided into three major groups, *EXO70.1*, *EXO70.2* and *EXO70.3*, which were further assigned to nine subgroups, from *EXO70A* to *EXO70I*, according to a phylogenetic tree (Figure 1b). The *EXO70A* subgroup belongs to the group *EXO70.1*, in which 43 (14.28%) *EXO70s* were included; the EXO70B, C, D, E, F, H, and I subgroups belong to the group *EXO70.2*, in which 229 (76.08%) *EXO70s* were included; the *EXO70G* subgroup belongs to the group *EXO70.3*, in which 29 (9.63%) *EXO70s* were included. The *EXO70I* subgroup has the most members (74, 24.58%), followed by *EXO70F* (48, 15.95%) and *EXO70A* (43, 14.28%) (Figure 1c). Based on the subgroups and genome allocation, the wheat *EXO70s* were designated [7,47]. For example, the *EXO70B1* from *T. dicoccoides* located on chromosome 1A was assigned *TdEXO70B1-1A*. In *Arabidopsis*, the *EXO70I* subgroup is missing, whereas the *EXO70I* subgroup in *Triticeae* species appeared to be the most divergent. However, our analysis led to a new insight: the *EXO70I* subgroup belongs to *EXO70.2*, rather than *EXO70.3* from other species [36] (Figure 1b,c).

a

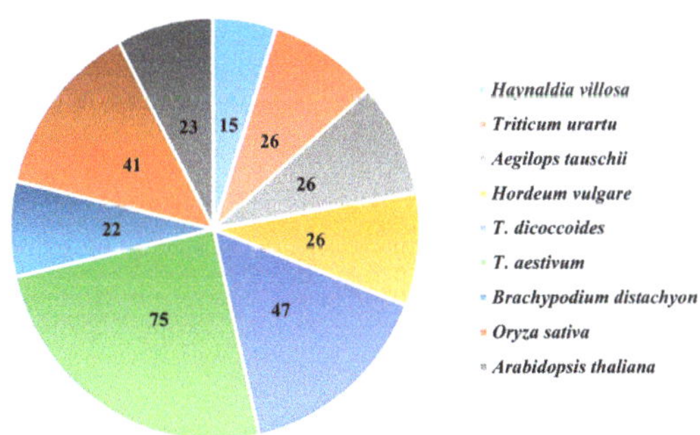

Figure 1. *Cont.*

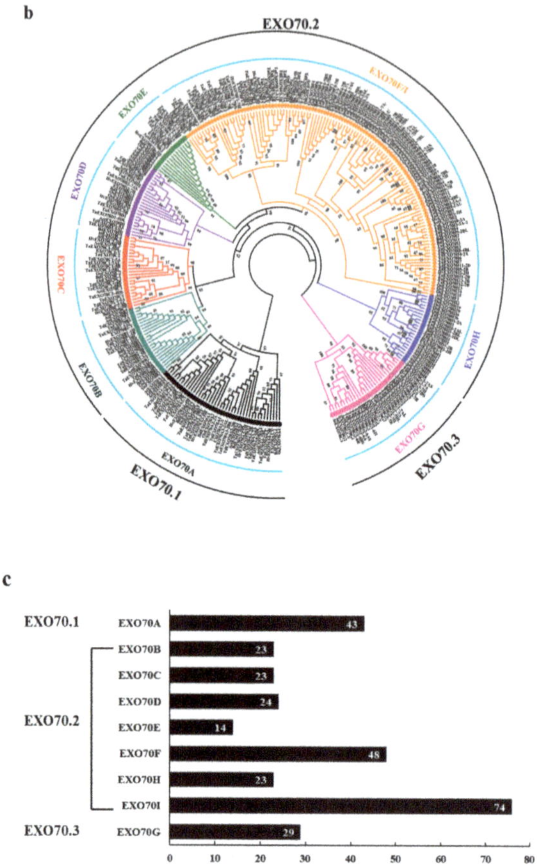

Figure 1. The number and phylogenetic relationships of *EXO70* family genes from *T. aestivum*, *T. urartu*, *Ae. tauschii*, *T. dicoccoides*, *H. vulgare*, *A. thaliana*, *Oryza sativa*, *Brachypodium distachyon* and *H. villosa*. (**a**) The total number of *EXO70* gene family in nine species. (**b**) The phylogenetic tree of nine *Triticeae* species. Species abbreviations: Ta, *Triticum aestivum*; Tu, *Triticum urartu*; Aet, *Aegilops tauschii*; Td, *Triticum dicoccoides*; Hv, *Hordeum vulgare*; At, *Arabidopsis thaliana*; Bd, *Brachypodium distachyon*; Os, *Oryza sativa*; -V, *Haynaldia villosa*. (**c**) The number of the *EXO70* gene family from nine species in each of the subgroups. The horizontal/longitudinal coordinate axis represents the number of genes and different subgroups, respectively.

The 15 cloned *EXO70s* from *H. villosa* were designated as *EXO70A1-V* to *EXO70I1-V* according to the phylogenetic relationship to wheat *EXO70s*. They belong to nine subgroups, three each to *EXO70A* and *F*, two each to *EXO70D* and *G* and one each to *EXO70B*, *C*, *E*, *H* and *I* (Figure 1). Their CDS length ranges from 801 bp (*EXO70H1-V*) to 2007 bp (*EXO70C1-V*) and their isoelectric point varies from 4.52 (*EXO70B1-V*) to 10.19 (*EXO70G1-V*) (Table 1).

Table 1. *EXO70* genes cloned from *H. villosa*.

Number	Name	ORF (bp)	AA (aa)	DL (aa)	PI	MW (KD)
1	EXO70A1-V	1914	637	268–623	7.71	71.48
2	EXO70A2-V	1944	647	273–631	8.78	73.07
3	EXO70A3-V	882	294	1–283	8.83	32.61
4	EXO70B1-V	1626	541	180–534	4.52	59.74
5	EXO70C1-V	2007	668	287–654	5.72	73.9
6	EXO70D1-V	1650	549	152–532	6.98	61.2
7	EXO70D2-V	1644	547	152–525	7.00	61.02
8	EXO70E1-V	1815	605	233–580	5.04	68.93
9	EXO70F1-V	1323	441	52–424	5.46	49.88
10	EXO70F2-V	1542	514	147–504	5.03	57.62
11	EXO70F3-V	1431	477	108–466	5.60	52.91
12	EXO70G1-V	807	268	1–232	10.19	30.95
13	EXO70G2-V	1437	478	85–439	9.42	53.8
14	EXO70H1-V	801	268	1–238	8.66	28.65
15	EXO70I1-V	1455	484	126–480	5.61	42.48

Abbreviations: ORF, open reading frame; AA, amino acids; DL, PFam03081 domain location; PI, protein isoelectric point; MW, protein molecular weight.

DNAMan was used to explore the amino acid sequence feature of *H. villosa EXO70s*. The pfam03081 domain at the C-terminus was relatively conservative, shown by the fact that all of the *EXO70s* had the same amino acid in the 11 sites (in the red rectangle in Figure 2a). R programming language was used to visualize the sequence similarities. The same subgroups shared more common sequences, e.g., *EXO70D1-V* and *EXO70D2-V* shared about 80% similarity, while different subgroups had low sequence similarities, e.g., *EXO70H1-V* and *EXO70I1-V* only had 21.46% similarity (Figure 2b).

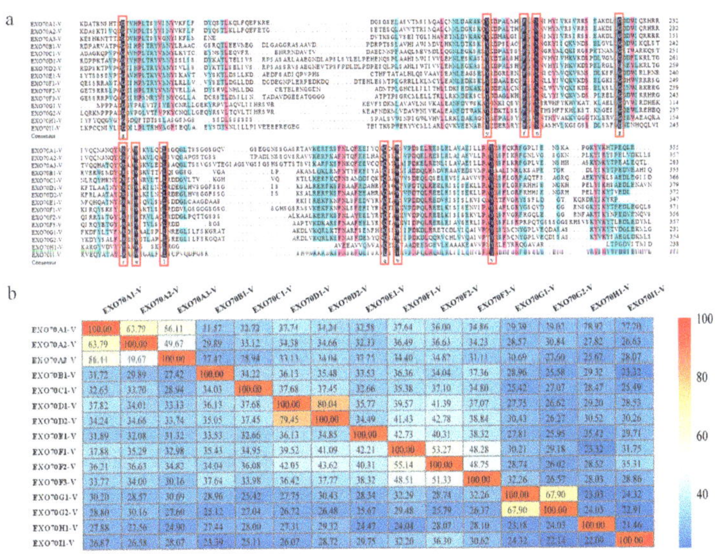

Figure 2. The amino acid sequence feature analysis of *EXO70* gene in *H. villosa*. (**a**) The amino acid sequence of pfam03081 domain for each *EXO70* was selected for multiple sequence alignment analysis by DNAMan. The 11 identical amino acids were indicated in red frame. The three colors of black, red and blue represent the level of similarity of amino acids, from high to low. (**b**) Sequence similarity analysis using the R programming language. The color scale bar represents sequence similarity between different genes. Red and yellow indicate that the sequence similarity was greater than 80% and 60%, respectively. Blue indicates that the sequence similarity was less than 40%.

2.2. The Number of EXO70s in Genomes of Triticeae Species

The diploid species *B. distachyon* and *Arabidopsis* have 22 and 23 *EXO70s*, respectively, while diploid rice has 41, nearly twice as many. We identified 22 *EXO70s* each in A and D of common wheat and B of the tetraploid wheat, 25 in A of tetraploid wheat, 26 in D of *Ae. Tauschii* and H of barley and 29 *EXO70s* in B of common wheat. For the A genome, tetraploid has three more *EXO70s* than hexaploidy wheat; for the B genome, hexaploidy has seven more than tetraploid wheat; for the D genome, *Ae. tauschii* has four more *EXO70s* than hexaploidy wheat. In common wheat, the chromosomes 5A, 5B and 5D had the same number of *EXO70s*, belonging to the same gene type and having the same gene order. The chromosomes 7A, 7B and 7D also have the same number of *EXO70s*; however, their gene type and gene order were not completely the same. For example, *TaEXO70D* is only present on 7A and 7B, but not on 7D; *TaEXO70G1* is only present on 7A and 7D, but not 7B. The gene number and order for *EXO70s* in the remaining five homoeologous were also different among corresponding homoeologous chromosomes, mainly due to their difference for subgroup *EXO70I* (Tables 2 and S2).

Table 2. Number of *EXO70* from different species in each of the chromosomes.

Chromosome	*T. aestivum*			*T. dicoccoides*		*Ae. tauschii*	*H. vulgare*	Total
	A	B	D	A	B	D	H	
Chr.1	1	1	3	1	1	1	1	9
Chr.2	5	8	4	8	6	6	7	44
Chr.3	4	5	3	4	3	4	4	27
Chr.4	1	3	2	2	2	2	2	14
Chr.5	3	3	3	2	2	3	3	19
Chr.6	1	2	0	1	1	1	1	7
Chr.7	7	7	7	7	7	9	7	51
Total	22	29	22	25	22	26	26	171
*Unknow		2					1	3

* The genes that were assigned to unknown chromosome.

The three analyzed diploid *Triticea* species, *H. vulgare*, *T. uratu* and *Ae. tauschii*, each had 26 *EXO70s* and had the same number as the *EXO70B, C, D* and H subgroups. They had the least *EXO70Hs* and the most *EXO70Is*. *T. uratu* only had one *EXO70G*, while the other two species had three. However, *T. uratu* had more *EXO70I* than the other two species (Table 3). The variation among different diploid species may be due either to the quality of the genome assembly or the duplication or deletion during evolution. The tetraploid *T. dicoccoides* and hexaploid *T. aestivum* have 47 and 75 *EXO70s*, which are about twice and three times the number in diploid species, respectively. In addition, the number of *EXO70s* from each subgroup is also exactly (*EXO70C, H*) or almost (*EXO70A, B, D, E, F, I*) 2 or 3 times that of the diploid species (Table 3). This indicated that, in polyploid wheat, the increased number of *EXO70s* is mostly due to the genome polyploidization.

We only identified 15 *EXO70s* in diploid *H. villosa*, which is much less than the average number for other diploid species. *H. villosa* had the same number of subgroups D and H, but fewer members of subgroups B, C, E, F and I. This is due either to the lack of the sequence of *H. villosa*, or the divergence of *H. villosa* from other *Triticea* species (Table 3).

Table 3. Numbers of *EXO70* paralogs encoded by the surveyed genomes in total and individual subgroups.

Genome	Total Number	Subgroup								
		A	B	C	D	E	F	G	H	I
H. villosa (VV)	15	3	1	1	2	1	3	2	1	1
H. vulgare (HH)	26	4	2	2	2	2	5	3	1	5
T. urartu (AA)	26	3	2	2	2	1	5	1	1	9
Ae. tauschii (DD)	26	4	2	2	2	1	4	3	1	7
T. dicoccoides (AABB)	22–25 (47)	3 (6)	2 (4)	2 (4)	2 (4)	1 (2)	4 (8)	3 (6)	1 (2)	7 (11)
T. aestivum (AABBDD)	22–29 (75)	4 (12)	2 (5)	2 (6)	2 (5)	1 (3)	5 (13)	3 (6)	1 (3)	14 (22)
B. distachyon (Bd)	22	4	2	2	2	1	4	3	1	3
Oryza sativa	41	4	3	2	2	1	5	3	5	16
Arabidopsis thaliana	23	3	2	2	3	2	1	2	8	0
Total	238 (301)	32 (43)	18 (23)	17 (23)	19 (24)	11 (14)	36 (48)	23 (29)	20 (23)	64 (74)

Numbers in brackets indicate number of copies for polyploid genomes.

2.3. The Chromosomal Distribution of EXO70s in Triticeae Species

The identified 174 *EXO70s* from four *Triticeae* species (*T. aestivum*, *Ae. tauschii*, *T. dicoccoides* and *H. vulgare*) were assigned to corresponding chromosomes (including three on unknown chromosome). With the exception of wheat 6D, all chromosomes have at least one *EXO70* gene (Figure 3). The *EXO70s* were not evenly distributed on difference chromosomes or in different homologous groups. In total, there were 9, 44, 27, 14, 19, 7 and 51 *EXO70s* in the homologous groups 1 to 7, with group 7 having the most *EXO70s* (29.31%), followed by group 2 (25.29%) (Table 2). The *EXO70s* in group 7 was dispersely distributed along the chromosome, while those in group 2 were clutered at the dital region of the long arm (Figure 3). We found that nine groups *EXO70s* were conversely present on the same homoeologous groups of the four species, such as the B1 in group1, F2/3 in group 2, E1 and F3/2 in group 3, C2/C1 in group 4, CI and D1 in group 5, F1 and F4 in group 7 (Table S2, Figure 3).

For homologous chromosomes from different genomes of wheat and its ancestral/related species, most of the corresponding *EXO70* orthologs were present at the syntenic genome regions. For example, *EXO70B1s* were on the long chromosome arm of homoeologous group 1 and *EXO70E1* on group 3 chromosomes (Figure 3). There are some exceptions. The *EXO70G1* is present on all group 6 chromosomes except for sub-genome 6D of the hexaploidy wheat. The *EXO70H1* is located on chromosomes 2A, 2B of common wheat and *T. dicoccoides*, 2D of *Ae. tauschii* and 2H of barley and on 1D of common wheat (Figure 3). We found that the *EXO70C2* is on the 4BS of common wheat and *T. dicoccoides* (Figure 3a,b), on the 4DS of common wheat and *Ae. tauschii* (Figure 3a,d), but on the 4AL of common wheat and *T. dicoccoides* (Figure 3a,b). This also supports the presence of an inter-arm translocation of 4A during the evolution from diploid to tetraploid wheat [48,49].

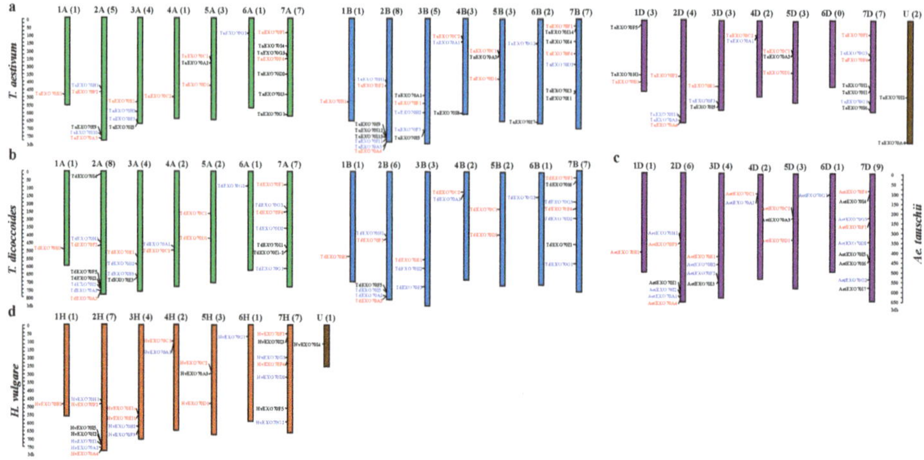

Figure 3. The chromosomal distribution of the *EXO70* gene family in four *Triticeae* species. (**a–d**) represent *T. aestivum*, *T. dicoccoides*, *A. tauschii* and *H. vulgare*, respectively. Chromosome numbers are indicated at the top of each bar and the number in parentheses corresponds to the number of *EXO70* genes present on that chromosome. The name of each gene is to the left of each chromosome. Gene names labeled with red, blue, or black indicate that they are conserved in four species, missing in one due to incomplete data, or missing in more than two, respectively.

2.4. The Diversification of Gene Structure of Triticeae EXO70s

The C-terminal PFam03081 domain, which may determine the function or structure of the proteins, is a specific characteristic of the *EXO70* superfamily [47]. All the predicted 200 and 15 homologous cloned EXO70 proteins possessed such a domain; however, their amino acid sequence length is different for different *EXO70s*, varying from 103aa to 669aa and with an average length of 345aa (Table S1).

The exon–intron structure of the 200 *Triticeae EXO70s* was visualized by using the online Gene Structure Display Server and compared among different subgroups or within each individual subgroup. The exon-intron structures for most of the *EXO70s* within the same subgroups were relatively conserved among *Triticeae* species, was similar to in *Arabidopsis* or rice [7,36]. Compared with other subgroups, the subgroup *EXO70A* with 30 genes has the most introns on average, e.g., *TuEXO70A3* has the most introns (20) and *HvEXO70A4-2H* has the fewest introns (five) (Figure 4a). The 170 *EXO70* genes in the remaining eight subgroups only have about one intron on average. Among them, 83 genes (41.50%) are intronless, including eight *EXO70Bs*, seven *EXO70Cs*, nine *EXO70Ds*, one *EXO70Es*, 18 *EXO70sF*, 12 *EXO70Gs*, five *EXO70Hs* and 23 *EXO70sI*; 60 genes (30.00%) have one intron (e.g., *AetEXO70B1* and *TaEXO70C1-5BL*), 19 genes (9.50%) have two introns (e.g., *TuEXO70C2* and *TaEXO70G2-6BS*) and eight genes (4.00%) have three to six introns (e.g., *AetEXO70I3* and *TaEXO70D1-5AL*). We also observed that genes that have a closer phylogenetic relationship in the same subgroup have a similar gene structure; however, within the same subgroup, some genes showed quite different gene structure. For instance, *TaEXO70D1-5AL* has six introns in *EXO70D* (Figure 4d) and the three copies of *TaEXO70I5* and *AetEXO70I3* have five introns, while the other genes in the same subgroup (*EXO70D* or *EXO70I*) only have one or two introns (Figure 4i). The most diversified gene structure of the *EXO70A* subgroup may be a clue that their diversified biological role is different from that of other subgroups.

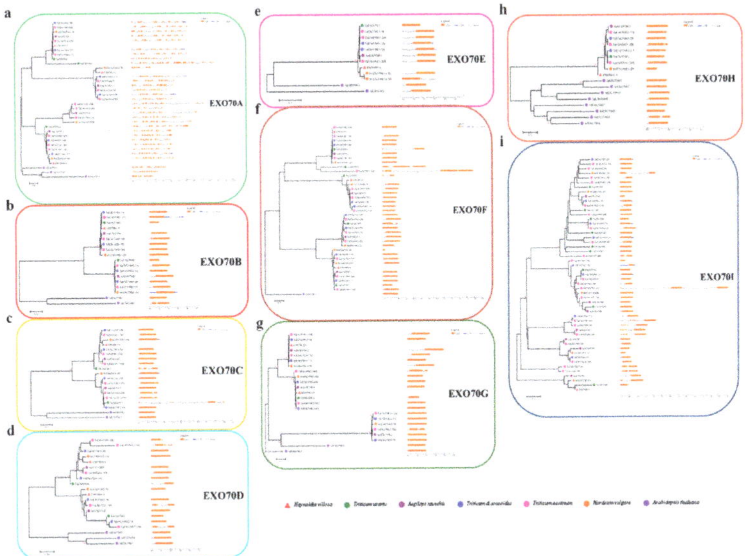

Figure 4. Phylogenetic analysis and exon–intron structures of *EXO70* gene family in common wheat and related *Triticeae* species. The phylogenetic analysis was performed using the sequences of the conserved domain of *EXO70*; proteins were aligned by ClustalW, constructed by MEGA6 using the N-J method, with 1000 bootstrap replicates; the branch length scale bar indicates the evolutionary distance. The left column identifies subgroups and is marked with different alternating background tones to make subgroup identification easier. Introns and exons are represented by black lines and colored boxes, respectively.

2.5. The Expression Pattern of TaEXO70 Genes

The expression patterns of different members in a gene family will help us predict their potential biological roles. To elucidate the potential roles of the identified *EXO70s*, their expression in different tissues or in responses to various biotic and abiotic stresses was investigated by in silico expression profiling or qRT-PCR analysis. The expression patterns of wheat *TaEXO70s* in different tissues (root, stem and leaf of seeding stage), under two biotic stresses (stripe rust pathogen *CYR31* and powdery mildew pathogen *E09*) and two abiotic stresses (drought and heat) [50,51] were first investigated using the wheat RNA-seq data from the publicly available databases. The expression level was measured as tags per million (TPM). To facilitate the portraits of transcript abundance, we assume the expression was high if $TPM \geq 2.5$; moderate if $2.5 > TPM \geq 1.5$; low if $1.5 > TPM > 0$; and undetectable if $TPM = 0$. All 75 *TaEXO70* genes exhibited significantly diverse expression patterns (Figure 5). Their expression can be classified into five groups. The group "e" includes 23 *TaEXO70s* (30.67%). Their transcript abundance was undetectable; among them 11 were from subgroup I and six were from subgroup C (Figure 5e). The group "d" includes 20 genes (26.67%). They had detectable but weak transcript abundance; among them six were from subgroup A and six were from subgroup I (Figure 5d). The group "c" has six genes (8.00%), whose expression is high in roots and stems, whereas it does not respond to the four stresses (Figure 5c). The group "b" has 15 genes (20.00%). These genes were found to be negligibly to moderately expressed in the three tissue types and in response to one or two stresses. Among them, nine were from the subgroup F (Figure 5b). The group "a" has ten genes (13.33%). Their expression was generally high, either in different tissues or in response to the four stresses and none of them were from the subgroup I (Figure 5a).

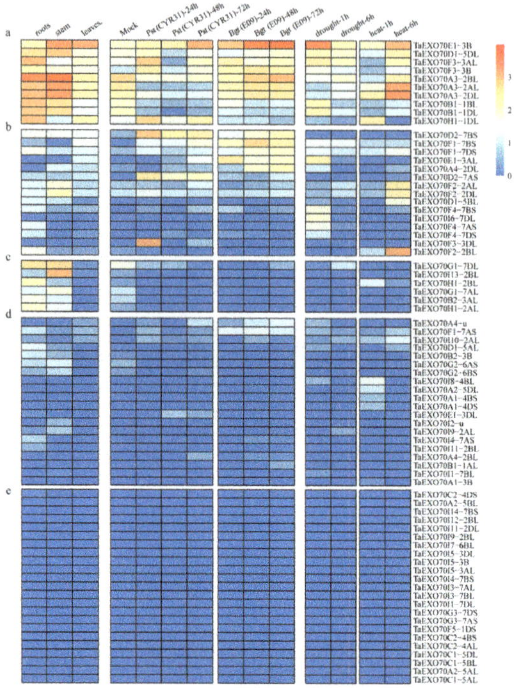

Figure 5. Heat map of the expression profiling of wheat *EXO70* genes in different tissues and under various stresses. The color scale bar represents the expression values of the genes. (**a**–**e**): Genes with different expression types. Abbreviations: *Bgt*, powdery mildew; *Pst*: Stripe rust.

Six genes were upregulated in response to both biotic and abiotic stresses. *TaEXO70E1-3B* was the only gene that was highly expressed in three tissues and upregulated in response to all four stresses. Its homologs on 3AL *TaEXO70E1-3AL* displayed moderate expression in three organs and were only responsive to *Bgt* inoculation and drought stress. The *TaEXO70D1-5DL* and *TaEXO70F3-3B* also displayed moderate expression in three organs and stresses. The *TaEXO70H1-1DL* expression was higher in root/leaf tissue and showed upregulation under both *Pst* infection and heat treatment. The *TaEXO70B1-1BL/1DL* expression was higher in root/stem tissue, and it was downregulated in response to *Bgt* and *Pst* inoculation but upregulated by drought treatment. *TaEXO70A3-2AL/2BL/2DL* had a similar expression pattern, being expressed in three tissues (stems > roots > leaves) and downregulated by *Pst* inoculation but upregulated by both drought and heat treatment. *TaEXO70D2-7BS* and *TaEXO70G1-7DL* were both only responsive to biotic stresses; however, they showed opposite patterns, with *TaEXO70D2-7BS* upregulated and *TaEXO70G1-7DL* downregulated. Twelve genes were only responsive to one of the four stresses; five were only upregulated in response to *Bgt* or *Pst* inoculation; and seven were only upregulated in response to drought or heat stress. Conservation roles were observed for all three homolog genes from different genomes, such as *TaEXO70A3(2AL/2BL/2DL)* and *TaEXO70F2(2AL/2BL/2DL)*.

2.6. Differential Expression of 15 EXO70 Genes from H. villosa

The expression patterns of 15 *EXO70*-Vs in different tissues and in response to different stresses or treatments in *H. villosa* were investigated by qRT-PCR. Different expression patterns were observed for the analyzed genes (Figure 6). Six of the genes (*EXO70D1-V*, *EXO70A2-V*, *EXO70F3-V*, *EXO70H1*-V, *EXO70I1*-V and *EXO70G2*-V) were not differentially expressed in all three tissues. Four genes

(*EXO70B1*-V, *EXO70G1*-V, *EXO70A1*-V, *EXO70C1*-V) showed more abundant transcript in stem; seven genes (*EXO70E1*-V, *EXO70A3*-V, *EXO70B1*-V, *EXO70G1*-V, *EXO70F1*-V, *EXO70D2*-V and *EXO70F2*-V) showed higher expression level in leaves (Figure 6a).

EXO70A2-V showed a similar expression level in all the three tissues; however, its expression was significantly increased in response to *Bgt* inoculation and treatments by chitin, flg22, heat stress, phytohormones, or H_2O_2. *EXO70G1-V* showed a similar expression level in the stems and leaves; its expression was also significantly increased by *Bgt* inoculation and treatment by chitin, flg22, four phytohormones, or H_2O_2. The expression of *EXO70H1-V* was strongly upregulated by *Bgt* inoculation and cold stress, and moderately upregulated by chitin, Flg22, ET and heat stress. *EXO70I1-V* only responded to abiotic stresses (drought, salt and cold) and ABA treatment. *EXOG2-V* was responsive to flg22 treatment and drought stress, and *EXO70A3-V* was only responsive to heat stress (Figure 6b,c). The divergence of expression patterns of different gene members indicated their clear-cut roles in the adaptation of *H. villosa* to various environments or stresses.

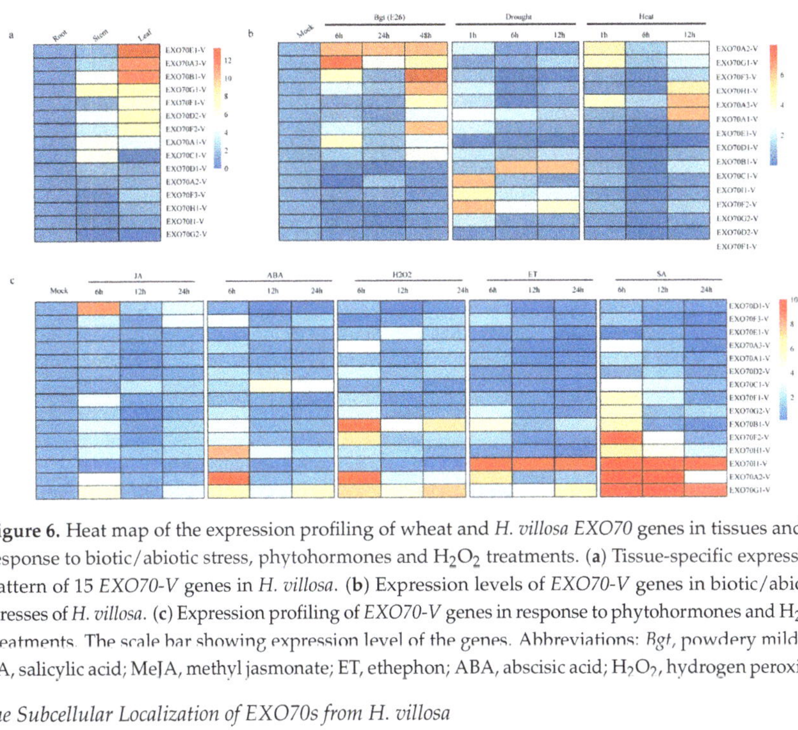

Figure 6. Heat map of the expression profiling of wheat and *H. villosa EXO70* genes in tissues and in response to biotic/abiotic stress, phytohormones and H_2O_2 treatments. (**a**) Tissue-specific expression pattern of 15 *EXO70-V* genes in *H. villosa*. (**b**) Expression levels of *EXO70-V* genes in biotic/abiotic stresses of *H. villosa*. (**c**) Expression profiling of *EXO70-V* genes in response to phytohormones and H_2O_2 treatments. The scale bar showing expression level of the genes. Abbreviations: *Bgt*, powdery mildew; SA, salicylic acid; MeJA, methyl jasmonate; ET, ethephon; ABA, abscisic acid; H_2O_2, hydrogen peroxide.

2.7. The Subcellular Localization of EXO70s from H. villosa

Knowledge of the subcellular localization of a plant protein can help us predict its potential role in the biological process. The subcellular localization of EXO70-Vs was investigated by transiently expressing the construct into leaves of *Nicotiana tabacum* via *Agrobacterium* method. Eleven EXO70-Vs generated fluorescence signals. Compared with the relatively even distribution of GFP signals in the cell (Figure 7a), the 11 confusion proteins had distinct localization patterns. EXO70A1, A3 and F1-V displayed weak signals on the plasma membrane (PM) (Figure 7b–d), while the PM signals for EXO70C1-V and EXO70D2-V were more intensive (Figure 7e,f). EXO70B1, E1 and F3-V displayed signals both in the PM and the nucleus (Figure 7g–i). EXO70D1-V and EXO70F2-V also produced signals in the PM; in addition, they also had small and discrete spot signals in the PM (Figure 7j,k). EXO70I1-V was the only one with no continuous PM localized signal; however, we observed discrete punctate signals along the PM (Figure 7l).

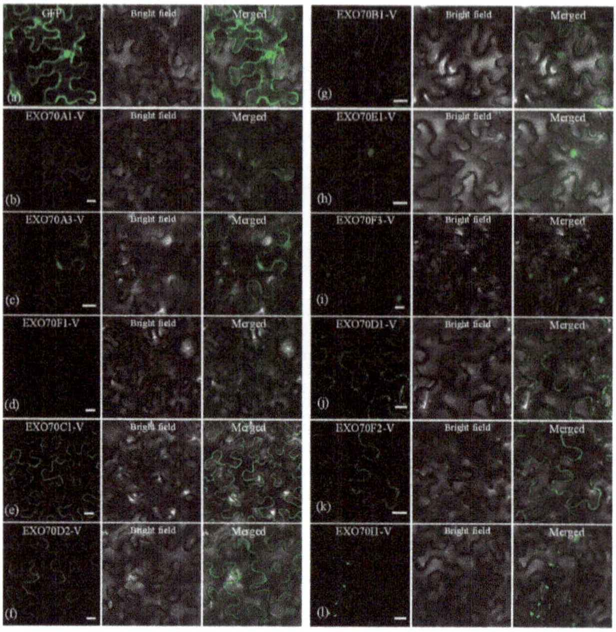

Figure 7. Subcellular localization of *H. villosa* EXO70-GFP-Vs proteins. *H. villosa* EXO70-GFP-Vs proteins were transiently expressed in *N. benthamiana* leaves carried out with injection of *Agrobacterium* and examined by a confocal microscope. Green fluorescence was observed 48 h after infection. Bar = 20 μm. (**a**) The empty GFP vector was used as the control. The green channel shows that GFP signals were localized in the nucleus and cytoplasmic and plasma membranes. (**b**–**l**) The subcellular localization pattern of EXO70A1-V to EXO70I1-V, respectively.

3. Discussion

3.1. Evolutionary Relationship of the EXO70 Gene Family in Wheat and Its Relatives

The evolutionary relationships of the *EXO70* gene family (between wheat, *T. urartu*, *Ae. tauschii*, *T. dicoccoides*, *H. vulgare* and *H. villosa*) have been speculated about based on the total number, classification, chromosomal distribution and structure. The surveyed diploid species of seven chromosome pairs except for *H. villosa* all possessed 26 *EXO70* genes, which suggested this gene family appeared before the divergence among Triticum species [52]. Allohexaploid wheat originated from two hybridizations between three diploid progenitors approximately 2.5–4.5 million years ago [53]. The number of *EXO70* genes in tetraploid and common wheat (a total of 47 and 75, respectively) is approximately twice and three times as many as diploids, implying they have undergone one and two rounds of polyploidization events [54]. Although the polyploidization event induced rapid and extensive genetic and epigenetic changes in the genome which were related to a large range of molecular and physiological adjustment [55] as well as a significant loss of gene family members upon domestication [53], by comparing with diploid species, the *EXO70* gene family did not go through a wide range of expansion or diminution in tetraploid wheat and allohexaploid wheat. This deduction is also supported by their chromosomal location analysis, which showed that 73.1% of genes have good collinearity among wheat, *Ae. tauschii*, *T. dicoccoides* and *H. vulgare*, and most of the same type of orthologous genes maintain the relative order of their ancestral genes (Figures 3 and 4). As for the small difference in the number of genes on individual subgroups among wheat and its relatives, this may be because of the quality of the genome assembling or the gene duplication of subgroup *EXO70I*.

Phylogenetic analysis showed that all three groups (*EXO70.1*, *EXO70.2* and *EXO70.3*) and nine subgroups (*EXO70A* to *EXO70I*) are represented in each of the six *Triticeae* species. A similar gene structure was found in the same subgroup; subgroup *EXO70A* consists of multiple introns, while the other eight subgroups had fewer or were intronless. The variable intron numbers confirmed the classifications of the *EXO70* genes. Additionally, *EXO70* subgroups diversified before the divergence within polyploid wheat and related species during the evolutionary process of the *EXO70* gene family; however, no new groups/subgroups have emerged.

EXO70I members were most represented in wheat and its five relatives (57), as well as in rice (16), but not in *Arabidopsis* [6,36]. *EXO70I* belongs to *EXO70.2*, not *EXO70.3*, which is different from what was found in previous studies [1,7,36]. This suggests that the *EXO70I* subgroup arose before the evolutionary divergence of rice from other *Triticeae* crops and disappeared during the evolution of *Arabidopsis*. The *EXO70I* subgroup underwent rapid divergence, producing a large number of members; this event can probably be explained by unequal cross-over or segmental chromosomal duplication [56,57]. During long-term natural selection, numerous *EXO70* genes diverged and evolved in order to respond to various conditions. The study of *Arabidopsis* showed that the duplicated gene loss process is non-random; those involved in DNA repair are more likely to be lost, while genes involved in signal transduction and transcription have been preferentially retained [56]. Therefore, the function of *EXO70I* subgroup in *Arabidopsis* was inclined to responses to DNA repair, and in grass species may participate in signal transduction. Research on a larger range of species is needed to figure out whether the *EXO70I* branch is unique to monocotyledons.

3.2. Diversification of Subcellular Localization Pattern of the EXO70 in H. villosa

Protein subcellular localization analysis provided important clues to their specialized biological functions [58]. The diversification of EXO70-V subcellular localization patterns implies functional differentiation. Except for EXO70I1-V, all *EXO70* genes showed plasma membrane (PM) signals. The EXO70D1/F2-V-GFP locates to the PM merged with some small, discrete punctate. At the same time, EXO70I1-V-GFP only gave rise to smaller fluorescent discrete punctate along with PM, which are similar to AtEXO70E2 in *Arabidopsis* protoplasts, which was a maker of a novel double-membraned structure termed EXPO (exocyst-positive organelles). AtEXO70E2 was involved in unconventional protein secretion for cytosolic proteins that lack a signal peptide, because of its ability to recruit several other exocyst complex subunits [8,9,59,60]. Therefore, *EXO70D1*, *F2* and *I1-V* might have the ability to recruit different partners, then form various complexes to execute different biological functions. AtEXO70A1 was distributed in different patterns in different systems' cytosol; it showed up in the nucleus and numerous small punctate structures in the BY-2 cell [36], at the apex of growing tobacco pollen tubes [61] and is strongly present in the cell plate [62]. In the study, EXO70A1/A3-V showed a weak PM signal, while EXO70A2-V was characterized by mis-localizations. This was probably because *AtEXO70A1* took part in a different vesicular transport process. Moreover, despite both AtEXO70B2 and AtEXO70H1 participating in the interaction between plants and pathogens, the signal of EXO70B2-GFP was mainly found in the cytoplasm, while EXO70H1-GFP was in vesicle-like structures in *Nicotiana benthamiana* leaf [31]. In our analysis, EXO70B1-V was present in the PM and nucleus (Figure 7g). It is likely that they went through different action sites to take part in the process of disease resistance. An exocyst is a tethering factor that mediates secretory vesicles to the plasma membrane before SNARE-mediated fusion [63,64]. EXPOs deliver cytosolic proteins to the cell surface [65,66] and therefore all of those were related to PM. The results explained why most genes had the PM location pattern.

3.3. Function Conserve or Differentiation of the EXO70 Gene Family in Common Wheat and H. villosa

An orthologous gene is one that diverged after evolution to give rise to different species; this gene generally maintains a similar function to that of the ancestral gene that it evolved from [67]. In our study, some *EXO70* orthologous genes from *H. villosa* exhibited a similar expression pattern to common wheat. For example, *EXO70A3-V* and *TaEXO70A3* showed a high expression level under heat stress; *EXO70B1-V* was preferred to *TaEXO70B1-3AL*, which was induced by *Bgt* at a late stage (48 h), but not by drought and heat stress; *EXO70E1-V* and *TaEXO70E1-3B* were upregulated by *Bgt* treatment; the expression of *EXO70H1-V* and *TaEXO70H1-1DL* was increased in response to heat (Figures 5 and 6). Therefore, it is reasonable to presume that some *EXO70* genes from *H. villosa* may have a similar function to the corresponding *EXO70* genes from common wheat. In plants, *AtEXO70B1*, *AtEXO70B2*, *AtEXO70H1* and *OsEXO70E1* are known for their roles in innate immunity [6,27–31, 34]. In the literature, genes such as *TaEXO70B1/B2* (with their homologous alleles), *EXO70B1-V*, *TaEXO70E1-3B*, *EXO70E1-V*, *TaEXO70H1-1DL* and *EXO70H1-V* were induced by *Pst/Bgt* treatment. Thus, we hypothesize that those genes also play an important role in plant defense responses and it is worth conducting further study to prove their function.

The long-term evolutionary fate of paralogous genes will still be determined by functions, with the genes that appeared to be sub-functionalized or neo-functionalized probably having higher rates of gene birth because of the increased adaptability. In contrast, the functional redundancy gene is unlikely to be stably maintained in the genome [55,57,68,69]. Paralogous/orthologous genes may diverge in expression to achieve more complex control of the same genetic network, balancing the relationship between internal growth and external environmental stimuli so that they "pay" the least and get the most [70]. In *Arabidopsis*, *EXO70C1/C2* were involved in pollen development and mainly localized pollen related tissue [21,22]. In common wheat, six *TaEXO70C* genes had an undetectable expression level in roots/stem/leaves (Figure 5a), but *EXO70C1-V* showed a moderate expression level in the stem and responded to drought and ABA treatment (Figure 6). Studies have shown that ABA has contributed to osmotic stress tolerance by regulating stomatal aperture and guard cells [71]. Therefore, *EXO70C1-V* perhaps plays an important role in drought tolerance. In rice, OsExo70F3 interacts with AVR-Pii and plays a crucial role in triggered immunity [35]. In our research, including the copy number, 13 *TaEXO70F* members were identified and *EXO70F1-V*, *EXO70F2-V* and *EXO70F3-V* were cloned from *H. villosa*. Expression studies have revealed that 11/13 common wheat varieties were induced by stress treatment (Figure 5), and three *EXO70-V* genes showed clearly diverse expression patterns, of which *EXO70F1-V* was only induced by SA, *EXO70F2-V* in response to drought, H_2O_2 and SA at an early stage, and *EXO70F3-V* in response to *Bgt* treatment at a late stage (48 h) (Figure 6). Research shows that both SA and H_2O_2 function as a key regulator against pathogens and stress tolerance [72–74]. Thus, we can infer the function of *EXO70F* genes not only in plant defense responses but also in abiotic stress.

In *N. benthamiana*, *EXO70D* and *EXO70G* mainly affect the size of the leaf [6]. In wheat, *TaEXO70D2-7BS* and *TaEXO70G1/G2* (with their homeoalleles) were induced by *Pst* and *Bgt* treatments. *EXO70D1-V* was upregulated by *Bgt* and MeJA treatments, *EXO70G1-V* had an increased expression level under phytohormones and H_2O_2 treatments, while *EXO70G2-V* was induced by drought and SA. MeJA is important for regulating the growth of plants and promotes plant resistance of various stresses [75]. This might suggest that *EXO70D1-V* plays an important role in plant growth and defense responses *EXO70G* are multifunctional. Of 22 *EXO70I* genes from common wheat, only five genes had distinct inducible expression. For instance, *TaEXO70I6-7DL* and *TaEXO70I8-4BL* were upregulated by drought and heat, respectively. *EXO70I1-V* was induced by ET and SA and maintained a high expression level. ET and SA regulate many diverse metabolic and developmental processes in plants, such as seed germination, abiotic stress response and pathogen defense [76,77]. Thus, we hypothesize that *EXO70I1-V* might play a vital role in growth or against multiple stresses. *EXO70* genes provide diverse expression patterns in different tissues and stresses, implying that *EXO70* genes may play an essential role in plant adaptation to a complicated and changeable environment.

4. Materials and Methods

4.1. Plant Materials

H. villosa (genome VV, accession no. 91C43), from the Cambridge Botanical Garden, Cambridge, UK, was used for gene cloning and expression analysis. Powdery mildew susceptible variety Sumai 3 was used for propagation of fresh spores of powdery mildew isolate *E26*. *Nicotiana benthamiana* plants were used for subcellular localization analysis. All the materials were grown in a greenhouse under a 14 h light/10 h dark cycle at 24 °C/18 °C, with 70% relative air humidity.

4.2. Plant Treatments

The seedlings of *H. villosa* were grown in liquid or soil until the three-leaf stage. For heat shock or drought stress treatment, the plants were transferred to 42 °C conditions, or dipped into 20% PEG 6000 and leaves were sampled at 0, 1, 6, or 12 h after treatment. For powdery mildew treatment, the plant was inoculated with pathogen isolate *E26* and the leaf tissues were sampled at 0, 24, 48 and 72 h after inoculation. For phytohormones and H_2O_2 treatments, the plants were sprayed with 5 mmol salicylic acid (SA), 0.1 mmol methyl jasmonate (MeJA), 0.1 mmol ethephon (ET), 0.2 mmol abscisic acid (ABA) and 7 mmol hydrogen peroxide (H_2O_2), respectively and all leaf tissues were collected at 0, 6, 12, or 24 h after spraying [78]. All the samples were rapidly frozen in liquid nitrogen, then stored in an ultra-freezer (−80 °C) before use.

4.3. RNA Isolation and Real-Time PCR Analysis

Total RNA was extracted using a Trizol Reagent kit (Invitrogen, CA, USA) according to the manufacturer's instructions and analyzed by gel electrophoresis. The first-strand cDNA was synthesized with random oligonucleotides using the HiScript® II Reverse Transcriptase system (Vazyme, Nanjing, China). qRT-PCR was carried out in a total volume of 20 μL containing 2 μL of cDNA, 0.4 μL gene-specific primers (10 μM), 10 μL SYBR Green Mix and 7.2 μL of RNase free ddH_2O, using the Roche LightCycler480 Real-time System (Roche, Basel, Swiss Confederation). The expression was represented in the form of relative fold change using the $2^{-\Delta\Delta CT}$ method [79]. Differentially expressed genes between each two samples pair were defined as two-fold up-regulated or two-fold down-regulated genes. Primers used for qRT-PCR are designed by Primer3 (Table S3). Three biological replications were performed. Heat map analysis of the expression data was performed using heat map drawing software MeV (version No. 4.7, Institute for Genomic Research, MD, USA)

4.4. Identification of EXO70 Gene Families in Wheat and Related Triticeae Species

We searched for the keywords 'exocyst subunit exo70 family protein' in the annotated proteins database of *Hordeum vulgare* (HH, $2n = 2x = 14$, accession No. FJWB02000000) and obtained entries containing gene ID and protein sequences [42]. Then we performed a Conserved Domains (CD) search (https://www.ncbi.nlm.nih.gov/Structure/cdd/wrpsb.cgi) and reserved the protein sequences that contain the typical pfam03081 domain [47].

The identified EXO70 protein sequences of *H. vulgare* were used as query sequences to blast (*E*-value $\leq 10^{-10}$) against protein database of the other species, including *Triticum urartu* (AA, $2n = 2x = 14$, accession No. NMPL02000000) [39], *T. aestivum* (AABBDD, $2n = 6x = 42$, accession No. NMPL02000000) [80], *Brachypodium distachyon* (BdBd, $2n = 2x = 14$, accession No. ADDN03000000) [81], Oryza sativa Japonica ($2n = 2x = 24$, accession No. AP008207–AP008218) [82], *T. dicoccoides* (AABB, $2n = 4x = 28$, accession No. FXXJ01000000) [43] and *Aegilops tauschii* (DD, $2n = 2x = 14$, accession No. AOCO02000000) [83]. After removing the redundant gene sequences for each the species, the alignment hits were validated by performing a CD search as described above.

4.5. Cloning and Protein Sequences Analysis of EXO70 Genes from *Haynaldia villosa*

According to the sequences obtained from the database of *H. vulgare*, primers (Table S3) for cloning the full-length cDNA of the *EXO70* gene from *Haynaldia villosa* were designed with online software Primer3 designing tool (v. 0.4.0, University of California, USA) [84] (Table S3). Mixed root, stem and leaf tissue cDNA of *H. villosa* served as a template for the isolation. This was performed at 95 °C for 30 s, followed by 35 cycles of 95 °C for 15 s, 58 °C for 15 s or 30 s and 72 °C for 3 min and then by 5 min at 72 °C in Phanta Max Super-Fidelity DNA polymerase (Vazyme, Nanjing, China). Before subcloning into their destination vectors, the PCR-amplified cDNA products were first cloned into the *pTOPO-Blunt* Vector (Aidlab, Beijing, China) as per the manufacturer's instructions. Multiple sequence alignments were conducted using DNAMan (Lynnon Corporation, Quebec, QC, Canada) software. The sequences similarity was visualized using the R programming language.

4.6. Subcellular Localization Assay

The ORFs of *EXO70-V* genes (without stop codon) were amplified from the *pTOPO-Blunt* Vector, then inserted into the *pCambia1305-GFP* vector, which contains a green fluorescent protein (GFP) reporter gene driven by the CaMV 35S promoter, using homologous cloning technology as per the manufacturer's instructions (Vazyme, Nanjing, China) (Table S3). Then it was introduced into *Agrobacterium tumefaciens* (strain GV3101) bacteria by a freeze–thaw procedure and grown in Luria-Bertani (LB) medium at 28 °C for 2 or 3 d.

Agrobacterium tumefaciens (strain GV3101) bacteria containing fusion constructed were grown in Luria-Bertani (LB) medium with both rifampicin and kanamycin (0.05 μg/mL) at 28 °C overnight. The bacterial cells were centrifuged and resuspended in an infiltration solution (10 mM MES pH 5.6, 0.1 mM Acetosyringone, 10 mM $MgCl_2$) to a final OD_{600} = 1.5. Bacterial suspensions were infiltrated into five- to six-week growing stage leaves of *N. benthamiana* by depressing the plunger of a 1-mL disposable needleless syringe into the abaxial side of leaves [85,86]. The fluorescence signals were observed 48–60 h after injection and images were captured using a confocal laser scanning microscope (LSM780; Carl Zeiss, Jena, Germany) according to the methods described by Wang et al. [87].

4.7. Phylogenetic Analysis of EXO70 Gene Family

Multiple sequence alignment was conducted by ClustalW which was integrated in Mega v6.0 [88]. Phylogenetic analysis was performed through online software PhyML 3.0 [89] using maximum-likelihood method with default parameter [90]. EXO70 proteins are rather diverse at their N-terminal and could not be aligned reliably, so we only used the conserved domain proteins to construct the phylogenetic tree and removed five genes where the length of the domain is fewer than 200 amino acids for further analysis.

4.8. Chromosomal Distribution and Exon-Intron Structure Analysis

Chromosomal information of predicted *EXO70* genes from each species was obtained after using cDNA sequences as a query sequence blasted to the genomic sequence to determine their chromosomal locations. Then we drew their locations onto the physical map of each chromosome using MapInspect tool (http://mapinspect.software.informer.com/).

The gff3 files of each species was downloaded from the Ensembl Plants FTP server (http://plants.ensembl.org/index.html) on 20 January 2018 for exon–intron structure analysis; the image of the exon–intron structure was obtained using the online Gene Structure Display Server (last accessed date on 20 January 2018) with the gff3 files for each species [91]. The corresponding evolutionary tree were constructed by Mega v6.0 [88], all sequences were aligned by ClustalW using the default parameters [92,93], used the Neighbor-Joining method with the pairwise deletion option, Poisson correction and bootstrap analysis conducted with 1000 replicates [26,94]

4.9. RNA-seq Expression Analysis

Publicly available RNA-seq data were retrieved from the expVIP [50] were used to analyze the expression pattern of predicted wheat *EXO70* genes in different tissues and stresses. The tissue-specific expression data were compiled from three wheat tissues (leaf, stem, root) collected from Chinese Spring at seeding development. The biotic stress expression data included two diseases (stripe rust and powdery mildew pathogen) and were collected from disease-resistant wheat varieties N9134 (at 7 days seedling stage). The abiotic stress expression data, including drought and heat treatments, were collected from heat-resistant wheat cultivar TAM107 (at 7 days seedling stage). The relative expression of each *TaEXO70* gene in different tissues and stresses was presented as a heat map, which was constructed by the heat map drawing software MeV (version No. 4.7, Institute for Genomic Research, MD, USA).

Supplementary Materials: Supplementary materials can be found at http://www.mdpi.com/1422-0067/20/1/60/s1.

Author Contributions: X.W., J.X., H.W. and J.Z. conceived and designed the study; J.Z., X.Z., W.W. and J.X. analyzed the data; H.Z., J.L. and M.L. collected the plant materials; J.Z. and X.Z. performed the experiments; J.Z., X.W. and J.X. wrote the manuscript. All authors reviewed and edited the manuscript.

Funding: This work was supported by the National Key Research and Development program (2016YFD0101004, 2016YFD0102001-004), the National Science Foundation of China (Nos. 31471490, 31661143005, 31290213), the Important National Science & Technology Specific Projects in Transgenic Research (2014ZX0800907B), the Chinese High Tech Program of China (No. 2011AA1001), the Program of Introducing Talents of Discipline to Universities (No. B08025), the 333 Talent Project of Jiangsu Province, the Shanghai Agriculture Applied Technology Development Program, China Grant (No. Z201502) and Development of Key Agricultural Varieties in Jiangsu Province (PZCZ201706).

Conflicts of Interest: The authors declare no conflict of interest.

Abbreviations

ABA	abscisic acid
DNA	deoxyribonucleic acid
EXPO	exocyst-positive organelle
ET	ethephon
GFP	green fluorescent protein
H_2O_2	hydrogen peroxide
MeJA	methyl jasmonate
PM	plasma membrane
Pm	powdery mildew
qRT-PCR	quantitative real time polymerase chain reaction
RNA	ribonucleic acid
SNARE	soluble NSF attachment protein receptor
SA	salicylic acid

References

1. Cvrckova, F.; Grunt, M.; Bezvoda, R.; Hala, M.; Kulich, I.; Rawat, A.; Zarsky, V. Evolution of the land plant exocyst complexes. *Front. Plant Sci.* **2012**, *3*, 159. [CrossRef] [PubMed]
2. He, B.; Guo, W. The exocyst complex in polarized exocytosis. *Curr. Opin. Cell Biol.* **2009**, *21*, 537–542. [CrossRef] [PubMed]
3. Ma, W.; Wang, Y.; Yao, X.; Xu, Z.; An, L.; Yin, M. The role of Exo70 in vascular smooth muscle cell migration. *Cell. Mol. Biol. Lett.* **2016**, *21*, 20. [CrossRef] [PubMed]
4. Elias, M. The exocyst complex in plants. *Cell Biol. Int.* **2003**, *27*, 199–201. [CrossRef]
5. Tu, B.; Hu, L.; Chen, W.; Li, T.; Hu, B.; Zheng, L.; Lv, Z.; You, S.; Wang, Y.; Ma, B.; et al. Disruption of OsEXO70A1 Causes Irregular Vascular Bundles and Perturbs Mineral Nutrient Assimilation in Rice. *Sci. Rep.* **2015**, *5*, 18609. [CrossRef] [PubMed]
6. Du, Y.; Overdijk, E.J.R.; Berg, J.A.; Govers, F.; Bouwmeester, K. Solanaceous exocyst subunits are involved in immunity to diverse plant pathogens. *J. Exp. Bot.* **2018**, *69*, 655–666. [CrossRef]

7. Synek, L.; Schlager, N.; Elias, M.; Quentin, M.; Hauser, M.T.; Zarsky, V. AtEXO70A1, a member of a family of putative exocyst subunits specifically expanded in land plants, is important for polar growth and plant development. *Plant J.* **2006**, *48*, 54–72. [CrossRef]
8. Wang, J.; Ding, Y.; Wang, J.; Hillmer, S.; Miao, Y.; Lo, S.W.; Wang, X.; Robinson, D.G.; Jiang, L. EXPO, an exocyst-positive organelle distinct from multivesicular endosomes and autophagosomes, mediates cytosol to cell wall exocytosis in Arabidopsis and tobacco cells. *Plant Cell* **2010**, *22*, 4009–4030. [CrossRef]
9. Ding, Y.; Wang, J.; Chun Lai, J.H.; Ling Chan, V.H.; Wang, X.; Cai, Y.; Tan, X.; Bao, Y.; Xia, J.; Robinson, D.G.; et al. Exo70E2 is essential for exocyst subunit recruitment and EXPO formation in both plants and animals. *Mol. Biol. Cell* **2014**, *25*, 412–426. [CrossRef]
10. Dellago, H.; Loscher, M.; Ajuh, P.; Ryder, U.; Kaisermayer, C.; Grillari-Voglauer, R.; Fortschegger, K.; Gross, S.; Gstraunthaler, A.; Borth, N.; et al. Exo70, a subunit of the exocyst complex, interacts with SNEV(hPrp19/hPso4) and is involved in pre-mRNA splicing. *Biochem. J.* **2011**, *438*, 81–91. [CrossRef]
11. Ren, J.; Guo, W. ERK1/2 regulate exocytosis through direct phosphorylation of the exocyst component Exo70. *Dev. Cell* **2012**, *22*, 967–978. [CrossRef]
12. Zhao, Y.; Liu, J.; Yang, C.; Capraro, B.R.; Baumgart, T.; Bradley, R.P.; Ramakrishnan, N.; Xu, X.; Radhakrishnan, R.; Svitkina, T.; et al. *Exo70* generates membrane curvature for morphogenesis and cell migration. *Dev. Cell* **2013**, *26*, 266–278. [CrossRef] [PubMed]
13. Zuo, X.; Zhang, J.; Zhang, Y.; Hsu, S.C.; Zhou, D.; Guo, W. Exo70 interacts with the Arp2/3 complex and regulates cell migration. *Nat. Cell Biol.* **2006**, *8*, 1383–1388. [CrossRef] [PubMed]
14. Liu, J.; Yue, P.; Artym, V.V.; Mueller, S.C.; Guo, W. The role of the exocyst in matrix metalloproteinase secretion and actin dynamics during tumor cell invadopodia formation. *Mol. Biol. Cell* **2009**, *20*, 3763–3771. [CrossRef] [PubMed]
15. Xiao, L.; Zheng, K.; Lv, X.; Hou, J.; Xu, L.; Zhao, Y.; Song, F.; Fan, Y.; Cao, H.; Zhang, W.; et al. Exo70 is an independent prognostic factor in colon cancer. *Sci. Rep.* **2017**, *7*, 5039. [CrossRef] [PubMed]
16. Li, S.; Chen, M.; Yu, D.; Ren, S.; Sun, S.; Liu, L.; Ketelaar, T.; Emons, A.M.; Liu, C.M. EXO70A1-mediated vesicle trafficking is critical for tracheary element development in Arabidopsis. *Plant Cell* **2013**, *25*, 1774–1786. [CrossRef] [PubMed]
17. Vukasinovic, N.; Oda, Y.; Pejchar, P.; Synek, L.; Pecenkova, T.; Rawat, A.; Sekeres, J.; Potocky, M.; Zarsky, V. Microtubule-dependent targeting of the exocyst complex is necessary for xylem development in Arabidopsis. *New Phytol.* **2017**, *213*, 1052–1067. [CrossRef]
18. Kulich, I.; Cole, R.; Drdova, E.; Cvrckova, F.; Soukup, A.; Fowler, J.; Zarsky, V. Arabidopsis exocyst subunits SEC8 and EXO70A1 and exocyst interactor ROH1 are involved in the localized deposition of seed coat pectin. *New Phytol.* **2010**, *188*, 615–625. [CrossRef] [PubMed]
19. Drdova, E.J.; Synek, L.; Pecenkova, T.; Hala, M.; Kulich, I.; Fowler, J.E.; Murphy, A.S.; Zarsky, V. The exocyst complex contributes to PIN auxin efflux carrier recycling and polar auxin transport in Arabidopsis. *Plant J.* **2013**, *73*, 709–719. [CrossRef] [PubMed]
20. Kalmbach, L.; Hematy, K.; De Bellis, D.; Barberon, M.; Fujita, S.; Ursache, R.; Daraspe, J.; Geldner, N. Transient cell-specific EXO70A1 activity in the CASP domain and Casparian strip localization. *Nat. Plants* **2017**, *3*, 17058. [CrossRef]
21. Lai, K.S. Analysis of EXO70C2 expression revealed its specific association with late stages of pollen development. *Plant Cell Tissue Organ Cult.* **2015**, *124*, 209–215. [CrossRef]
22. Synek, L.; Vukasinovic, N.; Kulich, I.; Hala, M.; Aldorfova, K.; Fendrych, M.; Zarsky, V. EXO70C2 Is a Key Regulatory Factor for Optimal Tip Growth of Pollen. *Plant Physiol.* **2017**, *174*, 223–240. [CrossRef] [PubMed]
23. Chen, C.; Liu, M.; Jiang, L.; Liu, X.; Zhao, J.; Yan, S.; Yang, S.; Ren, H.; Liu, R.; Zhang, X. Transcriptome profiling reveals roles of meristem regulators and polarity genes during fruit trichome development in cucumber (*Cucumis sativus* L.). *J. Exp. Bot.* **2014**, *65*, 4943–4958. [CrossRef] [PubMed]
24. Kulich, I.; Vojtikova, Z.; Glanc, M.; Ortmannova, J.; Rasmann, S.; Zarsky, V. Cell wall maturation of Arabidopsis trichomes is dependent on exocyst subunit EXO70H4 and involves callose deposition. *Plant Physiol.* **2015**, *168*, 120–131. [CrossRef] [PubMed]
25. Wang, Z.; Li, P.; Yang, Y.; Chi, Y.; Fan, B.; Chen, Z. Expression and Functional Analysis of a Novel Group of Legume-specific WRKY and Exo70 Protein Variants from Soybean. *Sci. Rep.* **2016**, *6*, 32090. [CrossRef] [PubMed]

26. Yang, Y.; Zhou, Y.; Chi, Y.; Fan, B.; Chen, Z. Characterization of Soybean WRKY Gene Family and Identification of Soybean WRKY Genes that Promote Resistance to Soybean Cyst Nematode. *Sci. Rep.* **2017**, *7*, 17804. [CrossRef] [PubMed]
27. Kulich, I.; Pecenkova, T.; Sekeres, J.; Smetana, O.; Fendrych, M.; Foissner, I.; Hoftberger, M.; Zarsky, V. Arabidopsis exocyst subcomplex containing subunit EXO70B1 is involved in autophagy-related transport to the vacuole. *Traffic* **2013**, *14*, 1155–1165. [CrossRef]
28. Stegmann, M.; Anderson, R.G.; Westphal, L.; Rosahl, S.; McDowell, J.M.; Trujillo, M. The exocyst subunit Exo70B1 is involved in the immune response of *Arabidopsis thaliana* to different pathogens and cell death. *Plant Signal. Behav.* **2013**, *8*, e27421. [CrossRef]
29. Zhao, T.; Rui, L.; Li, J.; Nishimura, M.T.; Vogel, J.P.; Liu, N.; Liu, S.; Zhao, Y.; Dangl, J.L.; Tang, D. A truncated NLR protein, TIR-NBS2, is required for activated defense responses in the exo70B1 mutant. *PLoS Genet.* **2015**, *11*, e1004945. [CrossRef]
30. Stegmann, M.; Anderson, R.G.; Ichimura, K.; Pecenkova, T.; Reuter, P.; Zarsky, V.; McDowell, J.M.; Shirasu, K.; Trujillo, M. The ubiquitin ligase PUB22 targets a subunit of the exocyst complex required for PAMP-triggered responses in Arabidopsis. *Plant Cell* **2012**, *24*, 4703–4716. [CrossRef]
31. Pecenkova, T.; Hala, M.; Kulich, I.; Kocourkova, D.; Drdova, E.; Fendrych, M.; Toupalova, H.; Zarsky, V. The role for the exocyst complex subunits Exo70B2 and Exo70H1 in the plant-pathogen interaction. *J. Exp. Bot.* **2011**, *62*, 2107–2116. [CrossRef] [PubMed]
32. Seo, D.H.; Ahn, M.Y.; Park, K.Y.; Kim, E.Y.; Kim, W.T. The N-Terminal UND Motif of the Arabidopsis U-Box E3 Ligase PUB18 Is Critical for the Negative Regulation of ABA-Mediated Stomatal Movement and Determines Its Ubiquitination Specificity for Exocyst Subunit Exo70B1. *Plant Cell* **2016**, *28*, 2952–2973. [CrossRef] [PubMed]
33. Hong, D.; Jeon, B.W.; Kim, S.Y.; Hwang, J.U.; Lee, Y. The ROP2-RIC7 pathway negatively regulates light-induced stomatal opening by inhibiting exocyst subunit Exo70B1 in Arabidopsis. *New Phytol.* **2016**, *209*, 624–635. [CrossRef] [PubMed]
34. Guo, J.; Xu, C.; Wu, D.; Zhao, Y.; Qiu, Y.; Wang, X.; Ouyang, Y.; Cai, B.; Liu, X.; Jing, S.; et al. Bph6 encodes an exocyst-localized protein and confers broad resistance to planthoppers in rice. *Nat. Genet.* **2018**, 50297–50306. [CrossRef] [PubMed]
35. Fujisaki, K.; Abe, Y.; Ito, A.; Saitoh, H.; Yoshida, K.; Kanzaki, H.; Kanzaki, E.; Utsushi, H.; Yamashita, T.; Kamoun, S.; et al. Rice Exo70 interacts with a fungal effector, AVR-Pii, and is required for AVR-Pii-triggered immunity. *Plant J.* **2015**, *83*, 875–887. [CrossRef] [PubMed]
36. Chong, Y.T.; Gidda, S.K.; Sanford, C.; Parkinson, J.; Mullen, R.T.; Goring, D.R. Characterization of the *Arabidopsis thaliana* exocyst complex gene families by phylogenetic, expression profiling, and subcellular localization studies. *New Phytol.* **2010**, *185*, 401–419. [CrossRef]
37. Marcussen, T.; Sandve, S.R.; Heier, L.; Spannagl, M.; Pfeifer, M.; Jakobsen, K.S.; Wulff, B.B.; Steuernagel, B.; Mayer, K.F.; Olsen, O.A. Ancient hybridizations among the ancestral genomes of bread wheat. *Science* **2014**, *345*, 1250092. [CrossRef]
38. Zimin, A.V.; Puiu, D.; Hall, R.; Kingan, S.; Clavijo, B.J.; Salzberg, S.L. The first near-complete assembly of the hexaploid bread wheat genome, *Triticum aestivum*. *GigaScience* **2017**, *6*, gix097. [CrossRef]
39. Ling, H.Q.; Ma, B.; Shi, X.; Liu, H.; Dong, L.; Sun, H.; Cao, Y.; Gao, Q.; Zheng, S.; Li, Y.; et al. Genome sequence of the progenitor of wheat A subgenome *Triticum urartu*. *Nature* **2018**, *557*, 424–428. [CrossRef]
40. Bettgenhaeuser, J.; Krattinger, S.G. Rapid gene cloning in cereals. *Theor. Appl. Genet.* **2018**. [CrossRef]
41. Luo, M.C.; Gu, Y.Q.; Puiu, D.; Wang, H.; Twardziok, S.O.; Deal, K.R.; Huo, N.; Zhu, T.; Wang, L.; Wang, Y.; et al. Genome sequence of the progenitor of the wheat D genome *Aegilops tauschii*. *Nature* **2017**, *551*, 498–502. [CrossRef] [PubMed]
42. Mascher, M.; Gundlach, H.; Himmelbach, A.; Beier, S.; Twardziok, S.O.; Wicker, T.; Radchuk, V.; Dockter, C.; Hedley, P.E.; Russell, J.; et al. A chromosome conformation capture ordered sequence of the barley genome. *Nature* **2017**, *544*, 427–433. [CrossRef] [PubMed]
43. Avni, R.; Nave, M.; Barad, O.; Baruch, K.; Twardziok, S.O.; Gundlach, H.; Hale, I.; Mascher, M.; Spannagl, M.; Wiebe, K.; et al. Wild emmer genome architecture and diversity elucidate wheat evolution and domestication. *Science* **2017**, *357*, 93–97. [CrossRef] [PubMed]

44. Xing, L.; Qian, C.; Cao, A.; Li, Y.; Jiang, Z.; Li, M.; Jin, X.; Hu, J.; Zhang, Y.; Wang, X.; et al. The Hv-SGT1 gene from Haynaldia villosa contributes to resistances towards both biotrophic and hemi-biotrophic pathogens in common wheat (*Triticum aestivum* L.). *PLoS ONE* **2013**, *8*, e72571. [CrossRef]
45. Cao, A.; Xing, L.; Wang, X.; Yang, X.; Wang, W.; Sun, Y.; Qian, C.; Ni, J.; Chen, Y.; Liu, D.; et al. Serine/threonine kinase gene *Stpk-V*, a key member of powdery mildew resistance gene *Pm21*, confers powdery mildew resistance in wheat. *Proc. Natl. Acad. Sci. USA* **2011**, *108*, 7727–7732. [CrossRef] [PubMed]
46. Xing, L.; Hu, P.; Liu, J.; Witek, K.; Zhou, S.; Xu, J.; Zhou, W.; Gao, L.; Huang, Z.; Zhang, R.; et al. *Pm21* from *Haynaldia villosa* Encodes a CC-NBS-LRR that Confers Powdery Mildew Resistance in Wheat. *Mol. Plant* **2018**, *11*, 874–878. [CrossRef] [PubMed]
47. Li, S.; van Os, G.M.; Ren, S.; Yu, D.; Ketelaar, T.; Emons, A.M.; Liu, C.M. Expression and functional analyses of *EXO70* genes in Arabidopsis implicate their roles in regulating cell type-specific exocytosis. *Plant Physiol.* **2010**, *154*, 1819–1830. [CrossRef] [PubMed]
48. Miftahudin; Ross, K.; Ma, X.F.; Mahmoud, A.A.; Layton, J.; Milla, M.A.; Chikmawati, T.; Ramalingam, J.; Feril, O.; Pathan, M.S.; et al. Analysis of expressed sequence tag loci on wheat chromosome group 4. *Genetics* **2004**, *168*, 651–663. [CrossRef]
49. Hao, M.; Luo, J.; Zhang, L.; Yuan, Z.; Zheng, Y.; Zhang, H.; Liu, D. In situ hybridization analysis indicates that 4AL-5AL-7BS translocation preceded subspecies differentiation of Triticum turgidum. *Genome* **2013**, *56*, 303–305. [CrossRef]
50. Borrill, P.; Ramirez-Gonzalez, R.; Uauy, C. expVIP: A Customizable RNA-seq Data Analysis and Visualization Platform. *Plant Physiol.* **2016**, *170*, 2172–2186. [CrossRef]
51. Pearce, S.; Vazquez-Gross, H.; Herin, S.Y.; Hane, D.; Wang, Y.; Gu, Y.Q.; Dubcovsky, J. WheatExp: An RNA-seq expression database for polyploid wheat. *BMC Plant Biol.* **2015**, *15*, 299. [CrossRef] [PubMed]
52. Li, X.; Gao, S.; Tang, Y.; Li, L.; Zhang, F.; Feng, B.; Fang, Z.; Ma, L.; Zhao, C. Genome-wide identification and evolutionary analyses of bZIP transcription factors in wheat and its relatives and expression profiles of anther development related TabZIP genes. *BMC Genom.* **2015**, *16*, 976. [CrossRef] [PubMed]
53. Brenchley, R.; Spannagl, M.; Pfeifer, M.; Barker, G.L.; D'Amore, R.; Allen, A.M.; McKenzie, N.; Kramer, M.; Kerhornou, A.; Bolser, D.; et al. Analysis of the bread wheat genome using whole-genome shotgun sequencing. *Nature* **2012**, *491*, 705–710. [CrossRef] [PubMed]
54. Feldman, M.; Levy, A.A. Allopolyploidy–a shaping force in the evolution of wheat genomes. *Cytogenet. Genome Res.* **2005**, *109*, 250–258. [CrossRef] [PubMed]
55. Feldman, M.; Levy, A.A. Genome evolution in allopolyploid wheat—A revolutionary reprogramming followed by gradual changes. *J. Genet. Genom.* **2009**, *36*, 511–518. [CrossRef]
56. Blanc, G.; Wolfe, K.H. Functional divergence of duplicated genes formed by polyploidy during *Arabidopsis* evolution. *Plant Cell* **2004**, *16*, 1679–1691. [CrossRef] [PubMed]
57. Zhang, J. Evolution by gene duplication: An update. *Trends Ecol. Evol.* **2003**, *18*, 292–298. [CrossRef]
58. Sperschneider, J.; Catanzariti, A.M.; DeBoer, K.; Petre, B.; Gardiner, D.M.; Singh, K.B.; Dodds, P.N.; Taylor, J.M. LOCALIZER: Subcellular localization prediction of both plant and effector proteins in the plant cell. *Sci. Rep.* **2017**, *7*, 44598. [CrossRef]
59. Lin, Y.S.; Ding, Y.; Wang, J.; Shen, J.B.; Kung, C.H.; Zhuang, X.H.; Cui, Y.; Yin, Z.; Xia, Y.J.; Lin, H.X.; et al. Exocyst-Positive Organelles and Autophagosomes Are Distinct Organelles in Plants. *Plant Physiol.* **2015**, *169*, 1917–1932.
60. Robinson, D.G.; Ding, Y.; Jiang, L. Unconventional protein secretion in plants: A critical assessment. *Protoplasma* **2016**, *253*, 31–43. [CrossRef]
61. Hala, M.; Cole, R.; Synek, L.; Drdova, E.; Kulich, I.; Pecenkova, T.; Hochholdinger, F.; Cvrckova, F.; Fowler, J.; Zarsky, V. Exocyst complex functions in plant development. *Comp. Biochem. Phys. A* **2008**, *150*, S188–S189. [CrossRef]
62. Fendrych, M.; Synek, L.; Pecenkova, T.; Toupalova, H.; Cole, R.; Drdova, E.; Nebesarova, J.; Sedinova, M.; Hala, M.; Fowler, J.E.; et al. The *Arabidopsis* exocyst complex is involved in cytokinesis and cell plate maturation. *Plant Cell* **2010**, *22*, 3053–3065. [CrossRef] [PubMed]
63. Wu, B.; Guo, W. The Exocyst at a Glance. *J. Cell Sci.* **2015**, *128*, 2957–2964. [CrossRef] [PubMed]
64. Lurick, A.; Kummel, D.; Ungermann, C. Multisubunit tethers in membrane fusion. *Curr. Biol.* **2018**, *28*, R417–R420. [CrossRef] [PubMed]

65. Chi, Y.; Yang, Y.; Li, G.; Wang, F.; Fan, B.; Chen, Z. Identification and characterization of a novel group of legume-specific, Golgi apparatus-localized WRKY and Exo70 proteins from soybean. *J. Exp. Bot.* **2015**, *66*, 3055–3070. [CrossRef] [PubMed]
66. Ebine, K.; Ueda, T. Roles of membrane trafficking in plant cell wall dynamics. *Front. Plant Sci.* **2015**, *6*, 878. [CrossRef] [PubMed]
67. Koonin, E.V. Orthologs, paralogs, and evolutionary genomics. *Annu. Rev. Genet.* **2005**, *39*, 309–338. [CrossRef] [PubMed]
68. Panchy, N.; Lehti-Shiu, M.; Shiu, S.H. Evolution of Gene Duplication in Plants. *Plant Physiol.* **2016**, *171*, 2294–2316. [CrossRef]
69. Roulin, A.; Auer, P.L.; Libault, M.; Schlueter, J.; Farmer, A.; May, G.; Stacey, G.; Doerge, R.W.; Jackson, S.A. The fate of duplicated genes in a polyploid plant genome. *Plant J.* **2013**, *73*, 143–153. [CrossRef] [PubMed]
70. Liang, Z.; Schnable, J.C. Functional Divergence Between Subgenomes and Gene Pairs After Whole Genome Duplications. *Mol. Plant.* **2018**, *11*, 388–397. [CrossRef]
71. Fernando, V.C.D.; Schroeder, D.F. Role of ABA in Arabidopsis Salt, Drought, and Desiccation Tolerance. In *Abiotic and Biotic Stress in Plants—Recent Advances and Future Perspectives*; IntechOpen: London, UK, 2016; Chapter 22.
72. Kumar, D. Salicylic acid signaling in disease resistance. *Plant Sci.* **2014**, *228*, 127–134. [CrossRef] [PubMed]
73. Maity, A.; Sharma, J.; Sarkar, A.; More, A.K.; Pal, R.K.; Nagane, V.P.; Maity, A. Salicylic acid mediated multi-pronged strategy to combat bacterial blight disease (*Xanthomonas axonopodis* pv. *punicae*) in pomegranate. *Eur. J. Plant Pathol.* **2018**, *150*, 923–937. [CrossRef]
74. You, J.; Chan, Z. ROS Regulation During Abiotic Stress Responses in Crop Plants. *Front. Plant Sci.* **2015**, *6*, 1092. [CrossRef] [PubMed]
75. Chung, H.S.; Howe, G.A. A critical role for the TIFY motif in repression of jasmonate signaling by a stabilized splice variant of the JASMONATE ZIM-domain protein JAZ10 in Arabidopsis. *Plant Cell.* **2009**, *21*, 131–145. [CrossRef] [PubMed]
76. Vlot, A.C.; Dempsey, D.A.; Klessig, D.F. Salicylic Acid, a multifaceted hormone to combat disease. *Annu. Rev. Phytopathol.* **2009**, *47*, 177–206. [CrossRef] [PubMed]
77. Bleecker, A.B.; Kende, H. Ethylene: A gaseous signal molecule in plants. *Annu. Rev. Cell Dev. Biol.* **2000**, *16*, 1–18. [CrossRef] [PubMed]
78. Zhu, Y.; Li, Y.; Fei, F.; Wang, Z.; Wang, W.; Cao, A.; Liu, Y.; Han, S.; Xing, L.; Wang, H.; et al. E3 ubiquitin ligase gene CMPG1-V from *Haynaldia villosa* L. contributes to powdery mildew resistance in common wheat (*Triticum aestivum* L.). *Plant J.* **2015**, *84*, 154–168. [CrossRef]
79. Livak, K.J.; Schmittgen, T.D. Analysis of relative gene expression data using real-time quantitative PCR and the $2^{-\Delta\Delta CT}$ Method. *Methods* **2001**, *25*, 402–408. [CrossRef]
80. Mayer, K.F.X.; Rogers, J.; Doležel, J.; Pozniak, C.; Eversole, K.; Feuillet, C.; Gill, B.; Friebe, B.; Lukaszewski, A.J.; Sourdille, P. A chromosome-based draft sequence of the hexaploid bread wheat (*Triticum aestivum*) genome. *Science* **2014**, *345*, 1251788.
81. International Brachypodium Initiative. Genome sequencing and analysis of the model grass *Brachypodium distachyon*. *Nature* **2010**, *463*, 763–768. [CrossRef]
82. Fujisawa, M.; Baba, T.; Nagamura, Y.; Nagasaki, H.; Waki, K.; Vuong, H.; Matsumoto, T.; Wu, J.Z.; Kanamori, H.; Katayose, Y. The map-based sequence of the rice genome. *Nature* **2005**, *436*, 793–800.
83. Zhao, G.; Zou, C.; Li, K.; Wang, K.; Li, T.; Gao, L.; Zhang, X.; Wang, H.; Yang, Z.; Liu, X. The *Aegilops tauschii* genome reveals multiple impacts of transposons. *Nat. Plants* **2017**, *3*, 946–955. [CrossRef] [PubMed]
84. Andreas, U.; Ioana, C.; Triinu, K.; Jian, Y.; Brant, C.F.; Maido, R.; Steven, G.R. Primer3—New capabilities and interfaces. *Nucleic Acids Res.* **2012**, *40*, e115.
85. Chen, H.; Zou, Y.; Shang, Y.; Lin, H.; Wang, Y.; Cai, R.; Tang, X.; Zhou, J.M. Firefly luciferase complementation imaging assay for protein-protein interactions in plants. *Plant Physiol.* **2008**, *146*, 368–376. [CrossRef] [PubMed]
86. Wu, C.; Tan, L.; van Hooren, M.; Tan, X.; Liu, F.; Li, Y.; Zhao, Y.; Li, B.; Rui, Q.; Munnik, T.; et al. *Arabidopsis* EXO70A1 recruits Patellin3 to the cell membrane independent of its role as an exocyst subunit. *J. Integr. Plant Biol.* **2017**, *59*, 851–865. [CrossRef]

87. Wang, Z.; Cheng, J.; Fan, A.; Zhao, J.; Yu, Z.; Li, Y.; Zhang, H.; Xiao, J.; Muhammad, F.; Wang, H.; et al. LecRK-V, an L-type lectin receptor kinase in *Haynaldia villosa*, plays positive role in resistance to wheat powdery mildew. *Plant Biotechnol. J.* **2018**, *16*, 50–62. [CrossRef]
88. Tamura, K.; Stecher, G.; Peterson, D.; Filipski, A.; Kumar, S. MEGA6: Molecular Evolutionary Genetics Analysis version 6.0. *Mol. Biol. Evol.* **2013**, *30*, 2725–2729. [CrossRef]
89. Lefort, V.; Longueville, J.E.; Gascuel, O. SMS: Smart Model Selection in PhyML. *Mol. Biol. Evol.* **2017**, *34*, 2422–2424. [CrossRef]
90. Guindon, S.; Dufayard, J.F.; Lefort, V.; Anisimova, M.; Hordijk, W.; Gascuel, O. New Algorithms and Methods to Estimate Maximum-Likelihood Phylogenies: Assessing the Performance of PhyML 3.0. *Syst. Biol.* **2010**, *59*, 307–321. [CrossRef]
91. Hu, B.; Jin, J.; Guo, A.Y.; Zhang, H.; Luo, J.; Gao, G. GSDS 2.0: An upgraded gene feature visualization server. *Bioinformatics* **2015**, *31*, 1296–1297. [CrossRef]
92. Thompson, J.D.; Gibson, T.J.; Higgins, D.G. Multiple sequence alignment using ClustalW and ClustalX. *Curr. Protoc. Bioinform.* **2002**. [CrossRef] [PubMed]
93. Larkin, M.A.; Blackshields, G.; Brown, N.P.; Chenna, R.; McGettigan, P.A.; McWilliam, H.; Valentin, F.; Wallace, I.M.; Wilm, A.; Lopez, R.; et al. Clustal W and Clustal X version 2.0. *Bioinformatics* **2007**, *23*, 2947–2948. [CrossRef] [PubMed]
94. Wang, M.; Yue, H.; Feng, K.; Deng, P.; Song, W.; Nie, X. Genome-wide identification, phylogeny and expressional profiles of mitogen activated protein kinase kinase kinase (MAPKKK) gene family in bread wheat (*Triticum aestivum* L.). *BMC Genom.* **2016**, *17*, 668. [CrossRef] [PubMed]

© 2018 by the authors. Licensee MDPI, Basel, Switzerland. This article is an open access article distributed under the terms and conditions of the Creative Commons Attribution (CC BY) license (http://creativecommons.org/licenses/by/4.0/).

International Journal of
Molecular Sciences

MDPI

Article

Molecular Identification and Characterization of Hydroxycinnamoyl Transferase in Tea Plants (*Camellia sinensis* L.)

Chi-Hui Sun [1], Chin-Ying Yang [2,*] and Jason T. C. Tzen [1,*]

1 Graduate Institute of Biotechnology, National Chung Hsing University, Taichung 40227, Taiwan; bettychi2121@gmail.com

2 Department of Agronomy, National Chung Hsing University, Taichung 40227, Taiwan

* Correspondence: emiyang@dragon.nchu.edu.tw (C.-Y.Y.); TCTZEN@dragon.nchu.edu.tw (J.T.C.T.); Tel.: +886-4-22840777 (ext. 608) (C.-Y.Y.); Fax: +886-4-22877054 (C.-Y.Y.)

Received: 2 November 2018; Accepted: 6 December 2018; Published: 7 December 2018

Abstract: Tea (*Camellia sinensis* L.) contains abundant secondary metabolites, which are regulated by numerous enzymes. Hydroxycinnamoyl transferase (HCT) is involved in the biosynthesis pathways of polyphenols and flavonoids, and it can catalyze the transfer of hydroxyconnamoyl coenzyme A to substrates such as quinate, flavanol glycoside, or anthocyanins, thus resulting in the production of chlorogenic acid or acylated flavonol glycoside. In this study, the *CsHCT* gene was cloned from the Chin-Shin Oolong tea plant, and its protein functions and characteristics were analyzed. The full-length cDNA of CsHCT contains 1311 base pairs and encodes 436 amino acid sequences. Amino acid sequence was highly conserved with other HCTs from *Arabidopsis thaliana*, *Populus trichocarpa*, *Hibiscus cannabinus*, and *Coffea canephora*. Quantitative real-time polymerase chain reaction analysis showed that *CsHCT* is highly expressed in the stem tissues of both tea plants and seedlings. The *CsHCT* expression level was relatively high at high altitudes. The abiotic stress experiment suggested that low temperature, drought, and high salinity induced *CsHCT* transcription. Furthermore, the results of hormone treatments indicated that abscisic acid (ABA) induced a considerable increase in the *CsHCT* expression level. This may be attributed to CsHCT involvement in abiotic stress and ABA signaling pathways.

Keywords: tea; hydroxycinnamoyl transferase; abiotic stress; ABA signaling; hormone

1. Introduction

After water, tea has been the most widely consumed beverage worldwide for several thousand years because of its unique aroma and taste. Tea plants (*Camellia sinensis* L.) contain abundant specialized secondary metabolites such as polyphenolic compounds, found in the largest proportions in tea plants alkaloids, terpenoids, and amino acids [1]. Tea polyphenols account for 30% of the dry weight of tea leaves. They can be roughly divided into the following five categories: flavanols, flavonols, flavones, proanthocyanidins, and phenolic acids [2,3].

Hydroxycinnamoyl transferase (HCT) catalyzes the transfer of hydroxycinnamoyl moiety to receptor substrates such as shikimic acid, quinic acid, anthocyanins, flavanol glycoside, polyamine, and long-chain fatty acids. Plants under environmental stresses can induce the related gene expression involved in the phenylpropanoid metabolic pathway to generate various secondary metabolites that resist or adapt to environmental stresses [4]. The HCT involved in the phenylpropanoid pathway catalyzes shikimic acid and quinic acid to participate in the upstream pathway of lignin biosynthesis. Lignins in plant cell walls provide a physical defense that protects polysaccharides in cell walls from degradation by microorganisms [5]. Low temperature, high salinity, drought, mechanical injury, abscisic acids (ABAs), salicylic acid (SA), and hydrogen peroxide can induce *HcHCT* expression in *Hibiscus cannabinus*. HcHCT increases abiotic stress tolerance in plants [6]. In *Cucumis sativus*, *HCT* expression was reduced with pectinase treatment. In addition, directing the phenylpropanoid pathway to generate H-lignin caused p-coumaraldehyde accumulation [7].

HCT not only participates in secondary metabolite acylation but also regulates hypersensitive responses (HRs) in plants. HCT1806 or HCT4918 in *Zea mays* interacts with Rp1-D21 translated from resistance genes, thereby inhibiting HR generation. When pathogens attack plants, effectors secreted by the pathogens can change the protein structure of HCT1806 or HCT4918, which influences how they interact with Rp1-D21 and causes plants to generate HR. This prevents the spread of pathogens in local cell necrosis [8]. HCT is specific to a wide range of substrates such as gentisate, 3-hydroxybenzoate, hydroxyanthranilate, and protocatechuate, and competes with shikimic acid or quinate acid for the binding site on the enzyme, which in turn produces other acylation products [9–11].

In this study, to clarify the molecular characteristics of HCT in tea plants, we analyzed highly conserved domains in the amino acid sequences of HCT in *Arabidopsis thaliana*, *Nicotiana tabacum*, *H. cannabinus*, *Theobroma cacao*, and *Fragaria vesca*, designed degenerate primers for use in polymerase chain reaction (PCR), and cloned the genetic sequence of *CsHCT* from a Chin-Shin Oolong tea plant. Quantitative real-time PCR (qRT-PCR) was used to analyze *CsHCT* expression levels in the tissues of tea plants and seedlings. The results demonstrated a high level of *CsHCT* expression in the stem tissues of tea plants and seedlings. The amount of *CsHCT* transcribed in tea plants at various altitudes and in different seasons was also measured, and the results indicated that CsHCT expression levels were relatively high at high altitudes and at low temperatures. Moreover, an abiotic stress experiment revealed that low-temperature, drought, and high-salinity stresses induced *CsHCT* transcription. In addition, *CsHCT* expression increased with ABA treatment. Thus, this study concluded that CsHCT may be involved response to abiotic stress and ABA signaling pathways in tea plants.

2. Methods

2.1. Plant Materials and Growth Conditions

This study used the tea [*C. sinensis* (L.) Kuntze] cultivar Chin-Shin Oolong in these experiments. These were obtained from a tea seed germination farm operated in Nantou in central Taiwan. The seedlings were 1–2 years old with a height of 40–50 cm. The tea plant (>10 years old) samples were obtained from a tea farmer in Nantou County. The samples were collected between 2015 and 2016 and consisted of buds and young leaf (YL), old leaf (OL), young stem (YS), and old stem (OS) tissues. They were obtained from high-mountain tea plantations in the Alishan area of Chiayi County. The tea plantations were located at altitudes between 700 and 1300 m, and all the tea plants were >10 years old.

2.2. Bioinformatics Analysis of the CsHCT Gene and Amino Acid Sequence

This study analyzed the highly conserved domains of the HCT protein sequence in *A. thaliana*, *N. tabacum*, *H. cannabinus*, *T. cacao*, and *F. vesca*, designed a degenerate primer, and used the cDNA of the Chin-Shin Oolong tea plant as the template to perform PCR for obtaining the gene fragment sequence of *CsHCT*. The SMARTerTM RACE cDNA amplification kit (Clontech Laboratories, Inc., Mountain View, CA, USA) was used to expand 5′-end and 3′-end cDNA sequences. The full-length cDNA sequence of *CsHCT* was obtained after sequencing.

In the bioinformatics analysis conducted on the amino acid sequence of CsHCT, the ExPASy Translate tool (https://web.expasy.org/translate/) (access on July, 2016) was used for estimating the amino acid sequence translated by a nucleotide. The ExPASy Compute pI/Mw tool (https://web.expasy.org/compute_pi/) (access on July, 2016) was used to estimate the protein molecular weight and isoelectric point. Multiple sequence alignments were performed using the EMMA function of the EMBOSS explorer (http://www.bioinformatics.nl/emboss-explorer/) (access on July, 2016) and the BLOSUM50 scoring matrix, and GeneDoc software was used to compare the results. Subsequently, Motif Scan (https://myhits.isb-sib.ch/cgi-bin/motif_scan) (access on July, 2016) was used to predict the structural and functional protein regions. This study used the National Center for Biotechnology Information (https://www.ncbi.nlm.nih.gov/) (access on July, 2016) and Phytozome v10.3 (https://phytozome.jgi.doe.gov/pz/portal.html) (access on July, 2016) websites to obtain the protein sequences of clade Vb of BAHD (BEAT, benzylalcohol-O-acetyltransferase; AHCT, anthocyanin O-hydroxycinnamoyltransferase; HCBT, anthranilate N-hydroxycinnamoyl-benzoyltransferase; and DAT, deacetylvindoline 4-O-acetyltransferase) acyltransferase from *A. thaliana*, *Oryza sativa*, *Populus trichocarpa*, *Coffea canephora*, and *H. cannabinus*. Sequence alignment was performed using the ClustalW model. A phylogenetic tree was constructed using the MEGA6 software, after which statistical analysis was conducted through the neighbor joining method. The 1000 iterations of the tree algorithm were performed using the bootstrap method. SignalP (http://www.cbs.dtu.dk/services/SignalP/) (access on July, 2016) was used to predict whether a protein signal peptide existed. The Hphob./Kyte and Doolittle method of the ExPASy ProtScale (https://web.expasy.org/protscale/) (access on July, 2016) was adopted for predicting whether the proteins were hydrophilic or hydrophobic. The subcellular localization of proteins was predicted using WoLF PSORT (https://www.genscript.com/wolf-psort.html) (access on July, 2016).

2.3. Abiotic Stress and Hormone Treatments on Tea Seedlings

The 1-year-old tea seedlings were treated with low temperature, high temperature, high salinity, and drought. Treatment conditions were as follows. The low- and high-temperature stresses were 5 °C and 35 °C, respectively. The seedlings were watered on optimum level and treated with the stresses for 12 h. Under the high-salinity stress, the seedlings were given 50 mL of 300 mM NaCl at 20 °C per day, whereas under the drought stress, they were not given additional water. The two treatments lasted for 5 days. In the control group, the seedlings were on optimum level watered at 20 °C for 5 days. After the treatments, samples were collected and preserved in a −80 °C environment for subsequent analysis. For hormone treatments, 100 μM solutions of ABA, SA, methyl jasmonate (MeJA), and 1-aminocyclopropane-1-carboxylic acid (ACC) solutions were prepared and sprayed on the YLs of the seedlings. After waiting for 6 h, the samples were collected and preserved in a −80 °C environment for subsequent analysis.

2.4. Extraction of Total RNA and qRT-PCR

In the experiment, 0.2 g of tea leaf samples was ground into powder in liquid nitrogen, and the total RNA was extracted using the Plant Total RNA Purification Kit (GeneMark, Taichung, Taiwan). The Moloney Murine Leukemia Virus (MMLV) first-strand synthesis kit (Gene DireX, Las Vegas, NV, USA) was used for reaction of 2 μg of total RNA. In the reaction solution, 1 μL of Oligo dT (1 μg/μL) was mixed with 1 μL of 10 mM dNTP to react for 10 min at 70 °C and for 5 min at 4 °C. After the reaction, 4 μL of 5× reaction buffer, 2 μL of 0.1 M Dithiothreitol (DTT), 1 μL of diethylpyrocarbonate (DEPC) H_2O, and 1 μL of MMLV reverse transcriptase were sequentially added into the solution for 1 h of reaction at 37 °C and 10 min of reaction at 65 °C, after which the reaction was terminated. Subsequently, 80 μL of DEPC H_2O was added into a 0.2-mL microcentrifuge tube to perform qRT-PCR on the obtained cDNA.

For qRT-PCR, the cDNA was amplified using the CFX ConnectTM Real-Time System (Bio-Rad, Hercules, CA, USA), and data were analyzed using Bio-Rad CFX Manager 3.1. The *18S rRNA* of the tea leaf samples were used as the internal control to normalize cDNA levels. The reaction conditions were as follows: 5 min at 94 °C; 15 s of 45 circulations at 94 °C, 60 °C, and 72 °C each; and finally, 10 min at 72 °C. Nonspecific products or primer dimers were identified based on their lower melting temperature than that of the specific amplicon. The primers used for qRT-PCR analyses were as follows: *CsHCT* forward sequence 5′-caaattaaccaaggaccaactcaac-3′ and reverse sequence 5′-tgtaattgaccatgttcccatcttc-3′; and *18S rRNA* forward sequence 5′-ccgctggcaccttatgagaa-3′ and reverse sequence 5′-tttcagccttgcgaccatact-3′. The qRT-PCR experiments were repeated at least 3 times each biologically independently, and the data shown are average values. Statistical analyses were performed using Statistical Analysis System (SAS) 9.4 software.

3. Results

3.1. Bioinformatics Analysis of the CsHCT Gene and Amino Acid Sequences of C. sinensis L.

C. sinensis L. contains numerous polyphenolic compounds that provide multiple health benefits. To understand the role of HCT in the reaction of acylated flavonol glycosides, the *CsHCT* gene was cloned from the Chin-Shin Oolong tea plant. The *CsHCT* gene has cDNA of length 1552 bp that includes 35-bp and 182-bp 5′ and 3′ untranslated regions (excluding a poly-A tail). The open reading frame includes 1311 nucleotide sequences, which can encode 436 amino acid sequences (GenBank accession number: MH271107).

Sequence alignment analysis was performed on the amino acid sequence of CsHCT of *C. sinensis* L. and the HCT sequences of *A. thaliana* (AtHCT), *H. cannabinus* (HcHCT), *P. trichocarpa* (PtHCT), and *C. canephora* (CcHCT). Using global alignment and the BLOSUM50 scoring matrix, the similarities between CsHCT and AtHCT, HcHCT, PtHCT, and CcHCT were found to be 79.5%, 81.5%, 81.9%, and 82.1%, respectively. CsHCT and the HCT of other higher plants exhibited the sequences HXXXD and DFGWG, which are the conserved sequences of BAHD acyltransferase (Figure 1). Motif scanning was then performed to predict the amino acid sequence of CsHCT, and the results indicated that the N-terminus of its protein possesses the predicted N-myristoylation, casein kinase II phosphorylation, and protein kinase C phosphorylation sites.

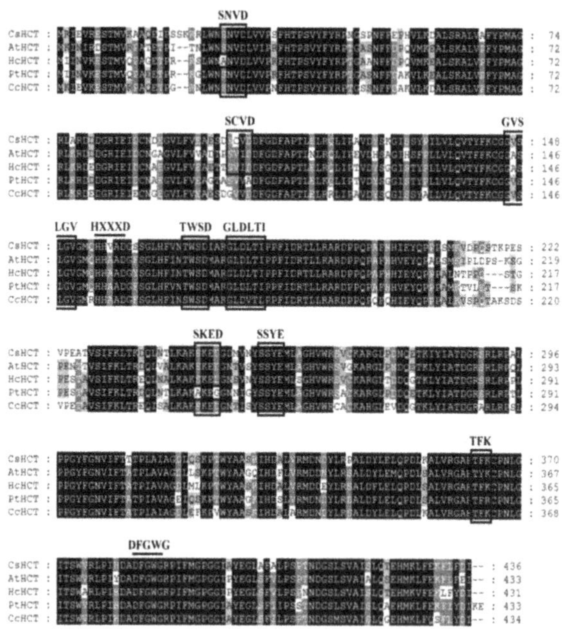

Figure 1. Alignment of deduced amino acid sequences of CsHCT with other putative HCTs. Black and gray shadings indicate conservation of 100% and at least 80%, respectively. Amino acid residues enclosed by squares correspond to consensus sequences of SXXD, SXXE, HXXXD, GVXXGV, TXXD, GLXXTI, DFGWG, etc. AtHCT, hydroxycinnamoyl-CoA shikimate/quinate hydroxycinnamoyl transferase was from *A. thaliana* (NP_199704); HcHcT, putative hydroxycinnamoyl-CoA shikimate/quinate hydroxycinnamoyl transferase was from *H. cannabinus* (AFN85668);

PtHCT, quinate O-hydroxycinnamoyl transferase/shikimate O-hydroxycinnamoyl transferase was from *P. trichocarpa* (ACC63882); and, CcHCT, hydroxycinnamoyl-CoA shikimate/quinate hydroxycinnamoyl transferase was from *C. canephora* (ABO47805). Alignment was performed using the ClustalW algorithm. Sequence identities with CsHCT were as follows: AtHCT, 79.5%; HcHCT, 81.5%; PtHCT, 81.9%; and, CcHCT, 82.1%.

Compute pI/Mw was used to analyze the amino acid sequence of CsHCT; its molecular weight and isoelectric point were predicted to be 48.53 kDa and 5.86, respectively. Analysis using SignalP failed to identify a signal peptide in CsHCT. The hydrophilicity and hydrophobicity of the protein were then analyzed using ProtScale, and the results demonstrated that the amino acid sequence did not possess an apparent hydrophobic end. The distribution of the hydropathy indices indicated that CsHCT is a hydrophilic protein (Figure 2A). Tuominen et al. (2011) reported that HCT is a member of clade Vb of the BAHD acyltransferase family. To understand the phylogenetic relationships between CsHCT and other clade Vb members, the MEGA6 software was used to perform neighbor joining. This generated a phylogenetic tree for the CsHCT of *C. sinensis* L. and the proteins of clade Vb members in *A. thaliana*, *O. sativa*, *P. trichocarpa*, *C. canephora*, and *H. cannabinus*. The results demonstrated that CsHCT and AtHCT had the closest phylogeny (Figure 2B); thus, they may have similar biochemical characteristics.

A

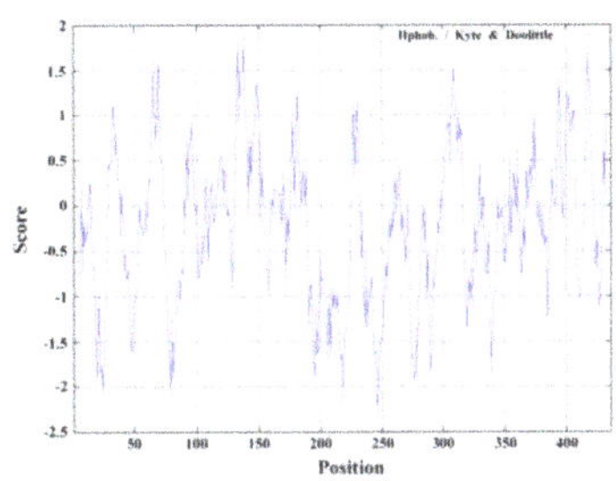

B

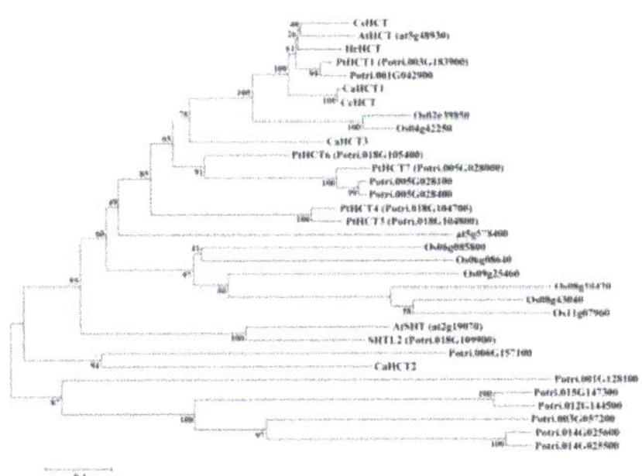

Figure 2. Hydropathy plot and phylogenetic tree analysis for CsHCT. (**A**) Hydropathy plot of CsHCT using the Kyte–Doolittle method with a window size of 436 amino acids. The window position values indicated on the x-axis of the graph reveal the average hydropathy of the entire window, with the corresponding amino acids as the middle element of that window. Plots above 0 (zero) in the graph indicate hydrophobic regions in the protein, and those below 0 (zero) indicate hydrophilic regions. (**B**) Unrooted phylogram of members of the HCT protein family. Phylogenetic tree of AtSHT (at2g19070), AtHCT (at5g48930), and HXXXD-type acyl transferase family protein (at5g57840) proteins in *Arabidopsis*; transferase family protein (Os02g39850; Os04g42250; Os06g08580; Os06g08640;

Os08g10420; Os08g43040; Os09g25460; and Os11g07960) in *O. sativa*; PtHCT1 (Potri.003G183900), PtHCT3 (Potri.018G104800), PtHCT4 (Potri.018G104700), PtHCT6 (Potri.018G105400), PtHCT7 (Potri.005G028000), SHTL2 (Potri.018G109900), Shikimate O-hydroxycinnamoyl transferase (Potri.005G028100 and Potri.005G028400), transferase family protein (Potri.006G157100; Potri.015G147300; Potri.003G057200; Potri.001G042900; Potri.001G128100; Potri.014G025500; and Potri.012G144500), anthranilate N-hydroxycinamoyl/benzoyltransferase-like protein (Potri.014G025600) in *P. trichocarpa*; CaHCT1 (CAJ40778), CaHCT2 (CAT00082), CaHCT3 (CAT00081), CcHCT (ABO77955) in *C. canephora*; and HcHCT (AFN85668) in *H. cannabinus*. The phylogenetic tree was constructed by the Neighbor Joining algorithm implemented in the MEGA 6 software package. The 1000 iterations of the tree algorithm were performed using the bootstrap method.

3.2. High Level of CsHCT Expression in the Stem Tissues of Tea Plants and Seedlings

To assess *CsHCT* expression in tissues of tea plants and seedlings, we selected YL, OL, YS, and OS tissues from 1-year-old tea seedlings and extracted the total RNA. Reverse transcriptase was then used to synthesize the first cDNA for qRT-PCR analysis. The results indicated that the *CsHCT* expression level was the highest in YS tissues, followed by OS tissues, buds, and leaves in tea seedlings (Figure 3A). The transcription level was the highest in OS tissues, followed by YS, YL, and OL tissues in tea plants (Figure 3B). Nonlignified YS tissues in tea seedlings exhibited the highest CsHCT expression level, but in tea plants, the CsHCT expression level was high in both OS and YS tissues. Moreover, CsHCT expression was evident in the buds and YL tissues of both tea plants and seedlings; however, expression levels remained lower than those in stem tissues (Figure 3).

A

B

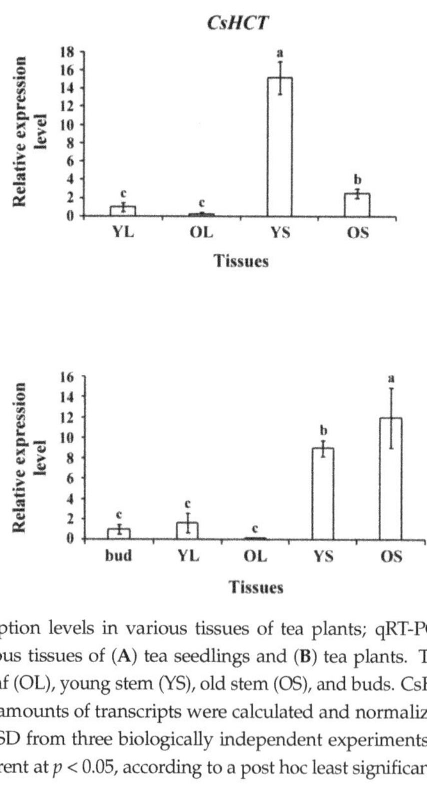

Figure 3. CsHCT transcription levels in various tissues of tea plants; qRT-PCR analysis of CsHCT transcription levels in various tissues of (**A**) tea seedlings and (**B**) tea plants. Total RNA was isolated from young leaf (YL), old leaf (OL), young stem (YS), old stem (OS), and buds. CsHCT transcription levels were determined. Relative amounts of transcripts were calculated and normalized to that of *18S rRNA*. Values represent means ± SD from three biologically independent experiments. Values with different letters are significantly different at $p < 0.05$, according to a post hoc least significant difference (LSD) test.

3.3. CsHCT Transcription Levels in Oolong Tea Plants during Four Growing Seasons and at Different Altitudes

Temperature affects gene expression related to the secondary metabolism of plants, which regulates the generation of secondary metabolites and enables plants to adapt to environmental changes. To evaluate the influence of temperature on *CsHCT* expression, we used qRT-PCR to analyze its transcription levels in one-tip-two-leaf tissues of oolong tea plants during four seasons and compared their *CsHCT* expression. Specifically, the tea samples were collected from Zhushan, Nantou County, Taiwan in April, June, September, and December. The results showed that the *CsHCT* expression level of tea samples from December was the highest, and that of the tea samples from June was the lowest (Figure 4A). According to the average monthly temperature in Zhushan between 2015 and 2016 released by the Central Weather Bureau, *CsHCT* expression levels were negatively correlated with temperature (Figure 4B). *CsHCT* expression levels were high in low-temperature seasons.

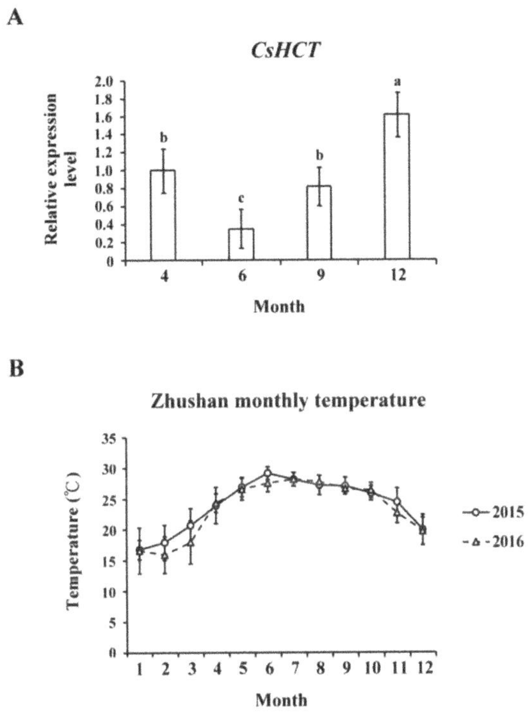

Figure 4. *CsHCT* transcription levels in four growing seasons. (**A**) The qRT-PCR analysis of CsHCT transcription levels in four growing seasons in the field. (**B**) Monthly average temperature in the Zhushan area. Total RNA was isolated from tissues of handpicked one-tip-two-leaf oolong tea samples collected in April, June, September, and December. Transcript levels of *CsHCT* were calculated and normalized to that of *18S rRNA*. Values represent means ± SD from five biologically independent experiments. Values with different letters are significantly different at $p < 0.05$, according to a post hoc least significant difference (LSD) test.

Altitude influences temperature, humidity, and sunlight intensity. Temperature decreases as altitude increases. To further verify *CsHCT* expression, the tea samples collected from the Alishan area of Nantou County from altitudes of 700, 1000, and 1300 m and bud tissues were analyzed using qRT-PCR to determine CsHCT expression. The results demonstrated that the *CsHCT* expression level was the highest at 1300 m and decreased as the altitude decreased (Figure 5).

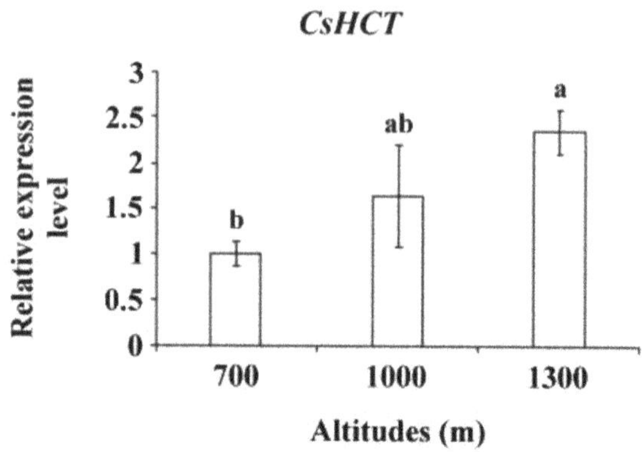

Figure 5. *CsHCT* expression levels at various altitudes; qRT-PCR analysis of *CsHCT* transcription levels in the Alishan area of Nantou County. Total RNA was isolated from the tissues of handpicked one-tip-two-leaf oolong tea samples collected at altitudes of 700, 1000, and 1300 m. Transcription levels of *CsHCT* were calculated and normalized to that of *18S rRNA*. Values represent means ± SD from three biologically independent experiments. Values with different letters are significantly different at $p < 0.05$, according to a post hoc LSD test.

3.4. Effects of Abiotic Stresses and Hormone Signaling on CsHCT Transcript Levels in Oolong Tea Seedlings

To investigate whether *CsHCT* expression is induced by abiotic stress, we exposed 1-year-old tea seedlings to stresses such as high temperature (35 °C), low temperature (5 °C), high salinity (300 mM NaCl), and drought; subsequently extracted the total RNA from one-tip-two-leaf tissues of oolong tea seedlings; and then performed qRT-PCR analysis. The results demonstrated that compared with the control group, *CsHCT* expression was higher at low-temperature stress and lower under high-temperature stress, thus verifying the association between *CsHCT* expression and temperature. Furthermore, *CsHCT* transcription levels also increased in response to the high-salinity and drought treatments (Figure 6A).

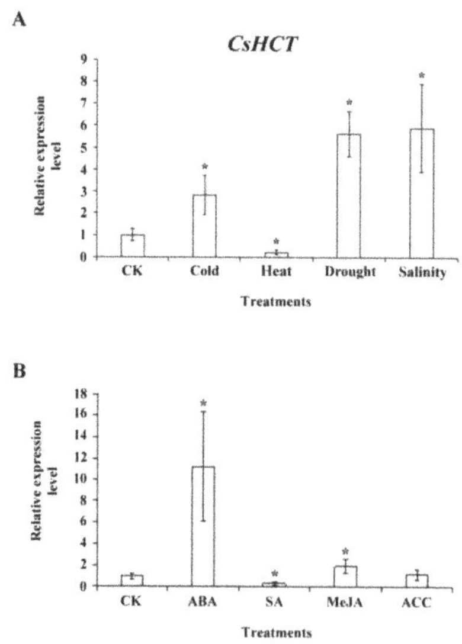

Figure 6. *CsHCT* expression levels under various abiotic stress and phytohormone treatment;

qRT-PCR analysis of *CsHCT* transcription levels in tea seedlings after (**A**) abiotic stress and (**B**) phytohormone treatment. Total RNA was isolated from YLs of tea seedlings after cold treatment (5 °C) for 12 h, heat treatment (35 °C) for 12 h, drought (no water) for 3 days, and salt treatment (300 mM NaCl) for 5 days. The phytohormone treatment included treatment with 100 μM ABA, SA, MeJA, and ACC for 6 h. *CsHCT* transcription levels were calculated and normalized to that of *18S rRNA*. Values represent means ± SD from four biologically independent experiments. * $p < 0.05$, versus value in control check (CK) treatment (Student's *t* test).

When a plant is under biotic or abiotic stresses, stress-related hormone signaling initiates the plant's defense mechanisms and increases its stress tolerance. To assess whether CsHCT expression is induced by various stress hormone signals, we treated the 1-year-old tea seedlings with ABA, SA, MeJA, and the ethylene precursor ACC for 6 h and subsequently collected the one-tip-two-leaf tissues for analysis of *CsHCT* expression through qRT-PCR. The results indicated that *CsHCT* expression was induced under ABA and MeJA treatment and that the expression level was the highest in the ABA group. By contrast, the *CsHCT* expression level in the SA group was significantly lower than that in the control (Figure 6B). Accordingly, this study determined that CsHCT expression can be induced by abiotic stresses such as low temperature, high salinity, and drought, and inferred that CsHCT may be involved in the ABA signaling pathways.

4. Discussion

Acyltransferase in higher plants can catalyze transfer of acyl group to donor substrate. Acyl esters are produced as a result of the acyl group transfer from the donor substrate to the acceptor substrate [12]. Acyltransferase can be divided into two protein families, namely BAHD acyltransferase and serine carboxypeptidase-like acyltransferase, according to different donor substrates. HCT is categorized into the clade Vb of BAHD acyltransferase and characterized by its ability to catalyze various substrates.

Amino acid analysis results demonstrated that the structure and function of the CsHCT and HCT sequences of other higher plants were highly conserved (Figure 1). The amino acid domain was highly conserved in clade Vb that possessed a particularly conserved sequence SXXDL in the BAHD acyltransferase family [13]. However, whether the amino acid domain affects the catalytic function of enzymes remains to be investigated and verified. BAHD acyltransferase acts mainly in the cytoplasm of plant cells. Its substrate, acyl coenzyme A thioesters, can perform biosynthesis in various cell organs and be transferred to the cytoplasm by the transporter on the cell membrane, which facilitates the catalytic reaction of the BAHD acyltransferase [14]. After analyzing the cellular localization of CsHCT, we inferred that CsHCT is primarily located in the cytoplasm, and the amino acid sequence does not possess a signal peptide or apparent hydrophobic end (Figure 2A). Therefore, this study inferred that CsHCT primarily reacts in the cytoplasm for catalytic reaction.

Our data determined that the *CsHCT* expression level in the YS tissues of tea seedlings was higher than that in the OS and bud tissues, whereas the *CsHCT* expression level in tea plants was higher in OS and YS tissues (Figure 3). Studies on *Trifolium pretense* have shown that *HCT1* is primarily expressed in stem and flower tissues, whereas *HCT2* is mainly expressed in leaf and flower tissues, which indicates that HCT1 and HCT2 have different catalytic functions in diverse plant tissues [15]. *P. trichocarpa* possesses seven *PtrHCTs* that can be expressed in the tissues of various plant parts and exhibit differences with respect to their relative performance. In particular, *PtrHCT1* and *PtrHCT6* are primarily expressed in stem tissues, whereas *PtrHCT3* has a higher level of expression in leaf tissues [16]. In this study, nonlignified YS tissues of tea seedlings were found to contain a relatively large amount of *CsHCT* transcripts. The *CsHCT* expression level in YS tissues was eight times higher than that in OS tissues. This indicated that secondary metabolites and the expression of related biosynthesis genes in tissues vary according to the growth stages of *C. sinensis* L.

When plants are under environmental stresses, the expression of genes related to the biosynthesis of secondary metabolites is induced, which results in the generation and accumulation of compounds such as phenylpropanoid, flavonoids, and anthocyanins that can increase plants' tolerance to stresses [4]. Our results demonstrated that the *CsHCT* expression level in *C. sinensis* L. was relatively high in winter and at high altitudes (Figures 4 and 5), indicating that *CsHCT* has a high level of expression in low temperatures. The *CsHCT* expression level increased under low-temperature stress and decreased under high-temperature stress (Figure 6A). Thus, *CsHCT* expression is induced in low temperatures and may be involved in the defense pathways against low-temperature stress. Research demonstrated that HCT expression is regulated by biotic and abiotic stresses, thereby increasing stress tolerance in plants [6]. Our data indicated that *CsHCT* expression in *C. sinensis* L. can be induced with low-temperature, high-salinity, and drought stresses, and the expression level was particularly high with ABA treatment (Figure 6B).

Phytohormone ABA involved in stress tolerance in plants can be divided into ABA-dependent and ABA-independent signaling pathways. ABA-dependent pathways transmit signals through ABA, thereby activating downstream transcription factors such as the ABRE-binding factor/ABA-responsive element-binding protein, myelocytomatosis, and myeloblastosis to regulate plants' stress tolerance [17,18]. This study demonstrated that *CsHCT* expression was induced by the abiotic stresses of low temperature, high salinity, and drought as well as ABA treatment. The signals of the three stresses may be transmitted through ABA-dependent pathways and may affect the expression of transcription factors of genes associated with the regulation of secondary metabolism. In this study, our results demonstrated the relationship between CsHCT expression and hormone signaling in oolong tea plants and may help improve the quality and possible health benefits of tea in the future.

5. Conclusions

Taken together, our study results indicated that *CsHCT* cloned from the Chin-Shin Oolong tea plant may be involved in tea plants' response to abiotic stresses (i.e., low temperature, high salinity, and drought) and various hormone signaling pathways that affect tea plants' secondary metabolic pathways.

Author Contributions: Conceptualization, C.-Y.Y.; Data curation, C.-H.S.; Formal analysis, C.-H.S.; Project administration, J.T.C.T. and C.-Y.Y.; Supervision, J.T.C.T. and C.-Y.Y.; Writing—original draft preparation, C.-Y.Y.

Funding: This work was supported by the National Science Council (NSC MOST 105-2311-B-005-007 and 106-2313-B-005-032-MY2 to Chin-Ying Yang).

Acknowledgments: We would like to thank Jian-Syun Chen for his generous supply of tea plants.

Conflicts of Interest: The authors declare that they have no conflicts of interest.

Abbreviations

ABA	Abscisic acid
ACC	1-aminocyclopropane-1-carboxylic acid
BAHD	BAHD actyltransferase
HCT	Hydroxycinnamoyl transferase
MeJA	Methyl jasmonate
qRT-PCR	quantitative real-time polymerase chain reaction
SA	Salicylic acid

References

1. Chen, Y.J.; Kuo, P.C.; Yang, M.L.; Li, F.Y.; Tzen, J.T.C. Effects of baking and aging on the changes of phenolic and volatile compounds in the preparation of old Tieguanyin oolong teas. *Food Res. Int.* **2013**, *53*, 732–743. [CrossRef]
2. Park, J.S.; Kim, J.B.; Hahn, B.S.; Kim, K.H.; Ha, S.H.; Kim, J.B.; Kim, Y.H. EST analysis of genes involved in secondary metabolism in Camellia sinensis (tea), using suppression subtractive hybridization. *Plant Sci.* **2004**, *166*, 953–961. [CrossRef]

3. Lin, L.Z.; Chen, P.; Harnly, J.M. New phenolic components and chromatographic profiles of green and fermented teas. *J. Agric. Food Chem.* **2008**, *56*, 8130–8140. [CrossRef] [PubMed]
4. Dixon, R.A.; Paiva, N.L. Stress-induced phenylpropanoid metabolism. *Plant Cell* **1995**, *7*, 1085–1097. [CrossRef] [PubMed]
5. Liu, Q.Q.; Luo, L.; Zheng, L.Q. Lignins: Biosynthesis and biological functions in plants. *Int. J. Mol. Sci.* **2018**, 19. [CrossRef] [PubMed]
6. Chowdhury, E.M.; Choi, B.S.; Park, S.U.; Lim, H.S.; Bae, H. Transcriptional analysis of hydroxycinnamoyl transferase (HCT) in various tissues of Hibiscus cannabinus in response to abiotic stress conditions. *Plant Omics J.* **2012**, *5*, 305–313.
7. Varbanova, M.; Porter, K.; Lu, F.C.; Ralph, J.; Hammerschmidt, R.; Jones, A.D.; Day, B. Molecular and biochemical basis for stress-induced accumulation of free and bound p-coumaraldehyde in cucumber. *Plant Physiol.* **2011**, *157*, 1056–1066. [CrossRef] [PubMed]
8. Wang, G.F.; He, Y.J.; Strauch, R.; Olukolu, B.A.; Nielsen, D.; Li, X.; Balint-Kurti, P.J. Maize homologs of hydroxycinnamoyltransferase, a key enzyme in lignin biosynthesis, bind the nucleotide binding leucine-rich repeat Rp1 proteins to modulate the defense response. *Plant Physiol.* **2015**, *169*, 2230–2243. [PubMed]
9. Sander, M.; Petersen, M. Distinct substrate specificities and unusual substrate flexibilities of two hydroxycinnamoyltransferases, rosmarinic acid synthase and hydroxycinnamoyl-CoA: Shikimate hydroxycinnamoyl-transferase, from Coleus blumei Benth. *Planta* **2011**, *233*, 1157–1171. [CrossRef] [PubMed]
10. Kang, S.; Kang, K.; Chung, G.C.; Choi, D.; Ishihara, A.; Lee, D.S.; Back, K. Functional analysis of the amine substrate specificity domain of pepper tyramine and serotonin N-hydroxycinnamoyltransferases. *Plant Physiol.* **2006**, *140*, 704–715. [CrossRef] [PubMed]
11. Eudes, A.; Pereira, J.H.; Yogiswara, S.; Wang, G.; Benites, V.T.; Baidoo, E.E.K.; Lee, T.S.; Adams, P.D.; Keasling, J.D.; Loque, D. Exploiting the substrate promiscuity of hydroxycinnamoyl-CoA: Shikimate hydroxycinnamoyl transferase to reduce lignin. *Plant Cell Physiol.* **2016**, *57*, 568–579. [CrossRef] [PubMed]
12. Moglia, A.; Acquadro, A.; Eljounaidi, K.; Milani, A.M.; Cagliero, C.; Rubiolo, P.; Genre, A.; Cankar, K.; Beekwilder, J.; Comino, C. Genome-wide identification of BAHD acyltransferases and in vivo characterization of HQT-like enzymes involved in caffeoylquinic acid synthesis in globe artichoke. *Front. Plant Sci.* **2016**, *7*, 1424. [CrossRef] [PubMed]
13. Tuominen, L.K.; Johnson, V.E.; Tsai, C.J. Differential phylogenetic expansions in BAHD acyltransferases across five angiosperm taxa and evidence of divergent expression among *Populus paralogues. BMC Genomics* **2011**, *12*, 236. [CrossRef] [PubMed]
14. Bontpart, T.; Cheynier, V.; Ageorges, A.; Terrier, N. BAHD or SCPL acyltransferase? What a dilemma for acylation in the world of plant phenolic compounds. *New Phytol.* **2015**, *208*, 695–707. [CrossRef] [PubMed]
15. Sullivan, M.L. A novel red clover hydroxycinnamoyl transferase has enzymatic activities consistent with a role in phaselic acid biosynthesis. *Plant Physiol.* **2009**, *150*, 1866–1879. [CrossRef] [PubMed]
16. Shi, R.; Sun, Y.H.; Li, Q.Z.; Heber, S.; Sederoff, R.; Chiang, V.L. Towards a systems approach for lignin biosynthesis in *Populus trichocarpa*: Transcript abundance and specificity of the monolignol biosynthetic genes. *Plant Cell Physiol.* **2010**, *51*, 144–163. [CrossRef] [PubMed]
17. Shinozaki, K.; Yamaguchi-Shinozaki, K.; Seki, M. Regulatory network of gene expression in the drought and cold stress responses. *Curr. Opin. Plant Biol.* **2003**, *6*, 410–417. [CrossRef]
18. Ishitani, M.; Xiong, L.M.; Stevenson, B.; Zhu, J.K. Genetic analysis of osmotic and cold stress signal transduction in Arabidopsis: Interactions and convergence of abscisic acid-dependent and abscisic acid-independent pathways. *Plant Cell* **1997**, *9*, 1935–1949. [CrossRef] [PubMed]

© 2018 by the authors. Licensee MDPI, Basel, Switzerland. This article is an open access article distributed under the terms and conditions of the Creative Commons Attribution (CC BY) license (http://creativecommons.org/licenses/by/4.0/).

International Journal of
Molecular Sciences

MDPI

Article

The WRKY Transcription Factor GmWRKY12 Confers Drought and Salt Tolerance in Soybean

Wen-Yan Shi [1,2,†], Yong-Tao Du [2,†], Jian Ma [3,†], Dong-Hong Min [1], Long-Guo Jin [2], Jun Chen [2], Ming Chen [2], Yong-Bin Zhou [2], You-Zhi Ma [2], Zhao-Shi Xu [2,*] and Xiao-Hong Zhang [1,*]

1 College of Life Sciences, College of Agronomy, Northwest A&F University, State Key Laboratory of Crop Stress Biology for Arid Areas, Yangling 712100, China; Shiwy12@126.com (W.-Y.S.); mdh2493@126.com (D.-H.M.)

2 Institute of Crop Science, Chinese Academy of Agricultural Sciences (CAAS), National Key Facility for Crop Gene Resources and Genetic Improvement, Key Laboratory of Biology and Genetic Improvement of Triticeae Crops, Ministry of Agriculture, Beijing 100081, China; duyongtao1994@126.com (Y.-T.D.); jinlongguo@caas.cn (L.-G.J.); chenjun@caas.cn (J.C.); chenming02@caas.cn (M.C.); zhouyongbin@caaas.cn (Y.-B.Z.); mayouzhi@caas.cn (Y.-Z.M.)

3 Faculty of Agronomy, Jilin Agricultural University, Changchun 130118, China; winter0106@163.com

* Correspondence: xuzhaoshi@caas.cn (Z.-S.X.); zhxh2493@126.com (X.-H.Z.); Tel.: +86-10-8210-6773 (Z.-S.X.)

† These authors contributed equally to this work.

Received: 14 November 2018; Accepted: 15 December 2018; Published: 17 December 2018

Abstract: WRKYs are important regulators in plant development and stress responses. However, knowledge of this superfamily in soybean is limited. In this study, we characterized the drought- and salt-induced gene *GmWRKY12* based on RNA-Seq and qRT-PCR. *GmWRKY12*, which is 714 bp in length, encoded 237 amino acids and grouped into WRKY II. The promoter region of *GmWRKY12* included ABER4, MYB, MYC, GT-1, W-box and DPBF *cis*-elements, which possibly participate in abscisic acid (ABA), drought and salt stress responses. *GmWRKY12* was minimally expressed in different tissues under normal conditions but highly expressed under drought and salt treatments. As a nucleus protein, *GmWRKY12* was responsive to drought, salt, ABA and salicylic acid (SA) stresses. Using a transgenic hairy root assay, we further characterized the roles of *GmWRKY12* in abiotic stress tolerance. Compared with control (Williams 82), overexpression of *GmWRKY12* enhanced drought and salt tolerance, increased proline (Pro) content and decreased malondialdehyde (MDA) content under drought and salt treatment in transgenic soybean seedlings. These results may provide a basis to understand the functions of *GmWRKY12* in abiotic stress responses in soybean.

Keywords: WRKY; stress responsive mechanism; drought tolerance; salt tolerance; transgenic hairy root assay; soybean

1. Introduction

Drought and salinity are the most important abiotic stress factors affecting plants growth and crop yield. On average, 1/3 of cultivable land suffers drought and salinization, which is equivalent to a loss of about 1,500,000 ha of crop land per year [1]. The damage caused by drought and salt are almost the sum of losses caused by other stress factors. Under limited land and water resources, it is necessary to breed new stress-resistant varieties to increase yield and ensure food security. Cultivation of stress-resistant crop varieties is also an important way to ensure high and stable yield of crops. Transgenic technology has become an important way to learn the function of genes in crops [2–4].

Being unable to move, plants encounter numerous biotic and abiotic stresses at different developmental stages which include drought, salinity, temperature changes, nutritional deficiency, pathogen invasion and competition from alien species. To overcome these unfavorable conditions, plants have evolved a complex and efficient signaling network, which can produce a series of responses to external stress signals and induce the expression of stress-related genes to protect the normal activities of the cells [5]. Inducible genes encoding proteins can be divided into three categories based on function: the first is functional genes, which are directly involved in stress response and are located downstream in the signaling network, such as HKT [6,7], SALT [8], NHX [9,10], CAX and CHX [11–13]. Another is transcription factors (TFs) that regulate the expression of functional genes in the middle of the signaling network, like DREB [14,15], MYB [16], WRKY [17,18], NAC [19,20], bZIP [21,22] and ERF [23,24]. The last group includes a variety of protein kinases, which conduct stress signals and are located upstream of the signaling network, such as GST [25], LEA [26] and FNS [27].

Among the three classes of stress-related genes, the TFs form a connecting link between the beginning and end of the signaling network; WRKYs are among the largest family of plant TFs. The WRKY domain is about 60 residues in length and is named by a conserved WRKY domain, containing the WRKYGQK heptapeptide at the N-terminus followed by a zinc-finger motif CX4-5CX22-23HXH or CX7CX23HXC [28,29]. Based on the number of WRKY domains and the structure of zinc finger motifs, WRKY TFs are divided into three groups. Group I includes two WRKY domains and either a CX4-5CX22-23HXH or CX7CX23HXC zinc-finger motif. Group II WRKY proteins contain a single WRKY domain and a CX4-5CX22-23HXH zinc-finger motif; due to differences in the primary amino acid sequence, Group II can be divided into five subgroups IIa-IIe [29,30]. Group III WRKY proteins have a single WRKY domain and a CX7CX23HXC zinc-finger motif.

As one of the members of the plant TF family, WRKY is heavily studied. Researchers have determined that WRKY TFs participate in various physiological and developmental processes [29], such as seed development [31], seed dormancy and germination [32], senescence [33], development [34], plant immune response [35], pathogen defense [18,36] and insect resistance [37,38]. Recent studies have revealed that WRKY proteins are involved in the signal transduction of plant hormones, like abscisic acid (ABA) [39,40], jasmonic acid (JA) [41] and gibberellin (GA) [39]. Numerous studies have demonstrated that WRKY TFs respond to abiotic stresses [42,43], such as salt [4], drought [44], cold [45] and heat [46–48]. There are 74 WRKY TF members in model plant *Arabidopsis* [49] and 18 WRKYs have been suggested to be induced by exposure to salt stress; overexpression of *WRKY25* or *WRKY33* was sufficient to increase *Arabidopsis* NaCl tolerance [50]. Overexpressing *TaWRKY2* and *TaWRKY19* exhibited salt and drought tolerance in transgenic *Arabidopsis* [51]. Moreover, researchers found that *OsWRKY11* directly bound to the promoter of a drought-responsive gene, RAB21, as well as enhanced heat and drought tolerance in transgenic rice seedlings [52,53]. Ectopic expression of *ZmWRKY33* and *ZmWRKY58* in *Oryza* and *Arabidopsis* improved drought and salt tolerance, respectively, in transgenic plants [54,55]. In addition, there is extensive cross-talk between responses to biotic/abiotic stresses and exogenous hormones, for example drought and salt stress with the plant hormones. *Arabidopsis WRKY46*, *WRKY54* and *WRKY70* are involved in Brassinosteroid-mediated drought response and plant growth [43]. Novel cotton WRKY-genes *GhWRKY25* and *GhWRKY6-like* confer tolerance to abiotic and biotic stresses in transgenic *Nicotiana* and enhanced salt tolerance by activating the ABA signaling pathway and scavenging reactive oxygen species [56]. SA-inducible poplar *PtrWRKY73* is also involved in disease resistance in *Arabidopsis* [37]. All of these studies illustrated that WRKY TFs play a significant role in plant developmental and physiological processes and abiotic and biotic stresses.

Soybean (*Glycine max*), is an important global cash crop, accounting for 59 percent of the world's oilseed production (http://soystats.com). Currently, due to its high protein content it is often treated as an important source of protein for both human consumption and as fodder. The demand for soybean is thus increasing rapidly and improving soybean yield has become a major research goal. Soybean productivity is greatly affected by growing environment, such as climatic and soil conditions (drought, salt, metallic pollution and fungus infection). Therefore, it is vital to cultivate soybean varieties that are resistant to stressors.

Recently, many studies based on biotechnological and RNA-Seq approaches have been conducted on soybean WRKY TFs. Researchers have identified 188 soybean WRKY genes genome-wide and 66 of the genes have been shown to respond rapidly and transiently to the imposition of salt stress [30]. In the latest version of the soybean genome (*Wm82.a2v1*), 176 GmWRKY proteins were confirmed and the expression of *GmWRKY47* and *GmWRKY58* decreased upon dehydration, while *GmWRKY92*, *GmWRKY144* and *GmWRKY165* increased under salt treatment [57]. *GmWRKY13* may function in plant growth and abiotic stress. *GmWRKY21* and *GmWRKY54* conferred tolerance to cold stress and salt and drought stress, respectively [58]. Here, based on RNA-Seq and several databases and bioinformatics methods, we identified *GmWRKY12*, which is associated with abiotic stress tolerance by quantitative RT-PCR. Overexpression of *GmWRKY12* could improve tolerance of soybean to drought and salt.

2. Results

2.1. Identification of GmWRKYs Up-Regulated under Drought/Salt Treatment

The GmWRKYs are distributed in different tissues or located upstream of soybean genes to bind the W-box consensus (TTGACY) in the promoters of target genes, initiating functions such as plant development, pathogen defense, insect resistance, response to biotic and abiotic stress and participating in signal transduction mediated by plant hormones [59,60]. In order to identify the function of genes or to explore whether GmWRKY mRNA expression goes up under biotic and abiotic stress, we conducted RNA-Seq (Tables S5 and S6). RNA-Seq data were used to screen GmWRKYs that are responsive to drought and salt. There were 105 GmWRKYs upregulated after drought treatment and fifty-three GmWRKYs were selected based on the rule that $\log_2$ (GH_treat/CK1_treat) >1 (Table 1). Nine GmWRKYs were selected from salt treatment RNA-Seq data based on the rule that $\log_2$ (NaCl_treat/CK2_treat) >1 (Table 2).

Table 1. Annotation of *Glycine max* WRKY transcription factors responding to drought stress (up-regulation).

Gene ID [a]	Name [b]	Chr	CDS (bp)	Protein (aa)	Group [c]
GLYMA_14G103100	*GmWRKY40*	14	849	282	IIb
GLYMA_18G056600	*GmWRKY62*	18	1689	542	IIb
GLYMA_17G042300	*GmWRKY6*	17	1173	390	IIe
GLYMA_04G054200	*GmWRKY50*	4	486	161	IIe
GLYMA_01G222300	*GmWRKY22*	1	738	245	IIc
GLYMA_02G293400	*GmWRKY31*	2	1278	425	IIa
GLYMA_04G218700	*GmWRKY21*	4	591	196	I
GLYMA_06G147100	*GmWRKY51*	6	591	196	III
GLYMA_01G224800	*GmWRKY12*	1	714	237	IIc
GLYMA_11G163300	*GmWRKY19*	11	1647	548	I
GLYMA_06G061900	*GmWRKY17*	6	885	294	IIb
GLYMA_10G011300	*GmWRKY54*	10	972	323	IIa
GLYMA_04G223300	*GmWRKY58*	4	954	317	III
GLYMA_18G213200	*GmWRKY57*	18	900	299	III
GLYMA_06G125600	*GmWRKY53*	6	1095	364	IIa
GLYMA_19G217800	*GmWRKY23*	19	873	290	IId
GLYMA_09G280200	*GmWRKY33*	9	1632	543	I
GLYMA_03G002300	*GmWRKY70*	3	747	248	IIc
GLYMA_13G310100	*GmWRKY36*	13	1845	614	IIc
GLYMA_14G200200	*GmWRKY49*	14	1728	575	IIc
GLYMA_16G026400	*GmWRKY60*	16	1122	373	IIc
GLYMA_16G0544001	*GmWRKY75*	16	588	195	IIb
GLYMA_04G223200	*GmWRKY55*	4	1020	339	IId
GLYMA_02G232600	*GmWRKY39*	2	1743	580	III
GLYMA_05G0290001	*GmWRKY72*	5	1785	594	I
GLYMA_03G220100	*GmWRKY41*	5	762	253	IIe

Table 1. *Cont.*

Gene ID [a]	Name [b]	Chr	CDS (bp)	Protein (aa)	Group [c]
GLYMA_08G021900	*GmWRKY46*	8	1080	356	III
GLYMA_15G003300	*GmWRKY27*	15	921	306	IIb
GLYMA_17G097900	*GmWRKY61*	17	1803	600	IIc
GLYMA_01G128100	*GmWRKY5*	1	1527	508	IId
GLYMA_12G212300	*GmWRKY16*	12	792	263	IIc
GLYMA_08G082400	*GmWRKY28*	8	881	293	III
GLYMA_07G227200	*GmWRKY3*	7	1602	533	IIc
GLYMA_03G256700	*GmWRKY43*	66	1089	362	IIe
GLYMA_15G168200	*GmWRKY42*	15	882	293	IIb
GLYMA_13G289400	*GmWRKY52*	13	798	265	IIc
GLYMA_08G011300	*GmWRKY25*	8	444	147	IId
GLYMA_09G061900	*GmWRKY47*	19	1573	296	IIc
GLYMA_17G222300	*GmWRKY30*	4	555	184	IIa
GLYMA_01G053800	*GmWRKY9*	1	1368	455	IIc
GLYMA_08G118200	*GmWRKY48*	7	789	262	IIc
GLYMA_01G056800	*GmWRKY32*	1	894	297	IId
GLYMA_08G218600	*GmWRKY56*	8	942	313	III
GLYMA_07G262700	*GmWRKY34*	7	1554	517	IIb
GLYMA_03G159700	*GmWRKY15*	1	1017	338	I
GLYMA_11G053100	*GmWRKY14*	11	963	320	I
GLYMA_05G096500	*GmWRKY11*	17	1050	334	I
GLYMA_17G222500	*GmWRKY63*	17	849	278	IIa
GLYMA_08G240800	*GmWRKY4*	2	1572	523	I
GLYMA_03G176600	*GmWRKY29*	5	1308	436	IIc
GLYMA_08G325800	*GmWRKY35*	8	1734	577	IIc
GLYMA_10G138300	*GmWRKY1*	14	1449	482	IIb
GLYMA_06G077400	*GmWRKY37*	6	903	300	III

[a]—The annotated GmWRKYs according to NCBI (https://www.ncbi.nlm.nih.gov/pubmed) and. PlantTFDB (http://planttfdb.cbi.pku.edu.cn/); [b]—The names of GmWRKYs are given according to SoyDB (http://soykb.org/); [c]—The grouping is according to [30,61].

Table 2. Annotation of *Glycine max* WRKY transcription factors responding to salt stress (up-regulation).

Gene ID [a]	Name [b]	Chr	CDS (pb)	Protein (aa)	Group [c]
GLYMA_11G053100	*GmWRKY14*	9	963	320	I
GLYMA_08G325800	*GmWRKY35*	8	1734	577	IIc
GLYMA_04G218700	*GmWRKY21*	10	591	196	I
GLYMA_14G200200	*GmWRKY49*	18	1728	575	IIc
GLYMA_07G227200	*GmWRKY3*	18	1602	533	IIc
GLYMA_02G115200	*GmWRKY28*	8	881	293	III
GLYMA_03G256700	*GmWRKY43*	16	1089	362	III
GLYMA_06G320700	*GmWRKY59*	6	2331	776	IIc
GLYMA_01G224800	*GmWRKY12*	7	714	237	IIc

[a]—The annotated GmWRKYs according to NCBI (https://www.ncbi.nlm.nih.gov/pubmed) and PlantTFDB (http://planttfdb.cbi.pku.edu.cn/). [b]—The names of GmWRKYs are given according to SoyDB (http://soykb.org/). [c]—The grouping is according to [30,61].

2.2. Tissue-Specific Expression Patterns of GmWRKYs

To thoroughly study GmWRKY expression profiles under normal conditions, hierarchical clustering was conducted using expression levels of fifty-three (drought-responsive) and nine (salt-responsive) GmWRKY genes in young leaf, flower, one cm pod, pod shell 10 days after flowering (DAF), pod shell 14 DAF, seed 10 DAF, seed 14 DAF, seed 21 DAF, seed 25 DAF, seed 28 DAF, seed 35 DAF, seed 42 DAF, root and nodule (Figures 1 and 2). Approximately 28% of GmWRKYs from different tissues were expressed at low levels or unexpressed (*GmWRKY3, 5, 6, 21, 22, 25, 29, 30, 31, 47, 50, 55, 63, 70* and *72*); by contrast, 45% of GmWRKYs were highly expressed in different tissues

(*GmWRKY4*, *9*, *11*, *14*, *16*, *17*, *19*, *28*, *32*, *34*, *35*, *36*, *37*, *39*, *41*, *42*, *46*, *48*, *49*, *52*, *56*, *57*, *60* and *61*). Among these GmWRKYs, *GmWRKY11* and *GmWRKY17* had the highest expression in four different tissues. *GmWRKY28*, *35*, *37*, *48* and *57* are highly expressed in nodule, seed 10 DAF, seed 42 DAF, root and flower. Within the nine GmWRKYs related to salt response, *GmWRKY3* and *GmWRKY21* had low expression and *GmWRKY14*, *28*, *35*, *49* and *59* were highly expressed in at least four different tissues. The analysis data are available in Tables S1 and S2.

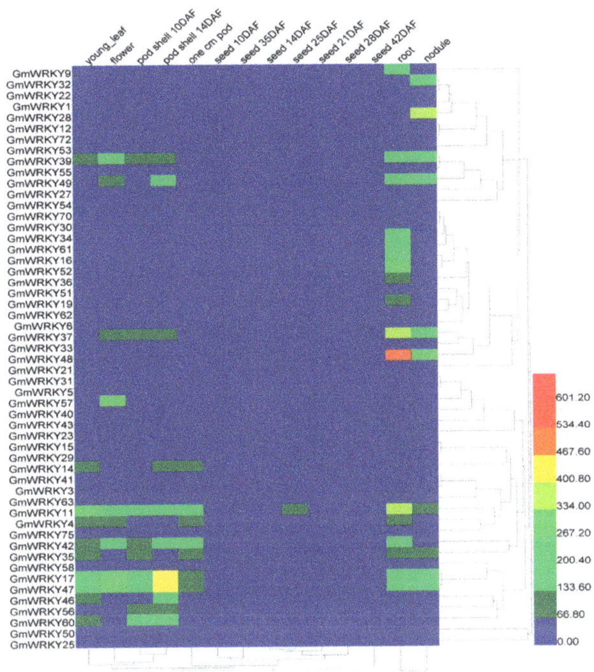

Figure 1. Expression pattern of fifty-three GmWRKYs in six different tissues (young leaf, flower, pod shell, seed, root and nodule). The fifty-three GmWRKYs were selected from drought treatment RNA-Seq data based on the rule that $\log_2$ (GH_treat/CK1_treat) >1. The tissue expression is from SoyDB (http://soykb.org/). The color legend refers to the different expression level under normal condition. "DAF" in the tissue label indicates days after flowering.

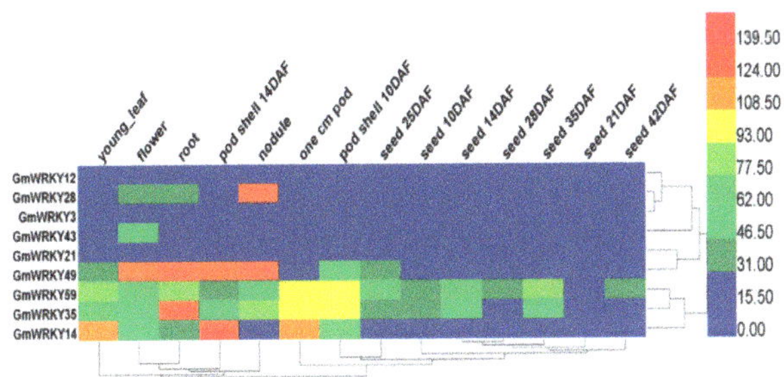

Figure 2. Expression pattern of nine GmWRKYs in six different tissues (young leaf, flower, pod shell, seed, root and nodule). The nine GmWRKYs were selected from salt treatment RNA-Seq data based on the rule that $\log_2$ (NaCl_treat/CK2_treat) >1. The tissue expression is from SoyDB (http://soykb.org/). The color legend refers to the different expression level under normal condition. "DAF" in the tissue label indicates days after flowering.

2.3. *GmWRKYs Responsive to Both Drought and Salt Treatments*

Based on RNA-Seq data and result of Venn method [62], seven *GmWRKY* genes were found to respond to both drought and salt treatments (*GmWRKY3, 12, 14, 21, 35, 43* and *49*) (Figure S1A). In order to confirm whether the seven GmWRKY genes are responsive to drought and salt, 10-day-old soybean seedlings were subjected to stress treatments. For drought treatment, soybean seedlings were put on filter paper to stimulate drought and then sampled 0.1 g of leaf on different periods (0, 0.5, 1, 2, 5, 8, 12 and 24 h); for salt treatment, the roots of soybean were soaked in 100 mM NaCl solution then sampled 0.1 g of leaf on different periods (0, 0.5, 1, 2, 5, 8, 12 and 24 h), all samples were submerged immediately in liquid nitrogen and stored at −80 °C for RNA extraction then quantitative real-time PCR (qRT-PCR) was conducted. Results confirmed that the seven *GmWRKY* genes were responsive to both treatments (Figure 3). Under drought treatment, the expression levels of *GmWRKY12* and *GmWRKY43* were gradually increased at 0, 0.5, 1, 2, 5, 8, 12 and 24 h. *GmWRKY12* was highly expressed after 12 h of drought treatment. While *GmWRKY14, GmWRKY21* and *GmWRKY35* had a tendency to rise first and then decrease, *GmWRKY49* was highly expressed at 2 h. Under drought conditions, the expression profiles of five *GmWRKY* genes were little changed at 0 to 5 h and then increased significantly at 12 h.

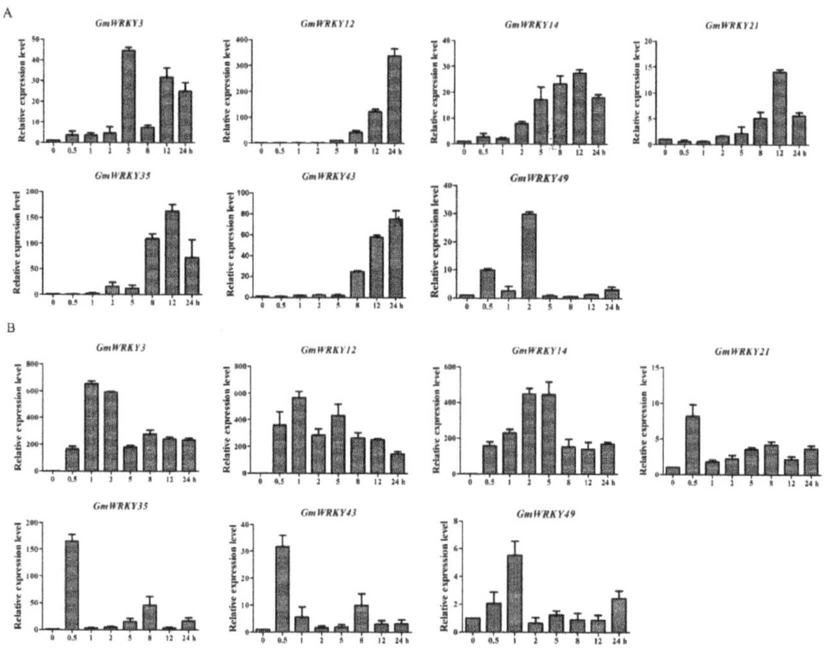

Figure 3. Quantitative RT-PCR of seven GmWRKYs under drought and salt treatment. (**A**) qRT-PCR of seven GmWRKYs under drought treatment. (**B**) qRT-PCR of seven GmWRKYs under salt treatment. The expression level of *GmActin* as a loading control. The data represent means ± SD of three biological replications.

Under salt treatment, the expression profile increased first and then decreased, meanwhile, there was a notable change at 0 to 0.5 h and *GmWRKY3, 12, 14* and *35* were highly expressed. *GmWRKY12*, which was 714 bp in length, encoded 237 amino acids and had low expression in different tissues under normal conditions but was highly expressed under drought and salt treatments was selected for further investigation (Figure S1B).

2.4. Multiple Sequence Alignment and Phylogenetic Analysis of GmWRKY12

Although WRKYGQK sequence is a conservative motif of WRKY proteins, WRKY variant domains, such as WRKYGEK, WRKYGKK, WQKYGQK, WSKYGQK and WRKYGM have been found in the genomes of *Arabidopsis* [28], rice [63], grape [64] and tomato [65]. This difference may be a variation of WRKY TFs developed over long-term evolution. The domain of these variations is unique and may represent a new type. Therefore, to identify conservation of *GmWRKY12*, *WRKY12* from 20 different species were selected for multiple sequence alignment (Figure 4A). Results showed that 20 species only harbored one WRKY variant WRKYGQK, with amino acid sequence similarity of 75%, which illustrated that *GmWRKY12* was highly conserved. To further evaluate the evolutionary relationship between *GmWRKY12* and *WRKY12* of 32 different species, a phylogenetic tree was constructed with the neighbor-joining method [66]. Phylogenetic results showed that the relationship between *GmWRKY12* and *VrWRKY12* (XP_014515898.1) was the closest (Figure 4B).

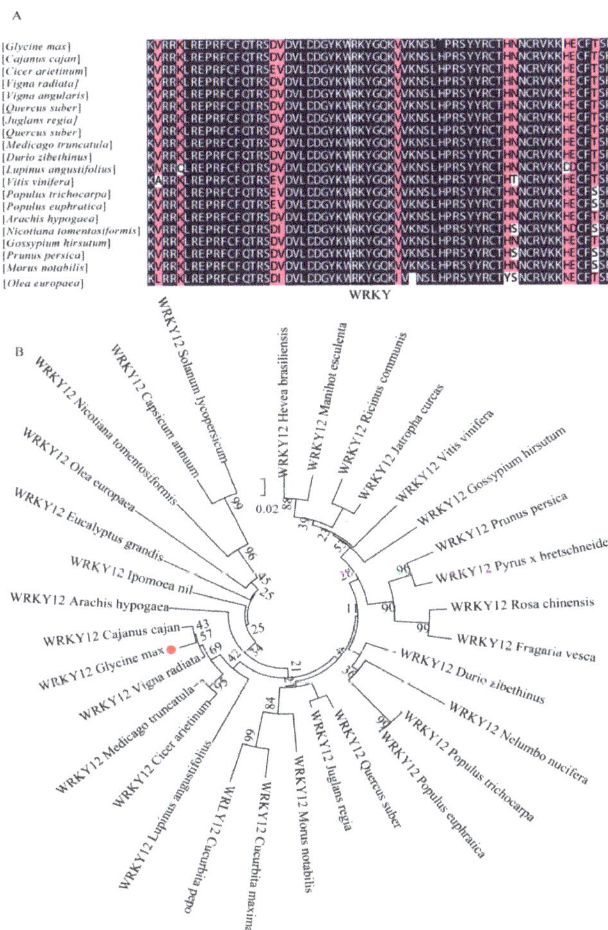

Figure 4. Multiple alignment and phylogenetic relationship of GmWRKY12 with different species. (**A**) Multiple alignment of GmWRKY12 with other WRKY12 proteins from other species. (**B**) Phylogenetic relationship of *GmWRKY12* in different species. The red dot in (**B**) means GmWRKY12. The number of nodes is the bootstrap value and the number on the branch is the evolutionary distance. Bootstrap replications are 1000.

2.5. Expression Patterns of *GmWRKY12* under Different Treatments

GmWRKY12 was responsive to drought and salt treatments (Figure 3). WRKY proteins are reported to be involved in signal transductions of plant hormones [39]. In order to identify whether *GmWRKY12* was responsive to other abiotic stresses, expression patterns were identified using qRT-PCR. Results indicated that *GmWRKY12* not only participated in drought and salt response but was also responsive to ABA and SA. Under low concentrations of SA, the expression profile of *GmWRKY12* was increased about 50-fold (Figure 5).

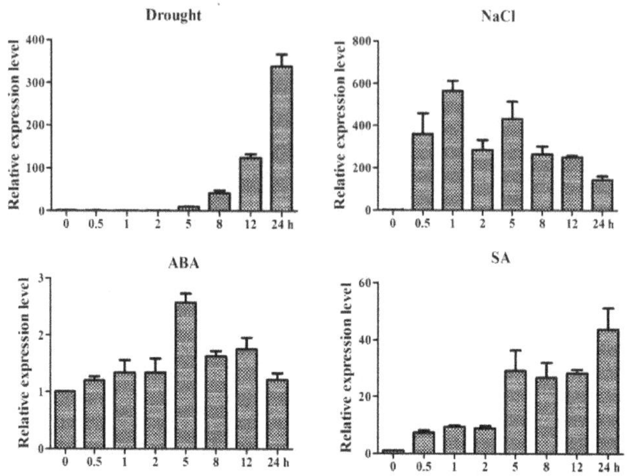

Figure 5. Expression patterns of *GmWRKY12* under drought, NaCl, exogenous ABA and SA. The ordinates are the relative expression level (fold) of *GmWRKY12* compared to non-stressed control. The horizontal ordinate is treatment time for 0, 0.5, 1, 2, 5, 8, 12 and 24 h. The expression level of *GmActin* as a loading control. All experiments were repeated three times. Error bars represent standard deviations (SDs). All data represent the means ± SDs of three independent biological replicates.

2.6. Cis-Acting Elements in Promoter

To further understand the regulatory mechanism of *GmWRKY12*, we isolated its promoter region. *Cis*-elements correlated to stress were present in the promoter region, including the ABA and wound responsive elements ABER4 and MYC, drought responsive element MYB, salt stress responsive element GT-1 and wound responsive element W-box. In addition, there was another element that participated in heat and GA response in the promoter region of *GmWRKY12* (Table 3). This analysis suggested that *GmWRKY12* may function in abiotic stress response.

Table 3. *Cis*-elements analysis of *GmWRKY12* promotor.

Cis-Elements	Numbers	Target Sequences	Functions
MYC	32	CANNTG	ABA and wound responsive element
W-box	21	TTGAC/TTTGACY/TGACY	SA responsive element
ABER4	18	ACGT	ABA responsive element
MYB	14	C/TAACNA/G	Drought responsive element
CCAATB	10	CCAAT	Heat-responsive element
GT-1	7	GAAAAA	Salt stress responsive element
DPBF	6	ACACNNG	Dehydration-responsive element
GARE	2	TAACAAR	GA-responsive element

"Numbers" corresponds to the number of *cis*-elements of each type present in the promoter.

2.7. GmWRKY12 was Located in the Nucleus

To investigate GmWRKY12 subcellular localization, GmWRKY12 were fused to the N-terminus of the humanized green fluorescent protein (hGFP) and co-transformed into wheat mesophyll protoplasts with the nucleus marker AT2G03340 (AtWRKY3)-mCherry [67,68]. The 35S::GFP vector was transformed as the control. Fluorescence of GmWRKY12 was specifically detected in the nucleus, whereas GFP fluorescence was distributed throughout the cell (Figure 6).

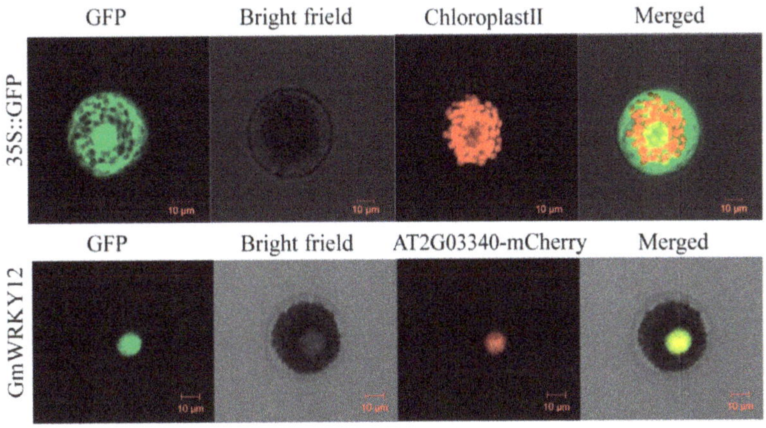

Figure 6. Co-localization of GmWRKY12. The recombinant plasmid of GmWRKY12-GFP and At2G03340-mCherry were co-transformed into wheat mesophyll protoplasts under the control of the CaMV 35S promoter. GmWRKY12 was localized in the nucleus of wheat mesophyll protoplasts protoplasts. Results were observed by a confocal laser scanning microscope (LSM700; CarlZeiss, Oberkochen Germany) after incubating in darkness at 22 °C for 18–20 h. Scale bars = 10 μm.

2.8. GmWRKY12 Improved Drought and Salt Tolerance of Soybean

We further used transgenic hairy root assays to investigate the roles of *GmWRKY12* in abiotic stress responses. Amplified cDNA sequence of *GmWRKY12* was constructed into pCAMBIA3301 to create an overexpression transgenic line and the control was pCAMBIA3301 plant expression vector with CaMV35S promoter. Two constructs were transferred into *Agrobacterium rhizogenes* strain K599 (NCPPB2659) [69] then transformed into soybean hairy roots as previously described [70,71]. After drought treatment for 20 days, both control and over-expression soybean seedlings had leaf shedding to different degrees, especially the old leaves of the plants (Figure 7A). However, compared with transgenic soybean seedlings, the control seedlings were severely wilted and almost 99% of the leaves had serious dehydration and drying. By contrast, there was slight shedding of the old leaves of transgenic soybean seedlings but the new leaves were still growing vigorously. Results of proline and malondialdehyde (MDA) content determination showed that overexpression of *GmWRKY12* increased proline content in transgenic lines, while the MDA content was decreased due to drought stress (Figure 7B,C). Fresh weight and main length of transgenic soybean hair roots under drought treatment were measured (Figure 8E,F), results showed overexpressed *GmWRKY12* in soybean roots enhanced drought tolerance of soybean by increasing the length of transgenic hair roots and the number of transgenic hair roots.

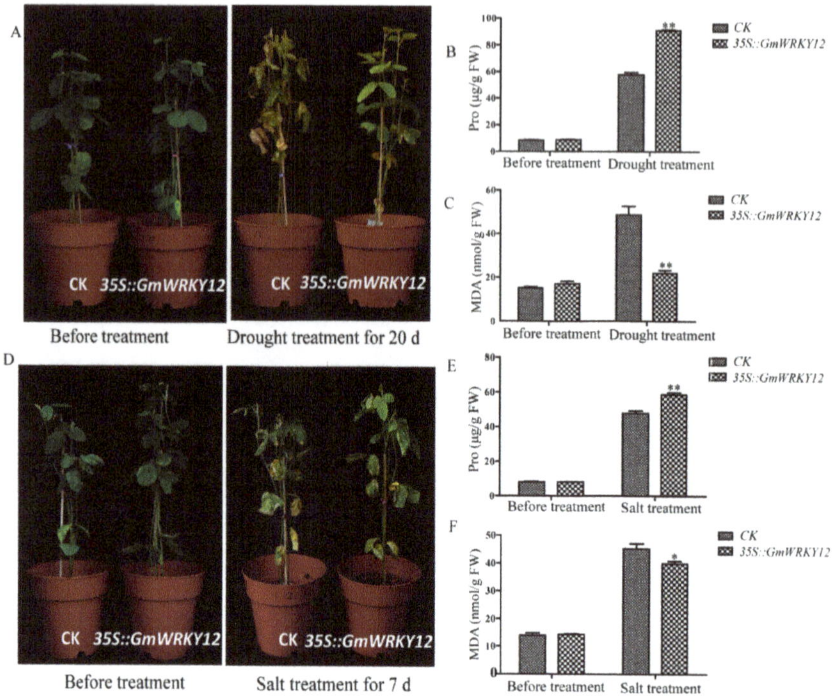

Figure 7. Phenotype identification of *GmWRKY12* under drought and salt treatments. (**A**) Images of drought stress resistance phenotypes of CK and *35S::GmWRKY12* soybean seedlings after drought treatment for 20 days. (**B**) Proline contents in CK and *35S::GmWRKY12* soybean seedlings under normal growth conditions and drought treatment. (**C**) MDA contents in in CK and *35S::GmWRKY12* soybean seedlings under normal growth conditions and drought treatment. (**D**) Images of salt stress resistance phenotypes of CK and *35S::GmWRKY12* soybean seedlings after 200 mM NaCl treatment for 7 days. (**E**) Proline contents in CK and *35S::GmWRKY12* soybean seedlings under normal growth conditions and salt treatment. (**F**) MDA contents in CK and *35S::GmWRKY12* soybean seedlings under normal growth conditions and salt treatment. All data represent the means ± SDs of three independent biological replicates. ANOVA tests demonstrated that there were significant differences (* $p < 0.05$, ** $p < 0.01$).

Meanwhile, under NaCl (200 mM) treatment, control and overexpression soybean seedlings had different degrees of leaf shedding (Figure 7D). Compared with the control, transgenic soybean seedlings were slightly wilted and slowly drying out, while the control seedlings were almost dry due to the osmotic stress. Results of Pro and MDA content in transgenic lines (Figure 7E,F) fresh weight and main length of transgenic soybean hair roots (Figure 8H,I) also showed that *GmWRKY12* improved salt tolerance of soybean. These results demonstrated that *GmWRKY12* confers stress tolerance in transgenic hairy roots.

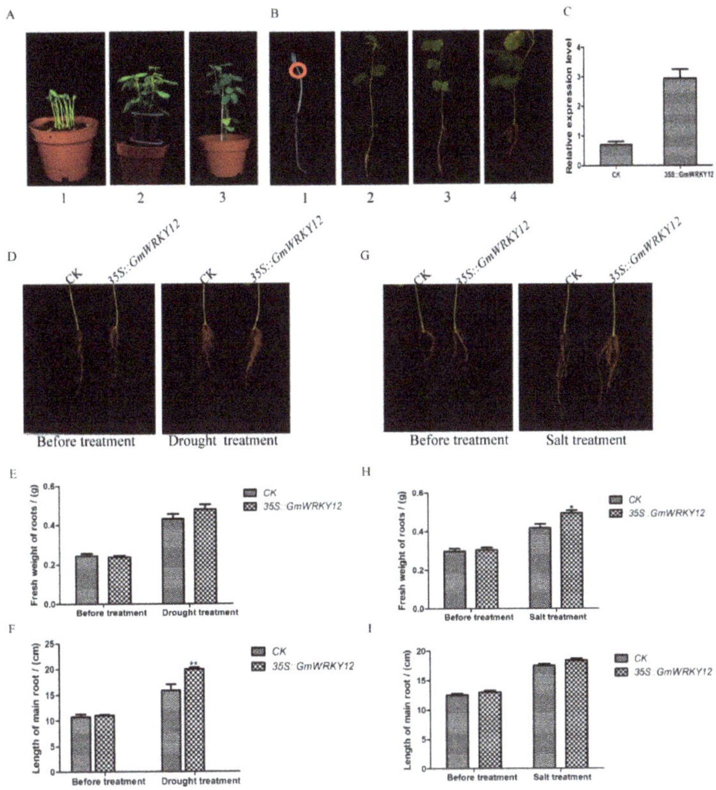

Figure 8. Different growth stage of transgenic soybean seedlings and phenotypes of transgenic soybean hair roots. (**A**) Images of different growth stage of transgenic soybean seedlings cultivated in flowerpot before any treatment. (**A1**) Soybean seedlings of 5-days-old without injected *A. rhizogenes* carrying *GmWRKY12*. (**A2**) Soybean seedlings which have injected *A. rhizogenes* carrying *GmWRKY12* for 7 days. (**A3**) Soybean seedlings which have injected *A. rhizogenes* carrying *GmWRKY12* for 14 days (The original main roots were removed by cutting from 1 cm below the infection site and the hairy roots of the seedlings were cultivated in nutritious soil with full water and grown with 16 h light (100 μM photons $m^{-2} \cdot s^{-1}$)/8 h dark at 25 °C). (**B**) Images of different growth stage of signal transgenic soybean seedling before any treatment. (**B1**) Soybean seedling of 5-days-old without injected *A. rhizogenes* carrying *GmWRKY12* and the red circle shows the inject site of *A. rhizogenes*. (**B2**) Soybean seedling which have injected *A. rhizogenes* carrying *GmWRKY12* for 7 days and new hair roots have generated. (**B3**) Soybean seedling which have injected *A. rhizogenes* carrying *GmWRKY12* for 14 days. (**B4**) Soybean seedling that have salt treatment for 7days. (**C**) Relative expression of CK and *35S::GmWRKY12* transgenic soybean hair roots under normal growth conditions. (**D**) Images of drought stress resistance phenotypes of CK and *35S::GmWRKY12* transgenic soybean hair roots after drought treatment for 20 days. (**E**) Fresh weight in CK and *35S::GmWRKY12* transgenic soybean hair roots under normal growth conditions and drought treatment. (**F**) Length in CK and *35S::GmWRKY12* transgenic soybean hair roots under normal growth conditions and drought treatment. (**G**) Images of salt stress resistance phenotypes of CK and *35S::GmWRKY12* transgenic soybean hair roots after 200 mM NaCl treatment for 7 days. (**H**) Fresh weight in CK and *35S::GmWRKY12* transgenic soybean hair roots under normal growth conditions and salt treatment. (**I**) Length in CK and *35S::GmWRKY12* transgenic soybean hair roots under normal growth condition and salt treatment. All data represent the means ± SDs of three independent biological replicates. ANOVA tests demonstrated that there were significant differences (* $p < 0.05$, ** $p < 0.01$).

3. Discussion

The WRKY transcription factor superfamily, as a recently described member of the TF family, has been studied by many researchers due to its numerous and diverse biological functions. Since the first reports of WRKY TFs [72], research conducted in different species [4,52,57,73,74] has shown that WRKY TFs play significant roles in plant development and stress responses. Recently, many studies of GmWRKY TFs have been based on biotechnological and RNA-Seq approaches [30,57]. However, these studies mainly reported genome-wide annotation of the WRKYs and structure analysis of some genes involved in response to abiotic and biotic stresses. Although these genes have been identified through biochemistry and bioinformatics approaches, knowledge about soybean stress tolerance was limited. In this study, based on qRT-PCR and RNA-Seq data, *GmWRKY12* was selected for investigation of stress tolerance in soybean (Figure S1).

According to classifications in the WRKY family [18,28,75], WRKY12 belongs to Group IIc and contains a single WRKY domain and a CX4-5CX22-23HXH zinc-finger motif. Recent studies have shown that the WRKYGQK heptapeptide, which can specifically recognize and bind to the W-box consensus sequence (TTGACY) in the promoters of target genes, can be replaced by WRKYGKK, WRKYGEK, WKKYEDK, or WKKYCEDK; variations of the WRKYGQK motif might change the DNA binding specificities to downstream target genes [75]. However, multiple sequence alignment results showed that WRKY12 in different species only harbor the same WRKYGQK heptapeptide, demonstrating that WRKY12 protein is evolutionarily conserved and can recognize and bind to downstream target genes (Figure 4A). The result was consistent with the results observed in other species [54,57,65,76,77]. Structural conservation determines functional specificity: in rice, *OsWRKY12* was related to normal plant growth and expression of *OsWRKY12* was low at the seedling stage but increased gradually with growth [78]; similar results were found in specific tissues in our study. *GmWRKY12* has low expression in young leaf, flower, one cm pod, pod shell 10 DAF, seed 10 DAF, seed 14 DAF, seed 21 DAF, seed 25 DAF, seed 28 DAF, seed 35 DAF, seed 42 DAF and root under normal conditions. At the pod shell 14 DAF and nodule stages, the expression levels gradually increase (Table S2), which may be because genes are differentially expressed at different growth stages, or may perform different activities, such as metabolism, nutrient absorption or material transformation. For example, at the nodule stage, plants are primarily vegetative, while at seed 42 DAF, plants are accumulating nutrients [57]. In addition, WRKY12 was related to plant flowering time: *Arabidopsis* plants overexpressing *MlWRKY12* showed early flowering phenotype [79]. WRKY12 and WRKY13 have opposite effects on flowering time in the action of GA [80]. Overexpression of three *Triticum* genes, *TaWRKY12*, *TaWRKY18* and *TaZFP2* induced the expression of some genes related to Pi absorption and transportation, enhancing the abilities of Pi uptake and Pi use efficiency in plants under low-Pi stress conditions [81]. Thus, *GmWRKY12*, like other WRKYs, is involved in plant growth and development.

There are many *cis*-acting elements in the *GmWRKY12* promoter region, such as MYC (ABA and wound responsive element), W-box (SA responsive element), ABER4 (ABA responsive element), MYB (drought responsive element), CCAATB (heat-responsive element), GT-1 (salt stress responsive element), DPBF (dehydration-responsive element) and GARE (GA-responsive element) (Table 3). The presence of these elements indicates that *GmWRKY12* may take part in various biotic and abiotic responses except for growth and development of plants. Research of tobacco transcription factors *NtWRKY12* and *TGA2.2* found that *NtWRKY12* alone was able to induce PR-1a::GUS expression to high levels, the PR-1a gene was salicylic acid-inducible to activate the expression of SA-inducible genes [82]. SA is an important endogenous molecule that activates plant hypersensitive response and systemic acquired resistance, which are often involved in disease resistance of plants [83]. As the closest orthologue of *AtWRKY12*, *BrWRKY12* from Chinese cabbage conferred enhanced resistance to *Pectobacterium carotovorum ssp. carotovorum* (*Pcc*) through transcriptional activation of defense-related genes [84]. Furthermore, *LrWRKY12* were responsive to SA and methyl jasmonate (MeJA) treatments and conferred more resistance to *B. cinerea* than in wild-type plants [85]. These results show that

WRKY12 plays an important role in disease defense of plants, mainly because WRKYGQK specifically binds to the W-box to induce expression of downstream target genes.

In addition to the significant roles of WRKY12 identified in development and disease defense of plants, WRKY12 also functions in plant stress responses. Under treatment with NaCl and PEG, the expression level of *THWRKY12* in *Tamarix* tissues was increased, the expression pattern of *THWRKY12* after ABA treatment was approximately the same as the expression level changes under NaCl and PEG treatment, showing that the gene may participate in regulating salt and drought tolerance through the signaling pathway regulated by ABA [86]. In our study, *GmWRKY12* was first screened following both drought and salt treatment using RNA-Seq. In order to confirm whether it was responsive to salt and drought stress, qRT-PCR was conducted and further showed that *GmWRKY12* was highly expressed under drought and salt treatment, which indicated that the gene was related to drought and salt tolerance (Figure 3). *Cis*-acting elements and expression pattern analysis of *GmWRKY12* also showed that it may participate in the ABA signaling pathway (Table 3 and Figure 5). However, compared to the high expression level under drought and salt treatment, on the condition of ABA, *GmWRKY12* had low expression. Resistance identification of *GmWRKY12* using a soybean hairy root assay further showed that *GmWRKY12* may be involved in regulating salt and drought tolerance by promoting the combination of *cis*-acting elements with drought and salt-related genes, thereby enhancing plant resistance (Figure 7). Similar results were also found in other studies [87–89].

4. Materials and Methods

4.1. Identification and Annotation of GmWRKYs Response to Drought/Salt Stress

Identification of the response of GmWRKYs to drought/salt stress was based on RNA-seq data collected from a set of drought and salt stress experiments (Tables S5 and S6). Seeds of Williams 82 were cultivated in a 10 × 10 cm flowerpot (vermiculite: nutritious soil is 1:3), fresh leaf of 10-day-old soybean seedlings were used for RNA-Seq.CK1_treat-Expression represented two independent replicates of plants sampled before any treatment; GH_treat-Expression related to drought treatment for 5 h (put on the filter paper to simulate drought) of soybean plants at room temperature; CK2_treat-Expression without NaCl treatment; and NaCl_treat-Expression salt treatment that soaking soybean roots with 100 mM NaCl solution for 1 h and then sampled for RNA-seq [57,68]. Both $\log_2$ (GH_treat/CK1_treat) >1, $\log_2$ (NaCl_treat/CK2_treat) >1 and up-regulated were treated as the rule to select GmWRKYs responding to drought/salt stress. Several databases: NCBI (https://www.ncbi.nlm.nih.gov/pubmed), PlantTFDB (http://planttfdb.cbi.pku.edu.cn/), Phytozome (https://phytozome.jgi.doe.gov/pz/portal.html) and SoyDB (http://soykb.org/), were used to annotate Gene ID, Name, Chromosomal localization, CDS, Protein and Group.

4.2. Tissue-Specific Expression Patterns of GmWRKYs

Data of six different tissues (young leaf, flower, pod shell, seed, root and nodule) from different growth periods was available from SoyBase (https://www.soybase.org/soyseq/). Heml1.0 software (http://www.patrick-wied.at/static/heatmapjs/) was used to perform hierarchical clustering of fifty-three and nine GmWRKYs under normal conditions. The analysis data are available in Tables S1 and S2.

4.3. RNA Extraction and qRT-PCR

Seeds of Williams 82 was cultivated in a 10 × 10 cm flowerpot (vermiculite: nutritious soil is 1:3), fresh leaf tissue of 10-day-old soybean seedlings were used for RNA extraction of different stress treatment. For drought treatment, soybean seedlings were dried on filter paper then sampled 0.1 g of leaf on different periods (0, 0.5, 1, 2, 5, 8, 12 and 24 h), for salt, ABA and SA treatment, the roots of soybean seedlings were soaked in 100 mM NaCl, 100 μmol·L^{-1} ABA and 100 μmol·L^{-1} SA solution, respectively [68]. Then sampled 0.1 g of leaf on different periods (0, 0.5, 1, 2, 5, 8, 12 and 24 h), all samples were submerged immediately in liquid nitrogen and stored at −80 °C for RNA extraction using RNA prep plant kit (TIANGEN, Beijing, China); cDNA was synthesized using a Prime Script First-Strand cDNA Synthesis Kit (TransGen, Beijing, China) following the manufacturer's instructions. cDNA of treatment for 0 h was used for screen one highly expressed gene from seven GmWRKYs that response to both drought and salt treatment (Figure S1B). qRT-PCR was performed with Super Real PreMix Plus (TransGen, Beijing, China) on an ABI Prism 7500 system (Applied Biosystems, Foster City, CA, USA). Specific primers of *GmWRKY3, 12, 14, 21, 28, 35, 43, 49 and* soybean actin primers are listed in Table S4. Three biological replicates were used for qRT-PCR analysis. The $2^{-\Delta\Delta Ct}$ method was used for quantification.

4.4. Gene Isolation and Phylogenetic Analysis of GmWRKY12

Venn2.0 (http://bioinfogp.cnb.csic.es/tools/venny/index.html) was used to screen GmWRKYs that respond to both drought and salt treatment, then qRT-PCR was used to find genes highly expressed under stresses. Full-length *GmWRKY12* was amplified by PCR with specific primers from soybean cDNA (*Williams 82*); primers of *GmWRKY12* are available in Table S4. PCR products were cloned into pLB vector (TIANGEN, Beijing, China) and sequenced for further study. The amino acid sequence of WRKY12 in different species were searched for in the NCBI database on account of the amino acid similarity between GmWRKY12 and WRKY12 in different species is more than 50%. DNAMAN was applied for multiple sequence alignment on the basis of the amino acid similarity between GmWRKY12 and WRKY12 in different species is more than 60%. Phylogenetic trees were constructed using MEGA 6.0 with the neighbor-joining method [66] and 1000 bootstrap replications. Information of WRKY12 in different species is listed in Table S3.

4.5. Co-Localization of GmWRKY

Seeds of Kenong199 were cultivated in a 10 × 10 cm flowerpot (vermiculite: nutritious soil is 1:3), fresh leaf tissue of 7-day-old wheat seedlings were used for preparation of wheat protoplasts. Amplified cDNA sequence of *GmWRKY12* was cloned into the N-terminus hGFP protein driven by the CaMV35S promoter. The cDNA coding sequences of AT2G03340 (AtWRKY3) which located in the nucleus [67] were fused to the N-terminus of the mCherry protein (WRKY25-RFP) under the control of the CaMV 35S promoter [68]. The recombinant plasmid of GmWRKY12-GFP and AtWRKY3-mCherry were co-transformed into wheat mesophyll protoplasts via the PEG4000-mediated method. The 35S::GFP vector was transformed as the control. Fluorescence was observed using a confocal laser scanning microscope (LSM700; CarlZeiss, Oberkochen, Germany) after incubating in darkness at 22 °C for 18–20 h. Primers are available in Table S4.

4.6. Cis-acting Elements in Promoter

The 2.0 kb promoter region upstream of the ATG start codon in the promoter of *GmWRKY12* was obtained from soybean genomic DNA in the Ensembl Plants website, *cis*-acting elements were analyzed by PLACE (http://www.dna.affrc.go.jp/PLACE/).

4.7. A. rhizogenes-mediated Drought and Salt Stress Assays

To generate a transgenic line of soybean the amplified cDNA sequence of *GmWRKY12* was constructed into pCAMBIA3301 for an overexpression transgenic line (*35S::GmWRLY12*) and the control was pCAMBIA3301 plant vector with CaMV35S promoter (CK) and two constructs transferred into *A. rhizogenes* strain K599 (NCPPB2659) [69]. Primers are available in Table S4. Williams 82 was cultivated in a 10 × 10 cm flowerpot (vermiculite: nutritious soil is 1:3) for stress experiments (Figure 8A1), soybean seeds were grown under a photoperiod of 16 h light (100 μM photons $m^{-2} \cdot s^{-1}$)/8 h dark at 25 °C. When plants displayed two cotyledons (Figure 8A1), *A. rhizogenes* strain K599 harboring pCAMBIA3301 (CK) and K599 harboring *35S::GmWRLY12* were injected at the cotyledonary node and/or hypocotyl (Figure 8B1). A plastic cup was used to surround the inoculated soybean seedlings to provide high humidity conditions. After 3 days, nutritious soil was prepared and used to fill the gaps in the plastic cup so that soybean seedlings could grow new roots (Figure 8A2); plants typically need two weeks to generate new roots (2–10 cm) at the infection site (Figure 8B2,B3). The original main roots were removed by cutting from 1 cm below the infection site and the hairy roots of the seedlings were cultivated in nutritious soil with full water and grown with 16 h light (100 μM photons $m^{-2}·s^{-1}$)/8 h dark at 25 °C for 5 days [70,71]. Each flowerpot cultivated 5 transgenic soybean seedlings and 5 replications of each stress treatment (Figure 8A3). Afterward, the transgenic soybean seedlings were subjected to natural dehydration and 200 mM NaCl for drought and salt stress assays [19,68]. For drought stress assay, both CK and transgenic soybean seedlings were grown without water for 20 days. For salt stress assay, CK and transgenic soybean seedlings were treated with 200 mM NaCl solution for 7 days. There are some Supplement Materials need to prepare for culturing *A. rhizogenes* strain K599 that harbored (*35S::GmWRLY12*) and the control (CK)., eg: Solidified LB medium with streptomycin sulfate (100 mg/L) and Kanamycin solution(100 mg/L) (10 g tryptone, 5 g yeast extract, 10 g NaCl, 15 g agar per liter), Liquid LB medium containing streptomycin sulfate (100 mg/L) and Kanamycin solution (100 mg/L) [69].

4.8. Measurements of Proline and MDA Contents

Both proline and MDA content were measured with the Pro and MDA assay kit (Comin, Beijing, China) based on the manufacturer's protocols; all measurements were from three biological replicates.

4.9. Measurements of Fresh Weight and Main Length

Transgenic soybean hair roots were used to measure the fresh weight and main length. All data represent the means ± SDs of three independent biological replicates.

5. Conclusions

In this study, using RNA-Seq, we identified 62 *GmWRKY* genes in the soybean genome that were differently expressed in six different tissues under normal condition. Seven GmWRKYs responded to both drought and salt treatment. Based on the qRT-PCR, GmWRKY12, a nucleus protein of 237 amino acids, belonging to WRKY Group II was identified. It was responsive to salt, drought and exogenous hormones ABA and SA. Results of Agrobacterium rhizogenes-mediated hairy roots assay showed that overexpressing *GmWRKY12* may improve tolerance to drought and salt in soybean. These results provided new insight into the roles of soybean WRKY genes in abiotic stress responses.

Supplementary Materials: Supplementary materials can be found at http://www.mdpi.com/1422-0067/19/12/4087/s1.

Author Contributions: Z.-S.X. coordinated the project, conceived and designed experiments and edited the manuscript; W.-Y.S. performed the experiments and wrote the first draft of the manuscript; Y.-T.D., J.M. and J.C. conducted the bioinformatic work and performed related experiments; D.-H.M. and L.-G.J. provided analytical tools and analyzed the data; M.C., Y.-B.Z. and X.-H.Z. contributed valuable discussion; and Y.-Z.M. coordinated the project. All authors have read and approved the final manuscript.

Funding: This research was financially supported by the National Key R & D Program of China (2016YFD0102000), the National Transgenic Key Project of the Ministry of Agriculture (2018ZX0800909B and 2016ZX08002-002), the National Natural Science Foundation of China (31871624) and the Major Research Development Program of Shaanxi Province (2018NY-026).

Acknowledgments: We are grateful to Hui Zhang (Institute of Crop Science, Chinese Academy of Agricultural Sciences) for providing the pCAMBIA3301 vector and pUC57-GmU6-sgRNA.

Conflicts of Interest: The authors declare that they have no competing interests.

Abbreviations

ABA	abscisic acid
ABRE	ABA-responsive element
JA	jasmonic acid
SA	salicylic acid
MeJA	methyl jasmonate
qRT-PCR	quantitative real-time PCR
Pro	proline
MDA	malondialdehyde
DAF	days after flowering
GFP	green fluorescent protein

References

1. Peng, X.; Ma, X.; Fan, W.; Man, S.; Cheng, L.; Alam, I.; Lee, B.H.; Qi, D.; Shen, S.; Liu, G. Improved drought and salt tolerance of *Arabidopsis thaliana* by transgenic expression of a novel DREB gene from *Leymus chinensis*. *Plant Cell Rep.* **2011**, *30*, 1493–1502.
2. Hong, C.; Cheng, D.; Zhang, G.; Zhu, D.; Chen, Y.; Tan, M. The role of *ZmWRKY4* in regulating maize antioxidant defense under cadmium stress. *Biochem. Bioph. Res. Commun.* **2016**, *482*, 1504–1510. [CrossRef]
3. Fu, J.; Liu, Q.; Wang, C.; Liang, J.; Liu, L.; Wang, Q. *ZmWRKY79* positively regulates maize phytoalexin biosynthetic gene expression and is involved in stress response. *J. Exp. Bot.* **2017**, *69*, 497–510. [CrossRef]
4. Liu, Q.L.; Xu, K.D.; Pan, Y.Z.; Jiang, B.B.; Liu, G.L.; Jia, Y.; Zhang, H.Q. Functional Analysis of a Novel Chrysanthemum WRKY Transcription Factor Gene Involved in Salt Tolerance. *Plant Mol. Biol. Rep.* **2014**, *32*, 282–289. [CrossRef]
5. Zhu, J.K. Cell signaling under salt, water and cold stresses. *Curr. Opin. Plant Biol.* **2001**, *4*, 401–406. [CrossRef]
6. Horie, T.; Hauser, F.; Schroeder, J.I. HKT transporter-mediated salinity resistance mechanisms in *Arabidopsis* and monocot crop plants. *Trends Plant Sci.* **2009**, *14*, 660–668. [CrossRef]
7. Takahashi, R.; Liu, S.; Takano, T. Cloning and functional comparison of a high-affinity K+ transporter gene *PhaHKT1* of salt-tolerant and salt-sensitive reed plants. *J. Exp. Bot.* **2007**, *58*, 4387–4395. [CrossRef] [PubMed]
8. Guan, R.; Qu, Y.; Guo, Y.; Yu, L.; Liu, Y.; Jiang, J.; Chen, J.; Ren, Y.; Liu, G.; Tian, L. Salinity tolerance in soybean is modulated by natural variation in *GmSALT3*. *Plant J.* **2015**, *80*, 937–950. [CrossRef]
9. Yokoi, S.; Quintero, F.J.; Cubero, B.; Ruiz, M.T.; Bressan, R.A.; Hasegawa, P.M.; Pardo, J.M. Differential expression and function of *Arabidopsis thaliana* NHX Na^+/H^+ antiporters in the salt stress response. *Plant J.* **2010**, *30*, 529–539. [CrossRef]
10. Rodríguez-Rosales, M.P.; Gálvez, F.J.; Huertas, R.; Aranda, M.N.; Baghour, M.; Cagnac, O.; Venema, K. Plant NHX cation/proton antiporters. *Plant Signal. Behav.* **2009**, *4*, 265–276. [CrossRef]
11. Shigaki, T.; Hirschi, K. Characterization of CAX-like genes in plants: Implications for functional diversity. *Gene* **2000**, *257*, 291–298. [CrossRef]
12. Shigaki, T.; Hirschi, K.D. Diverse functions and molecular properties emerging for CAX cation/H^+ exchangers in plants. *Plant Biol.* **2006**, *8*, 419–429. [CrossRef] [PubMed]
13. Qi, X.; Li, M.W.; Xie, M.; Liu, X.; Ni, M.; Shao, G.; Song, C.; Kay-Yuen, Y.A.; Tao, Y.; Wong, F.L. Identification of a novel salt tolerance gene in wild soybean by whole-genome sequencing. *Nat. Commun.* **2014**, *5*. [CrossRef] [PubMed]
14. Wang, J.W.; Yang, F.P.; Chen, X.Q.; Liang, R.Q.; Zhang, L.Q.; Geng, D.M.; Zhang, X.D.; Song, Y.Z.; Zhang, G.S. Induced expression of DREB transcriptional factor and study on its physiological effects of drought tolerance in transgenic wheat. *Acta Genetica Sinica* **2006**, *33*, 468–476. [CrossRef]
15. Morran, S.; Eini, O.; Pyvovarenko, T.; Parent, B.; Singh, R.; Ismagul, A.; Eliby, S.; Shirley, N.; Langridge, P.; Lopato, S. Improvement of stress tolerance of wheat and barley by modulation of expression of DREB/CBF factors. *Plant Biotechnol. J.* **2011**, *9*, 230–249. [CrossRef] [PubMed]
16. Dubos, C.; Stracke, R.; Grotewold, E.; Weisshaar, B.; Martin, C.; Lepiniec, L. MYB transcription factors in *Arabidopsis*. *Trends Plant Sci.* **2010**, *15*, 573–581. [CrossRef] [PubMed]
17. Eulgem, T.; Rushton, P.J.; Robatzek, S.; Somssich, I.E. The WRKY superfamily of plant transcription factors. *Trends Plant Sci.* **2000**, *5*, 199–206. [CrossRef]

18. Eulgem, T.; Somssich, I.E. Networks of WRKY transcription factors in defense signaling. *Curr. Opin. Plant Biol.* **2007**, *10*, 366–371. [CrossRef] [PubMed]
19. Hao, Y.J.; Wei, W.; Song, Q.X.; Chen, H.W.; Zhang, Y.Q.; Wang, F.; Zou, H.F.; Lei, G.; Tian, A.G.; Zhang, W.K. Soybean NAC transcription factors promote abiotic stress tolerance and lateral root formation in transgenic plants. *Plant J.* **2011**, *68*, 302–313. [CrossRef]
20. Jin, H.X.; Huang, F.; Cheng, H.; Song, H.N.; Yu, D.Y. Overexpression of the GmNAC2 Gene, an NAC Transcription Factor, Reduces Abiotic Stress Tolerance in Tobacco. *Plant Mol. Biol. Rep.* **2013**, *31*, 435–442. [CrossRef]
21. Guerinot, M.L. The ZIP family of metal transporters. *BBA Biomembr.* **2000**, *1465*, 190–198. [CrossRef]
22. Liao, Y.; Zou, H.F.; Wei, W.; Hao, Y.J.; Tian, A.G.; Huang, J.; Liu, Y.F.; Zhang, J.S.; Chen, S.Y. Soybean *GmbZIP44*, *GmbZIP62* and *GmbZIP78* genes function as negative regulator of ABA signaling and confer salt and freezing tolerance in transgenic *Arabidopsis*. *Planta* **2008**, *228*, 225–240. [CrossRef]
23. Xu, Z.S.; Chen, M.; Ma, L.C.; Ma, Y.Z. Functions and application of the AP2/ERF franscription factor family in crop improvement. *Bull Bot.* **2011**, *61*, 570–585.
24. Xu, Z.S.; Chen, M.; Li, L.C.; Ma, Y.Z. Functions of the ERF transcription factor family in plants. *Botany* **2008**, *86*, 969–977. [CrossRef]
25. Fukuda, A.; Okada, Y.; Suzui, N.; Fujiwara, T.; Yoneyama, T.; Hayashi, H. Cloning and characterization of the gene for a phloem-specific glutathione S -transferase from rice leaves. *Physiol. Plant.* **2010**, *120*, 595–602. [CrossRef] [PubMed]
26. Hundertmark, M.; Hincha, D.K. LEA (Late Embryogenesis Abundant) proteins and their encoding genes in *Arabidopsis thaliana*. *BMC Genomics* **2008**, *9*. [CrossRef]
27. Yan, J.; Wang, B.; Jiang, Y.; Cheng, L.; Wu, T. GmFNSII-Controlled Soybean Flavone Metabolism Responds to Abiotic Stresses and Regulates Plant Salt Tolerance. *Plant Cell Physiol.* **2014**, *55*, 74–86. [CrossRef]
28. Fei, C.; Yue, H.; Vannozzi, A.; Wu, K.; Cai, H.; Yuan, Q. The WRKY transcription factor family in model plants and crops. *Crit. Rev. Plant Sci.* **2018**, *36*, 1–25.
29. Rushton, P.J.; Somssich, I.E.; Ringler, P.; Shen, Q.J. WRKY transcription factors. *Plant Signal. Behav.* **2010**, *15*, 247–258. [CrossRef]
30. Yu, Y.; Wang, N.; Hu, R.; Xiang, F. Genome-wide identification of soybean WRKY transcription factors in response to salt stress. *Springerplus* **2016**, *5*. [CrossRef]
31. Lagacã, M.; Matton, D.P. Characterization of a WRKY transcription factor expressed in late torpedo-stage embryos of Solanum chacoense. *Planta* **2004**, *219*, 185–189. [CrossRef] [PubMed]
32. Zentella, R.; Zhang, Z.; Park, M.; Thomas, S.; Endo, A.; Murase, K.; Fleet, C.; Jikumaru, Y.; Nambara, E.; Kamiya, Y. Global analysis of della direct targets in early gibberellin signaling in *Arabidopsis*. *Plant Cell* **2007**, *19*, 3037–3057. [CrossRef] [PubMed]
33. Silke, R.; Imre, E.S. Targets of *AtWRKY6* regulation during plant senescence and pathogen defense. *Genes Dev* **2002**, *16*, 1139–1149.
34. Johnson, C.S.; Kolevski, B.; Smyth, D.R. Transparent TESTA glabra2, a trichome and seed coat development gene of *Arabidopsis*, encodes a WRKY transcription factor. *Plant Cell* **2002**, *14*, 1359–1375. [CrossRef] [PubMed]
35. Pandey, S.P.; Somssich, I.E. The role of WRKY transcription factors in plant immunity. *Plant Physiol.* **2009**, *150*, 1648–1655. [CrossRef] [PubMed]
36. Birkenbihl, R.P.; Diezel, C.; Somssich, I.E. *Arabidopsis* WRKY33 is a key transcriptional regulator of hormonal and metabolic responses toward Botrytis cinerea infection. *Plant Physiol.* **2012**, *159*, 266–285. [CrossRef] [PubMed]
37. Duan, Y.; Jiang, Y.; Ye, S.; Karim, A.; Ling, Z.; He, Y.; Yang, S.; Luo, K. *PtrWRKY73*, a salicylic acid-inducible poplar WRKY transcription factor, is involved in disease resistance in *Arabidopsis thaliana*. *Plant Cell Rep.* **2015**, *34*, 831–841. [CrossRef]
38. Qiu, Y.P.; Yu, D.Q. Over-expression of the stress-induced *OsWRKY45* enhances disease resistance and drought tolerance in *Arabidopsis*. *Environ. Exp. Bot.* **2009**, *65*, 35–47. [CrossRef]
39. Zhang, L.; Gu, L.; Ringler, P.; Smith, S.; Rushton, P.J.; Shen, Q.J. Three WRKY transcription factors additively repress abscisic acid and gibberellin signaling in aleurone cells. *Int. J. Exp. Plant Biol.* **2015**, *236*, 214–222. [CrossRef]
40. Zou, X.; Seemann, J.R.; Neuman, D.; Shen, Q.J. A WRKY gene from creosote bush encodes an activator of the abscisic acid signaling pathway. *J. Biol. Chem.* **2004**, *279*, 55770–55779. [CrossRef]

41. Shimono, M.; Sugano, S.; Nakayama, A.; Jiang, C.J.; Ono, K.; Toki, S.; Takatsuji, H. Rice WRKY45 plays a crucial role in benzothiadiazole-inducible blast resistance. *Plant Cell* **2007**, *19*, 2064–2076. [CrossRef] [PubMed]
42. Chen, L.; Song, Y.; Li, S.; Zhang, L.; Zou, C.; Yu, D. The role of WRKY transcription factors in plant abiotic stresses. *BBA Gene Regul. Mech.* **2012**, *1819*, 120–128. [CrossRef] [PubMed]
43. Chen, J.; Nolan, T.; Ye, H.; Zhang, M.; Tong, H.; Xin, P.; Chu, J.; Chu, C.; Li, Z.; Yin, Y. *Arabidopsis WRKY46, WRKY54* and *WRKY70* transcription factors are involved in brassinosteroid-regulated plant growth and drought response. *Plant Cell* **2017**, *29*, 1425–1439. [CrossRef] [PubMed]
44. Erpen, L.; Devi, H.S.; Grosser, J.W.; Dutt, M. Potential use of the DREB/ERF, MYB, NAC and WRKY transcription factors to improve abiotic and biotic stress in transgenic plants. *Plant Cell Tissue Organ* **2018**, *132*, 1–25. [CrossRef]
45. Taji, T.; Ohsumi, C.; Iuchi, S.; Seki, M.; Kasuga, M.; Kobayashi, M.; YamaguchiShinozaki, K.; Shinozaki, K. Important roles of drought- and cold-inducible genes for galactinol synthase in stress tolerance in *Arabidopsis thaliana*. *Plant J.* **2010**, *29*, 417–426. [CrossRef]
46. Wang, C.T.; Ru, J.N.; Liu, Y.W.; Li, M.; Zhao, D.; Yang, J.F.; Fu, J.D.; Xu, Z.S. Maize WRKY Transcription Factor *ZmWRKY106* Confers Drought and Heat Tolerance in Transgenic Plants. *Int. J. Mol. Sci.* **2018**, *19*. [CrossRef] [PubMed]
47. He, G.H.; Xu, J.Y.; Wang, Y.X.; Liu, J.M.; Li, P.S.; Ming, C.; Ma, Y.Z.; Xu, Z.S. Drought-responsive WRKY transcription factor genes *TaWRKY1* and *TaWRKY33* from wheat confer drought and/or heat resistance in *Arabidopsis*. *BMC Plant Biol.* **2016**, *16*. [CrossRef]
48. Wang, C.T.; Ru, J.N.; Liu, Y.W.; Yang, J.F.; Li, M.; Xu, Z.S.; Fu, J.D. The Maize WRKY transcription factor *ZmWRKY40* confers drought resistance in transgenic *Arabidopsis*. *Int. J. Mol. Sci.* **2018**, *19*. [CrossRef]
49. Ulker, B.; Somssich, I.E. WRKY transcription factors: From DNA binding towards biological function. *Curr. Opin. Plant Biol.* **2004**, *7*, 491–498. [CrossRef]
50. Jiang, Y.; Qiu, Y.; Hu, Y.; Yu, D. Heterologous expression of *AtWRKY57* confers drought tolerance in *Oryza sativa*. *Front. Plant Sci.* **2016**, *7*. [CrossRef]
51. Niu, C.F.; Wei, W.; Zhou, Q.Y.; Tian, A.G.; Hao, Y.J.; Zhang, W.K.; Ma, B.; Lin, Q.; Zhang, Z.B.; Zhang, J.S. Wheat WRKY genes *TaWRKY2* and *TaWRKY19* regulate abiotic stress tolerance in transgenic *Arabidopsis* plants. *Plant Cell Environ.* **2012**, *35*, 1156–1170. [CrossRef] [PubMed]
52. Wu, X.; Shiroto, Y.; Kishitani, S.; Ito, Y.; Toriyama, K. Enhanced heat and drought tolerance in transgenic rice seedlings overexpressing OsWRKY11 under the control of HSP101 promoter. *Plant Cell Rep.* **2009**, *28*, 21–30. [CrossRef] [PubMed]
53. Lee, H.; Cha, J.; Choi, C.; Choi, N.; Ji, H.S.; Park, S.R.; Lee, S.; Hwang, D.J. Rice *WRKY11* plays a role in pathogen defense and drought tolerance. *Rice* **2018**, *11*. [CrossRef] [PubMed]
54. Cai, R.; Zhao, Y.; Wang, Y.; Lin, Y.; Peng, X.; Li, Q.; Chang, Y.; Jiang, H.; Xiang, Y.; Cheng, B. Overexpression of a maize *WRKY58* gene enhances drought and salt tolerance in transgenic rice. *Plant Cell Tissue Organ* **2014**, *119*, 565–577. [CrossRef]
55. Li, H.; Gao, Y.; Xu, H.; Dai, Y.; Deng, D.; Chen, J. *ZmWRKY33*, a WRKY maize transcription factor conferring enhanced salt stress tolerances in *Arabidopsis*. *Plant Growth Regul.* **2013**, *70*, 207–216. [CrossRef]
56. Ullah, A.; Sun, H.; Hakim; Yang, X.; Zhang, X. A novel cotton WRKY-gene, *GhWRKY6*-like, improves salt tolerance by activating the ABA signalling pathway and scavenging of reactive oxygen species. *Physiol. Plant* **2017**, *162*, 439–454. [CrossRef] [PubMed]
57. Song, H.; Wang, P.; Hou, L.; Zhao, S.; Zhao, C.; Xia, H.; Li, P.; Zhang, Y.; Bian, X.; Wang, X. Global analysis of WRKY genes and their response to dehydration and salt stress in soybean. *Front. Plant Sci.* **2016**, *7*. [CrossRef]
58. Zhou, Q.Y.; Tian, A.G.; Zou, H.F.; Xie, Z.M.; Lei, G.; Huang, J.; Wang, C.M.; Wang, H.W.; Zhang, J.S.; Chen, S.Y. Soybean WRKY-type transcription factor genes, *GmWRKY13*, *GmWRKY21* and *GmWRKY54*, confer differential tolerance to abiotic stresses in transgenic *Arabidopsis* plants. *Plant Biotechnol. J.* **2010**, *6*, 486–503. [CrossRef]
59. Li, J.; Wang, J.; Wang, N.; Guo, X.; Gao, Z. *GhWRKY44*, a WRKY transcription factor of cotton, mediates defense responses to pathogen infection in transgenic *Nicotiana benthamiana*. *Plant Cell Tissue Organ* **2015**, *121*, 127–140. [CrossRef]

60. Liu, X.; Song, Y.; Xing, F.; Wang, N.; Wen, F.; Zhu, C. *GhWRKY25*, a group I WRKY gene from cotton, confers differential tolerance to abiotic and biotic stresses in transgenic *Nicotiana benthamiana*. *Protoplasma* **2015**, *253*, 1–17. [CrossRef]
61. Wang, Y.; Jiang, L.; Chen, J.; Tao, L.; An, Y.; Cai, H. Overexpression of the alfalfa *WRKY11* gene enhances salt tolerance in soybean. *PLoS ONE* **2018**, *13*, e0192382. [CrossRef]
62. Mehdi, P.; Vijayaraj, N.; Youping, D. GeneVenn—A web application for comparing gene lists using Venn diagrams. *Bioinformation* **2007**, *1*, 420–422.
63. Berri, S.; Abbruscato, P.; FaivreRampant, O.; Brasileiro, A.C.; Fumasoni, I.; Satoh, K.; Kikuchi, S.; Mizzi, L.; Morandini, P.; Pè, M.E. Characterization of WRKY co-regulatory networks in rice and *Arabidopsis*. *BMC Plant Biol.* **2009**, *9*, 1–22. [CrossRef] [PubMed]
64. Guo, C.; Guo, R.; Xu, X.; Gao, M.; Li, X.; Song, J.; Zheng, Y.; Wang, X. Evolution and expression analysis of the grape (*Vitis vinifera* L.) *WRKY* gene family. *J. Exp. Bot.* **2014**, *65*, 1513–1528. [CrossRef] [PubMed]
65. Huang, S.; Gao, Y.; Liu, J.; Peng, X.; Niu, X.; Fei, Z.; Cao, S.; Liu, Y. Genome-wide analysis of WRKY transcription factors in *Solanum lycopersicum*. *Mol. Genet. Genomics* **2012**, *287*, 495–513. [CrossRef] [PubMed]
66. Tamura, K.; Stecher, G.; Peterson, D.; Filipski, A.; Kumar, S. MEGA6: Molecular evolutionary genetics analysis version 6.0. *Mol. Biol. Evol.* **2013**, *30*, 2725–2729. [CrossRef] [PubMed]
67. Miller, M.J.; Barrett-Wilt, G.A.; Zhihua, H.; Vierstra, R.D. Proteomic analyses identify a diverse array of nuclear processes affected by small ubiquitin-like modifier conjugation in *Arabidopsis*. *Proc. Natl. Acad. Sci. USA* **2011**, *107*, 16512–16517. [CrossRef]
68. Du, Y.T.; Zhao, M.J.; Wang, C.T.; Gao, Y.; Wang, Y.X.; Liu, Y.W.; Chen, M.; Chen, J.; Zhou, Y.B.; Xu, Z.S. Identification and characterization of *GmMYB118* responses to drought and salt stress. *BMC Plant Biol.* **2018**, *18*. [CrossRef]
69. Kereszt, A.; Li, D.; Indrasumunar, A.; Nguyen, C.D.; Nontachaiyapoom, S.; Kinkema, M.; Gresshoff, P.M. *Agrobacterium rhizogenes*-mediated transformation of soybean to study root biology. *Nat. Protoc.* **2007**, *2*, 948–952. [CrossRef]
70. Cao, D.; Hou, W.S.; Song, S.K.; Sun, H.B.; Wu, C.X.; Gao, Y.S.; Han, T.F. Assessment of conditions affecting *Agrobacterium rhizogenes*-mediated transformation of soybean. *Plant Cell Tissue Organ* **2009**, *96*, 45–52. [CrossRef]
71. Wang, F.; Chen, H.W.; Li, Q.T.; Wei, W.; Li, W.; Zhang, W.K.; Ma, B.; Bi, Y.D.; Lai, Y.C.; Liu, X.L. *GmWRKY27* interacts with *GmMYB174* to reduce expression of *GmNAC29* for stress tolerance in soybean plants. *Plant J.* **2015**, *83*, 224–236. [CrossRef] [PubMed]
72. Ishiguro, S.; Nakamura, K. Characterization of a cDNA encoding a novel DNA-binding protein, SPF1, that recognizes SP8 sequences in the 5′ upstream regions of genes coding for sporamin and β-amylase from sweet potato. *Mol. Genet. Genomics* **1994**, *244*, 563–571. [CrossRef]
73. Wei, K.F.; Chen, J.; Chen, Y.F.; Wu, L.J.; Xie, D.X. Molecular phylogenetic and expression analysis of the complete WRKY transcription factor family in Maize. *DNA Res.* **2012**, *19*, 153–164. [CrossRef] [PubMed]
74. Wu, K.L. The WRKY family of transcription factors in rice and *Arabidopsis* and their origins. *DNA Res.* **2005**, *12*, 9–26. [CrossRef]
75. Jue, D.; Sang, X.; Liu, L.; Shu, B.; Wang, Y.; Liu, C.; Xie, J.; Shi, S. Identification of WRKY Gene Family from *Dimocarpus longan* and its expression analysis during flower induction and abiotic stress responses. *Int. J. Mol. Sci.* **2018**, *19*. [CrossRef] [PubMed]
76. Kim, C.Y.; Vo, K.T.X.; Cong, D.N.; Jeong, D.H.; Lee, S.K.; Kumar, M.; Kim, S.R.; Park, S.H.; Kim, J.K.; Jeon, J.S. Functional analysis of a cold-responsive rice WRKY gene, *OsWRKY71*. *Plant Biotechnol. Rep.* **2016**, *10*, 13–23. [CrossRef]
77. Ding, M.; Chen, J.; Jiang, Y.; Lin, L.; Cao, Y.; Wang, M.; Zhang, Y.; Rong, J.; Ye, W. Genome-wide investigation and transcriptome analysis of the WRKY gene family in *Gossypium*. *Mol. Genet. Genomics* **2015**, *290*, 151–171. [CrossRef]
78. Shi, J.N.; Li-Yun, L.I.; Wen-Jing, X.U.; Guan, M.L.; Xue-Jiao, L.I.; Niu, D.D.; Lan, J.P.; Dou, S.J.; Liu, L.J.; Liu, G.Z. Expression analysis of eight WRKY transcription factors in rice leaf growth and disease resistance response. *Acta Phytopathol. Sin.* **2014**, *44*, 54–64.
79. Yu, Y.; Hu, R.; Wang, H.; Cao, Y.; He, G.; Fu, C.; Zhou, G. *MlWRKY12*, a novel Miscanthus transcription factor, participates in pith secondary cell wall formation and promotes flowering. *Int. J. Exp. Plant Biol.* **2013**, *212*, 1–9. [CrossRef]

80. Li, W.; Wang, H.; Yu, D. *Arabidopsis* WRKY Transcription Factors WRKY12 and WRKY13 oppositely regulate flowering under short-day conditions. *Mol. Plant* **2016**, *9*, 1492–1503. [CrossRef]
81. Li, X.J.; Guo, C.J.; Lu, W.J.; Duan, W.W.; Zhao, M.; Ma, C.Y.; Gu, J.T.; Xiao, K. expression pattern analysis of Zinc finger protein genes in wheat (*Triticum aestivum* L.) under phosphorus deprivation. *J. Integr. Agric.* **2014**, *13*, 1621–1633. [CrossRef]
82. Van Verk, M.C.; Neeleman, L.; Bol, J.F.; Linthorst, H.J. Tobacco transcription factor *NtWRKY12* interacts with TGA2.2 in vitro and in vivo. *Front. Plant Sci.* **2011**, *2*, 1085–1091. [CrossRef] [PubMed]
83. Thurow, C.; Schiermeyer, A.; Krawczyk, S.; Butterbrodt, T.; Nickolov, K.; Gatz, C. Tobacco bZIP transcription factor TGA2.2 and related factor TGA2.1 have distinct roles in plant defense responses and plant development. *Plant J.* **2010**, *44*, 100–113. [CrossRef] [PubMed]
84. Kim, H.S.; Park, Y.H.; Nam, H.; Lee, Y.M.; Song, K.; Choi, C.; Ahn, I.; Park, S.R.; Lee, Y.H.; Hwang, D.J. Overexpression of the Brassica rapa transcription factor WRKY12 results in reduced soft rot symptoms caused by *Pectobacterium carotovorum* in *Arabidopsis* and Chinese cabbage. *Plant Biol.* **2015**, *16*, 973–981. [CrossRef] [PubMed]
85. Cui, Q.; Yan, X.; Gao, X.; Zhang, D.M.; He, H.B.; Jia, G.X. Analysis of WRKY transcription factors and characterization of two *Botrytis cinerea*-responsive LrWRKY genes from Lilium regale. *Plant Physiol. Biochem.* **2018**, 525–536. [CrossRef] [PubMed]
86. Wang, J.X.; Tang, Y.K.; Liu, Q.; Liang, C.Q.; Wang, Y.C. Cloning and expression analysis of *THWRKY12* gene from *Tamarix hispida*. *Chin. J. Agric. Biotechol.* **2013**, *5*, 55–57.
87. Baranwal, V.K.; Negi, N.; Khurana, P. Genome-wide identification and structural, functional and evolutionary analysis of WRKY components of *Mulberry*. *Sci. Rep.* **2016**, *6*. [CrossRef]
88. Tao, X.; Chen, C.; Li, C.; Liu, J.; Liu, C.; He, Y. Genome-wide investigation of WRKY gene family in pineapple: Evolution and expression profiles during development and stress. *BMC Genomics* **2018**, *19*. [CrossRef]
89. Tan, C.K.; Carey, A.J.; Cui, X.; Webb, R.I.; Ipe, D.; Crowley, M.; Cripps, A.W.; Benjamin, B.W., Jr.; Ulett, K.B.; Schembri, M.A. Genome-wide mapping of cystitis due to *Streptococcus agalactiae* and *Escherichia coli* in mice identifies a unique bladder transcriptome that signifies pathogen-specific antimicrobial defense against urinary tract infection. *Infect. Immun.* **2012**, *80*, 3145–3160. [CrossRef]

© 2018 by the authors. Licensee MDPI, Basel, Switzerland. This article is an open access article distributed under the terms and conditions of the Creative Commons Attribution (CC BY) license (http://creativecommons.org/licenses/by/4.0/).

International Journal of
Molecular Sciences

MDPI

Article

Genome-Wide Characterization and Identification of Trihelix Transcription Factor and Expression Profiling in Response to Abiotic Stresses in Rice (*Oryza sativa* L.)

Jiaming Li [1], Minghui Zhang [2], Jian Sun [1], Xinrui Mao [1], Jing Wang [3], Jingguo Wang [1], Hualong Liu [1], Hongliang Zheng [1], Zhen Zhen [2], Hongwei Zhao [1] and Detang Zou [1,*]

1 College of Agriculture, Northeast Agricultural University, Harbin 150030, China; simons2016@163.com (J.L.); sunjian8416@163.com (J.S.); mxr1025559316@163.com (X.M.); wangjg@neau.edu.cn (J.W.); liuhualongneau@163.com (H.L.); zhenghongliang008@163.com (H.Z.); hongweizhao_cool@126.com (H.Z.)

2 College of Life Science, Northeast Agricultural University, Harbin 150030, China; zhangmh@neau.edu.cn (M.Z.); nneehhzz@126.com (Z.Z.)

3 Agriculture Technology and Popularization Center, Jixi 158100, China; 13895943955@163.com

* Correspondence: zoudtneau@126.com; Tel.: +86-451-55190635

Received: 29 October 2018; Accepted: 6 January 2019; Published: 10 January 2019

Abstract: Trihelix transcription factors play a role in plant growth, development and various stress responses. Here, we identified 41 trihelix family genes in the rice genome. These *OsMSLs* (Myb/SANT-LIKE) were located on twelve chromosomes. Synteny analysis indicated only six duplicated gene pairs in the rice trihelix family. Phylogenetic analysis of these *OsMSLs* and the trihelix genes from other species divided them into five clusters. *OsMSLs* from different groups significantly diverged in terms of gene structure and conserved functional domains. However, all *OsMSLs* contained the same five *cis*-elements. Some of these were responsive to light and dehydration stress. All *OsMSLs* expressed in four tissues and six developmental stages of rice but with different expression patterns. Quantitative real-time PCR analysis revealed that the *OsMSLs* responded to abiotic stresses including drought and high salt stress and stress signal molecule including ABA (abscisic acid), hydrogen peroxide. *OsMSL39* were simultaneously expressed under all treatments, while *OsMSL28* showed high expression under hydrogen peroxide, drought, and high salt treatments. Moreover, *OsMSL16/27/33* displayed significant expression under ABA and drought treatments. Nevertheless, their responses were regulated by light. The expression levels of the 12 chosen *OsMSLs* differed between light and dark conditions. In conclusion, our results helped elucidate the biological functions of rice trihelix genes and provided a theoretical basis for further characterizing their biological roles in responding to abiotic stresses.

Keywords: rice; trihelix transcription factor; phylogenetic analysis; stress response; light

1. Introduction

Transcription factors are ubiquitous in plants. They play crucial roles in various growth and development processes and respond to abiotic stresses [1]. Previous studies reported more than 60 transcription factor families in plants [2,3]. However, little is known about several important transcription factor families. Trihelix transcription factors occur only in plants. They were first identified and isolated from pea (*Pisum sativum*) in the 1990s. They bind to the core sequence of 5′-G-Pu-(T/A)-A-A-(T/A)-3′ of the promoter region of rbcS-3A gene to regulate light-dependent expression [4]. They were initially called GT factors because they bind to light-responsive GT elements.

The DNA-binding domain of the GT factors has a typical tandem trihelix (helix-loop-helix-loop-helix) structure which was later renamed the trihelix transcription factor. Subsequent research revealed that the trihelix structure of the GT factors resembles the solution structure of the Myb/SANT-LIKE DNA-binding domain [5]. GT factors evolved from Myb/SANT-LIKE proteins in plants. Gaps between helix pairs created different recognition sequences between GT factors and Myb/SANT-LIKE proteins [5,6]. According to databases like Pfam, the Myb/SANT-LIKE domain represents the trihelix conserved domain.

Trihelix is a family of transcription factors that have only recently received attention. However, the trihelix genes have been systematically studied mainly in dicotyledonous plants such as *Arabidopsis*, tomato and chrysanthemum, while almost no research has been carefully carried out in a monocotyledonous plant. In *Arabidopsis*, 30 GT family members were identified and divided into the GT-1, GT-2, GTγ, SH4, and SIP1 subfamilies named after their founding members [7]. The 96 trihelix proteins of tomato (*Solanum lycopersicum*) were classified into six subfamilies (clades GT-1, GT-2, SH4, SIP1, GTγ, and GTδ). The GTδ subfamily is apparently missing in *Arabidopsis* [8]. Most of the trihelix gene subfamily structures vary substantially, especially at the C-terminus. The exceptions are GT1 and GT2.

Earlier studies identified the trihelix family genes as a class of light regulators. Nevertheless, the roles of GT factors in light regulation must be systematically established. In *Arabidopsis*, the GT1 subfamily genes may participate in salt stress and pathogen responses and their expression was induced by light in 3-d seedlings [9]. In contrast, the rice GT-1 gene *RML1* (*OsMSL21* in the present study) was repressed by light in etiolated seedlings [10]. The trihelix transcription factors in soybean, *GmGT-2A* and *GmGT-2B*, were induced by ABA (abscisic acid), drought, high salt levels, and cold in soybean seedlings [11]. Loss-of-function analysis of *GTL1* revealed that *gtl1* mutants had fewer stomata than wild type plants. In this way, the former had comparatively lower water loss and higher drought tolerance than the latter [12]. The expression of the rice GTγ clade gene *OsGTγ-1* increased 2.5 to 10 times in response to salt stress and was also upregulated by ABA treatment [13]. On the other hand, the expression of several trihelix genes in Chrysanthemum was downregulated by ABA [14]. Trihelix transcription factors are also associated with plant morphogenesis. The trihelix transcription factor *PETAL LOSS* (*PTL*) determines the number of petals per flower and sepal fusion in *Arabidopsis*. The rice SH4 clade gene (*SH4*) promotes the abscission layer development and function in mature seed peduncles [15]. However, the function of the SH4 clade has not yet been investigated. The *Arabidopsis* SIP1 genes *ASIL1* and *ASIL2* downregulated the LEA (Late Embryogenesis Abundant) genes in *Arabidopsis* seedlings [7]. The trihelix genes also have multiple functions throughout plant development. The molecular mechanisms of their stress responses and their involvement in the signaling pathway require elucidation.

Rice (*Oryza sativa* L.) is both a major global cereal crop and an important tool in plant research. In this study, we identified 41 rice trihelix genes by the Myb/SANT-LIKE domain using HMM-search in silico. We analyzed their chromosomal distributions, gene synteny, phylogenetic analysis, gene structures, motif compositions, *cis*-elements, and expression patterns in different tissues, developmental stages, and environmental stress responses. The aim of this study was to analyze the structure and function of rice trihelix genes and phylogenetic relationship between rice trihelix proteins and other species including dicotyledonous and monocotyledonous plant. To establish the role of the trihelix genes' response to stress, we evaluated their response to abiotic stress factors including drought and high salt, and to stress signal molecules, such as abscisic acid and hydrogen peroxide. Our results provide a theoretical basis for the functional analysis of the rice trihelix family genes especially in abiotic stress responses.

2. Results

2.1. Identification of Trihelix Genes in Rice

The HMM (Hidden Markov Model) for the Myb/SANT-LIKE domain identified 117 gene candidates and a rice-specific Myb/SANT-LIKE domain was built using them. The HMM profile search was performed on the whole rice genome with the rice-specific Myb/SANT-LIKE domain and 79 new candidate genes were found. Only genes with E-value < 0.01 were classified in the trihelix family. Putative genes were verified in the Pfam and InterPro databases to confirm the existence of the complete Myb/SANT-LIKE domains. Finally, 41 trihelix genes were identified.

All trihelix genes mapped onto the rice chromosomes, they were named *OsMSL01-OsMSL41* according to the gene distribution order on the chromosomes. *OsMSL25* and *OsMSL34* have two alternative splicing. "MSL" stands for "Myb/SANT-LIKE". The characteristics of OsMSLs including the gene MSU_Locus ID, the chromosomes locations, the lengths of the CDS (coding sequence) and amino acid sequences, the number of exons, the protein sizes, and the isoelectric points are summarized in Table 1. OsMSL19 was the smallest protein with 266 amino acids, whereas OsMSL12 was the largest with 882 amino acids. The protein MW (Molecular Weight) ranged from 28.62 kDa to 97.37 kDa. Their predicted isoelectric points varied from 4.45 (OsMSL09) to 11.38 (OsMSL17). Twenty-nine of the trihelix transcription factors were localized in the nucleus, ten in the chloroplast, and two in the peroxisome (OsMSL04 and OsMSL29).

Table 1. Detailed information of all trihelix family genes identified in the rice genome.

Gene Name	Gene Locus	Chr	ORF (bp)	Exon No.	Length (aa)	MW (kD)	pI	Localization
OsMSL01	LOC_Os01g11200.1	Chr1	828	3	275	31.8	8.57	Nucleus
OsMSL02	LOC_Os01g27590.1	Chr1	1131	2	376	41.08	8.6	Nucleus
OsMSL03	LOC_Os01g34400.1	Chr1	1227	3	408	46.06	9.22	Chloroplast
OsMSL04	LOC_Os01g36850.1	Chr1	1170	2	389	57.61	9.46	Peroxisome
OsMSL05	LOC_Os01g48320.1	Chr1	999	1	332	35.92	9.62	Nucleus
OsMSL06	LOC_Os01g52090.1	Chr1	969	1	322	35.76	5.54	Nucleus
OsMSL07	LOC_Os01g62410.1	Chr1	1764	7	587	64.1	8.5	Nucleus
OsMSL08	LOC_Os02g01380.1	Chr2	1113	1	370	40.5	5.33	Nucleus
OsMSL09	LOC_Os02g07800.1	Chr2	1308	2	435	46.24	4.45	Chloroplast
OsMSL10	LOC_Os02g08450.1	Chr2	1968	4	655	74.35	7.09	Chloroplast
OsMSL11	LOC_Os02g31160.1	Chr2	1125	2	374	39.55	5.17	Nucleus
OsMSL12	LOC_Os02g33610.1	Chr2	2649	18	882	97.37	8.97	Chloroplast
OsMSL13	LOC_Os02g33770.1	Chr2	1233	1	410	46.67	6.28	Nucleus
OsMSL14	LOC_Os02g35690.1	Chr2	1260	1	419	44.44	6.11	Nucleus
OsMSL15	LOC_Os02g43300.1	Chr2	1887	3	628	67.74	4.87	Nucleus
OsMSL16	LOC_Os03g44130.1	Chr3	1104	3	367	41.74	6.61	Chloroplast
OsMSL17	LOC_Os03g46350.1	Chr3	1038	2	345	42.49	11.38	Nucleus
OsMSL18	LOC_Os04g21860.1	Chr4	1254	1	417	47.15	9.1	Nucleus
OsMSL19	LOC_Os04g30890.1	Chr4	801	1	266	28.62	9.64	Nucleus
OsMSL20	LOC_Os04g36790.1	Chr4	1254	1	417	43.93	6.53	Nucleus
OsMSL21	LOC_Os04g40930.1	Chr4	1158	5	385	41.93	5.82	Nucleus
OsMSL22	LOC_Os04g45750.1	Chr4	1587	2	528	57.46	5.74	Nucleus
OsMSL23	LOC_Os04g57530.1	Chr4	1173	2	390	41.44	9.01	Nucleus
OsMSL24	LOC_Os05g03740.1	Chr5	1002	1	333	36.95	5.85	Nucleus
OsMSL25	LOC_Os05g48690.1	Chr5	1041	1	346	37.44	9.92	Nucleus
OsMSL26	LOC_Os06g32944.1	Chr6	876	3	291	32.82	8.22	Nucleus
OsMSL27	LOC_Os07g02500.1	Chr7	1104	3	367	41.74	6.61	Chloroplast
OsMSL28	LOC_Os07g10950.1	Chr7	1437	8	478	54.23	9.06	Chloroplast
OsMSL29	LOC_Os08g08130.1	Chr8	1170	2	389	44.28	6.2	Peroxisome
OsMSL30	LOC_Os08g12950.1	Chr8	1254	1	417	47.15	8.92	Nucleus
OsMSL31	LOC_Os08g37810.1	Chr8	948	1	315	35.06	7.09	Nucleus
OsMSL32	LOC_Os08g44690.1	Chr8	912	1	303	34.83	9.29	Nucleus
OsMSL33	LOC_Os09g03570.1	Chr9	1932	7	643	73.57	9.06	Chloroplast
OsMSL34	LOC_Os09g38570.1	Chr9	1011	1	336	36.34	6.58	Nucleus
OsMSL35	LOC_Os10g33030.1	Chr10	1104	3	367	41.7	6.61	Chloroplast
OsMSL36	LOC_Os10g41460.1	Chr10	1011	1	336	35.64	8.9	Nucleus
OsMSL37	LOC_Os11g06410.1	Chr11	1492	2	483	55.06	6.24	Nucleus
OsMSL38	LOC_Os11g17954.1	Chr11	1545	4	514	58.1	8.3	Nucleus
OsMSL39	LOC_Os11g38660.1	Chr11	1938	5	645	72.97	7.34	Chloroplast
OsMSL40	LOC_Os12g06640.1	Chr12	1299	1	432	48.77	6.13	Nucleus
OsMSL41	LOC_Os12g10550.1	Chr12	888	2	295	33.5	7.74	Nucleus

2.2. Chromosomal Distributions and Synteny Analysis of Rice Trihelix Genes

The extraction of the chromosomal information of the *OsMSLs* identified their chromosomal locations. As shown in Figure 1, all *OsMSLs* had precise positions in the chromosomes. Each rice chromosome contains ≥1 *OsMSL*. The *OsMSLs* are unevenly and non-randomly distributed on 12 chromosomes. Chr2 (chromosomal 2) contains the largest number of *OsMSLs* (eight) whereas Chr6 contains only one. The first four chromosomes contain 23 trihelix genes while chromosomes 5–12 have on average only 2–3 genes per chromosome. Therefore, *OsMSLs* are distributed mainly on the first four rice chromosomes. Although Chr2 is relatively short, it contains the most *OsMSLs*. Chr1 is the longest in rice and also contains numerous *OsMSLs*. Chr10, the shortest chromosome, contains two *OsMSLs*. In contrast, Chr6 is longer than Chr10 but contains only one *OsMSL*. There is no apparent correlation between the chromosome length and *OsMSL* gene distribution. Moreover, only *OsMSL12* and *OsMSL13* form gene clusters on Chr2.

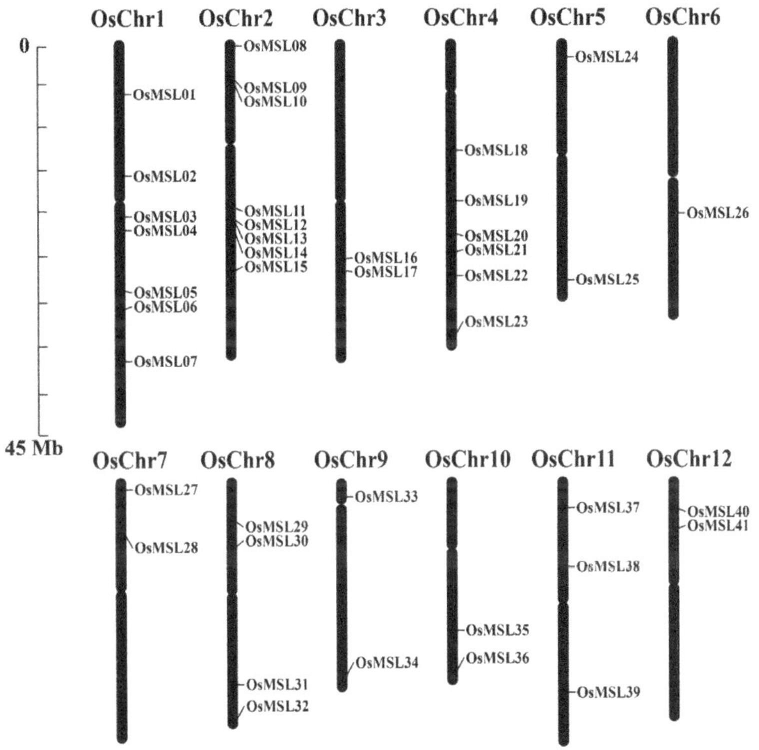

Figure 1. Chromosomal locations of rice trihelix genes. Black bars represent the chromosomes. Chromosome numbers are shown at the tops of the bar. Trihelix genes are labeled at the right of the chromosomes. Scale bar on the left indicates the chromosome lengths (Mb).

Synteny was also used to analyze rice trihelix gene duplication. Chromosomal region within 200 kb containing two or more genes is defined as a tandem duplication event [16]. As shown in Figure 1, four rice trihelix genes (*OsMSL09/10* and *OsMSL12/13*) were clustered into two tandem duplication event regions on rice chromosomal 2. Besides the tandem duplication events, segmental duplications were also investigated by BLASTP and MCScanX methods [17]. Four segmental duplication events with eight rice trihelix genes were also identified, which are located on duplicated segments on chromosomes 1, 2, 4, 5, and 11 (Figure 2). This finding is consistent with the highly divergent, non-conservative evolution of *OsMSLs*.

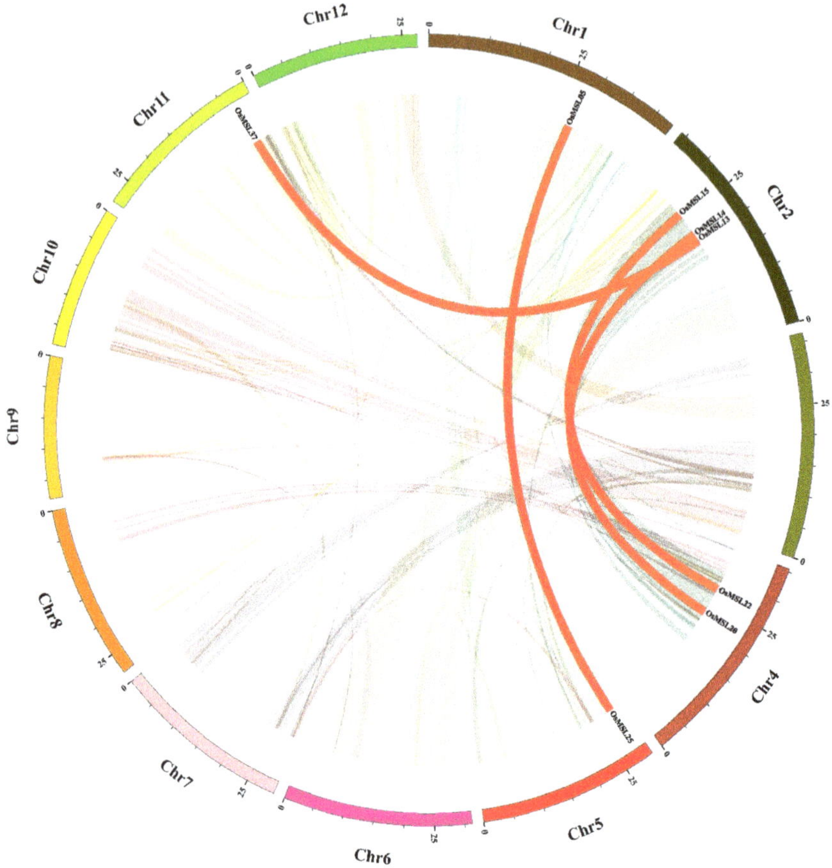

Figure 2. Schematic representations of segmental duplications of rice trihelix genes. Different color lines indicate all synteny blocks in rice genome between each chromosome, and the thick red lines indicate duplicated trihelix gene pairs. The chromosome number is indicated at the bottom of each chromosome. Scale bar marked on the chromosome indicating chromosome lengths (Mb).

To further understand the gene duplication mechanisms of the rice trihelix family, we constructed four comparative syntenic maps of rice associated with four representative species, including one dicots (*Arabidopsis*) (Figure 3A) and three monocots (*Brachypodium distachyon*, wheat and maize) (Figure 3B). A total of 23 rice trihelix genes showed a syntenic relationship with those in maize, followed by wheat (21), *Brachypodium distachyon* (19) and *Arabidopsis* (2), indicating that in comparison with monocotyledonous plants, rice trihelix genes show a high evolution divergence with dicotyledonous plants. Congruously, previous research reported that 14 pairs of orthologous trihelix genes were found between tomato and *Arabidopsis* [8]. Some *OsMSLs* were found to be associated with at least three syntenic gene pairs, such as *OsMSL14*, *OsMSL17*, and *OsMSL21*. These genes may have played a crucial role in the trihelix gene family during evolution. To better understand the evolutionary constraints acting on the trihelix gene family, the Ka/Ks ratios of the trihelix gene pairs were calculated (Tables S1–S5). All segmental and tandem duplicated *OsMSL* gene pairs, and the majority of orthologous trihelix gene pairs had Ka/Ks < 1, suggesting that the rice trihelix gene family might have experienced a strong purifying selective pressure during evolution.

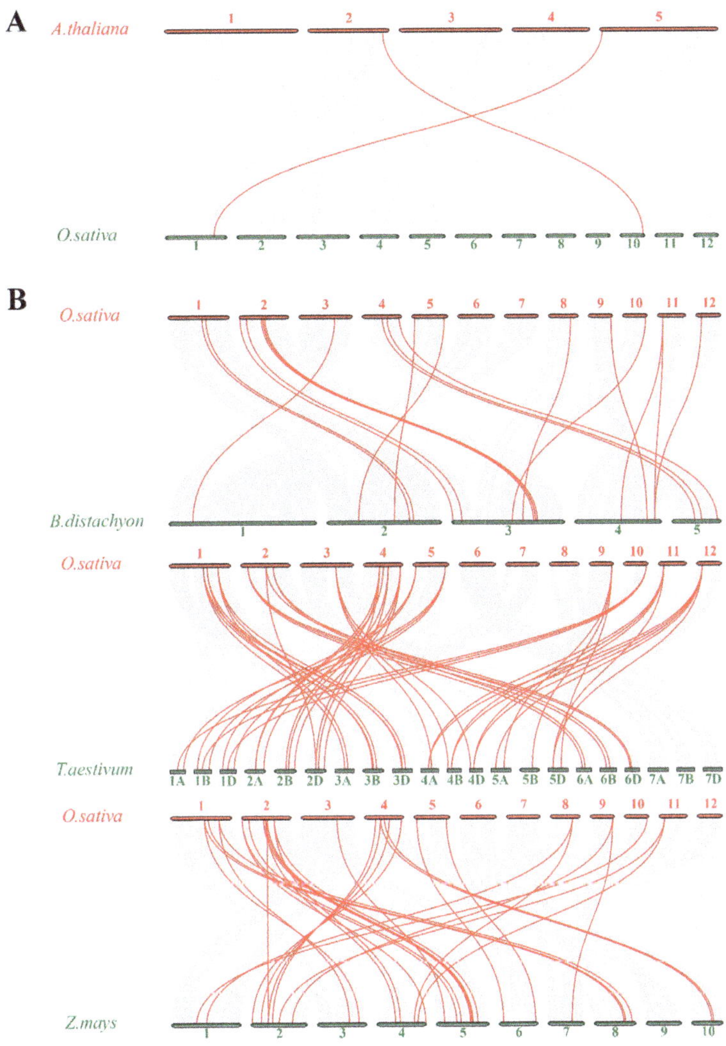

Figure 3. Synteny analysis of trihelix genes between rice and (**A**) dicotyledonous plant *Arabidopsi thaliana*, (**B**) monocotyledonous plant *Brachypodium distachyon*, wheat and maize. Gray lines in the background indicate the collinear blocks within rice and other plant genomes, while the red lines highlight the syntenic trihelix gene pairs. The species names with the prefixes '*A. thaliana*', '*B. distachyon*', '*T. aestivum*', '*Z. mays*' and '*O. sativa*' indicate *Arabidopsi thaliana, Brachypodium distachyon, Triticum aestivum, Zea mays* and *Oryza sativa*, respectively. Red or green bars represent the chromosomes. The chromosome number is labeled at the top or bottom of each chromosome.

2.3. Phylogenetic Analysis, Gene Structure, and Motif Composition of Trihelix Genes

To better understand the phylogenetic relationships of trihelix genes, a maximum likelihood phylogenetic tree was built based on the multiple sequence alignment of Myb/SANT-LIKE domains among rice and other species which include dicotyledonous plants such as *Arabidopsis*, soybean, tomato, chrysanthemum and monocotyledonous plant such as maize, wheat, wild rice, *Brachypodium distachyon*. As shown in Figure 4, OsMSLs were divided into five subfamilies named SIP1, GTγ, GT, SH4, and GTδ according to the characteristics of their trihelix DNA binding domains. Some genes that have been classified previously such as *SlGT-4/7/12/18/36* in tomato [8], *CmTH2/6/12/17/19/20* in chrysanthemum [14], *GmGT-2A* and *GmGT-2B* in soybean [11] was as a classified marker. The GT clade was the largest subfamily, containing 28 trihelix genes, whereas the SH4 clade was the smallest, consisting of 13 members, indicating that trihelix genes were distributed unevenly in the different clades. All clades consisted of genes both from dicot and monocot species. There is a similar classification in rice which was previously named GTδ in tomato and two tomato trihelix genes *SlGT-4* and *SlGT-12* have been found in this subfamily. To demonstrate the evolutionary relationships among *OsMSLs*, we constructed an unrooted phylogenetic tree using the full-length amino acid sequences of the OsMSLs. Of the 43 transcripts of the 41 rice trihelix genes, nine belonged to SIP1, 10 belonged to GTγ, 11 belonged to GT, five belonged to SH4, and eight belonged to GTδ (Figure 5A). Most of the duplicated genes were present in the GTδ classification. The phylogenetic tree of the all MSLs between rice and *Arabidopsis* was constructed and is shown in Figure S1. However, we found that the GTδ subfamily does not contain *Arabidopsis* trihelix genes.

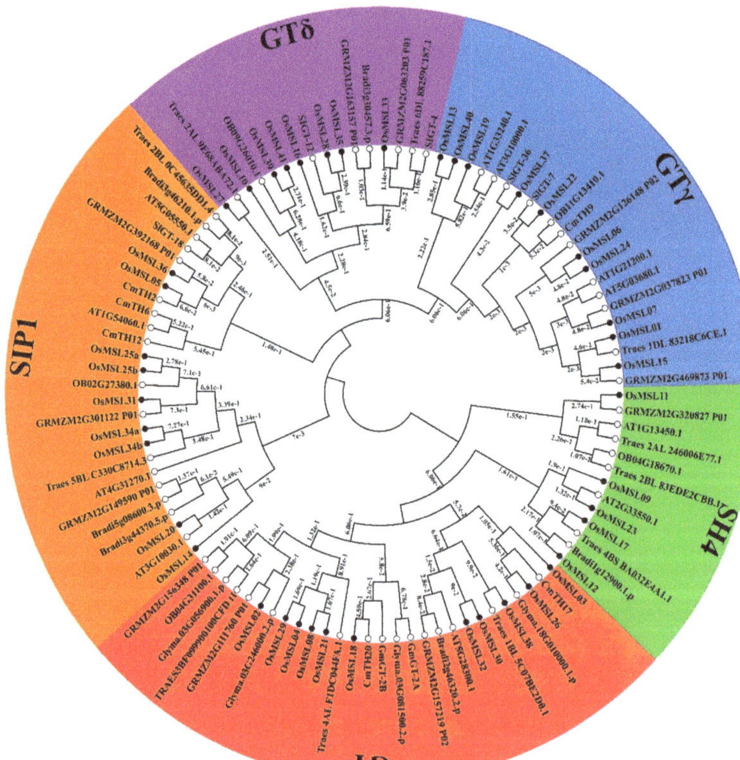

Figure 4. Phylogenetic relationships among 105 trihelix proteins in rice, *Arabidopsis*, soybean, maize, tomato, wheat, chrysanthemum, wild rice and *Brachypodium distachyon*. The maximum likelihood tree was created using MEGA v. 7.0 (bootstrap value = 1000) and the bootstrap value of each branch is displayed. Forty-three OsMSL proteins are marked with black circles and other species are marked with white circles. The phylogenetic tree was clustered into SIP1, GTγ, GT, SH4, and GTδ.

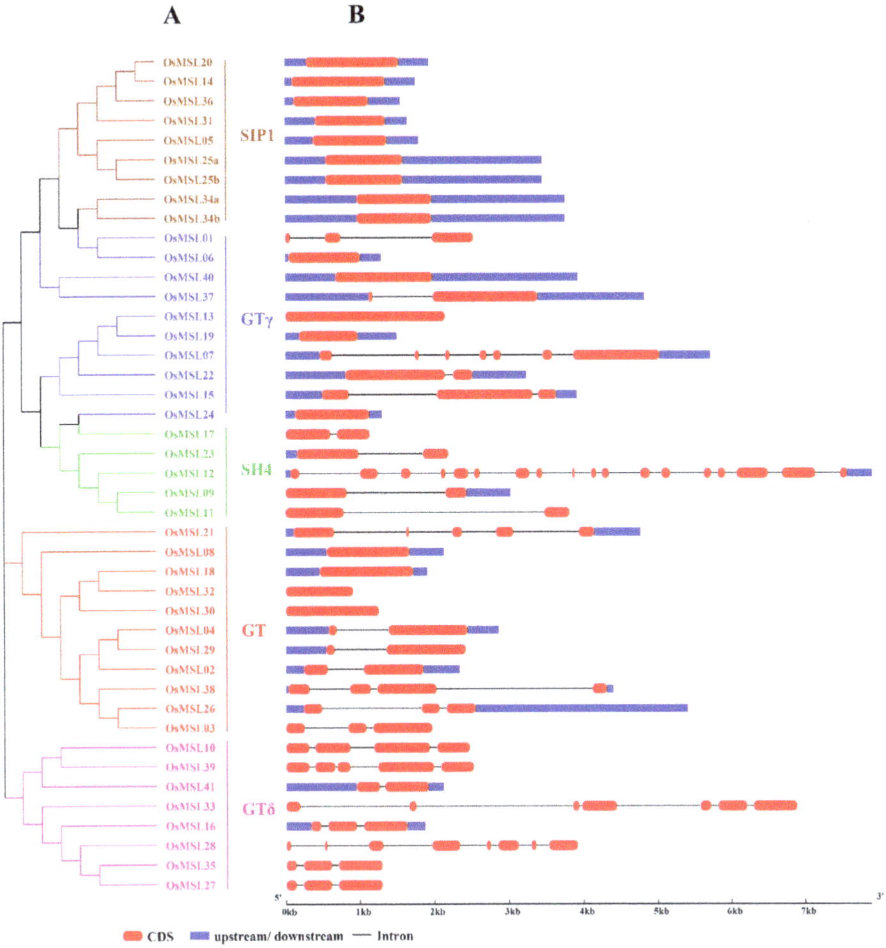

Figure 5. Phylogenetic analysis and gene structure of the rice trihelix family. (**A**) Phylogenetic analysis of the rice trihelix family. The phylogenetic tree was constructed based on the full-length amino acid sequences of the rice trihelix proteins by using MEGA v. 7.0 with the maximum-likelihood method. Bootstrap = 1,000. SIP1, GTγ, GT, SH4, and GTδ are marked with different colors. (**B**) Gene structures of the rice trihelix family. These were analyzed by the Gene Structure Display Server (GSDS v. 2.0). Exons, introns, and untranslated regions are marked by round red rectangles, black lines, and blue rectangles, respectively. The scale bar at the bottom estimates the lengths of the exons, introns, and untranslated regions.

To identify the differences between the rice trihelix family genes, we analyzed the *OsMSL* gene structure by comparing each coding sequence with its corresponding genomic sequence. As shown in Figure 5B, the number of *OsMSL* exons is discontinuously distributed from 1 through 18. Combining the gene structure with the phylogenetic tree, we found that the *OsMSL* exon-intron distribution is related to its classification. Closely related genes usually have homologs. Therefore, their gene structures are similar. For example, the *OsMSL* genome sequences in the SIP1 subfamily have no introns and only one exon. Therefore, the evolution of this gene subfamily is relatively conservative. The genes in the GTδ subfamily have no UTR region and only exons and introns except for *OsMSL16* and *OsMSL41*. In contrast, the structures of the various genes in the GTγ, GT, and SH4 subfamilies are relatively different. These results indicate that although the *OsMSLs* are subdivided into five families, their genes are relatively conservative.

To determine the functions of the trihelix family genes, the OsMSL motif composition was analyzed by amino acid sequence in the MEME program. Ten motifs with $E < 1.8 \times 10^{-45}$ were identified. These resemble the MSLs in chrysanthemum. The genes for each subfamily were classified [14]. As shown in Figure 6, except for OsMSL01 and OsMSL09, most trihelix family genes contain motif 1 (Myb-type DNA-binding domain) located at the *N*-terminus of the amino acid sequence. Motifs 2, 6, and 8 are various trihelix DNA binding domains (WWW, WWF, and WWI). These determine OsMSL classification, structure, and function [18]. As the gene structure analysis indicated, the gene motifs and distribution patterns are closely related to their subfamilies. SIP1 contains motif 8, GTγ contains motif 6, and only OsMSL06 contains an extra motif 8. Both GT and SH4 contain motif 2 but that in SH4 is longer than that in GT. OsMSL09 and OsMSL12 in SH4 also contain an additional motif 8. OsMSL02, OsMSL03, OsMSL04, OsMSL26, OsMSL29, and OsMSL38 in the GT subfamily also contain motif 6. Motif 2 with other functional domains and conservative sequences are contained in the rice-specific GTδ subfamily. Although their functions have yet to be elucidated, they may indicate that the GTδ gene in rice has multiple functions.

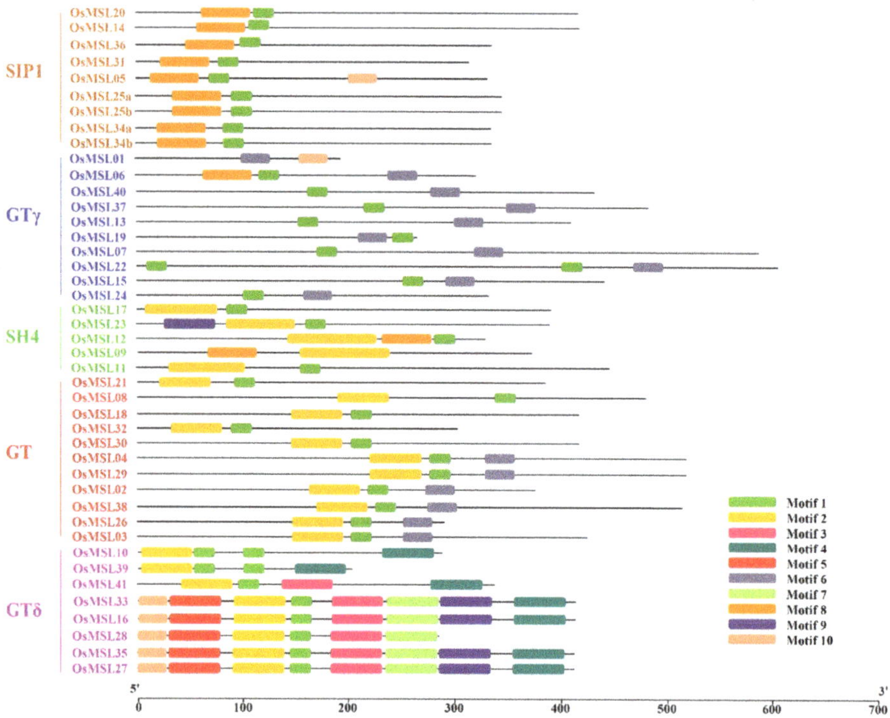

Figure 6. Motif composition of rice trihelix proteins. Motif analysis was performed using the MEME program as described in the methods section. The trihelix proteins are listed on the left. Boxes of different colors represent the various motifs. Their location in each sequence is marked. Motif sequences are shown in Figure S2. The scale bar at the bottom indicates the lengths of the trihelix protein sequences.

2.4. Cis-Element Analysis of Rice Trihelix Genes

To understand the genetic functions, metabolic networks, and regulatory mechanisms of rice trihelix genes, the shared *cis*-elements in the promoter regions of the *OsMSLs* were analyzed. The 1500-bp upstream *OsMSL* sequence was obtained and identified as a hypothetical promoter. The potential shared *OsMSL cis*-element was scanned and screened out and its distribution and function were analyzed. Two dehydration-responsive- and three light-responsive *cis*-elements common to all *OsMSLs* were identified and labeled by different colors in the promoter sequence (Figure 7).

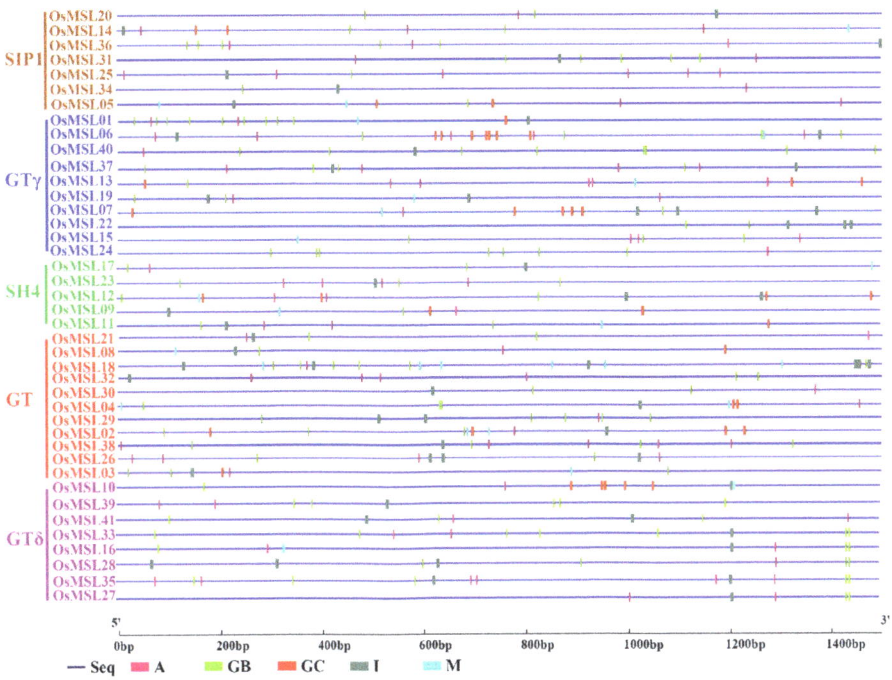

Figure 7. Predicted *cis*-elements in the promoter regions of the rice trihelix genes. All promoter sequences (−1500 bp) were analyzed. The trihelix genes are shown on the left side of the figure. The scale bar at the bottom indicates the length of promoter sequence. Green bar (GB): GATABOX element; purple bar (A): ACGTATERD1 element; red bar (GC): GT1CONSENSUS element; gray bar (I): INRNTPSADB element; blue bar (M): GT1GMSCAM4 element

As shown in Figure 7, A (ACGTATERD1) and M (GT1GMSCAM4) element are two dehydration-responsive elements and M is a core element. Therefore, *OsMSLs* probably participate in dehydration (including drought and salt) stress responses. GB (GATABOX element), GC (GT1CONSENSUS), and I (INRNTPSADB) are three light-responsive elements. They indicate that the *OsMSLs* family potentially consists of light-inducible/repressible genes. Light responsiveness is typical of the GT factor (now known as the trihelix family gene) and was confirmed in our *cis*-element study. To verify whether *OsMSLs* are regulated by light under both normal- and stress conditions, a dark treatment was added to the *OsMSL* expression analysis.

2.5. Expression Profiles of Trihelix Genes in Rice Tissues and Developmental Stages

The expression profiles of the various rice tissues including the root, stem, leaf, and sheath of four-leaf rice seedlings were investigated (Figure 8). As shown in Figure 8A, all 43 transcripts expressed in all tissues but their expression levels varied greatly in each tissue. Particularly, there are more highly expressing trihelix genes in the leaves and sheaths than the other organs. There are almost no genes with low expression levels and most of the genes remained at high expression levels in the leaves. In contrast, comparatively fewer genes with extremely high expression exist in the stems and none of them express at an extremely high level in the roots.

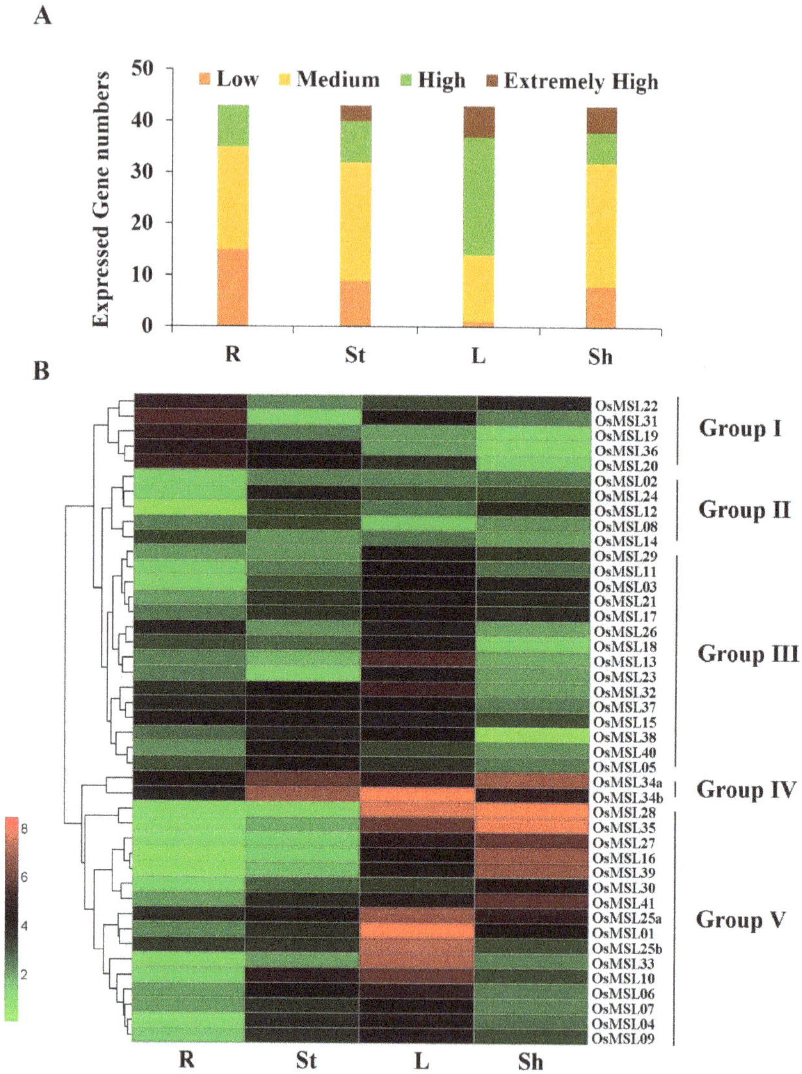

Figure 8. Expression of the rice trihelix gene family in various tissues (R: Root; St: Stem; L: Leaf; Sh: Sheath). (**A**) Numbers of expressed genes in each tissue. Expression data of the rice trihelix gene family were retrieved from the Expression Atlas database. Extremely high: Expression value > 6, high: 6 ≥ expression value > 4, medium: 4 ≥ expression value > 2, low: 2 ≥ expression value > 0; (**B**) Expression patterns of the trihelix genes in various rice tissues. Heatmaps were created in HemI v.1.0 and based on the expression data. Expression levels are depicted by different colors on the scale. Green and red represent low and high expression levels, respectively.

The *OsMSLs* were clustered into groups I to V according to their expression characteristics in different tissues (Figure 8B). *OsMSLs* in groups I and II expressed at lower levels in all tissues, but members in group I displayed relatively higher expression in root tissue. For instance, although *OsMSL19* maintained very low expression levels in most tissues, its expression displayed a distinct enrichment in root. In contrast, the genes in group II, especially *OsMSL02*, *OsMSL08*, and *OsMSL12*, remained low in the various tissues. On the other hand, in *Arabidopsis*, the At5g63420 gene, the orthology of *OsMSL12*, is highly expressed specifically in the seeds, suggesting that also *OsMSL12* could show a dominant expression level in seeds rather than in any of these four tissues [19]. Contrary to the group I, most of the genes in group III expressed at higher levels in the stems or leaves and at lower levels in the roots. However, the transcription levels of *OsMSL34a* and *OsMSL34b* in group IV were high in all four tissues. Group V members also displayed tissue specific expression, but expressed at higher levels than group I and III. Besides, the genes in group V, especially *OsMSL16*, *OsMSL27*, *OsMSL28*, *OsMSL35*, and *OsMSL39*, expressed extremely highly in the leaves and sheaths, whereas their expression levels in the roots and stems were lower than those of genes in group I and III, respectively. In conclusion, *OsMSLs* displayed tissue expression specificity, indicating their potential roles in different mechanisms.

We also investigated the *OsMSL* expression profiles at different rice developmental stages including 7 (S1), 20 (S2), 40 (S3), 80 (S4), 100(S5), and 140(S6) days after sowing. As shown in Figure 9A, the numbers of genes with high- and extremely high expression levels are greater in S2 and S3 than in the other stages. In contrast, no genes expressed at high levels during S4, S5, or S6. The genes were subdivided into five groups according to their expression characteristics at different stages (Figure 9B). *OsMSLs* in group I, were at comparatively higher transcription levels during the S2 and S3 stages except for *OsMSL28*, which displayed high expression levels at S1 stage as well. In general, the expression levels of the genes in groups II and III were similar to those in group I, but their expression levels in S2 and S3 were lower than in group I. Notably, the group IV genes had lower expression levels than the other four clusters at almost all developmental stages. Contrary to groups II and III, high gene expression levels were observed mainly at S1 in group V. Besides, with the exception of *OsMSL35* and *OsMSL06*, all other genes in group V expressed at significantly lower levels by S2. In conclusion, *OsMSLs* play a potential role at early developmental stages.

2.6. Quantitative Real-Time PCR Analysis of Rice Trihelix Genes in Responses to Different Treatments under Normal and Dark Conditions

A prediction of the *cis*-elements of the *OsMSLs* suggested that they may participate in rice dehydration stress tolerance and light-mediated signaling pathways. To verify this hypothesis, we subjected rice seedlings to ABA, hydrogen peroxide, drought, and high salt then carefully selected 12 genes expressing positively in the leaves (Figures 10 and 11). On the whole, although trihelix genes were induced by multiple treatments, their expression levels under one treatment were much higher than other treatments. For instance, six *OsMSLs* (*OsMSL25a*, *OsMSL25b*, *OsMSL28*, *OsMSL34a*, *OsMSL34b*, and *OsMSL35*) were induced by multiple treatments, but the transcript levels significantly increased under hydrogen peroxide treatment compared to other treatments. Several genes were induced after being repressed, such as *OsMSL41*, which remained downregulated up until 12 h of ABA treatment. Some trihelix genes showed high transcript levels under multiple treatments. For example, *OsMSL39* simultaneously responded to all treatments. *OsMSL28* was significantly induced by three tested treatments except ABA treatment. In conclusion, all 12 genes are induced by various abiotic stress or stress signaling molecules, but the expression levels are different and there was no significant correlation between gene expression and its classification.

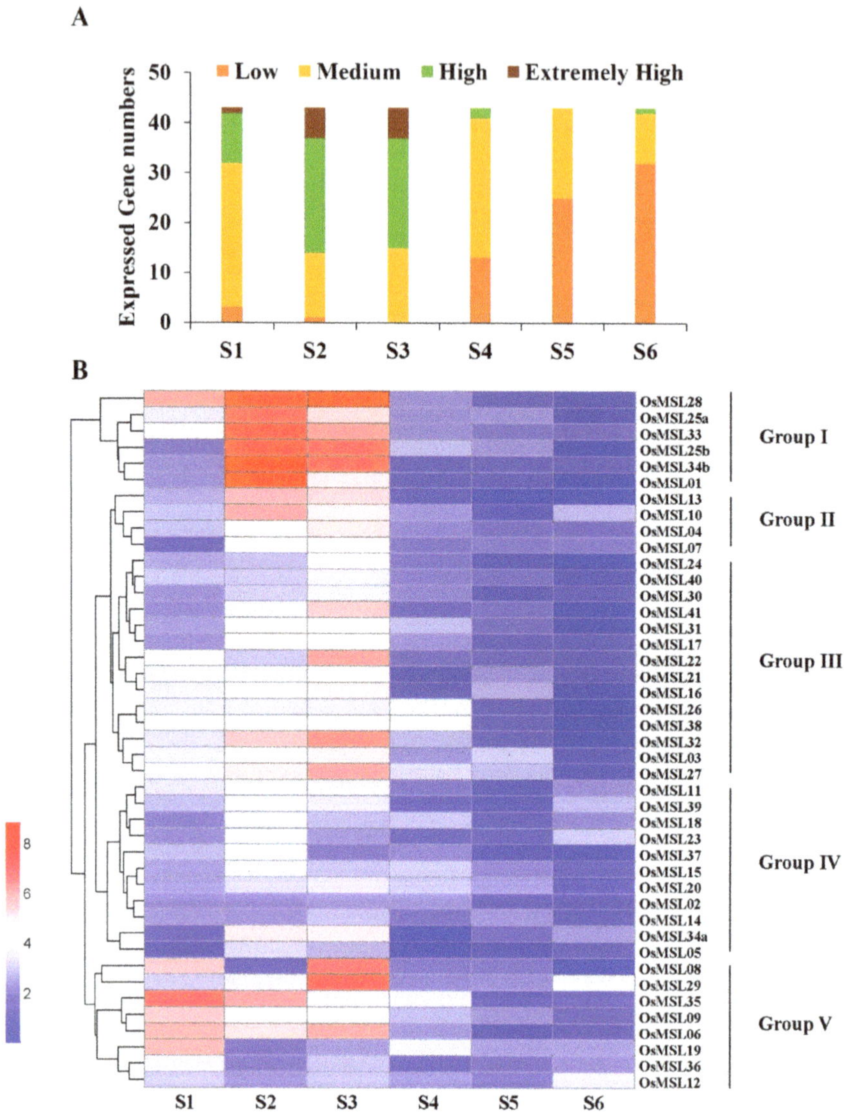

Figure 9. Expression of the rice trihelix gene family at different developmental stages (S1: 7 days after sowing; S2: 20 days after sowing; S3: 40 days after sowing; S4: 80 days after sowing; S5: 100 days after sowing; S6: 140 days after sowing). (**A**) Numbers of expressed genes in various developmental stages. Expression data of the rice trihelix gene family genes were retrieved from the Expression Atlas database. Extremely high: Expression value > 6, high: 6 ≥ expression value > 4, medium: 4 ≥ expression value > 2, low: 2 ≥ expression value > 0; (**B**) Expression patterns of trihelix genes in various rice developmental stages. Heatmaps were created in HemI v.1.0 and based on the expression data. Expression levels are depicted by different colors on the scale. Blue and red represent low and high expression levels, respectively.

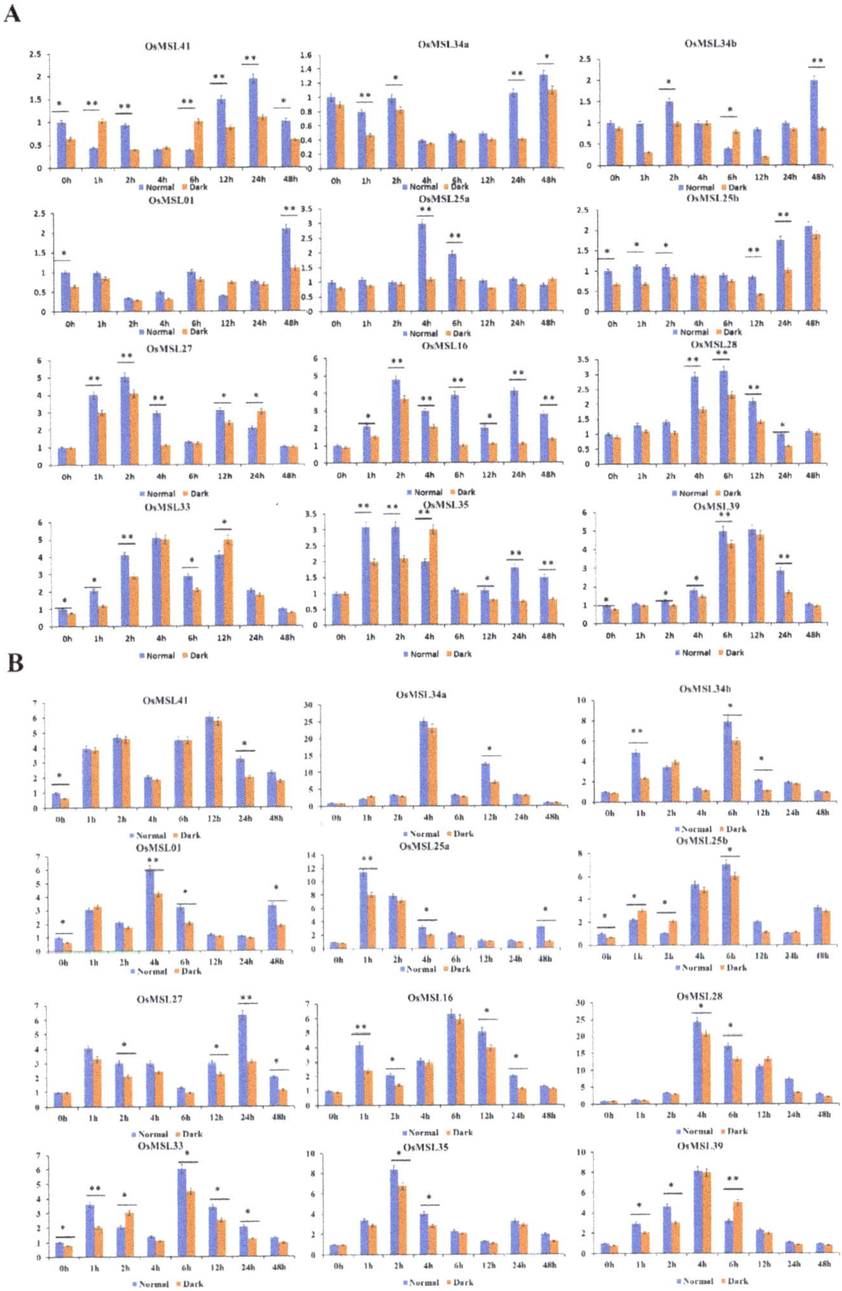

Figure 10. Expression profiles of 12 selected *OsMSLs* in response to (**A**) ABA (abscisic acid), (**B**) hydrogen peroxide treatments. Data were normalized to the β-actin gene. Vertical bars indicate standard deviations. Asterisks indicate corresponding genes significantly upregulated or downregulated compared with the control. (* $p < 0.05$; ** $p < 0.01$; Student's t-test).

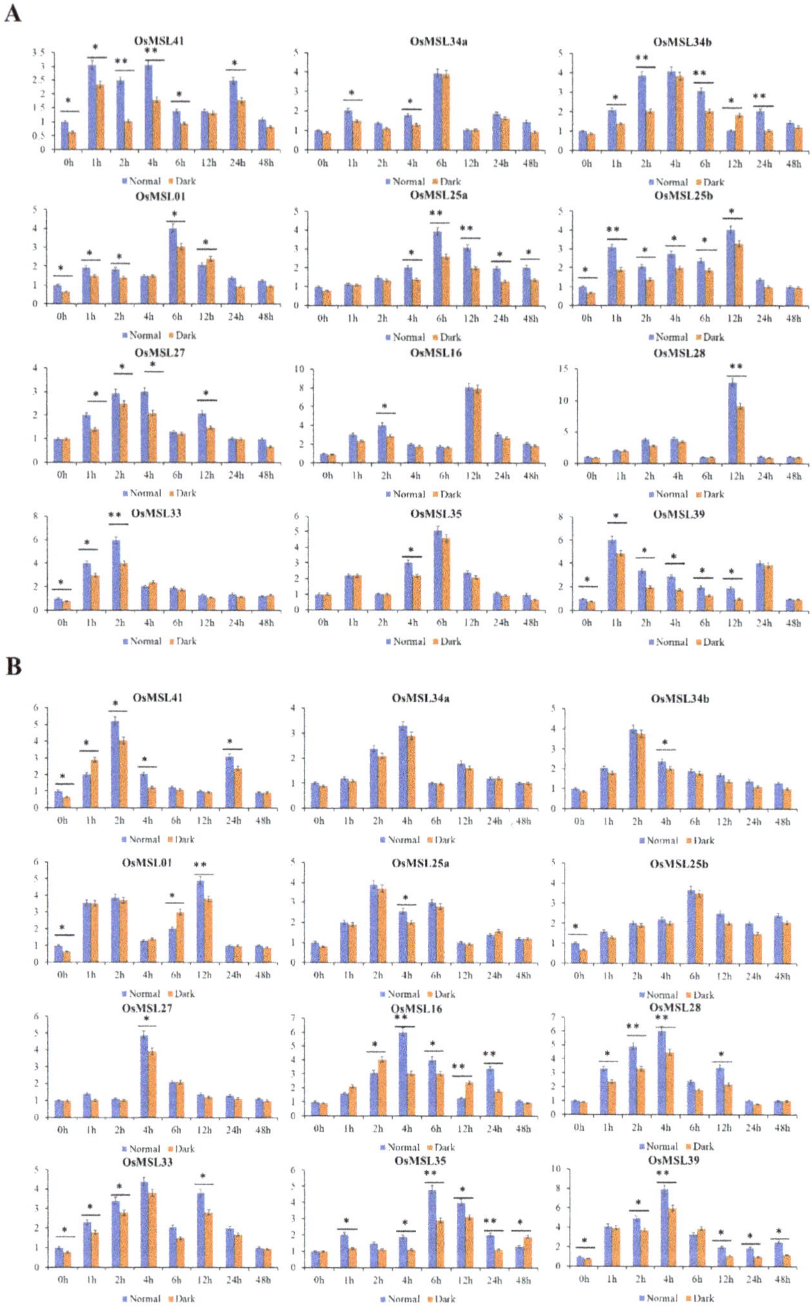

Figure 11. Expression profiles of 12 selected OsMSLs in response to (**A**) drought, (**B**) high salt stress, treatments. Data were normalized to the β-actin gene. Vertical bars indicate standard deviations. Asterisks indicate corresponding genes significantly upregulated or downregulated compared with the control. (* $p < 0.05$; ** $p < 0.01$; Student's *t*-test).

To determine whether light influences trihelix genes expression, the expression levels of 12 *OsMSLs* from each treatment group under light and dark conditions were investigated (Figures 10 and 11). Overall, some *OsMSLs* were induced under normal conditions. For example, at 0 h, *OsMSL01*, *OsMSL25b*, *OsMSL33*, *OsMSL39*, and *OsMSL41* expression levels significantly differed between the two conditions. Several genes began to be light-regulated expressed under different treatments. For instance, after ABA exposure, the expression levels of *OsMSL16*, *OsMSL25a*, *OsMSL25b*, *OsMSL28*, *OsMSL34a*, and *OsMSL39* at different time points under normal conditions were significantly higher than those at the same intervals under dark conditions. In contrast, some genes were repressed by light at different points, such as the expression levels of *OsMSL41* at 2 h under ABA treatment, *OsMSL28* at 6 h under hydrogen peroxide treatment, *OsMSL34b* at 12 h under drought stress and *OsMSL16* at 12 h under high salty stress were higher in the dark than under normal conditions. Therefore, the relationship between trihelix genes and light is complex and trihelix gene expression is not directly regulated by light but may be controlled by multiple regulatory mechanisms.

3. Discussion

The trihelix DNA binding domain is usually associated with that of Myb/SANT LIKE. There are three strongly conserved, regularly spaced tryptophan (W) residues in each repeating Myb α-helix. The residues of the Myb α-helix regions are also strongly conserved between the GT factors and the Myb/SANT-LIKE proteins. Individual helices are longer, so the trihelix domains of the GT factors and the Myb/SANT LIKE proteins are related [20]. In contrast, other amino acids at this location have longer Myb repeat sequences than the helix-turn-helix structure formed by the Myb-type DNA-binding domain. Therefore, the trihelix family genes have functions and target gene sequences differing from those of the MYB transcription factor family [20]. In the present study, 41 rice trihelix genes were identified using the Myb-type (Myb/SANT-LIKE) DNA-binding domain. However, in a previous study, only 31 rice trihelix genes were identified [21]. Because the repeated search method was performed in the present study, the trihelix gene could be identified in the rice genome more comprehensively. Based on the wide ranges of *OsMSL* protein MW, isoelectric point, and subcellular localization, we speculated that *OsMSLs* are not conservatively evolved. Besides, in *Arabidopsis*, the chimeric trihelix gene At4g17060 was misannotated in the genome [22]. Its ortholog LOC_Os10g41460 (*OsMSL36*), was identified in this study.

In general, gene families expand by tandem and segmental duplications [23]. Evolutionary conservatism increased with the number of duplicated genes in a gene family [16]. Chromosomal distribution and gene duplication analyses indicated that there are six pairs of duplicated genes in rice among total 41 trihelix gene, including two pairs of tandem genes and four pairs of segmental genes. However, in *Arabidopsis*, fifteen pairs of duplicated trihelix genes were previously detected among 34 trihelix genes [24]. These results suggested that *OsMSLs* are less conserved, and most genes may not originate from the same ancestor. On the other hand, these results demonstrate that the rice trihelix family has a high degree of evolutionary divergence and is non-conservative. These properties may account for the substantial differences among the rice trihelix proteins.

We conducted a phylogenetic analysis to elucidate the evolutionary relationships within the rice and other species trihelix gene family. In a previous study, no GTδ was designated for rice or *Arabidopsis* [18] because the *Arabidopsis* family Myb/SANT-LIKE DNA-binding domain or protein sequence was aligned with the rice genome in the attempt to identify the potential rice trihelix gene. However, there is a relatively long evolutionary distance between rice and *Arabidopsis*. Therefore, this method may overlook the specific trihelix genes in rice. Misclassification may have resulted in deviations because of the influence of the *Arabidopsis* trihelix genes. In the present study, a class of rice trihelix genes and some tomato trihelix genes previously assigned to this subfamily have been found in the GTδ clade. Subsequent investigation revealed that this subfamily has a high structure similarity. For this reason, the evolutionary relationships of its members are more conservative than those in other subgroups. In previous studies, the GT clade was divided into the GT1 and

GT2 subfamilies cited [13,14,24]. According to our evolutionary analysis, there are two GT subfamily clusters (Figure 3A). The difference between GT1 and GT2 is smaller than those among SIP1, SH4, GTγ, and GTδ. The genes in GT1 each have one trihelix DNA binding domain with three conserved tryptophans. Those in GT2 each have an additional trihelix DNA binding domain with two tryptophans and one phenylalanine [25]. Consequently, we classified both the GT1 and GT2 clades in the GT subfamily.

Although the evolution of the trihelix family was not conservative, our gene structure and conservative functional domain analyses indicated that the genes within the same subfamilies (especially SIP1 and GTδ) were still relatively conserved. In fact, most duplicated genes occurred in GTδ. The MEME analysis revealed that the functional domain distribution of each *OsMSL* was related to its classification (Figure 4). Therefore, these conserved functional domains play central roles in group-specific functions. In contrast, the gene structures among the various groups differed greatly from the conserved functional domains. For this reason, they may have different downstream regulatory genes and participate in different signaling pathways.

The distribution and type of *cis*-elements on the gene promoter may determine *OsMSL* functions. In this study, we identified five *cis*-elements shared by all genes out of a large number among 41 *OsMSLs*. The results showed that *OsMSLs* were mainly involved in abiotic stress and light-induced responses. To date, little functional analysis has been conducted on the trihelix transcription factors in plants. Previous studies showed that they participate in responses to pathogens and abiotic stress, light induction, and nitrogen metabolism [18]. The light-induced process is a major feature of the trihelix genes. Light induces massive reprogramming of the plant transcriptome and upregulates or downregulates gene expression and its corresponding signaling pathway [26]. Light signaling coordinates the induction or repression of specific downstream genes like bHLH [27], bZIP [28], R2R3-MYB [29], FAR1 [30], and FHY3 [31]. Studies on light regulatory mechanisms in plants focused on the long-term effects of light exposure. However, little attention has been paid to the transient light-responsive processes of transcription factors in plant stress reactions. In the present study, the *OsMSL* expression profiles disclosed that their responses to light are transient and change with the processing time. Therefore, *OsMSLs* may be regulated by light in response to abiotic stress in the same way that phototropism, chloroplast movement, and stomatal opening participate in rapid light-responsive processes and are not under extensive transcriptional regulation. This mechanism substantially differs from that observed in relation to gene expression changes in response to the long-term effects of light on photoperiod.

Gene expression specificity in plant tissues and developmental stages may indicate possible gene functions. Previous studies showed that certain trihelix genes in tomato and chrysanthemum exhibited stable expression levels in all tissues [8,14]. However, most *OsMSLs* do not maintain stable expression levels in different tissues. *OsMSL* expression levels may vary substantially among tissues. For example, the expression levels of *OsMSL16*, *OsMSL27*, *OsMSL28*, *OsMSL35*, and *OsMSL39* were extremely high in the leaves and sheaths but comparatively low in the roots and stems. Subcellular localization revealed that these genes are expressed in the chloroplasts, and these are present only in the rice leaves and sheaths. The expression patterns of *OsMSL16*, *OsMSL27*, *OsMSL28*, *OsMSL35*, and *OsMSL39* in GTδ were similar. Therefore, the different *OsMSLs* within the same group explain the parallel functions of the rice trihelix family. *OsMSL02*, *OsMSL08*, and *OsMSL14* in Group II were expressed at low levels in all four tissues. Either they are inducible or they are only upregulated under special conditions [32].

High *OsMSL* expression was observed mainly in 7–40 days seedlings (from germination to early tillering). Therefore, they indicate that *OsMSLs* contribute primarily to the early stages of rice growth, as do many other genes. For example, *ZFP182* overexpression enhanced salt, drought, and cold tolerance in transgenic rice seedlings [33]. Loss of the ABA transporter *OsPM1* in 35 days rice seedlings conferred greater drought sensitivity than that seen in the WT [34]. The expression patterns of *OsMSL34a* and *OsMSL34b* disclosed that they were, in fact, different variable splicing forms of the same gene with very different expression levels. Since *OsMSL34b* might be the primary variable splicing form of the gene, *OsMSL34a* was downregulated to some extent.

In the previous study on rice trihelix genes, their responses to various plant hormones were highlighted [21], and the present study focused on abiotic stress and stress signaling molecules. Other than plant growth and development, MSLs participate in stress responses [18]. Although no *cis*-elements related to the ABA signaling pathway were found in the *OsMSL* promoter region, ABA nonetheless, induced the rice trihelix genes. ABA accumulates when plants are subjected to a water deficit. It regulates the expression of drought stress-related genes and modulates the molecular, cellular, and physiological mechanisms for adaptation to environmental stress [35]. In chrysanthemum, however, ABA downregulated the trihelix genes but others were upregulated after prolonged ABA exposure [14]. In contrast, *GmGT-2A* and *GmGT-2B* in soybean were upregulated by ABA [11]. In the present study, twelve *OsMSLs* from all subfamilies have been induced by ABA. These results suggest that the signaling mechanisms of the trihelix family genes vary with species. Whether the ABA presence has a negative regulation on the trihelix gene requires further experimental verification.

ROS (reactive oxygen species) are produced in response to most environmental stress. Excessive ROS accumulation may irreversibly damage cells [36,37]. In previous studies, trihelix family genes had at least one response during plant osmotic stress defense [18]. It follows that trihelix family genes may also participate in ROS scavenging and enhance plant tolerance to various stresses. However, little is known about the mechanism of the peroxide reaction mediated by the trihelix family. We performed a quantitative PCR analysis on 12 *OsMSLs* subjected to hydrogen peroxide. All 12 *OsMSLs* responded to hydrogen peroxide stress. Therefore, they may help improve the permeation tolerance by increasing the ROS scavenging capacity in rice.

The trihelix transcription factors bind to GT elements on the light-regulating genes [25]. In darkness, photo regulatory genes are repressed and their associated trihelix family genes are also affected. Nevertheless, certain trihelix family genes are downregulated in response to light exposure, apparently because they must be repressed to be able to downregulate target whose expression is light-dependent. Certain constitutively expressed trihelix genes occur in *Arabidopsis*. Their expression is ubiquitous and indifferent to the light regime. They are concentrated mainly in GT1 and GT2 [38]. In the present study, however, the expression of the 12 rice trihelix genes changed direction at least once after light or dark treatment. Therefore, they may be inducible rather than constitutive. It remains to be determined whether *OsMSLs* are regulated by one wavelength or a wide light spectrum. Further investigation of the light path and its components is necessary. The results of our study have helped initiate the research of the rice trihelix transcription factors. In future research, the relationships among the *OsMSLs*, ABA-mediated dehydration stress tolerance, and ROS scavenging ability under light regulation should be explored.

4. Materials and Methods

4.1. Identification and Sequence Analysis of Trihelix Transcription Factor Family in Rice

The rice trihelix transcription factors were identified according to a previously described method with minor changes [39]. The Hidden Markov Model of Myb/SANT-LIKE domain (PF13837) was downloaded from the Pfam database (http://pfam.xfam.org/) [40]. The entire rice amino acid, genome, and CDS sequence assembly and corresponding annotation were downloaded from the EnsemblPlants database (http://plants.ensembl.org/index.html) [41]. The candidate proteins were sought by the HMMSEARCH program (https://www.ebi.ac.uk/Tools/hmmer/search/hmmsearch) base on the Bio-Linux system (Dr Tracey Timms-Wilson, Centre for Ecology & Hydrology (CEH), Oxfordshire, UK). The domain sequences of these candidate proteins were extracted and used to build a rice-specific Hidden Markov Model. All rice proteins were detected by the rice-specific Hidden Markov Model. Those with E-value < 0.01 were selected. The trihelix proteins were verified using the Pfam and InterPro databases (http://www.ebi.ac.uk/interpro/) [42]. Proteins obtained by the domain and database screening confirmation were considered trihelix family members. The corresponding CDS and gene sequences were extracted according to their protein identifications.

The MEME program (http://meme-suite.org/) identified conserved motifs of the trihelix family proteins with the following parameters: Any number of repetitions; minimum seven motifs; maximum 49 motifs; optimum 10–200 amino acids; expected E-value $< 1 \times 10^{-48}$. The trihelix family gene structures were displayed by comparing the coding and genomic sequences with the Gene Structure Display Server tools (http://gsds.cbi.pku.edu.cn/) [43]. The chromosomal locations of the trihelix family genes were mapped onto the rice linkage map with an online tool according to their TIGR numbers [44]. The isoelectric points and molecular weights of trihelix family proteins were estimated with ExPASy (http://expasy.org/) [45].

4.2. Phylogenetic Analysis

A multiple alignment was performed with the full-length amino acid sequences of the rice trihelix family proteins using MEGA v. 7.0 (https://www.megasoftware.net/) [46]. Unrooted trees were constructed by the maximum-likelihood (ML) method with the following parameters: Poisson correction; pairwise deletion; 1000 bootstrap replicates.

4.3. Gene Duplication and Ka/Ks Analysis

Synteny blocks of the rice genome were downloaded from the Plant Genome Duplication Database (PGDD, http://chibba.agtec.uga.edu/duplication/) [47]. Duplicated *OsMSLs* pairs were connected by solid lines.

4.4. Cis-Element Analysis of Trihelix Transcription Factor Family

Promoters of the trihelix family genes were downloaded from the Phytozome database (https://phytozome.jgi.doe.gov/pz/portal.html#) [48]. The PLACE database (https://sogo.dna.affrc.go.jp/) was used to analyze the *cis*-regulatory elements of the trihelix family gene promoters [49].

4.5. Plant Growth Conditions and Treatments

Nipponbare rice seeds (*O. sativa* L. *ssp. japonica*) were surface-sterilized with 10% sodium hypochlorite solution for 30 min then sown on 1/2 MS (Murashige & Skoog) solid medium and cultured in a light incubator. After 2 weeks, the seedlings were at the two true leaf stage. They were transplanted into Hoagland's nutrient solution and cultured in an artificial climate chamber under controlled conditions (14 h light at 28 °C/10 h dark at 22 °C; relative humidity 70%). The rice seedlings were subjected to various stresses at the three-leaf stage (4 weeks) [50].

For the drought, salt, and hydrogen peroxide stress treatments, the rice seedlings were transferred to Hoagland's nutrient solution containing 20% polyethylene glycol (PEG)-6000 (*w*/*v*), 150 mM NaCl, or 2% hydrogen peroxide (*v*/*v*), respectively. For the ABA treatment, the rice seedlings were cultured on 1/2 MS solid medium containing 10 µM ABA. The control group was maintained on normal nutrient solution or medium. All other culture conditions were the same as described above. Treated rice tissues were harvested at 0 h, 1 h, 2 h, 4 h, 6 h, 12 h, 24 h, and 48 h. The samples were immediately placed in liquid nitrogen and stored at −80 °C until use. Untreated material was used as a control. The experimental procedure was repeated at least three times.

4.6. Expression Analysis of Trihelix Transcription Factor Family

Total RNA was extracted from rice tissues by the TRIzol method (Thermo Fisher Scientific, Waltham, MA, USA) and treated with DNase to eliminate any DNA contamination. RNA quality was assessed by electrophoresis and stored at −80 °C until use. First-strand cDNA (10 µL) was synthesized according to the instructions for the PrimeScript™ RT Master Mix (Takara Biomedical Technology (Beijing) Co., Ltd., Beijing, China). Primers were designed with Primer Premier v. 5.0

(PREMIER Biosoft International, Palo Alto, CA, USA) and were based on the trihelix gene family transcript sequences. Gene specific primers for quantitative real-time PCR are listed in Table S6. Primer amplification specificity was verified in the rice genome database using Blast from NCBI (https://www.ncbi.nlm.nih.gov/) [51]. Rice β-actin was the internal reference gene. Quantitative real-time PCR was performed in the ABI 7300 Real Time PCR System (Applied Biosystems, Foster City, CA, USA) using SYBR Green chemistry and reaction mix consists of 10 μL SYBR qPCR Master Mix (Vazyme Biotech Co.,Ltd., Nanjing, China), 0.4 μL upstream and downstream primers respectively, 0.4 μL ROX, 2 μL cDNA (10 times dilution) and 6.8 μL ddH_2O to 20 μL. The PCR reaction protocol was 95 °C for 5 min; 95 °C for 10 s; 60 °C for 20 s; 72 °C for 20 s; 45 cycles. Gene expression levels were calculated by the $2^{-\Delta\Delta CT}$ method: $\Delta\Delta CT = (CT_{target} - CT_{actin})$ at time x − $(CT_{target} - CT_{actin})$ at time 0 [52]. The test was repeated three times. Expression data for the rice trihelix family genes were retrieved from the Expression Atlas database (https://www.ebi.ac.uk/gxa/home) [53]. Heatmaps were created in HemI v.1.0 (The CUCKOO Workgroup, Hubei, China) and based on the expression data [54].

5. Conclusions

Trihelix transcription factors participate in many plant biological processes but have not been systematically studied. Here, 41 rice trihelix transcription factors were identified by bioinformatics analysis. Gene synteny analysis showed that *OsMSLs* are less conserved, and most genes may not originate from the same ancestor. Phylogenetic analysis categorized them into five subfamilies. The gene structures and conserved functional domains of the rice trihelix transcription factors varied greatly but they shared five cis-elements governing light dehydration stress responses. Expression pattern analysis revealed that the trihelix transcription factors had the highest expression levels in the early rice growth stages and most of the strongly upregulated genes were localized in the leaves and sheaths. Real-time quantitative PCR analysis of the trihelix family genes subjected to various stressors or ABA revealed that they were induced in response to drought, high salt, hydrogen peroxide, and ABA. However, these responses were also regulated by light and individual *OsMSL* expressions differed with the presence or absence of light. Our study helped elucidate the biological functions of the trihelix transcription factors in rice.

Supplementary Materials: Supplementary materials can be found at http://www.mdpi.com/1422-0067/20/2/251/s1.

Author Contributions: Conceptualization, J.L.; data curation, X.M.; formal analysis, J.S.; funding acquisition, D.Z.; investigation, J.W. (Jing Wang) and J.W. (Jingguo Wang); project administration, H.Z. (Hongwei Zhao); resources, H.Z. (Hualong Liu); software, H.Z. (Hongliang Zheng); supervision, Z.Z.; writing—original draft, J.L.; Writing—review & editing, M.Z. All authors read and approved the final manuscript.

Funding: This research was funded by 1. National Science and Technology Major Project (2018ZX0800912B-002). 2. Youth Science Foundation of Heilongjiang Province (QC2017015). 3. National Natural Science Foundation (31701507). 4. Natural Science Foundation of Heilongjiang Province of China (No. C2015003).

Acknowledgments: We thank Editage for its linguistic assistance during the preparation of this manuscript.

Conflicts of Interest: The authors declare no conflict of interest.

References

1. Lindemose, S.; O'Shea, C.; Jensen, M.K.; Skriver, K. Structure, Function and Networks of Transcription Factors Involved in Abiotic Stress Responses. *Int. J. Mol. Sci.* **2013**, *14*, 5842–5878. [CrossRef] [PubMed]
2. Jin, J.P.; Tian, F.; Yang, D.C.; Meng, Y.Q.; Kong, L.; Luo, J.C.; Gao, G. PlantTFDB 4.0: Toward a central hub for transcription factors and regulatory interactions in plants. *Nucleic. Acids Res.* **2017**, *45*, D1040–D1045. [CrossRef] [PubMed]
3. Riechmann, J.L.; Heard, J.; Martin, G.; Reuber, L.; Jiang, C.; Keddie, J.; Adam, L.; Pineda, O.; Ratcliffe, O.J.; Samaha, R.R. Arabidopsis transcription factors: Genome-wide comparative analysis among eukaryotes. *Science* **2000**, *290*, 2105. [CrossRef] [PubMed]

4. Green, P.J.; Kay, S.A.; Chua, N.H. Sequence-specific interactions of a pea nuclear factor with light-responsive elements upstream of the rbcS-3A gene. *EMBO J.* **1987**, *6*, 2543–2549. [CrossRef] [PubMed]
5. Nagano, Y. Several Features of the GT-Factor Trihelix Domain Resemble Those of the Myb DNA-Binding Domain. *Plant Physiol.* **2000**, *124*, 491–493. [CrossRef] [PubMed]
6. Qin, Y.; Ma, X.; Yu, G.H.; Wang, Q.; Wang, L.; Kong, L.R.; Kim, W.; Wang, H.W. Evolutionary History of Trihelix Family and Their Functional Diversification. *DNA Res.* **2014**, *21*, 499–510. [CrossRef]
7. Gao, M.J.; Lydiate, D.J.; Li, X.; Lui, H.; Gjetvaj, B.; Hegedus, D.D.; Rozwadowski, K. Repression of Seed Maturation Genes by a Trihelix Transcriptional Repressor in Arabidopsis Seedlings. *Plant Cell* **2009**, *21*, 54–71. [CrossRef] [PubMed]
8. Yu, C.Y.; Cal, X.F.; Ye, Z.B.; Li, H.X. Genome-wide identification and expression profiling analysis of trihelix gene family in tomato. *Biochem. Biophys. Res. Commun.* **2015**, *468*, 653–659. [CrossRef] [PubMed]
9. Murata, J.; Takase, H.; Hiratsuka, K. Characterization of a Novel GT-box Binding Protein from Arabidopsis. *Plant Biotechnol.* **2002**, *19*, 103–112. [CrossRef]
10. Wang, R.H.G.; Han, B. Transcript abundance of rml1, encoding a putative GT1-like factor in rice, is up-regulated by Magnaporthe grisea and down-regulated by light. *Gene* **2004**, *324*, 105–115. [CrossRef] [PubMed]
11. Xie, Z.M.; Zou, H.F.; Lei, G.; Wei, W.; Zhou, Q.Y.; Niu, C.F.; Liao, Y.; Tian, A.G.; Ma, B.; Zhang, W.K. Soybean Trihelix Transcription Factors GmGT-2A and GmGT-2B Improve Plant Tolerance to Abiotic Stresses in Transgenic Arabidopsis. *PLoS ONE* **2009**, *4*, e6898. [CrossRef]
12. Chan, Y.Y.; Pence, H.E.; Jing, B.J.; Miura, K.; Gosney, M.J.; Hasegawa, P.M.; Mickelbart, M.V. The Arabidopsis GTL1 Transcription Factor Regulates Water Use Efficiency and Drought Tolerance by Modulating Stomatal Density via Transrepression of SDD1. *Plant Cell* **2010**, *22*, 4128.
13. Fang, Y.; Xie, K.; Xin, H.; Hu, H.; Xiong, L. Systematic analysis of GT factor family of rice reveals a novel subfamily involved in stress responses. *Mol. Genet. Genom.* **2010**, *283*, 157–169. [CrossRef] [PubMed]
14. Song, A.P.; Wu, D.; Fan, Q.Q.; Tian, C.; Chen, S.M.; Guan, Z.Y.; Xin, J.J.; Zhao, K.K.; Chen, F.D. Transcriptome-Wide Identification and Expression Profiling Analysis of Chrysanthemum Trihelix Transcription Factors. *Int. J. Mol. Sci.* **2016**, *17*, 198. [CrossRef] [PubMed]
15. Li, C.; Zhou, A.; Sang, T. Rice Domestication by Reducing Shattering. *Science* **2006**, *311*, 1936–1939. [CrossRef] [PubMed]
16. Holub, E.B. The arms race is ancient history in Arabidopsis, the wildflower. *Nat. Rev. Genet.* **2001**, *2*, 516–527. [CrossRef]
17. Wang, Y.; Tang, H.; DeBarry, J.D.; Tan, X.; Li, J.; Wang, X.; Lee, T.-H.; Jin, H.; Marler, B.; Guo, H. MCScanX: A toolkit for detection and evolutionary analysis of gene synteny and collinearity. *Nucleic Acids Res.* **2012**, *40*, e49. [CrossRef] [PubMed]
18. KaplanLevy, R.N.; Brewer, P.B.; Quon, T.; Smyth, D.R. The trihelix family of transcription factors—Light, stress and development. *Trends Plant Sci.* **2012**, *17*, 163–171. [CrossRef] [PubMed]
19. Markus, S.; Davison, T.S.; Henz, S.R.; Pape, U.J.; Monika, D.; Martin, V.; Bernhard, S.L.; Detlef, W.; Lohmann, J.U. A gene expression map of Arabidopsis thaliana development. *Nat. Genet.* **2005**, *37*, 501–506.
20. Nagata, T.; Niyada, E.; Fujimoto, N.; Nagasaki, Y.; Noto, K.; Miyanoiri, Y.; Murata, J.; Hiratsuka, K.; Katahira, M. Solution structures of the trihelix DNA-binding domains of the wild-type and a phosphomimetic mutant of Arabidopsis GT-1: Mechanism for an increase in DNA-binding affinity through phosphorylation. *Proteins* **2010**, *78*, 3033–3047. [CrossRef] [PubMed]
21. Jianhui, J.; Yingjun, Z.; Hehe, W.; Liming, Y. Genome-wide analysis and functional prediction of the Trihelix transcription factor family in rice. *Hereditas* **2015**, *37*, 1228. (In Chinese)
22. Geraldo, N.; Baurle, I.; Kidou, S.; Hu, X.; Dean, C. FRIGIDA Delays Flowering in Arabidopsis via a Cotranscriptional Mechanism Involving Direct Interaction with the Nuclear Cap-Binding Complex. *Plant Physiol.* **2009**, *150*, 1611. [CrossRef]
23. Cannon, S.B.; Mitra, A.; Baumgarten, A.; Young, N.D.; May, G. The roles of segmental and tandem gene duplication in the evolution of large gene families in Arabidopsis thaliana. *BMC Plant Biol.* **2004**, *4*, 10. [CrossRef] [PubMed]
24. Ali, M.A.; Yasmeen, E.; Riaz, M.; Azeem, F.; Sultan, S.; Abbas, A.; Riaz, K. Genome-wide analysis of trihelix Transcription factor gene family in Arabidopsis thaliana. *Pak. J. Agric. Sci.* **2016**, *53*, 439–448.

25. Lam, E. Domain analysis of the plant DNA-binding protein GT1a: Requirement of four putative alpha-helices for DNA binding and identification of a novel oligomerization region. *Mol. Cell. Biol.* **1995**, *15*, 1014–1020. [CrossRef] [PubMed]
26. Jiao, Y.; Lau, O.S.; Deng, X.W. Light-regulated transcriptional networks in higher plants. *Nat. Rev. Genet.* **2007**, *8*, 217–230. [CrossRef] [PubMed]
27. Toledoortiz, G.; Huq, E.; Quail, P.H. The Arabidopsis Basic/Helix-Loop-Helix Transcription Factor Family. *Plant Cell* **2003**, *15*, 1749–1770. [CrossRef]
28. Jakoby, M.; Weisshaar, B.; Dröge-Laser, W.; Vicente-Carbajosa, J.; Tiedemann, J.; Kroj, T.; Parcy, F. bZIP transcription factors in Arabidopsis. *Trends Plant Sci.* **2002**, *7*, 106–111. [CrossRef]
29. Ballesteros, M.L.; Bolle, C.; Lois, L.M.; Moore, J.M.; Viellecalzada, J.P.; Grossniklaus, U.; Chua, N.H. LAF1, a MYB transcription activator for phytochrome A signaling. *Gene Dev.* **2001**, *15*, 2613–2625. [CrossRef]
30. Hudson, M.; Ringli, C.; Boylan, M.T.; Quail, P.H. The FAR1 locus encodes a novel nuclear protein specific to phytochrome A signaling. *Gene Dev.* **1999**, *13*, 2017–2027. [CrossRef]
31. Wang, H.; Deng, X.W. Arabidopsis FHY3 defines a key phytochrome A signaling component directly interacting with its homologous partner FAR1. *EMBO J.* **2002**, *21*, 1339–1349. [CrossRef]
32. Tang, Y.; Qin, S.; Guo, Y.; Chen, Y.; Wu, P.; Chen, Y.; Li, M.; Jiang, H.; Wu, G. Genome-Wide Analysis of the AP2/ERF Gene Family in Physic Nut and Overexpression of the JcERF011 Gene in Rice Increased Its Sensitivity to Salinity Stress. *PLoS ONE* **2016**, *11*, e0150879. [CrossRef] [PubMed]
33. Huang, J.; Sun, S.; Xu, D.; Lan, H.; Sun, H.; Wang, Z.; Bao, Y.; Wang, J.; Tang, H.; Zhang, H. A TFIIIA-type zinc finger protein confers multiple abiotic stress tolerances in transgenic rice (Oryza sativa L.). *Plant Mol. Biol.* **2012**, *80*, 337. [CrossRef] [PubMed]
34. Yao, L.; Cheng, X.; Gu, Z.; Huang, W.; Li, S.; Wang, L.; Wang, Y.F.; Xu, P.; Ma, H.; Ge, X. The AWPM-19 Family Protein OsPM1 Mediates Abscisic Acid Influx and Drought Response in Rice. *Plant Cell* **2018**, *30*, 1258–1276. [CrossRef]
35. Yamaguchi-Shinozaki, K.; Shinozaki, K. Transcriptional regulatory networks in cellular responses and tolerance to dehydration and cold stresses. *Annu. Rev. Plant Biol.* **2006**, *57*, 781–803. [CrossRef] [PubMed]
36. Apel, K.; Hirt, H. Reactive oxygen species: Metabolism, oxidative stress, and signal transduction. *Annu. Rev. Plant Biol.* **2004**, *55*, 373–399. [CrossRef] [PubMed]
37. Miller, G.; Suzuki, N.; Ciftci-Yilmaz, S.; Mittler, R. Reactive oxygen species homeostasis and signalling during drought and salinity stresses. *Plant Cell Environ.* **2010**, *33*, 453–467. [CrossRef] [PubMed]
38. Gilmartin, P.M.; Memelink, J.; Hiratsuka, K.; Kay, S.A.; Chua, N.H. Characterization of a gene encoding a DNA binding protein with specificity for a light-responsive element. *Plant Cell* **1992**, *4*, 839–849. [CrossRef]
39. Lozano, R.; Hamblin, M.T.; Prochnik, S.; Jannink, J.L. Identification and distribution of the NBS-LRR gene family in the Cassava genome. *BMC Genom.* **2015**, *16*, 360. [CrossRef] [PubMed]
40. El-Gebali, S.; Mistry, J.; Bateman, A.; Eddy, S.R.; Luciani, A.; Potter, S.C.; Qureshi, M.; Richardson, L.J.; Salazar, G.A.; Smart, A. The Pfam protein families database in 2019. *Nucleic Acids Res.* **2018**, *28*, 263–266. [CrossRef]
41. Kersey, P.J.; Allen, J.E.; Allot, A.; Barba, M.; Boddu, S.; Bolt, B.J.; Carvalho-Silva, D.; Christensen, M.; Davis, P.; Grabmueller, C. Ensembl Genomes 2018: An integrated omics infrastructure for non vertebrate species. *Nucleic Acids Res.* **2018**, *46*, D802–D808. [CrossRef] [PubMed]
42. Mitchell, A.L.; Attwood, T.K.; Babbitt, P.C.; Blum, M.; Bork, P.; Bridge, A.; Brown, S.D.; Chang, H.-Y.; El-Gebali, S.; Fraser, M.I. InterPro in 2019: Improving coverage, classification and access to protein sequence annotations. *Nucleic Acids Res.* **2018**. [CrossRef] [PubMed]
43. Hu, B.; Jin, J.; Guo, A.Y.; Zhang, H.; Luo, J.; Gao, G. GSDS 2.0: An upgraded gene feature visualization server. *Bioinformatics* **2014**, *31*, 1296. [CrossRef] [PubMed]
44. Kurata, N.; Yamazaki, Y. Oryzabase. An integrated biological and genome information database for rice. *Plant Physiol.* **2006**, *140*, 12–17. [CrossRef] [PubMed]
45. Artimo, P.; Jonnalagedda, M.; Arnold, K.; Baratin, D.; Csardi, G.; De Castro, E.; Duvaud, S.; Flegel, V.; Fortier, A.; Gasteiger, E. ExPASy: SIB bioinformatics resource portal. *Nucleic Acids Res.* **2012**, *40*, W597–W603. [CrossRef] [PubMed]
46. Kumar, S.; Stecher, G.; Tamura, K. MEGA7: Molecular evolutionary genetics analysis version 7.0 for bigger datasets. *Mol. Biol. Evol.* **2016**, *33*, 1870–1874. [CrossRef] [PubMed]

47. Lee, T.H.; Tang, H.; Wang, X.; Paterson, A.H. PGDD: A database of gene and genome duplication in plants. *Nucleic Acids Res.* **2013**, *41*, 1152–1158. [CrossRef] [PubMed]
48. Goodstein, D.M.; Shu, S.; Howson, R.; Neupane, R.; Hayes, R.D.; Fazo, J.; Mitros, T.; Dirks, W.; Hellsten, U.; Putnam, N. Phytozome: A comparative platform for green plant genomics. *Nucleic Acids Res.* **2011**, *40*, D1178–D1186. [CrossRef]
49. Higo, K.; Ugawa, Y.; Iwamoto, M.; Korenaga, T. Plant cis-acting regulatory DNA elements (PLACE) database: 1999. *Nucleic Acids Res.* **1999**, *27*, 297–300. [CrossRef] [PubMed]
50. Ren, J.; Gao, F.; Wu, X.; Lu, X.; Zeng, L.; Lv, J.; Su, X.; Luo, H.; Ren, G. Bph32, a novel gene encoding an unknown SCR domain-containing protein, confers resistance against the brown planthopper in rice. *Sci. Rep.* **2016**, *6*, 37645. [CrossRef] [PubMed]
51. Coordinators, N.R. Database resources of the national center for biotechnology information. *Nucleic Acids Res.* **2017**, *45*, D12.
52. Livak, K.J.; Schmittgen, T.D. Analysis of relative gene expression data using real-time quantitative PCR and the 2(-Delta Delta C(T)) Method. *Methods* **2001**, *25*, 402–408. [CrossRef] [PubMed]
53. Papatheodorou, I.; Fonseca, N.A.; Keays, M.; Tang, Y.A.; Barrera, E.; Bazant, W.; Burke, M.; Füllgrabe, A.; Fuentes, A.M.-P.; George, N. Expression Atlas: Gene and protein expression across multiple studies and organisms. *Nucleic Acids Res.* **2017**, *46*, D246–D251. [CrossRef] [PubMed]
54. Deng, W.; Wang, Y.; Liu, Z.; Cheng, H.; Xue, Y. HemI: A Toolkit for Illustrating Heatmaps. *PLoS ONE* **2014**, *9*, e111988. [CrossRef] [PubMed]

© 2019 by the authors. Licensee MDPI, Basel, Switzerland. This article is an open access article distributed under the terms and conditions of the Creative Commons Attribution (CC BY) license (http://creativecommons.org/licenses/by/4.0/).

International Journal of *Molecular Sciences*

MDPI

Article

Identification, Characterization, and Expression Patterns of TCP Genes and microRNA319 in Cotton

Zujun Yin [1], Yan Li [1], Weidong Zhu [1], Xiaoqiong Fu [1], Xiulan Han [2], Junjuan Wang [1], Huan Lin [1] and Wuwei Ye [1,*]

1 Research Base, Zhengzhou University, State Key Laboratory of Cotton Biology, Institute of Cotton Research of Chinese Academy of Agricultural Sciences, Anyang 455000, Henan, China; yinzujun@caas.cn (Z.Y.); lycaas@163.com (Y.L.); weidongzhu17@163.com (W.Z.); m13663726262@163.com (X.F.); wangjj@cricaas.com.cn (J.W.); linhuan0829@163.com (H.L.)

2 State Key Laboratory of Crop Biology, College of Agronomy, Shandong Agricultural University, Taian 271000, Shandong, China; genetics@sdau.edu.cn

* Correspondence: yewuwei@caas.cn; Tel.: +86-372-2562283

Received: 17 October 2018; Accepted: 6 November 2018; Published: 20 November 2018

Abstract: The TEOSINTE BRANCHED 1, CYCLOIDEA, and PROLIFERATING CELL FACTORS (TCP) gene family is a group of plant-specific transcription factors that have versatile functions in developmental processes and stress responses. In this study, a total of 73 *TCP* genes in upland cotton were identified and characterizated. Phylogenetic analysis classified them into three subgroups: 50 belonged to PCF, 16 to CIN, and 7 to CYC/TB1. *GhTCP* genes are randomly distributed in 22 of the 26 chromosomes in cotton. Expression patterns of *GhTCPs* were analyzed in 10 tissues, including different developmental stages of ovule and fiber, as well as under heat, salt, and drought stresses. Transcriptome analysis showed that 44 *GhTCP* genes exhibited varied transcript accumulation patterns in the tested tissues and 41 *GhTCP* genes were differentially expressed in response to heat, salt, and drought stresses. Furthermore, three *GhTCP* genes of the CIN clade were found to contain miR319-binding sites. An anti-correlation expression of *GhTCP21* and *GhTCP54* was analyzed with miR319 under salt and drought stress. Our results lay the foundation for understanding the complex mechanisms of GhTCP-mediated developmental processes and abiotic stress-signaling transduction pathways in cotton.

Keywords: upland cotton; TCP genes; abiotic stress; miR319; target genes

1. Introduction

Transcription factors are essential for the control of gene expression. Gene expression can be regulated by transcription factors that either activate or repress transcription, so they are vital for many cell biological process [1]. The TEOSINTE BRANCHED 1, CYCLOIDEA, and PROLIFERATING CELL FACTORS (TCP) gene family is a small group of transcription factors exclusive to higher plants [2]. This class of transcription factors has many functions in regulating diverse plant growth and development processes by controlling cell proliferation [3]. They are characterized by a highly conserved 59-amino-acid basic helix–loop–helix (bHLH) motif at the N-terminus designated as the TCP domain [4]. This domain is responsible for DNA binding, nuclear targeting, and is involved in protein–protein interactions [5]. Based on variation in the TCP domain, TCP family members can be classified into two classes: Class I (also known as the PCF or TCP-P class) and class II (also known as the TCP-C class) [6,7]. Class II is further subdivided into the CINCINNATA (CIN) and CYC/TB1 subgroups. In addition to the TCP domain, several class II members possess an 18–20-residue arginine-rich motif [8]. This so-called R domain was predicted to form a hydrophilic α-helix or a coiled-coil structure that mediates protein–protein interactions [2].

It has been reported that many TCP transcription factors participate in the regulation of diverse physiological and biological processes, such as phytohormone biosynthesis and signal transduction, branching, leaf morphogenesis flower development and senescence, pollen development, and regulation of the circadian clock in various plants [9–11]. In *Arabidopsis thaliana* seeds, *TCP14* was expressed in the vascular tissues of embryos. It promotes germination through antagonism of abscisic acid signaling [12]. *TCP1* is expressed in restricted areas of the flower meristem, leaf vasculature, and at the junctions of roots and hypocotyls. It mediates the expression of a key brassinosteroid (BR) biosynthetic gene by directly associating with the two GGNCCC motifs in the promoter region of *DWARF4* (*DWF4*) [13]. *DWF4* encodes a 22-hydroxylase and is responsible for multiple 22-hydroxylation steps during BR biosynthesis [14]. The expression levels of *DWF4* were positively correlated with *TCP1* abundance in *planta*. In Arabidopsis flowers, the gynoecium and silique development was modulated by *TCP15* through partly regulating auxin biosynthesis. The ectopic expression of Arabidopsis *TCP15* represses style and stigma development, thus producing gynoecia with decreased stigmatic tissue and/or carpel fusion defects in apical parts [15]. TCP17 and its two closely related homologs, TCP5 and TCP13, play an important role in mediating shade-induced hypocotyl elongation by up-regulating auxin biosynthesis via a PHYTOCHROME INTERACTING FACTORS (PIF)-dependent and a PIF-independent pathway [16]. In rice (*Oryza sativa*), *OsTCP19* was upregulated under salt and water-deficit stress. Overexpression of *OsTCP19* in Arabidopsis caused upregulation of *INDOLE-3-ACETIC ACID3*(*IAA3*), *ABSCISIC ACID INSENSITIVE 3*(*ABI3*), and *ABI4*, and downregulation of *LIPOXYGENASE2* (*LOX2*), thus leading to developmental abnormalities, such as less lateral roots [17]. MicroRNAs (miRNAs) are a class of smallnon-coding RNAs generated from single-strand hairpin RNA precursors. They regulate gene expression by binding to complementary sequences within target mRNAs [18]. Considerable progress has been made in identifying the targets of plant miRNAs. In Arabidopsis, five CIN-like *TCP* genes (*TCP2*, *TCP3*, *TCP4*, *TCP10*, and *TCP24*) were targeted by miR319 and have been implicated in regulating leaf morphogenesis [19]. Knockdown of a subset of Class II TCP transcription factors by overexpression of miR319 increases tolerance to dehydration and salinity stress in bentgrass (*Agrostis stolonifera*) [20]. Accumulated functional characterization of TCPs indicated their diverse function in a developmental-, tissue-, and signal-dependent context. In addition to their importance as transcriptional regulators of cell-cycle genes, TCPs have other functions with comparable impact on plant development. Characterization of TCPs and their signaling pathway will be beneficial to unravel their exact role in the control of plant development and evolution.

Allotetraploid upland cotton (*Gossypiumhirsutum* L.) accounts for more than 90% of cultivated cotton worldwide, is the main source of renewable textile fibers, and is also grown to produce oilseed. It has proven to be difficult to sequence, owing to its complex allotetraploid (A_tD_t) genome [21]. Recently, its whole genome was sequenced by integrating whole-genome shotgun reads, bacterial artificial chromosome-end sequences, and genotype-by-sequencing genetic maps [22,23]. Repeated sequences account for 67.2% of the A_tD_t genome, and transposable elements originating from D_t were more active than those from A_t [23]. Availability of the genome information can provide a great opportunity to identify and characterize TCP genes in this plant species for the first time. In this study, we identified and characterized 73 non-redundant TCP transcription factors in the *G. hirsutum* genome. Detailed information regarding their genomic structures, chromosomal locations, and a phylogenetic tree were also provided. Using RNA-seq data, we investigated their transcript profiles in different tissues, including different developmental stages of ovule and fiber, as well as their response to heat, drought, and salt stress. Furthermore, the miR319-targeted TCP genes were characterized.

2. Results

2.1. Identification and Characterization of TCP Proteins in G. hirsutum

To identify the *TCP* genes in the *G. hirsutum* genome, protein sequences of Arabidopsis and rice TCPs serve as BLAST search queries, and multiple-alignment was performed. A total of 73 *TCP* genes were identified. All candidate TCP genes were confirmed to encode the conserved TCP domain using the InterProScan database and NCBI's CDD, the Conserved Domain Database [24]. Seven *GhTCP* genes were found to possess the R domain. These characteristic features suggested that they were members of the TCP gene family. Detailed characteristics of the TCP transcription factors in *G. hirsutum* are offered in Table S1. The GhTCP proteins are different in their length, molecular weight (Mw), and theoretical isoelectric point (pI). The mean length and Mw of these proteins was 347 amino acids and 37.58 kDa, respectively. The pI varied from pH 5.80 (GhTCP9) to 10.07 (GhTCP38) with an average of pH 8.09. All the GhTCP proteins were predicted to localize in the nucleus. Proteins were localized at their appropriate subcellular compartment to perform their desired function [3,25].

Unrooted phylogenetic trees were constructed based on the multiple sequence alignment of 73 GhTCP protein sequences and their Arabidopsis and rice homologs. The TCP transcription factors from the three species were distributed in almost all clades, indicating that the TCP family diversified before divergence of these plants. The phylogenetic tree placed the GhTCPs into two classes (Figure 1), as was also found for all species so far. Class I was named the TCP-P or PCF class, and class II was named the TCP-C class. The class II genes were further divided into two groups: CYC/TB1 and CIN. In *G. hirsutum*, CYC/TB1 and CIN were a larger family: For CYC/TB1, approximately twice the size of those of Arabidopsis and rice; and for CIN, approximately five times the size. Seven GhTCP genes belonged to the CYC/TB1 group–in Arabidopsis and rice, three of 24 AtTCPs and three of 21 OsTCPs were grouped into this subfamily. Fifty GhTCPs belonged to the PCF group, and13 AtTCPs and10 OsTCPs were also grouped into this subfamily. CYC/TB-type proteins were divided into two subgroups. One group contained four *G. hirsutum* TCPs, but only one Arabidopsis TCP and none from rice, which indicated that this group was either acquired after the divergence of monocots and dicots or was lost in rice. In *G. hirsutum*, the number of TCP genes was significantly higher than those in tomato, *Citrullus lanatus*, *Arabidopsis*, rice, and *Prunus mume* (Figure 2).

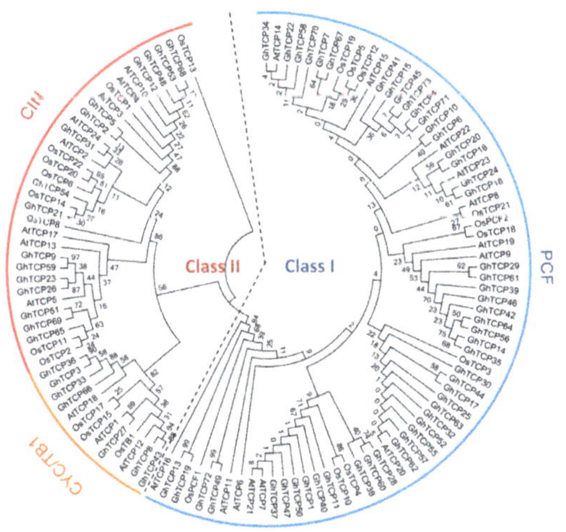

Figure 1. Phylogenetic analysis of TCP proteins from *G. hirsutum*, *Arabidopsis*, and rice. The deduced full-length amino acid sequences were aligned using ClustalX 2.0 and the phylogenetic tree was constructed using MEGA 6.0 by the Neighbor-Joining (NJ) method with 1000 bootstrap replicates. The three subclasses are indicated with different colors.

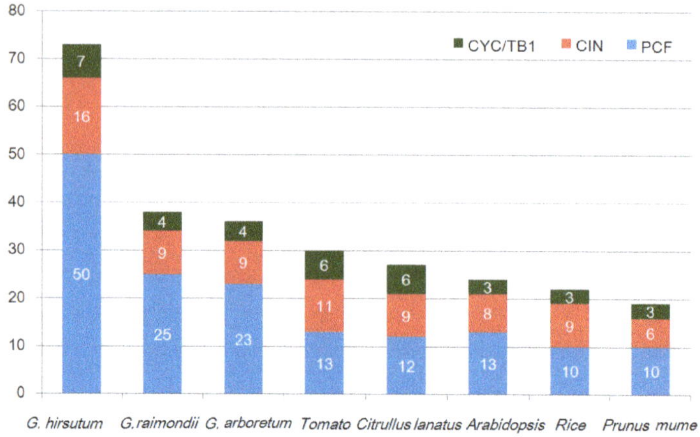

Figure 2. TCP family members of *G. hirsutum*, *G. raimondii*, *G. arboretum*, tomato, *Citrullus lanatus*, Arabidopsis, rice, and *Prunus mume*. Different colors represent the different subclasses, and the number of genes in each subclass is shown. Green: *CYC/TB1* genes; Red: *CIN* genes; Blue: *PCF* genes.

2.2. Genomic Distribution, Gene Structural Organization, and Domain Analysis of GhTCP Genes

The complete genome sequences provided an overview of the chromosomal distribution of these TCP genes. Among the 73 *G. hirsutum* TCPs, 67 members were located on the 22 chromosomes, and the other six were located at six unmapped scaffolds. *GhTCP* genes were unevenly distributed on 22 of the 26 *G. hirsutum* chromosomes, with the number of *TCP* genes per chromosome in the range of 0–8 (Figure S1). Chromosomes, A12 and D11, contained eight and seven genes, respectively, while chromosomes A02, A06, D03, D06, and D13 had no TCP genes.

To better understand the gene structures of GhTCP family genes, we analyzed their exon–intron organization. Overall, 88% of the GhTCPs contained only one exon (Figure S2). Seven GhTCP genes contained one intron and two exons: *GhTCP3*, *GhTCP23*, *GhTCP26*, *GhTCP56*, *GhTCP64*, and *GhTCP66*. Only *GhTCP33* in the CYC/TB1 group possessed four introns and five exons. Losses or gains of exons were identified during the evolution of the PCF group genes. *GhTCP19* comprised seven introns and eight exons, whereas *GhTCP13* consisted of four introns and five exons. Comparing their structural patterns showed the loss of an exon in the middle of the *GhTCP13* sequence. Two PCF class genes contained one intron and two exons, and the remaining PCF class genes contained only one exon. Analysis of the pattern of exon–intron junctions can provide important understanding into the evolution of gene families. Our results suggested that TCP genes maintained a relatively constant exon–intron composition during evolution of the *G. hirsutum* genome.

The conserved motif of TCP proteins in *G. hirsutum* was investigated using Clustal X. The sequences were found to encode a putative TCP-domain protein that contained a bHLH-type motif at the N-terminus (Figure S3). The components of the loop, and helixes I and II, were quite different between class I and II proteins. Within the TCP domain, several putative residues involved in DNA binding were located in the basic region and several putative hydrophobic residues located in helixes I and II. In the basic region, the CIN and CYC/TB1 type proteins contained an insertion of four amino acids. The R domain, an arginine-rich motif of 18–20 residues, was absent from all class I proteins and was mainly present in CYC/TB1 group proteins.

2.3. Expression Analysis of GhTCP Genes in Different Tissues and under Various Stress Conditions

To provide reliable information on the growth and developmental functions of TCP genes in *G. hirsutum*, their transcript accumulation patterns in mature leaves, stem, root, torus, petal, stamen, pistil, cylycle, ovules, and fibers of *G. hirsutum* was investigated (Figure 3). We obtained transcriptome data from the NCBI Sequence Read Archive (accession number RJNA248163). Some TCP genes with close phylogenetic relationships showed similar or divergent expression patterns. For instance, the paralogous pair, *GhTCP2* and *GhTCP5*, was expressed highly in both the torus and petal, at moderate levels in the ovules at 20 days post anthesis (DPA), and at low levels in mature leaves. Most CYC/TB1-type genes were only weakly or not expressed in all tissues, suggesting that they were primarily expressed in other organs not tested or under special conditions. In contrast, *GhTCP31*, *GhTCP40*, and *GhTCP47* were constitutively expressed at very high levels in all tissues tested, indicating that these genes played regulatory roles during multiple development stages. Some TCP genes exhibited tissue-specific expression. *GhTCP31* and *GhTCP40* were highly expressed only in reproductive organs: Torus, petal, stamen, pistil, and calycles.

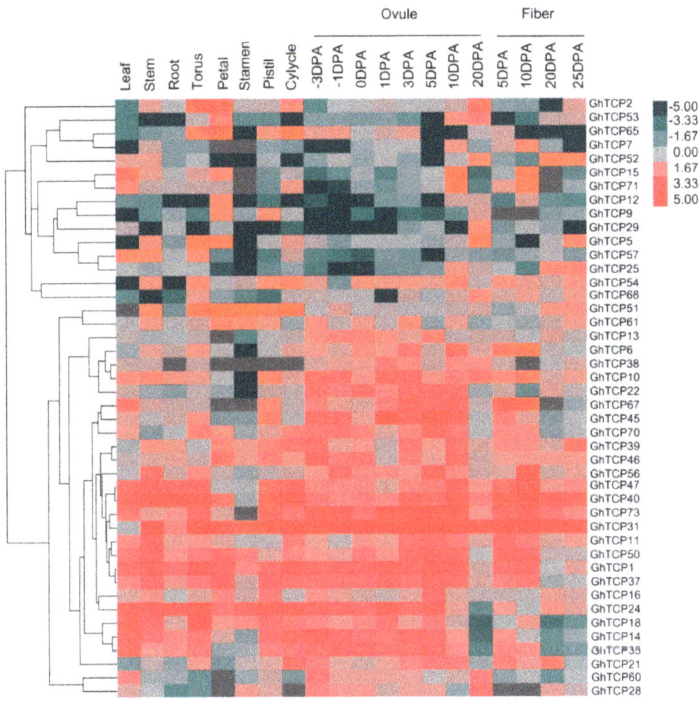

Figure 3. Heat map representation of GhTCP gene expression in different tissues. The tissues used for expression profiling are indicated at the top. The genes are shown on the left of the expression bars and the phylogenetic relationship is shown. The −3 to 20 days post anthesis (DPA) indicate −3, −1, 0, 1, 3, 5, 10, and 20 days after pollination.

To predict possible functions of *TCP* genes in environmental adaptation, we investigated the transcriptional profile of *TCP* genes under various stress conditions, including heat, salt, and drought stresses. In total, 41 genes exhibited variations in expression (Figure 4). Of the three treatments, heat stress caused relatively more fluctuations in the transcript abundance of *TCPs* than did salt or drought stress. Under heat stress conditions, 18 *TCP* genes were downregulated and eight were upregulated. In response to salt treatment, the expression of five GhTCPs (*TCP7*, *14*, *25*, *33*, and *35*) increased instantly, and then decreased slowly during continued salt stress. Six GhTCP genes were selected at random for quantitative RT-PCR (qRT-PCR) analysis to determine the relative expression under salt and drought stresses (Figure 5). The qRT-PCR results indicated that these *GhTCP* genes showed similar expression patterns to the transcriptome sequencing results.

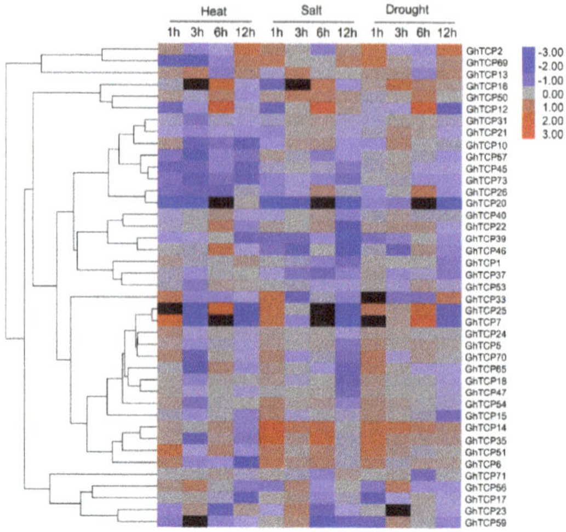

Figure 4. Expression of GhTCP genes under heat, salt, and drought stresses. The genes are shown on the left of the expression bars and the phylogenetic relationship is shown. The abiotic stresses used for expression profiling are indicated at the top. The 1, 3, 6, and 12 h indicate hours after treatment.

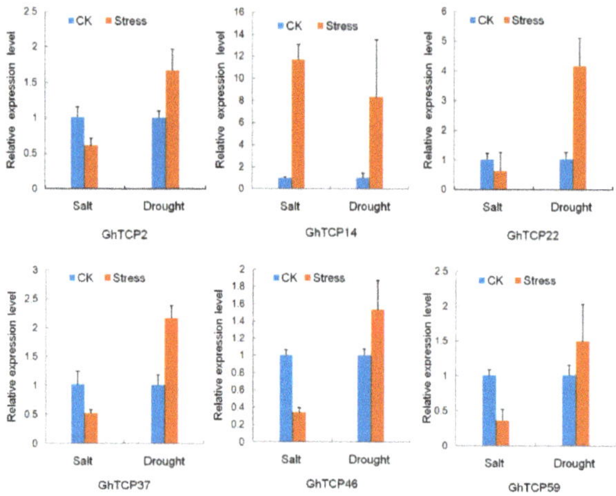

Figure 5. Relative expression levels of six *GhTCP* genes under salt and drought stress. QRT-PCR analyses were performed using RNA generated from cotton leaves after NaCl and Polyethylene Glycol (PEG) treatment. Error bars represent standard error of the mean.

2.4. Target Sites of miR319 in GhTCP Genes

The miRNAs can cause endonucleolytic cleavage of mRNA by extension, which often perfect complementarity to mRNAs. In plants, miR319 was one of the first characterized and conserved miRNA families, which has been demonstrated to target *TCP* genes. In *G. hirsutum*, miR319 had only 1-nt mismatch compared with sequences in Arabidopsis (Figure 6). The predicted hairpin structures of the miR319 precursor had 191 nt. The miR319 sequence was located at the 3′-end of the pre-miRNAs and began with a 5′-uridine. Using a set of strict standards, *GhTCP21*, *GhTCP31*, and *GhTCP54* were predicted as targets of miR319. These three miR319 target sites were all located in the coding regions, and all miR319-targeted genes belonged to the CIN clade. Similarly, there were five and three *TCP* genes containing miR319-binding sites in Arabidopsis and *P. mume*, respectively, and they also belonged to the CIN group [26]. This suggests that miR319 held homologous target interactions during the evolution and diversification of plants.

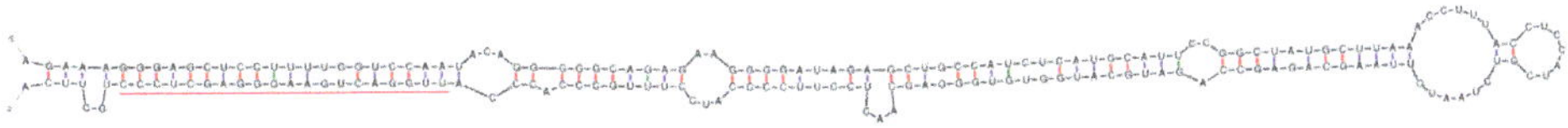

Figure 6. Mature and predicted fold-back structures of miR319 precursors in *G. hirsutum*. Sequences of mature miR319 are underlined.

Degradome sequencing had been widely used to identify plant miRNA cleavage sites. In this study, the *GhTCP* mRNA degradation sites were determined by BLASTing the sequenced degraded fragments against the *G. hirsutum* TCP genes. The degradome sequencing data are available at Gene Expression Omnibus (GEO accession number GSE69820). Using PairFinder software, miR319 was found to cleave *GhTCP21* and *GhTCP54* mRNA transcripts (Figure 7). The miR319–mRNA pair was at the cleavage site of the two TCP genes. There were both 67 raw reads at the position, with abundance at the position equal to the maximum on the transcript, and with only one maximum on the transcript. The 5′-ends of the mRNA fragments mapped to the nucleotide that paired to the tenth nucleotide of the miR319 sequence. To research the biological function of miR319, a negative-correlation expression test was undertaken for miR319 and its target *GhTCP21* and *GhTCP54* mRNAs using qRT-PCR. In response to salt and drought treatments, miR319 showed different degrees of upregulation. The transcriptome sequencing analysis showed that expression of *GhTCP21* and *GhTCP54* was downregulated.

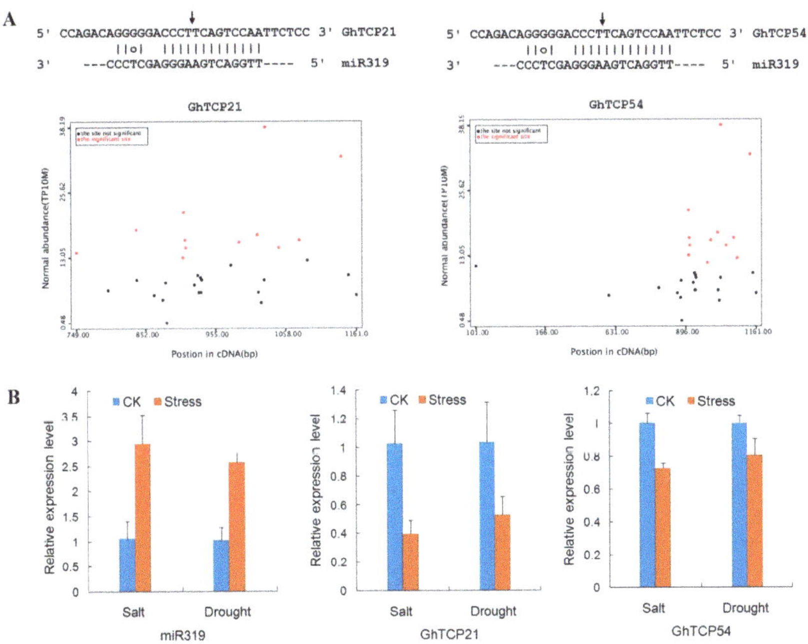

Figure 7. miR319 and its target genes, *GhTCP21* and *GhTCP54*, in *G. hirsutum*. (**A**) Target plot (t-plot) for *GhTCP21* and *GhTCP54*, which were targeted by miR319. Arrows indicate the signatures corresponding to the miRNA cleavage site. Partial mRNA sequences of target genes aligned with the miRNAs show perfect matches (straight lines) and G-U wobbles (circles). (**B**) Relative expression levels of miR319, *GhTCP21*, and *GhTCP54* under salt and drought stresses. QRT-PCR analyses were performed using RNA generated from cotton leaves after NaCl and Polyethylene Glycol (PEG) treatment. Error bars represent standard error of the mean.

3. Discussion

A number of TCP proteins had been recently identified in various plants due to completion of their whole-genome sequence, including *Arabidopsis*, rice, tomato (*Solanum lycopersicum*), and watermelon (*Citrullus lanatus*), *Orchis italica*, and *Populus euphratica* [27–31]. The allotetraploid, *G. hirsutum*, is not only the world's most important fiber crop, but is also a model polyploid crop. Despite being among the largest and most diverse gene families, the TCP gene family has not been systematically identified in the *G. hirsutum* genome. In this study, we identified 73 *TCP* genes in the sequenced genome of *G. hirsutum*. We analyzed their phylogenetic relationship, genomic distribution, conserved protein motif, and exon–intron organization. Over 80% of *GhTCP* genes were intronless, which was quite similar to the structure of *G. raimondii* and *G. arboretum TCP* genes [32,33]. Generally speaking, most *GhTCPs* within the same subclade showed similar gene structure in terms of numbers and lengths of introns and exons. Furthermore, similar to the exon–intron organization, members of the same subclade also showed similar motif composition, indicating their functional similarities. Additionally, some motifs were only present at specific subclades, such as the R domain, suggesting that they can have subclade-specific functions.

The *GhTCP* genes possessed an expanded family, with approximately three-fold size compared with *Arabidopsis*, tomato, and rice, and approximately two-fold compared with *G. arboreum* and *G. raimondii*. This suggests that although plant *TCP* genes may derive from a common ancestor, many had undergone distinct patterns of differentiation with the divergence of different lineages. Based mainly on amino acid sequence differences, especially in the basic region of the TCP domain, the TCP transcription factors are divided into three groups. There were 50 *GhTCP* genes in the PCF group, 16 in the CIN group, and seven in the CYC/TB1 group. The numbers of genes in each group were approximately twice those in *G. arboretum* and *G. raimondii*. According to a recent study, all tetraploid cotton species (A_tD_t) evolved from A-genome diploid, *G. arboretum*, and D-genome diploid, *G. raimondii*, at around 1–2 Mya [34]. In addition, previous studies indicated that gene duplication contributed to increasing the number of gene family members on various scales, including whole-genome duplication [35]. The expansion of regulatory genes is rarely achieved simply through single gene duplication alone, implying that genome duplication contributed to the amplification of the TCP gene family in *G. hirsutum*.

TCP transcription factors was involved in the regulation of cell growth and proliferation, which performed diverse functions in multiple aspects of plant growth and development [3]. We determined the spatial and temporal expression profiles of *G. hirsutum* TCP genes in 10 tissues, which included different developmental stages of ovule and fiber, using transcriptome analysis. The expression in different tissues varied widely among *GhTCP* genes and different organs for individual TCP genes. This implies functional divergence of *GhTCP* genes during different plant developmental processes. *GhTCP15* and *GhTCP71* were relatively highly expressed in ovules and fibers at 10 DPA. Previous study demonstrated that *GbTCP* was preferentially expressed in elongating *G. barbadense* fiber during5 to 15 DPA [36]. Overexpression of *GbTCP* enhanced root hair initiation and elongation in *Arabidopsis* and regulated branching. Both *GbTCP* in *G. barbadense* and *GhTCP71* are orthologs of *AT1G69690* in *Arabidopsis* (named *AtTCP15*), compared with which they had only one amino acid difference within the TCP domain [37]. *AtTCP15* was expressed in trichomes and rapidly dividing tissues and vascular tissue, and the protein promoted mitotic cell division, but inhibited endo-reduplication by modulating the expression of several key cell-cycle genes [15,38]. In our study, *GhTCP14* was also expressed predominantly in fiber cells, especially at the initiation and elongation stages of development as previously reported. Induced expression of *GhTCP14* can increases the density and length of root hairs and trichomes and affects gravitropism of *Arabidopsis* [39]. These results suggested that cotton fiber and *Arabidopsis* root hair elongation may have a similar regulatory mechanism for *TCP* genes.

Many *Arabidopsis* TCP genes with similar functions tended to cluster in the same clade, implying that TCP genes within the same clade may have similar functions in *G. hirsutum*. In *Arabidopsis*, some angiosperm members of the CIN-like clade involved in leaf and flower morphogenesis are targeted by miR319–for example, *AtTCP2, 3, 4, 10,* and *24* [40]. Loss of function of these genes results in enlarged leaves, due to an excess of cells that are smaller in size, while their gain of function leads to smaller leaves [41,42]. The miR319, previously known as "miR-JAW", was first described in *Arabidopsis* because its involvement in the control of leaf morphogenesis [43]. Several studies had reported the involvement of miR319 in plants in response to stress conditions via downregulation of its target genes [44]. Transgenic creeping bentgrass overexpressing a rice miR319, *Osa-miR319a*, exhibited enhanced salt and drought tolerance [45]. In this study, we observed upregulation of miR319 and downregulation of the targets in both salt and drought treatments. To understand the responses of *GhTCP* genes to stresses, the expression profiles were investigated in response to abiotic stresses, such as heat, salinity, and drought. In total, 40 *GhTCP* genes exhibited variations in expression. It is noteworthy that some genes showed instantaneous upregulation, and decreased slowly during continued stress. For example, *GhTCP6, 14, 35,* and *51* exhibited their highest expression at 3 h of dehydration and salinity treatment. However, no significantly upregulated expression was found at late time points. It is plausible to postulate that these genes might be the part of a stress-signaling system. The functions of these stress-responsive *GhTCP* genes in abiotic stress resistance will be further characterized in future work.

In this study, a total of 73 non-redundant TCP encoding genes were identified in *G. hirsutum*. Our results provided evidence for the relationship between structure and function in the *G. hirsutum* TCP gene family, and laid the foundation for further identification of the functions of the *GhTCP* gene family and their relationship with miR319.

4. Materials and Methods

4.1. Plant Materials and Treatments

The *G. hirsutum* L. accession TM-1 was used in this study. The seeds were provided by the National Mid-term Genebank of the Institute of Cotton Research in China. Cotton seeds were sterilized, and germinated in vermiculite under greenhouse conditions: 30/22 °C day/night temperature, 55–70% relative humidity, and a 14/10 h light/dark cycle under 450 μmol $m^{-2}{\cdot}s^{-1}$ light intensity. At the two-leaf stage, healthy seedlings were placed in pots containing aerated nutrient solution. Plants were cultured under normal conditions for 10 d to ensure full establishment before starting the drought and salt stress treatments. The pH was maintained close to 6.9 by adding H_2SO_4 or KOH as required. The roots of cotton seedlings were irrigated with 20% PEG to test the response to drought. The seedlings were treated with 150 Mm NaCl solution to test the response to salt. After exposing the seedlings to drought and salt stress for 24 h, leaves were harvested directly into liquid nitrogen and stored at −80 °C for subsequent use.

4.2. Sequence Retrieval and TCP Gene Identification

To identify TCPs in *G. hirsutum*, multiple database searches were performed. The completed genome sequence and protein sequences of this species were downloaded from the CottonGen database (http://www.cottongen.org) and the Cotton Genome Project (http://cgp.genomics.org.cn/page/species/index.jsp). A local protein database was constructed using the protein sequences. The TCP proteins from *Arabidopsis* and rice were used as query sequences, and were collected from published literature and downloaded from The *Arabidopsis* Information Resource (TAIR release 10, http://www.arabidopsis.org) and the Rice Genome Annotation Project (ftp://ftp.plantbiology.msu.edu), respectively. The BLASTP (http://cgp.genomics.org.cn/) was used to do the BLAST search. The e-value was set at 1e-10. The candidate *TCP* genes were further aligned to remove redundant sequences. To verify the reliability of the initial results, all non-redundant candidate TCP sequences were analyzed to confirm the presence of the conserved TCP domain using the InterProScan database (https://www.ebi.ac.uk/) and the NCBI's CDD (http://www.ncbi.nlm.nih.gov/Structure/cdd/cdd.shtml). Based on the results, the sequences that did not include the TCP domain were eliminated.

4.3. Analysis of Protein Features and Chromosomal Locations

The Mw and pI of each TCP protein were obtained using the online ExPASy program (http://web.expasy.org/compute_pi/). Protein pI was calculated using pK values of amino acids. Protein Mw was calculated by the addition of average isotopic masses of amino acids in the protein and the average isotopic mass of one water molecule. The subcellular localization of each GhTCP protein was analyzed using the CELLO v2.5 server (http://cello.life.nctu.edu.tw/). Through BLASTN (http://cgp.genomics.org.cn/) searches against the *G. hirsutum* whole genome, some information was obtained about the physical locations of each *GhTCP* genes on chromosomes.

4.4. Phylogenetic Analysis and Gene Structure

To analyzethe phylogenetic relationships between *TCP* genes in *G. hirsutum* and other species, the protein sequences of the identified *GhTCP* genes, Arabidopsis *TCP* genes, and rice *TCP* genes, were used to generate aphylogenetic tree. The ClustalX program was used to align the TCP domains. Phylogenetic trees were constructed by MEGA6.0 using the NJ and Minimal Evolution (ME) methods. For both methods, the bootstrap test of phylogeny was performed with 1000 replications. The exon/intron structures for each *GhTCP* gene was determined by aligning the CDS sequences to their corresponding genomic DNA sequences. The structures were shown using the Gene Structure Display Server 2.0 (http://gsds.cbi.pku.edu.cn/).

4.5. Expression Analyses of the TCP Genes and Search for miR319 Targets

Expression data for *GhTCP* genes were obtained from transcriptome data. RNA-seq data were obtained from the NCBI Sequence Read Archive (SRA: PRJNA248163). The expression pattern of *GhTCP* genes was analyzed in leaves, roots, and stems of 2-week-old plants; petals, torus, pistils, stamens, and lower sepals dissected from whole mature flowers; ovules from −3, −1, 0, 1, 3, 5, 10, and 20 days after pollination; fibers from 5, 10, 20, and 25 days; and true leaves of seedlings treated with salt, PEG, and heat. Gene expression levels were calculated according to Fragments Per Kilobase Million (FPKM) values and the default empirical abundance threshold of FPKM > 1 was used to identify the expressed gene.

Degradome sequencing data were used to find miR319 that caused TCP transcript degradation (GEO: GSE69820). We matched the degraded fragments to the GhTCP gene sequences, identified the cDNA sequences expressed, and then calculated normalized expression numbers of each degraded site along every cDNA, blast with miR319 sequences. A t-plot figure was constructed to show the tag distributions. PairFinder software was used to identify the sliced targets for miRNAs.

4.6. RNA Extraction and qRT-PCR Analysis

Total RNA was extracted with TRIzol Reagent (Invitrogen, 15596-026, Dalian, China) according to the manufacturer's instructions. For the first-strand cDNA synthesis experiment of miR319, a One Step PrimeScript® miRNA cDNASynthesis Kit (Takara, Dalian, China) was used. For each sample, 4 μg of total RNA was converted to cDNA in a 20-μL reaction system, which contained 10 μL of 2× miRNA reaction buffer mix, 2 μL of 0.1% BSA, and 2 μL of miRNA PrimeScript®RT Enzyme Mix. qRT-PCR was performed using SYBR® Premix Ex TaqTM II (Takara) and undertaken with a 7500 Fast Real-Time PCR system (Applied Biosystems Inc., Foster City, CA, USA). The specific miR319 and TCP genes primers used are given in Table S2. The reactions were incubated in a 96-well plate at 95 °C at 30 s, followed by 40 cycles of 95 °C at 15 s and 60 °C at 30 s. The 25-μL reaction solutions contained 12.5 μL of SYBR®*Premix* Ex Taq™II (2×), 1 μL of PCR forward primer (10 μM), 1 μL of PCR reverse primer (10 μM) and 2 μL of five fold diluted cDNA template. All reactions were performed with three replicates. Relative expression levels were calculated by the comparative threshold cycle ($2^{-\Delta\Delta T}$) method.

Supplementary Materials: Supplementary materials can be found at http://www.mdpi.com/1422-0067/19/11/3655/s1.

Author Contributions: Z.Y. and W.Y. conceived and designed the experiments; Z.Y., W.Z., Y.L., J.W. and H.L. performed the experiments; Z.Y. and X.F. analyzed the data; Z.Y. and Y.L. wrote the paper. Y.L. and X.F. reviewed and edited the manuscript. All authors read and approved the manuscript.

Funding: This work was supported by the National Natural Science Foundation of China (31201247), Young Elite Scientists Sponsorship Program by CAST (2016QNRC001), and State Key Laboratory of Crop Biology Open Fund (2015KF12).

Conflicts of Interest: The authors declare no conflict of interest.

References

1. Lee, T.I.; Young, R.A. Transcription of eukaryotic protein-coding genes. *Annu. Rev. Genet.* **2000**, *34*, 77–137. [CrossRef] [PubMed]
2. Cubas, P.; Lauter, N.; Doebley, J.; Coen, E. The TCP domain: A motif found in proteins regulating plant growth and development. *Plant J.* **1999**, *18*, 215–222. [CrossRef] [PubMed]
3. Martin-Trillo, M.; Cubas, P. TCP genes: A family snapshot ten years later. *Trends Plant Sci.* **2010**, *15*, 31–39. [CrossRef] [PubMed]
4. Aggarwal, P.; Das Gupta, M.; Joseph, A.P.; Chatterjee, N.; Srinivasan, N.; Nath, U. Identification of specific DNA binding residues in the TCP family of transcription factors in Arabidopsis. *Plant Cell* **2010**, *22*, 1174–1189. [CrossRef] [PubMed]
5. Kosugi, S.; Ohashi, Y. DNA binding and dimerization specificity and potential targets for the TCP protein family. *Plant J.* **2002**, *30*, 337–348. [CrossRef] [PubMed]
6. Navaud, O.; Dabos, P.; Carnus, E.; Tremousaygue, D.; Herve, C. TCP transcription factors predate the emergence of land plants. *J. Mol. Evol.* **2007**, *65*, 23–33. [CrossRef] [PubMed]
7. Viola, I.L.; Reinheimer, R.; Ripoll, R.; Manassero, N.G.; Gonzalez, D.H. Determinants of the DNA binding specificity of class I and class II TCP transcription factors. *J. Biol. Chem.* **2012**, *287*, 347–356. [CrossRef] [PubMed]
8. Sarvepalli, K.; Nath, U. CIN-TCP transcription factors: Transiting cell proliferation in plants. *IUBMB Life* **2018**, *70*, 718–731. [CrossRef] [PubMed]
9. Kosugi, S.; Ohashi, Y. PCF1 and PCF2 specifically bind to cis elements in the rice proliferating cell nuclear antigen gene. *Plant Cell* **1997**, *9*, 1607–1619. [CrossRef] [PubMed]
10. Li, C.; Potuschak, T.; Colon-Carmona, A.; Gutierrez, R.A.; Doerner, P. Arabidopsis TCP20 links regulation of growth and cell division control pathways. *Proc. Nat. Acad. Sci. USA* **2005**, *102*, 12978–12983. [CrossRef] [PubMed]
11. Danisman, S. TCP Transcription Factors at the Interface between Environmental Challenges and the Plant's Growth Responses. *Front. Plant Sci.* **2016**, *7*, 1930. [CrossRef] [PubMed]
12. Tatematsu, K.; Nakabayashi, K.; Kamiya, Y.; Nambara, E. Transcription factor AtTCP14 regulates embryonic growth potential during seed germination in Arabidopsis thaliana. *Plant J.* **2008**, *53*, 42–52. [CrossRef] [PubMed]
13. Gao, Y.; Zhang, D.; Li, J. TCP1 Modulates DWF4 Expression via Directly Interacting with the GGNCCC Motifs in the Promoter Region of DWF4 in Arabidopsis thaliana. *J. Genet. Genom.* **2015**, *42*, 383–392. [CrossRef] [PubMed]
14. Kim, H.B.; Kwon, M.; Ryu, H.; Fujioka, S.; Takatsuto, S.; Yoshida, S.; An, C.S.; Lee, I.; Hwang, I.; Choe, S. The regulation of DWARF4 expression is likely a critical mechanism in maintaining the homeostasis of bioactive brassinosteroids in Arabidopsis. *Plant Physiol.* **2006**, *140*, 548–557. [CrossRef] [PubMed]
15. Lucero, L.E.; Uberti-Manassero, N.G.; Arce, A.L.; Colombatti, F.; Alemano, S.G.; Gonzalez, D.H. TCP15 modulates cytokinin and auxin responses during gynoecium development in Arabidopsis. *Plant J.* **2015**, *84*, 267–282. [CrossRef] [PubMed]
16. Zhou, Y.; Zhang, D.Z.; An, J.X.; Yin, H.J.; Fang, S.; Chu, J.F.; Zhao, Y.D.; Li, J. TCP Transcription Factors Regulate Shade Avoidance via Directly Mediating the Expression of Both PHYTOCHROME INTERACTING FACTORs and Auxin Biosynthetic Genes. *Plant Physiol.* **2018**, *176*, 1850–1861. [CrossRef] [PubMed]

17. Mukhopadhyay, P.; Tyagi, A.K. OsTCP19 influences developmental and abiotic stress signaling by modulating ABI4-mediated pathways. *Sci. Rep.* **2015**, *5*, 9998. [CrossRef] [PubMed]
18. Carrington, J.C.; Ambros, V. Role of microRNAs in plant and animal development. *Science* **2003**, *301*, 336–338. [CrossRef] [PubMed]
19. Schommer, C.; Palatnik, J.F.; Aggarwal, P.; Chetelat, A.; Cubas, P.; Farmer, E.E.; Nath, U.; Weigel, D. Control of jasmonate biosynthesis and senescence by miR319 targets. *PLoS Biol.* **2008**, *6*, e230. [CrossRef] [PubMed]
20. Zhou, M.; Li, D.; Li, Z.; Hu, Q.; Yang, C.; Zhu, L.; Luo, H. Constitutive expression of a miR319 gene alters plant development and enhances salt and drought tolerance in transgenic creeping bentgrass. *Plant Physiol.* **2013**, *161*, 1375–1391. [CrossRef] [PubMed]
21. Chen, Z.J.; Scheffler, B.E.; Dennis, E.; Triplett, B.A.; Zhang, T.; Guo, W.; Chen, X.; Stelly, D.M.; Rabinowicz, P.D.; Town, C.D.; et al. Toward sequencing cotton (Gossypium) genomes. *Plant Physiol.* **2007**, *145*, 1303–1310. [CrossRef] [PubMed]
22. Zhang, T.; Hu, Y.; Jiang, W.; Fang, L.; Guan, X.; Chen, J.; Zhang, J.; Saski, C.A.; Scheffler, B.E.; Stelly, D.M.; et al. Sequencing of allotetraploid cotton (Gossypium hirsutum L. acc. TM-1) provides a resource for fiber improvement. *Nat. Biotechnol.* **2015**, *33*, 531–537. [CrossRef] [PubMed]
23. Li, F.; Fan, G.; Lu, C.; Xiao, G.; Zou, C.; Kohel, R.J.; Ma, Z.; Shang, H.; Ma, X.; Wu, J.; et al. Genome sequence of cultivated Upland cotton (Gossypium hirsutum TM-1) provides insights into genome evolution. *Nat. Biotechnol.* **2015**, *33*, 524–530. [CrossRef] [PubMed]
24. Marchler-Bauer, A.; Zheng, C.; Chitsaz, F.; Derbyshire, M.K.; Geer, L.Y.; Geer, R.C.; Gonzales, N.R.; Gwadz, M.; Hurwitz, D.I.; Lanczycki, C.J.; et al. CDD: Conserved domains and protein three-dimensional structure. *Nucleic Acids Res.* **2013**, *41*, D348–352. [CrossRef] [PubMed]
25. Qin, L.J.; Guo, X.Z.; Feng, X.Z.; Weng, L.; Yan, J.; Hu, X.H.; Luo, D. Cloning of LjCYC1 gene and nuclear localization of LjCYC1 protein in Lotus japonicus. *Zhi Wu Sheng Li Yu Fen Zi Sheng Wu Xue Xue Bao* **2004**, *30*, 523–532. [PubMed]
26. Zhou, Y.; Xu, Z.; Zhao, K.; Yang, W.; Cheng, T.; Wang, J.; Zhang, Q. Genome-Wide Identification, Characterization and Expression Analysis of the TCP Gene Family in Prunus mume. *Front. Plant Sci.* **2016**, *7*, 1301. [CrossRef] [PubMed]
27. Parapunova, V.; Busscher, M.; Busscher-Lange, J.; Lammers, M.; Karlova, R.; Bovy, A.G.; Angenent, G.C.; de Maagd, R.A. Identification, cloning and characterization of the tomato TCP transcription factor family. *BMC Plant Biol.* **2014**, *14*, 157. [CrossRef] [PubMed]
28. De Paolo, S.; Gaudio, L.; Aceto, S. Analysis of the TCP genes expressed in the inflorescence of the orchid Orchis italica. *Sci. Rep.* **2015**, *5*, 16265. [CrossRef] [PubMed]
29. Ma, X.; Ma, J.; Fan, D.; Li, C.; Jiang, Y.; Luo, K. Genome-wide Identification of TCP Family Transcription Factors from Populus euphratica and Their Involvement in Leaf Shape Regulation. *Sci. Rep.* **2016**, *6*, 32795. [CrossRef] [PubMed]
30. Shi, P.; Guy, K.M.; Wu, W.; Fang, B.; Yang, J.; Zhang, M.; Hu, Z. Genome-wide identification and expression analysis of the ClTCP transcription factors in Citrullus lanatus. *BMC Plant Biol.* **2016**, *16*, 85. [CrossRef] [PubMed]
31. Yao, X.; Ma, H.; Wang, J.; Zhang, D. Genome-Wide Comparative Analysis and Expression Pattern of TCP Gene Families in Arabidopsis thaliana and Oryza sativa. *J. Integrat. Plant Biol.* **2007**, *49*, 32795. [CrossRef]
32. Ma, J.; Wang, Q.; Sun, R.; Xie, F.; Jones, D.C.; Zhang, B. Genome-wide identification and expression analysis of TCP transcription factors in Gossypium raimondii. *Sci. Rep.* **2014**, *4*, 6645. [CrossRef] [PubMed]
33. Ma, J.; Liu, F.; Wang, Q.; Wang, K.; Jones, D.C.; Zhang, B. Comprehensive analysis of TCP transcription factors and their expression during cotton (Gossypium arboreum) fiber early development. *Sci. Rep.* **2016**, *6*, 21535. [CrossRef] [PubMed]
34. Paterson, A.H.; Wendel, J.F.; Gundlach, H.; Guo, H.; Jenkins, J.; Jin, D.; Llewellyn, D.; Showmaker, K.C.; Shu, S.; Udall, J.; et al. Repeated polyploidization of Gossypium genomes and the evolution of spinnable cotton fibres. *Nature* **2012**, *492*, 423–427. [CrossRef] [PubMed]
35. Kurosaki, M.; Bolis, M.; Fratelli, M.; Barzago, M.M.; Pattini, L.; Perretta, G.; Terao, M.; Garattini, E. Structure and evolution of vertebrate aldehyde oxidases: From gene duplication to gene suppression. *Cell Mol. Life Sci.* **2013**, *70*, 1807–1830. [CrossRef] [PubMed]

36. Hao, J.; Tu, L.; Hu, H.; Tan, J.; Deng, F.; Tang, W.; Nie, Y.; Zhang, X. GbTCP, a cotton TCP transcription factor, confers fibre elongation and root hair development by a complex regulating system. *J. Exp. Bot.* **2012**, *63*, 6267–6281. [CrossRef] [PubMed]
37. Kieffer, M.; Master, V.; Waites, R.; Davies, B. TCP14 and TCP15 affect internode length and leaf shape in Arabidopsis. *Plant J.* **2011**, *68*, 147–158. [CrossRef] [PubMed]
38. Li, Z.Y.; Li, B.; Dong, A.W. The Arabidopsis transcription factor AtTCP15 regulates endoreduplication by modulating expression of key cell-cycle genes. *Mol. Plant* **2012**, *5*, 270–280. [CrossRef] [PubMed]
39. Wang, M.Y.; Zhao, P.M.; Cheng, H.Q.; Han, L.B.; Wu, X.M.; Gao, P.; Wang, H.Y.; Yang, C.L.; Zhong, N.Q.; Zuo, J.R.; et al. The cotton transcription factor TCP14 functions in auxin-mediated epidermal cell differentiation and elongation. *Plant Physiol.* **2016**, *162*, 1669–1680. [CrossRef] [PubMed]
40. Mendez-Vigo, B.; de Andres, M.T.; Ramiro, M.; Martinez-Zapater, J.M.; Alonso-Blanco, C. Temporal analysis of natural variation for the rate of leaf production and its relationship with flowering initiation in Arabidopsis thaliana. *J. Exp. Bot.* **2010**, *61*, 1611–1623. [CrossRef] [PubMed]
41. Koyama, T.; Furutani, M.; Tasaka, M.; Ohme-Takagi, M. TCP transcription factors control the morphology of shoot lateral organs via negative regulation of the expression of boundary-specific genes in Arabidopsis. *Plant Cell* **2007**, *19*, 473–484. [CrossRef] [PubMed]
42. Efroni, I.; Blum, E.; Goldshmidt, A.; Eshed, Y. A protracted and dynamic maturation schedule underlies Arabidopsis leaf development. *Plant Cell* **2008**, *20*, 2293–2306. [CrossRef] [PubMed]
43. Palatnik, J.F.; Allen, E.; Wu, X.; Schommer, C.; Schwab, R.; Carrington, J.C.; Weigel, D. Control of leaf morphogenesis by microRNAs. *Nature* **2003**, *425*, 257–263. [CrossRef] [PubMed]
44. Thiebaut, F.; Rojas, C.A.; Almeida, K.L.; Grativol, C.; Domiciano, G.C.; Lamb, C.R.; de Engler, J.A.; Hemerly, A.S.; Ferreira, P.C. Regulation of miR319 during cold stress in sugarcane. *Plant Cell Environ.* **2012**, *35*, 502–512. [CrossRef] [PubMed]
45. Zhou, M.; Luo, H. Role of microRNA319 in creeping bentgrass salinity and drought stress response. *Plant Signal. Behav.* **2014**, *9*, e28700. [CrossRef] [PubMed]

© 2018 by the authors. Licensee MDPI, Basel, Switzerland. This article is an open access article distributed under the terms and conditions of the Creative Commons Attribution (CC BY) license (http://creativecommons.org/licenses/by/4.0/).

International Journal of
Molecular Sciences

MDPI

Article

Mining *Late Embryogenesis Abundant* (LEA) Family Genes in *Cleistogenes songorica*, a Xerophyte Perennial Desert Plant

Blaise Pascal Muvunyi, Qi Yan, Fan Wu, Xueyang Min, Zhuan Zhuan Yan, Gisele Kanzana, Yanrong Wang * and Jiyu Zhang *

State Key Laboratory of Grassland Agro-ecosystems; Key Laboratory of Grassland Livestock Industry Innovation, Ministry of Agriculture; College of Pastoral Agriculture Science and Technology, Lanzhou University; Lanzhou 730000, China; muvunyi14@lzu.edu.cn (B.P.M.); yanq16@lzu.edu.cn (Q.Y.); wuf15@lzu.edu.cn (F.W.); minxy15@lzu.edu.cn (X.M.); yanzhzh16@lzu.edu.cn (Z.Z.Y.); giskanzana@gmail.com (G.K.)

* Correspondence: yrwang@lzu.edu.cn (Y.W.); zhangjy@lzu.edu.cn (J.Z.); Tel.: +86-931-891-4051 (Y.W.); +86-138-933-29958 (J.Z.)

Received: 19 September 2018; Accepted: 23 October 2018; Published: 1 November 2018

Abstract: Plant growth and development depends on its ability to maintain optimal cellular homeostasis during abiotic and biotic stresses. *Cleistogenes songorica*, a xerophyte desert plant, is known to have novel drought stress adaptation strategies and contains rich pools of stress tolerance genes. Proteins encoded by *Late Embryogenesis Abundant* (LEA) family genes promote cellular activities by functioning as disordered molecules, or by limiting collisions between enzymes during stresses. To date, functions of the *LEA* family genes have been heavily investigated in many plant species except perennial monocotyledonous species. In this study, 44 putative *LEA* genes were identified in the *C. songorica* genome and were grouped into eight subfamilies, based on their conserved protein domains and domain organizations. Phylogenetic analyses indicated that *C. songorica* Dehydrin and LEA_2 subfamily proteins shared high sequence homology with stress responsive Dehydrin proteins from Arabidopsis. Additionally, promoter regions of *CsLEA_2* or *CsDehydrin* subfamily genes were rich in G-box, drought responsive (MBS), and/or Abscisic acid responsive (ABRE) *cis*-regulatory elements. In addition, gene expression analyses indicated that genes from these two subfamilies were highly responsive to heat stress and ABA treatment, in both leaves and roots. In summary, the results from this study provided a comprehensive view of *C. songorica* *LEA* genes and the potential applications of these genes for the improvement of crop tolerance to abiotic stresses.

Keywords: *Cleistogenes songorica*; LEA proteins; gene expression analysis; abiotic stresses

1. Introduction

Abiotic stresses from increasing temperature or salinity can disrupt optimal plant performance and cause significant crop yield losses [1]. To maintain proper homeostasis for normal growth, plants have evolved multiple ways to combat harsh environments by mobilizing a wide spectrum of stress responsive genes [2]. For example, proteins encoded by the *late embryogenesis abundant* (LEA) family are known to play defensive roles in plants during abiotic stresses [3,4]. The *LEA* family genes were first studied in cotton seed at the late phases of seed development [4]. The *LEA* family genes were later identified in various tissues of many other plant species and the proteins encoded by these genes were shown to be important during cold, drought and/or high salinity stresses [5,6]. LEA proteins are not plant specific, they are also found in invertebrates, fungi and bacteria [7,8]. Typical LEA proteins are highly hydrophilic due to high contents of charged amino acid residues, as well as amino acids like threonine, serine and alanine residues in their sequences [9]. It was suggested that LEA proteins have molecular shield functions [10] and are capable of abating protein aggregation and preventing enzyme degradation [11], thereby promoting proper cellular homeostasis during stresses [8,12]. LEA proteins are also flexible proteins which can undergo conformational changes and interact with other macromolecules including proteins, membranes and/or nucleic acids during different adverse stress conditions [10].

LEA proteins are divided into at least eight distinct subfamilies, based on their conserved protein domains in the Pfam database: LEA (1–6), Dehydrin and Seed Maturation Protein (SMP) [13]. Motif structures within subfamily genes are mostly conserved, except the genes in the *LEA_2/LEA5C* subfamily [9]. In addition, proteins in the LEA_2/LEA5C subfamily are known to have other non-canonical LEA protein properties like high hydrophobicity and at least one atypical LEA domain known as Water stress and Hypersensitive response (WHy) domain. The presence of atypical LEA domain(s) in LEA proteins indicate that these proteins may function differently from typical LEA proteins [14–16]. Proteins in the Dehydrin subfamily are featured with at least one K-segment, a 15 amino acid residue rich in lysine (i.e., EKKGIMDKIKEKLPG) and can function like chaperones to protect peripheral membrane and proteins during dehydration [2,17–20]. Numerous earlier studies have demonstrated that the *LEA* family genes are potential abiotic stress responsive genes, important for enhancing plant stress tolerance. For instance, transgenic Arabidopsis [21,22], maize [23], alfalfa [24] bacteria [25], yeast and tobacco [26] expressing different *LEA* genes exhibited improved abiotic stress tolerance compared with their respective wild-type plant, bacteria or yeast.

C. songorica is a xerophyte C4 desert plant distributed widely in the wild lands in the northwest part of China, with an annual precipitation of about 100 mm [27]. Previous genome-wide surveys of *LEA* family genes were done for multiple plant species except perennial monocotyledonous species. In this study, we investigated the *LEA* family genes in *C. songorica* and analyzed the responses of four selected *LEA* genes to heat stress or abscisic acid (ABA) treatment in leaves and shoots. Results from this study provide new information on the evolution of the LEA family proteins, protein structures and potential applications of these genes for the improvement of crop tolerance against abiotic stresses.

2. Results

2.1. Identification of CsLEA Genes and Phylogenetic Analysis

A total of 44 putative *C. songorica* LEA proteins were identified in this study (Table S1). These proteins were named from CsLEA1 to CsLEA44 and grouped into eight different subfamilies (Figure 1): CsLEA_1, CsLEA_2/LEA5C (Battaglia classification), CsLEA_3, CsLEA_4, CsLEA_5, CsLEA_6, SMP and Dehydrin, based on their Pfam conserved protein domains and their homology with the published LEA proteins of *A. thaliana* [28]. Two atypical LEA stress related domains, Water Stress and Hypersensitive response (WHy) and LEA14-like desiccation related protein (COG5608), were detected in the proteins from CsLEA_2 subfamily (Figure S1).

2.2. Structures, Physiochemical Properties and Subcellular Localizations of CsLEA Proteins

Most proteins within the same family exhibited similar structures and properties (Table S1). Over one third of the CsLEA proteins were classified as unstable proteins with an instability index value higher than 40. Furthermore, this property varied significantly among the proteins within the Dehydrin subfamily, ranging from 3.83 to 56.11. All the proteins in the LEA_5 subfamily showed instability index values greater than 47 and were considered as the most unstable and/or potentially disordered proteins [29–31]. The GRAVY (grand average of hydropathicity index) values of more than 90% CsLEA proteins were below 0, stressing that CsLEA proteins are likely to have low hydrophobicity features. CsLEA_2 subfamily proteins were the most hydrophobic proteins, while Cs_LEA5 proteins were the highest hydrophilic proteins. These results are consistent with previous studies [28,32] and reinforce the structural disordered properties of LEA proteins by which they are capable of interacting with other molecules and mitigating the collision of enzymes during plant stress conditions [11]

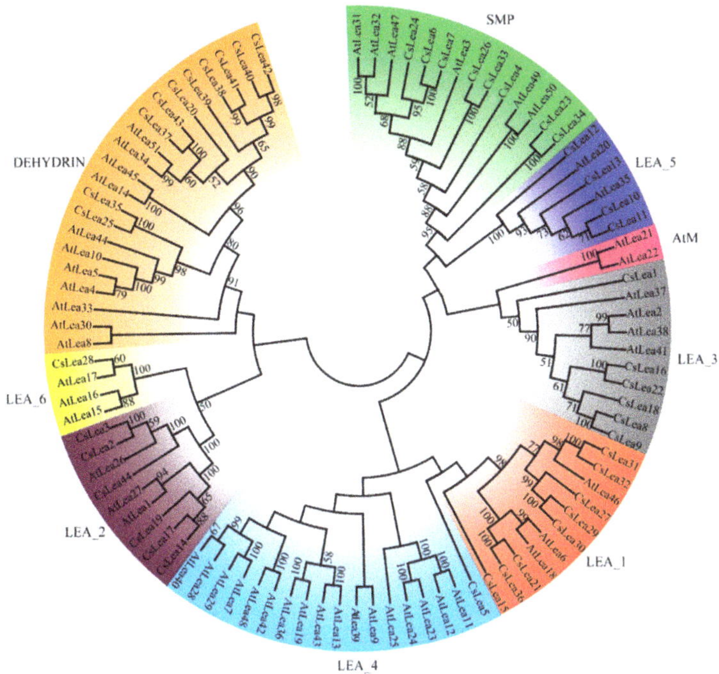

Figure 1. Phylogenetic analysis of *C. songorica* LEA proteins. Full-length amino acid sequences of the 44 CsLEA proteins were analyzed using the unrooted method in the ClustalW software.

2.3. Gene and Motif Structure Analyses

LEA genes within the same subfamily showed similar exon and intron architectures (Figure 2, right panel). Further investigation of structures of paired genes at the short-end branches in the phylogenetic tree revealed that six of them (e.g., CsLEA 43–37, 35–25, 2–3, 8–9, 23–34 and 6–7) might have experienced exon-intron gain/loss events during their evolutionary history. Similar situations have been reported for Brassica *LEA* genes [21,33]. In total, 18 motifs were identified in 43 CsLEA proteins (Figure 2, left panel). No motif was found in CsLEA28 protein. Except for LEA_2 and LEA_3 subfamily proteins, motif structures and compositions were nearly identical among the proteins in the same subfamily, but differed significantly between the proteins belonging to different subfamilies, implying functional specificities of different CsLEA subfamily proteins [14,34].

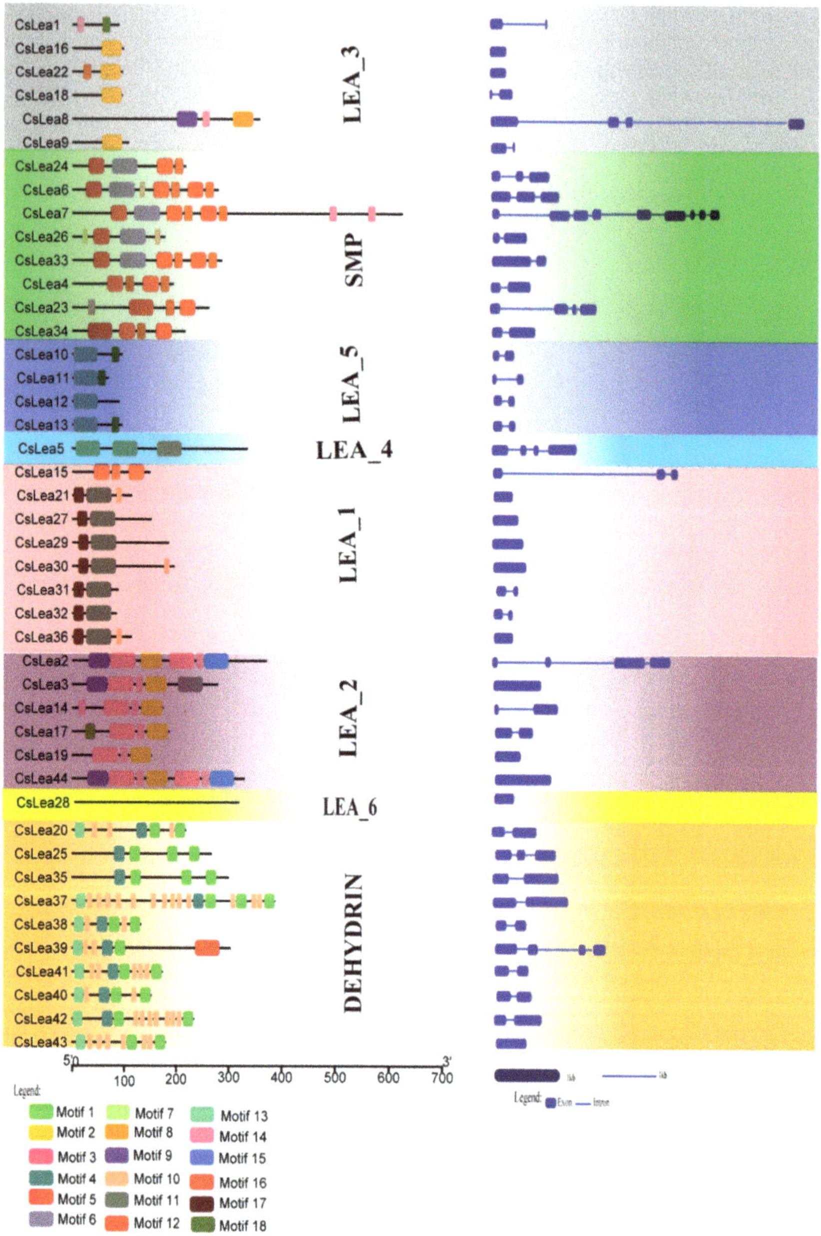

Figure 2. Motif structure and exon–intron organizations of the 44 CsLEA genes. The 18 motifs discovered in this study are shown on these *CsLEA* genes (**left**). The blue boxes represent exons and the blue lines represent introns (**right**).

2.4. *Sequence Alignment of* C. Songorica *and Arabidopsis Dehydrin Proteins*

Multiple sequence alignment using *C. songorica* Dehydrin protein sequences and their Arabidopsis counterparts revealed the conservation stress response related segments including Y, K and S segments (YKS) in *C. songorica* Dehydrin proteins (Figure 3).

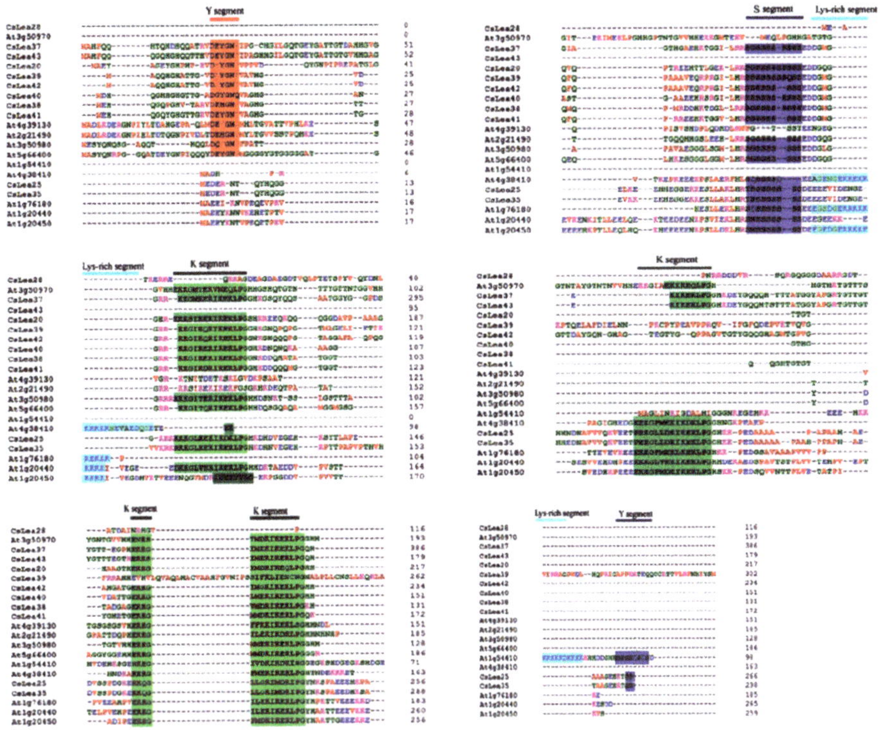

Figure 3. Multiple sequence alignment using *C. songorica* and Arabidopsis Dehydrin protein sequences. The identified Y segment, K segment and S segment are indicated by different colors. Y segment = red, K segment – green and S segment – purple.

2.5. *Cis-Regulatory Element in* C. Songorica *LEA Gene Promoters*

Cis-regulatory elements control expression patterns of stress responsive genes in various tissues and organs. These elements are located upstream of gene coding sequences and provide binding sites for transcription factors (TFs) [35]. More than three G-box *cis*-elements were recorded for each *Dehydrin*, *LEA_2* and *SMP* genes and nearly more than two MBS elements were detected for each *Dehydrin* and *LEA_2* subfamily gene (Table 1).

Table 1. Stress-related *cis*-regulatory elements in 44 *C. songorica LEA* gene promoters.

CsLEA Subfamilies	Gene Names	Functional *cis*-Element Names and Sequences			
		MBS (CGGTC)	G-Box (GTGCAT/CACGAC)	ABRE (GACACGTACGT)	CGTCA Motif
		Functions			
		Drought Responsive (MYB Binding Site)	Light Responsive	Abscisic Acid Responsive	MeJA Responsive
LEA_1	*CsLEA29*	3	1	1	0
	CsLEA30	2	3	1	4
	CsLEA31	4	1	1	0
	CsLEA32	2	2	1	0
	CsLEA36	4	5	2	2
LEA_2	*CsLEA2*	2	1	0	3
	CsLEA3	0	3	2	2
	CsLEA14	2	6	1	0
	CsLEA17	3	9	3	0
	CsLEA19	5	0	0	1
	CsLEA44	0	0	0	2
LEA_3	*CsLEA1*	4	2	0	2
	CsLEA16	1	2	0	3
	CsLEA22	1	3	1	3
	CsLEA18	8	3	3	0
	CsLEA8	0	7	1	2
	CsLEA9	3	0	0	0
LEA_4	*CsLEA5*	1	5	4	1
LEA_5	*CsLEA10*	0	2	0	1
	CsLEA11	0	3	0	3
	CsLEA12	2	7	1	1
	CsLEA13	2	1	3	5
LEA_6	*CsLEA28*	0	9	5	3
SMP	*CsLEA24*	0	4	1	0
	CsLEA6	2	0	2	0
	CsLEA7	2	6	0	0
	CsLEA26	1	5	1	1
	CsLEA33	0	5	1	4
	CsLEA4	2	1	0	0
	CsLEA23	2	1	0	0
	CsLEA34	1	4	0	1
	CsLEA42	0	3	1	2
	CsLEA43	0	0	1	5
Dehydrin	*CsLEA20*	4	5	0	0
	CsLEA25	4	2	0	1
	CsLEA35	3	8	2	1
	CsLEA37	0	6	3	4
	CsLEA38	2	10	2	3
	CsLEA39	3	1	0	1
	CsLEA41	2	0	1	2
	CsLEA40	0	2	0	1
	CsLEA42	0	3	1	2
	CsLEA43	0	0	1	5

2.6. Chromosomal Mapping of CsLEA Genes

C. songorica genome has in total twenty chromosomes. Positions of the 44 *CsLEA* genes on 15 different *C. songorica* chromosomes were estimated (Figure 4). Genes from the same subfamily were mostly found on different chromosomes, suggesting a strategy to exert their functions across the whole *C. songorica* genome. However, genes in the *LEA_5* and *Dehydrin* subfamily were mostly found in clusters on the 14th, 15th and 18th chromosome.

2.7. Gene Expression Analysis qRT-PCR Validation

The expression levels of the *CsLEA 14*, *CsLEA 19*, *CsLEA 37* and *CsLEA 38* genes were induced after 24 h of ABA or heat treatment but were not tissue specific (Figure 5). To validate results from expression profile analysis, qRT-PCR was carried out for *CsLEA 14*, *CsLEA 19* (from the *LEA_2* subfamily) and for *CsLEA37* and *CsLEA 38* (from the *Dehydrin* subfamily) as these genes showed a relatively high number of stress related *cis* acting elements and motifs.

We carried out qRT-PCR analyses using *CsLEA14* and *CsLEA19* (*LEA_2* subfamily), and *CsLEA37* and *CsLEA 38* (*Dehydrin* family). Results showed that after 24 h heat treatment the expression levels of these four genes up-regulated by 156.5 (*CsLEA 19*), 95.8 (*CsLEA14*), 52.6 (*CsLEA38*) and 14.6 fold (*CsLEA37*) in *C. songorica* leaves compared with the untreated plant leaf samples. The expression levels of these four genes were slightly up-regulated after the ABA treatment, especially *CsLEA 38* (14.6 fold, Figure 6).

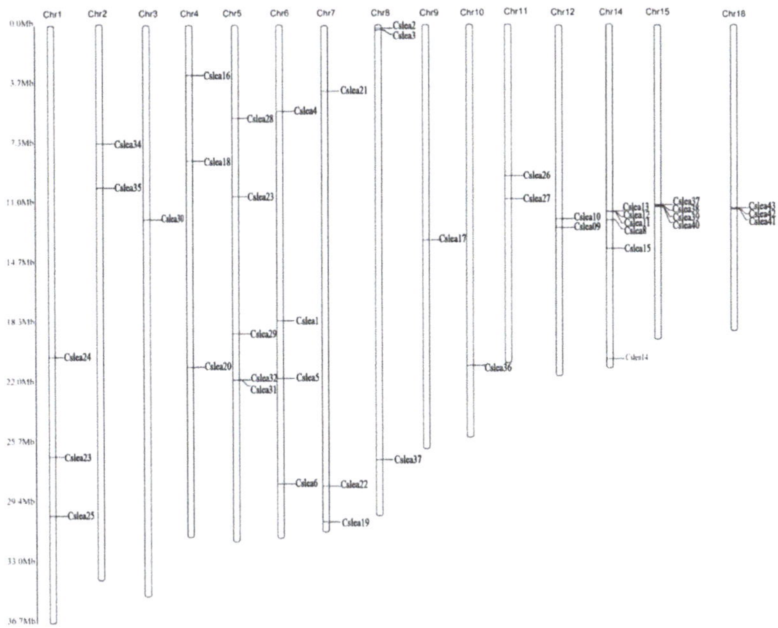

Figure 4. Locations of the 44 *CsLEA* genes on 15 chromosomes of *C. songorica*.

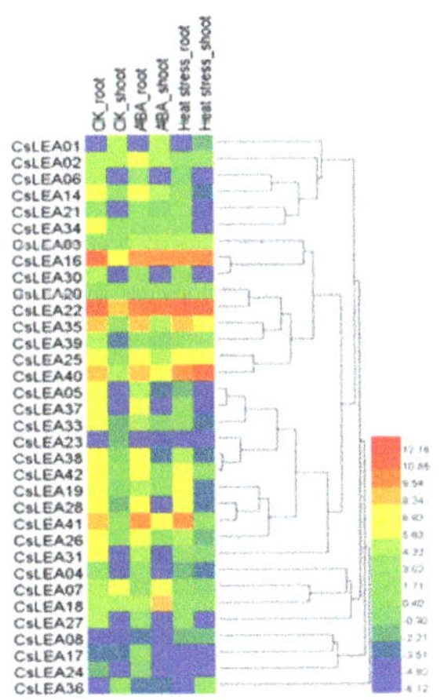

Figure 5. Hierarchical clustering of *CsLEA* gene expression profiles in root and shoot tissues after 24 h heat or ABA treatment. The log transformed values for the relative expressions of *CsLEA* genes were used for the hierarchical clustering analysis. The blue scale means low transcript expression and the red scale means high transcript expression.

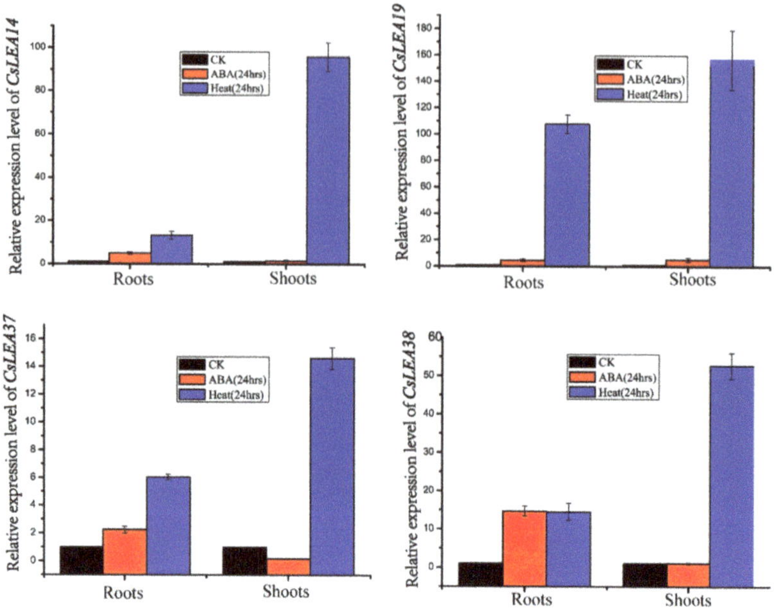

Figure 6. Hierarchical clustering of *CsLEA* gene expression profiles in root and shoot tissues after 24 h heat or ABA treatment. The log transformed values for the relative expressions of *CsLEA* genes were used for the hierarchical clustering analysis. The blue scale means low transcript expression and the red scale means high transcript expression.

3. Discussion

3.1. Phylogeny Analysis and Protein Sequence Analysis

Genome surveys of LEA subfamily proteins (e.g., LEA_1 to LEA_6, SMP and DEHYDRIN) were done for *A. thaliana* [28], rice [36] and maize [37]. *C. songorica* is a perennial monocotyledonous desert plant with a high tolerance for drought stress. Investigation of stress responsive proteins encoded by *LEA* family genes in a desert plant should benefit crop improvement for drought and other abiotic stresses. Phylogenetic analysis grouped the CsLEA proteins into eight different subfamilies. Several *C. songorica* and Arabidopsis Dehydrin subfamily proteins were clustered together with high bootstrap values, which implied potential significant functional similarities between *C. songorica* and Arabidopsis DEHYDRIN proteins. Many of the Arabidopsis Dehydrin proteins are known as stress regulatory proteins such as RAB18 (*AtLea51*) and COR47 (*AtLea4*), two ABA and cold inducible proteins [38,39], ERD14 (*AtLea4*) and ERD10 (*AtLea5*), two disordered chaperon proteins [40] and Dehydrin Xero2 (*AtLea33*), a disordered cold responsive protein with membrane binding activity [41].

Sequence alignment using CsDehydrin proteins and their Arabidopsis counterparts revealed that all *C. songorica* Dehydrin proteins contained YKS segments but lacked the lysine rich segment. The K segment is critical for the formation of structural disordered alpha-helical compounds that can enhance bindings between proteins and their targeted molecules [10,42]. The presence of the K segment in *C.songorica* DEHYDRIN proteins emphasizes their role in limiting aggregation of molecules and thence promoting proper cellular homeostasis during dehydration stresses [11]. Additionally, the detection of the S segment in the *C. songorica* DEHYDRIN protein also suggests their implication in enhancing plant tolerance against abiotic stresses through protein phosphorylation, as previous studies indicated that the S segment participates in calcium binding through protein phosphorylation [43]. The findings above support the fact that disordered LEA proteins are flexible proteins, capable of adjusting their conformation to maintain proper cellular homeostasis during detrimental stress conditions [43,44].

3.2. Protein Domain Analysis

Two additional non-LEA conserved protein domains that are associated with stress response were detected in the CsLEA_2 subfamily proteins (Supplementary Figure S1). At least one WHy (Water stress and Hypersensitive response) [15,16] and one COG5608 domain (LEA14-like desiccation related protein) was spotted within each single protein sequence from the CsLEA_2 subfamily. The COG5608 domain was previously detected in the Arabidopsis LEA14 protein, a well characterized abiotic stress marker protein, which suggests that the CsLEA_2 subfamily proteins may function similarly to Arabidopsis LEA14 protein (*Atlea1*). On the other hand, a thorough functional characterization of the WHy domain has only been elucidated in a few bacterial genes, *dwhy1* [15] and *drwh* [45]. Studies in vivo indicated that *dwhy1* confers cold and freeze damage resistance. Furthermore, in *E. coli*, the function of *drwh* is related to oxidative stress tolerance and salinity stresses. Silencing this gene triggered reduced activity of antioxidant enzymes such as lactate dehydrogenase (LDH) malate dehydrogenase (MDH) [15,45]. All the *CsLea_2* subfamily genes contained the WHy domain which could explain the functional importance of this domain during low water availability in a typical desert grass, *C. songorica*.

3.3. *C. songorica* Gene Promoter and Gene Expression Analysis

ABA and stress responsive *cis*-regulatory elements, such as ABRE, MBS/MYB and G-Box were found to be abundant in the promoters of *C. songorica* Dehydrin and the *LEA_2* subfamilies genes. These regulatory elements are known to provide binding sites for transcription factors like ABEF (a member of the bZIPTFs family), BHLH and ERF for the transcription of downstream stress responsive genes [46]. Expression analysis of *CsLEA37* and *CsLEA38* (*DEHYDRIN* genes), and *CsLEA14* and *CsLEA19* (*LEA_2* subfamily genes) with qRT-PCR indicated that the expression levels of these four genes in root and shoot tissues were significantly up-regulated after the drought or ABA treatment.

The relevant role of DEHYDRIN or the LEA_2 subfamily proteins during plant stress tolerance has been reported in earlier studies using various transgenic plants. For example, transgenic tobacco plant overexpressing the *CaLEA6* gene showed an enhanced dehydration and salt tolerance [47]. Additionally, sweet potato plants overexpressing the *IbLEA14* gene exhibited an improved salinity and dehydration tolerance [48]. A Foxtail millet plant overexpressing the *SiLEA4* gene displayed salt and drought resilience [49]. For DEHYDRIN proteins, transgenic Arabidopsis plant expressing a *Dehydrin* gene from an olive showed an enhanced osmotic stress tolerance [50]. Similarly, a wheat *Dhn-5* gene increased salinity and dehydration stress tolerance in transgenic Arabidopsis plants [51]. *C. songorica DEHYDRIN* and *LEA_2* gene transcripts accumulation during water deficit and ABA treatment, reinforcestheir functional importance under detrimental stress conditions.

4. Materials and Methods

4.1. Mining LEA Genes in the C. Songorica Genome

C. songorica LEA genes were mined based on their protein sequence homology with the previously published *A. thaliana* [28], *Oryza sativa* (rice) [36] and *Zea mays* (maize) [37] LEA protein sequences. The published full length *A. thaliana*, rice and maize LEA protein sequences or coding sequences were retrieved from (https://phytozome.jgi.doe.gov/pz/portal.html) [28,32]. The obtained LEA protein sequences were used as queries to blast search the whole *C. songorica* genome sequence retrieved from the BMK cloud: http://www.biocloud.net/ using a local blast tool [52,53].

The resulting non-redundant sequences were further examined with the Hidden Markov Model available in the Pfam database (http://pfam.sanger.ac.uk/search) [13] and then submitted to the SMART database (http://smart.embl-heidelberg.de/) [54] and the NCBI Conserved Domain Search database (http://www.ncbi.nlm.nih.gov/Structure/cdd/wrpsb.cgi) [55], respectively, to confirm CsLEA Pfam domain families. The obtained LEA nucleotide and protein sequences were then submitted to the Genbank to obtain respective accession numbers (Table S1).

4.2. *Multiple Sequence Alignment and Phylogenetic Analysis of CsLEA Family Proteins.*

The alignment of the *C. songorica* and Arabidopsis LEA protein sequences was performed using the ClustalW software [56] in the MEGA 6 program with a default parameter setting. After sequence alignment and pair-wise deletion of gaps as previously described [57], a phylogenetic tree was constructed using the Neighbor Joining (NJ) algorithm with bootstrap analysis of 1000 trials [57,58]. Multiple sequence alignment and sequence homology analysis of *C. songorica* and Arabidopsis Dehydrin proteins were performed using the ClustalW algorithm embedded in the DNAMAN version 6 program as instructed (Lynnon Corporation, Quebec, Canada).

4.3. *In Silico Analyses of CsLEA Proteins.*

Determination of GRAVY (grand average of hydropathicity index) values and pI (theoretical isoelectric point) were carried out by the ProtParam Tool (web.expasy.org/protparam/) [59]. Protein Prowler Subcellular Localization Predictor version 1.2 (http://bioinf.scmb.uq.edu.au/pprowler_webapp_1--2/) 53 and TargetP1.1 (http://www.cbs.dtu.dk/services/TargetP/) servers [60] were used to predict the subcellular locations of *C. songorica* LEA proteins. All of the prediction servers were run under the default settings. To determine the conserved motifs in different *C. songorica* and Arabidopsis LEA proteins, protein sequences were analyzed using MEME (The Multiple Expectation Maximization for Motif Elicitation) platform (http://alternate.meme-suite.org/) [61]. MEME parameters were then customized to detect a maximum of 40 motifs with a width covering 6 to 50 amino acid residues.

4.4. *Analysis of* Cis-*Regulatory Elements and Motifs*

Sequences of 2000 bp from promoters of the 44 identified *C. songorica LEA* genes were analyzed for potential *cis*-regulatory elements and motifs by querying them through the PlantCARE database (http://bioinformatics.psb.ugent.be/webtools/plantcare/html/) [62]. Stress- and ABA-related *cis*-regulatory elements, including MYB binding site (drought responsive) [63], G-box (light inducible) [64], ABRE (Abscisic acid responsive) [65] and CGTCA-motif (Methyl jasmonate responsive) [66] were recorded.

4.5. *Plant Material Preparation and Transcriptomic Data Analysis*

C. songorica seeds were sown in vermiculite medium supplied with 1/4 diluted Hoagland's nutrient solution, pH 5.8. Growth chamber conditions were set at 75–80% relative humidity, 30/28 °C (day/night), and 16/8 h (day/night) light at 200 mmol photons m^{-2} s^{-1}. One-month old seedlings were treated with 40 °C or with 100 µM ABA. Root and shoot tissue samples were collected at 0 and 24 h post the treatment and kept at −80 °C till RNA extraction. Three root and shoot samples were collected from each treatment. Total RNA was isolated from the samples using the Shengong RNA isolation kit as instructed (Shengong Ltd., Shanghai, China). RNA pools were constructed following Illumina sequencing guidelines and then sequenced conferring to RNA-seq procedure. In total 24 million 250-bp raw reads were produced from the 12 samples. To eliminate adapter sequences from raw reads, the FASTX version 0.0.13 toolkit (http://hannonlab.cshl.edu/fastxtoolkit/) was used. Additionally, the FastQC server tool (http://www.bioinformatics.babraham.ac.uk/projects/fastqc/) was utilized to assess the quality of sequences. The resulting clean reads were aligned with the *C. songorica* genome by means of Tophat v.2.0.10 server (http://tophat.cbcb.umd.edu/) [67] and the produced alignment files were used as Cufflinks inputs to create transcriptome assemblies [68]. The *C. songorica* gene expression levels were estimated based on fragments per kilobase of exon model per million mapped reads (FPKM) for root and shoot tissues. Following this, a random sampling model built on read count for each individual gene was applied to determine differentially expressed [69]. Gene expression levels were normalized with Pearson coefficients to generate hierarchical clustering with average linkage using HemI toolkit [70].

4.6. Gene Expressions Analysis.

The isolated RNA was then used to synthesize first strand cDNAs using an oligo dT primer and the cDNA synthesis kit (Shengong Ltd, Shanghai, China). The resulting cDNA samples were individually diluted to 100 ng/μL prior to qPCR using gene specific primers (Table S2). For each PCR reaction, three biological replicates with three technical replicates were each used. qPCR reactions were 40 cycles of 95 °C for 5 s, 60 °C for 15 s, and 72 °C for 34 s using a SYBR Green Master (Shengong Ltd, Shanghai, China). The relative gene expression levels were determined using the comparative $\Delta\Delta Ct$ method [71]. The expression level of the *C. songorica GADPH* gene was used as an internal control. Two-way analysis of variance and Duncan's multiple range test (DMRT) were used for multiple mean comparisons. SPSS (IBM Corp. 2013, IBM SPSS Statistics for Windows, Version 21.0, Armonk, NY, USA) was used to determine the significant differences between means ($p < 0.005$).

5. Conclusions

At a glance, this study methodically investigated at a genome wide level LEA proteins from a monocot perennial desert plant, *Cleistogenes songorica*. A total of 44 genes discovered were classified into eight different subfamilies and were found to be patchily spread over the *C. songorica* chromosomes. Analysis of the physio-chemical properties, motif and gene structure, homology and phylogenetic relationships detected that they were mostly similar within the same groups, but greatly differed among different subfamilies. Our study particularly explored CsLEA_2 and CsDEHYDRIN subfamilies proteins and elucidated their striking links with the regulatory mechanisms of plant abiotic stress tolerance. This study delivers a comprehensive summary of the evolution of the *C. songorica LEA* genes and some groundbreaking insights to the functional roles of this family that can be a critical foundation for crop abiotic tolerance improvement.

Supplementary Materials: Supplementary materials can be found at http://www.mdpi.com/1422-0067/19/11/3430/s1.

Author Contributions: Conceptualization, Y.W., J.Z., B.P.M. and X.M.; methodology, Y.W., J.Z., B.P.M., Z.Y.Y. and X.M.; software, B.P.M., Q.Y., Z.Y.Y. and F.W.; validation, B.P.M., X.M. and G.K.; formal analysis, B.P.M.; investigation, Q.Y. and J.Z.; resources, Y.W.; data curation, B.P.M., F.W.; writing-original draft preparation, B.P.M.; writing-review and editing, J.Z. and B.M.; visualization, G.K.; supervision, Y.W., project administration, Y.W., and G.K.; funding acquisition, Y.W.

Funding: This work was supported by the Program for Changjiang Scholars and Innovative Research Team in University (IRT_17R50), the National Natural Science Foundation of China (31572453), the Open Project Program of State Key Laboratory of Grassland Agro-ecosystems (SKLGAE201702) and the 111 project (B12002).

Acknowledgments: We thank Xin Shun Ding for his suggestions during manuscript preparation. We sincerely thank the anonymous reviewers for their critical comments and detailed suggestions for revision.

Conflicts of Interest: The authors declare no conflicts of interest.

References

1. Shahbaz, M.; Ashraf, M. Critical reviews in plant sciences improving salinity tolerance in cereals improving salinity tolerance in cereals. *Crit. Rev. Plant Sci.* **2013**, *32*, 237–249. [CrossRef]
2. Verma, G.; Dhar, Y.V.; Srivastava, D.; Kidwai, M.; Chauhan, P.S.; Bag, S.K.; Asif, M.H.; Chakrabarty, D. Genome-wide analysis of rice dehydrin gene family: Its evolutionary conservedness and expression pattern in response to peg induced dehydration stress. *PLoS ONE* **2017**, *12*. [CrossRef] [PubMed]
3. Huang, Z.; Zhong, X.-J.; He, J.; Jin, S.-H.; Guo, H.-D.; Yu, X.-F.; Zhou, Y.-J.; Li, X.; Ma, M.-D.; Chen, Q.-B.; et al. Genome-wide identification, characterization, and stress-responsive expression profiling of genes encoding lea (late embryogenesis abundant) proteins in moso bamboo (phyllostachys edulis). *PLoS ONE* **2016**, *11*, e0165953. [CrossRef] [PubMed]

4. Dure, L., 3rd; Greenway, S.C.; Galau, G.A. Developmental biochemistry of cottonseed embryogenesis and germination: Changing messenger ribonucleic acid populations as shown by in vitro and in vivo protein synthesis. *Biochemistry* **1981**, *20*, 4162–4168. [CrossRef] [PubMed]
5. Wang, F.; Zhu, H.; Cheng, W.; Liu, Y.; Cheng, X.; Sun, J.; Gill, S.S.; Tuteja, N. Polyamines and fabiotic stress tolerance in plants. *Plant Signal Behav.* **2010**, *5*, 26–33.
6. Thomashow, M.F. Plant cold acclimation: Freezing tolerance genes and regulatory mechanisms. *Annu. Rev. Plant Physiol. Plant Mol. Biol.* **1999**, *50*, 571–599. [CrossRef] [PubMed]
7. Kikawada, T.; Nakahara, Y.; Kanamori, Y.; Iwata, K.-I.; Watanabe, M.; McGee, B.; Tunnacliffe, A.; Okuda, T. Dehydration-induced expression of lea proteins in an anhydrobiotic chironomid. *Biochem. Biophys. Res. Commun.* **2006**, *348*, 56–61. [CrossRef] [PubMed]
8. Hand, S.C.; Menze, M.A.; Toner, M.; Boswell, L.; Moore, D. Lea proteins during water stress: Not just for plants anymore. *Annu. Rev. Physiol.* **2011**, *73*, 115–134. [CrossRef] [PubMed]
9. Battaglia, M.; Olvera-Carrillo, Y.; Garciarrubio, A.; Campos, F.; Covarrubias, A.A. The enigmatic lea proteins and other hydrophilins. *Plant Physiol.* **2008**, *148*, 6–24. [CrossRef] [PubMed]
10. Boucher, V.; Buitink, J.; Lin, X.; Boudet, J.; Hoekstra, F.A.; Hundertmark, M.; Renard, D.; Leprince, O. Mtpm25 is an atypical hydrophobic late embryogenesis-abundant protein that dissociates cold and desiccation-aggregated proteins. *Plant Cell Environ.* **2010**, *33*, 418–430. [CrossRef] [PubMed]
11. Chakrabortee, S.; Tripathi, R.; Watson, M.; Schierle, G.S.; Kurniawan, D.P.; Kaminski, C.F.; Wise, M.J.; Tunnacliffe, A. Intrinsically disordered proteins as molecular shields. *Mol. Biosyst.* **2012**, *8*, 210–219. [CrossRef] [PubMed]
12. Olveracarrillo, Y.; Campos, F.; Reyes, J.L.; Garciarrubio, A.; Covarrubias, A.A. Functional analysis of the group 4 late embryogenesis abundant proteins reveals their relevance in the adaptive response during water deficit in arabidopsis. *Plant Physiol.* **2010**, *154*, 373–390. [CrossRef] [PubMed]
13. Finn, R.D.; Bateman, A.; Clements, J.; Coggill, P.; Eberhardt, R.Y.; Eddy, S.R.; Heger, A.; Hetherington, K.; Holm, L.; Mistry, J.; et al. Pfam: The protein families database. *Nucleic Acids Res.* **2014**, *42*, D222–D230. [CrossRef] [PubMed]
14. Bies-Etheve, N.; Gaubier-Comella, P.; Debures, A.; Lasserre, E.; Jobet, E.; Raynal, M.; Cooke, R.; Delseny, M. Inventory, evolution and expression profiling diversity of the lea (late embryogenesis abundant) protein gene family in arabidopsis thaliana. *Plant Mol. Biol.* **2008**, *67*, 107–124. [CrossRef] [PubMed]
15. Ciccarelli, F.D.; Bork, P. *The Why Domain Mediates the Response to Desiccation in Plants and Bacteria*; Oxford University Press: Oxford, UK, 2005; pp. 1304–1307.
16. Jaspard, E.; Hunault, G. Comparison of amino acids physico-chemical properties and usage of late embryogenesis abundant proteins, hydrophilins and why domain. *PLoS ONE* **2014**, *9*, e109570. [CrossRef] [PubMed]
17. Mouillon, J.-M.; Gustafsson, P.; Harryson, P. Structural investigation of disordered stress proteins. Comparison of full-length dehydrins with isolated peptides of their conserved segments. *Plant Physiol.* **2006**, *141*, 638–650. [CrossRef] [PubMed]
18. Tolleter, D.; Jaquinod, M.; Mangavel, C.; Passirani, C.; Saulnier, P.; Manon, S.; Teyssier, E.; Payet, N.; Avelange-Macherel, M.H.; Macherel, D. Structure and function of a mitochondrial late embryogenesis abundant protein are revealed by desiccation. *Plant Cell* **2007**, *19*, 1580–1589. [CrossRef] [PubMed]
19. Koag, M.-C.; Wilkens, S.; Fenton, R.D.; Resnik, J.; Vo, E.; Close, T.J. The k-segment of maize dhn1 mediates binding to anionic phospholipid vesicles and concomitant structural changes. *Plant Physiol.* **2009**, *150*, 1503–1514. [CrossRef] [PubMed]
20. Rahman, L.N.; Chen, L.; Nazim, S.; Bamm, V.V.; Yaish, M.W.; Moffatt, B.A.; Dutcher, J.R.; Harauz, G. Interactions of intrinsically disordered thellungiella salsuginea dehydrins tsdhn-1 and tsdhn-2 with membranes—synergistic effects of lipid composition and temperature on secondary structure. *Biochem. Cell Biol.* **2010**, *88*, 791–807. [CrossRef] [PubMed]
21. Liang, J.; Zhou, M.; Zhou, X.; Jin, Y.; Xu, M.; Lin, J. Jclea, a novel lea-like protein from jatropha curcas, confers a high level of tolerance to dehydration and salinity in arabidopsis thaliana. *PLoS ONE* **2014**, *8*, e83056. [CrossRef] [PubMed]
22. Zhang, J.; Kong, L.; Liu, Z.; Jahufer, Z.; Duan, Z.; Huo, Y.; Di, H.; Wang, Y. Stress-induced expression in arabidopsis with a dehydrin lea protein from cleistogenes songorica, a xerophytic desert grass. *Plant Omics* **2015**, *8*, 485–492.

23. Waie, B.; Rajam, M.V. Effect of increased polyamine biosynthesis on stress responses in transgenic tobacco by introduction of human s -adenosylmethionine gene. *Plant Sci.* **2003**, *164*, 727–734. [CrossRef]
24. Zhang, J.; Duan, Z.; Zhang, D.; Zhang, J.; Di, H.; Wu, F.; Wang, Y. Co-transforming bar and cslea enhanced tolerance to drought and salt stress in transgenic alfalfa (medicago sativa l.). *Biochem. Biophys. Res. Commun.* **2016**, *472*, 75–82. [CrossRef] [PubMed]
25. Gao, J.; Lan, T. Functional characterization of the late embryogenesis abundant (lea) protein gene family from pinus tabuliformis (pinaceae) in *Escherichia coli*. *Sci. Rep.* **2016**, *6*. [CrossRef] [PubMed]
26. Liu, Y.; Wang, L.; Xing, X.; Sun, L.; Pan, J.; Kong, X.; Zhang, M.; Li, D. Zmlea3, a multifunctional group 3 lea protein from maize (zea mays l.), is involved in biotic and abiotic stresses. *Plant Cell Physiol.* **2013**, *54*, 944–959. [CrossRef] [PubMed]
27. Zhang, J.; John, U.P.; Wang, Y.; Li, X.; Gunawardana, D.; Polotnianka, R.M.; Spangenberg, G.C.; Nan, Z. Targeted mining of drought stress-responsive genes from est resources in *Cleistogenes songorica*. *J. Plant Physiol.* **2011**, *168*, 1844–1851. [CrossRef] [PubMed]
28. Hundertmark, M.; Hincha, D.K. Lea (late embryogenesis abundant) proteins and their encoding genes in *Arabidopsis thaliana*. *BMC Genomics* **2008**, *9*. [CrossRef] [PubMed]
29. Baker, J.; Dennsteele, C.V.; Iii, L.D. Sequence and characterization of 6 lea proteins and their genes from cotton. *Plant Mol. Biol.* **1988**, *11*, 277–291. [CrossRef] [PubMed]
30. Galau, G.A.; Wang, H.Y.; Hughes, D.W. Cotton Lea5 and Lea14 encode atypical late embryogenesis-abundant proteins. *Plant Physiol.* **1993**, *101*, 695–696. [CrossRef] [PubMed]
31. He, S.; Tan, L.; Hu, Z.; Chen, G.; Wang, G.; Hu, T. Molecular characterization and functional analysis by heterologous expression in e. Coli under diverse abiotic stresses for oslea5, the atypical hydrophobic lea protein from oryza sativa l. *Mol. Genet. Genom.* **2012**, *287*, 39–54. [CrossRef] [PubMed]
32. Lan, T.; Gao, J.; Zeng, Q.Y. Genome-wide analysis of the lea (late embryogenesis abundant) protein gene family in *Populus trichocarpa*. *Tree Genet. Genom.* **2013**, *9*, 253–264. [CrossRef]
33. Cheng, F.; Wu, J.; Wang, X. Genome triplication drove the diversification of brassica plants. *Hortic. Res.* **2014**, *1*. [CrossRef] [PubMed]
34. Liang, Y.; Xiong, Z.; Zheng, J.; Xu, D.; Zhu, Z.; Xiang, J.; Gan, J.; Nadia, R.; Yin, Y.; Li, M. Genome-wide identification, structural analysis and new insights into late embryogenesis abundant (lea) gene family formation pattern in *Brassica napus*. *Sci. Rep.* **2016**, *6*. [CrossRef] [PubMed]
35. Yamaguchi-Shinozaki, K.; Shinozaki, K. Organization of cis-acting regulatory elements in osmotic- and cold-stress-responsive promoters. *Trends Plant Sci.* **2005**, *10*, 88–94. [CrossRef] [PubMed]
36. Wang, X.-S.; Zhu, H.-B.; Jin, G.-L.; Liu, H.-L.; Wu, W.-R.; Zhu, J. Genome-scale identification and analysis of lea genes in rice (oryza sativa l.). *Plant Sci.* **2007**, *172*, 414–420. [CrossRef]
37. Li, X.; Cao, J. Late embryogenesis abundant (lea) gene family in maize: Identification, evolution, and expression profiles. *Plant Mol. Biol. Rep.* **2015**, *34*, 15–28. [CrossRef]
38. Lång, V.; Palva, E.T. The expression of a rab-related gene, rab18, is induced by abscisic acid during the cold acclimation process of arabidopsis thaliana (l.) heynh. *Plant Mol. Boil.* **1992**, *20*, 951–962. [CrossRef]
39. Zhu, Y.; Wang, B.; Tang, K.; Hsu, C.C.; Xie, S.; Du, H.; Yang, Y.; Tao, W.A.; Zhu, J.K. An Arabidopsis nucleoporin NUP85 modulates plant responses to ABA and salt stress. *PLoS Genet.* **2017**, *13*, e1007124. [CrossRef] [PubMed]
40. Kovacs, D.; Kalmar, E.; Torok, Z.; Tompa, P. Chaperone activity of erd10 and erd14, two disordered stress-related plant proteins. *Plant Physiol.* **2008**, *147*, 381–390. [CrossRef] [PubMed]
41. Eriksson, S.K.; Kutzer, M.; Procek, J.; Grobner, G.; Harryson, P. Tunable membrane binding of the intrinsically disordered dehydrin lti30, a cold-induced plant stress protein. *Plant Cell* **2011**, *23*, 2391–2404. [CrossRef] [PubMed]
42. Candat, A.; Macherel, D. The ubiquitous distribution of late embryogenesis abundant proteins across cell compartments in Arabidopsis offers tailored protection against abiotic stress. *Plant Cell* **2014**, *26*, 3148–3166. [CrossRef] [PubMed]
43. Hanin, M.; Brini, F.; Ebel, C.; Toda, Y.; Takeda, S.; Masmoudi, K. Plant dehydrins and stress tolerance: Versatile proteins for complex mechanisms. *Plant Signal Behav.* **2011**, *6*, 1503–1509. [CrossRef] [PubMed]
44. Close, T.J. Dehydrins: A commonalty in the response of plants to dehydration and low temperature. *Physiologia Plantarum* **1997**, *100*, 291–296. [CrossRef]

45. Jiang, S.; Wang, J.; Liu, X.; Liu, Y.; Guo, C.; Zhang, L.; Han, J.; Wu, X.; Xue, D.; Gomaa, A.E.; et al. Drwh, a novel why domain-containing hydrophobic lea5c protein from deinococcus radiodurans, protects enzymatic activity under oxidative stress. *Sci. Rep.* **2017**, *7*. [CrossRef] [PubMed]
46. Franco-Zorrilla, J.M.; López-Vidriero, I.; Carrasco, J.L.; Godoy, M.; Vera, P.; Solano, R. DNA-binding specificities of plant transcription factors and their potential to define target genes. *Proc. Nati. Acad. Sci. USA* **2014**, *111*, 2367–2372. [CrossRef] [PubMed]
47. Kim, H.S.; Lee, J.H.; Kim, J.J.; Kim, C.H.; Jun, S.S.; Hong, Y.N. Molecular and functional characterization of CaLea6, the gene for a hydrophobic LEA protein from capsicum annuum. *Gene* **2005**, *344*, 115–123. [CrossRef] [PubMed]
48. Park, S.-C.; Kim, Y.-H.; Jeong, J.C.; Kim, C.Y.; Lee, H.-S.; Bang, J.-W.; Kwak, S.-S. Sweetpotato late embryogenesis abundant 14 (iblea14) gene influences lignification and increases osmotic- and salt stress-tolerance of transgenic calli. *Planta* **2011**, *233*, 621–634. [CrossRef] [PubMed]
49. Wang, M.; Li, P.; Li, C.; Pan, Y.; Jiang, X.; Zhu, D.; Zhao, Q.; Yu, J. SiLEA14, a novel atypical lea protein, confers abiotic stress resistance in foxtail millet. *BMC Plant Biol.* **2014**, *14*. [CrossRef] [PubMed]
50. Chiappetta, A.; Muto, A.; Bruno, L.; Woloszynska, M.; Lijsebettens, M.V.; Bitonti, M.B. A dehydrin gene isolated from feral olive enhances drought tolerance in arabidopsis transgenic plants. *Front. Plant Sci.* **2015**, *6*. [CrossRef] [PubMed]
51. Brini, F.; Hanin, M.; Lumbreras, V.; Amara, I.; Khoudi, H.; Hassairi, A.; Pages, M.; Masmoudi, K. Overexpression of wheat dehydrin dhn-5 enhances tolerance to salt and osmotic stress in *Arabidopsis thaliana*. *Plant Cell Rep.* **2007**, *26*, 2017–2026. [CrossRef] [PubMed]
52. Finn, R.D.; Mistry, J.; Tate, J.; Coggill, P.; Heger, A.; Pollington, J.E.; Gavin, O.L.; Gunasekaran, P.; Ceric, G.; Forslund, K.; et al. The pfam protein families database. *Nucleic Acids Res.* **2010**, *38*, D211–D222. [CrossRef] [PubMed]
53. Altschul, S.F.; Gish, W.; Miller, W.; Myers, E.W.; Lipman, D.J. Basic local alignment search tool. *J. Mol. Biol.* **1990**, *215*, 403–410. [CrossRef]
54. Letunic, I.; Copley, R.R.; Schmidt, S.; Ciccarelli, F.D.; Doerks, T.; Schultz, J.; Ponting, C.P.; Bork, P. Smart 4.0: Towards genomic data integration. *Nucleic Acids Res.* **2004**, *32*, D142–D144. [CrossRef] [PubMed]
55. Marchler-Bauer, A.; Derbyshire, M.K.; Gonzales, N.R.; Lu, S.; Chitsaz, F.; Geer, L.Y.; Geer, R.C.; He, J.; Gwadz, M.; Hurwitz, D.I.; et al. Cdd: Ncbi's conserved domain database. *Nucleic Acids Res.* **2015**, *43*, D222–D226. [CrossRef] [PubMed]
56. Feng, D.-F.; Doolittle, R.F. Progressive sequence alignment as a prerequisitetto correct phylogenetic trees. *J. Mol. Evol.* **1987**, *25*, 351–360. [CrossRef] [PubMed]
57. Tamura, K.; Stecher, G.; Peterson, D.; Filipski, A.; Kumar, S. Mega6: Molecular evolutionary genetics analysis version 6.0. *Mol. Biol. Evol.* **2013**, *30*, 2725–2729. [CrossRef] [PubMed]
58. Higgins, D.G.; Sharp, P.M. Clustal: A package for performing multiple sequence alignment on a microcomputer. *Gene* **1988**, *73*, 237–244. [CrossRef]
59. Gasteiger, E.; Gattiker, A.; Hoogland, C.; Ivanyi, I.; Appel, R.D.; Bairoch, A. Expasy: The proteomics server for in-depth protein knowledge and analysis. *Nucleic Acids Res.* **2003**, *31*, 3784–3788. [CrossRef] [PubMed]
60. Emanuelsson, O.; Brunak, S.; von Heijne, G.; Nielsen, H. Locating proteins in the cell using targetp, signalp and related tools. *Nat. Protoc.* **2007**, *2*, 953–971. [CrossRef] [PubMed]
61. Bailey, T.L.; Boden, M.; Buske, F.A.; Frith, M.; Grant, C.E.; Clementi, L.; Ren, J.; Li, W.W.; Noble, W.S. Meme suite: Tools for motif discovery and searching. *Nucleic Acids Res.* **2009**, *37*, W202–W208. [CrossRef] [PubMed]
62. Rombauts, S.; Dehais, P.; Van Montagu, M.; Rouze, P. Plantcare, a plant cis-acting regulatory element database. *Nucleic Acids Res.* **1999**, *27*, 295–296. [CrossRef] [PubMed]
63. Yue, R.; Lu, C.; Sun, T.; Peng, T.; Han, X.; Qi, J.; Yan, S.; Tie, S. Identification and expression profiling analysis of calmodulin-binding transcription activator genes in maize (*Zea mays* L.) under abiotic and biotic stresses. *Front. Plant Sci.* **2015**, *6*. [CrossRef] [PubMed]
64. Petrov, V.; Vermeirssen, V.; De, C.I.; Van, B.F.; Minkov, I.; Vandepoele, K.; Gechev, T.S. Identification of cis-regulatory elements specific for different types of reactive oxygen species in arabidopsis thaliana. *Gene* **2012**, *499*, 52–60. [CrossRef] [PubMed]

65. Passricha, N.; Saifi, S.; Ansari, M.W.; Tuteja, N. Prediction and validation of cis-regulatory elements in 5′ upstream regulatory regions of lectin receptor-like kinase gene family in rice. *Protoplasma* **2017**, *254*, 669–684. [CrossRef] [PubMed]
66. Acharya, B.R.; Jeon, B.W.; Zhang, W.; Assmann, S.M. Open stomata 1 (OST1) is limiting in abscisic acid responses of arabidopsis guard cells. *New Phytol.* **2013**, *200*, 1049–1063. [CrossRef] [PubMed]
67. Trapnell, C.; Pachter, L.; Salzberg, S.L. Tophat: Discovering splice junctions with RNA-seq. *Bioinformatics* **2009**, *25*, 1105–1111. [CrossRef] [PubMed]
68. Trapnell, C.; Roberts, A.; Goff, L.; Pertea, G.; Kim, D.; Kelley, D.R.; Pimentel, H.; Salzberg, S.L.; Rinn, J.L.; Pachter, L. Differential gene and transcript expression analysis of rna-seq experiments with tophat and cufflinks. *Nat. Protoc.* **2012**, *7*. [CrossRef] [PubMed]
69. Wang, L.; Feng, Z.; Wang, X.; Wang, X.; Zhang, X. Degseq: An r package for identifying differentially expressed genes from rna-seq data. *Bioinformatics* **2010**, *26*, 136–138. [CrossRef] [PubMed]
70. Deng, W.; Wang, Y.; Liu, Z.; Cheng, H.; Xue, Y. Hemi: A toolkit for illustrating heatmaps. *PLoS ONE* **2014**, *9*, e111988. [CrossRef] [PubMed]
71. Fujisawa, M.; Takita, E.; Harada, H.; Sakurai, N.; Suzuki, H.; Ohyama, K.; Shibata, D.; Misawa, N. Pathway engineering of brassica napus seeds using multiple key enzyme genes involved in ketocarotenoid formation. *J. Exp. Bot.* **2009**, *60*, 1319–1332. [CrossRef] [PubMed]

© 2018 by the authors. Licensee MDPI, Basel, Switzerland. This article is an open access article distributed under the terms and conditions of the Creative Commons Attribution (CC BY) license (http://creativecommons.org/licenses/by/4.0/).

International Journal of
Molecular Sciences

MDPI

Article

Transcriptomic Analysis of *Betula halophila* in Response to Salt Stress

Fenjuan Shao [1], Lisha Zhang [1], Iain W. Wilson [2] and Deyou Qiu [1,*]

1 State Key Laboratory of Tree Genetics and Breeding, Key Laboratory of Tree Breeding and Cultivation of State Forestry Administration, The Research Institute of Forestry, Chinese Academy of Forestry, Beijing 100091, China; shaofenjuan@caf.ac.cn (F.S.); zlsxxb@aliyun.com (L.Z.)

2 CSIRO Agriculture and Food, Canberra, ACT 2601, Australia; Iain.Wilson@csiro.au

* Correspondence: qiudy@caf.ac.cn; Tel.: +86-10-6288-9641; Fax: +86-10-6287-2015

Received: 9 October 2018; Accepted: 25 October 2018; Published: 31 October 2018

Abstract: Soil salinization is a matter of concern worldwide. It can eventually lead to the desertification of land and severely damage local agricultural production and the ecological environment. *Betula halophila* is a tree with high salt tolerance, so it is of importance to understand and discover the salt responsive genes of *B. halophila* for breeding salinity resistant varieties of trees. However, there is no report on the transcriptome in response to salt stress in *B. halophila*. Using Illumina sequencing platform, approximately 460 M raw reads were generated and assembled into 117,091 unigenes. Among these unigenes, 64,551 unigenes (55.12%) were annotated with gene descriptions, while the other 44.88% were unknown. 168 up-regulated genes and 351 down-regulated genes were identified, respectively. These Differentially Expressed Genes (DEGs) involved in multiple pathways including the Salt Overly Sensitive (SOS) pathway, ion transport and uptake, antioxidant enzyme, ABA signal pathway and so on. The gene ontology (GO) enrichments suggested that the DEGs were mainly involved in a plant-type cell wall organization biological process, cell wall cellular component, and structural constituent of cell wall molecular function. Kyoto Encyclopedia of Genes and Genomes (KEGG) pathway enrichment showed that the top-four enriched pathways were 'Fatty acid elongation', 'Ribosome', 'Sphingolipid metabolism' and 'Flavonoid biosynthesis'. The expression patterns of sixteen DEGs were analyzed by qRT-PCR to verify the RNA-seq data. Among them, the transcription factor AT-Hook Motif Nuclear Localized gene and dehydrins might play an important role in response to salt stress in *B. halophila*. Our results provide an important gene resource to breed salt tolerant plants and useful information for further elucidation of the molecular mechanism of salt tolerance in *B. halophila*.

Keywords: *Betula halophile*; salt stress; transcriptomes

1. Introduction

Soil salinization is worldwide problem that can alter the soil osmotic potential to the point where it inhibits the uptake of water by plants, severely impacting agricultural production and the ecological environment. It has been reported that more than 6% of the world's land is affected by salt [1–3], and increased salinization may lead to the loss of 30% arable land in the next 25 years. It has been reported that more than 6% of the world's land is affected by salt [1–3], and increased salinization may lead to the loss of 30% arable land in the next 25 years and up to 50% by 2050 [1–3]. Therefore, soil salinization is a serious threat to the growth and development of plants. At present, it is particularly urgent to search for salinity resistant varieties of plants and screen for salt tolerant gene alleles or transform them genetically to enable plants to grow and reproduce with increasing salinity stress [4]. Moreover, understanding the mechanism of salt tolerance in plants can provide valuable information for effective engineering strategies.

In plants, the salt resistance mechanism is very complicated and involves a complex of processes at the molecular, cellular, metabolic, physiological, and whole-plant levels. Once the plant is under salt stress, multiple signal transduction pathways are activated to cope with salt stress [2,4–6]. In recent years, although extensive studies among ion uptake and transport, osmotic regulation, hormone metabolism, antioxidant metabolism, and stress signaling have made significant progress [4–11], the molecular mechanisms involved in salt tolerance remain to be elucidated. In addition, next-generation high-throughput sequencing based RNA-seq analysis has been widely used to uncover expression patterns under abiotic stress, and it provides a comprehensive means of identifying and studying the differential expression genes [12–15].

Betula halophila is a haloduric species in China, belonging to the family Betulaceae. It was first discovered in a swamp with extremely high salinity in Xinjiang province [16] in 1956 by Professor Renchang Qin. *B. halophila* is a critically endangered plant, which has high salt tolerance, and high ecological and economic value in promoting the afforestation of saline soil in arid and semi-arid areas16. Thus *B. halophila* is a potent source of salt tolerant genes. However, to the best of our knowledge, there is no published information on genes associated with salt tolerance in *B. halophila*. Understanding the molecular mechanisms of salt tolerance are potentially important for breeding salt tolerant varieties. With the aim of identifying the genes in response to increasing salt concentration and potentially the molecular mechanisms of salt tolerance in *B. halophila*, we constructed transcriptome libraries from the leaves of control *B. halophila* plants and plants subjected to salt treatment. The aims were to detect salt responsive genes from *B. halophila* and explore their roles in response to salt stress. Our results provide insight into the molecular mechanisms of salt tolerance in *B. halophila*. A better understanding of these tolerance mechanisms can be used to breed crops with improved yield performance under salinity stress.

2. Results

2.1. Transcriptome Sequencing and Assembly

In order to explore the salt tolerant genes of *B. halophila*, six cDNA libraries were constructed from leaves of eight-months-old control plants (untreated) and plants treated with 200 mM of NaCl for 24 h, and sequenced using the Illumina deep sequencing platform. A total of 68,348,352, 75,684,144, 70,510,750, 86,101,536, 77,402,824, and 84,837,142 raw reads were generated by Illumina sequencing, respectively, after adapter sequence and low quality sequences had been removed, a total of 66,834,236, 73,359,338, 69,082,978, 84,155,542, 75,577,004, and 82,931,756 clean reads were obtained, and the Q20 percentage (proportion of nucleotides with a quality value larger than 20) for each data was 96.5%, 96.48%, 96.25%, 96.54%, 96.49%, and 96.45%, respectively. The GC (%) ratio for each library was 46.88%, 47.4%, 47.29%, 47.28%, 47.33% and 47.29% (Table 1). Transcriptome assembly was accomplished based on the left.fq and right.fq using Trinity with the min_kmer_cov set to 2 by default and all other parameters set to default [17]. As a result, a total of 117,091 unigenes with lengths ranging from 201 bp to 15,762 bp were obtained. The size distribution of the unigenes is shown in Figure 1. The size distribution showed that the unigenes ranged from 200 bp to 1 kbp was the majority (Figure 1). The average length, median length, and N50 of the assembled unigenes were 956 bp, 631 bp and 1408 bp, separately. The total length of 117,091 unigenes was 110 Mb, which suggests that most of the sequencing data had been successfully assembled into relatively long unigenes.

Table 1. Summary of the sequencing data of the Betula halophila transcriptome.

Sample	Raw Reads	Clean Reads	Clean Bases	Error (%)	Q20 (%)	Q30 (%)	GC Content(%)
CK_1	68,348,352	66,834,236	10.03G	0.03	96.5	94.17	46.88
CK_2	75,684,144	73,359,338	11G	0.03	96.48	94.14	47.4
CK_3	70,510,750	69,082,978	10.36G	0.03	96.25	93.83	47.29
SC_1	86,101,536	84,155,542	12.62G	0.03	96.54	94.23	47.28
SC_2	77,402,824	75,577,004	11.34G	0.03	96.49	94.16	47.33
SC_3	84,837,142	82,931,756	12.44G	0.03	96.45	94.12	47.29

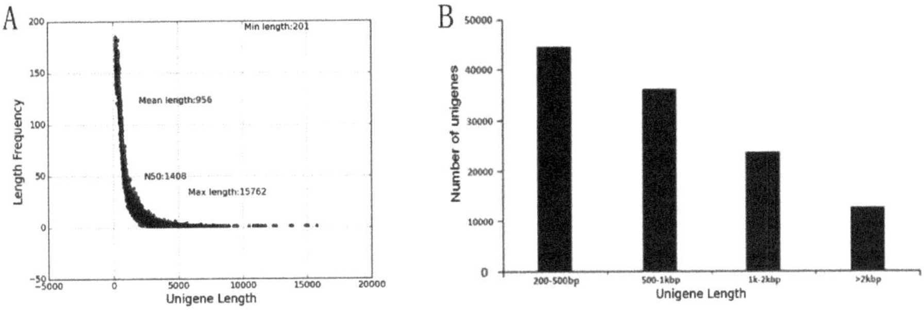

Figure 1. Length distribution of the assembled unigenes. (**A**) The number of contigs with same length. (**B**) The number of contigs with length 200–500 bp, 500 bp−1 kb, 1–2 kb and larger than 2 kb are shown.

2.2. Functional Annotation and Classification of the Unigenes

Functional annotation of unigenes were performed to search for homologues against the NCBI non-redundant protein sequence database (*N*r), NCBI nucleotide sequences (*N*t), Pfam (Protein family), Kyoto Encyclopedia of Genes and Genomes (KEGG), swiss-prot sequence databases (SwissProt), Gene ontology (GO), and Eukaryotic Orthologous Groups (KOG) using the Basic Local Alignment Search Tool (BLAST) [18]. An e-value cut-off of 10^{-5} was applied to the homologue recognition. The results were shown in Table 2. 64551 (55.12%) total unigenes were annotated in at least one database and 8973 unigenes (7.66%) were annotated in all databases. 51,105 (43.64%) total unigenes were annotated in the *N*r protein database.

Table 2. Summary of function annotation of the *Betula halophila* transcriptome.

	Number of Unigenes	Percentage (%)
Annotated in NR	51,105	43.64
Annotated in NT	45,933	39.22
Annotated in KO	18,876	16.12
Annotated in SwissProt	40,624	34.69
Annotated in PFAM	40,661	34.72
Annotated in GO	41,116	35.11
Annotated in KOG	15,572	13.29
Annotated in all Databases	8973	7.66
Annotated in at least one Database	64,551	55.12
Total Unigenes	117,091	100

The GO analysis indicated that a total of 41,116 unigenes were summarized into the three main GO categories (biological process, cellular component, and molecular function) and 56 sub-categories (Figure 2). In the biological process category, genes involved in cellular process, metabolic process, and single-organism process were dominant. As for the cellular component category, genes involved in cell, cell part, and organelle were highly represented. The molecular function category mainly included genes involved in binding and catalytic activity.

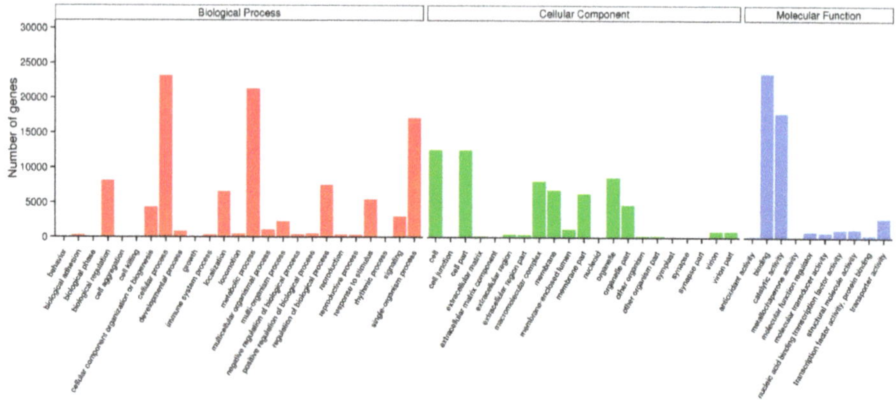

Figure 2. Gene ontology (GO) classification of unigenes. The GO terms are summarized into three main categories: biological process, cellular component, and molecular function.

The KOG analysis showed that all of the 15,572 unigenes were divided into 26 different functional classes, which were represented by A to Z (Figure 3). Among the 26 categories, the largest group was 'Post-translational modification, protein turnover, chaperon' (2104, 13.51%) followed by 'General function prediction' (1906, 12.24%), 'Translation, ribosomal structure, and biogenesis' (1552, 9.97%), 'RNA processing and modification' (1272, 8.17%) and 'Signal Transduction' (1248, 8.01%). The smallest group was 'Cell motility' (7, 0.04%) and 'Unnamed protein' (2, 0.01%).

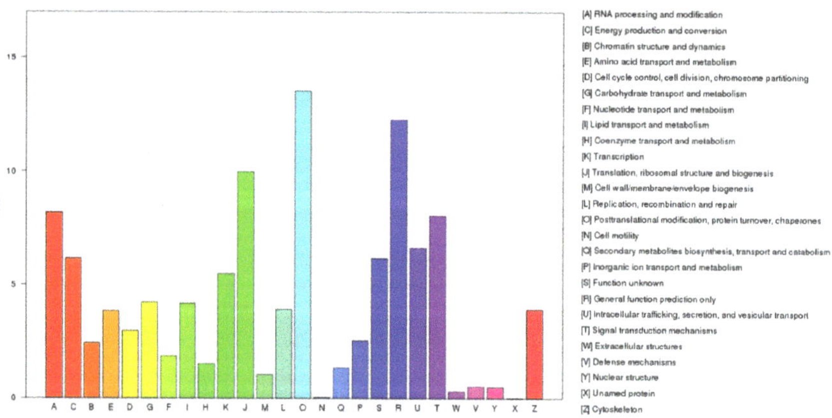

Figure 3. Eukaryotic Orthologous Groups (KOG) classification of the unigenes.

The KEGG pathway analysis revealed that 18876 (16.12%) of the unigenes could be mapped to the KEGG database and referred to 129 pathways (Figure 4). The pathway involved the highest number of unigenes was 'Translation' (1915, 10.14%), followed by 'Folding, sorting, and degradation' (1465, 7.76%), 'Carbohydrate metabolism' (1388, 7.35%) and 'Overview' (1012, 5.36%). These results are very important for studying the mechanism in *B. halophila* response to salt.

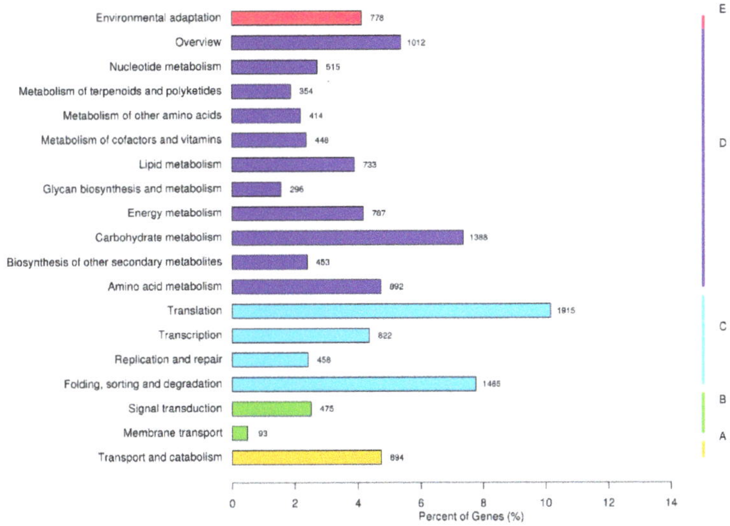

Figure 4. Kyoto Encyclopedia of Genes and Genomes (KEGG) classification of KO annotated unigenes.

2.3. *Differential Expression Genes in B. halophila Response to Salt*

To obtain the differential expression genes' response to salt in *B. halophila*, we compared the differentially expressed tags of two libraries. As a results, a total of 519 differentially expressed genes (DEGs) with q value < 0.05 and |log2 (fold change)| >1 were identified in the two libraries (Table S1). As shown in Figure 5a, there were more down-regulated genes (351) than up-regulated genes (168). Among these DEGs, 332 DEGs were present in both libraries, (Figure 5b). 66 DEGs were only detected in the salt stress library (Figure 5b) and 121 DEGs were only detected in the control library. In this study, the transcription factor AT-Hook Motif Nuclear Localized gene (AHL) was the most up-regulate gene in leaves after the salt stress. Conversely, a dehydrin (DHNs) was the most down-regulated gene. These results suggest that the two genes may have a high correlation with salt resistance of *B. halophila*. The GO and KEGG classification of the 519 DEGs were analyzed (Figure S1). GO enrichment and KEGG enrichment were performed for further analysis of the functions of 519 DEGs.

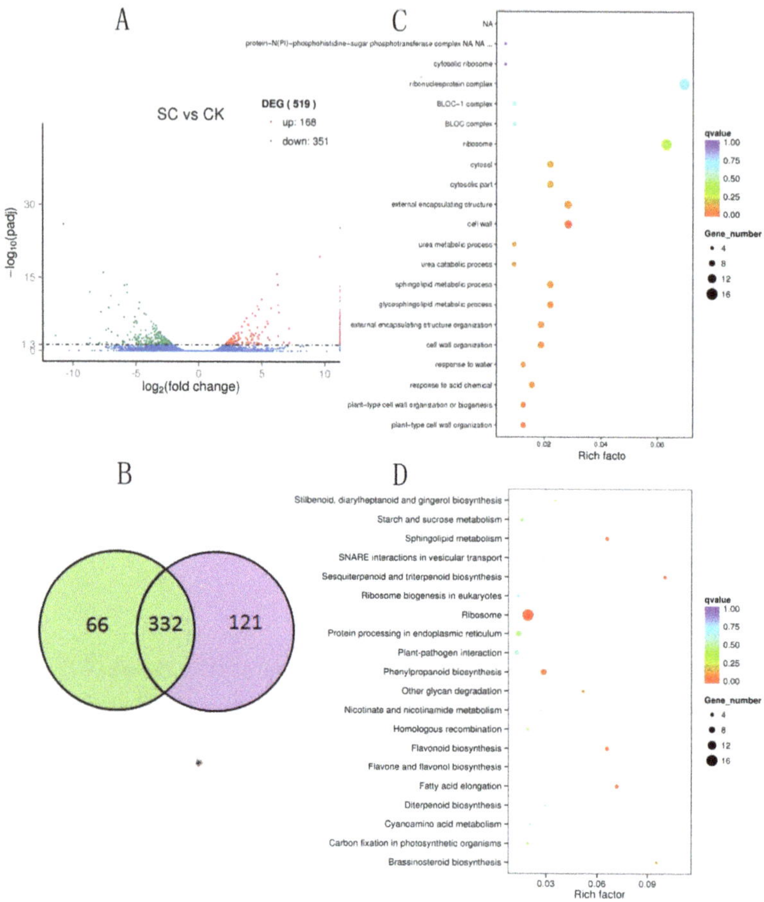

Figure 5. A. Up-regulated and down-regulated differentially expressed genes in SC vs. CK; **B**. Venn diagrams showing unique and shared differentially expressed genes (DEGs) in SC (green) vs. CK (purple); **C**. Scatterplot of GO category enrichment of DEGs in SC vs. CK; **D**. Scatterplot of enriched KEGG pathways for DEGs in SC vs. CK. Rich factor is the ratio of the differentially expressed gene number to the total gene number in a certain pathway. The size and color of dot represent the gene number and the range of the q value, respectively.

2.4. GO category Enrichment of DEGs Under Salt Stress

To characterize the function of the DEGs under salt stress, the GO category enrichment analysis was performed using Fisher's exact test with p value ≤ 0.05 as the cutoff. GO category enrichment analysis for 519 DEGs under salt stress showed that these DEGs were mainly involved in a plant-type cell wall organization biological process, plant-type cell wall organization or biogenesis biological process, cell wall cellular component and structural constituent of cell wall molecular function (Figure 5c, Table S2). For the up-regulated DEGs, metalloendopeptidase activity molecular function was most highly enriched (Table S3). For down-regulated DEGs (Figure 5c), in the BP category, 'plant-type cell wall organization biological process', 'plant-type cell wall organization or biogenesis biological process', 'cell wall organization biological process', and 'external encapsulating structure organization biological process' were most highly enriched. In the CC category, 'cell wall cellular component', 'cytosolic part cellular component', 'cytosol cellular component', and 'external encapsulating structure cellular component' were the main enriched terms. In MF, the most enriched term was structural constituent of cell wall molecular function (Table S4).

2.5. KEGG Enrichment of DEGs under Salt Stress

The KEGG pathway enrichment analysis were performed to identify the candidate pathways involved in salt stress using KOBAS 2.0 [19]. The results showed that genes with KO number within 519 DEGs under salt stress were enriched in 48 KEGG pathways (Table S5). The top-four enriched pathways for DEGs in SC vs CK were 'Fatty acid elongation', 'Ribosome', 'Sphingolipid metabolism' and 'Flavonoid biosynthesis'. For up-regulated DEGs (Table S6), the most highly enriched pathways were 'Sphingolipid metabolism', 'Other glycan degradation', 'Brassinosteroid biosynthesis' and 'Citrate cycle (TCA cycle)' (Figure 5d). For down-regulated DEGs (Table S7), 'Ribosome', 'Fatty acid elongation', 'Flavonoid biosynthesis' and 'Phenylpropanoid biosynthesis' were the top-four enriched pathways (Figure 5d).

2.6. qRT-PCR Analysis

In order to validate the RNA-seq data and confirm the differential expression genes, we performed qRT-PCR on sixteen candidate DEGs associated with salt stress. The results revealed that these DEGs include AHL, dehydrin-1, highly ABA-induced PP2C gene, proline transporter 2-like, sodium/hydrogen exchanger 4, serine/threonine-protein kinase, heat shock protein, Cation/H(+) antiporter, wall-associated receptor kinase-like and dehydrin-3 were up-regulated in the leaves with salt treatment (Figure 6), whereas major allergen variant Cor, dehydrin-2, phosphoglycerate kinase, potassium transporter, peroxidase 4 and WRKY transcription factor were down- regulated in the leaves with salt treatment. The results indicated that these sixteen candidate DEGs had the same expression patterns compared with the sequencing data, suggesting the reliability of the RNA-seq data.

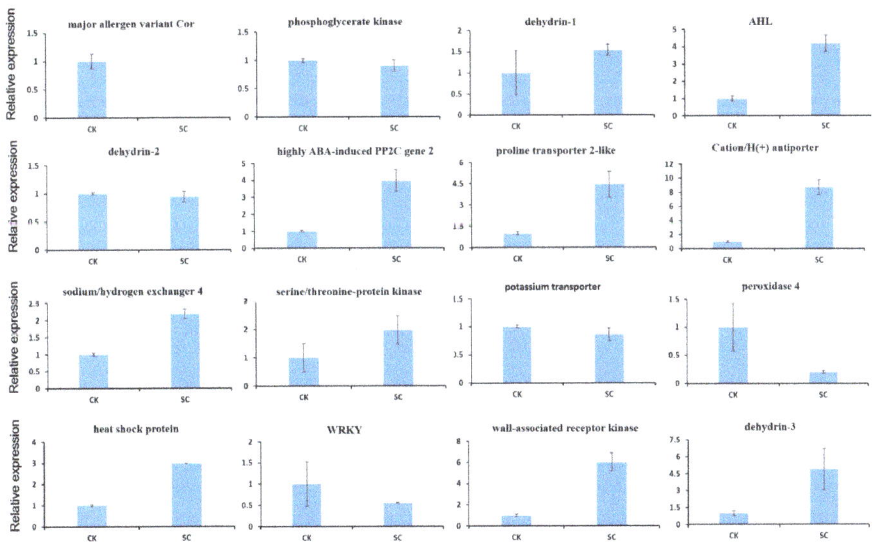

Figure 6. qRT-PCR validation of sixteen selected DEGs in leaves. Fold changes of the DEGs are shown. The expression levels in CK were arbitrarily set to 1. Error bars represent the standard deviations of three technical PCR replicates.

3. Discussion

B. halophila is a plant with high salt tolerance, so it is important to discover the salt tolerant genes of *B. halophila* for breeding salinity resistant varieties of trees. However, to our best knowledge, there is no report about genes associated with salt tolerance in *B. halophila*. In this study, we analyzed the transcriptomic data from the leaves of wild type *B. halophila* plants and plants with salt treatment. As a result, approximately 460 M raw reads were generated and were further assembled into 117,091 unigenes, among these unigenes, 64551 unigenes (55%) were annotated with gene descriptions, while the other 45% were unknown. This is the first report of transcriptome data from *B. halophila*. This transcriptome data provides an important genus resource for insight into the molecular mechanism of salt tolerance and facilitates discovery of novel genes responsive to salt stress in *B. halophila*.

In plants, salt stress responsive mechanisms are very complicated, which involve a complex interaction of physiological processes, metabolic pathways, and regulation at the molecular and cellular levels. Although plant response to salt stress has been extensively studied at different levels, the mechanisms underlying salinity tolerance are far from being completely understood. In addition, salt stress responsive mechanisms in different plants are also different. At present, the main mechanisms for which plants respond to salt stress include ion homeostasis and compartmentalization, ion transport and uptake, biosynthesis of osmoprotectants and compatible solutes, activation of antioxidant enzyme, and synthesis of antioxidant compounds [4–7,11]. In this study, 168 up-regulated genes and 351 down-regulated genes were identified in *B. halophila* under salt stress, respectively. These DEGs include dehydrin proteins, sodium/hydrogen exchanger, potassium transporter, sarcoplasmic/endoplasmic reticulum calcium ATPase, Ca^{2+} antiporter/cation exchanger, Nodulin MtN21/EamA-like transporter, heat shock protein, phosphoenolpyruvatecarboxykinase, NADH dehydrogenase, highly ABA-induced PP2C gene, homeobox-leucine zipper protein, phosphoglycerate kinase, WRKY transcription factor, AP2/ERF and B3 domain-containing transcription factor, flavonoid 3′,5′-hydroxylase, which is consistent with the other plants that are reported to be responsive to salt stress [20–24].

The analysis of GO enrichments suggested that the 519 DEGs response to salt stress was mainly involved in plant-type cell wall organization biological process, plant-type cell wall organization or biogenesis biological process, cell wall cellular component and structural constituent of cell wall molecular function. KEGG pathway enrichment results showed that the top-four enriched pathways for DEGs was 'Fatty acid elongation', 'Ribosome', 'Sphingolipid metabolism', and 'Flavonoid biosynthesis'. The expression patterns of sixteen of these DEGs were analyzed by qRT-PCR to verify the RNA-seq results. It revealed that the qRT-PCR results were consistent with RNA-seq data.

Based on the functional annotations of the 519 DEGs and the physiological evidence of *B. halophila* in response to salt stress [25], the possible mechanism of salt tolerance in the leaves of *B. halophila* was summarized in Figure 7. The possible salt tolerance mechanism is coordinately linked with ion homeostasis, osmotic protection, antioxidant regulation, ABA signal pathway, transcription factors and chaperons. When the plant is treatment with 200mM NaCl, multiple signal pathways are activated to cope with salt stress such as the SOS pathway, antioxidant pathway and ABA signal pathway and so on. Meanwhile, the osmoprotectants such as proline and polyols were accumulated to protect the cell. In addition, the transcription factors (WRKY, ERF, ZIP and AHL) and (LEAs, HSPs and AQPs) were activated to regulate the genes involved in the above pathways [2–6]. Overall, the salt tolerance mechanism in *B. halophila* is a complex network that involved the interactions at multiple levels. This information will be useful in elucidating the salt tolerance mechanisms in *B. halophila*.

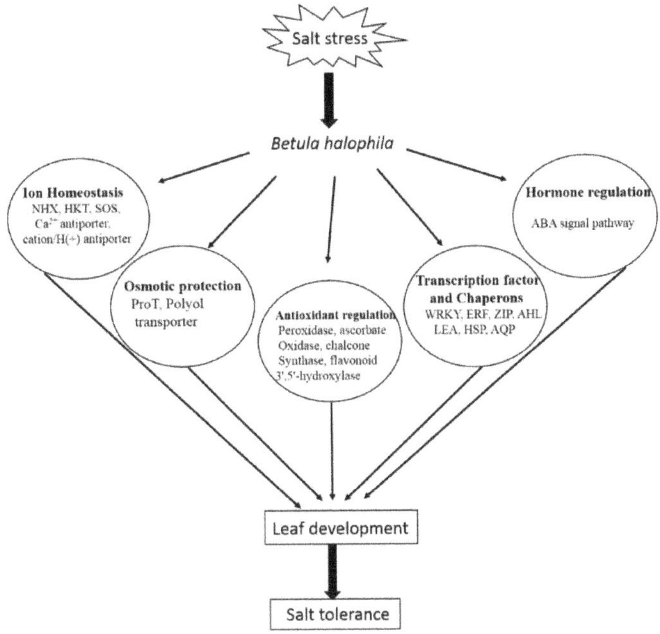

Figure 7. The possible mechanism of salt tolerance in the leaves of *B. halophila*.

In the present study, we observed that one dehydrin (DHNs) which is the most down-regulated gene among these DEGs and two other dehydrins showed a distinct salt responsive expression, suggesting that these dehydrin proteins may play different roles in response to salt stress in *B. halophila*. Dehydrins, also known as group 2 LEA (Late Embryogenesis Abundant) proteins, play a fundamental role in plant response to abiotic stresses [26–28]. Their expression is often induced under salinity, dehydration, cold and frost stress. Dehydrins are divided into five structural subgroups: Kn, SKn, KnS, YxKn and YxSKn [28]. The three dehydrins protein features of *B. halophila* were all SK3 subclass. It has been shown that SK3 dehydrins play an important protective role in plant stress tolerance, including drought, cold, and salinity [27]. For example, the expression of the durum wheat DHN-5 in A. thaliana led to an increase in salt and osmotic stress tolerance [28]. Rab16A in salt-tolerant Indica rice variety Pokkali can enhance tolerance to drought and salt stress in tobacco plants [29]. Similarly, overexpression of the wheat dehydrin PMA80 (as well as the LEAI protein PMA1959) enhances rice tolerance to drought and salt stress [30]. Although experimental evidence suggests that dehydrins have diverse roles (membrane protection, cryoprotection of enzymes, and protection from reactive oxygen species) in response to stresses [27–31], further efforts are still needed to precisely confirm the roles of these dehydrins and explore the regulatory mechanism underlying these functions in plant adaptive response to abiotic stresses.

In addition, our results indicated that the transcription factor AT-Hook Motif Nuclear Localized gene (AHL) was the most up-regulated gene in leaves after salt stress, implying that it might play an important role in response to salt stress in *B. halophila*. Previous studies showed that the AHL genes regulate diverse aspects of growth and development in plants. Such as the homeostasis of phytohormones [32], and defense responses [33–40]. However, there is no report about the function of AHL genes associated with salt stress. Further studies are still needed to understand the function of AHL genes in salt stress.

Therefore, our results provide a list of candidate genes for further investigation to determine whether they have a role in allowing *B. halophila* to tolerate high salt levels, and may be helpful in the understanding of the molecular mechanisms of salt stress response in *B. halophila*.

4. Materials and Methods

4.1. Plant Materials

The seeds of *B. halophila* were obtained from Xinjiang Academy of Forestry. After germination, the seedlings of *B. halophila* were grown in the greenhouse in Chinese Academy of Forestry. Leaves were collected from eight-months-old plants, the fourth or fifth leaf from top to bottom was used for sampling and RNA extraction. Three independent biological replicates were performed for each experiment. All samples were frozen and stored in liquid nitrogen until use.

4.2. Salt stress Treatment

Based on *Betula halophila* physiological response to salt stress as described by Zhang et al. [25], plantlets were treated with 200 mM of NaCl for 24h and then leaves were collected from stressed plants, plantlets treated with water were used as controls. Three independent biological replicates were performed for each experiment. All samples were frozen and stored in liquid nitrogen until use.

4.3. Library Construction and Sequencing for RNA-seq

In order to construct cDNA libraries, total RNAs were extracted from the control and the NaCl treated plant using Trizol RNA extraction kit (Life Technology, Beijing, China) according to the manufacturer's instruction. Six samples were sequenced by Novogene (Tianjin, China) using Illumina HiSeq2500 system.

4.4. Transcriptome Assembly and Bioinformatics Analysis

Transcriptome assembly was accomplished based on the left.fq and right.fq using Trinity [17] with min_kmer_cov set to 2 by default and all other parameters set default. In brief, the left files (read1 files) from all libraries/samples were pooled into one big left.fq file, and right files (read2 files) into one big right.fq file. Gene function was annotated based on Nr (NCBI non-redundant protein sequences), Nt (NCBI non-redundant nucleotide sequences), Pfam (Protein family), KOG/COG (Clusters of Orthologous Groups of proteins), Swiss-Prot (A manually annotated and reviewed protein sequence database), KO (KEGG Ortholog database), GO (Gene Ontology) databases. Gene expression levels were estimated by RSEM [41] for each sample. Clean data were mapped back onto the assembled transcriptome. Read counts for each gene were obtained from the mapping results. For differential expression analysis, prior to differential gene expression analysis, for each sequenced library, the read counts were adjusted by edgeR program package [42] through one scaling normalized factor. Differential expression analysis of six samples was performed using the DEGseq R package [43].

All transcripts were searched against the latest versions (as of August 2018) of Nr (nonredundant) database (http://www.ncbi.nlm.nih.gov/) and the Swiss-Prot database (http://www.gpmaw.com/html/swiss-prot.html) using the BLAST program with an $e < 10^{-5}$. The transcripts with the top hits were selected as unigenes. Open reading frames (ORFs) were predicted using the GetORF program contained in the EMBOSS software package. The Blast2GO program was used for GO annotation (http://www.geneontology.org), and the unigenes were aligned to the eggNOG (evolutionary genealogy of genes: non-supervised orthologous groups) database (http://www.ncbi.nlm.nih.gov/COG/) to identify functional categories. The KEGG database (http://www.genome.jp/kegg/) was used for pathway annotation. All searches were conducted using an e-value cut-off of 10^{-5}. GO terms were downloaded from the GO Analysis Toolkit and Database for Agriculture Community (AGRI go, http://bioinfo.cau.edu.cn/agriGO/download.php). All the genes identified with significant differential expression ($p < 0.05$) and FC >2 in this study were used as inputs to carry out GO enrichment analysis. Gene Ontology (GO) enrichment analysis of the differentially expressed genes (DEGs) was implemented by the GOseq R packages based Wallenius non-central hyper-geometric distribution [44] that can adjust for gene length bias in DEGs. KEGG pathway enrichment analysis used KOBAS [19] software to test the statistical enrichment of the differential expression genes in the KEGG pathways. In the scatterplot, the rich factor is the ratio of the differentially expressed gene number to the total gene number in a certain pathway.

4.5. Quantitative RT-PCR

The candidate DEGs in response to salt stress were selected to validate the reliability of the RNA-seq data using quantitative RT-PCR following the previously reported procedures [45,46]. Gene-specific primers were listed in Table S8. *BhActin* was used as a reference gene. Three independent biological replicates were performed. The results from gene-specific amplification were analyzed using the comparative *Cq* method, which uses an arithmetic formula, $2^{-\Delta\Delta Cq}$, to achieve results for relative quantification [47]. *Cq* represents the threshold cycle.

5. Conclusions

We sequenced and comparatively analyzed the transcriptomes from the leaves of wild type *B. halophila* plants and plants with salt treatment. This work enabled us to characterize gene expression profiles and identify functional genes related to salt tolerance. A total of 519 genes were differentially expressed under salt stress. These DEGs appear to be involved in many aspects, such as the SOS pathway, ion transport and uptake, antioxidant enzyme, ABA signal pathway and so on. It has been shown that one gene encoding the AT-Hook Motif Nuclear Localized transcription factor and three genes encoding dehydrins, might play important roles in response to salt stress in *B. halophila*. The results provide good candidate genes to breed salt tolerant plants, and will be helpful in understanding of the molecular mechanisms of salt stress in *B. halophila*.

Supplementary Materials: Supplementary materials can be found at http://www.mdpi.com/1422-0067/19/11/3412/s1. The sequencing data have been submitted to the SRA database under the accession number SRP146369.

Author Contributions: F.S. analyzed the data. F.S. and I.W.W. wrote the manuscript. L.Z. performed qRT-PCR. F.S. and D.Q. designed the experiment. All authors have read and approved the version of manuscript.

Funding: This work was supported by the National Key R & D Program of China (grant number 2016YFC0503103).

Conflicts of Interest: The authors declare that they have no competing interests.

Abbreviations

AHL	AT-Hook Motif Nuclear Localized gene
BLAST	the Basic Local Alignment Search Tool
DEGs	differentially expressed genes
DHNs	dehydrins
GO	Gene ontology
KEGG	Kyoto Encyclopedia of Genes and Genomes
KOG	Eukaryotic Orthologous Groups
LEA	Late Embryogenesis Abundant
Nr	NCBI non-redundant protein sequence database
Nt	NCBI nucleotide sequences
ORFs	Open reading frames
Pfam	Protein family
SOS	Salt Overly Sensitive
SwissProt	Swiss-prot sequence data bases

References

1. Flowers, T.J. Improving crop salt tolerance. *J. Exp. Bot.* **2004**, *55*, 307–319. [CrossRef] [PubMed]
2. Munns, R.; Tester, M. Mechanisms of salinity tolerance. *Annu. Rev. Plant Biol.* **2008**, *59*, 651–681. [CrossRef] [PubMed]
3. Wang, M.C.; Peng, Z.Y.; Li, C.L.; Li, F.; Liu, C.; Xia, G.M. Proteomic analysis on a high salt tolerance introgression strain of Triticumaestivum/Thinopyrumponticum. *Proteomics* **2008**, *8*, 1470–1489. [CrossRef] [PubMed]
4. Roy, S.; Chakraborty, U. Salt tolerance mechanisms in Salt Tolerant Grasses (STGs) and their prospects in cereal crop improvement. *Bot. Stud.* **2014**, *55*, 31. [CrossRef] [PubMed]
5. Gupta, B.; Huang, B. Mechanism of Salinity Tolerance in Plants: Physiological, Biochemical, and Molecular Characterization. *Int. J. Genom.* **2014**, *2014*, 701596. [CrossRef] [PubMed]
6. Ji, H.; Pardo, J.M.; Batelli, G.; Van Oosten, M.J.; Bressan, R.A.; Li, X. The Salt Overly Sensitive (SOS) pathway: Established and emerging roles. *Mol. Plant* **2013**, *6*, 275–286. [CrossRef] [PubMed]
7. Barragán, V.; Leidi, E.O.; Andrés, Z.; Rubio, L.; De Luca, A.; Fernández, J.A.; Cubero, B.; Pardo, J.M. Ion exchangers NHX1 and NHX2 mediate active potassium uptake into vacuoles to regulate cell turgor and stomatal function in Arabidopsis. *Plant Cell* **2012**, *24*, 1127–1142. [CrossRef] [PubMed]
8. Dugasa, M.T.; Cao, F.; Ibrahim, W.; Wu, F. Genotypic difference in physiological and biochemical characteristics in response to single and combined stresses of drought and salinity between the two wheat genotypes (Triticumaestivum) differing in salt tolerance. *Physiol. Plant* **2018**. [CrossRef] [PubMed]
9. James, R.A.; Blake, C.; Byrt, C.S.; Munns, R. Major genes for Na+ exclusion, Nax1 and Nax2 (wheatHKT1;4 and HKT1;5), decrease Na+ accumulation in bread wheat leaves under saline and waterlogged conditions. *J. Exp. Bot.* **2011**, *62*, 2939–2947. [CrossRef] [PubMed]
10. Xue, H.W.; Chen, X.; Mei, Y. Function and regulation of phospholipid signalling in plants. *Biochem. J.* **2009**, *421*, 145–156. [CrossRef] [PubMed]
11. Apel, K.; Hirt, H. Reactive oxygen species: Metabolism, oxidative stress, and signal transduction. *Annu. Rev. Plant Biol.* **2004**, *55*, 373–399. [CrossRef] [PubMed]
12. Wang, X.C.; Zhao, Q.Y.; Ma, C.L.; Zhang, Z.H.; Cao, H.L.; Kong, Y.M.; Yue, C.; Hao, X.Y.; Chen, L.; Ma, J.Q.; et al. Global transcriptome profiles of Camellia sinensis during cold acclimation. *BMC Genom.* **2013**, *14*, 415. [CrossRef] [PubMed]
13. Zhang, Q.; Cai, M.; Yu, X.; Wang, L.; Guo, C.; Ming, R.; Zhang, J. Transcriptome dynamics of Camellia sinensis in response to continuous salinity and drought stress. *Tree Genet. Genomes* **2017**, *13*, 78. [CrossRef]
14. Zeng, A.; Chen, P.; Korth, K.L.; Ping, J.; Thomas, J.; Wu, C.; Srivastava, S.; Pereira, A.; Hancock, F.; Brye, K.; et al. RNA sequencing analysis of salt tolerance in soybean (Glycine max). *Genomics* **2018**. [CrossRef] [PubMed]
15. Yu, J.; Chen, S.; Zhao, Q.; Wang, T.; Yang, C.; Diaz, C.; Sun, G.; Dai, S. Physiological and Proteomic Analysis of Salinity Tolerance in Puccinelliatenuiflora. *J. Proteome Res.* **2011**, *10*, 3852–3870. [CrossRef] [PubMed]
16. Wei, G.S.; Jian, D.Y.; Fu, C.Z.; Yu, F.Z.; Xue, W.D. Research on introduction and salt tolerance of Betulahalophila. *J. Gansu Agric. Univ.* **2011**, *5*, 101–105.
17. Grabherr, M.G.; Haas, B.J.; Yassour, M.; Levin, J.Z.; Thompson, D.A.; Amit, I.; Adiconis, X.; Fan, L.; Raychowdhury, R.; Zeng, Q.; et al. Full-length transcriptome assembly from RNA-Seq data without a reference genome. *Nat. Biotechnol.* **2011**, *29*, 644–652. [CrossRef] [PubMed]
18. Altschul, S.F.; Madden, T.L.; Schäffer, A.A.; Zhang, J.; Zhang, Z.; Miller, W.; Lipman, D.J. Gapped BLAST and PSI-BLAST: A new generation of protein database search programs. *Nucleic Acids Res.* **1997**, *25*, 3389–3402. [CrossRef] [PubMed]
19. Xie, C.; Mao, X.; Huang, J.; Ding, Y.; Wu, J.; Dong, S.; Kong, L.; Gao, G.; Li, C.Y.; Wei, L. KOBAS 2.0: A web server for annotation and identification of enriched pathways and diseases. *Nucleic Acids Res.* **2011**, *39*, W316–W322. [CrossRef] [PubMed]
20. Zhang, J.; Feng, J.; Lu, J.; Yang, Y.; Zhang, X.; Wan, D.; Liu, J. Transcriptome differences between two sister desert poplar species under salt stress. *BMC Genom.* **2014**, *15*, 337. [CrossRef] [PubMed]
21. Zhou, Y.; Yang, P.; Cui, F.; Zhang, F.; Luo, X.; Xie, J. Transcriptome Analysis of Salt Stress Responsiveness in the Seedlings of Dongxiang Wild Rice (OryzarufipogonGriff.). *PLoS ONE* **2016**, *11*, e0146242. [CrossRef]

22. Wang, W.S.; Zhao, X.Q.; Li, M.; Huang, L.Y.; Xu, J.L.; Zhang, F.; Cui, Y.R.; Fu, B.Y.; Li, Z.K. Complex molecular mechanisms underlying seedling salt tolerance in rice revealed by comparative transcriptome and metabolomic profiling. *J. Exp. Bot.* **2016**, *67*, 405–419. [CrossRef] [PubMed]
23. Villarino, G.H.; Hu, Q.; Scanlon, M.J.; Mueller, L.; Bombarely, A.; Mattson, N.S. Dissecting Tissue-Specific Transcriptomic Responses from Leaf and Roots under Salt Stress in Petunia hybrida Mitchell. *Genes* **2017**, *8*, 195. [CrossRef] [PubMed]
24. Villarino, G.H.; Bombarely, A.; Giovannoni, J.J.; Scanlon, M.J.; Mattson, N.S. Transcriptomic analysis of Petunia hybrida in response to salt stress using high throughput RNA sequencing. *PLoS ONE* **2014**, *9*, e94651. [CrossRef] [PubMed]
25. Zhang, H.B.; Zeng, Y.L.; Lan, H.Y.; Zhang, F.C. Physiological response of Betulahalophila (Betaluceae) to salt stress. *Acta Bot. Yunnanica* **2009**, *31*, 260–264. [CrossRef]
26. Liu, Y.; Song, Q.; Li, D.; Yang, X.; Li, D. Multifunctional Roles of Plant Dehydrins in Response to Environmental Stresses. *Front. Plant Sci.* **2017**, *8*, 1018. [CrossRef] [PubMed]
27. Kumar, M.; Lee, S.C.; Kim, J.Y.; Kim, S.J.; Aye, S.S.; Kim, S.R. Over-expression of dehydrin gene, OsDhn1, improves drought and salt stress tolerance through scavenging of reactive oxygen species in rice (*Oryza sativa* L.). *J. Plant Biol.* **2014**, *57*, 383–393. [CrossRef]
28. Brini, F.; Hanin, M.; Lumbreras, V.; Amara, I.; Khoudi, H.; Hassairi, A.; Pagès, M.; Masmoudi, K. Overexpression of wheat dehydrin DHN5 enhances tolerance to salt and osmotic stress in Arabidopsis thaliana. *Plant Cell Rep.* **2007**, *26*, 2017–2026. [CrossRef] [PubMed]
29. Roy Choudhury, A.; Sengupta, D.N. Transgenic tobacco plants overexpressing the heterologous lea gene Rab16A from rice during high salt and water deficit display enhanced tolerance to salinity stress. *Plant Cell Rep.* **2007**, *26*, 1839–1859. [CrossRef] [PubMed]
30. Cheng, Z.; Targolli, J.; Huang, X.; Wu, R. Wheat LEA genes, PMA80 and PMA1959, enhance dehydration tolerance of transgenic rice (Oryza sativa L.). *Mol. Breed.* **2002**, *10*, 71–82. [CrossRef]
31. Perdiguero, P.; Collada, C.; Soto, A. Novel dehydrins lacking complete K-segments in Pinaceae. The exception rather than the rule. *Front. Plant Sci.* **2014**, *5*, 682. [CrossRef] [PubMed]
32. Close, T.J. Dehydrins: A commonalty in the response of plants to dehydration and low temperature. *Physiol. Plant* **1997**, *100*, 291–296. [CrossRef]
33. Liu, C.C.; Li, C.M.; Liu, B.G.; Ge, S.J.; Dong, X.M.; Li, W.; Zhu, H.; Wang, B.; Yang, C. Genome-wide identification and characterization of a dehydrin gene family in poplar (Populustrichocarpa). *Plant Mol. Biol. Rep.* **2012**, *30*, 848–859. [CrossRef]
34. Du, H.; Liu, H.; Xiong, L. Endogenous auxin and jasmonic acid levels are differentially modulated by abiotic stresses in rice. *Front. Plant Sci.* **2013**, *4*, 397. [CrossRef] [PubMed]
35. Kim, H.B.; Oh, C.J.; Park, Y.C.; Lee, Y.; Choe, S.; An, C.S.; Choi, S.B. Comprehensive analysis of AHL homologous genes encoding AT-hook motif nuclear localized protein in rice. *BMB Rep.* **2011**, *44*, 680–685. [CrossRef] [PubMed]
36. Matsushita, A.; Furumoto, T.; Ishida, S.; Takahashi, Y. AGF1, an AT-hook protein, is necessary for the negative feedback of AtGA3ox1 encoding GA3-oxidase. *Plant Physiol.* **2007**, *143*, 1152–1162. [CrossRef] [PubMed]
37. Endt, D.V.; Silva, M.S.; Kijne, J.W.; Pasquali, G.; Memelink, J. Identification of a bipartite jasmonate-responsive promoter element in the Catharanthusroseus ORCA3 transcription factor gene that interacts specifically with AT-hook DNA-binding proteins. *Plant Physiol.* **2007**, *144*, 1680–1689. [CrossRef] [PubMed]
38. Street, I.H.; Shah, P.K.; Smith, A.M.; Avery, N.; Neff, M.M. The AT-hook-containing proteins SOB3/AHL29 and ESC/AHL27 are negative modulators of hypocotyl growth in Arabidopsis. *Plant J.* **2008**, *54*, 1–14. [CrossRef] [PubMed]
39. Lim, P.O.; Kim, Y.; Breeze, E.; Koo, J.C.; Woo, H.R.; Ryu, J.S.; Park, D.H.; Beynon, J.; Tabrett, A.; Buchanan-Wollaston, V.; et al. Overexpression of a chromatin architecture-controlling AT-hook protein extends leaf longevity and increases the post-harvest storage life of plants. *Plant J.* **2007**, *52*, 1140–1153. [CrossRef] [PubMed]
40. Lu, H.; Zou, Y.; Feng, N. Overexpression of AHL20 negatively regulates defenses in Arabidopsis. *J. Int. Plant Biol.* **2010**, *52*, 801–808. [CrossRef] [PubMed]
41. Li, B.; Dewey, C. RSEM: Accurate transcript quantification from RNA-Seq data with or without a reference genome. *BMC Bioinform.* **2011**, *12*, 323. [CrossRef] [PubMed]

42. Robinson, M.D.; McCarthy, D.J.; Smyth, G.K. edgeR: A Bioconductor package for differential expression analysis of digital gene expression data. *Bioinformatics* **2010**, *26*, 139–140. [CrossRef] [PubMed]
43. Wang, L.; Feng, Z.; Wang, X.; Wang, X.; Zhang, X. DEGseq: An R package for identifying differentially expressed genes from RNA-seq data. *Bioinformatics* **2010**, *26*, 136–138. [CrossRef] [PubMed]
44. Young, M.D.; Wakefield, M.J.; Smyth, G.K.; Oshlack, A. Gene ontology analysis for RNA-seq: Accounting for selection bias. *Genome Biol.* **2010**, *11*, R14. [CrossRef] [PubMed]
45. Shao, F.; Lu, Q.; Wilson, I.W.; Qiu, D. Genome-wide identification and characterization of the SPL gene family in Ziziphusjujuba. *Gene* **2017**, *627*, 315–321. [CrossRef] [PubMed]
46. Lu, Q.; Shao, F.; Macmillan, C.; Wilson, I.W.; Van der Merwe, K.; Hussey, S.G.; Myburg, A.A.; Dong, X.; Qiu, D. Genomewide analysis of the lateral organ boundaries domain gene family in Eucalyptus grandis reveals members that differentially impact secondary growth. *Plant Biotechnol. J.* **2018**, *16*, 124–136. [CrossRef] [PubMed]
47. Livak, K.J.; Schmittgen, T.D. Analysis of relative gene expression data using real-time quantitative PCR and the 2 (-Delta Delta, C.(T.)) method. *Methods* **2001**, *25*, 402–408. [CrossRef] [PubMed]

© 2018 by the authors. Licensee MDPI, Basel, Switzerland. This article is an open access article distributed under the terms and conditions of the Creative Commons Attribution (CC BY) license (http://creativecommons.org/licenses/by/4.0/).

International Journal of
Molecular Sciences

MDPI

Article

Maize WRKY Transcription Factor ZmWRKY106 Confers Drought and Heat Tolerance in Transgenic Plants

Chang-Tao Wang [1,†], Jing-Na Ru [2,†], Yong-Wei Liu [3], Meng Li [1], Dan Zhao [1], Jun-Feng Yang [4], Jin-Dong Fu [2,*] and Zhao-Shi Xu [2,*]

1 Beijing Advanced Innovation Center for Food Nutrition and Human Health/Beijing Key Lab of Plant Resource Research and Development, Beijing Technology and Business University, Beijing 100048, China; wangct@th.btbu.edu.cn (C.-T.W.); limeng@th.btbu.edu.cn (M.L.); zhaodanustb@126.com (D.Z.)

2 Institute of Crop Science, Chinese Academy of Agricultural Sciences (CAAS)/National Key Facility for Crop Gene Resources and Genetic Improvement, Key Laboratory of Biology and Genetic Improvement of Triticeae Crops, Ministry of Agriculture, Beijing 100081, China; rujingna1993@163.com

3 Institute of Genetics and Physiology, Hebei Academy of Agriculture and Forestry Sciences/Plant Genetic Engineering Center of Hebei Province, Shijiazhuang 050051, China; liuywmail@126.com

4 Hebei Wangfeng Seed Industry Co., Ltd., Xingtai 054900, China; Yangjunfenghb@163.com

* Correspondence: fujindong@caas.cn (J.-D.F.); xuzhaoshi@caas.cn (Z.-S.X.); Tel.: +86-10-82106773 (Z.-S.X.)

† These authors contributed equally to this work.

Received: 12 September 2018; Accepted: 1 October 2018; Published: 6 October 2018

Abstract: WRKY transcription factors constitute one of the largest transcription factor families in plants, and play crucial roles in plant growth and development, defense regulation and stress responses. However, knowledge about this family in maize is limited. In the present study, we identified a drought-induced WRKY gene, *ZmWRKY106*, based on the maize drought *de novo* transcriptome sequencing data. ZmWRKY106 was identified as part of the WRKYII group, and a phylogenetic tree analysis showed that ZmWRKY106 was closer to OsWRKY13. The subcellular localization of ZmWRKY106 was only observed in the nucleus. The promoter region of *ZmWRKY106* included the C-repeat/dehydration responsive element (DRE), low-temperature responsive element (LTR), MBS, and TCA-elements, which possibly participate in drought, cold, and salicylic acid (SA) stress responses. The expression of *ZmWRKY106* was induced significantly by drought, high temperature, and exogenous abscisic acid (ABA), but was weakly induced by salt. Overexpression of *ZmWRKY106* improved the tolerance to drought and heat in transgenic *Arabidopsis* by regulating stress-related genes through the ABA-signaling pathway, and the reactive oxygen species (ROS) content in transgenic lines was reduced by enhancing the activities of superoxide dismutase (SOD), peroxide dismutase (POD), and catalase (CAT) under drought stress. This suggested that *ZmWRKY106* was involved in multiple abiotic stress response pathways and acted as a positive factor under drought and heat stress.

Keywords: WRKY; ZmWRKY106; drought tolerance; thermotolerance; maize

1. Introduction

Changing environmental factors, such as abiotic stresses, influence plant growth and development [1]. Among them, drought and heat stresses seriously threaten crop productivity and quality. Plants must respond appropriately to changing environmental challenges to survive. Thus, it is important to explore the stress response mechanisms of plants and to enhance their tolerance to drought and heat to increase crop productivity without expanding cultivated land [2].

Environmental stresses initiate transcription factor (TF)-mediated expression of a variety of genes in plants, including bZIP, AP2/EREBP, MYB/MYC, NAC and WRKY [1,3–6]. WRKY TFs are identified by their conserved DNA-binding WRKY domains (WRKYGQK) in N-termini, and a zinc -finger motif (C-X4-5-C-X22-23-H-X1-H or C-X7-C-X23-H-X1-C) in C-termini [7,8]. It has been reported that WRKYs participated in defense responses by binding to the W-box located in the promoters of plant defense-related genes [9–11]. Another study found that 15 WRKY rice genes were induced by infection with the pathogen *Magnaporthe grisea* [12]. Statistically, 13 rice WRKY genes have regulated resistance against pathogens [13–17].

Recently, studies have revealed the involvement of the WRKY family in plant responses to abiotic stresses [18,19]. For example, *ABO3/WRKY63* took part in responses to abscisic acid (ABA) and drought stress in *Arabidopsis*. *AtWRKY57*-overexpressing *Arabidopsis* exhibited improved tolerance to drought by combining the promoter sequences of *NCED3* through the ABA pathway [20]. Overexpression of *OsWRKY30* enhanced resistance to drought stress in rice by the phosphorylation process of mitogen-activated protein kinases (MAPKs) [11]. *TaWRKY2* is a nuclear-located protein, and overexpression of *TaWRKY2* in *Arabidopsis* led to enhanced tolerance to drought and salt stresses by improving the expressions of *STZ* and *RD29B*; moreover, the exogenous expression of *TaWRKY19* in *Arabidopsis* not only conferred resistance to salt and drought, but also improved freezing tolerance [21]. In addition, *CmWRKY10*-overexpression in chrysanthemum revealed enhanced resistance to drought stress by regulating stress-related genes [19].

Maize (*Zea mays*) is a major food and economic crop. A few studies on the genome-wide analysis of WRKYs in maize have been reported in recent years. Wei et al. (2012) identified 136 WRKY proteins encoded by 119 *ZmWRKY* genes in maize, and Zhang et al. (2017) identified three additional new *ZmWRKY* genes and analyzed the gene expression profiles of *ZmWRKYs* using data from microarray, three RNA-seq studies, and the results of RT-PCR, which improved knowledge of WRKYs in maize [22,23]. In this paper, we performed drought-treated *de novo* transcriptome sequencing of maize (SRP144573) to investigate potential drought-tolerant WRKY genes in the maize genome. We identified a drought-responsive WRKY gene, *ZmWRKY106*, (Gene ID: GRMZM2G013391), which was named by Wei et al. (2012) and Zhang et al. (2017) [22,23]. The exogenous expression of *ZmWRKY106* in *Arabidopsis* led to enhanced tolerance of drought and heat.

2. Results

2.1. De Novo Transcriptome Sequencing Analysis

To find maize stress-responsive genes under drought stress, three-leaf seedlings were dehydrated on filter paper for 4 h, and then were collected for transcriptome sequencing analysis. The results showed that the transcription levels of many genes had changed after drought treatment (Figure 1A). Gene ontology (GO) analyses were used to classify the differentially expressed genes (DEGs) into functional groups. Almost 30 functionally enriched GO terms were identified for DEGs, and the results are shown in Supplementary Figure S1A. Among the predominantly enriched GO terms, signaling process was the most enriched term related to biological process. To further understand which pathways the stress-responsive genes may be involved in, the DEGs were analyzed against the Kyoto Encyclopedia of Genes and Genomes (KEGG) pathway database. The top 20 enriched pathways were identified, and the "plant hormone signal transduction" pathway enriched the most DEGs under drought treatment (Supplementary Figure S1B). DEGs including many transcription factors that play vital roles in plant growth, development, morphogenesis, and abiotic stress responses through regulating the expression of downstream genes [1,4,18]. Among these transcription factors, WRKYs play important roles in response to biotic and abiotic stresses [10,19]. We searched for *ZmWRKYs* among the DEGs, and found 14 *ZmWRKYs* induced by drought treatment (Figure 1B). We chose the gene GRMZM2G013391 named *ZmWRKY106* for further study.

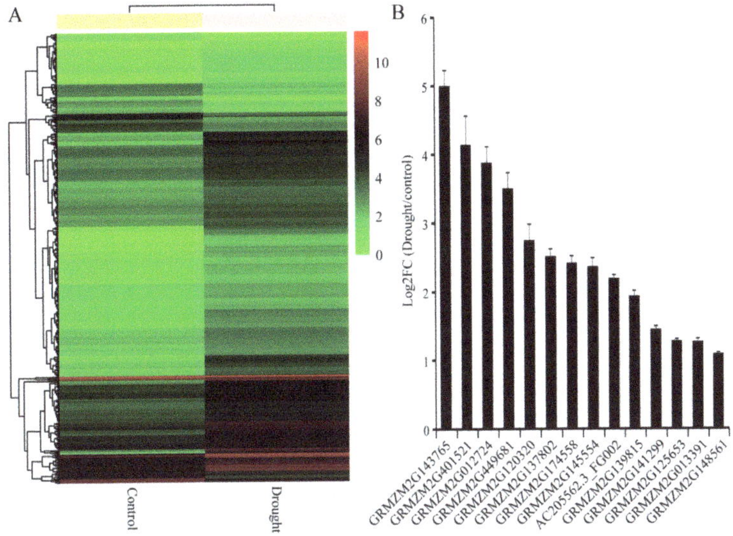

Figure 1. *De novo* transcriptome sequencing analysis of maize under drought stress. (**A**) Cluster analysis of the differentially expressed genes (DEGs) under drought treatment. (**B**) Transcription levels of the 14 differentially expressed *ZmWRKYs* under drought treatment. Error bar represent standard deviations (SD). The data represent means ± SD of three biological replications.

2.2. Phylogenetic Analysis of Maize *ZmWRKY106*

After selecting from the drought-treated maize *de novo* transcriptome data, we got a putative WRKY gene *ZmWRKY106* encoding 277 amino acids. The BLASTp online tool was used to search for the homologous amino acid sequences of *ZmWRKY106* in rice and *Arabidopsis*. The amino acid sequence alignment and phylogeny analysis of ZmWRKY106 orthologs are shown in Figure 2. ZmWRKY106 shared a mean identity of 28.47% with its rice, *Arabidopsis*, and barley orthologs and had a conserved signature WRKYGQK at the N-terminus followed by a C2H2 zinc-finger motif (C-X5-C-X23-H-X1-H), which characterized group II (Figure 2A). The sequences outside the conserved domain/motif were very different. The results of phylogenesis showed that ZmWRKY106 was closer to OsWRKY13, with a 61% bootstrap rate, followed by HvWRKY39, with a frequency of 100% (Figure 2B). However, the identity of ZmWRKY106 with other orthologs was lower than that with OsWRKY13, which indicated that ZmWRKY106 may have an extensive difference from other members.

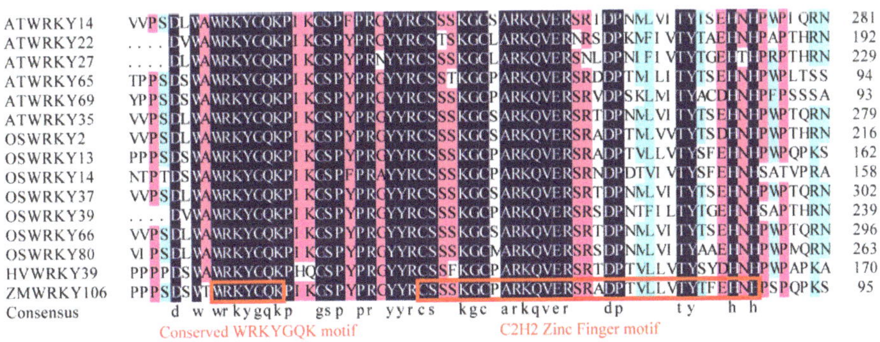

(**A**)

Figure 2. *Cont.*

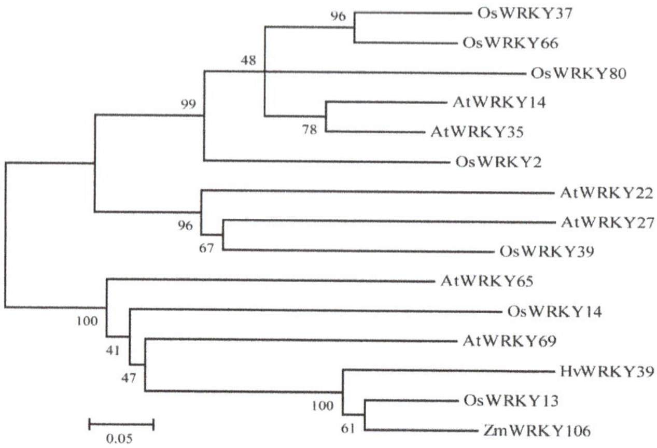

(B)

Figure 2. Multiple alignment and phylogenetic relationships of ZmWRKY106 with other orthologs in rice, *Arabidopsis*, and barley. The phylogenetic tree was produced using the aligned file with 1000 bootstraps in the MEGA 5.0 program. (**A**) Multiple alignment of ZmWRKY106 homologous proteins in rice, *Arabidopsis*, and barley. The different background colors represent the similar degree of amino acid sequences. (**B**) Phylogenetic relationship of ZmWRKY106 and other orthologs in different species. The first red box indicates the WRKYGQK motif, and the second indicates the conserved C2H2 zinc-finger motif.

2.3. *ZmWRKY106 Was Localized in the Nucleus*

The transient expression vector p16318h-ZmWRKY106 was transformed to maize mesophyll protoplasts by the PEG-mediated method to determine the cell localization. After incubation in darkness for 18 h, the fluorescence signals were monitored by a confocal laser scanning microscope. As shown in Figure 3, relative to the control distributed throughout the cell, the p16318h-ZmWRKY106 fusion protein was specifically detected in the nucleus.

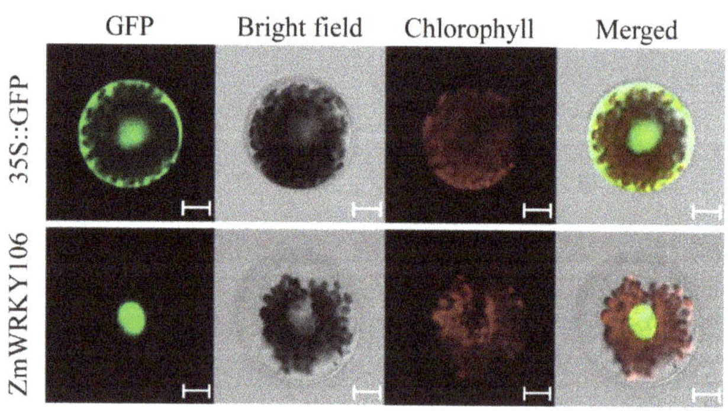

Figure 3. Subcellular localization of *ZmWRKY106*. The p16318hGFP and p16318hGFP-ZmWRKY106 constructs were transiently expressed in maize protoplasts. The green indicates green fluorescent, and the red indicates chloroplast autofluorescence. Results were observed after transformation for 18 h with confocal microscopy. Scale bars = 10 μm.

2.4. *ZmWRKY106 Promoter Domain Contained Various Stress-Related Cis-Elements*

To further understand the regulation mechanism of *ZmWRKY106*, we isolated the promoter region upstream of the *ZmWRKY106* ATG start codon. Types of *cis*-elements correlated to stress were present in the promoter region, including the C-repeat/DRE element referred to cold and dehydration response, low-temperature responsive element LTR and the drought-induced element MBS. In addition, there was another TCA-element that participated in salicylic acid (SA) response in the promoter region of *ZmWRKY106* (Table 1). This analysis suggested that *ZmWRKY106* may function in abiotic stress response.

Table 1. Putative *cis*-elements in the *ZmWRKY106* promoter.

Elements	Sequence	Function
C-repeat/DRE	TGGCCGAC	involved in cold- and dehydration-responsiveness
LTR	CCGAAA	involved in low-temperature responsiveness
MBS	TAACTG	MYB binding site involved in drought-inducibility
TCA-element	TCAGAAGAGG	involved in SA responsiveness

DRE—dehydration responsive element; LTR—low-temperature responsive; SA—salicylic acid.

2.5. *ZmWRKY106* Was Involved in Abiotic Stress Responses

To explore the possible signal pathways which *ZmWRKY106* may be involved in, we performed qRT-PCR to investigate the expression patterns of *ZmWRKY106* in maize treated with drought, high-salt, high-temperature, and ABA treatments. *ZmWRKY106* was remarkably induced by drought, high temperature and ABA, but was weakly induced by salt (Figure 4). For dehydration treatment, the transcript of *ZmWRKY106* was rapidly up-regulated more than 10-fold after 1 h of dehydration stress (Figure 4A). *ZmWRKY106* was slightly induced by salt at a maximum level of about 1.5-fold (Figure 4B). High temperature also significantly affected the expression of *ZmWRKY106*. Under high-temperature stress, the transcription level of *ZmWRKY106* increased gradually, peaked at 7.6-fold after 2 h of stress, and then rapidly declined to a constitutive level. With exogenous ABA treatment, the transcription level of *ZmWRKY106* was increased more than three-fold at 6 h after treatment.

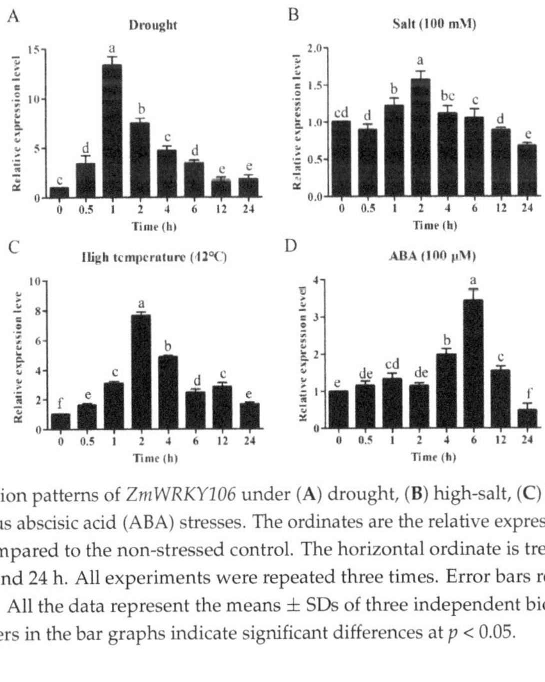

Figure 4. Expression patterns of *ZmWRKY106* under (**A**) drought, (**B**) high-salt, (**C**) high-temperature, and (**D**) exogenous abscisic acid (ABA) stresses. The ordinates are the relative expression level (fold) of *ZmWRKY106* compared to the non-stressed control. The horizontal ordinate is treatment time for 0, 0.5, 1, 2, 4, 6, 12 and 24 h. All experiments were repeated three times. Error bars represent standard deviations (SDs). All the data represent the means ± SDs of three independent biological replicates. The different letters in the bar graphs indicate significant differences at $p < 0.05$.

2.6. ZmWRKY106 Enhanced Drought Tolerance in Transgenic Arabidopsis

To investigate the function of *ZmWRKY106*, the pBI121-*ZmWRKY106* recombinant was transformed into wild-type (WT) *Arabidopsis* (Columbia-0). T_3 generation transgenic lines with relatively high expressions were selected by qRT-PCR for further analysis. The expression levels of transgenic lines are exhibited in Supplementary Figure S2. On MS medium, no significant differences in seed germination rates were observed between transgenic and WT plants. In the presence of 4% PEG6000, the germination rate of transgenic seeds was nearly 9% higher than WT after four days. Moreover, the germination was suppressed under 8% PEG6000, but transgenic seeds showed a higher germination rate than WT seeds (Figure 5A). For root growth assays, as shown in Figure 5B, *ZmWRKY106* transgenic lines had similar phenotypes to WT on MS medium. When supplemented with PEG6000, the growth of all transgenic and WT plants was repressed; however, transgenic plants showed clear differences compared to WT ones, with significantly longer total root lengths than those of WT under both PEG treatments. These results showed that *ZmWRKY106* transgenic lines had a stronger capacity to resist drought.

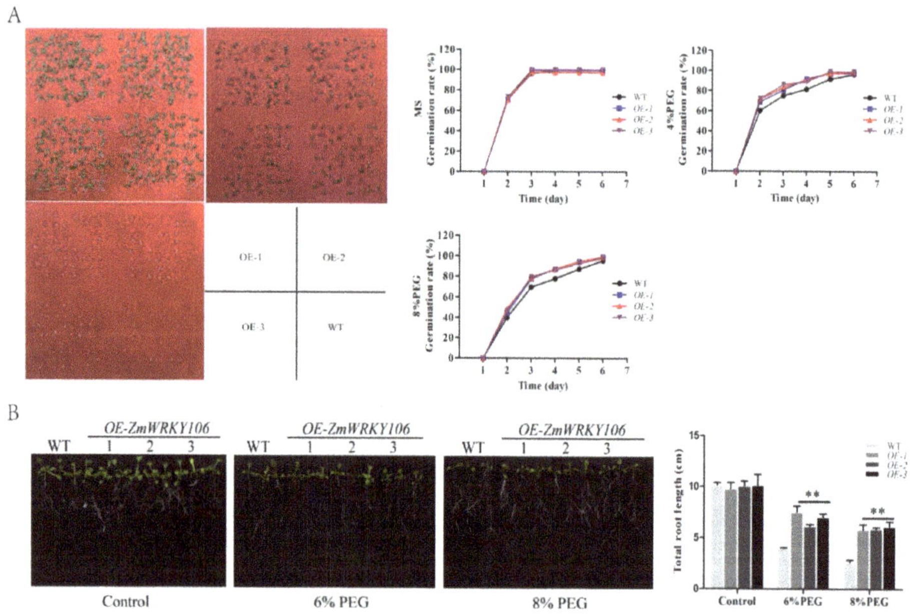

Figure 5. Phenotypes of *ZmWRKY106* transgenic *Arabidopsis* under drought treatment. (**A**) Seed germinations of wild-type (WT) and *ZmWRKY106*-overexpressing lines. (**B**) Root lengths of WT and *ZmWRKY106* transgenic plants. Five-day-old seedlings were transferred to MS medium supplemented with or without PEG6000 for seven days, and then root lengths were measured. All the data represent the means ± SDs of three independent biological replicates and asterisks (**) represent the significant differences at $p < 0.01$ (Student's *t*-test).

2.7. ZmWRKY106 Enhanced Heat Tolerance in Transgenic Arabidopsis

Under high temperature, the expression of *ZmWRKY106* was up-regulated. Following this result, we observed the phenotypes among WT and transgenic lines under 45 °C (Figure 6). The survival rates of transgenic and WT plants were 100% under normal conditions, while higher a survival rate was exhibited in OE lines than WT after heat treatment for 5 h. *ZmWRKY106*-overexpressing lines had a survival rate of more than 30%, compared to less than 20% for WT plants after heat treatment. This suggested that *ZmWRKY106* may improve thermotolerance of transgenic plants.

Figure 6. Survival rates of WT and *ZmWRKY106* transgenic lines under heat stress. Five-day-old seedlings were placed at 45 °C for 5 h and then resumed growth at 22 °C. The data represent the means ± SDs of three independent biological replicates. Asterisks (**) represent the significant differences ($p < 0.01$) compared with the control (Student's *t*-test).

2.8. *ZmWRKY106* Regulated the Expression of Stress-Related Genes

To understand the molecular mechanisms of *ZmWRKY106* in stress responses, expression of stress-responsive genes, including *RD29A*, *HSP90*, *DREB2A*, *CuZnSOD*, *NCED3*, and *NCED6*, was examined using qRT-PCR under normal and drought conditions. The results showed that the expression levels of *HSP90* and *NCED3* were low in both WT and OE lines under normal conditions, while the expression levels in OE lines remained higher than WT plants after treatment (Figure 7B,E). Meanwhile, *CuZnSOD* and *NCED6* in OE lines were up-regulated after 4 h of stress treatment, and sharply increased to the maximum (Figure 7D,F). The expression levels of *RD29A* and *DREB2A* in OE lines were remarkably higher following all treatments (Figure 7A,C). Because the expressions of ABA and stress-related genes were altered in transgenic lines, we conjectured that *ZmWRKY106* may play a role in the abiotic stress response by regulating stress-related genes through the ABA-signaling pathway.

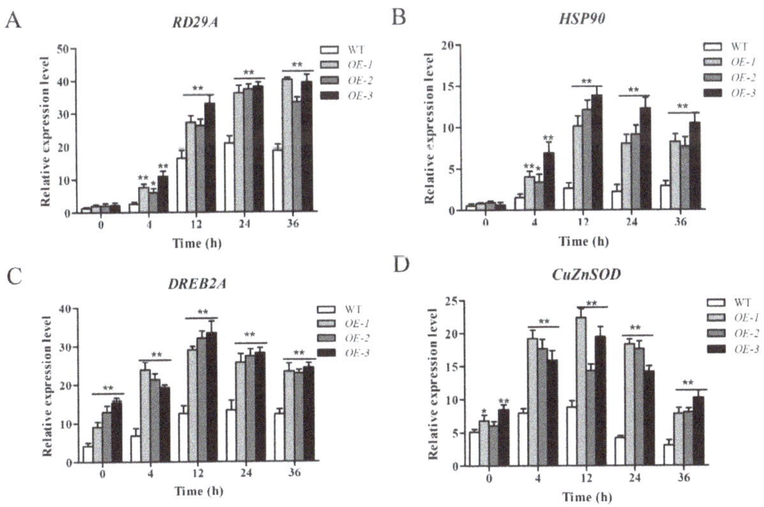

Figure 7. *Cont.*

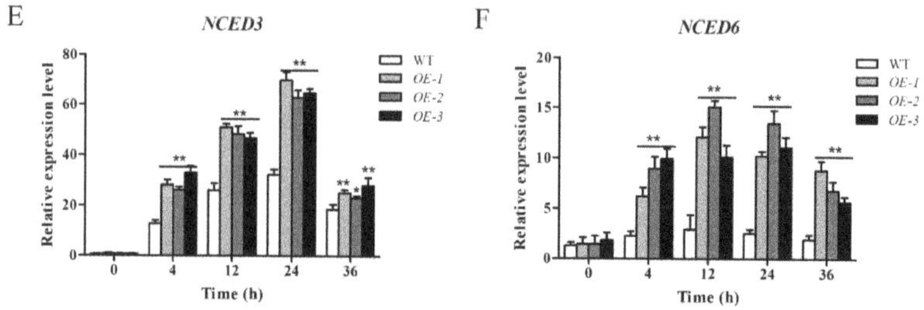

Figure 7. The relative expression of stress-related genes. (**A**) *RD29A*, (**B**) *HSP90*, (**C**) *DREB2A*, (**D**) *CuZnSOD*, (**E**) *NCED3*, and (**F**) *NCED6* were examined under control and drought conditions for various time points (4, 12, 24 and 36 h). Values are means ± SDs of three replicates, and asterisks (* or **) represent the significant differences at $p < 0.05$ or $p < 0.01$, respectively (Student's *t*-test).

2.9. Overexpression of ZmWRKY106 Reduced Reactive Oxygen Species (ROS) Content and Enhanced the Activities of Superoxide Dismutase (SOD), Peroxide Dismutase (POD), and Catalase (CAT) under Drought Treatment

The ROS content and the enzyme activities were assessed in transgenic lines and WT plants at 0, 4, 12 and 24 h after drought treatment (Figure 8). As shown in Figure 8A, the ROS accumulation in transgenic lines was less than that in WT plants at all times, while the ROS content was increased in WT plants and remained at a higher level during the whole experiment. The activities of SOD, POD and CAT were increased in OE lines compared to WT lines (Figure 8B–D). The activity of SOD was almost unchanged in WT before and after drought treatment, whereas in OE lines the activity of SOD was greater, and reached a maximum at 24 h after drought treatment (Figure 8B). Increases of POD activity were observed in both WT and transgenic lines, but the increases in WT were smaller, and there was consistently higher POD activity in transgenic lines than in WT lines (Figure 8C). In the case of CAT, the CAT activity of WT lines remained consistent at 0.016 U and had almost no significant change during the stress treatment; however, the CAT concentration in the OE lines remained significantly higher compared to that in WT lines under drought stress (Figure 8D). In a word, overexpression of *ZmWRKY106* reduced ROS content by enhancing the activities of SOD, POD and CAT to resist drought stress.

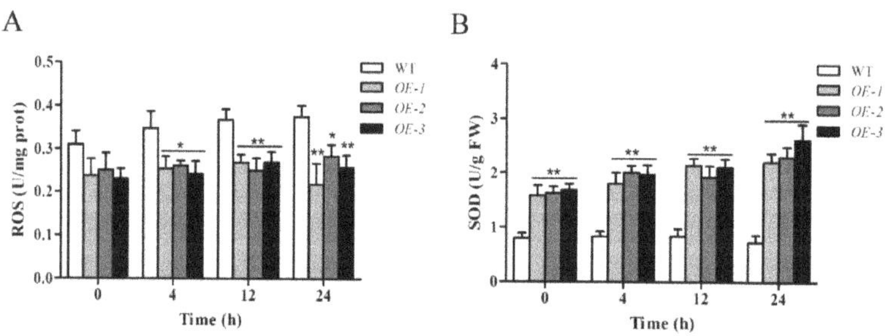

Figure 8. *Cont.*

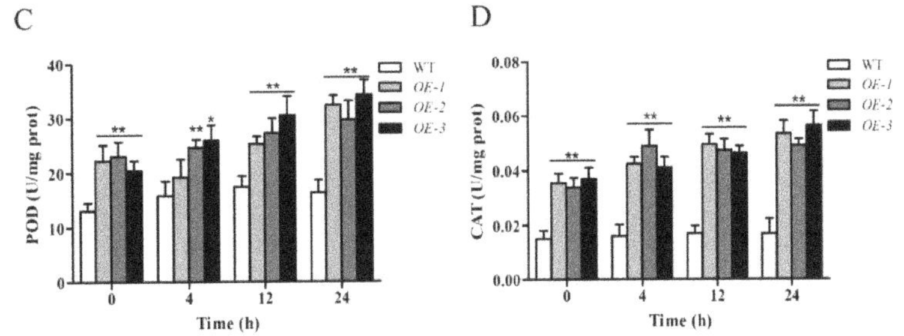

Figure 8. (**A**) The reactive oxygen species (ROS) content and the activities of (**B**) superoxide dismutase (SOD), (**C**) peroxide dismutase (POD), and (**D**) catalase (CAT) under different conditions at different time points (0, 4, 12, and 24 h). Values are means ± SDs of three replicates, and asterisks (* or **) represent the significant differences at $p < 0.05$ or $p < 0.01$, respectively (Student's *t*-test).

3. Discussion

Biotic and abiotic stresses seriously affect plant growth and development. Under adverse environments, transcriptome changes are the earliest responses, and transcriptional regulation plays a crucial role in plant defense responses [15]. Thus far, many TFs have been identified as participating in plant defense responses, including MYB, bZIP, and WRKY proteins. There are many more biotic stress-related genes in WRKYs than in other TFs, and an increasing number of studies have revealed that WRKY TFs play positive or negative roles in plants' disease prevention [24]. For example, *AtWRKY46*, coordinated with *AtWRKY70* and *AtWRKY53*, positively regulated basal resistance to *Pseudomonas syringae* [25]; *OsWRKY6* played a positive role in plant defense response by activating the expression of defense-related genes [26]; *GhWRKY44* was induced by pathogen injection, and overexpression of *GhWRKY44* led to enhanced resistance against bacterial and fungal pathogens [27]. These results all suggest that the WRKY family plays an important role in responding to biotic stresses [28].

However, knowledge about the role of WRKYs in abiotic stresses is limited [29,30]. Maize is a major food and economic crop and plays an important role in basic and applied biological research. So far, known research about WRKYs has been mostly related to defense response in dicotyledon plants such as *Arabidopsis*, tomato, and tobacco, but little information about the role of maize WRKYs has been reported [31–33]. It is rather crucial to elucidate the functional maize WRKY protein in abiotic stress response. In maize, Wei et al. [22] have identified 136 WRKY proteins encoded by 119 WRKY genes, numbered them, and performed a phylogenetic tree analysis of the maize WRKYs with orthologs in *Arabidopsis*, rice, and barley, which improved knowledge of WRKYs in maize. In addition, Zhang et al. [23] identified three new additional *ZmWRKY* genes, analyzed the gene expression profiles of *ZmWRKYs* using data from various studies, and found that ten genes, including *ZmWRKY9*, *ZmWRKY25*, *ZmWRKY47*, *ZmWRKY97*, *ZmWRKY80*, *ZmWRKY39*, *ZmWRKY106*, *ZmWRKY53*, *ZmWRKY36* and *ZmWRKY113*, were responsive under drought treatment in at least in three studies, which provided the basis for cloning functional *ZmWRKY* genes. In this study, we revealed the function of *ZmWRKY106* in abiotic stress responses. Our study showed that ZmWRKY106 belongs to group II, shares a mean identity with its rice, *Arabidopsis* and barley orthologs, and is closer to *OsWRKY13* (Figure 2).

Increasing evidence has indicated that WRKYs play an important role in abiotic stress response, for example, *GmWRKY21* improved freezing tolerance in transgenic *Arabidopsis*, and *GmWRKY54* played a positive role in response to salt and drought stresses, whereas *GmWRKY13* markedly increased sensitivity to salt and mannitol [34]. Overexpression of *AtWRKY25* and *AtWRKY33* in *Arabidopsis* led to enhanced resistance to salt and hypersensitivity to ABA [35]. In rice, *OsWRKY11* enhanced heat and drought tolerance [29]. In barley, *Hv-WRKY38* played key roles in the response to cold and drought stresses, and enhanced drought tolerance in turf and forage grass [36,37]. In this study, expression profiles analysis revealed that *ZmWRKY106* was induced significantly by drought, high temperature and ABA (Figure 4), possibly related to various stress-related cis-elements in its promoter region (Table 1). Under drought treatment, the transgenic seeds of *ZmWRKY106* germinated faster than WT seeds, and roots of OE lines were remarkably longer than those of WT lines (Figure 5). Meanwhile, overexpression of *ZmWRKY106* reduced ROS content and enhanced the activities of SOD, POD and CAT under drought treatment (Figure 8). Furthermore, the survival rates of OE lines were higher than those of WT lines (Figure 6). These results all showed that *ZmWRKY106* exhibited drought tolerance and thermotolerance.

ABA is a major phytohormone referred to plant response under drought stress, and there exist ABA-dependent and ABA-independent pathways in drought stress response. In our study, the expression levels of six stress-related genes were assessed under normal and drought conditions (Figure 7). *DREB2A* is a well-known marker gene in ABA-independent stress responses [38]. ABRE and DRE/CRT motifs were found in the promoters of many stress-inducible genes, such as *RD29A*, which contained several DREs and one ABRE in the promoter domain, and was strongly induced by cold, drought and salt stresses [39–41]. *HSP90* played a major role in stress signal transduction, and overexpression of HSP90 affected the phenotype of transgenic plants [42–45]. In our study, the expressions of *RD29A*, *HSP90*, and *DREB2A* genes were all up-regulated in *ZmWRKY106* transgenic lines (Figure 7A–C), suggesting that *ZmWRKY106* may play a positive role in drought and heat response. *CmWRKY10* acted as a positive factor in response to drought stress by regulating the expression of *DREB1A*, *DREB2A*, *CuZnSOD*, *NCED3A*, and *NCED3B*, which proved that *CmWRKY10* enhanced the drought tolerance through the ABA-dependent pathway [19]. These genes could play key roles in the physiological process of abiotic stress response [46,47]. We found that the expressions of ABA-related genes were higher in transgenic lines, which indicated that overexpression of *ZmWRKY106* led to enhanced tolerance of drought stress through the ABA-dependent pathway (Figure 7D–F). These results all indicated that *ZmWRKY106* may play a role in the abiotic stress response by regulating stress-related genes through the ABA-signaling pathway (Figure 7). Nevertheless, the role and regulation mechanisms of *ZmWRKY106* in maize still need further research.

4. Materials and Methods

4.1. De Novo Transcriptome Sequencing

Three-leaf stage untreated maize seedlings and seedlings dehydrated on filter paper for 4 h were collected for RNA-seq analysis. The detailed process of RNA-seq was undertaken as previously described [18]. The transcriptome data are available in the National Center for Biotechnology Information (NCBI) under accession number SRP144573.

4.2. Plant Materials and Stress Treatments

The seeds of maize (X178) used in this study were provided by Zhuan-Fang Hao (Institute of Crop Science, Chinese Academy of Agricultural Sciences, Beijing, China). The maize seeds were sown as previously described [48]. Three-leaf stage maize seedlings were exposed to drought, salt, high-temperature and ABA treatments. For dehydration treatment, seedlings were quickly cleaned and then transferred on to filter paper to rapidly dry in air as previously described [49]. Seedlings were placed in 42 °C chambers for high-temperature treatment. For salt and ABA treatments, seedlings were exposed to water solutions supplemented with 100 mM NaCl, and 100 μM ABA, respectively. The samples were collected at 0, 0.5, 1, 2, 4, 6, 12 and 24 h after treatment. Harvested seedlings were dropped immediately into liquid nitrogen and stored at −80 °C for RNA extraction.

4.3. RNA Extraction and Quantitative Real-Time PCR (qRT-PCR)

Total RNAs were extracted from maize tissue using an RNAprep plant kit (Tiangen, Beijing, China), and cDNA was synthesized as previously described [18]. The qRT-PCR was performed with

SuperReal PreMix Plus (Tiangen, Beijing, China) by an ABI Prism 7500 system (Applied Biosystems, Foster City, CA, USA). The specific primers of *ZmWRKY106* are listed in Supplementary Table S1. Each PCR was repeated three times and data were analyzed, as previously described [50].

4.4. Gene Isolation and Sequence Analysis

The full length of the *ZmWRKY106* gene was amplified by PCR with specific primers from maize cDNA. The primers of *ZmWRKY106*-F and *ZmWRKY106*-R are listed in Supplementary Table S1. The PCR products were cloned into pLB vector (Tiangen, China) and sequenced. The homologs of *ZmWRKY106* in different species were searched for in the NCBI database. Sequence alignments of *ZmWRKY106* orthologs were performed by ClustalX software. The phylogenetic tree was constructed using the neighbor-joining method by the MEGA 5.0 program with bootstrap analysis of 1000 replicates [51].

4.5. Subcellular Localization

The coding region of *ZmWRKY106* was fused to the subcellular localization vector p16318h with green fluorescent protein (GFP) tags containing the CaMV35S promoter. The specific primers are listed in Supplementary Table S1. For transient expression assays, the p16318h-*ZmWRKY106* reconstruction plasmid was transformed to maize mesophyll protoplasts by the PEG-mediated method, while the p16318hGFP vector was transformed as control [52]. The fluorescence signals were observed by a confocal laser scanning microscope (LSM700; CarlZeiss, Oberkochen, Germany) after incubation in darkness at 22 °C for 18 h.

4.6. Cis-Acting Elements in Promoter

The 2.0 kb promoter region upstream of *ZmWRKY106* was obtained from maize genomic DNA on the EnsemblPlants website (available online: http://plants.ensembl.org/index.html). Putative cis-acting elements in the promoter region were analyzed using the PLACE database [53].

4.7. Generation of Transgenic Arabidopsis and Its Phenotype under Stress Treatment

Plant expression vector pBI121-*ZmWRKY106* was constructed as previously described [54], and was transformed to wild-type (WT) *Arabidopsis* using the *Agrobacterium*-mediated floral dip method. Columbia-0 (WT) was used for exogenous expression of *ZmWRKY106*. The transformed seeds were selected on MS medium containing 50 mM Kanamycin at 22 °C with a photoperiod of 16 h light/8 h dark (60% humidity) to obtain the positive plants. Three T_3 generation overexpression lines (*OE-ZmWRKY106-1*, *OE-ZmWRKY106-2*, *OE-ZmWRKY106-3*) with higher expression levels of *ZmWRKY106* were selected by qRT-PCR for further analysis. *Arabidopsis* seeds were grown as described previously [50]. Four-week-old seedlings of transgenic and WT *Arabidopsis* were collected at 0, 4, 12, 24 and 36 h after drought treatment (dried on filter paper) to examine the expression of stress-related genes by qRT-PCR. The specific primers of stress-related genes are listed in Supplementary Table S1. Three biological replicates were performed for qRT-PCR.

For the germination assay, WT and transgenic *Arabidopsis* seeds were placed on MS medium and MS medium supplemented with 4% (*w*/*v*) or 8% PEG6000. When the radicle had emerged from the seed coat, we considered the seed germinated. Seed germination was followed for five days, and the germination rate was analyzed. For the root growth assay, five-day-old seedlings were transferred to MS medium with or without 6% and 8% PEG6000 for seven days, and then root lengths were measured. Each treatment contained three independent replicates.

For high-temperature stress assay, five-day-old seedlings were placed at 45 °C for 5 h and then resumed growth at 22 °C as described previously [18]. After growing under normal conditions for seven days, we took photos and analyzed the survival rate. Each treatment contained three independent replicates. Values are means ± SD and statistically significant differences were based on the Student's test.

4.8. Measurements of Reactive Oxygen Species (ROS) Content and Enzyme Activity

To better understand the function of *ZmWRKY106* under drought treatment, we assessed the activities of superoxide dismutase (SOD), peroxide dismutase (POD), catalase (CAT) and the ROS content in WT and transgenic lines at 0, 4, 12 and 24 h after drought stress. The ROS content and the activities of SOD, POD, and CAT were measured as previously described [19]. To obtain reproducible results, each experiment was repeated three times. Values are means ±SD and statistically significant differences were based on the Student's test.

5. Conclusions

We identified a drought-induced WRKYII gene *ZmWRKY106* based on maize drought *de novo* transcriptome sequencing data (SRP144573). *ZmWRKY106* was only observed in the nucleus. The expression of *ZmWRKY106* was induced significantly by drought, high-temperature, and exogenous abscisic acid (ABA) treatments, but was induced weakly by salt. Further research revealed that overexpression of *ZmWRKY106* could improve tolerance to drought and heat in transgenic *Arabidopsis* by regulating stress-related genes through the ABA-signaling pathway, and could reduce the ROS content in transgenic lines by enhancing the activities of SOD, POD and CAT under drought stress. These results may provide a basis for understanding the functions of *ZmWRKY106* in abiotic stress response in maize.

Supplementary Materials: Supplementary materials can be found at http://www.mdpi.com/1422-0067/19/10/3046/s1.

Author Contributions: Conceptualization, C.-T.W., J.-N.R., and J.-F.Y.; methodology, C.-T.W., J.-N.R., and Y.-W.L.; validation, C.-T.W., J.-N.R., and Y.-W.L.; formal analysis, C.-T.W. and J.-N.R.; investigation, C.-T.W. and J.-N.R.; data, C.-T.W. and J.-N.R.; writing—original draft preparation, C.-.T.W. and J.-N.R.; writing—review and editing, Z.-S.X. and J.-D.F.; supervision, J.-F.Y. and Y.-W.L.; project administration, Z.-S.X. and J.-D.F.; funding acquisition, M.L. and D.Z.

Funding: This study was financially supported by the Funding Project for Beijing Advanced Innovation Center for Food Nutrition and Human Health, and the Open Research Fund Program of Beijing Key Lab of Plant Resource Research and Development, Beijing Technology and Business University.

Acknowledgments: We thank Li-Na Ning for critically reading the manuscript.

Conflicts of Interest: The authors declare no conflict of interest.

Abbreviations

ABA	abscisic acid
ABRE	ABA-responsive element
DRE	dehydration responsive element
GFP	green fluorescent protein
LTR	low-temperature responsive
MAPKs	mitogen-activated protein kinases
qRT-PCR	quantitative real-time PCR
RT-PCR	reverse transcription PCR
SA	salicylic acid
TF	transcription factor
WT	wild type

References

1. Xu, Z.S.; Chen, M.; Li, L.C.; Ma, Y.Z. Functions of the ERF transcription factor family in plants. *Botany* **2008**, *86*, 969–977. [CrossRef]
2. Mehta, R.H.; Ponnuchamy, M.; Kumar, J.; Reddy, N.R. Exploring drought stress-regulated genes in senna (*Cassia angustifolia* Vahl.): A transcriptomic approach. *Funct. Integr. Genom.* **2017**, *17*, 1–25. [CrossRef] [PubMed]

3. Rushton, P.J.; Somssich, I.E. Transcriptional control of plant genes responsive to pathogens. *Curr. Biol.* **1998**, *1*, 311–315. [CrossRef]
4. Tuteja, N. Abscisic acid and abiotic stress signaling. *Plant Signal Behav.* **2007**, *2*, 135–138. [CrossRef] [PubMed]
5. Singh, D.; Laxmi, A. Transcriptional regulation of drought response: A tortuous network of transcriptional factors. *Front. Plant Sci.* **2015**, *6*, 895. [CrossRef] [PubMed]
6. Gahlaut, V.; Jaiswal, V.; Kumar, A.; Gupta, P.K. Transcription factors involved in drought tolerance and their possible role in developing drought tolerant cultivars with emphasis on wheat (*Triticum aestivum* L.). *Theor. Appl. Genet.* **2016**, *129*, 2019–2042. [CrossRef] [PubMed]
7. Eulgem, T.; Rushton, P.J.; Robatzek, S.; Somssich, I.E. The WRKY superfamily of plant transcription factor. *Trends Plant Sci.* **2000**, *5*, 199–206. [CrossRef]
8. Jiang, W.; Wu, J.; Zhang, Y.; Yin, L.; Lu, J. Isolation of a *WRKY30* gene from *Muscadinia rotundifolia* (Michx) and validation of its function under biotic and abiotic stresses. *Protoplasma* **2015**, *252*, 1361–1374. [CrossRef] [PubMed]
9. Ciolkowski, I.; Wanke, D.; Birkenbihl, R.P.; Somssich, I.E. Studies on DNA-binding selectivity of WRKY transcription factors lend structural clues into WRKY-domain function. *Plant Mol. Biol.* **2008**, *68*, 81–92. [CrossRef] [PubMed]
10. Verk, M.C.V.; Pappaioannou, D.; Neeleman, L.; Bol, J.F.; Linthorst, H.J.M. A novel WRKY transcription factor is required for induction of PR-1a gene expression by salicylic acid and bacterial elicitors. *Plant Physiol.* **2008**, *146*, 1983–1995. [CrossRef] [PubMed]
11. Shen, H.; Liu, C.; Zhang, Y.; Meng, X.; Zhou, X.; Chu, C.; Wang, X.P. OsWRKY30 is activated by MAP kinases to confer drought tolerance in rice. *Plant Mol. Biol.* **2012**, *80*, 241–253. [CrossRef] [PubMed]
12. Ryu, H.S.; Han, M.; Lee, S.K.; Cho, J.I.; Ryoo, N.; Heu, S.; Lee, Y.H.; Bhoo, S.H.; Wang, G.L.; Hahn, T.R.; et al. A comprehensive expression analysis of the WRKY gene superfamily in rice plants during defense response. *Plant Cell Rep.* **2006**, *25*, 836–847. [CrossRef] [PubMed]
13. Cheng, H.T.; Li, H.B.; Deng, Y.; Xiao, J.H.; Li, X.H.; Wang, S.P. The WRKY45-2-WRKY13-WRKY42 transcriptional regulatory cascade is required for rice resistance to fungal pathogen. *Plant Physiol.* **2015**, *167*, 1087–1099. [CrossRef] [PubMed]
14. Choi, C.; Hwang, S.H.; Fang, I.R.; Kwon, S.I.; Park, S.R.; Ahn, I.; Kim, J.B.; Hwang, D.J. Molecular characterization of *Oryza sativa* WRKY6, which binds to W-box-like element 1 of the *Oryza sativa pathogenesis–related* (*PR*) 10a promoter and confers reduced susceptibility to pathogens. *New Phytol.* **2015**, *208*, 846–859. [CrossRef] [PubMed]
15. Wang, H.H.; Meng, J.M.; Peng, X.X.; Tang, X.K.; Zhou, P.L.; Xiang, J.H.; Deng, X.B. Rice WRKY4 acts as a transcriptional activator mediating defense responses toward *Rhizoctonia solani*, the causing agent of rice sheath blight. *Plant Mol. Biol.* **2015**, *89*, 157–171. [CrossRef] [PubMed]
16. Hwang, S.H.; Kwon, S.I., Jang, J.Y.; Fang, I.L.; Lee, H.; Choi, C.; Park, S.; Ahn, I.; Bae, S.C.; Hwang, D.J. OsWRKY51, a rice transcription factor, functions as a positive regulator in defense response against *Xanthomonas oryzae pv. oryzae*. *Plant Cell Rep.* **2016**, *35*, 1975–1985. [CrossRef] [PubMed]
17. Peng, X.; Wang, H.; Jang, J.C.; Xiao, T.; He, H.; Jiang, D.; Tang, X. OsWRKY80-OsWRKY4 module as a positive regulatory circuit in rice resistance against *Rhizoctonia solani*. *Rice* **2016**, *9*, 63–76. [CrossRef] [PubMed]
18. He, G.H.; Xu, J.Y.; Wang, Y.X.; Liu, J.M.; Li, P.S.; Chen, M.; Ma, Y.Z.; Xu, Z.S. Drought-responsive WRKY transcription factor genes *TaWRKY1* and *TaWRKY33* from wheat confer drought and/or heat resistance in *Arabidopsis*. *BMC Plant Biol.* **2016**, *16*, 116. [CrossRef] [PubMed]
19. Jaffar, M.A.; Song, A.P.; Faheem, M.; Chen, S.M.; Jiang, J.F.; Liu, C.; Fan, Q.Q.; Chen, F.D. Involvement of *CmWRKY10* in drought tolerance of chrysanthemum through the ABA–Signaling pathway. *Int. J. Mol. Sci.* **2016**, *17*, 693. [CrossRef] [PubMed]
20. Jiang, Y.J.; Liang, G.; Yu, D.Q. Activated expression of *WRKY57* confers drought tolerance in *Arabidopsis*. *Mol. Plant* **2012**, *5*, 1375–1388. [CrossRef] [PubMed]
21. Niu, C.F.; Wei, W.; Zhou, Q.Y.; Tian, A.G.; Hao, Y.J.; Zhang, W.K.; Ma, B.; Lin, Q.; Zhang, Z.B.; Zhang, J.S.; et al. Wheat WRKY genes *TaWRKY2* and *TaWRKY19* regulate abiotic stress tolerance in transgenic *Arabidopsis* plants. *Plant Cell Environ.* **2012**, *35*, 1156–1170. [CrossRef] [PubMed]
22. Wei, K.F.; Chen, J.; Chen, Y.F.; Wu, L.J.; Xie, D.X. Molecular phylogenetic and expression analysis of the complete WRKY transcription factor family in Maize. *DNA Res.* **2012**, *19*, 153–164. [CrossRef] [PubMed]

23. Zhang, T.; Tan, D.F.; Zhang, L.; Zhang, X.Y.; Han, Z.X. Phylogenetic analysis and drought-responsive expression profiles of the WRKY transcription factor family in maize. *Agri Gene* **2017**, *3*, 99–108. [CrossRef]
24. Pandey, S.P.; Somssich, I.E. The role of WRKY transcription factors in plant immunity. *Plant Physiol.* **2009**, *150*, 1648–1655. [CrossRef] [PubMed]
25. Hu, Y.R.; Dong, Q.Y.; Yu, D.Q. *Arabidopsis* WRKY46 coordinates with WRKY70 and WRKY53 in basal resistance against pathogen *Pseudomonas syringae*. *Plant Sci.* **2012**, *185–186*, 288–297. [CrossRef] [PubMed]
26. Huang, S.H.; Yie, S.W.; Hwang, D.J. Heterologous expression of *OsWRKY6* gene in *Arabidopsis* activates the expression of defense related genes and enhances resistance to pathogens. *Plant Sci.* **2011**, *181*, 316–323.
27. Li, J.; Wang, J.; Wang, N.X.; Guo, X.Q.; Gao, Z. GhWRKY44, a WRKY transcription factor of cotton, mediates defense responses to pathogen infection in transgenic *Nicotiana benthamiana*. *Plant Cell Tissue Organ Cult.* **2015**, *121*, 127–140. [CrossRef]
28. Ulker, B.; Somssich, I.E. WRKY transcription factors: From DNA binding towards biological function. *Curr. Opin. Plant Biol.* **2004**, *7*, 491–498. [CrossRef] [PubMed]
29. Wu, X.L.; Shiroto, Y.; Kishitani, S.; Ito, Y.; Toriyama, K. Enhanced heat and drought tolerance in transgenic rice seedlings overexpressing *OsWRKY11* under the control of *HSP101* promoter. *Plant Cell Rep.* **2009**, *28*, 21–30. [CrossRef] [PubMed]
30. Yan, Y.; Jia, H.H.; Wang, F.; Wang, C.; Liu, S.C.; Guo, X.Q. Overexpression of *GhWRKY27a* reduces tolerance to drought stress and resistance to *Rhizoctonia solani* infection in transgenic *Nicotiana benthamiana*. *Front. Physiol.* **2015**, *6*, 265. [CrossRef] [PubMed]
31. Rizhsky, L.; Liang, H.; Mittler, R. The combined effect of drought stress and heat shock on gene expression in tobacco. *Plant Physiol.* **2002**, *130*, 1143–1151. [CrossRef] [PubMed]
32. Li, S.J.; Fu, Q.T.; Chen, L.G.; Huang, W.D.; Yu, D.Q. *Arabidopsis* thaliana WRKY25, WRKY26, and WRKY33 coordinate induction of plant thermotolerance. *Planta* **2011**, *233*, 1237–1252. [CrossRef] [PubMed]
33. Atamian, H.S.; Eulgem, T.; Kaloshian, I. SlWRKY70 is required for Mi-1-mediated resistance to aphids and nematodes in tomato. *Planta* **2012**, *235*, 299–309. [CrossRef] [PubMed]
34. Zhou, Q.Y.; Tian, A.G.; Zou, H.F.; Xie, Z.M.; Lei, G.; Huang, J.; Wang, C.M.; Wang, H.W.; Zhang, J.S.; Chen, S.Y. Soybean WRKY-type transcription factor genes, *GmWRKY13*, *GmWRKY21*, and *GmWRKY54*, confer differential tolerance to abiotic stresses in transgenic *Arabidopsis* plants. *Plant Biotechnol. J.* **2008**, *6*, 486–503. [CrossRef] [PubMed]
35. Jiang, Y.; Deyholos, M.K. Functional characterization of *Arabidopsis* NaCl-inducible WRKY25 and WRKY33 transcription factors in abiotic stresses. *Plant Mol. Biol.* **2009**, *69*, 91–105. [CrossRef] [PubMed]
36. Marè, C.; Mazzucotelli, E.; Crosatti, C.; Francia, E.; Stanca, A.M.; Cattivelli, L. Hv-WRKY38: A new transcription factor involved in cold- and drought-response in barley. *Plant Mol. Biol.* **2004**, *55*, 399–416. [CrossRef] [PubMed]
37. Xiong, X.; James, V.A.; Zhang, H.N.; Altpeter, F. Constitutive expression of the barley HvWRKY38 transcription factor enhances drought tolerance in turf and forage grass (*Paspalum notatum* Flugge). *Mol. Breed.* **2010**, *25*, 419–432. [CrossRef]
38. Shinozaki, K.; Yamaguchi-Shinozaki, K. Gene networks involved in drought stress response and tolerance. *J. Exp. Bot.* **2007**, *58*, 221–227. [CrossRef] [PubMed]
39. Qin, Y.X.; Wang, M.C.; Tian, Y.C.; He, W.X.; Han, L.; Xia, G.M. Over-expression of *TaMYB33* encoding a novel wheat MYB transcription factor increases salt and drought tolerance in *Arabidopsis*. *Mol. Biol. Rep.* **2012**, *39*, 7183–7192. [CrossRef] [PubMed]
40. Xiong, L.; Schumaker, K.S.; Zhu, J.K. Cell signaling during cold, drought, and salt stress. *Plant Cell* **2000**, *14*, S165. [CrossRef]
41. Msanne, J.; Lin, J.; Stone, J.M.; Awada, T. Characterization of abiotic stress-responsive *Arabidopsis* thaliana *RD29A* and *RD29B* genes and evaluation of transgenes. *Planta* **2011**, *234*, 97–107. [CrossRef] [PubMed]
42. Pratt, W.B.; Galigniana, M.D.; Harrell, J.M.; Defranco, D.B. Role of hsp90 and the hsp90-binding immunophilins in signalling protein movement. *Cell. Signal.* **2004**, *16*, 857–872. [CrossRef] [PubMed]
43. Yamada, K.; Fukao, Y.; Hayashi, M.; Fukazawa, M.; Suzuki, I.; Nishimura, M. Cytosolic HSP90 regulates the heat shock response that is responsible for heat acclimation in *Arabidopsis thaliana*. *J. Biol. Chem.* **2007**, *282*, 37794–37804. [CrossRef] [PubMed]
44. Queitsch, C.; Sangster, T.A.; Lindquist, S. Hsp90 as a capacitor of phenotypic variation. *Nature* **2000**, *417*, 618–624. [CrossRef] [PubMed]

45. Sangster, T.A.; Bahrami, A.; Wilczek, A.; Watanabe, E.; Schellenberg, K.; Mclellan, C.; Kelley, A.; Kong, S.W.; Queitsch, C.; Lindquist, S. Phenotypic diversity and altered environmental plasticity in *Arabidopsis thaliana* with reduced *Hsp90* levels. *PLoS ONE* **2007**, *2*, e648. [CrossRef] [PubMed]
46. Seki, M.; Kamei, A.; Yamaguchi-Shinozaki, K.; Shinozaki, K. Molecular responses to drought, salinity and frost: Common and different paths for plant protection. *Curr. Opin. Biotechnol.* **2003**, *14*, 194–199. [CrossRef]
47. Shinozaki, K.; Yamaguchi-shinozaki, K.; Seki, M. Regulatory network of gene expression in the drought and cold stress responses. *Curr. Opin. Biol.* **2003**, *6*, 410–417. [CrossRef]
48. Song, W.; Zhao, H.; Zhang, X.; Lei, L.; Lai, J. Genome-Wide identification of VQ motif-containing proteins and their expression profiles under abiotic stresses in Maize. *Front. Plant Sci.* **2016**, *6*, 1177. [CrossRef] [PubMed]
49. Liu, W.X.; Zhang, F.C.; Zhang, W.Z.; Song, L.F.; Wu, W.H.; Chen, Y.F. *Arabidopsis* Di19 functions as a transcription factor and modulates *PR1*, *PR2*, and *PR5* expression in response to drought stress. *Mol. Plant* **2013**, *6*, 1487–1502. [CrossRef] [PubMed]
50. Feng, Z.J.; Cui, X.Y.; Cui, X.Y.; Chen, M.; Yang, G.X.; Ma, Y.Z.; He, G.Y.; Xu, Z.S. The soybean GmDi19-5 interacts with GmLEA3.1 and increases sensitivity of transgenic plants to abiotic stresses. *Front. Plant Sci.* **2015**, *6*, 179. [CrossRef] [PubMed]
51. Tamura, K.; Peterson, D.; Peterson, N.; Stecher, G.; Nei, M.; Kumar, S. MEGA5: Molecular evolutionary genetics analysis using maximum likelihood, evolutionary distance, and maximum parsimony methods. *Mol. Biol. Evol.* **2011**, *28*, 2731–2739. [CrossRef] [PubMed]
52. Toppo, S.; Vanin, S.; Bosello, V.; Tosatto, S.C. Evolutionary and structural insights into the multifaceted glutathione peroxidase (Gpx) superfamily. *Antioxid. Redox Signal.* **2008**, *10*, 1501–1514. [CrossRef] [PubMed]
53. Higo, K.; Ugawa, Y.; Iwamoto, M.; Korenaga, T. Plant cis–acting regulatory DNA elements (PLACE) database. *Nucleic Acids Res.* **1999**, *27*, 297–300. [CrossRef] [PubMed]
54. Xu, Z.S.; Xia, L.Q.; Chen, M.; Cheng, X.G.; Zhang, R.Y.; Li, L.C.; Li, Y.X.; Zhao, Y.X.; Lu, Y.; Ni, Z.Y.; et al. Isolation and molecular characterization of the *Triticum aestivum* L. ethylene-responsive factor 1 (*TaERF1*) that increases multiple stress tolerance. *Plant Mol. Biol.* **2007**, *65*, 719–732. [CrossRef] [PubMed]

© 2018 by the authors. Licensee MDPI, Basel, Switzerland. This article is an open access article distributed under the terms and conditions of the Creative Commons Attribution (CC BY) license (http://creativecommons.org/licenses/by/4.0/).

International Journal of
Molecular Sciences

MDPI

Article

Genotyping-by-Sequencing Enhances Genetic Diversity Analysis of Crested Wheatgrass [*Agropyron cristatum* (L.) Gaertn.]

Kiran Baral [1], Bruce Coulman [1], Bill Biligetu [1,*] and Yong-Bi Fu [2,*]

1 Department of Plant Sciences, University of Saskatchewan, 51 Campus Drive, Saskatoon, SK S7N 5A8, Canada; kiran.baral@usask.ca (K.B.); bruce.coulman@usask.ca (B.C.)

2 Plant Gene Resources of Canada, Saskatoon Research and Development Centre, Agriculture and Agri-Food Canada, 107 Science Place, Saskatoon, SK S7N 0X2, Canada

* Correspondence: bill.biligetu@usask.ca (B.B.); yong-bi.fu@agr.gc.ca (Y.-B.F.); Tel.: +1-306-966-4007 (B.B.); +1-306-385-9298 (Y.-B.F.)

Received: 3 August 2018; Accepted: 28 August 2018; Published: 31 August 2018

Abstract: Molecular characterization of unsequenced plant species with complex genomes is now possible by genotyping-by-sequencing (GBS) using recent next generation sequencing technologies. This study represents the first use of GBS application to sample genome-wide variants of crested wheatgrass [*Agropyron cristatum* (L.) Gaertn.] and assess the genetic diversity present in 192 genotypes from 12 tetraploid lines. Bioinformatic analysis identified 45,507 single nucleotide polymorphism (SNP) markers in this outcrossing grass species. The model-based Bayesian analysis revealed four major clusters of the samples assayed. The diversity analysis revealed 15.8% of SNP variation residing among the 12 lines, and 12.1% SNP variation present among four genetic clusters identified by the Bayesian analysis. The principal coordinates analysis and dendrogram were able to distinguish four lines of Asian origin from Canadian cultivars and breeding lines. These results serve as a valuable resource for understanding genetic variability, and will aid in the genetic improvement of this outcrossing polyploid grass species for forage production. These findings illustrate the potential of GBS application in the characterization of non-model polyploid plants with complex genomes.

Keywords: genotyping-by-sequencing; *Agropyron*; genetic diversity; genetic structure; SNP

1. Introduction

Genotyping-by-sequencing (GBS) is a powerful genomic approach for identification of genetic variation on a genome-wide scale for genetic diversity analysis of non-model plants [1–3]. This approach produces high-density, low-cost genotypic information without the requirement for a reference genome sequence [4]. The detailed GBS approach in plant diversity analysis is described in Peterson et al. [3]. In brief, the GBS analysis involves five major steps: (1) genome complexity reduction with restriction enzyme; (2) barcoding the seared genomic DNAs with indexed adaptors; (3) high-throughput sequencing of barcoded DNA fragments; (4) identification of genetic variants through a bioinformatics analysis of de-multiplexed reads; and (5) a genetic diversity analysis of sequenced samples based on sample-by-variant matrix. The GBS application, despite being a powerful approach, has certain limitations, including many missing data points, uneven genome coverage, complex bioinformatics, and issues related to polyploidy [5–8]. To overcome these limitations, a GBS-based pipeline, called Haplotag, was developed by Tinker et al. [9], which can generate tag-level haplotype and single nucleotide polymorphism (SNP) data for polyploid organisms. This approach has been successfully applied in the study of diploid and polyploid genomes in oat (*Avena sativa*) [10–12] and genetic diversity analysis of northern wheatgrass (*Elymus lanceolatus* ssp. *Lanceolatus*) [13].

Crested wheatgrass [CWG; *Agropyron cristatum* (L.) Gaertn.] is one of the perennial species of the genus *Agropyron* that comprises 10–15 species in a polyploid series of diploid (2n = 2x = 14), tetraploid (2n = 4x = 28) and hexaploid (2n = 6x = 42) forms with the P genome [14,15]. *Agropyron* species are native to temperate-frigid grassland and sandy soils of Eurasia [14,16,17], and were first introduced to Canada in 1911 [16]. CWG is the most important commercial species of the crested wheatgrass complex in Canadian grasslands [18]. It is characterized by an extensive root system, making it drought tolerant and winter hardy. CWG is considered an important pasture grass for early spring grazing, providing highly palatable and nutritious forage [19]. This species is easy to establish, has strong competitive ability, tolerates insect predation, provides high forage yield, and can be managed for multiple harvests in a season [16,19,20]. It performs well on marginal lands and semi-desert environments to moist moderately saline soils [19,20]. Due to these features, this species can be used for land reclamation of abandoned croplands, burnt and degraded areas, as well as in erosion control [21]. It has persisted as a high yielding species compared to native forage species, even in 20- to 40-year-old pastures, despite heavy grazing and trampling [19,22]. In addition, CWG is also known to possess traits of interest, including disease resistance, tolerance to abiotic stress, and high yield, which have been utilized in wheat and barley breeding [23–27]. The palatability and nutrient content of CWG declines after anthesis, and it becomes less desirable for summer grazing [19]. Thus, a goal of present CWG breeding programs is to develop later maturing cultivars that would maintain nutritive value into the summer grazing season. Development of high forage-quality, late-maturing CWG cultivars is limited by the relatively long varietal development process, few studies to assess genetic variability of the germplasm, and lack of an effective marker system for marker-assisted and/or genomic selection/breeding. Recent RNA-seq studies in CWG have identified flowering time related genes and flowering related differentially expressed genes [28,29]. This emphasizes the need for genetic diversity studies of CWG for the management and utilization of proper genetic resources in a breeding program as exogamous perennial forage species are often morphologically comparable, though they are genetically highly heterogeneous and heterozygous [30,31]. An adequate level of genetic diversity is crucial for both germplasm adaptation and the long-term sustainability of plant communities [32].

Attempts have been made to assess genetic variability within and among the genus *Agropyron* using molecular markers like amplified fragment length polymorphism (AFLP) [18] and simple sequence repeat (SSR) markers [31,33,34]. The revealed variabilities have allowed for better understanding of the extent of diversity present in the genus. However, these marker systems are unable to provide high resolution of genetic diversity and population structure information to understand the ancestry and microevolution of the populations. Research is needed to assess molecular characteristics of CWG for plant breeding. The molecular characterization is now more feasible than before with the advanced sequencing technology and reduced cost to acquire informative markers such as SNPs in non-model polyploid CWG plants. Recent GBS studies in polyploid plants [10,13] demonstrate the likelihood that GBS will unveil genetic variability on a genome-wide scale in CWG plants, and characterize CWG germplasm for breeding and genetic research.

This study was conducted with the objective to apply GBS in combination with the Universal Network Enabled Analysis Kit (UNEAK) [35] and the Haplotag pipelines to (1) identify genome-wide SNP markers; (2) assess the genetic diversity present in 12 lines of *A. cristatum*; and (3) assess whether the GBS application is useful in the genetic diversity analysis of complex polyploid plants.

2. Results

2.1. SNP Discovery and Characterization

The Miseq run of 192 genotypes from 12 CWG lines (Table 1) generated approximately 87.8 million raw forward (R1) sequence reads of 250 bp. The number of raw forward sequence reads per sample ranged from 190,606 to 775,160 with an average of 457,279. Combined UNEAK and Haplotag analysis at the 20%, 30%, 40%, and 50% level of missing data generated 227; 1,884; 10,738; and 45,507 SNPs, respectively across the 192 genotypes. In addition, this analysis also generated many metagenomic files associated with the SNP discovery, which are described and accessible in the online Supplementary Materials. The distribution of the minor allele frequency in 45,507 SNPs' data ranged from 0.025 to 0.5, and exhibited a steady decline of minor alleles with increased occurrence of frequencies from 0.075 to 0.5 (Figure 1A). Likewise, there were more SNPs at the higher percentages of missing data (Figure 1B).

Table 1. List of the 12 crested wheatgrass (*A. cristatum*) lines used in the study.

Lines	CN Number [a]	Alternative Identification [a]	Origin	Type
Kirk	CN108662	PI 536010	Canada	Cultivar
AC-Goliath	CN108673		Canada	Cultivar
NewKirk		FOR552	Canada	Cultivar
Vysokij 9	CN30995	PI 370654	Siberia, Former Soviet Union, Omsk region	Genebank line
Karabalykskij 202	CN31068	PI 326204	Kazakhstan, Former Soviet Union, Kustanai region	Genebank line
PGR 16830	CN43478		Kazakhstan	Genebank line
S8959E		FOR917	Siberia/Canada	Breeding line
S9491		S9491	Canada	Breeding line
S9514		S9514	Canada	Breeding line
S9516		S9516	Canada	Breeding line
S9544		S9544	Canada	Breeding line
S9556		S9556	Canada	Breeding line

[a] CN number is the line identification in Plant Gene Resources of Canada, Agriculture, and Agri-Food Canada (AAFC), while the alternative identifications, including FOR or S, are from the joint forage breeding program of the University of Saskatchewan and AAFC, and PI is from plant inventory book, National Germplasm Resources Laboratory, USA.

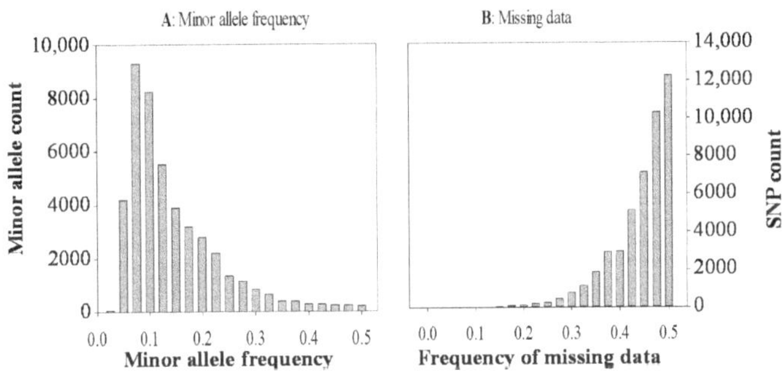

Figure 1. The minor allele frequency distribution (**A**) and the frequency of missing data (**B**) for 45,507 SNP markers in 192 genotypes of 12 crested wheatgrass lines.

2.2. Genetic Structure and Relationship

The genetic structure estimated for 192 genotypes from 12 CWG lines without consideration of prior population information in the STRUCTURE [36] analysis revealed four optimal clusters (Figure 2A) with strong support from change in LnP(K) variance (Figure 2B) and the largest delta K value (Figure 2C). Cluster 1 (red in color) consisted of 17 genotypes (16 from Vysokij 9 and one from S8959E). Cluster 2 (green in color) had 22 genotypes (16 from S9491 and 6 from S9514). Cluster 3 (blue in color) was the largest cluster, with 95 genotypes from seven lines. Cluster 4 (yellow in color), with 58 genotypes from five lines, was the second largest cluster. The neighbor-joining (NJ) tree was in agreement with clusters obtained from the STRUCTURE analysis (Figure 3). However, there existed some discrepancies, as some members of cluster 4 (yellow in color) were spread into cluster 2 (green in color) and cluster 3 (blue in color).

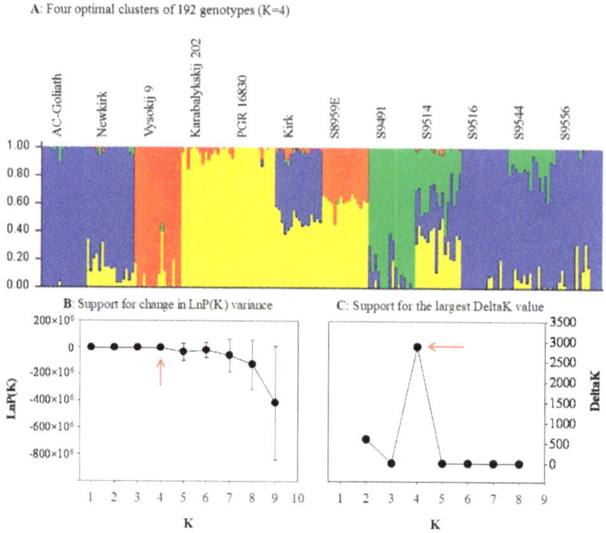

Figure 2. Four genetic clusters of 192 genotypes of the 12 crested wheatgrass lines inferred by STRUCTURE based on 45,507 SNP markers. (**A**) The mixture coefficients of 192 genotypes with K = 4, presented in the original order of genotypes from 12 lines (see Table 1 for line label); (**B**) support from the LnP(K) estimation; (**C**) support from the estimation of the largest value of the delta K = mean (|Ln″(K)|)/sd (LnP(K)).

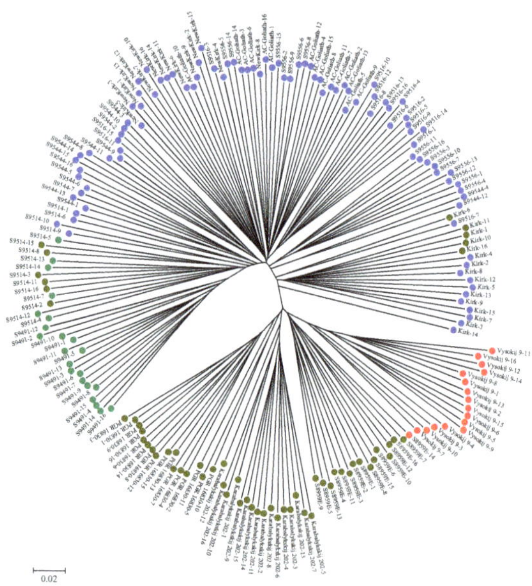

Figure 3. Genetic relationship of 192 genotypes of the 12 crested wheatgrass lines as revealed by neighbor-joining clustering with the 45,507 SNP markers. Each genotype is numbered after its line label. Each node for a genotype is represented with colored circle followed by genotype name. Red, green, blue, and yellow represent plants in Clusters 1, 2, 3, and 4, inferred from the STRUCTURE analysis (Figure 2A), respectively.

The principal coordinates analysis (PCoA) revealed that the genetic relationship of 192 genotypes (Figure 4A) was not in accordance to the Bayesian inferences from the STRUCTURE analysis. The clusters II, III, and IV identified by the Bayesian inferences appeared to overlap and became undistinguishable with PCoA. However, the PCoA plot was able to distinguish four lines Karabalykskij 202 (from Kazakhstan), PGR 16,830 (from Kazakhstan), Vysokij 9 (from Russia) and S8,959E (selected from Vysokij 9) from the rest of the lines (Figure 4B). We also observed lines S9,516, S9,544 and S9,556 from cluster 3 (blue in color from the model-based Bayesian analysis) were more dispersed than other breeding lines and cultivars, likely indicating the larger genetic diversity present in those breeding lines (Figure 4B).

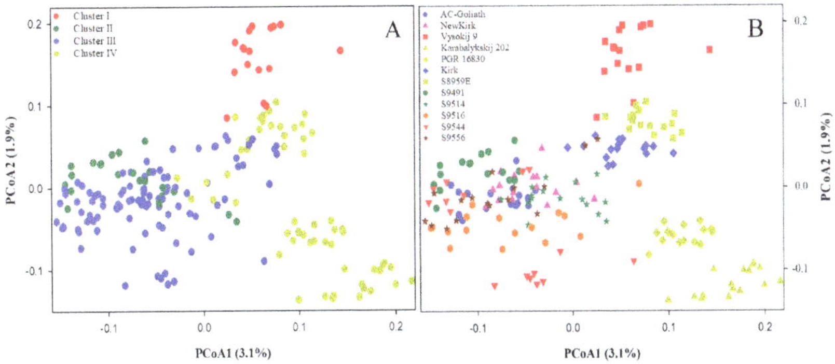

Figure 4. Genetic relationship of 192 genotypes of the 12 crested wheatgrass lines as revealed by principal coordinates analysis (PCoA) with the 45,507 SNP markers. Two panels are identical, but in the left panel (**A**) each genotype is labelled with colored circles representing the clusters obtained from the STRUCTURE analysis, while the right panel (**B**) labels genotypes for 12 lines.

2.3. Genetic Differentiation

The analysis of molecular variance (AMOVA) revealed that most of the SNP variations were present within the lines (84.2%), while much smaller variations reside among lines (15.8%) or among the four Bayesian clusters (12.07%) (Table 2). Line-specific F_{st} was also estimated from AMOVA for each line as the weighted variation among individual plants within a line to observe the extent of inbreeding. They were obtained in the range of 0.56 (in line S9491) to 0.64 (in the cultivar Kirk) with mean of 0.60 (Figure 5B). The pairwise genetic distance among the 12 lines ranged from 0.055 (between AC-Goliath and S9544) to 0.32 (between Karabalykskij 202 and S9491) with an average distance of 0.15.

Table 2. Results of the analysis of molecular variance for two models of genetic structure (12 lines and four clusters from the STRUCTURE analysis) based on 45,507 SNP markers.

Model/Source of Variation	df	Sum of Squares	Variance Explained	Variance (%) [a]
12 lines				
Among lines	11	101,048.8	246.0	15.8
Within lines	372	488,598.0	1313.4	84.2
Four clusters from STRUCTURE				
Among clusters	3	54,736.5	193.3	12.1
Within clusters	380	534,910.3	1407.7	87.9

[a] These variances were statistically significant from zero at $P < 0.0001$.

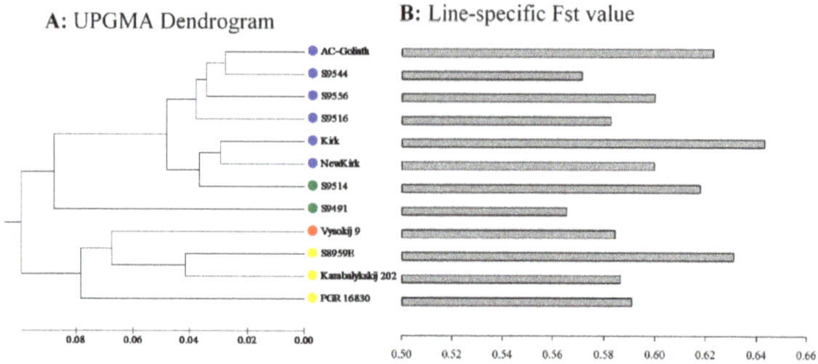

Figure 5. Genetic diversity and genetic relationships of the 12 crested wheatgrass lines. Left panel (**A**) shows their genetic relationship in the unweighted pair group method with arithmetic mean (UPGMA) dendrogram based on the Phi statistics obtained from the AMOVA. The right panel (**B**) displays the line-specific F_{st} values for the 12 lines.

The dendrogram based on AMOVA showed the grouping of the 12 CWG lines into three genetically distinct clusters at the Phi statistic of 0.08 or more (Figure 5A). The dendrogram grouped the lines from Kazakhstan and Russia in one distinct cluster. The second distinct cluster consisted of the single line S9491. The largest of all is the third cluster, with seven lines consisting of cultivars and breeding lines from Canada.

2.4. Effects of Missing Data on Diversity Analysis

The optimal numbers of genetic clusters inferred from STRUCTURE analyses with respect to the extent of missing data from M20%, M30%, M40%, and M50% datasets provided 4, 6, 6, and 4 optimal clusters, respectively (Figure 6A). Comparing the proportions of SNP variance residing among the 12 lines inferred from the AMOVA analysis showed 24.6%, 20.3%, 17.8%, and 15.8% for M20%, M30%, M40%, and M50%, respectively (Figure 6B).

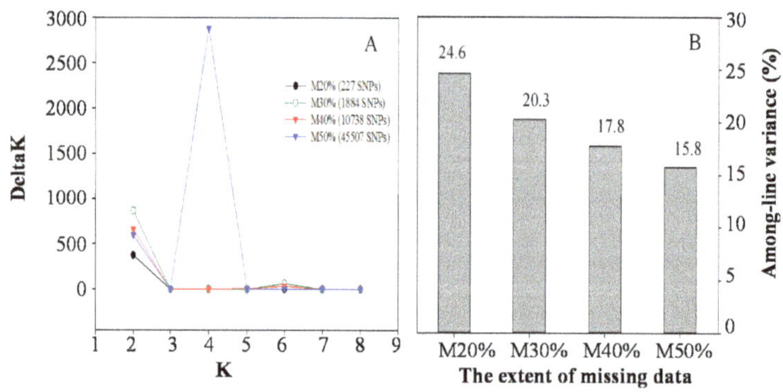

Figure 6. The impact of missing SNP data on the inferences of STRUCTURE and AMOVA analysis. The left panel (**A**) shows the four optimal clusters obtained from the STRUCTURE analyses at the missing level of M20% and M50%, and six clusters at M30% and M40%. The right panel (**B**) shows the SNP variances, ranging from 24.6 to 15.78%, inferred from AMOVA analyses residing among 12 lines at the increasing level of missing values from M20% to M50%, respectively.

3. Discussion

This study utilized the gd-GBS application, in combination with Haplotag pipeline, for the first time in CWG, to generate a data matrix of 192 genotypes × 45,507 SNP markers, and captured genome-wide genetic variants to evaluate the genetic diversity present in tetraploid CWG. The diversity analysis revealed 15.8% of SNP variation residing among the 12 lines and the model-based Bayesian analysis identified four major clusters of the assayed samples. These research outputs are not only useful for understanding the genetic diversity of CWG and for its breeding, but also are encouraging for molecular characterization of non-model polyploid plants.

The revealed patterns of genetic diversity are interesting. First, the model-based Bayesian approach in the STRUCTURE identified four major clusters of the assayed genotypes, while the distance-based approaches like PCoA and UPGMA identified three major clusters; however, the neighbor-joining analysis was in accordance with the result from STRUCTURE analysis. Following the pedigree of the assayed genotypes (Table S1), we could infer that the model-based Bayesian analysis and neighbor-joining analysis were able to genetically infer population substructure—an outcome of probable processes such as genetic drift, migration, mutation, and selection—more distinctly than distance-based approaches. Results also showed most of the genotypes grouped together within their lines, revealing that different lines were distinct. The STRUCTURE analysis (Figure 2A), neighbor-joining analysis (Figure 3), PCoA (Figure 4B), and UPGMA dendrogram (Figure 5A) revealed the genetic distinctness of lines Karabalykskij 202, PGR 16830, S8959E, and Vysokij 9. S8959E is a breeding line in the Saskatoon program, but it is a selection from Russian genebank line Vysokij 9. Although it has been recurrently selected for vigorous growth and plant type, it has not been interpollinated with any other lines, explaining its distinctness from other Canadian cultivars/breeding lines. However, STRUCTURE revealed all genotypes, except one (S8959E-14; Figure 2A) from line S8959E, showing high affinity with the line from Kazakhstan. This is also supported by UPGMA clustering (Figure 5A), while neighbor-joining analysis revealed the relatedness of lines from Russia. These findings will serve as valuable information for the genetic improvement of CWG for forage production.

Our analysis showed high within-line genetic variation (Table 2) of assayed CWG lines, which is in agreement with studies on highly outcrossing species [37]. Overall, our genetic diversity results are in accordance with diversity studies of CWG reported by Mellish et al. [18] using AFLP markers and Che et al. [31] and Che et al. [33,34] using SSR markers. The somewhat higher among population variation (15.8%) observed in the present study may partly be due to narrower genetic base of eight of the breeding lines/cultivars relative to the three genebank lines and one line of Russian origin (S8959E). Most of the Canadian cultivars and breeding lines shared one or more common parents in their genetic background (Table S1), and they have gone through many cycles of recurrent selection for vigor and yield. Thus, there has probably been a slight reduction in heterozygosity as indicated by the generally higher inbreeding coefficients (Figure 5B). The distinctness of the lines S8959E, Vysokij 9, Karabalykskij 202, and PGR 16830 can be attributed to their Asian origin and absence of interpollination with Canadian cultivars/lines and selection under Canadian conditions, except for the recurrent selection of line S8959E, mentioned above. Thus, the cultivars/breeding lines likely have reduced the within-line variation, while diverging more from the unselected Asian lines, explaining some increase of the among-line variation. Further research is needed on the utilization of the genetic variability of these lines with focus on morpho-physiological studies, adaptation, and their utilization in breeding programs. Likewise, the distinctness of the line S9491 in the UPGMA analysis (Figure 5A) is attributed to its synthesis from seven different lines/cultivars from breeding programs in Saskatoon and Logan, Utah, USA. The line S9514 was directly selected from S9491, which explains why these two lines clustered (green cluster) together in the STRUCTURE analysis (Figure 2) and neighbor-joining analysis (Figure 3). However, the Canadian cultivar "Kirk" developed partly from a plant introduction from a botanical garden in Finland (University of Turku) in 1968 showed shared pedigree with some or all of the Kazakhstan lines based on model-based Bayesian clustering (Figure 2A) and neighbor-joining analysis (Figure 3). While the origin of the plant introduction from the University of Turku remains unknown, it can be reasoned that this original introduction may have common genetic background with some of the Kazakhstan lines based on Bayesian clustering.

It was observed that the extent of reduction in heterozygosity, as explained by F_{st}, was more in cultivars than most of the breeding lines. Two cultivars "AC-Goliath" and "Kirk" had lower diversity as indicated by higher inbreeding coefficient (F_{st} values) (Figure 5B), perhaps because of being synthesized from the interpollination of fewer genotype than many of the breeding lines. Also, most of the breeding lines included cultivars "Kirk", "AC-Goliath", and other sources, in their pedigrees. The cultivar "Newkirk" was selected from progenies of crosses between "Kirk" and "AC-Goliath". However, the inbreeding coefficient of "Newkirk" was lower than the parental cultivars, indicating a higher level of heterozygosity. The three breeding lines S9516, S9544, and S9556 showed high within-line genetic diversity according to greater dispersal of these lines on PCoA (Figure 4B), higher within line variation (92.2%) as explained by a separate AMOVA, and lower line-specific F_{st} (Figure 5B). This greater genetic diversity could be attributed to inclusion of diverse germplasm sources during their synthesis (Table S1). The high within-line variability suggests that there is sufficient genetic variation in all lines in this study to make progress from selection. Inclusion of germplasm from the Asian lines in the breeding program to interpollinate with Canadian cultivars/breeding lines will increase diversity.

Our gd-GBS application has identified thousands of genome-wide SNP markers to assess the extent of genetic diversity in the non-model polyploid CWG with no prior genomic information. These results demonstrated the technical feasibility and effectiveness of GBS to sample genome-wide genetic variability in other perennial grass species with complex genomes. High resolution plant genetic diversity analysis, with 45,000 SNP markers spread over a genome, is more informative than with relatively few markers, like AFLP and SSR used in previous studies [1,12,18,38–40]. Also, the experimental cost for sampling genome-wide variants in this study was roughly $12,000, suggesting the feasibility of a wider application of GBS to characterize other perennial polyploid grass species. The results of the present study, along with those published in northern wheatgrass and wild oat [12,13], demonstrate the utility of GBS in molecular characterization of non-model plants with complex ploidy and genetic structures.

4. Materials and Methods

4.1. Plant Materials

The study material comprised 12 tetraploid CWG lines consisting of six breeding lines, three cultivars, and three genebank accessions (Table 1). These accessions were acquired from USDA-ARS plant germplasm system, Plant Gene Resources of Canada (PGRC), and the joint forage breeding program of the University of Saskatchewan and Agriculture and Agri-Food Canada (AAFC). For ease of interpretation, all the acquired material will be referred to as lines, rather than accessions, in this study. Seeds of each line were grown for six weeks in the greenhouse at the Saskatoon Research and Development Centre, AAFC, under the following growth conditions: 16 h photoperiod at 22 °C and 8 h dark at 16 °C. Young leaf tissues were collected from 16 randomly selected plants for each of the lines and stored at −80 °C prior to DNA extraction. A total of 192 genotypes from the 12 tetraploid lines, listed in Table 1, were used for bioinformatics and genetic diversity analyses.

4.2. Genotyping-by-Sequencing

For each of the 192 genotypes, DNA was extracted from 0.1 g finely ground tissue following the protocols of NucleoSpin® Plant II Kit (Macherey-Nagel, Bethlehem, PA, USA), and was eluted in a 1.5 mL Eppendorf tube with Elution Buffer. NanoDrop 8000 (Thermo Fisher Scientific, Waltham, MT, USA) was used to measure the quality of the DNA by comparing the 260 and 280 nm absorptions. DNA samples were further quantified through the Quant-iTTM PicoGreen®dsDNA assay kit (Invitrogen, Carisbad, CA, USA) and diluted to 60 ng/μL with 1× TE buffer prior to sequencing analysis.

A genetic diversity-focused GBS (gd-GBS) protocol by Peterson et al. [3] was used for the preparation of multiplexed GBS libraries. In brief, for each library, 200 ng purified genomic DNA was first digested with the restriction enzyme combination *Pst*I and *Msp*I (New England Biolabs, Whitby, ON, Canada). Ligation of customized adapters onto the 5′ and 3′ ends of the restriction fragments by T4 ligase was subsequently carried out. Then, the ligation fragments were purified by an AMPure XP kit (Beckman Coulter, Brea, CA, USA). Following the purification, Illumina TruSeq HT multiplexing primers were added through PCR amplification. The amplicon fragments were further quantified, concentrated, and pooled to form 4 subgroups of 12 samples each. The samples in the subgroups were pre-selected using a Pippin Prep instrument (Sage Science, Beverly, MA, USA) for an insert size range of 250–450 bp, before pooling the samples into a library. Each pooled library was diluted to 6 pM, and denatured with 5% of sequencing-ready Illumina PhiX Library Control (Illumina, San Diego, CA, USA) that can serve for calibration. Sequencing was completed using an Illumina MiSeq Instrument with paired-ends of 250 bp in length. MiSeq runs generated 384 FASTQ sequence files from 192 genotypes of 12 lines (one forward and one reverse for each of 192 genotypes). All the raw pair-end sequencing data in FASTQ format were deposited into the National Center for Biotechnology Information (NCBI) Sequence Read Archive (SRA) with accession number SRP115373 as part of the larger sequencing effort to enhance crested wheatgrass breeding [41]. The sequencing information for all 192 assayed samples is described in the online Supplementary Material, Section A.

4.3. Bioinformatics Analysis

Bioinformatic analysis began with sequence (FASTQ) data cleaning, using Trimmomatic version 0.36 [42] to remove any sequenced-through Illumina adapters, low quality sequence (sliding window of 10 bases, average Phred of 20), and fragments under 64 bases long.

As the UNEAK-GBS pipeline [35] only considers sequences of 64 bp (after barcode removal) with an intact 5-base *Pst*I residue (TGCAG) at the beginning, each FASTQ file of 250 bp was first split into three fragment sets with a custom Perl script *fastq184CutandCode-Pst.pl*. The first set comprised the first 64 bases with the *Pst*I residual restriction site, and the next two sets each with 59 base portions and an added 5-base *Pst*I residue. The script also provided an arbitrary barcode sequence (CATCAT) at the start of each sequence fragment, since the UNEAK pipeline expects to deconvolute barcoded sequence reads which are not already separated by sample. The three 70-base-long fragments formed, thereafter, were independent, as their relationship was not preserved. Each fragment set was recognized by the UNEAK-GBS pipeline [35], and was passed into UNEAK as an independent dataset.

Each fragment set (70 bases long) was analyzed with UNEAK and the Haplotag pipelines [9], resulting in the analysis of a total of 177 bases of genetic sequence. Online Supplementary Material, Section B, describes the procedures to run UNEAK. Two types of meta data files—a single mergedAll.txt (all tags observed more than 10 times) and a set of individual tagCount files (one per sample) needed for the Haplotag pipeline—were generated from the UNEAK run.

Haplotag was run with the parameters and filtering threshold settings described in the HTinput.txt file, and generated a matrix of samples by SNP loci (online Supplementary Material, Section B). A set of tag-level haplotypes ("HTgenos") are first generated by Haplotag, followed by a set of SNP data derived from these haplotypes ("HTSNPgenos"). These two data types are technically redundant, so choosing one of them relies on the implementation and preference of software. In the present study, most (97.5%) haplotypes were found to contain only a single SNP; thus, we decided to analyze the SNP dataset for simplicity and compatibility with downstream analysis software.

The character by Taxa (CbyT) program supplied by N. Tinker was used to generate a filtered SNP file. In brief, Haplotag generated three separate "HTSNPGenos" files, which were merged before running CbyT. The "minimum presence" value in CbyT was set to 80%, 70%, 60%, and 50% for 20%, 30%, 40%, and 50% missing data, respectively. A SNP-by-sample matrix in the output files was used in further analyses. Additional descriptions of the SNP data matrix and the custom Perl and Shell scripts are available in the online Supplementary Material, Section A. Analyses from FASTQ file separation to

SNP generation were conducted using Microsoft Windows 7 64-bit OS with an Intel (R) Xeon (R) CPU E5-2623 v3 @ 3.00 GHz (8 threads) and 32 GB RAM.

4.4. Genetic Diversity Analysis

The diversity analysis was based on 45,507 SNP markers, with 50% or less missing values in 192 genotypes from 12 CWG lines. Data analysis began with calculation of the minor allele frequency and the extent of missing SNP data with Microsoft Excel®. Thereafter, diversity analyses at the individual and line levels were carried out.

Three types of diversity analysis were performed at individual genotype level. First, genetic structure of 192 CWG genotypes was examined using a model-based Bayesian method implemented in the program STRUCTURE version 2.2.3 [36,43]. Linux server with 60 core parallel computing was used to run the STRUCTURE program, where each population subgroup (K = 1–9) was run 20 times, using an admixture model with 10,000 replicates each for burn-in and during the analysis. Based on (1) a plot of likelihood of these models, (2) the rate of change in the second derivative (ΔK) between successive K values [44], and (3) the consistency of group configuration across 20 runs, the final population subgroups were determined. For a given population subgroup (K) with 20 runs, the run having the highest likelihood value was chosen to assign the posterior membership coefficients to each sample. These posterior membership coefficients were used to create a graphical bar plot. The size and formation of each optimal cluster with respect to population were evaluated. Second, a neighbor-joining (NJ) analysis of the 192 genotypes was conducted using MEGA version 7.0.14 [45] based on the dissimilarity matrix obtained from R routine AveDissR [46,47], and a radiation tree was displayed. Third, a PCoA of all 192 genotypes was also done using the R routine AveDissR [46,47] to assess genetic distinctness and redundancy, and to assess the genotype associations, plots of the first two resulting principal components were generated. For comparison, the resulting NJ trees and PCoA plots were individually labeled for the inferred structures.

Genetic variation present among the 12 lines was evaluated with AMOVA using Arlequin version 3.5 [48] on 45,507 markers. In addition, the pairwise genetic distances were computed and line-specific F_{st} values (inbreeding coefficient) for each line [49] were generated to infer the reduction in heterozygosity. To inspect the genetic variation among the clusters identified from the STRUCTURE analysis, additional AMOVA was performed. Unweighted pair group method, with arithmetic mean (UPGMA) dendrogram based on pairwise genetic distances among the 12 lines obtained from AMOVA, were generated using MEGA version 7.0.14 [45], to evaluate line differentiation and distinctness.

To estimate the influence of missing SNP data on the genetic diversity analysis, four datasets of 272; 1884; 10,738; and 45,507 SNPs representing 20%, 30%, 40%, and 50% of missing SNPs (M20%, M30%, M40%, and M50%) were attained for the 192 genotypes, respectively. For each dataset, the among-line variance from AMOVA and the optimal number of genetic clusters from STRUCTURE were obtained and compared among the four datasets of varying percentages of missing data.

5. Conclusions

With the application of GBS, it has been possible to generate 45,507 SNP markers for a diversity analysis of crested wheatgrass. The variation residing among these 12 lines of CWG was found to be 15.8%. Further analysis grouped the assayed samples into four genetic clusters, and revealed the genetic distinctness of two cultivars each from Kazakhstan and Russia, respectively. These results can enhance parental selection for increased genetic variation and improved offspring performance in crested wheatgrass breeding. The findings in this study can also aid in the application of GBS in the characterization of non-model plants with complex genomes.

Supplementary Materials: Supplementary materials can be found at http://www.mdpi.com/1422-0067/19/9/2587/s1.

Author Contributions: Y.-B.F., B.B. and B.C. conceived the project; Y.-B.F. designed research; B.C. prepared the study material; Y.-B.F. conducted sequencing; K.B. and Y.-B.F. performed data analysis; K.B. wrote the manuscript; B.C., Y.-B.F. and B.B. made revisions to the manuscript. All authors read and approved the final manuscript.

Funding: The work was financially supported by the Beef Cattle Research Council of Canada and Agriculture and Agri-Food Canada (AAFC) Growing Forward 2 Funds (FGR.08.13).

Acknowledgments: The author would like to thank Gregory Peterson and Carolee Horbach for their technical assistance; Isobel Parkin for the access to and the use of the Illumina MiSeq instrument; Compute Canada and Westgrid for providing the high-performance computing service as well as of their technical support; and Helen Booker, Bunyamin Tar'an and two anonymous journal reviewers for their helpful comments on the early version of the manuscript.

Conflicts of Interest: The authors declare no conflict of interest.

References

1. Fu, Y.B.; Peterson, G.W. Genetic diversity analysis with 454 pyrosequencing and genomic reduction confirmed the eastern and western division in the cultivated barley gene pool. *Plant Genome* **2011**, *4*, 226–237. [CrossRef]
2. Peterson, B.; Weber, J.N.; Kay, E.H.; Fisher, H.S.; Hoekstra, H.E. Double digest RADseq: An inexpensive method for de novo SNP discovery and genotyping in model and non-model species. *PLoS ONE* **2012**, *7*, e37135. [CrossRef] [PubMed]
3. Peterson, G.W.; Dong, Y.; Horbach, C.; Fu, Y.B. Genotyping-by-sequencing for plant genetic diversity analysis: A lab guide for SNP genotyping. *Diversity* **2014**, *6*, 665–680. [CrossRef]
4. Poland, J.A.; Rife, T.W. Genotyping-by-sequencing for plant breeding and genetics. *Plant Genome* **2012**, *5*, 92–102. [CrossRef]
5. Poland, J.A.; Brown, P.J.; Sorrells, M.E.; Jannink, J.L. Development of high-density genetic maps for barley and wheat using a novel two-enzyme genotyping-by-sequencing approach. *PLoS ONE* **2012**, *7*, e32253. [CrossRef] [PubMed]
6. Huang, Y.F.; Poland, J.A.; Wight, C.P.; Jackson, E.W.; Tinker, N.A. Using genotyping-by-sequencing (GBS) for genomic discovery in cultivated oat. *PLoS ONE* **2014**, *9*, e102448. [CrossRef] [PubMed]
7. Fu, Y.B.; Peterson, G.W.; Dong, Y. Increasing genome sampling and improving SNP genotyping for genotyping-by-sequencing with new combinations of restriction enzymes. *G3* **2016**, *6*, 845–856. [CrossRef] [PubMed]
8. Fu, Y.B.; Yang, M.H. Genotyping-by-sequencing and its application to oat genomic research. In *Oat—Methods and Prtocols*; Gasparis, S., Ed.; Springer Science+Business Media: New York, NY, USA, 2017; pp. 169–187. [CrossRef]
9. Tinker, N.A.; Bekele, W.A.; Hattori, J. Haplotag: Software for haplotype-based genotyping-by-sequencing analysis. *G3* **2016**, *6*, 857–863. [CrossRef] [PubMed]
10. Yan, H.; Bekele, W.A.; Wight, C.P.; Peng, Y.; Langdon, T.; Latta, R.G.; Fu, Y.B.; Diederichsen, A.; Howarth, C.J.; Jellen, E.N.; et al. High-density markers profiling confirms ancestral genomes of Avena species and identifies D-genome chromosomes of hexaploid oat. *Theor. Appl. Genet.* **2016**, *129*, 2133–2149. [CrossRef] [PubMed]
11. Bekele, W.A.; Wight, C.P.; Chao, S.; Howarth, C.J.; Tinker, N.A. Haplotype based genotyping-by-sequencing in oat genome research. *Plant Biotechnol. J.* **2018**, *16*, 1452–1463. [CrossRef] [PubMed]
12. Al-Hajaj, N.; Peterson, G.W.; Horbach, C.; Al-Shamaa, K.; Tinker, N.A.; Fu, Y.B. Genotyping-by-sequencing empowered genetic diversity analysis of Jordanian oat wild relative *Avena sterilis*. *Genet. Resour. Crop Evol.* [CrossRef]
13. Li, P.; Bhattarai, S.; Peterson, G.P.; Coulman, B.E.; Schellenberg, M.P.; Biligetu, B.; Fu, Y.B. Genetic diversity of northern wheatgrass (*Elymus lanceolatus* ssp. *lanceolatus*) as revealed by genotyping-by-sequencing. *Diversity* **2018**, *10*, 23. [CrossRef]
14. Dewey, D.R. The genomic system of classification as a guide to intergeneric hybridization with the perennial *Triticeae*. In *Gene Manipulation in Plant Improvement, Proceedings of the 6th Stadler Genetics Symposium*; Gustafson, J.P., Ed.; Columbia University Press: New York, NY, USA, 1984; pp. 209–279.
15. Asay, K.H.; Jensen, K.B.; Hsiao, C.; Dewey, D.R. Probable origin of standard crested wheatgrass, *Agropyron desertorum* Fisch Ex Link, Schultes. *Can. J. Plant Sci.* **1992**, *72*, 763–772. [CrossRef]

16. Rogler, G.A.; Lorenz, R.L. Crested wheatgrass-early history in the United States. *J. Range Manag.* **1983**, *36*, 91–93. [CrossRef]
17. Chen, S.Y.; Ma, X.; Zhang, X.Q.; Huang, L.K.; Zhou, J.N. Genetic diversity and relationships among lines of five crested wheatgrass species (Poaceae: *Agropyron*) based on gliadin analysis. *Genet. Mol. Res.* **2013**, *12*, 5704–5713. [CrossRef] [PubMed]
18. Mellish, A.; Coulman, B.E.; Ferdinandez, Y. Genetic relationships among selected crested wheatgrass cultivars and species determined on the basis of AFLP markers. *Crop Sci.* **2002**, *42*, 1662–1668. [CrossRef]
19. Looman, J.; Heinrichs, D. Stability of crested wheatgrass pastures under long-term pasture use. *Can. J. Plant Sci.* **1973**, *53*, 501–506. [CrossRef]
20. Asay, K.H.; Jensen, K.B. Wheatgrass. In *Cool-Season Forage Grasses*; Moser, L.E., Buxton, D., Casler, M.D., Eds.; Agron Monogr ASA, CSSA, SSSA: Madison, WI, USA, 1996; pp. 691–724.
21. Zlatnik, E. *Agropyron cristatum*. In Fire Effects Information System, [Online]. U.S. Department of Agriculture, Forest Service, Rocky Mountain Research Station, Fire Sciences Laboratory (Producer), 1999; p. 8. Available online: https://www.fs.fed.us/database/feis/plants/graminoid/agrcri/all.html (accessed on 24 July 2018).
22. Hull, G.J.; Klomp, A.C. Longevity of crested wheatgrass in the sagebrush-grass type in southern Idaho. *J. Range Manag.* **1966**, *19*, 5–11. [CrossRef]
23. Sharma, H.C.; Gill, B.S.; Uyemoto, J.K. High levels of resistance in *Agropyron* species to barley yellow dwarf and wheat streak mosaic viruses. *J. Phytopath.* **1984**, *110*, 143–147. [CrossRef]
24. Dong, Y.S.; Zhou, R.H.; Xu, S.J.; Li, L.H.; Cauderon, Y.; Wang, R.R. Desirable characteristics in perennial *Triticeae* collected in China for wheat improvement. *Hereditas* **1992**, *116*, 175–178. [CrossRef]
25. Wu, J.; Yang, X.M.; Wang, H.; Li, H.J.; Li, L.H.; Li, X.Q.; Liu, W.H. The introgression of chromosome 6P specifying for increased numbers of florets and kernels from *Agropyron cristatum* into wheat. *Theor. Appl. Genet.* **2006**, *114*, 13–20. [CrossRef] [PubMed]
26. Ochoa, V.; Madrid, E.; Said, M.; Rubiales, D.; Cabrera, A. Molecular and cytogenetic characterization of a common wheat-*Agropyron cristatum* chromosome translocation conferring resistance to leaf rust. *Euphytica* **2015**, *201*, 89–95. [CrossRef]
27. Zhang, J.; Liu, W.; Han, H.; Song, L.; Bai, L.; Gao, Z.; Zhang, Y.; Yang, X.; Gao, L.A.; Li, L. De novo transcriptome sequencing of *Agropyron cristatum* to identify available gene resources for the enhancement of wheat. *Genomics* **2015**, *106*, 129–136. [CrossRef] [PubMed]
28. Zeng, F.; Biligetu, B.; Coulman, B.E.; Schellenberg, M.P.; Fu, Y.B. RNA-Seq analysis of gene expression for floral development in crested wheatgrass (*Agropyron cristatum* L.). *PLoS ONE* **2017**, *12*. [CrossRef] [PubMed]
29. Zeng, F.; Biligetu, B.; Coulman, B.E.; Schellenberg, M.P.; Fu, Y.B. RNA-Seq analysis of plant maturity in crested wheatgrass (*Agropyron cristatum* L.). *Genes.* **2017**, *8*, 291. [CrossRef] [PubMed]
30. Forster, J.W.; Jones, E.S.; Kölliker, R.; Drayton, M.C.; Dumsday, J.; Dupal, M.P.; Guthridge, K.M.; Mahoney, N.L.; van Zijil de Jong, E.; Smith, K.F. Development and implementation of molecular markers for forage crop improvement. In *Molecular Breeding of Forage Crops*; Spangenberg, G., Ed.; Kluwer Academic Press: Dordrecht, The Netherlands, 2001; pp. 101–133.
31. Che, Y.H.; Li, H.J.; Yang, Y.P.; Yang, X.M.; Li, X.Q.; Li, L.H. On the use of SSR markers for the genetic characterization of the *Agropyron cristatum* (L.) Gaertn. in Northern China. *Genet. Resour. Crop Evol.* **2008**, *55*, 389–396. [CrossRef]
32. Rogers, D.L.; Montalvo, A.M. *Genetically Appropriate Choices for Plant materials to Maintain Biological Diversity*; Report to the USDA Forest Service; University of California: Rocky Mountain Region, Lakewood, CO, USA, 2004; p. 343.
33. Che, Y.H.; Yang, Y.P.; Yang, X.M.; Li, X.Q.; Li, L.H. Genetic diversity between ex situ and in situ samples of *Agropyron cristatum* (L.) Gaertn. based on simple sequence repeat molecular markers. *Crop Past. Sci.* **2011**, *62*, 639–644. [CrossRef]
34. Che, Y.H.; Yang, Y.P.; Yang, X.M.; Li, X.Q.; Li, L.H. Phylogenetic relationship and diversity among *Agropyron* Gaertn. germplasm using SSRs markers. *Plant Syst. Evol.* **2015**, *301*, 163–170. [CrossRef]
35. Lu, F.; Lipka, A.E.; Glaubitz, J.; Elshire, R.; Cherney, J.H.; Casler, M.D.; Buckler, E.S.; Costich, D.E. Switchgrass genomic diversity, ploidy, and evolution: Novel insights from a network-based SNP discovery protocol. *PLoS Genet.* **2013**, *9*, e1003215. [CrossRef] [PubMed]
36. Pritchard, J.; Stephens, M.; Donnelly, P. Influence of population structure using multilocus genotype data. *Genetics* **2000**, *155*, 945–959. [PubMed]

37. Hamrick, J.L.; Godt, M.J.W. Allozyme diversity in plant species. In *Plant Population Genetics, Breeding and Genetic Resources*; Brown, A.H.D., Clegg, M.T., Kahler, A.L., Weir, B.S., Eds.; Sinauer Associates: Sunderland, MA, USA, 1989; pp. 43–63.
38. Fu, Y.B.; Coulman, B.E.; Ferdinandez, Y.S.N.; Cayouette, J.; Peterson, P.M. Genetic diversity of fringed brome (*Bromus ciliatus*) as determined by amplified fragment length polymorphism. *Can. J. Bot.* **2005**, *83*, 1322–1328. [CrossRef]
39. Biligetu, B.; Schellenberg, M.P.; Fu, Y.B. Detecting genetic diversity of side-oats grama grass populations using AFLP Marker. *Can. J. Plant Sci.* **2013**, *93*, 1105–1114. [CrossRef]
40. Fu, Y.B.; Phan, A.T.; Coulman, B.E.; Richards, K.W. Genetic diversity in natural populations and corresponding seed collections of little bluestem as revealed by AFLP markers. *Crop Sci.* **2004**, *44*, 2254–2260. [CrossRef]
41. Li, P.; Biligetu, B.; Coulman, B.E.; Schellenberg, M.P.; Fu, Y.B. Genotyping-by-sequencing data of 272 crested wheatgrass (*Agropyron cristatum*) genotypes. *Data Brief* **2017**, *15*, 401–406. [CrossRef] [PubMed]
42. Bolger, A.M.; Lohse, M.; Usadel, B. Trimmomatic: A flexible trimmer for Illumina sequence data. *Bioinformatics* **2014**, *30*, 2114–2120. [CrossRef] [PubMed]
43. Falush, D.; Stephens, M.; Pritchard, J.K. Inference of population structure using multilocus genotype data: Dominant markers and null alleles. *Mol. Ecol. Notes* **2007**, *7*, 574–578. [CrossRef] [PubMed]
44. Evanno, G.; Regnaut, S.; Goudet, J. Detecting the number of clusters of individuals using the software STRUCTURE: A simulation study. *Mol. Ecol.* **2005**, *14*, 2611–2620. [CrossRef] [PubMed]
45. Kumar, S.; Stecher, G.; Tamura, K. MEGA7: Molecular evolutionary genetics analysis version 7.0 for bigger datasets. *Mol. Biol. Evol.* **2016**, *33*, 1870–1874. [CrossRef] [PubMed]
46. Yang, M.H.; Fu, Y.B. AveDissR: An R function for assessing genetic distinctness and genetic redundancy. *Appl. Plant Sci.* **2017**, *5*, 1700018. [CrossRef] [PubMed]
47. R Development Core Team. *R: A Language and Environment for Statistical Computing*; R Foundation for Statistical Computing: Vienna, Austria, 2016; ISBN 3-900051-07-0. Available online: http://www.r-project.org/ (accessed on 3 August 2018).
48. Excoffier, L.; Lischer, H.E.L. Arlequin suite ver 3.5. 5: A new series of programs to perform population genetics analyses under Linux and Windows. *Mol. Ecol. Resour.* **2010**, *10*, 564–567. [CrossRef] [PubMed]
49. Weir, B.S.; Hill, W.G. Estimating F-statistics. *Ann. Rev. Genet.* **2002**, *36*, 721–775. [CrossRef] [PubMed]

© 2018 by the authors. Licensee MDPI, Basel, Switzerland. This article is an open access article distributed under the terms and conditions of the Creative Commons Attribution (CC BY) license (http://creativecommons.org/licenses/by/4.0/).

International Journal of
Molecular Sciences

MDPI

Article

Assessment of Genetic Diversity, Population Structure, and Evolutionary Relationship of Uncharacterized Genes in a Novel Germplasm Collection of Diploid and Allotetraploid *Gossypium* Accessions Using EST and Genomic SSR Markers

Allah Ditta [1,2,†], Zhongli Zhou [1,†], Xiaoyan Cai [1], Xingxing Wang [1], Kiflom Weldu Okubazghi [1,3], Muhammad Shehzad [1], Yanchao Xu [1], Yuqing Hou [1], Muhammad Sajid Iqbal [1], Muhammad Kashif Riaz Khan [2], Kunbo Wang [1,*] and Fang Liu [1,*]

1 State Key Laboratory of Cotton Biology/Institute of Cotton Research, Chinese Academy of Agricultural Sciences, Anyang 455000, Henan, China; adbotanist@yahoo.com (A.D.); zhonglizhou@163.com (Z.Z.); cxycri@163.com (X.C.); wxx1991@126.com (X.W.); wediweldu81@yahoo.com (K.W.O.); mshehzad534@gmail.com (M.S.); xuyanchao2016@163.com (Y.X.); houyuqing18@163.com (Y.H.); sajidses@hotmail.com (M.S.I.)

2 Nuclear Institute for Agriculture and Biology (NIAB), Jhang Road, Faisalabad 38000, Punjab, Pakistan; mkrkhan@gmail.com

3 Hamelmalo Agricultural College, P.O. Box 397, Keren, Eritrea

* Correspondence: wkbcri@163.com (K.W.); liufcri@163.com (F.L.)

† These authors contributed equally.

Received: 23 June 2018; Accepted: 13 August 2018; Published: 14 August 2018

Abstract: This study evaluated the genetic diversity and population structures in a novel cotton germplasm collection comprising 132 diploids, including *Glossypium klotzschianum* and allotetraploid cotton accessions, including *Glossypium barbadense*, *Glossypium darwinii*, *Glossypium tomentosum*, *Glossypium ekmanianum*, and *Glossypium stephensii*, from Santa Cruz, Isabella, San Cristobal, Hawaiian, Dominican Republic, and Wake Atoll islands. A total of 111 expressed sequence tag (EST) and genomic simple sequence repeat (gSSR) markers produced 382 polymorphic loci with an average of 3.44 polymorphic alleles per SSR marker. Polymorphism information content values counted 0.08 to 0.82 with an average of 0.56. Analysis of a genetic distance matrix revealed values of 0.003 to 0.53 with an average of 0.33 in the wild cotton collection. Phylogenetic analysis supported the subgroups identified by STRUCTURE and corresponds well with the results of principal coordinate analysis with a cumulative variation of 45.65%. A total of 123 unique alleles were observed among all accessions and 31 identified only in *G. ekmanianum*. Analysis of molecular variance revealed highly significant variation between the six groups identified by structure analysis with 49% of the total variation and 51% of the variation was due to diversity within the groups. The highest genetic differentiation among tetraploid populations was observed between accessions from the Hawaiian and Santa Cruz regions with a pairwise F_{ST} of 0.752 ($p < 0.001$). DUF819 containing an uncharacterized gene named yjcL linked to genomic markers has been found to be highly related to tryptophan-aspartic acid (W-D) repeats in a superfamily of genes. The RNA sequence expression data of the yjcL-linked gene Gh_A09G2500 was found to be upregulated under drought and salt stress conditions. The existence of genetic diversity, characterization of genes and variation in novel germplasm collection will be a landmark addition to the genetic study of cotton germplasm.

Keywords: novel accessions; PIC; PCR; EST-gSSRs; genes; genetic distance

1. Introduction

The leading natural fiber in the world is a product of cotton crops. Cotton is placed in the taxonomic order with the genus *Gossypium* and has broad phenotypic diversity, which includes more than 50 species [1–3]. There are now 7 tetraploid and 46 diploid cotton species after molecular confirmation and taxonomic designated two new tetraploid ones, i.e., *Gossypium ekmanianum* (AD6) and *Gossypium stephensii* (AD7) [3–6]. Among those, four are cultivated throughout the world: two of these species are diploids ($2n = 2x = 26$) and two are allotetraploids ($2n = 4x = 52$). Global cotton production is manifested from the two allotetraploid species *Gossypium hirsutum* and *Gossypium barbadense* [7–9].

Data on allotetraploid cotton evolution indicates that the seven tetraploid cottons evolved about 1.5 million years ago by hybridization of the Old world cotton *Gossypium herbaceum* (A_1 genome) and the New world cotton *Gossypium raimondii* (D_5 genome) as a consequence of subsequent diploidization and domestication [3,4,8,10–12]. *Gossypium hirsutum*, also called "Upland cotton", represents 90% of global cotton fiber production [13], while *Gossypium barbadense* (also known as Pima) is valued for its extra-long staple fiber source, is domesticated in North-West South America, has its native origin in Egypt, and contributes around 8% of total world lint [9]. Wild *Gossypium darwinii* originated from Galapagos Island and, relative to *G. barbadense*, also has good fiber fineness characteristics and is a rich source of resistance to *fusarium* and *verticillum* wilts [14]. The D-genome *Gossypium klotzschianum*, having glabrous seed coverings, evolved through long-distance dispersals, is endemic to Galapagos Island, and is considered a New-World D-genome diploid along with *G. raimondii*. *Gossypium tomentosum* is drought-tolerant, native to a Hawaiian Island, and has a more diffuse population structure falling typically as scattered individuals and small populations on several islands. *G. tomentosum* (AD3), *G. darwinii* (AD5), *G. ekmanianum* (AD6), and *G. stephensii* (AD7) are wild and are not grown commercially [1,2,15,16]. Wendel and Percy analyzed 58 *G. darwinii* accessions from six islands using 17 isozyme markers and identified a high genetic diversity level within these accessions and relationships with *G. barbadense* and *G. hirsutum* genomes. This classic study suggested that *G. darwinii* and *G. barbadense* are separated and each has a distinct genome [17].

The genetic diversity of different plant species is an essential element for crop production in agriculture, including cotton. Genetic variation in the *Gossypium* species is widespread, covering large geographic and ecological niches. It is a vital source of conserved genetic diversity in situ in Mexico for cotton origin [18,19] and is preserved ex situ within worldwide cotton germplasm collections and materials of breeding programs. The productivity of cotton and future efforts to improve cotton depend to a large extent on the elucidation of genetic diversity in cotton genetic stocks and their effective utilization in cotton improvement programs [20].

The narrow genetic background of Upland cotton has become a major concern as low genetic diversity gives rise to stagnant yield and quality of breeding. The elite breeding programs cannot make robust inferences without using the unexploited standing genetic variation of archaic cultivars typically associated with wild accessions [14,21,22]. The characterization of genetic diversity between and within groups enables us to find heterozygous groups, understand population structures, and isolate a core set of lines for genetic analysis studies in cotton. A multitude of studies indicate the extensive usage of model-based structure analysis for investigating genetic diversity in cotton [22,23]. Genetic diversity estimates have been established using genotypic data and DNA-based molecular markers [24–28]. Molecular markers are more reliable since they can directly determine allelic diversity and give robust estimates of genetic distances.

The DNA-based markers used for determining genetic diversity in cotton include restriction fragment length polymorphisms (RFLPs) [29], random amplified polymorphic DNA (RAPD) [30–32], amplified fragment length polymorphisms (AFLPs) [33], simple sequence repeat (SSR) [9,34–36], expressed sequence tags (ESTs) [37], inter-simple sequence repeat (ISSR) [38,39], and single nucleotide polymorphisms (SNPs) [40]. Compared with other biomarkers, SSR has advantages that include more reproducibility, co-dominant inheritance, distribution throughout the genome, and its being highly transferable, informative, and reliable [41].

Although data from several studies implicates the marker-based estimation of genetic diversity in cotton, the majority of those remain bound to the number of accessions included or the number of markers used to describe genetic diversity [42]. Recently, an effort has been made by Kirungu et al. [43] to explore the important genes linked to SSR markers by constructing a genetic linkage map between *Gossypium davidsonii* and *G. klotzschianum*. Similarly, a study of gene diversity, their functionality, and especially the diagnosis of uncharacterized domains of proteins in developing the evolutionary relationship among cotton accessions will be fruitful for exploring the mystery of cotton evolution. Among all protein domains with a unique structure and functions, nearly more than 20% are currently described as "domains of unknown function" (DUFs). They are often overlooked as irrelevant as many of them are found in only a few genomes. Approximately 2700 DUFs exist in bacteria as compared to eukaryotes, which have only 1500. More than 800 DUFs have been found to be common in bacteria and eukaryotes, and about 300 of these are also present in archaea. Evolutionary conservation suggests that many of these DUFs are important in biology as they mostly represent single-domain proteins, clearly establishing the biological importance of DUFs [44].

The importance of prioritizing DUFs has been recognized in various experimental and/or computational characterization efforts [45–48]. We identified DUF819 (PF005684), which is not only highly conserved but also plays an important role against biotic and abiotic stress, among four sequenced cotton species by using the WDR (PF00400) superfamily as reference-genome-sequenced proteins. Genome-wide characterization of WD-repeats, also known as tryptophan-aspartic acid or the W-D superfamily, has only been conducted in Arabidopsis and Cucumber [49,50] till now. Therefore, a comprehensive study comprising a wide collection of germplasms, more efficient genotyping, and collective genomic platforms is required to measure the overall genetic diversity in diploid and allotetraploid cotton, which will help overcome the future challenges of the gene pool's disastrous escape.

The objectives of this study were to explore the genetic diversity and evolutionary relationship among the domains of uncharacterized proteins in natural diploid and allotetraploid cotton germplasm resources and to analyze the population structures to maximize estimations about the accessions of cotton present in a wild nursery of China for their efficient utilization in cotton-breeding programs.

2. Results

2.1. SSR Marker Analysis

Among a total of 853 SSR primer pairs used for genotyping 132 accessions, 205 primer pairs were found to be polymorphic with a polymorphism rate of 24%. Accessions with more than 5% missing data were removed and 94 SSRs were dropped; the selected 111 SSR primer pairs can be scored confidently and read clearly on PCR products. Data for monomorphic loci were also excluded from the analysis. Data generated from the selected 111 SSR primer pairs was analyzed. Among 132 accessions, a total of 382 SSR alleles were detected as marker loci with an average of 3.44 alleles per SSR ranging from 2 to 8. All 382 SSR loci were found to be polymorphic. The average polymorphism information content (PIC) value for SSRs was 0.555 with a range of 0.078 to 0.821, and the major allele frequency was 0.738 ranging from 0.541 to 0.959 for the complete panel. Seventy-six (68.468%) SSR markers in total were found to be highly informative with a PIC value $\geq$0.50, 29 (26.126%) were moderately informative with PICs value $\geq$0.25 and <0.50, and 6 (5.405%) were least informative with a PIC value <0.25. A summary of marker statistics for *G. hirsutum* accessions is listed in Supplementary Table S2.

2.2. Unique Alleles

Among the 382 alleles detected in the studied accessions, 123 alleles were found and were termed as unique alleles (Supplementary Table S3). A high percentage (17.51%) of unique alleles was observed in *G. ekmanianum* genotypes (Table 1). Twenty-five unique alleles were found in two accessions of *G. hirsutum*. Nineteen, 13, 18, 13, and 4 unique alleles were observed in *G. barbadense*, *G. tomentosum*,

G. darwinii, *G. klotzschianum*, and *G. stephensii*, respectively. *G. ekmanianum* had the highest number (31) of unique alleles, which were collected from the Dominican Republic, National Plant Germplasm System (NPGS) USA (Supplementary Table S3). These unique alleles are an important genetic resource for cotton and have never been studied before.

Table 1. Summary of unique (present in one accession) and rare alleles (present in <5% accessions) observed in a combined Panel of 132 accessions.

Panel	Total Alleles	Total Lines	Unique Alleles	Rare Alleles (Freq < 5%)
Combined Panel	382	132	123 (32.19%)	108 (28.27%)
Gossypium barbadense	258	20	19 (7.36%)	12 (4.65%)
Gossypium darwinii	309	59	18 (5.83%)	57 (18.44%)
Gossypium tomentosum	205	32	13 (6.34%)	15 (7.31%)
Gossypium hirsutum	143	2	25 (17.48%)	22 (15.38%)
Gossypium ekmanianum	177	10	31 (17.51%)	2 (1.13%)
Gossypium stephensii	125	4	4 (3.2%)	0
Gossypium klotzschianum	90	5	13 (14.44%)	0

2.3. Common Alleles

Common alleles were estimated to understand the phenomenon of cotton evolution and the gene flow mechanism. All six species of cotton considered in this study have common alleles at 114 loci, keeping *G. hirsutum* as fixed. The SSR marker DPL0330-A showed the maximum number of common alleles (124) among all tetraploid cottons except for *G. klotzschianum*, which is diploid, while DPL0249-C showed the minimum number of common alleles, which were only found in *G. barbadense*, *G. klotzschianum*, and *G. hirsutum*. The number of common alleles ranged from 21 to 123 in all six species of *Gossypium*. In this investigation, a total of 459 common loci were observed. Eighty-eight, 101, 82, 91, 67, and 30 loci having common alleles specific to *G. hirsutum* were observed in *G. barbadense*, *G. darwinii*, *G. tomentosum*, *G. ekmanianum*, *G. stephensii*, and *G. klotzschianum*, respectively (Supplementary Table S4). These *G. hirsutum*-specific alleles were amplified by 85 out of 111 SSR markers. The presence of *G. hirsutum*-specific alleles in all six species of *Gossypium* indicated a high level of natural introgression. The level of introgression was found to vary among these wild-type accessions [36,51].

2.4. Analysis of Population Structures

Based on the ΔK value, the analysis of population structures divided 97 out of 132 accessions into six subpopulations (Figures 1–3). Group 1 contained five diploid (D3) accessions collected from Santa Cruz Island. Group 2 had 32 accessions of *G. tomentosum* (AD3) obtained from a Hawaiian Island. Group 3 was composed of 14 accessions and was demarcated with the accession of *G. darwinii* (AD5) collected from Isabella Island but also including one from the China Wild Cotton Germplasm Nursery. Group 4 had 10 accessions of *G. ekmanianum* (AD6) collected from the Dominican Republic, NPGS, USA. Group 5 contained 19 accessions of *G. darwinii* that were collected from San Cristobal. Group 6 had 17 accessions of *G. barbadense* (AD2); of them, two were collected from the China Wild Cotton Germplasm Nursery (Supplementary Table S5). Based on a phylogenetic analysis using the Unweighted Pair-Group Method using Arithmetic average (UPGMA), the same accessions were placed under discriminating subgroups having significant genetic distance in accordance with the geographical locations of the collection. The results were further validated by using Shannon's information index to determine the genetic diversity among six populations. It was found that population 3 (Isabella) had the highest degree of heterozygosity with 55.5% polymorphic loci (Figure 4).

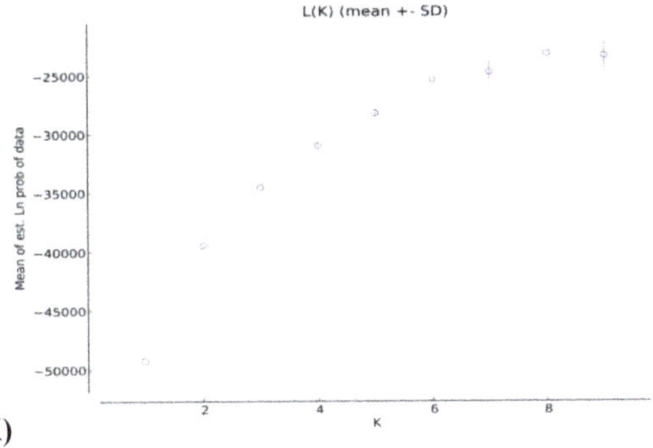

L(K)

Figure 1. K means for 132 accessions.

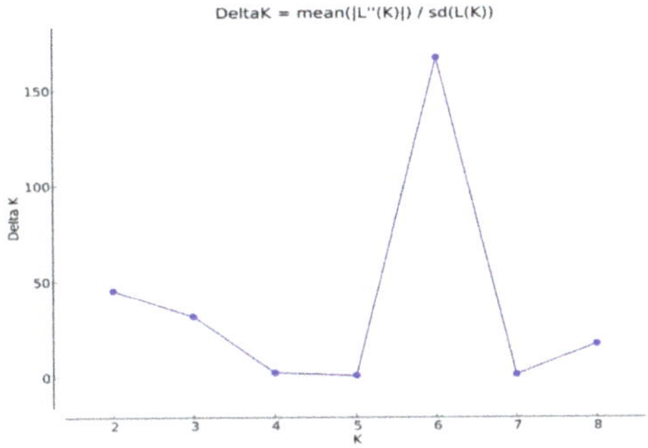

Figure 2. Delta K for 132 accessions.

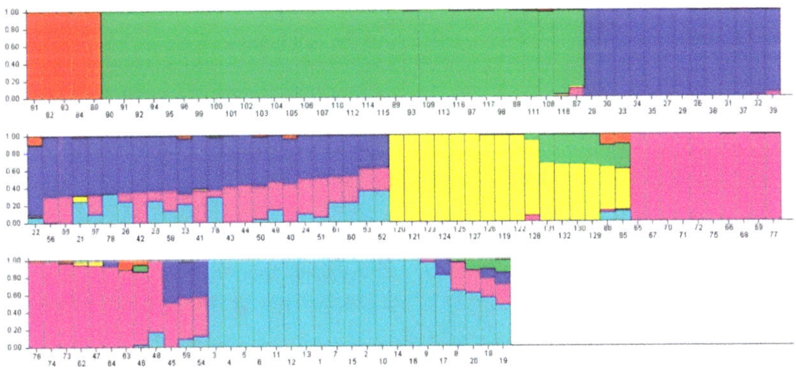

Figure 3. Q plot showing clustering of 132 accessions in 6 subpopulations based on an analysis of genotypic data using STRUCTURE software ver. 2.2. Each accession is indicated by vertical bars. The color subsections within each vertical bar represent the membership coefficient (Q) of the accession to different colors. Six groups were identified. The identified groups are I (red), II (lime), III (Blue), IV (yellow), V (Fuchsia), and VI (Aqua) colors in regular patterns.

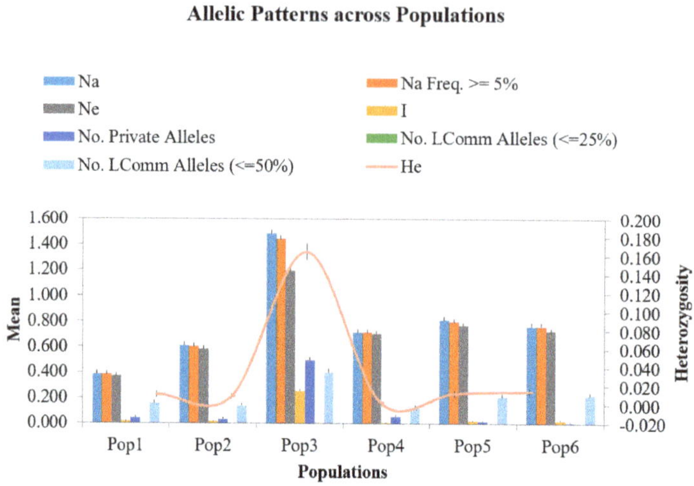

Figure 4. Alleleic patterns across populations.

2.5. Genetic Diversity and Cluster Analysis of Phylogenetic Tree

A total of 382 alleles, generated by 111 EST-SSRs, were used to run UPGMA for generating the dendrogram. Based on Nei's criteria [52], the genetic distance among wild cotton accessions ranged from 0.003 to 0.529 with an average of 0.325. The highest genetic distance (0.529) was between D3k-21-3 and AD5-lz. The phylogenetic tree was in agreement with the structure results with the exception that *G. hirsutum* and *G. stephensii* sit in different clusters in the phylogenetic tree but in the structure analysis these were grouped together. In order to see how the results correspond to each other between the STRUCTURE and phylogenetic analyses, the dendrogram was manually edited to show the STRUCTURE grouping (Figure 5 and Supplementary Figure S1). Six groups identified in the structure analysis were also clustered together in the phylogenetic tree analysis. Overall, there was good agreement between the two estimates. The clustering pattern also showed agreement with relationships based on pedigree studies [53]. The first two axes of the principal coordinate analysis (PCoA) accounted for 42.2% of the variation (Figure 6). This indicates a high level of genetic diversity in the *Gossypium* germplasm with continuous variation between and within the subgroups. Analysis of molecular variance (AMOVA) revealed highly significant variation between the six groups identified by the structure analysis, with 49% of the total variation contributing to between-group differences. However, a larger amount of variation (51%) was due to diversity within the groups having different populations (Table 2). Pairwise F_{ST} analysis revealed that accessions from Pop 3 (Isabella region) are closer to accessions from the San Cristobal (Pop 5) and Santa Cruz regions (Pop 6) as compared with the Hawaiian accessions. The highest genetic differentiation was observed among tetraploid populations between accessions from the Hawaiian (Pop 2) and Santa Cruz (Pop 6) regions with a pairwise F_{ST} of 0.752 ($p < 0.001$) (Table 3).

A cluster analysis clearly discriminated diploid wild-type cotton from other tetraploid wild-types. These accessions were collected from different locations, namely the Galapagos Islands, Hawaii, the Dominican Republic, Wake Atoll, and the Wild Cotton Germplasm Nursery of China. The dendrogram was truncated at a genetic distance level of (0.05) and divided 132 cotton genotypes into seven clusters (Supplementary Figure S1).

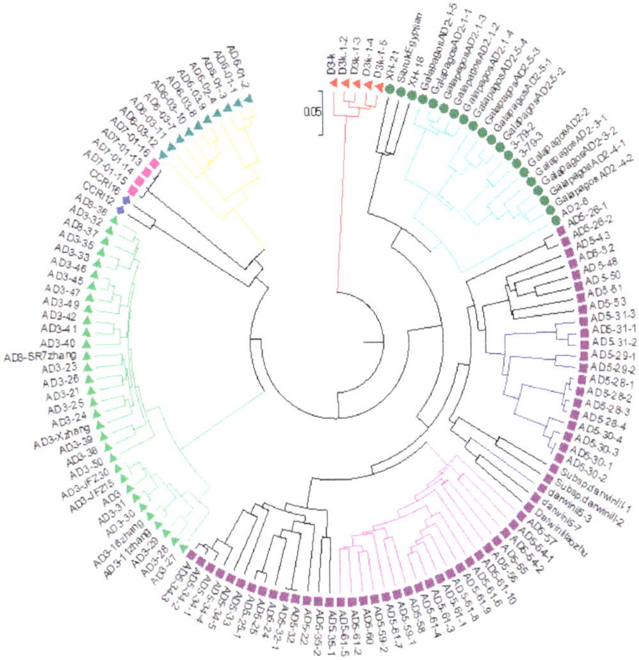

Figure 5. Dendrogram of 132 wild cotton accessions by Unweighted Pair-Group Method using Arithmetic average (UPGMA) analysis. Colors in the dendrogram lines correspond to *Gossypium* accession populations as identified by structure analysis while the colors in the circle represent the seven species. A membership threshold of 70% was used to assign accessions to different clusters in this dendrogram based on structure analysis.

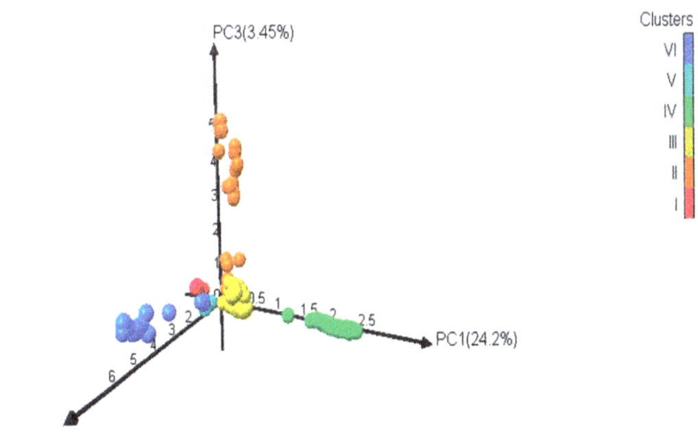

Figure 6. Three-dimensional principal coordinate analysis (PC_OA) of a *Gossypium* accessions diversity panel genotyped with expressed sequence tags (EST) and Genomic simple sequence repeats (SSRs). The different colors in the figure correspond to six clusters: Red (Cluster I), orange (Cluster II), yellow (Cluster III), Bright green (Cluster IV), Sky blue (Cluster V), Blue (Cluster VI).

Table 2. Analysis of molecular variance for wild cotton accessions among and within six populations as identified by STRUCTURE.

Source of Variation	*df*	Sum of Squares	Mean Squares	Estimated Variation	Percentage of Variation
Among Pops	5	4201.563	840.313	38.134 **	49%
Within Pops	126	4932.945	39.150	39.150	51%
Total	131	9134.508		77.284	100%

(PhiPT < 0.493; ** significance at $p < 0.001$).

Table 3. Pairwise Fst estimates for the five groups corresponding to six regions of accession collections as identified by STRUCTURE.

Populations with Origin	Pop1 (Santa Cruz)	Pop2 (Hawaiian)	Pop3 (Isabella)	Pop4 (Dominican Republic)	Pop5 (San Cristobal)
Pop2 (Hawaiian)	0.869				
Pop3 (Isabella)	0.697	0.651			
Pop4 (Dominican Republic)	0.757	0.689	0.544		
Pop5 (San Cristobal)	0.810	0.749	0.301	0.638	
Pop6 (Santa Cruz)	0.813	0.752 **	0.413	0.626	0.517

(** significance at $p < 0.001$).

2.6. Phylogenetic Analysis of Mined Genes and Functional Annotation of DUF819 (PF005684)

The study was extended to dissect the evolutionary relationship among the uncharacterized genes because these are considered to be highly conserved and have an important role in biology. DUF819 is a family containing proteins (PF005684) found in 532 species with a total of 756 sequences. A total 1517 genes were found among *Gossypium arboreum* (258), *G. raimondii* (258), *G. hirsutum* (513), and *G. barbadense* (488) from the cotton functional genomic database (www.cottonfgd.org). These were the best-fit matched homologue genes having the highest similarity with the four cotton species. Out of 1517, only 116 genes differentially expressed in experiments and belonging to PF00400 and PF05684 were identified among *G. arboreum*, *G. raimondii*, *G. hirsutum*, and *G. barbadense*. A total of 24 uncharacterized genes identified to diagnose their evolutionary relationship with these cotton species. A phylogenetic tree consisting of 115 genes out of 116 expressed genes, including uncharacterized genes, in different experiments was constructed for the sorted PF00400 (105) and PF005684 (11) (Figure 7). The protein sequence of one gene remained unaligned during a ClustalW alignment. The PF00400 belonging to the superfamily WDR was used as a reference because 13 out of 24 uncharacterized genes were linked to this protein domain. The remaining 11 uncharacterized genes were named yjcL. The 11 yjcL (PF005684) genes were found to be more closely related to Cotton_A04_RF178 and then to GOBAR_DD_12SPA2, whose functions are known. Cotton_A04_RF178 is a well-known E3 ubiquitous protein having an important role in stress response in plants and animals [54]. It is predicted that, as these genes make a very close cluster with a well-known gene playing a crucial role in plant survival, they may have same function because they can be assigned to proteins by using the bioinformatics tools in comparative genomics [55]. The yjcL of A_ and D_, which are subgenomes of *G. hiursutum* and *G. barbadense*, make close groups with the yjcL of *G. arboreum* and *G. raimondii*, respectively. This indicates that yjcL genes may flow from *G. arboreum* and *G. raimondii* to *G. barbadense* and *G. hirsutum* in equal proportion. Moreover, the uncharacterized genes in *G. hirsutum*, *G. raimondii*, and *G. arboreum* indicated as all2124 are grouped close to Gh_A03ago, which has the known function of a protein related to F-box/WD repeats. Similarly, these results can also be validated by predicting that Gh_A01all2124, Gorai-all2124, Gorai-alr3466, and Cott_A_all2124 perform the same function as that determined for Gh_A03ago. GOBAR_AA12SPAC343 is separate from but close to the gene pak1ip1 with the known function of p21-activated protein kinase-interacting protein 1-like. The gene SPBC1711 located on the A_ and D_ genome of *G. hirsutum*, *G. arboreum*, and *G. raimondii* makes a close group with the gene named RUP2 in the A03 and A09 chromosomes of *G. hirsutum*. RUP2 is a gene composed of WD protein domains. Meanwhile, another gene, SPAC3H5, originating in *G. hirsutum* and *G. arboreum* was found to be lying nearest to the DDB2 gene of known function in *G. hirsutum*. These genes were distributed throughout the 26 chromosomes. The maximum number of genes (11) was found on chromosome D11, and the minimum number (4) was found on chromosome 2 (Supplementary Table S6). The coding DNA sequences (CDS) were characterized and the GC content percentage ranged from 38.9 to 52.7 with its length ranging from 417 to 3221 (bp). The maximum number of exons was noted to be 24 in the Gh_D11G0779 homologue of GOBAR_DD21377, indicating the highest intron disruption (Supplementary Table S6). The majority of these mined genes, especially the uncharacterized ones, have a single protein domain, which means that these genes are highly conserved. We analyzed the features of these genes and the results showed several categories related to stress and fiber development in upland cotton. We further analyzed the genes through annotations and Gene Ontology (GO) terms that were associated with the mined genes, which describe the genes in relation to cellular components (CCs), molecular function (MF), and biological process (BP) [56]. In cellular components, functions such as microtubule organizing center (11%), microtubule-associated complex (10%), membrane coat (13%), coated membrane (13%), cytoskeleton part (10%), microtubule cytoskeleton (%), cytoskeleton (10%), and protein complex (23%) were observed. Similarly, 14 molecular functions and 5 biological processes were observed (Figure 8). Finally, we carried out RNA sequence expression to validate our results. The 65 genes with differential expression in *G. hirsutum* were selected to construct a heat map. The genes were both up and downregulated in cold, hot, polyethylene glycol (PEG), and salt treatments and different developmental stages of different tissue organs, such as calyx, leaf, petal, pistil, root, stamen, stem, and torus tissue (Figure 9). The genes were categorized into two main groups. Group 1 comprised 34 genes that were significantly expressed; i.e., with fragments per kilobase of transcript per million mapped reads (FPKM) value of more than 1. Among the 34 upregulated genes, SPA2 (protein SPA1-RELATED 2) with Gene ID Gh_D12G2294 has five Go functions: protein kinase activity (GO:0004672 = MF), protein binding (GO:0005515 = MF), ATP binding (GO:0005524 = MF), protein phosphorylation (GO:0006468 = BP), and transferase activity transferring phosphorus-containing groups (GO:0016772 = MF). CDC40 (Pre-mRNA-processing factor 17) with Gene ID Gh_A05G0018 depicts three GO functions: mRNA splicing via spliceosome (GO:0000398 = BP), protein binding (GO:0005515 = MF), and catalytic step 2 spliceosome (GO:0071013 = CC). Two Guanine nucleotide-binding protein subunit beta-2s with different Gene IDs were found to have two similar GO functions, namely protein binding (GO: 0005515 = MF) and signal transduction (GO:0007165 = BP). All remaining genes were found to be associated in molecular function with protein binding with GO:0005515. The yjcL-linked gene ID Gh_A09G2500 showed significant expression against drought and salt stress and fell into group 1. The other two Gene IDs associated with SPAC3H5, an uncharacterized WD-repeat-containing protein (GO:0005515 = MF), also indicated significant expression. Group 2 has 31 genes that exhibited the differential expression of both up and downregulation (Supplementary Table S7). Among these, only the Gh_D07G1711 gene showed three GO functions: mRNA splicing via spliceosome (GO:0000398 = BP), protein binding (GO:0005515 = MF), and catalytic step 2 spliceosome (GO:0071013 = CC). All others were associated with WDR25 (WD-repeat containing protein 25) with the GO function GO:0005515 = MF except for three genes with the IDs Gh_D11G0109, Gh_A11G2961, and Gh_D09G0432, which are linked to the uncharacterized gene yjcL and have no GO functions. In the second group, four genes, namely Gh_D07G2259, Gh_A10G2180, Gh_D09G0432, and Gh_D02G1696, were relatively downregulated while all other genes showed differential expression. Gh_D04G1713 showed upregulated expression in petal and stamen tissues but relative downregulation in other tissues and under stress treatments.

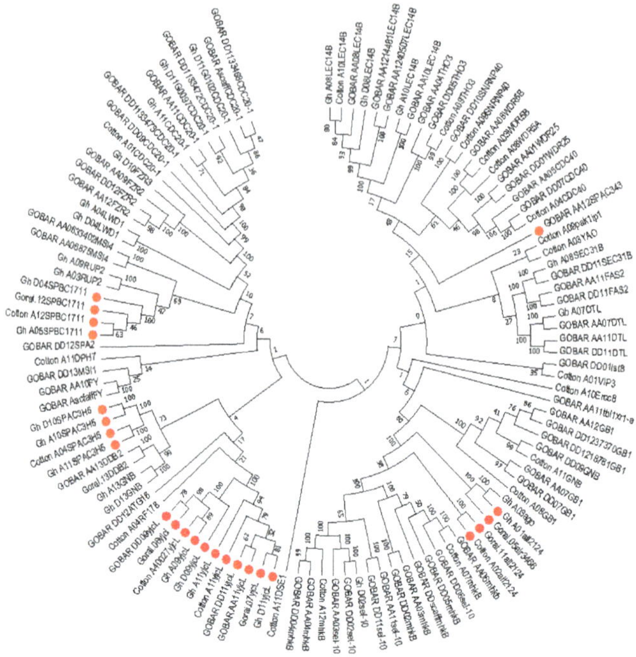

Figure 7. Evolutionary relationship of 115 genes belonging to protein domains of DUF819 (PF005684) and WDR (PF00400) in *Gossypium arboreum*, *G. raimondii*, *G. hirsutum*, and *G. barbadense*. The phylogenetic tree was constructed using MEGA software ver. 7.0 by the neighbor-joining method. The parameters were 1000 bootstraps and pairwise deletion. The 24 uncharacterized genes are indicated by red dots in four *Gossypium* species.

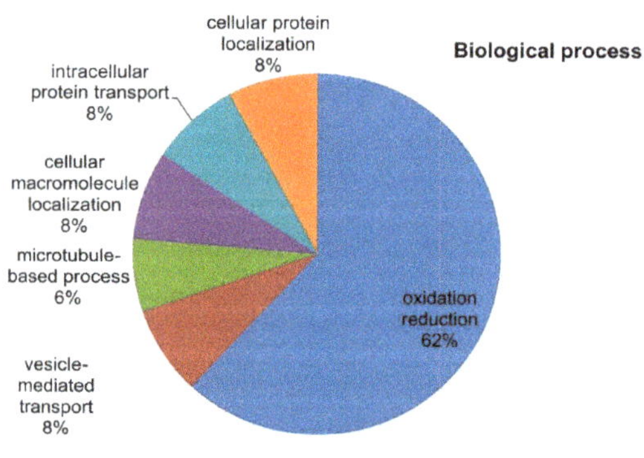

Figure 8. *Cont.*

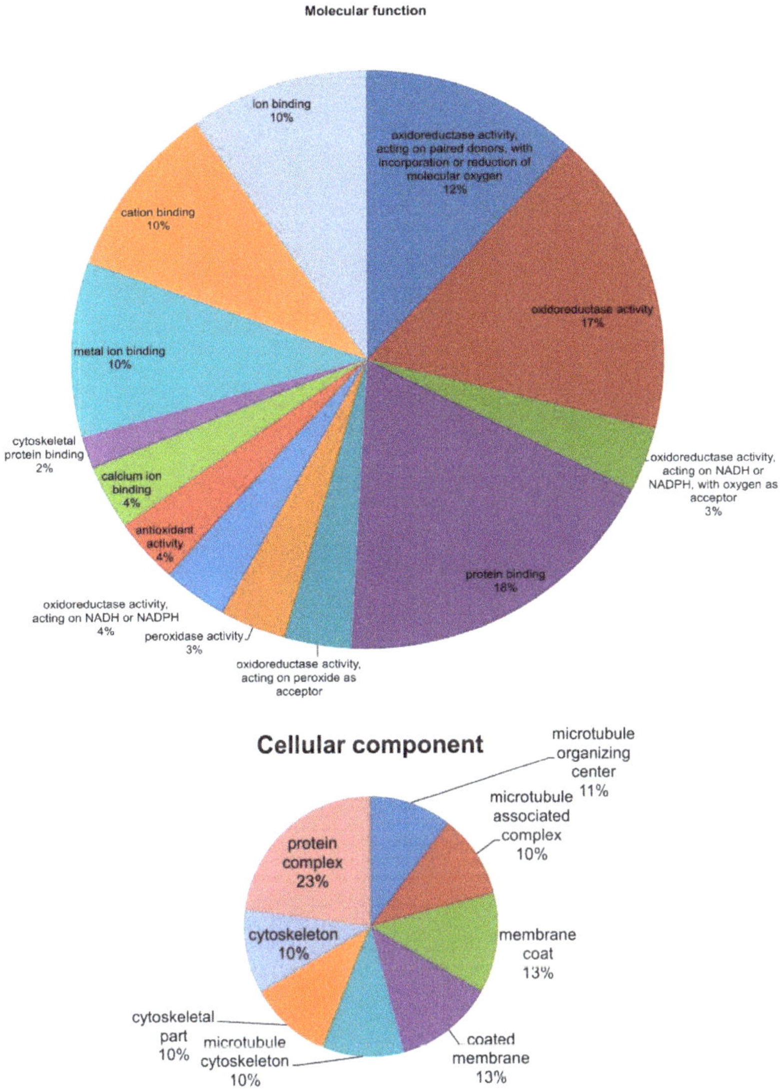

Figure 8. Genes were analyzed using the Agrigo v 2.0 software. Gene Ontology (GO) annotation results for *Gossypium hirsutum* mined genes. GO functional classification of genes mined with protein sequences predicted for their involvement in biological processes (BPs), molecular functions (MFs), and cellular component (CCs).

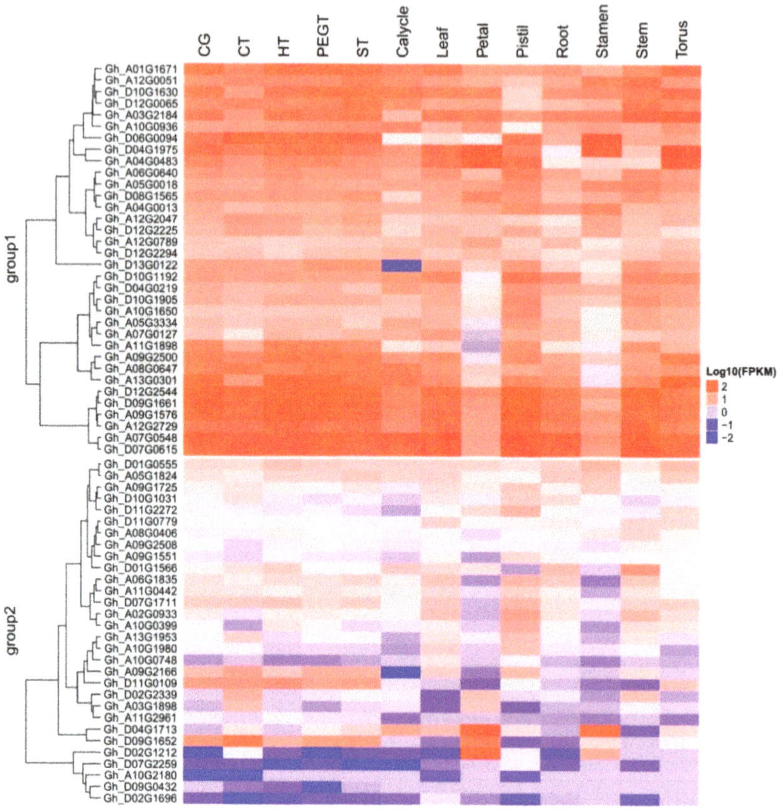

Figure 9. RNA sequence data analysis of 65 differentially expressed genes in eight different cotton tissues with reference to a control group (CG) under different cold (CT), heat (HT), PEG (PEGT), and salt treatments (ST) listed at the top of the figure. The names of genes are listed to the left of the figure. The heat map was generated from the log10 (FPKM) of the expression values by using R software. The *Y* axis represents the relative expression ($2^{-\Delta\Delta Ct}$).

3. Discussion

In this study, 111 SSR primer pairs generated 382 polymorphic loci in the 132 tested accessions. An average of 3.44 alleles amplified per marker was observed for all accessions, ranging from 2 to 8 alleles. This value is comparable with the findings of Dahab et al. [57] on allele number using 70 SSR markers on *Gossypium hirsutum*. Consistent results were found with 3.93 alleles per locus by Bardak and Bolek [58] for assessing genetic diversity in diploid and tetraploid cottons using Simple Sequence Repeat (SSR) and Inter Simple Sequence Repeat (ISSR) markers. Wendel and Percy [17] detected 3.47 enzymes per locus by studying 17 enzymes encoding 59 loci in a collection of 58 accessions of *darwinii* from six islands. However, other studies showed a variable allele number per locus. For example, 2.13, 2.20, 5.46, and 7.64 alleles per marker have been found in several genetic diversity assessments for the cotton germplasm [26,59–61]. The semi-wild accessions retaining a diverse germplasm showed high allele numbers in the majority of studies consistent with a recent study because these accessions have not yet been exposed to extensive human selection pressure for accumulating a particular type of alleles [26,62,63]. The number of alleles observed per marker is contingent on the selection of markers, the collection of germplasm to be genotyped, and the platform used for the resolution of amplified products [64].

Our study results determined an average PIC value of 0.555 with a range of 0.078 to 0.821, which completely corresponds to the literature-cited average PIC value for cotton SSRs, which ranges from 0.122 [65] to 0.71 [26]. Higher PIC values in cotton as shown in the current study suggest that these accessions can be useful for improving cotton [57]. The unique alleles identified in this study had percentages of 7.36, 5.83, 6.43, 17.36, 17.51, 3.2, and 14.44 in *G. barbadense*, *G. darwinii*, *G. tomentosum*, *G. hirsutum*, *G. ekmanianum*, *G. stephensii* and *G. klotzschianum*, respectively. These are higher than the percentages reported in the earlier report [65]. It is an interesting finding that all the unique alleles were found in the newly collected accessions along with a few from *G. hirsutum*. The unique alleles may be related to unique characteristics, such as extra-long fibers in *G. barbadense* and drought and salt tolerance in *G. tomentosum* and *G. darwinii*, and be similar to other wild accessions.

The common alleles were estimated to understand the gene flow mechanism of the *Gossypium* species during evolution. The results are supported by previous studies and the hypothesis that *G. hirsutum* has a single evolutionary lineage because all of the species in this study have common alleles with reference to the fixed alleles of *G. hirsutum*. *G. tomentosum* is considered to be a sister of *G. hirsutum*, while *G. barbadense* originates from a geographically overlapping region. *G. darwinii* has the same origin, the Galapagos Island, as *G. barbadense*. The *G. klotzschianum* diploid cotton is considered endemic to the New World and has an origin similar to that of *G. raimondii*. The two new species *G. ekmanianum* and *G. stephensii* are sister clades even though they make distinct groups with *G. hirsutum*, but *G. ekmanianum* and *G. stephensii* are monophyletic to *G. hirsutum* which strengthens the hypothesis of gene flow from different species to *G. hirsutum* having a single-lineage evolution [1,13,15]. The present analysis with a high level of natural introgression among the wild accessions shows consistency with the results of Yu et al. [51] and Hinze et al. [36] who described the distribution of introgression within the *G. hirsutum* and *G. barbadense* genomes using the chromosome positions as markers.

Populations from different islands are isolated distinctly, indicating general correspondence to Wedel and Percy's investigations [17]. Due to good agreement with Wendel and Percy [17], an exploration that occurred 30 years prior to the present study, this novel study will also be helpful in understanding the basis of the hybridization and domestication phenomenon in cotton evolution. Our findings also suggest that a wild germplasm has higher genetic diversity than that in cultivated cotton.

A phylogenetic tree constructed based on genotypic data completely validated the distinct clustering of the accessions detected. The results are quite congruent to prior taxonomic studies [2,3,66]. The average genetic distance (GD = 0.325) revealed the overall level of genetic diversity to be high among semi-wild and cultivated accessions; this finding is similar to earlier reports [26,67,68]. However, this estimate may be inflated since data from monomorphic SSR loci were excluded in the current study.

An evolutionary relationship among the genes was also developed. It has been estimated that the majority of genes are linked to responses towards biotic and abiotic stress conditions. For example, the damage-specific DNA binding protein 2 (DDB2), Autophagy-related protein 16 (ATG16), WD repeat-containing protein LWD1, Denticleless protein homolog (DTL), Protein FIZZY-RELATED FZR2, FZR3, Pre-mRNA-processing factor 17 (CDC40), Diphthine methyltransferase (DPH7), Chromatin assembly factor 1 subunit FAS2, DNA excision repair protein ERCC-8, Flowering time control protein FY, Guanine nucleotide-binding protein subunit beta-like protein (GB1), Myosin heavy chain kinase B (mhkB), WD-40 repeat-containing protein MSI1, MSI4, F-box/WD repeat-containing protein sel-10, U5 small nuclear ribonucleoprotein 40 kDa protein (SNRNP40), THO complex subunit 3 (THO3), WD repeat-containing protein VIP3, WDR25, WDR5A, WDR5B, and U3 snoRNP-associated protein-like YAO belong to the superfamily of WDR and are involved in repairing damaged DNA under various stress conditions [69,70]. Similarly, RUP2 has been found to play a very crucial role in vegetative development and flowering in Arabidopsis [49]. F-box/WD repeat-containing protein 7, named as "ago" and associated with Gh_A03G1152, supports plants against disease and repairs damaged DNA [71]. SEC31 homolog B transports proteins and is situated at Golgi-associated endoplasmic reticulum exit sites. CDC20-1 is a known component of the anaphase promoting complex/cyclosome (APC/C), a cell-cycle-regulated E3 ubiquitin–protein ligase complex that controls progression through mitosis and the G1 phase of the cell cycle. This protein is involved in the pathway protein ubiquitination, which is part of protein modification. The intron-containing CDC20 gene copies provide conserved and redundant functions for cell-cycle progression in plants and are required for meristem maintenance, plant growth, and male gametophyte formation [69]. These results also support our hypothesis that yjcL (PF005684) has the same functions as these WDR family genes.

The current study is highly associated with the pedigree information recently provided by Gallagher et al. [3] after molecular confirmation of newly designated species of *Gossypium*. Genetic diversity within the group was highest for the Isabella group and lowest for the Hawaiian group (Table 2 and Figure 1). The genetic differentiation between groups was further validated by AMOVA, with 49% of the variation among populations and 51% of the variation within populations (** significant $p > 0.001$) being explained by the population structure of the wild cotton germplasm (Table 3). Such higher variation may be due to the complete study of seven different species of diploid and tetraploid cotton. This also indicates the presence of a great genetic difference among tetraploid and diploid cottons as well as a good level of genetic diversity within each group, which can be used in further hybridization breeding programs in cotton to broaden the narrow genetic base of *G. hirsutum*, which is becoming a serious threat due to limited allelic availability [72]. The F_{ST} values for the diploid and tetraploid cottons observed in this study (0.301–0.869) are very high, indicating high genetic distance and diversity. PCoA plots separated tetraploid cottons from diploid plants, supporting the AMOVA results. All these results are in good agreement with Noormohammadi et al.'s genetic diversity analysis [68] between diploid and tetraploid accessions.

Thus, our results could help breeders to determine the selection of appropriate parental combinations in germplasm enhancement programs and conserve genetic diversity and the evolutionary relationship among the genes of uncharacterized functions. The presence of profound population differentiation could pose a challenge to successful Genome-Wide Association Mapping (GWAS) studies in the Upland cotton germplasm for traits that are associated with population structures. The power of structure-based association studies to detect the effects of a single gene would be reduced if a large fraction of variation was explained by the population structures [22,73]. In such cases, alternative association mapping populations would be more useful.

4. Materials and Methods

This study was conducted at the Institute of Cotton Research (ICR, Anyang, China), Chinese Academy of agricultural Sciences (CAAS), Anyang, China. The cotton accessions were obtained from six islands, namely Santa Cruz, San Cristobal, Isabella of Galapagos Island, a Hawaiian Island, Dominican Republic, and Wake Atoll. The screening of this unique collection was carried out using microsatellite markers for the detection of a polymorphism among these accessions.

4.1. Plant Material and DNA Extraction

We sampled a total of 132 accessions belonging to different species, including five *G. klotzschianum* (D3), two *G. hirsutum* (AD1), 20 *G. barbadense* (AD2), 32 *G. tomentosum* (AD3), 59 *G. darwinii* (AD5), 10 *G. ekmanianum* (AD6), and four *G. stephensii* (AD7), from six islands and the Wild Cotton Germplasm Nursery of China. Among the 132 accessions, 32, 25, 16, 32, 10, 4, and 12 were obtained from Santa Cruz, Isabella, San Cristobal, a Hawaiian island, the Dominican Republic, Wake Atoll (NPGS, USA), and the Wild Cotton Germplasm Nursery of China, respectively (Supplementary Table S1). Seedlings of these accessions were grown at the wild cotton germplasm nursery of China, Sanya Hainan during October 2015, 2016, and 2017, respectively. When the plants were about 30–35 days old, fresh leaves were sampled and immediately frozen at −80 °C for later DNA extraction. Total genomic DNA was extracted from the frozen leaves by the cetyltrimethylamonium bromide (CTAB) method as described by Zhang and Stewart [74] with slight modifications. DNA was quantified using Nanodrop at a 260/280 nm absorbance ratio and the quality was checked by 1% (*w*/*v*) agarose gel electrophoresis.

4.2. SSR Marker Selection and Genotyping

A total of 853 randomly selected SSR markers, including 200 DPL, 310 MonCGR, 48 NAU, 41 MUCS, and 254 SWU, were surveyed for their polymorphisms in 132 genotypes belonging to seven cotton species. Then, 111 EST and genomic SSR (based on D-genome) polymorphic primers from the Cotton Marker Database (CMD; http://www.cottonmarker.org/) were used in the SSR analysis. The reaction contained 5 μL 2× Taq Master Mix (containing buffer, dNTPs, and Taq DNA Polymerase), 2 μL primers, 1 μL DNA, and 2 μL H_2O. The PCR reaction was performed using a together TP 600 thermal cycler (TAKARA Bio Inc., Kusatsu, Japan) and then followed by silver staining according to a previous method described by Zhang et al. [75]. The PCR temperature program was two cycles of 95 °C for 3 min pre-denaturing followed by 30 cycles of 94 °C for 45 s denaturing, 57 °C for 36 s annealing, 72 °C for 1 min extension, with a final step of 1 cycle at 72 °C for 5 min extension. To confirm that the observed amplicons were amplified from genomic DNA and not a primer artifact, genome DNA was omitted from the control reaction. No amplification products were detected without genomic DNA in any PCR.

4.3. Analysis of Genotypic Data and Genetic Diversity

Pairwise genetic distances between accessions were calculated using the Powermarker software package ver.3.25 by Nei et al. [52] D_A distance. The dendrogram was constructed on the basis of the distance matrix. We estimated the similarity between genotypes for each accession by awarding a score to each microsatellite (i.e., 0 when an allele was absent, 1 when the allele was present). The cluster analysis was carried out using the unweighted pair group method using arithmetic average (UPGMA) and the dendrogram resulting from these calculations was plotted using MEGA 6.0 to visualize and edit the dendrogram. The basic summary statistics for biallelic data were calculated using the POWERMARKER software package version 3.25 [76]. The polymorphism information content (PIC) of an SSR marker was determined according to the method described by Anderson et al. [77] based on the allele frequency of all genotypes.

$$\text{PIC} = 1 - \sum_{i=1}^{n} pi2 \tag{1}$$

where P_{ij} is the frequency of the allele for locus i and the summation covered n patterns.

A PIC value of 1 indicates that the marker can differentiate each line, and 0 indicates a monomorphic marker. The informative potential of a marker is high if its PIC value is more than 0.5, moderate if its PIC is between 0.5 and 0.25, and only slightly informative if its PIC value is below 0.25. Other statistics calculated were the number of alleles and availability and gene diversity for each marker. Further analysis of genetic structure was done by means of Principal co-ordinate analysis (PC_OA) using XLSTAT, 2014 [78] and a three-dimensional diagram was constructed. Dominant data (0, 1 binary data) were used for the PCoA analysis.

4.4. Analysis of Genetic Structure

The STRUCTURE software version 2.3.4 [79] was employed to define 132 accessions into clusters consisting of genotypes by using co-dominant genotypic data. The admixture model was used to estimate a mixed group by using correlated allele frequencies between populations as described by Falush et al. [80]. The optimum number of subpopulations was calculated based on the recommendation of Evanno et al. [81] by defining the values for K = 2 to K =10 with a burn length of 10,000 and a run length of 100,000 each in 10 runs. The results were uploaded in a Zip file to the STRUCTURE harvester software for finding the ΔK [82]. Grouping and subgrouping of accessions was done if the probability of membership was more than 70% [83]. The accessions with membership <70% were placed into the mixed subgroup.

4.5. Gene Mining and Phylogenetic Analysis of DUF819 Proteins

The complete sequence of Markers SWU15000–SWU15194 mapped for chromosome 6 was downloaded from the Genome database of *G. raimondii* [84] and blastx was used to find the homologue similarity of genes in the genome sequence of *G. raimondii*, *G. arboreum*, *G. hirsutum*, and *G. barbadense*. The mining of genes from the marker regions has been done extensively; see for instance Kirungu et al. [43]. Similarly, the same has been applied by Magwanga et al. [85]. The uncharacterized gene named yjcL of DUF819 (PF008654) was selected for the evolutionary study of genes in sequenced *Gossypium* species. The full-length sequences of DUF819 (PF005684) were downloaded from the pfam database (http://pfam.xfam.org/). The dendrogram was constructed by using Molecular Evolutionary Genetics version 7.0 [86]. The functional description related to domains of uncharacterized proteins has been predicted using the protein sequence of 116 genes downloaded from the Cotton Functional Genomics database (www.cottonfgd.org) [87]. The evolutionary relationship among all selected genes was summed up to provide a clear picture of functions with reference to upregulated genes of the superfamily WDR.

Thus, 115 genes out of 116 were grouped into 6 major clusters and their evolutionary history was inferred using the Neighbor-Joining method [88]. The optimal tree with the sum of branch length equal to 41.56 is shown. The percentage of replicate trees in which the associated taxa clustered together in the bootstrap test (1000 replicates) is shown next to the branches [89]. The tree is drawn to scale, with branch lengths in the same units as those of the evolutionary distances used to infer the phylogenetic tree. The evolutionary distances were computed using the Poisson correction method [90] and the units for the number of amino acid substitutions per site. The analysis involved 115 amino acid sequences. All ambiguous positions were removed for each sequence pair. There were a total of 1466 positions in the final dataset. Evolutionary analyses were conducted in MEGA7 (Figure 7).

The explored genes were analyzed for their gene features, protein characteristics, and RNA expression using the cotton functional genome database (https://cottonfgd.org/search/), While GO functional classification was done using Agrigo ver. 2.0 software acquiring *Gossypium hirsutum* as the reference genome. The analysis of RNA expression data inferred was then carried out to construct a heatmap using the R statistical software package.

5. Conclusions

SSR markers can be used to describe the degree of differentiation between populations and to control the conservation of genetic resources. The study concludes that the evaluated cotton accessions have a broad genetic basis. The recurrent use of these accessions as parents will produce significant results. The genetic diversity and evolutionary relationship recognized among the uncharacterized genes and population structures established in this study would be informative to select parental accessions for breeding and genetic analysis as well as for efficient management and conservation of allotetraploid cotton genetic diversity. We identified DUF819 (PF005684), which is not only highly conserved but also plays an important role against biotic and abiotic stress, among four sequenced cotton species by using the WDR (PF00400) superfamily as reference genome-sequenced proteins. Additionally, the current diversity panel of semi-wild cottons will be invaluable as a community resource for measuring linkage disequilibrium (LD) and for fine-scale mapping of traits through LD mapping or a Genome-Wide Association Study (GWAS) that can be streamlined for genomics-assisted plant breeding programs. Our findings suggest that allotetraploid cotton species, including *G. barbadense* (AD2), *G. tomentosum* (AD3), *G. darwinii* (AD5), *G. ekmanianum* (AD6), and *G. stephensii* (AD7), are a rich source for the creation of genetic diversity in upland cotton.

Supplementary Materials: Supplementary materials can be found at http://www.mdpi.com/1422-0067/19/8/2401/s1.

Author Contributions: K.W., A.D., and F.L. designed the experiments. A.D. and Z.Z. conceived the experiments and analyzed the results. A.D. and Z.Z. carried out the majority of the experiments and contributed equally. A.D. carried out all computational analyses. F.L., X.C., X.W., K.W.O., M.S., Y.X., Y.H., M.K.R.K., and M.S.I. participated in the mapping experiments. A.D. drafted the manuscript and K.W. revised the manuscript. All authors read and approved the final manuscript.

Funding: This research program was financially sponsored by the National key research and development plan (2016YFD0100306, 2016YFD0100203) and National Natural Science Foundation of China (31530053, 31671745).

Acknowledgments: We are indebted to give appreciation to State Key Laboratory of Cotton Biology, Institute of Cotton Research (ICR), Chinese Academy of Agricultural Sciences (CAAS); Anyang, China for providing umbrella for research program. We are also grateful to National Plant Germplasm System (NPGS), USA for providing germplasm. We express profound sense of reverence to Kunbo Wang and Fang Liu (Institute of Cotton Research), for timely guidance and provision of material whenever we needed during research work. To the entire research team, friends, and any other person who contributed, we have deep gratitude for you so much.

Conflicts of Interest: The authors declare no conflict of interest.

Abbreviations

PIC	Polymorphism information content
MAF	Major Allele frequency
GD	Genetic distance
Pop	population
AMOVA	Analysis of molecular variance
PCA	Principal coordinate analysis
PCR	Polymerase chain reaction
DNA	Deoxyribonucleic acid
NPGS	National Plant Germplasm System
EST	Expressed sequence tags
gSSR	Genomic SSR
SSR	Simple sequence repeats
RFLP	Restriction fragment polymorphism
RAPD	Random amplified polymorphic DNA
AFLP	Amplified fragment length polymorphism
SNP	Single nucleotide polymorphism
G	*Gossypium*
WCGN	Wild cotton Germplasm Nursery of China

References

1. Wendel, J.F.; Brubaker, C.; Alveraz, I.; Cronn, R.; Stewart, J.M. Evolution and natural history of cotton genus. *Genet. Genom.* **2009**, *3*, 3–22. [CrossRef]
2. Wendel, J.F.; Grover, C.E. *Taxonomy and Evolution of Cotton Genus, Gossypium*; Issue Agronmonogr; Cotton, American Society of Agronomy, Inc.: Madison, WI, USA; Crop Science Society of America, Inc.: Madison, WI, USA; Soil Science Society of America, Inc.: Madison, WI, USA, 2015; Volume 57. [CrossRef]
3. Gallagher, J.P.; Grover, C.E.; Rex, K.; Moran, M.; Wendel, J.F. A New Species of Cotton from Wake Atoll, *Gossypium stephensii* (Malvaceae). *Syst. Bot.* **2017**, *42*, 115–123. [CrossRef]
4. Grover, C.E.; Zhu, X.; Grupp, K.K.; Jareczek, J.J.; Gallagher, J.P.; Szadkowski, E.; Seijo, J.G.; Wendel, J.F. Molecular confirmation of species status for the allopolyploid cotton species, *Gossypium ekmanianum* Wittmack. *Genet. Resour. Crop Evol.* **2015**, *62*, 103–114. [CrossRef]
5. Chandrakanth, K. *Wild Crop Relatives: Genomic and Breeding Resources*; Springer: Berlin/Heidelberg, Germany, 2011; pp. 109–122.
6. Krapovickas, A.; Seijo, G. *Gossypium ekmanianum* (Malvaceae), algodon silvestre de la Republica Dominicana. *Bonplandia* **2008**, *17*, 55–63.
7. Wendel, J.F.; Brubaker, C.L.; Percival, A.E. Genetic diversity in *Gossypium hirsutum* and the origin of Upland cotton. *Am. J. Bot.* **1992**, *79*, 1291–1310. [CrossRef]

8. Wendel, J.F.; Cronn, R.C. Polyploidy and the evolutionary history of cotton. *Adv. Agron.* **2003**, *78*, 139–186.
9. Wang, Q.; Fang, L.; Chen, J.; Hu, Y.; Si, Z.; Wang, S.; Chang, L.; Guo, W.; Zhang, T. Genome-Wide Mining, Characterization, and Development of Microsatellite Markers in *Gossypium* Species. *Sci. Rep.* **2015**, *5*, 10638. [CrossRef] [PubMed]
10. Wendel, J.F.; Brubaker, C.L. RFLP diversity in *Gossypium hirsutum* L. and new insights into the domestication of cotton. *Am. J. Bot.* **1993**, *80*, 71.
11. Brubaker, C.L.; Paterson, A.H.; Wendel, J.F. Comparative genetic mapping of allotetraploid cotton and its diploid progenitors. *Genome* **1999**, *42*, 184–203. [CrossRef]
12. Senchina, D.S.; Alvarez, I.; Cronn, R.C.; Liu, B.; Rong, J.; Noyes, R.D.; Paterson, A.H.; Wing, R.A.; Wilkins, T.A.; Wendel, J.F. Rate variation among nuclear genes and the age of polyploidy in Gossypium. *Mol. Biol. Evol.* **2003**, *20*, 633–643. [CrossRef] [PubMed]
13. Campbell, B.T.; Saha, S.; Percy, R.; Frelichowski, J.; Jenkins, J.N.; Park, W.; Mayee, C.D.; Gotmare, V.; Dessauw, D.; Giband, M.; et al. Status of the global cotton germplasm resources. *Crop Sci.* **2010**, *50*, 1161–1179. [CrossRef]
14. Chen, H.; Khan, M.K.R.; Zhou, Z.; Wang, X.; Cai, X.; Ilyas, M.K.; Wang, C.; Wang, Y.; Li, Y.; Liu, F.; et al. A high density SSR genetic map constructed from F_2 population of *Gossypium hirsutum* and *Gossypium darwinii*. *Gene* **2015**, *574*, 273–286. [CrossRef] [PubMed]
15. Dejoode, D.; Wendel, J. Genetic diversity and origin of the hawaiian-islands cotton, *Gossypium tomentosum*. *Am. J. Bot.* **1992**, *79*, 1311–1319. [CrossRef]
16. Wendel, J.F.; Rowley, R.; Stewart, J. Genetic diversity in and phylogenetic-relationships of the brazilian endemic cotton, *Gossypium mustelinum* (*Malvaceae*). *Plant Syst. Evol.* **1994**, *192*, 49–59. [CrossRef]
17. Wendel, J.F.; Percy, R. Allozyme diversity and introgression in the galapagos-islands endemic *Gossypium darwinii* and its relationship to continental *Gossypium barbadense*. *Biochem. Syst. Ecol.* **1990**, *18*, 517–528. [CrossRef]
18. Ulloa, M.; Stewart, J.M.; Garcia-C, E.A.; Goday, A.S.; Gaytan-M, A.; Acosta, N.S. Cotton Genetic Resources in the Western States of Mexico: In Situ Conservation Status and Germplasm Collection for ex situ Preservation. *Genet. Resour. Crop Evol.* **2006**, *53*, 653–668. [CrossRef]
19. Abdurakhmonov, I.Y.; Buriev, Z.T.; Shermatov, S.E.; Abdullaev, A.A.; Urmonov, K.; Kushanov, F.; Egamberdiev, S.S.; Shapulatov, U.; Abdukarimov, A.; Saha, S.; et al. Genetic Diversity in *Gossypium* genus. In *Genetic Diversity in Plants*; Caliskan, M., Ed.; InTech: Rijeka, Croatia, 2012; pp. 313–338. [CrossRef]
20. Van Esbroeck, G.A.; Bowman, D.T. Cotton germplasm diversity and its importance to cultivar development. *J. Cotton Sci.* **1998**, *2*, 121–129.
21. Bolek, Y.; El-Zik, K.M.; Pepper, A.E.; Bell, A.A.; Magill, C.W.; Thaxton, P.M.; Reddy, O.U.K. Mapping of verticillium wilt resistance genes in cotton. *Plant Sci.* **2005**, *168*, 1581–1590. [CrossRef]
22. Tyagi, P.; Gore, M.A.; Bowman, D.T.; Campbell, B.T.; Udall, J.A.; Kuraparthy, V. Genetic diversity and population structure in the US Upland cotton (*Gossypium hirsutum* L.). *Theor. Appl. Genet.* **2014**, *127*, 283–295. [CrossRef] [PubMed]
23. Iqbal, M.A.; Rehman, M.U. Identification of Marker-Trait Associations for Lint Traits in Cotton. *Front. Plant Sci.* **2017**, *8*, 86. [CrossRef] [PubMed]
24. Kalivas, A.; Xanthopoulos, F.; Kehagia, O.; Tsaftaris, A.S. Agronomic characterization, genetic diversity and association analysis of cotton cultivars using simple sequence repeat molecular markers. *Genet. Mol. Res.* **2011**, *10*, 208–217. [CrossRef] [PubMed]
25. Sun, G.; He, S.; Pan, Z.; Du, X. Homologous simple sequence repeats (SSRs) analysis in tetraploid (AD1) and diploid (A2, D5) genomes of *Gossypium*. *Hereditas (Beijing)* **2015**, *37*, 192–203.
26. Kiflom, W.O.; Xiao, L.; Cai, X.; Wang, X.; Chen, H.; Zhou, Z.; Wang, C.; Wang, Y.; Liu, F.; Wang, K. Genome wide assessment of genetic diversity and fiber quality traits characterization in *Gossypium hirsutum* races. *J. Integr. Agric.* **2017**, *16*, 2402–2412. [CrossRef]
27. Khan, M.K.R.; Haodong, C.; Zhongli, Z.; Ilyas, M.K.; Xingxing, W.; Cai, X.; Chunying, W.; Fang, L.; Kunbo, W. Genome Wide SSR High Density Genetic Map Construction from an Interspecific Cross of Gossypium hirsutum X Gossypium tomentosum. *Front. Plant Sci.* **2016**, *7*, 436. [CrossRef] [PubMed]
28. Ramakrishnan, M.; Ceasar, S.A.; Duraipandiyan, V.; Dhabi, N.A.A.; Ignacimuthu, S. Assessment of genetic diversity, population structure and relationships in Indian and non-Indian genotypes of finger millet (*Eleusine coracana* (L.) Gaertn) using genomic SSR markers. *SpringerPlus* **2016**, *5*, 120. [CrossRef] [PubMed]

29. Van Becelaere, G.; Lubbers, E.L.; Paterson, A.H.; Chee, P.W. Pedigree vs. DNA marker-based genetic similarity estimates in cotton. *Crop Sci.* **2005**, *45*, 2281–2287. [CrossRef]
30. Iqbal, M.J.; Aziz, N.; Saeed, N.A.; Zafar, Y. Genetic diversity evaluation of some elite cotton varieties by RAPD analysis. *Theor. Appl. Genet.* **1997**, *94*, 139–144. [CrossRef] [PubMed]
31. Bakht, J.; Iqbal, M.; Shafi, M. Genetic diversity and phylogenetic relationship in different genotypes of cotton for future breeding. *Int. Quart. J. Biol. Sci.* **2017**, *5*, 25–29.
32. Rahman, M.; Yasmin, T.; Tabbasam, N.; Ullah, I.; Asif, M.; Zafar, Y. Studying the extent of genetic diversity among *Gossypium arboretum* L. genotypes/cultivars using DNA fingerprinting. *Genet. Resour. Crop. Evol.* **2008**, *55*, 331–339. [CrossRef]
33. Shaheen, N.; Pearce, S.R.; Khan, M.A.; Mahmood, T.; Yasmin, G.; Hayat, M.Q. AFLP mediated genetic diversity of *Malvaceae* species. *J. Med. Plant Res.* **2010**, *4*, 148–154.
34. Liu, S.; Cantrell, R.G.; McCarty, J.C.J.; Stewart, J.M. Simple sequence repeat-based assessment of genetic diversity in cotton race stock accessions. *Crop Sci.* **2000**, *40*, 1459–1469. [CrossRef]
35. Liu, D.; Guo, X.; Lin, Z.; Nie, Y.; Zhang, X. Genetic diversity of Asian cotton (*Gossypium arboretum* L.) in china evaluated by microsatellite analysis. *Genet. Res. Crop Evol.* **2006**, *53*, 1145–1152. [CrossRef]
36. Hinze, L.L.; Gazave, E.; Gore, M.A.; Fang, D.D.; Scheffler, B.E.; Yu, J.Z.; Jones, D.C.; Frelichowski, J.; Percy, R.G. Genetic Diversity of the Two Commercial Tetraploid Cotton Species in the *Gossypium* Diversity Reference Set. *J. Hered.* **2016**, *107*, 274–286. [CrossRef] [PubMed]
37. Zhang, Y.; Wang, X.F.; Li, Z.K.; Zhang, G.Y.; Ma, Z.Y. Assessing genetic diversity of cotton cultivars using genomic and newly developed expressed sequence tag-derived microsatellite markers. *Genet. Mol. Res.* **2011**, *10*, 1462–1470. [CrossRef] [PubMed]
38. Dongre, A.; Bhandarkar, M.; Banerjee, S. Genetic diversity in tetraploid and diploid cotton (*Gossypium* spp.) using ISSR and microsatellite DNA markers. *Indian J. Biotechnol.* **2007**, *6*, 349–353.
39. Liu, B.; Wendel, J.F. Intersimple sequence repeat (ISSR) polymorphisms as a genetic marker system in cotton. *Mol. Ecol. Notes* **2001**, *1*, 205–208. [CrossRef]
40. Meng, K.; Wei, S.J.; Wang, Y.Q.; Zhou, D.U.; Ma, L.; Fang, D.; Yang, W.H.; Ma, Z.Y. Development of a core set of SNP markers for the identification of upland cotton cultivars in China. *J. Integr. Agric.* **2016**, *15*, 954–962. [CrossRef]
41. Park, Y.J.; Lee, J.K.; Kim, N.S. Simple Sequence Repeat Polymorphisms (SSRPs) for Evaluation of Molecular Diversity and Germplasm Classification of Minor Crops. *Molecules* **2009**, *14*, 4546–4569. [CrossRef] [PubMed]
42. Campbell, B.T.; Williams, V.E.; Park, W. Using molecular markers and field performance data to characterize the pee dee cotton germplasm resources. *Euphytica* **2009**, *169*, 285–301. [CrossRef]
43. Kirungu, J.N.K.; Deng, Y.; Cai, X.; Mangwanga, R.O.; Zhou, Z.; Wang, X.; Wang, Y.; Zhang, Z.; Wang, K.; Liu, F. Simple Sequence Repeat (SSR) Genetic Linkage Map of D Genome Diploid Cotton Derived from an Interspecific Cross between *Gossypium davidsonii* and *Gossypium klotzschianum*. *Int. J. Mol. Sci.* **2018**, *19*, 204. [CrossRef] [PubMed]
44. Goodacre, N.F.; Gerloff, D.L.; Uetz, P. Protein Domains of Unknown Function Are Essential in Bacteria. *mBio* **2013**, *5*, e00744-13. [CrossRef] [PubMed]
45. Mulder, N.J.; Kersey, P.; Pruess, M.; Apweiler, R. In silico characterization of proteins: UniProt, InterPro and Integr8. *Mol. Biotechnol.* **2008**, *38*, 165–177. [CrossRef] [PubMed]
46. Bateman, A.; Coggill, P.; Finn, R.D. DUFs: Families in search of function. *Acta Crystallogr. Sect. F Struct. Biol. Cryst. Commun.* **2010**, *66*, 1148–1152. [CrossRef] [PubMed]
47. Littler, E. Combinatorial domain hunting: Solving problems in protein expression. *Drug Discov. Today* **2010**, *15*, 461–467. [CrossRef] [PubMed]
48. Hauser, R.; Pech, M.; Kijek, J.; Yamamoto, H.; Titz, B.; Naeve, F.; Tovchigrechko, A.; Yamamoto, K.; Szaflarski, W.; Takeuchi, N.; et al. RsfA (YbeB) proteins are conserved ribosomal silencing factors. *PLoS Genet.* **2012**, *8*, e1002815. [CrossRef] [PubMed]
49. Gruber, H.; Heijde, M.; Heller, W.; Albert, A.; Seidlitz, H.K.; Ulm, R. Negative feedback regulation of UV-B-induced photomorphogenesis and stress acclimation in Arabidopsis. *Proc. Natl. Acad. Sci. USA* **2010**, *107*, 20132–20137. [CrossRef] [PubMed]
50. Li, Q.; Zhao, P.; Li, J.; Zhang, C.; Wang, L.; Ren, Z. Genome-wide analysis of the WD-repeat protein family in Cucumber and Arabidopsis. *Mol. Genet. Genom.* **2013**, *289*, 103–124. [CrossRef] [PubMed]

51. Yu, J.; Fang, D.; Kohel, R.; Ulloa, M.; Hinze, L.; Percy, R.; Zhang, J.; Chee, P.; Schefer, B.; Jones, D. Development of a core set of SSR markers for the characterization of *Gossypium* germplasm. *Euphytica* **2012**, *187*, 203–213. [CrossRef]
52. Nei, M.; Tajima, F.; Tateno, Y. Accuracy of estimated phylogenetic trees from molecular data. *J. Mol. Evol.* **1983**, *19*, 153–170. [CrossRef] [PubMed]
53. Smith, C.W.; Cothren, J.T. *Cotton: Origin, History, Technology, and Production*; John Wiley & Sons: Hoboken, NJ, USA, 1999; p. 43.
54. Shu, K.; Yang, W. E3 Ubiquitin Ligases: Ubiquitous Actors in Plant Development and Abiotic Stress Responses. *Plant Cell Physiol.* **2017**, *58*, 1461–1476. [CrossRef] [PubMed]
55. Matteo, P.; Edward, M.M.; Michael, J.T.; David, E.; Todd, O.Y. Assigning protein functions by comparative genome analysis: Protein phylogenetic profiles. *Proc. Natl. Acad. Sci. USA* **1999**, *96*, 4285–4288.
56. Tian, T.; Liu, Y.; Yan, H.; You, Q.; Yi, X.; Du, Z.; Xu, W.; Su, Z. agriGO v2.0: A GO analysis toolkit for the agricultural community, 2017 update. *Nucleic Acids Res.* **2017**, *45*, W12–W129. [CrossRef] [PubMed]
57. Dahab, A.A.; Saeed, M.; Mohamed, B.B.; Ashraf, M.A.; Puspito, A.N.; Bajwa, K.S.; Shahid, A.A.; Husnain, T. Genetic diversity assessment of cotton (*Gossypium hirsutum* L.) genotypes from Pakistan using simple sequence repeat markers. *Aust. J. Crop Sci.* **2013**, *7*, 261–267, ISSN 1835-2707.
58. Bardak, A.; Bolek, Y. Genetic Diversity of Diploid and Tetraploid Cottons Determined By Ssr and Issr Markers. *Turk. J. Field Crops* **2012**, *2*, 139–144.
59. Bertini, C.H.C.D.; Schuster, I.; Sediyama, T.; Barros, E.G.; Moreira, M.A. Characterization and genetic diversity analysis of cotton cultivars using microsatellites. *Genet. Mol. Biol.* **2006**, *29*, 321–329. [CrossRef]
60. Qin, H.; Chen, M.; Yi, X.; Bie, S.; Zhang, C.; Zhang, Y.; Lan, J.; Meng, Y.; Yuan, Y.; Jiao, C. Identification of Associated SSR Markers for Yield Component and Fiber Quality Traits Based on Frame Map and Upland Cotton Collections. *PLoS ONE* **2015**, *10*, e0118073. [CrossRef] [PubMed]
61. Baytar, A.A.; Erdogan, O.; Frary, A.; Frary, A.; Doganlar, S. Molecular diversity and identification of alleles for Verticillium wilt resistance in elite cotton (*Gossypium hirsutum* L.) germplasm. *Euphytica* **2017**, *213*, 31. [CrossRef]
62. Iqbal, M.J.; Reddy, O.U.K.; El-Zik, K.M.; Pepper, A.E. A genetic bottleneck in the evolution under domestication of upland cotton *Gossypium hirsutum* L. examined using DNA fingerprinting. *Theor. Appl. Genet.* **2001**, *103*, 547–554. [CrossRef]
63. Rungis, D.; Llewellyn, D.; Dennis, E.S.; Lyon, B.R. Simple sequence repeat (SSR) markers reveal low levels of polymorphism between cotton (*Gossypium hirsutum* L.) cultivars. *Aust. J. Agric. Res.* **2005**, *56*, 301–307. [CrossRef]
64. Lacape, J.M.; Dessauw, D.; Rajab, M.; Noyer, J.L.; Hau, B. Microsatellite diversity in tetraploid Gossypium germplasm: Assembling a highly informative genotyping set of cotton SSRs. *Mol. Breed.* **2007**, *19*, 45–58. [CrossRef]
65. Abdurakhmonov, I.Y.; Kohel, R.J.; Yu, J.Z.; Pepper, A.E.; Abdullaev, A.A.; Kushanov, F.N.; Salakhutdinov, L.B.; Buriev, Z.T.; Saha, S.; Scheffler, B.E.; et al. Molecular diversity and association mapping of fiber quality traits in exotic *G. hirsutum* L. germplasm. *Genomics* **2008**, *92*, 478–487. [CrossRef] [PubMed]
66. Grover, C.E.; Gallagher, J.P.; Jareczek, J.J.; Page, J.T.; Udal, J.A.; Gore, M.A.; Wendel, J.F. Re-evaluating the phylogeny of allopolyploid *Gossypium* L. *Mol. Phylogenet. Evol.* **2015**, *92*, 45–52. [CrossRef] [PubMed]
67. Noormohammadi, Z.; Farahani, Y.H.A.; Sheidai, M.; Baraki, S.G.; Alishah, O. Genetic diversity analysis in Opal cotton hybrids based on SSR, ISSR, and RAPD markers. *Genet. Mol. Res.* **2013**, *12*, 256–269. [CrossRef] [PubMed]
68. Noormohammadi, Z.; Sheidai, M.; Foroutan, M.; Alishah, O. Networking and Bayesian analyses of genetic affinity in cotton germplasm. *Nucleus* **2015**, *58*, 33–45. [CrossRef]
69. Wu, J.F.; Wang, Y.; Wu, S.H. Two New Clock Proteins, LWD1 and LWD2, Regulate Arabidopsis Photoperiodic Flowering. *Plant Physiol.* **2008**, *148*, 948–959. [CrossRef] [PubMed]
70. Biedermann, S.; Hellmann, H. The DDB1a interacting proteins ATCSA-1 and DDB2 are critical factors for UV-B tolerance and genomic integrity in Arabidopsis thaliana. *Plant J.* **2010**, *62*, 404–415. [CrossRef] [PubMed]
71. Strohmaier, H.; Spruck, C.H.; Kaiser, P.; Won, K.A.; Sangfelt, O.; Reed, S.I. Human F-box protein hDcd4 targets cyclin E for proteolysis and is mutated in a breast cancer cell line. *Nature* **2001**, *413*, 316–322. [CrossRef] [PubMed]
72. Brown, W.L. Genetic diversity and genetic vulnerability: An appraisal. *Econ. Bot.* **1983**, *37*, 4–12. [CrossRef]

73. Flint-Garcia, S.A.; Anne-Ce line, T.; Yu, J.; Pressoir, G.; Romero, S.M.; Mitchell, S.E.; Doebley, J.; Kresovich, S.; Goodman, M.M.; Buckler, E.S. Maize association population: A high resolution platform for quantitative trait locus dissection. *Plant J.* **2005**, *44*, 1054–1064. [CrossRef] [PubMed]
74. Zhang, J.; Stewart, J.M. Economical and rapid method for extraction cotton genomic DNA. *J. Cotton Sci.* **2000**, *4*, 193–201.
75. Zhang, J.; Guo, W.; Zhang, T. Molecular linkage map of allotetraploid cotton (*Gossypium hirsutum* L. X *Gossypium barbadense* L.) with a haploid population. *Theor. Appl. Genet.* **2002**, *105*, 1166–1174. [CrossRef] [PubMed]
76. Liu, K.; Muse, S.V. PowerMarker: An integrated analysis environment for genetic marker analysis. *Bioinformatics* **2005**, *21*, 2128–2129. [CrossRef] [PubMed]
77. Anderson, J.A.; Churchill, G.A.; Autrique, J.E.; Tanksley, S.D.; Sorrellis, M.E. Optimizing parental selection for genetic linkage maps. *Genome* **1993**, *36*, 181–186. [CrossRef] [PubMed]
78. XLSTAT. *Data Analysis and Statistical Solutions for Microsoft Excell*; Addinsoft: Paris, France, 2014.
79. Pritchard, J.K.; Stephens, M.; Donnelly, P. Inference of population structure using multilocus genotype data. *Genetics* **2000**, *155*, 945–959. [PubMed]
80. Falush, D.; Stephens, M.; Pritchard, J.K. Inference of population structure using multilocus genotype data: Linked loci and correlated allele frequencies. *Genetics* **2003**, *164*, 1567–1587. [PubMed]
81. Evanno, G.; Regnaut, S.; Goudet, J. Detecting the number of clusters of individuals using the software STRUCTURE: A simulationstudy. *Mol. Ecol.* **2005**, *14*, 2611–2620. [CrossRef] [PubMed]
82. Dent, A.E.; Bridgett, M.V. STRUCTURE HARVESTER: A website and program for visualizing STRUCTURE output and implementing the Evanno method. *Conserv. Genet. Resour.* **2012**, *4*, 359–361. [CrossRef]
83. Liu, K.J.; Goodman, M.; Muse, S.; Smith, J.S.; Buckler, E.; Doebley, J. Genetic structure and diversity among maize inbred lines as inferred from DNA microsatellites. *Genetics* **2003**, *165*, 2117–2128. [PubMed]
84. Wang, K.; Wang, Z.; Li, F.; Ye, W.; Wang, J.; Song, G.; Yue, Z.; Cong, L.; Shang, H.; Zhu, S.; et al. The draft genome of diploid *Gossypium raimondii*. *Nat. Genet.* **2012**, *44*, 1098–1103. [CrossRef] [PubMed]
85. Magwanga, R.O.; Lu, P.; Nyangasi Kirungu, J.; Diouf, L.; Dong, Q.; Hu, Y.; Cai, X.; Xu, Y.; Hou, Y.; Zhou, Z.; et al. GBS Mapping and Analysis of Genes Conserved between Gossypium tomentosum and Gossypium hirsutum Cotton Cultivars that Respond to Drought Stress at the Seedling Stage of the BC 2 F 2 Generation. *Int. J. Mol. Sci.* **2018**, *19*, 1614. [CrossRef] [PubMed]
86. Kumar, S.; Stecher, G.; Tamura, K. MEGA7: Molecular Evolutionary Genetics Analysis version 7.0 for bigger database. *Mol. Biol. Evol.* **2016**, *33*, 1870–1874. [CrossRef] [PubMed]
87. Priyam, A.; Woodcroft, B.J.; Rai, V.; Munagala, A.; Moghul, I.; Ter, F.; Gibbins, M.A.; Moon, H.K.; Leonard, G.; Rumpf, W.; et al. Sequenceserver: A modern graphical user interface for custom BLAST databases. *bioRxiv* **2015**. [CrossRef]
88. Saitou, N.; Nei, M. The neighbor joining method: A new method for reconstructing phylogenetic trees. *Mol. Bio. Evol.* **1987**, *4*, 406–425.
89. Felsenstein, J. Confidence limits on phylogenies: An approach using the bootstrap. *Evolution* **1985**, *39*, 783–791. [CrossRef] [PubMed]
90. Zuckerkandl, E.; Pauling, L. Evolutionary divergence and convergence in proteins. In *Evolving Genes and Proteins*; Bryson, V., Vogel, H.J., Eds.; Academic Press: New York, NY, USA, 1965; pp. 97–166.

© 2018 by the authors. Licensee MDPI, Basel, Switzerland. This article is an open access article distributed under the terms and conditions of the Creative Commons Attribution (CC BY) license (http://creativecommons.org/licenses/by/4.0/).

International Journal of
Molecular Sciences

Article

Substantial Epigenetic Variation Causing Flower Color Chimerism in the Ornamental Tree *Prunus mume* Revealed by Single Base Resolution Methylome Detection and Transcriptome Sequencing

Kai-Feng Ma [1], Qi-Xiang Zhang [1,2,*], Tang-Ren Cheng [1], Xiao-Lan Yan [3], Hui-Tang Pan [1] and Jia Wang [1]

1 Beijing Key Laboratory of Ornamental Plants Germplasm Innovation & Molecular Breeding, National Engineering Research Center for Floriculture, Beijing Laboratory of Urban and Rural Ecological Environment, Key Laboratory of Genetics and Breeding in Forest Trees and Ornamental Plants of Ministry of Education, School of Landscape Architecture, Beijing Forestry University, Beijing 100083, China; makaifeng@bjfu.edu.cn (K.-F.M.); chengtangren@163.com (T.-R.C.); htpan2000@163.com (H.-T.P.); wangjia8248@163.com (J.W.)

2 Beijing Advanced Innovation Center for Tree Breeding by Molecular Design, Beijing Forestry University, Beijing 100083, China

3 Mei Research Center of China, Wuhan 430074, China; important12@sina.com

* Correspondence: zqxbjfu@126.com; Tel.: +86-10-6233-6321

Received: 4 June 2018; Accepted: 2 August 2018; Published: 7 August 2018

Abstract: Epigenetic changes caused by methylcytosine modification participate in gene regulation and transposable element (TE) repression, resulting in phenotypic variation. Although the effects of DNA methylation and TE repression on flower, fruit, seed coat, and leaf pigmentation have been investigated, little is known about the relationship between methylation and flower color chimerism. In this study, we used a comparative methylomic–transcriptomic approach to explore the molecular mechanism responsible for chimeric flowers in *Prunus mume* "Danban Tiaozhi". High-performance liquid chromatography-electrospray ionization mass spectrometry revealed that the variation in white (WT) and red (RT) petal tissues in this species is directly due to the accumulation of anthocyanins, i.e., cyanidin 3,5-*O*-diglucoside, cyanidin 3-*O*-glucoside, and peonidin 3-*O*-glucoside. We next mapped the first-ever generated methylomes of *P. mume*, and found that 11.29–14.83% of the genomic cytosine sites were methylated. We also determined that gene expression was negatively correlated with methylcytosine level in general, and uncovered significant epigenetic variation between WT and RT. Furthermore, we detected differentially methylated regions (DMRs) and DMR-related genes between WT and RT, and concluded that many of these genes, including differentially expressed genes (DEGs) and transcription factor genes, are critical participants in the anthocyanin regulatory pathway. Importantly, some of the associated DEGs harbored TE insertions that were also modified by methylcytosine. The above evidence suggest that flower color chimerism in *P. mume* is induced by the DNA methylation of critical genes and TEs.

Keywords: DNA methylation; flower color chimera; bisulfate sequencing; transcriptome; comparative epigenomes; transposon; ornamental *Prunus mume*

1. Introduction

Prunus mume Sieb. et Zucc. ($2n = 2x = 16$), a well-known ornamental tree, is widely grown for its fruits and its abundant, colorful flowers with their unique fragrance [1]. Following the domestication of this species more than 3000 years ago in China, the cultivation of *P. mume* has spread widely to other countries in East Asia. Its petal color, which ranges from white to pale yellow, pink, red, and reddish-purple, determines the desirability and economic value of individual plants and is one of the central ornamental features attracting viewers and admirers. In the 1940s, varieties with a novel characteristic, flower color chimerism, were discovered and have since served as important materials for landscaping applications and genetic improvement [2]. Drawing on our breeding experience with these varieties, we recognize five types of flowers according to petal color patterns and arrangements on individual trees: (i) bicolored flowers, which are mostly white with some red-spotted or streaked petals; (ii) pure white flowers and (iii) pure red flowers, which are both found together in the same cluster; (iv) white flowers from branches bearing white flowers only; and (v) red flowers from branches bearing red flowers only. Although graft-propagated branches bearing chimeric flowers (i, ii, and iii) produce chimeric individuals, branches with only single-color flowers (iv and v) generate single-color clones. However, at present, little information is available on the genetic mechanisms of *P. mume* floral chimerism.

At the cytological level, the formation of plant chimeras is believed to be linked to genetic changes in primordial cells located in the apical meristem that then proliferate mechanically [3]. The resulting somatic cell lines contain pigments, including flavonoid, carotenoid, and betalain secondary metabolites, that can be directly visualized [4–7]. At the molecular level, chimeric variation was initially explained by the action of transposable elements (TEs, or transposons), or, more specifically, the *Activator/Dissociation* (*Ac/Ds*) system that regulates the mixture of purple and yellow pigments in maize kernels by activating or repressing a C group gene [8]. Much later, the insertions of TEs *Tpn1*, *Tpn2*, and *Tpn3* into structural genes *dihydroflavonolreductase-B* (*DFR-B*), *chalcone isomerase* (*CHI*), and *chalcone synthase-D* (*CHS-D*), respectively, were revealed to be the critical factors giving rise to the corresponding variegated color mutants *flecked*, *specked*, and *r-1* [9–11]. The integration of *Tpn4*, an *En/Spm*-related transposon, into the *purple-mutable* (*pr-m*) gene encoding a vacuolar Na^+/H^+ exchanger was found to be responsible for a mutation giving rise to purple flowers with blue sectors [12,13]. A similar functional mechanism has been linked to tobacco flower color [14]. Other examples include *Gret1* activation regulating the expression of *VvmybA1* to produce colorless grape skin [15]; *Tam1* transposon insertion in *Glycine max* [16]; and the insertion of either *Ty1dic1* or *Retdic1* in *AA5GT*, whose disruption prevents glycosylation at the 5 position of anthocyanins in *Dianthus caryophyllus* [17].

Other studies have uncovered evidence supporting a relationship between chimeric petals and the expression of structural/biosynthetic genes encoding enzymes and other regulatory factors involved in floral pigment biosynthesis and metabolism. For instance, *CHS*, *cinnamate-4-hydroxylase* (*C4H*), *flavanone 3-hydroxylase* (*F3H*), *DFR*, *anthocyanidin synthase* (*ANS*), and *UDP-glucose: flavonoid 3-O-glucosyltransferase* (*UFGT*) genes have been found to exhibit differential expression patterns between red and white flower petal tissues of individual higher plants [7,18–21]. The alternative splicing of *ANS* results in red flower petals [22], while the sequence-specific silencing of *CHS* generates white sectors in *Petunia hybrida* "Red Star" flowers [23]. Flower color variegation was also observed when the *regulator involved in anthocyanin transport* (*Riant*) gene, encoding a GST protein, was expressed while harboring an insertion–deletion polymorphism in exon 3 [24], and a TT2-like R2R3 MYB has been shown to regulate anthocyanin biosynthesis in flowering *P. persica* "Genpei" [25]. At the same time, epigenetic modification, such as the use of hypomethylated promoters of *A1*, *DFR-B*, and *OgCHS* genes driving the brick-red pelargonidin pigmentation of flower tissue [11,26,27], has been introduced to reveal genetic variation in variegated flowers.

With reference to the previous example, DNA methylation indeed appears to be one of the best-studied epigenetic modifications regulating eukaryotic growth and development [28–32] that also leads to morphological abnormalities in plants [33,34]. For instance, the extensive methylation and transcriptional silencing of a *Lcyc* gene leads to a fundamental change in floral symmetry, from bilateral to radial flowers [35], while methylated genes encoding MYB transcription factors are inversely associated with red and green-skinned fruits of apple and pear cultivars [36–38]. Importantly, DNA methylation modification is related to the silencing or reactivation of TEs that generally remain inactive [39–42]. However, little research has focused on the relationship between methylated TEs and chimeric traits.

In this study, we first assumed that the flower color chimerism of *P. mume* is associated with DNA methylation modification of structural genes or regulators, as well as methylated TEs, through the color regulation pathway. We performed transcriptome sequencing (RNA-seq) and advanced single base resolution methylome detection, which is a technique that has been used to elucidate fruit ripening in tomato [43], dynamic changes during seed development in soybean [44], photoperiodic sensitivity in cotton [45], and drought stress in cotton, apple, and rice [46–48], to examine three issues: (i) the methylome landscape of *P. mume*; (ii) differentially methylated region (DMR)-related genes contributing to pigment variation; and (iii) the question of whether TEs with DNA methylation modification contribute to bicolored flower formation.

2. Results

2.1. Variation in Pigmentation in White (WT) and Red (RT) Petal Tissues

Three types of anthocyanins, namely, cyanidin 3,5-*O*-diglucoside (Cy3,5G; 0.077 mg/g fresh weight), cyanidin 3-*O*-glucoside (Cy3G; 0.103 mg/g fresh weight), and peonidin 3-*O*-glucoside (Pn3G; 0.124 mg/g fresh weight), were detected in red petal tissue (RT) samples by high performance liquid chromatography-electrospray ionization-mass spectrometry (HPLC-ESI-MS). In contrast, no compounds similar to these secondary metabolite products were detected at an absorption wavelength of 520 nm in white petal tissue (WT) samples (Figure 1a,b and Figure S1).

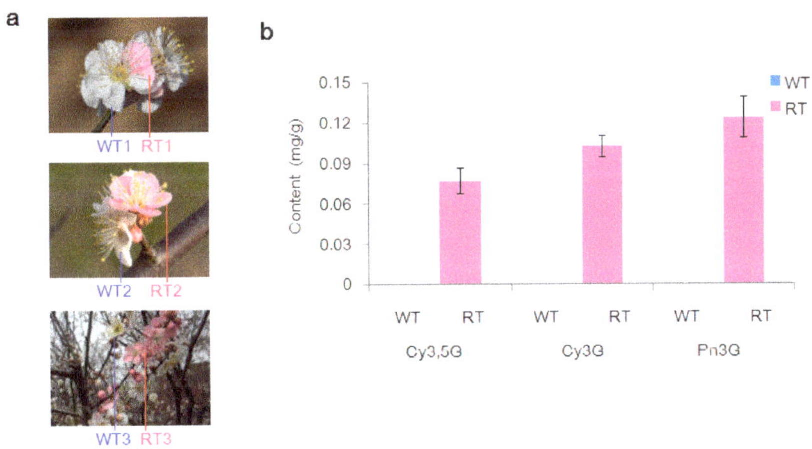

Figure 1. Color phenotypes and petal-tissue anthocyanin contents of flowers of *Prunus mume* "Danban Tiaozhi" collected in this study. (**a**) Examples of the six types of sampled petals. Samples WT1 and RT1 were respectively collected from white and red petals of bicolored flowers; WT2 and RT2 were collected from flowers with only white or red petals, respectively; WT3 and RT3 were collected from flowers on branches with only white flowers and only red flowers, respectively. (**b**) Anthocyanin content (mg/g fresh weight) of white petal tissue (WT) and red petal tissue (RT) samples. Cy3,5G, cyanidin 3,5-*O*-diglucoside; Cy3G, cyanidin 3-*O*-diglucoside; Pn3G, peonidin 3-*O*-glucoside.

2.2. Genome Methylation Landscape of *Prunus mume*

We used the BS-seq method, the "gold standard" of DNA methylation detection, to reveal the methylomes of six petal tissue groups from a single ornamental tree. In total, 40.0–50.9 million clean reads were generated, which corresponded to 11.6–14.6 Gb and greater than 41-fold coverage of the genome (estimated size = 280 Mb). To confirm the quality of the sequences, we also calculated QC20 (>97%) and QC30 (>92%) values, the bisulfite conversion rate (>99%), and GC content (21.37–21.54%). Approximately 59% of clean reads could be mapped to the reference genome, with a duplication rate of approximately 11.13–19.09% (Table S1). As revealed by sequencing coverage statistics, the maximum coverage was obtained at a sequencing depth of approximately 30× to 40× of the reference genome (Figure 2a), with each chromosome sequenced at a depth of around 24× to 34.2× (Figure S2). The coverage of cytosine sites, 8.8–10.95% of which were methylated, was reliably about 10.5× to 13.8×, with differing levels of mCG, mCHG, and mCHH sites (Figure S3, Table S2).

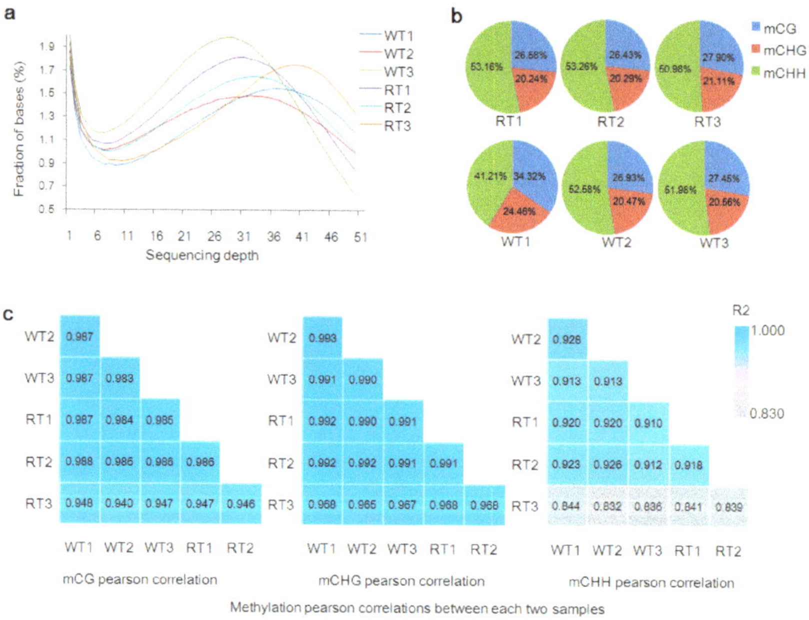

Figure 2. Genome sequencing coverage and methylation percentages and correlations between samples. WT1, WT2, and WT3 represent white petal tissue samples 1, 2, and 3, respectively. RT1, RT2, and RT3 represent red petal tissue samples 1, 2 and 3, respectively. (**a**) Distribution of the coverage of the six sequenced genomes. The *x*-axis represents sequencing depth, and the *y*-axis indicates the percentage of bases covered. (**b**) Percentage of methylated sites by sequence context (mCG, mCHG, and mCHH) relative to total methylated sites within each sample genome. (**c**) Pairwise correlations of methylation levels by sequence context (mCG, mCHG and mCHH) between sequenced genomes.

To assess the influence of non-methylated cytosine on BS-seq library construction and control for methylcytosine sequencing preference [49], methylated loci mapped to unique sites in the reference genome were detected after M-bias assessment (Figure S4). The percentage of methylated cytosine to total cytosine, mC/C, was 13.72%, and the ratios of mCG/CG, mCHG/CHG and mCHH/CHH sites were 38.04%, 20.42% and 9.21%, respectively (Table 1). The relative proportion of mCG, mCHG, and mCHH sites throughout the genome was 28.28% (26.43–34.32%), 21.19% (20.24–24.46%), and 50.53% (41.42–53.26%), respectively (Figure 2b). Meanwhile, there existed a tendency toward mCG and mCHG sites with highly methylation levels; 70–100% were methylated with a large proportion. However, the mCHH context showed the opposite trend with a low methylation level; 0–30% were methylated with a large proportion (Figure S5). An examination of logo plots exploring methylation preferences at sites and nearby regions revealed no significant variations in mC preference (Figure S6). Chromosome 3 had the highest number of methylated sites of any chromosome (Figure S7).

Table 1. Methylation of C contexts mapping to the reference genome. mC/C: methylated cytosine to total cytosine.

	C Site	mC (mC/C)	CG	mCG (mCG/CG)	CHG	mCHG (mCHG/CHG)	CHH	mCHH (mCHH/CHH)
WT1	82368087	9302812 11.29%	8335974	3193016 38.30%	11664623	2275905 19.51%	62367490	3833891 6.14%
WT2	82368087	11706910 14.21%	8335974	3153580 37.83%	11664623	2396865 20.54%	62367490	6156465 9.87%
WT3	82368087	10982193 13.33%	8335974	3015038 36.16%	11664623	2258108 19.35%	62367490	5709047 9.15%
WT-average	82368087	10663972 12.94%	8335974	3120545 37.43%	11664623	2310293 19.80%	62367490	5233134 8.39%
RT1	82368087	11609228 14.09%	8335974	3086648 37.02%	11664623	2350640 20.15%	62367490	6171940 9.89%
RT2	82368087	11991003 14.55%	8335974	3170368 38.03%	11664623	2434099 20.86%	62367490	6386536 10.24%
RT3	82368087	12222792 14.83%	8335974	3411185 40.92%	11664623	2580335 22.12%	62367490	6231272 9.99%
RT-average	82368087	11941008 14.49%	8335974	3222734 38.66%	11664623	2455025 21.04%	62367490	6263249 10.04%
Average	82368087	11302490 13.72%	8335974	3171639 38.04%	11664623	2382659 20.42%	62367490	5748192 9.21%

"WT1", "WT2" and "WT3" represent the white petal tissues; "RT1", "RT2" and "RT3" represent the red petal tissues.

Methylation levels of CG, CHG, and CHH sites were highly correlated ($R^2 > 0.83$) among different samples of the same petal color (Figure 2c), thus demonstrating that RT1–RT3 and WT1–WT3 could be analyzed as replicates of red petal tissue (RT) and WT samples, respectively. mCG, mCHG, and mCHH sites on chromosomes had similar distribution trends, with the most highly methylated regions in WT and RT corresponding to low-density genes (Figure 3a).

2.3. Levels of mC Variation in Different Colored Petal Tissues

The status and level of DNA methylation can vary between different individuals, tissues, and even genomic regions. We therefore explored variation in methylation between WT and RT samples in chromosomes as well as genomic regions. As shown in Figure 3b, the majority of strongly and weakly methylated regions occurred at similar chromosomal positions between the two samples. Calculated mCG/CG, mCHG/CHG, and mCHH/CHH ratios in the RT sample were 2.3–85%, 1.7–59%, and 0.4–4.3%, respectively. Many mC loci in RT were different from those in WT, where the mCG/CG, mCHG/CHG, and mCHH/CHH ratios were 2.8–83%, 1.7–61%, and 0.6–5%, respectively.

Although methylation levels in different genomic functional domains were higher in WT than in RT, they followed similar trends, with the highest and lowest methylation context ratios corresponding to mCG/CG and mCHH/CHH, respectively (Figure 3c). In the context of CG, methylation levels in exon and intron regions were higher than in promoter regions, but lower than in TE regions. For CHG and CHH, mCHG/CHG and mCHH/CHH ratios were lower in exon and intron regions than in promoter and transcriptional end site (TES) regions, although the difference in the mCHH/CHH ratio across various regions was not very pronounced. The highest methylation levels within gene bodies were at CG sites, whereas methylation levels in the ~2-kb upstream and downstream regions were highest at the CHG sites (Figure 3d). The lowest methylation levels, including at CG and CHG sites, were observed within regions adjacent to transcriptional start sites (TSSs) and TESs.

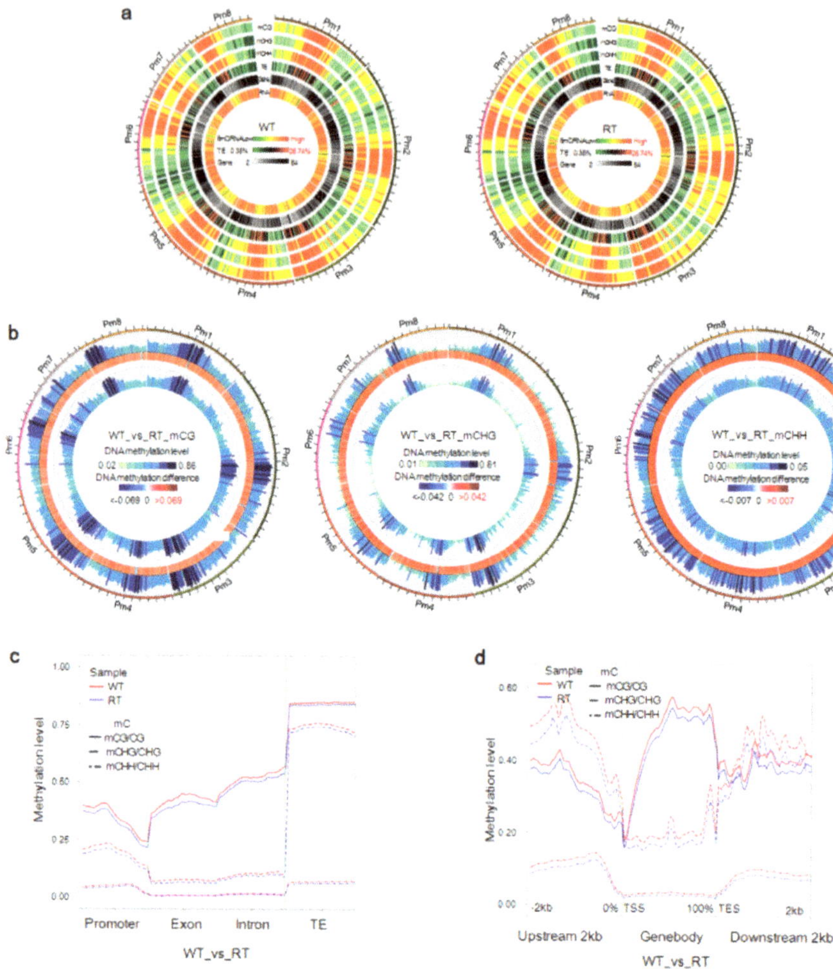

Figure 3. Epigenomic landscape and distribution of DNA methylation in white petal tissues (WT) and red petal tissues (RT) of *Prunus mume*. (**a**) Circos plots of chromosomes in *P. mume* WT and RT samples. The track order (from outside to inside) is as follows: (1–3) methylation density by sequence context (mCG, mCHG, and mCHH); (4) transposable element (TE) density; (5) gene density; and (6) number of reads generated by transcriptome sequencing. (**b**) Circos plots of differences in DNA methylation levels by methylation context (mCG, mCHG, and mCHH) in WT (outer track) and RT (inner track) genomes. The middle track in each plot indicates the degree of DNA methylation-level differences. (**c**) Distribution of DNA methylation levels within gene functional regions. The *x* and *y*-axes indicate gene functional domains and methylation levels, respectively. (**d**) Distribution of DNA methylation levels within gene-body domains and 2-kb upstream and downstream regions (TSS, transcriptional start site; TES, transcriptional end site). The *x* and *y*-axes indicate gene functional domains and methylation levels, respectively.

2.4. DMR-Related Genes and Bicolored Flowers on Individual Trees

DMRs can be used to identify methylation differences between individuals or developmental stages and their involvement in gene transcriptional regulation [50,51]. In the present study, 13,468 DMRs—10,121 hypermethylated and 3347 hypomethylated—were predicted between WT and RT types using DSS software. These DMRs, whose lengths were normally distributed (Figure S8a), were methylated at levels of approximately 38.5% (CG), 31.5% (CHG), and 8.4% (CHH) (Figure S8b). DMR chromosomal distributions and levels of significance are displayed in Figure 4a and Figure S8c. In each gene functional region (except for other regions in CG and CHG contexts), the number of DMRs between WT and RT exhibiting hypermethylation was higher than those with hypomethylation. In the CHH context, many more hypermethylated DMRs were detected than hypomethylated ones, and the TSS and TES regions contained few DMRs (Figure 4b). Heat maps of methylation levels of CG, CHG, and CHH DMRs revealed that variation was present in the methylation of WT and RT samples (Figure 4c). As shown in Figure 4d, these DMRs were overlapped with 4376 gene bodies and 4622 gene promoters. A total of 80 genes containing mCG, mCHG, and mCHH sites in their transcribed region were detected; similarly, 57 gene promoters were predicted containing all three types of methylated sites.

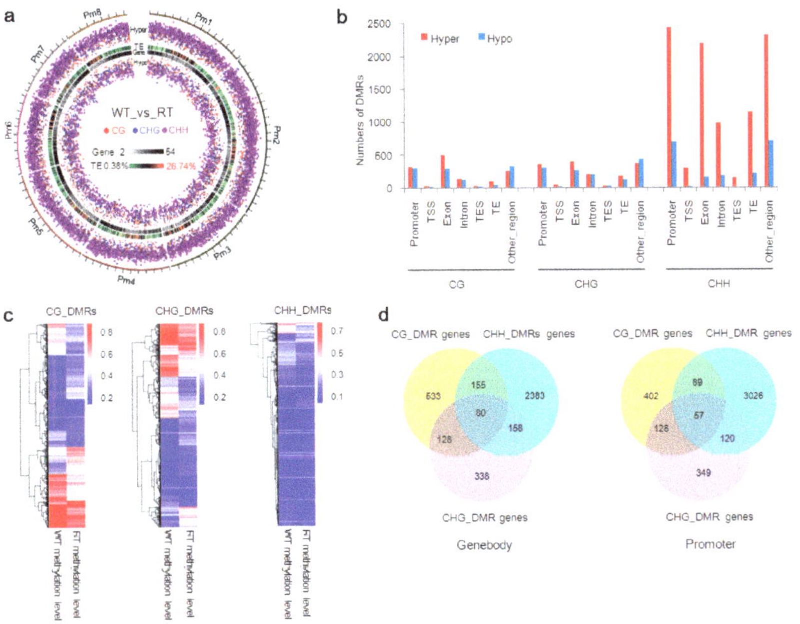

Figure 4. Distribution, methylation level, and predicted genes of differentially methylated regions (DMRs) in white petal tissues (WT) and red petal tissues (RT) samples. (**a**) Circos plots of CG, CHG, and CHH DMRs and transposable element (TE) and gene densities on each chromosome of *Prunus mume*. Track order (outside to inside) is as follows: scatter plot of hypermethylation (Hyper); TE density (TE); gene density (Gene); and scatter plot of hypermethylation (Hypo). Red, blue, and purple dots indicate CG, CHG, and CHH DMRs, respectively. (**b**) Distribution of CG, CHG, and CHH DMRs within gene functional regions. TSS, transcriptional start site; TES, transcriptional end site. (**c**) Heat maps of methylation levels of CG, CHG, and CHH DMRs. (**d**) Venn diagrams of predicted genes linked with CG, CHG, and CHH DMRs. "Genebody" and "promoter" indicate predicted genes anchored within gene body and promoter regions, respectively.

Following DMR detection, Gene Ontology (GO) and Kyoto Encyclopedia of Genes and Genomes (KEGG) annotations were performed to explore the functions of DMR-related genes. These analyses uncovered the enrichment of 896 CG–DMR-anchored genes (543 hypermethylated and 361 hypomethylated), 704 CHG–DMR-anchored genes (432 hypermethylated and 288 hypomethylated), and 3777 CHH–DMR-anchored genes (2531 hypermethylated and 292 hypomethylated). In addition, 696, 654, and 3292 genes with promoters overlapping with CG, CHG, and CHH DMRs, respectively, were found to be enriched. The identified genes have important molecular functions in various biological processes, especially phenylalanine metabolism and the biosynthesis of phenylpropanoids, carotenoids, flavonoids, and plant hormones. Since these are critical processes in flower color formation, their over-representation among DMR-anchored genes suggests that variation in methylation levels of DMRs affected the color of WT and RT samples (Figures S9–S12).

2.5. Analysis of Differential Transcription between WT and RT

To reveal expression differences between WT and RT, we compared the transcriptomes of the six petal tissue samples. RNA-seq generated 44,202,732–55,656,090 clean reads per sample, of which 38,737,776–49,062,666 (88%; 85% unique) were mapped to the reference genome. Approximately half of the unique reads were mapped to the positive-sense strand of chromosomes (Figure S13, Table S3), and 294 transcription factors were detected (Table S4). Gene expression levels were distributed similarly, and significantly positively correlated among the sample genomes ($R^2 > 0.93$) (Figure S14a,b). In the next step, the six samples were divided into two groups—WT (WT1, WT2, and WT3) and RT (RT1, RT2, and RT3)—for further analysis.

In this analysis, 16,383 expressed genes, 221 WT-specific and 765 RT-specific, were detected using a FPKM > 1 threshold (Figure S14c), with FPKM standing for fragments per kilobase of exon per million fragments mapped. Screening with DESeq yielded 958 upregulated and 1134 downregulated differentially expressed genes (DEGs) widely distributed across WT and RT genomes (Figure 5a,b). These DEGs belonged to 55 functional groups, including 36 biological process, 14 cellular components, and five molecular function categories. Many of the enriched genes take part in metabolic processes (Figure S14d). KEGG pathway analysis was used to further explore DEGs associated with anthocyanin biosynthesis and metabolic regulation (Figure 5c). We found that the genes were enriched mainly in pathways controlling plant hormone signal transduction (ko pmum04075), biosynthesis of secondary metabolites (ko pmum01110), metabolism (ko pmum01100), and flavonoid biosynthesis (ko pmum00941). Structural genes, including *Pm020453* (*YUCCA8*), *Pm004176* (*ANGLT*), *Pm031359* (*UGT79B6*), *Pm006139* (*GSTF1*), *Pm011195* (*GSTXC*), *Pm012985* (*GSTF7*), and *Pm025127* (*GSTX6*), were downregulated, but other critical genes, such as *Pm013782* (*DFRA*), *Pm018402* (*DFRA*), *Pm023202* (*DFRA*), *Pm017146* (*FLRT*), and *Pm008680* (*UGFGT*), were upregulated (Supplementary Data 1). In addition, 189 transcription factor genes, including *MYB*, *bHLH*, and *WD*, were detected using iTAK software (Supplementary Data 2), which suggests that transcription factors take part in the regulation of bicolored flowers.

2.6. Correlation between Gene Expression Levels and DNA Methylation

The DNA methylation of genes is correlated with changes in their expressions. To reveal the potential roles of expression level differences in WT versus RT regulated by DNA methylation, we analyzed the distribution of gene methylation and expression levels within each chromosome and gene functional region. As shown in Figure 3a, highly expressed genes were distributed in regions of low methylcytosine density. This negative correlation between gene expression levels and DNA methylation in general was apparent on scatter plots and heat maps (Figure S15).

To analyze the relationship between levels of expression and methylation within gene bodies (including 2-kb upstream and downstream domains), we divided genes into high, medium, low, and no expression groups [high expression (FPKM $\geq$ FPKM_75%); medium expression (FPKM_25% $\leq$ FPKM < FPKM_75%); low expression (1 $\leq$ FPKM < FPKM_25%); and no expression

(FPKM < 1). FPKM stands for fragments per kilobase of exon per million fragments mapped, and FPKM_25% and FPKM_75% refer to values at the boundary of the 25th and 75th percentiles of expression levels, respectively]. And their methylation levels were plotted as a function of location [48]. As shown in Figure 6, the non-expressed genes displayed high methylation levels at CG sites within domains that are 2-kb upstream or downstream of the gene body and at CHG sites in all of the gene regions. Within the gene body and 2-kb downstream regions, non-expressed genes showed high methylation levels at CHH sites, but gene expression was positively correlated with CG methylation within the gene body. Similarly, a positive correlation was detected between gene expression and CHH methylation levels within 2-kb upstream of the TSS. Interestingly, two peaks in CG and CHG methylation levels were observed in the 2-kb upstream and downstream regions of highly expressed genes.

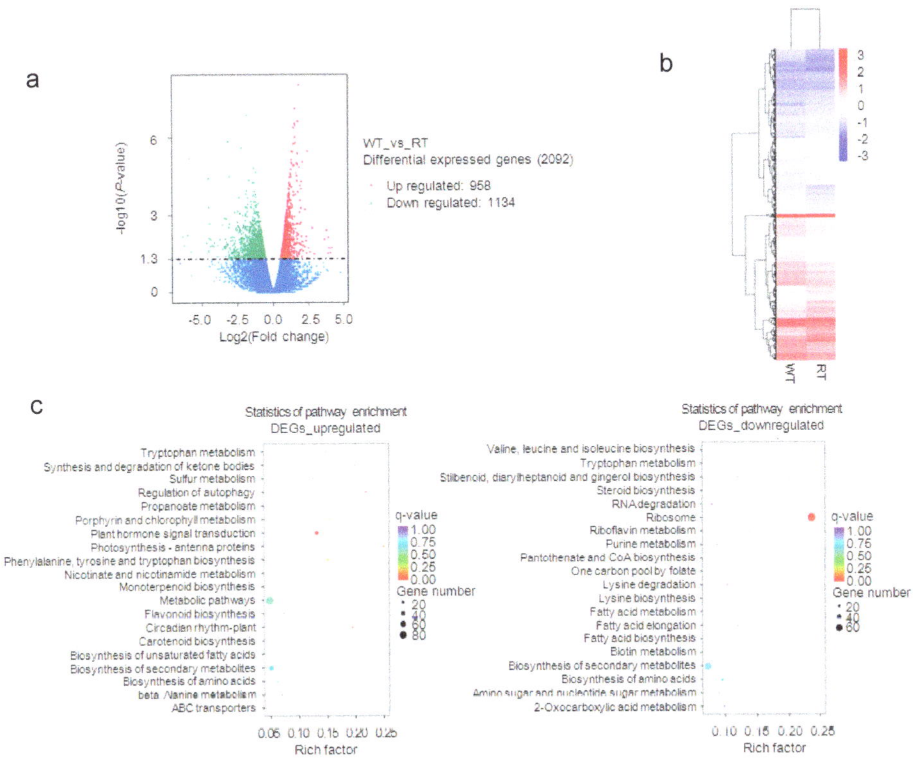

Figure 5. Analysis of differentially expressed genes (DEGs). (**a**) Detection and distribution of genes differentially expressed between white petal tissues (WT) and red petal tissues (RT). Green and red dots indicate upregulated and downregulated DEGs, respectively. (**b**) Cluster dendrogram of DEGs. (**c**) Kyoto Encyclopedia of Genes and Genomes (KEGG) pathway enrichment of upregulated and downregulated DEGs.

Another statistical approach, projecting methylation levels onto a coordinate system of expression levels (*x*-axis) and frequency (*y*-axis), was used to further explore gene expression and DNA methylation (Figure 7). We divided the methylation levels of gene bodies and promoters into five groups according to Xu et al. [48], and found that DNA methylation and gene expression levels were negatively correlated in most genes. In contrast, genes whose promoters lacked mCHH sequences were only weakly expressed (Figure 7); this situation may have been responsible for the positive correlation between gene expression and methylation levels within 2-kb upstream regions (Figure 6).

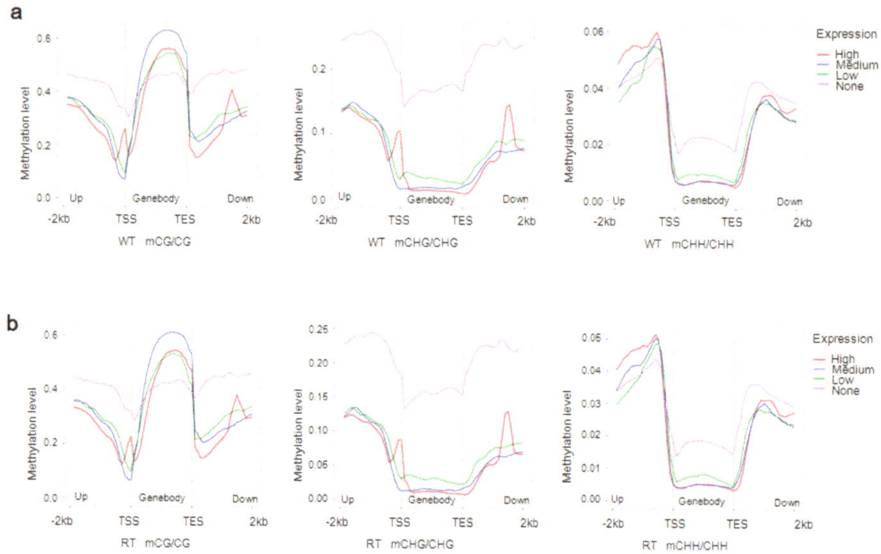

Figure 6. Distributions of methylation levels within gene bodies and their 2-kb upstream and downstream regions. Genes were classified into four groups according to their expression levels: no expression (pink; FPKM <1); low expression (green; 1≤ FPKM < FPKM_25%); medium expression (blue; FPKM_25% ≤ FPKM < FPKM_75%); and high expression (red; FPKM ≥ FPKM_75%). FPKM stands for fragments per kilobase of exon per million fragments mapped, and FPKM_25% and FPKM_75% refer to values at the boundary of the 25th and 75th percentiles of expression levels, respectively. Next, the different gene regions (gene body and 2-kb upstream and downstream) were divided into 50 bins, and the methylation levels of each were averaged. WT and RT indicate white petal tissues and red petal tissues, respectively. The *x* and *y*-axes represent gene body regions and DNA methylation levels by sequence context (mCG, mCHG, and mCHH), respectively. (**a**,**b**) Distributions of methylation levels within gene bodies and 2-kb upstream and downstream regions in samples WT [(**a**); FPKM_25% = 4.60, FPKM_75% = 45.31] and RT [(**b**); FPKM_25% = 4.91, FPKM_75% = 45.62].

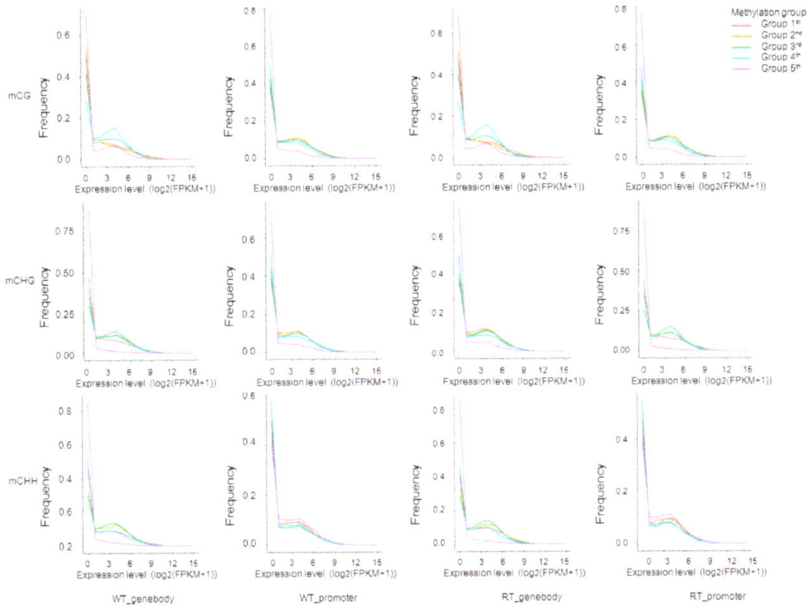

Figure 7. Frequencies of expressed genes anchored by methylation sites within gene bodies and promoters. For each methylation sequence context (mCG, mCHG and mCHH), expressed genes were divided into five groups according to methylation level: group 1 (red; methylation level < level_20%); group 2 (yellow-green; level_20% ≤ methylation level < level_40%); group 3 (green; level_40% ≤ methy_level < level_60%); group 4 (blue; level_60% ≤ methy_level < level_80%); group 5 (pink; methy_level ≥ level_80%). Level_20%, _40%, _60%, and _80% represent values at the boundaries of the 20th, 40th, 60th, and 80th percentiles of methylation levels. WT and RT indicate white petal tissues and red petal tissues, respectively. The x and y-axes represent gene expression levels and gene frequencies, respectively.

2.7. DEGs with Methylation Modification and DMR-Related Gene Expression

Analysis of DEG DNA methylation levels revealed large differences between WT and RT (Figure 8). In WT samples, high methylation levels were observed within DEGs, including downregulated and upregulated ones (i.e., involving promoters), at CG, CHG, and CHH sites. In contrast, DEGs in RT samples exhibited low methylation levels, with especially low levels observed at CHG and CHH sites within gene body domains. We also observed that DEG expression level differences between WT and RT samples were positively correlated with ratios of mCG/CG ($r = 0.037$, $p < 0.01$), mCHG/CHG ($r = 0.046$, $p < 0.01$), and mCHH/CHH ($r = 0.037$, $p < 0.01$) in gene body regions. However, in 2-kb downstream regions, a negative correlation ($r = -0.038$, $p < 0.01$) was observed between CHG methylation and expression level differences between WT and RT (Figure S16).

Next, an expression analysis of DMR-related genes (located in both gene body and promoter regions) identified 1154 mCG–DMR-related genes, 852 mCHG–DMR-related genes, and 4282 mCHH–DMR-related genes. Hypomethylated DMR-related genes were mostly associated with highly expressed genes, whereas the hypermethylation of DMR-related genes was associated with decreased expression levels (Figure 9). Methylation levels of DMR-related genes were negatively correlated with their expression levels with one exception: hypomethylation at CG sites within gene body regions was positively correlated with gene expression (Figure 10).

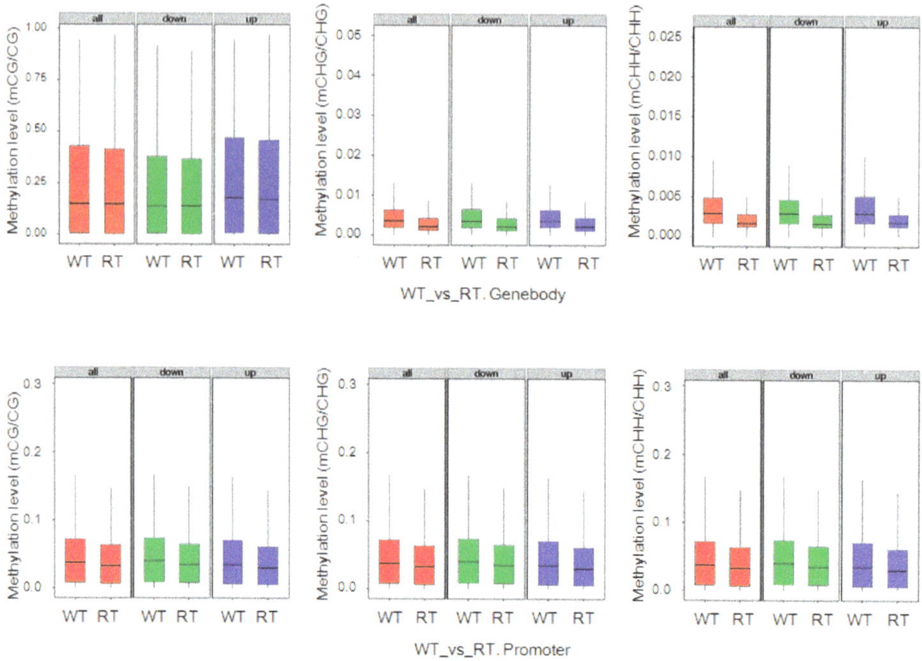

Figure 8. Methylation levels of differentially expressed genes (DEGs) by sequence context (mCG, mCHG, and mCHH) and gene region in white petal tissues (WT) and red petal tissues (RT). The terms "all", "up", and "down" indicate all, upregulated, and downregulated DEGs, respectively. DNA methylation levels within gene bodies and promoters of genes differentially expressed between WT and RT are shown ("WT vs. RT. Genebody" and "WT vs. RT. Promoter", respectively).

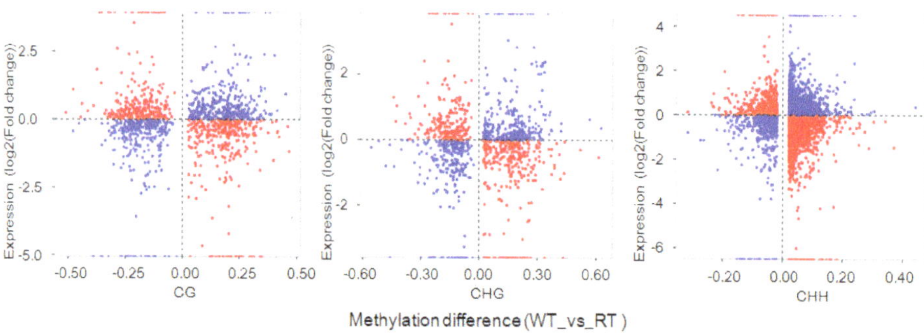

Figure 9. Distributions of differential expression of differentially methylated region (DMR)-related genes between white petal tissues (WT) and red petal tissues (RT). CG, CHG, and CHH refer to CG-DMR, CHG-DMR, and CHH-DMR-related genes, respectively. The *x*-axis indicates methylation-level differences, and the *y*-axis indicates gene differential expression levels. Red dots represent hypermethylated (hypomethylated) DMR-related genes associated with downregulated (upregulated) expression. Blue dots represent hypomethylated (hypermethylated) DMR-related genes associated with upregulated (downregulated) expression.

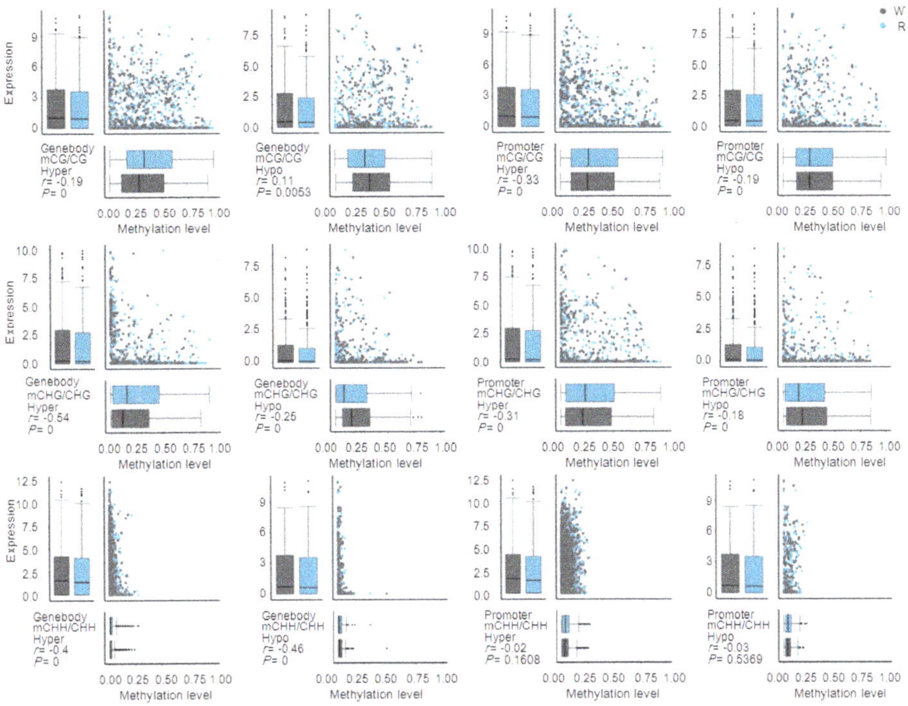

Figure 10. Combined maps of correlations between differentially methylated region (DMR)-related genes and expression levels. Each of the 12 subfigures is divided into four sections: upper left, box plot of DMR-related gene expression; upper right, scatter plot of DMR-related gene expression vs. methylation level; bottom left, comparison and correlation statistics; bottom right, box plot of DMR-related gene methylation level. WT and RT indicate white petal tissues (gray) and red petal tissues (blue), respectively.

2.8. Gene Ontology (GO) and Kyoto Encyclopedia of Genes and Genomes (KEGG) Enrichment of DMR-Related Genes Associated with DEGs

To investigate the effect of DNA methylation on genes with unique functions, we performed GO and KEGG enrichment analyses of gene sets generated from an association analysis between DMR-related genes and 506 DEGs. We first identified 285 DMR-related genes associated with gene-body domains of downregulated or upregulated DEGs. Of these 285, 126, and 125 hypermethylated DMR-related genes were associated with upregulated and downregulated DEGs, respectively, whereas 27 and 18 hypomethylated DMR-related genes were associated with upregulated and downregulated DEGs, respectively. In addition, six upregulated and five downregulated DEGs overlapped with both hypermethylated and hypomethylated DMR-related genes (Figure 11). We also discovered that 282 associated DEGs, 61 of which also overlapped with DMR-related genes within gene body regions, were hypermethylated (213) or hypomethylated (83) within promoter domains. A total of 43 and 40 DMR-related genes that were hypomethylated within promoter regions were associated with downregulated and upregulated DEGs, respectively; these numbers were larger than the number of genes hypomethylated within promoter domains (Figure S17).

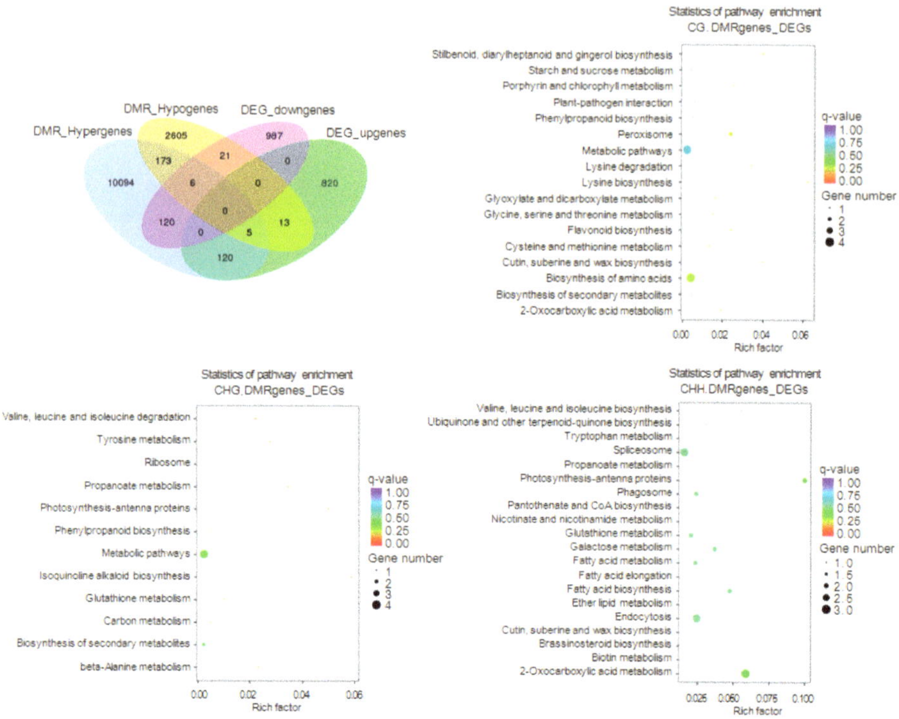

Figure 11. Venn diagram and KEGG pathway enrichment of differentially methylated region (DMR)-related differentially expressed genes (DEGs). DEGs are anchored by DMRs within gene body regions. DMR_Hypergenes and DMR_Hypogenes indicate hypermethylated and hypomethylated DMR-related genes, respectively. DEG_downgenes and DEG_upgenes indicate downregulated and upregulated DEGs, respectively. CG.DMRgenes, CHG.DMRgenes, and CHH.DMRgenes refer to CG-DMR, CHG-DMR, and CHH-DMR-related genes, respectively.

GO enrichment analysis was then carried out to functionally classify the DEGs associated with DMR-related genes into biological process, cellular component, and molecular function categories (Figure S18). We also performed a KEGG analysis on the associated DEGs, which revealed the enrichment of the following pathways that play important roles in flower color formation: flavonoid biosynthesis (ko pmum00941), the biosynthesis of secondary metabolites (ko pmum01110), phenylpropanoid biosynthesis (ko pmum00940), and plant hormone signal transduction (ko pmum04075) (Figure 11, Figures S17, S19 and S20). Examination of methylation level modifications of key structural and transcription factor genes differentially expressed between WT and RT indicated that *Pm031359* (*UGT79B6*) was hypermethylated at CHH sites in exon regions, and *Pm027422* (*YUCCA2*) and *Pm008425* (*MYB305*) were hypermethylated at CHH sites in the promoters of downregulated genes. In upregulated DEGs, *Pm008680* (*UFGT*) and *Pm019063* (*CHI*) were hypomethylated at CG and CHH sites within exons and CG sites within promoters. Hypermethylation was also detected in promoter and intron regions of the *Pm013782* (*DFRA*) gene (Supplementary Data 1).

2.9. Detection of Methylated TEs Regulating Candidate Gene Expression

At the global genome level, TE density was heaviest in genomic regions with a high methylation density and lightest in areas of low gene density and low gene expression (Figure 3a). Most long terminal repeat (LTR) TEs were located in 2-kb upstream and downstream regions of gene body domains, while *Helitron* TEs were mainly scattered throughout the gene body and long interspersed nuclear elements (LINE) were approximately evenly distributed (Figure 12a). Class-II TEs in intragenic regions were the most heavily methylated of the intergenic and intragenic class-I and class-II TEs. In addition, more methylated sites were found within TEs in the WT sample than in the RT sample. An analysis of the methylation sequence context of TEs revealed that mCG and mCHG accounted for a large proportion of methylated sites, especially within intragenic areas (Figure 12b).

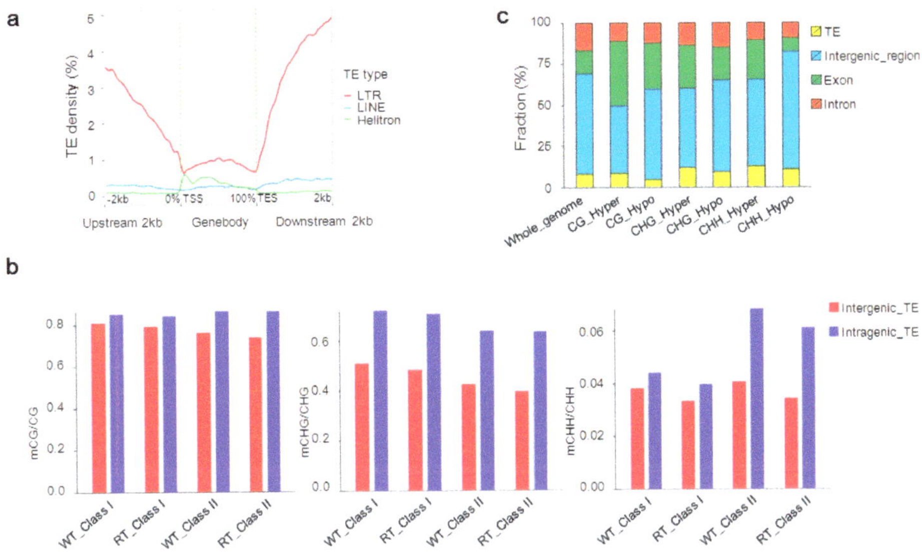

Figure 12. Methylation of transposable elements (TEs) in the genomes of *Prunus mume*. (**a**) Distribution of TEs density within gene body and 2-kb upstream and downstream domains. (**b**) Distribution of methylated TEs according to sequence context (mCG, mCHG, and mCHH) in intergenic (red) and intragenic (blue) domains. WT and RT indicate white petal tissues and red petal tissues, respectively. Class I and class II refer to class-I and class-II TEs, respectively. (**c**) Genome-wide length percentages of differentially methylated regions (DMRs) within TEs and other different functional regions (leftmost bar) and the methylome-wide percentage distributions of DMRs with different methylation types in different functional regions.

A total of 1833 DMRs (1437 hypermethylated and 396 hypomethylated) comprising 158 mCG, 302 mCHG, and 1373 mCHH sequences were detected within TEs between WT and RT samples. Hypermethylated mCG, hypomethylated mCG, hypermethylated mCHG, hypomethylated mCHG, hypermethylated mCHH, and hypomethylated mCHH within TEs accounted for 8.3%, 4.8%, 11.6%, 9.2%, 12.6% and 10.7%, respectively, of DMRs in genomic functional regions (Figure 12c). Among these DMRs, 197 were located in 196 TEs inserted into 106 DEGs associated with DMR-related genes (Supplementary Data 3). The gene *Pm008680* (*UGFGT*), with three hypomethylated-mCHH *Copia* TE insertions, was upregulated in WT relative to RT, as was *Pm013782* (*DFRA*) carrying three inserted hypermethylated TEs (two hypermethylated-mCHG *L1* and one hypermethylated-mCHG *Copia* elements). In contrast, the *Pm031359* (*UGT79B6*) gene, harboring a hypermethylated-mCHH *Copia* TE insertion in its 2-kb upstream region, and the *Pm031359* (*UGT79B6*) gene, with a hypermethylated-mCHG *Helitron* TE insertion in its 2-kb downstream region, were both downregulated (Supplementary Data 1).

3. Discussion

Since variation in petal color is highly prized in ornamental plant species, *P. mume* trees bearing chimeric red and white flowers have been selected during the process of genetic improvement. Floral chimerism is caused by the spatially and temporally restricted deposition of plant pigments, for instance, secondary metabolites of anthocyanins [7,25,37,52,53]. In our study, we detected Cy3G, Cy35G, and Pn3G in RT samples by HPLC-MS. The implied function of these compounds as determinants of color in red petal tissues suggests that genes related to the anthocyanin regulatory pathway, as revealed by an association analysis between transcriptomes and methylomes, are the molecular basis of chimerism in *P. mume* "Danban Tiaozhi".

3.1. *Genes within Anthocyanin Regulation Pathway Were Differentially Expressed*

Numerous molecular studies have revealed the processes controlling the genetics of plant organ coloration, which include differences in expressions of structural and transcription factor genes. For example, Han et al. [54] have reported that the downregulation of *CHI* and *DFR* by *anthocyanidin reductase* (*ANR*) results in yellow-skinned apple fruits. In bicolored flowers of *Petunia hybrida*, the enhancement of *CHS* expression induces blue or red coloration, respectively, in blue–white or red–white sectors of variegated flowers [5,23]. RNA-Seq technology has revealed that the significant upregulation of annotated anthocyanin biosynthetic genes *CHS*, *F3H*, *F3H*, *DFR*, *leucoanthocyanidin dioxygenase* (*LDOX*), *ANS*, and/or *UF3GT* is responsible for purple flesh coloration in a *Dioscoreaalata* cultivar, bicolored tepals in lily, and variegated petals in *P. mume* "Fuban Tiaozhi" [7,20,53]. In our study, we found that the anthocyanin biosynthetic gene homologs *Pm020453* (*YUCCA8*), *Pm004176* (*ANGLT*), and *Pm031359* (*UGT79B6*) were upregulated in red petal tissues, but other critical genes, such as *Pm013782* (*DFRA*), *Pm018402* (*DFRA*), *Pm023202* (*DFRA*), *Pm017146* (*FLRT*), and *Pm008680* (*UGFGT*) were downregulated. These differences between *P. mume* "Danban Tiaozhi" and "Fuban Tiaozhi" may be due to the temporal and tissue-specific nature of transcriptome expression [7], which has different development stages, and its genotypes show a variety of gene expression [8]. We also discovered that *GST* genes, which encode proteins transporting cyanidins and/or anthocyanins to the tonoplast [55], were upregulated in red tissues. This result is similar to the finding of a previous study that the *Riant* gene encoding a GST protein induces red flower coloration in peach [24].

The identification and functional characterization of flavonoid-related R2R3-MYB transcription factors, which show active or repressive effects on anthocyanin biosynthetic genes, is important for revealing plant pigmentation [56]. A study of blood-fleshed *P. persica* uncovered a mechanism whereby *BLOOD* (*BL*) was the key gene for the blood-fleshed trait via its activation of *PpMYB10.1*, and the silencing of *BL* reduced anthocyanin pigmentation in maturing fruits [6]. The MYB10 promoter is more variable in *Malus* × *domestica* "Honeycrisp" than in "Royal Gala", which results in a more variable color pattern in the peel of the first cultivar [36]. A transcriptomic comparison of red and green-colored leaves of *P. persica* has identified a MYB transcription regulator, *PpMYB10.4*, whose transient expression induces anthocyanin accumulation [57].

In "Lollypop" Asiatic lilies, the transcriptional profiling of *LhMYB12* has revealed that the presence of bicolored tepals is controlled by the transcriptional regulation of anthocyanin biosynthetic genes [53]. Differential expression of *Peace* (*peach anthocyanin colour enhancement*, a *R2R3 MYB-like* gene) determines the pattern of flower coloration in variegated petals within individual trees of flowering *P. persica* "Genpei" [25]. In addition, jasmonate has been reported to regulate WD-repeat/bHLH/MYB complex-mediated anthocyanin accumulation in *Arabidopsis thaliana* [58]. Using iTAK software, we detected 189 differentially expressed transcription factor genes, including *MYB*, *bHLH*, and *WD*, suggesting that transcription factors may play important roles in the formation of chimeras in *P. mume* "Danban Tiaozhi".

3.2. Methylcytosine Modification Affected the Expression of Anthocyanin-Related Genes

Cytosine methylation is a common form of DNA modification that is closely interwoven with the process of gene transcription. Although important for the epigenetic regulation of endogenous genes, the extent to which this type of DNA modification regulates genomes remains elusive [59,60]. DNA methylation occurs at higher levels in heterochromatin than euchromatin, and performs specific functions across different species [43,48,61,62]. In this study, we mapped the first-reported methylomes of *P. mume* flower petals using the single base resolution technique. We also compared methylomes and transcriptomes to elucidate the relationship of methylcytosine and gene expression. We discovered that 11.29–14.83% of cytosine sites within genomes were methylated, and the patterns of methylation (mCG, mCHG and mCHH) were similar to those reported in *Arabidopsis* [60,63], rice [64,65], maize [66], soybean [44,67], tomato [43], apple [48], and cotton [45,47]; specifically, the highest and lowest levels of methylation were at the CG and CHH sites, respectively, with significant differences observed within both WT and RT tissues. We also found that the mCG/CG ratios were highest in exon and intron regions; in contrast, mCHG/CHG and mCHH/CHH ratios were lower in the exon and intron regions than in the promoter and TE regions.

Gene transcription is influenced or regulated by DNA methylation [59,68], with methylcytosine levels negatively correlated with gene expression in general [69]. In our study, we observed a similar pattern. We also found that methylcytosine levels of RT samples were lower than those of WT samples, which suggests that anthocyanidin-related genes are suppressed by the high levels of methylcytosine in WT genomes. However, a contradictory result was seen in gene body regions, where methylcytosine levels of hypomethylated CG-DMR-related genes were positively correlated with gene expression. This latter observation is consistent with the finding that genes methylated in transcribed regions are highly expressed and constitutively active in *A. thaliana* [60], and that CG methylation is often linked to increased gene expression [68,70,71].

Next, we searched for DMRs and investigated the expression of DMR-related genes (especially DEGs). We focused on DMRs in functional regions, as gene body methylation is conserved across species among constitutively expressed genes [60,64,71], and because methylation within promoters is highly tissue-specific in nature and strongly associated with transcriptional repression in plants [60,69,72]. We detected 13,468 DMRs associated with 4376 gene bodies (285 DEGs) and 4622 gene promoters (282 DEGs). These associated DEGs were enriched in KEGG terms such as flavonoid biosynthesis, the biosynthesis of secondary metabolites, phenylpropanoid biosynthesis, plant hormone signal transduction, and transcription factor activity. In previous studies, gene methylation has been observed to influence floral pigmentation. For instance, red pigmentation is observed in the flowers of the transgenic petunia line 17-R upon hypomethylation of the 35S promoter driving the *A1* gene [26], while mosaic red anthocyanin in lip crests, sepals, and petals of yellow flowers of *Oncidium* "Gower Ramsey" may be attributed to activation of the *OgCHS* gene resulting from the demethylation of the five-upstream promoter region [27]. As a third example, methylation levels of *MYB* genes are associated with the formation of red-skinned pears and apples [37,38]. Thus, we suggest that the methylation or demethylation of genes participating in the anthocyanin regulation pathway is responsible for flower color chimerism in *P. mume* "Danban Tiaozhi".

3.3. TEs with Methylcytosine Affected the Expression of Anthocyanin-Related Genes

Various evidence supports the role of TE insertions in gene expression changes and phenotypic variation in higher plants. In anthocyanin biosynthetic genes, for example, TE-induced insertions cause null mutations that result in variations in seed, peel, and flower coloration. The insertion and excision of the *Ds* transposon has given rise to variation in the size and intensity of colored spots in maize kernels, and a *Candystripe 1* insertion in the second intron domain of the *y-candystripe* allele has altered the pigmentation of the sorghum grain pericarp from solid red to variegated [8,73,74]. Similarly, the *TRANSPARENT TESTA8* (*BrTT8*) locus, encoding a bHLH protein, lost its function with the insertion of a *Helitron* transposon, resulting in yellow seeds in *Brassicarapa* [75]. The golden pigmentation in hulls and internodes of *Oryza sativa* mutants is due to the complete suppression of the *OsCHI* gene following insertion of a *Dasheng* retrotransposon into its 5′ untranslated region (UTR), while a retrotransposon insertion in the upstream sequence of the pigmentation-related gene *VvmybA1* is regarded as the molecular basis for white-skinned coloration in grape cultivars [15,76]. TE-mediated insertional mutations have also caused alterations in seed coat and flower color in both *Ipomoea purpurea* and *Glycine max* [77–80]. Transformed tobacco plants carrying an inserted *Tag1* element between the *CaMV* 35S promoter and the maize *R* gene have variegated flowers, and the insertion of either *Ty1dic1* or *Retdic1* transposons can disrupt the *AA5GT* (*acyl-glucose-dependent anthocyanin 5-O-glucosyltransferase*) gene to prevent glycosylation of the 5′ position of anthocyanins in *Dianthus caryophyllus* [14,17].

One unanswered question concerns how TEs are activated or repressed to ensure a stable phenotype. Emerging evidence is demonstrating that TEs are silenced or reactivated by epigenetic mechanisms such as DNA methylation modification [39,41,42,68,81,82]. In the model plant *A. thaliana*, the imprinted gene *FWA* is a flowering-time modifier, with its silencing dependent on the cytosine methylation of a SINE retro element in the promoter region [83]. Transposons in *Arabidopsis* are heavily methylated at both CG and non-CG sites, whereas non-CG methylation is rarely found in active genes [70]. In rice, the essential *chloroplast protease 5* (*OsClpP5*) gene with the insertion of an epigenetically silenced *autonomous DNA-based active rice transposon 1* (*aDart1*) may induce leaves to show variegation [84]. Similarly, in our study, we detected 197 DMRs associated with 196 TEs inserted into 106 DEGs. We annotated these DEGs with GO and KEGG functional terms and pathways. The genes *Pm008680* (*UGFGT*), *Pm031359* (*UGT79B6*), *Pm031359* (*UGT79B6*), *Pm013782* (*DFRA*), and *Pm011195* (*GSTXC*), all participating directly in the color regulation pathway and anthocyanin transport, were enriched. This result suggests that the insertion of methylcytosine-modified transposons affects the expression of anthocyanin genes, resulting in chimerism.

4. Materials and Methods

4.1. Plant Materials

White and red petal tissues were collected from fresh flower blossoms of the flower color chimera *P. mume* "Danban Tiaozhi" on a fine day between 9:00–11:00. The collected tissues were carefully sorted under subzero temperatures with fine-tipped tweezers into six groups according to their color and origin (Figure 1a). Tissues in the first two groups came from individual flowers having both white (WT1) and red (RT1) petal tissues. The second two groups individually consisted of white (WT2) and red (RT2) petal tissues from single-color (white or red) flowers that were located together in the same branches. The final two groups comprised petal tissues derived from separate branches bearing either white (WT3) or red (RT3) flowers only. To facilitate the use of these materials for biochemical detection and a variation analysis of single base resolution genomes, all of the tissues were harvested from a single ornamental tree, snap-frozen in liquid nitrogen, and stored at −80 °C.

4.2. Qualitative and Quantitative Analysis of Floral Pigments

Previously frozen petal tissues were ground into fine powder in liquid nitrogen. Next, 200.0-mg portions of individual petal tissue samples were placed in centrifuge tubes containing extraction reagent (70:27:2:1 (*v*/*v*/*v*/*v*) methanol-water-formic acid-trifluoroacetic acid) [85] and vortexed for 1 min at room temperature followed by ultrasonic extraction in a KQ 2200B ultrasonic cleaner (Jiangsu, China) at 4 °C for 15 min. After storage overnight at −20 °C, the mixtures were centrifuged (A-14C, Sartorius, Goettingen, Germany) for 10 min at 12,000 rpm and 4 °C. Supernatants were collected and filtered through 0.22-μm nylon membranes [86] for high-performance liquid chromatography-electrospray ionization-mass spectrometry (HPLC-ESI-MS) analysis.

Preliminary HPLC analysis [86] was performed on a Waters 2695 system (USA) equipped with a W2996 photodiode array and a C18 column (5 μm, 4.6 × 250 mm i.d.; WondaCract ODS-2, Shimadzu,

Shanghai, China). The major parameters were set as follows: column temperature = 30 °C, absorption spectrum = 200 to 600 nm, injection volume = 10 μL, and flow rate = 0.8 mL/min. Gradient separation was carried out using a two-solvent system, 0.5% formic acid in water (phase A) and 0.1% formic acid in acetonitrile (phase B), as follows: 0 min, 5% B; 5 min, 10% B; 20 min, 20% B; 30 min, 25% B; 33 min, 10% B; 35 min, 5% B; and 50 min, 5% B. Quantification of WT and RT samples was performed using three replicates (WT1, WT2 and WT3, and RT1, RT2 and RT3, respectively) and an external standard. The external standard was cyanidin 3-*O*-glucoside chloride (Sigma-Aldrich, St. Louis, MO, USA), which was dissolved in the same extraction reagent (70:27:2:1 (*v*/*v*/*v*/*v*) methanol-water-formic acid-trifluoroacetic acid) with a concentration of 0.125 mg/mL.

Following preliminary HPLC identification, anthocyanins were detected using an HPLC system (Agilent 1200LC, Santa Clara, CA, USA) equipped with a diode array detector at 520 nm and the same C18 column used above and coupled to an electrospray ionization-mass spectrometer (ESI-MS) (6310 MSD Trap VL, Agilent, USA). ESI-MS was performed with the following settings: positive ionization mode (ESI, m/z 50–1000 mass units), gas temperature = 350 °C, flow rate = 8.0 L/min, nebulizer pressure = 35 psi, and capillary exit voltage = 120.4 V.

4.3. Bisulfite Sequencing and DMR Analysis

4.3.1. Extraction of DNA and BS-Seq

WT1, WT2, WT3, RT1, RT2, and RT3 samples were separately ground into fine powder in liquid nitrogen. Genomic DNA was extracted using a DNeasy Plant Mini kit (QianGen, Shanghai, China) following the manufacturer's instructions and then checked on 0.1% agarose gels and a NanoPhotometer spectrophotometer (Implen, Westlake Village, CA, USA). DNA concentrations were determined with a Qubit DNA Assay kit on a Qubit 2.0 fluorometer (Life Technologies, Carlsbad, CA, USA).

Next, 5.2 μg of qualified DNA spiked with 26 ng of lambda DNA (negative control) was sheared in a Covaris S220 ultrasonicator into random 200–300-bp fragments. The resulting fragments were then subjected to end repair, adenylation, and methyl-treated adapter ligation. Two bisulfite treatments with an EZ DNA Methylation-Gold kit (Zymo Research, Irvine, CA, USA) were applied to these fragments to transform non-methylated cytosines into uracil for subsequent base pairing with thymine by PCR. After PCR amplification using KAPA HiFi HotStart Uracil + ReadyMix (2×), the generated BS-seq library was quantified on a Qubit 2.0 fluorometer (Life Technologies) and by quantitative PCR, and insert size was assayed on an Agilent Bioanalyzer 2100 system. Sequencing of the BS-seq library, which generated 125/150-bp paired-end reads, was performed on an Illumina Hiseq 2500 platform followed by Illumina CASAVA pipeline analysis.

4.3.2. Quality Assessment of Sequencing Data

The sequenced paired-end reads (raw reads or raw data) were checked for quality using FastQC (fastqc_v0.11.5) and stored in the FASTQ file format (Babraham Bioinformatics, http://www.bioinformatics.babraham.ac.uk/projects/fastqc/). The resulting files were pre-processed with Trimmomatic v0.36 software [87] using the following parameters: SLIDINGWINDOW: 4:15; LEADING:3, TRAILING:3; ILLUMINACLIP: adapter.fa: 2:30:10; and MINLEN:36. Reads passing these filtering steps were counted as clean reads for use in subsequent analyses. Finally, basic quality statistics on the clean reads were obtained using FastQC.

All the sequencing data (accession number: CRA000731) are available in the database of Genome Sequence Archive (GSA, http://bigd.big.ac.cn/).

4.3.3. Reference Genome Preparation and Mapping of Clean Reads

To enable the mapping of clean reads, we first prepared a reference genome of *P. mume* [1]. Then, we performed reverse complementation process (C to T, and G to A) using Bismarkv 0.16.3 software [88] and Bowtie2 [89]. In addition, a gene annotation file in gene transfer format, a Gene Ontology (GO) annotation file, a description file, and a gene region file in browser extensible data format were generated for subsequent annotation and function analyses.

The quality-checked clean reads generated by BS-seq were aligned against the two strands of the converted reference genome (X700-dovetail). The best unique alignment of these sequence reads was selected from the two sets of pairwise comparisons. To infer cytosine methylation states and positions, the sequences were then compared against the normal genomic sequence. Identical sequences aligned to a unique genomic region were regarded as duplicates and used to estimate sequencing depth and coverage. To allow their visualization in the IGV browser, sequences were transformed into bigWig format (non-overlap) [88,90]. The bisulfite non-conversion rate was defined as the number of sequenced cytosines at all of the cytosine reference positions divided by the number in the lambda genome.

4.3.4. Evaluation of Methylation Level and Distribution

Methylation level was calculated as ML(mC) = reads(mC)/(reads(mC) + reads(C)), where mC is methylcytosine, C is non-methylated cytosine, and ML(mC) is the methylcytosine level. As recommended in a previous study [91], the parameter ML(mC) was corrected to ML(corrected) according to the following formula: ML(corrected) = (ML(mC) − r)/(1 − r), where r represents the bisulfite non-conversion rate. Methylcytosine sequence contexts—mCG, mCHG, and mCHH (where H represents A, T, or C)—were analyzed. Methylation level densities and methylcytosine distributions in each chromosome and gene functional region (promoter, exon, intron, and 2-kb upstream and downstream regions) were also analyzed [43,92,93]. Differences in global methylation levels and methylcytosine distributions in gene structural regions (including 2-kb upstream and downstream) were compared between samples [72]. To focus on petal color variation, WT1, WT2, and WT3 samples were merged together as three biological replicates of WT; similarly, RT1, RT2, and RT3 served as the three RT biological replicates.

4.3.5. Correlation Analysis and DMR Detection

A correlation analysis was carried out based on Pearson's coefficient [94]. DSS software was used to identify DMRs and differentially methylated loci between WT and RT samples [95–97]. Information from neighboring cytosine sites (i.e., spatial correlation) and site read depths were analyzed to improve the accuracy of long cytosine reads. Variance among biological replicates was analyzed using a beta-binomial distribution model. DMRs were annotated, and DMR-related genes were defined as those having coding regions (from the transcriptional start site (TSS) to the transcriptional end site (TES)) or promoter regions (i.e., upstream 2-kb from the TSS) that overlapped with the distribution of DMRs.

4.3.6. GO and KEGG Enrichment Analysis of DMR-Related Genes

GO enrichment analysis of DMR-related genes was performed using the GOseq R package [98], which also corrects for gene length bias. A GO term was considered to be significantly enriched in DMR-related genes at a corrected p-value threshold of 0.05. KEGG analysis (http://www.genome.jp/kegg/), an approach for understanding high-level functions and relationships in biological systems, was applied to uncover the pathway enrichment of DMR-associated genes [99]. KOBAS software [100] was used to test for statistical enrichment of DMR-related genes, which were then subdivided into all, hypermethylation, and hypomethylation categories and assigned to KEGG pathways.

4.4. Transcriptome Sequencing and Differentially Expressed Gene (DEG) Analysis

4.4.1. RNA Isolation and Sequencing

RNA was isolated from WT1, WT2, WT3, RT1, RT2, and RT3 samples using an RNeasy Plant Mini kit (QianGen). Extracted RNA was checked for degradation and contamination using 1% agarose gels and a NanoPhotometer spectrophotometer (Implen, CA, USA), respectively, and then quantified with a Qubit RNA Assay kit on a Qubit 2.0 fluorometer (Life Technologies). RNA integrity was assessed on a Bioanalyzer 2100 system (Agilent) using the supplied RNA Nano 6000 assay kit.

For sequencing library construction, first and second-strand cDNA synthesis was carried out using 3 μg of RNA and a NEBNext Ultra RNA Library Prep Kit for Illumina (NEB, Ipswich, MA, USA), following the manufacturer's instructions. After purification using an AMPure XP system (Beckman Coulter, Beverly, MA, USA), the synthesized strands were subjected to three-end adenylation and the addition of a poly-A tail and a NEBNext adapter. Next, 150–200-bp adapter-ligated fragments were preferentially selected using AMPure XP beads and amplified by PCR. The quality of the enriched cDNA library was assessed on an Agilent Bioanalyzer 2100 system. Clusters were generated from the qualified library using a cBot Cluster Generation System with a TruSeq PE Cluster kit v3-cBot-HS (Illumina, San Diego, CA, USA) and then sequenced on an Illumina Hiseq platform to generate 125-bp/150-bp paired-end reads (raw reads).

4.4.2. Mapping and DEG Analysis

After quality control to remove low-quality reads and adapter contaminants from the raw reads, the remaining clean reads were aligned to the *P. mume* reference genome [1] using HISAT2.0.4 software with default parameters [101]. Cufflinks v2.1.1 was then used to assemble and identify known and novel transcripts, and HTSeq v0.6.1 (-m union) was used to estimate gene expression levels based on fragments per kilobase of transcript sequence per millions of base pairs (FPKM) [102]. Differential expression analysis of WT and RT groups, with three biological replicates per group, was performed in DESeq v1.10.1, a program providing statistical routines for the determination of differential expression in digital gene expression data with a negative binomial distribution (Kij~NB(μij,σij2)) [103,104]. Significant DEGs were identified using an adjusted *p*-value cutoff of 0.05 and then subjected to GO [98] and KEGG [100] annotation.

4.5. Identification of Transcription Factors and TEs

Transcription factors and transcriptional regulatory factors were predicted and classified from the assembled transcriptome data according to Pérez-Rodríguez et al. [105] and Jin et al. [106]. In addition, the DEG sequences were screened for transcription factors using the iTAK tool (http://bioinfo.bti.cornell.edu/tool/itak), a program to identify plant transcription factors and transcriptional regulators and protein kinases, which currently provides both online and standalone versions with a hidden Markov model [107].

TEs were identified from the reference genome using RepeatMasker (http://www.repeatmasker.org/). TE density based on the length ratio of each bin within every chromosome was displayed as a Circos plot [48]. TEs annotated within gene set domains (including 2-kb upstream and downstream sequences) of DMR-related genes overlapping with DEGs were used for DNA methylation analysis. TEs were classified, and their distributions and mC contexts were analyzed.

4.6. Methylation Modification of Gene Expression

After the completion of independent methylome and transcriptome analyses, the correlation between methylation level and gene expression was progressively analyzed on four different levels. First, gene methylation level and gene expression densities were mapped onto the chromosomes of *P. mume*, and relationships between methylation levels and the expressions of promoters or gene coding regions, upstream (2-kb) and downstream (2-kb) regions, TSSs, and TESs were explored from a global perspective [48,90,108]. Second, CG, CHG, and CHH methylation modification patterns of DEGs (including promoters and 2-kb upstream and downstream regions) identified from the transcriptome data were analyzed [109,110]. Third, we attempted to relate the expression of DMR-related genes to different methylation modification patterns [111,112]. Finally, we identified the set of DMR-related genes that overlapped with DEGs, and subjected them to GO and KEGG enrichment analyses.

5. Conclusions

Flower color chimerism has since served as important material for landscaping application and genetic improvement. In our study, we detected the specific color substances, i.e., cyanidin 3,5-*O*-diglucoside, cyanidin 3-*O*-glucoside, and peonidin 3-*O*-glucoside, in red petal tissues of *P. mume* "Danban Tiaozhi". Simultaneously, we investigated the molecular mechanism of chimeric flowers by using a comparative methylomic–transcriptomic approach. We mapped the first-ever generated methylomes of *P. mume*, and determined that gene expression was negatively correlated with methylcytosine level in general and uncovered significant epigenetic variation between WT and RT. We also detected DMRs and DMR-related genes between WT and RT, and concluded that many of these genes, including DEGs and transcription factor genes, are critical participants in the anthocyanin regulatory pathway. Importantly, some of the associated DEGs harbored TE insertions that were also modified by methylcytosine. It suggests that flower color chimerism in *P. mume* is induced by the DNA methylation of critical genes and TEs.

Supplementary Materials: The following are available online at http://www.mdpi.com/1422-0067/19/8/2315/s1.

Author Contributions: K.-F.M., Q.-X.Z. designed the experiments; T.-R.C., X.-L.Y., H.-T.P., and J.W. collected the plant materials and performed the qualitative and quantitative analysis on the content of anthocyanins; K.-F.M. did the experiment in molecular biology and analyzed the data profiles; K.-F.M. and Q.-X.Z. wrote the manuscript; T.-R.C., X.-L.Y., H.-T.P., and J.W. provided suggestions for manuscript revision.

Funding: This research was supported by the Fundamental Research Funds for the Central Universities (Nos. 2016ZCQ02 and BLX2013010), the National Natural Science Foundation of China (No. 31501787), and the Special Fund for Beijing Common Construction Project.

Conflicts of Interest: The authors declare no conflicts of interest.

References

1. Zhang, Q.X.; Chen, W.B.; Sun, L.D.; Zhao, F.Y.; Huang, B.Q.; Yang, W.R.; Tao, Y.; Wang, J.; Yuan, Z.Q.; Fan, G.Y.; et al. The genome of *Prunus mume*. *Nat. Commun.* **2012**, *3*, 1318. [CrossRef] [PubMed]
2. Chen, J.Y. *Chinese Mei Flowers*; Hainan Publishing House: Haikou, China, 1996; pp. 12–58. (In Chinese)
3. Marcotrigiano, M. Chimeras and variegation: Patterns of deceit. *Hortscience* **1997**, *32*, 773–784.
4. Suzuki, M.; Miyahara, T.; Tokumoto, H.; Hakamatsuka, T.; Goda, Y.; Ozeki, Y.; Nobuhiro Sasaki, N. Transposon-mediated mutation of CYP76AD3 affects betalain synthesis and produces variegated flowers in four o'clock (*Mirabilis jalapa*). *J. Plant Physiol.* **2014**, *171*, 1586–1590. [CrossRef] [PubMed]
5. Wang, C.; Chin, Y.; Lin, C.; Chen, P.; To, K. Transforming the snapdragon aurone biosynthetic genes into petunia alters coloration patterns in transgenic flowers. *Adv. Biosci. Biotechnol.* **2015**, *6*, 702–722. [CrossRef]
6. Zhou, H.; Linwang, K.; Wang, H.L.; Gu, C.; Dare, A.P.; Epley, R.V.; He, H.P.; Allan, A.C.; Han, Y.P. Molecular genetics of blood-fleshed peach reveals activation of anthocyanin biosynthesis by NAC transcription factors. *Plant J.* **2015**, *82*, 105–121. [CrossRef] [PubMed]
7. Wu, X.; Gong, Q.; Ni, X.; Zhou, Y.; Gao, Z. UFGT: The key enzyme associated with the petals variegation in Japanese apricot. *Front. Plant Sci.* **2017**, *8*, 108. [CrossRef] [PubMed]
8. McClintock, B. Chromosome organization and gene expression. *Cold Spring Harb. Symp. Quant. Biol.* **1951**, *16*, 13–47. [CrossRef] [PubMed]
9. Inagaki, Y.; Hisatomi, Y.; Iida, S. Somatic mutations caused by excision of the transposable element, *Tpn1*, from the *DFR* gene for pigmentation in sub-epidermal layer of periclinally chimeric flowers of Japanese morning glory and their germinal transmission to their progeny. *Theor. Appl. Genet.* **1996**, *92*, 499–504. [CrossRef] [PubMed]
10. Abe, Y.; Hoshino, A.; Iida, S. Appearance of flower variegation in the mutable speckled line of the Japanese morning glory is controlled by two genetic elements. *Genes Genet. Syst.* **1997**, *72*, 57–62. [CrossRef]
11. Iida, S.; Morita, Y.; Choi, J.; Park, K.; Hoshino, A. Genetics and epigenetics in flower pigmentation associated with transposable elements in morning glories. *Adv. Biophys.* **2004**, *38*, 141–159. [CrossRef]
12. Fukada-Tanaka, S.; Inagaki, Y.; Yamaguchi, T.; Saito, N.; Iida, S. Colour-enhancing protein in blue petals. *Nature* **2000**, *407*, 581. [CrossRef] [PubMed]

13. Yamaguchi, T.; Fukadatanaka, S.; Inagaki, Y.; Saito, N.; Yonekurasakakibara, K.; Tanaka, Y.; Kusumi, T.; Iida, S. Genes en-coding the vacuolar Na^+/H^+ exchanger and flower coloration. *Plant Cell Physiol.* **2001**, *42*, 451–461. [CrossRef] [PubMed]
14. Liu, D.; Galli, M.; Crawford, N.M. Engineering variegated floral patterns in tobacco plants using the *Arabidopsis* transposable element *Tag1*. *Plant Cell Physiol.* **2001**, *42*, 419–423. [CrossRef] [PubMed]
15. Kobayashi, S.; Goto-Yamamoto, N.; Hirochika, H. Retrotransposon-induced mutations in grape skin color. *Science* **2004**, *304*, 982. [CrossRef] [PubMed]
16. Zabala, G.; Vodkin, L. A putative autonomous 20.5 kb-CACTA transposon insertion in an *F3′H* allele identifies a new CACTA transposon subfamily in *Glycine max*. *BMC Plant Biol.* **2008**, *8*, 124. [CrossRef] [PubMed]
17. Nishizaki, Y.; Matsuba, Y.; Okamoto, E.; Okamura, M.; Yoshihiro Ozeki, Y.; Sasaki, N. Structure of the acyl-glucose-dependent anthocyanin 5-*O*-glucosyltransferase gene in carnations and its disruption by transposable elements in some varieties. *Mol. Genet. Genom.* **2011**, *286*, 383–394. [CrossRef] [PubMed]
18. Fukada-Tanaka, S.; Hoshino, A.; Hisatomi, Y.; Habu, Y.; Hasebe, M.; Iida, S. Identification of new chalcone synthase genes for flower pigmentation in the Japanese and common morning glories. *Plant Cell Physiol.* **1997**, *38*, 754–758. [CrossRef] [PubMed]
19. Chen, Y.N.; Mao, Y.; Liu, H.L.; Yu, F.X.; Li, S.X.; Yin, T.M. Transcriptome analysis of differentially expressed genes relevant to variegation in peach flowers. *PLoS ONE* **2014**, *9*, e90842. [CrossRef] [PubMed]
20. Wu, Z.G.; Jiang, W.; Mantri, N.; Bao, X.Q.; Chen, S.L.; Tao, Z.M. Transciptome analysis reveals flavonoid biosynthesis regulation and simple sequence repeats in yam (*Dioscorea alata* L.) tubers. *BMC Genom.* **2015**, *16*, 346. [CrossRef] [PubMed]
21. Zhang, Y.; Cheng, Y.; Ya, H.; Xu, S.; Han, J. Transcriptome sequencing of purple petal spot region in tree peony reveals differentially expressed anthocyanin structural genes. *Front Plant Sci.* **2015**, *8*, 964. [CrossRef] [PubMed]
22. Hassani, D.; Liu, H.L.; Chen, Y.N.; Wan, Z.B.; Zhuge, Q.; Li, S.X. Analysis of biochemical compounds and differentially expressed genes of the anthocyanin biosynthetic pathway in variegated peach flowers. *Genet. Mol. Res.* **2015**, *14*, 13425–13436. [CrossRef] [PubMed]
23. Koseki, M.; Goto, K.; Masuta, C.; Kanazawa, A. The star-type color pattern in *Petunia hybrida* 'Red Star' flowers is induced by sequence-specific degradation of *Chalcone Synthase* RNA. *Plant Cell Physiol.* **2005**, *46*, 1879–1883. [CrossRef] [PubMed]
24. Cheng, J.; Liao, L.; Zhou, H.; Gu, C.; Wang, L.; Han, Y.P. A small indel mutation in an anthocyanin transporter causes variegated colouration of peach flowers. *J. Exp. Bot.* **2015**, *66*, 7227–7239. [CrossRef] [PubMed]
25. Uematsu, C.; Katayama, H.; Makino, I.; Inagaki, A.; Arakawa, O.; Martin, C. Peace, a MYB-like transcription factor, regulates petal pigmentation in flowering peach 'Genpei' bearing variegated and fully pigmented flowers. *J. Exp. Bot.* **2014**, *65*, 1081–1094. [CrossRef] [PubMed]
26. Meyer, P.; Heidmann, I.; Niedenhof, I. Differences in DNA-methylation are associated with a paramutation phenomenon in transgenic petunia. *Plant J.* **1993**, *4*, 89–100. [CrossRef] [PubMed]
27. Liu, X.J.; Chuang, Y.N.; Chiou, C.Y.; Chin, D.C.; Shen, F.Q.; Yeh, K.W. Methylation effect on chalcone synthase gene expression determines anthocyanin pigmentation in floral tissues of two *Oncidium* orchid cultivars. *Planta* **2012**, *236*, 401–409. [CrossRef] [PubMed]
28. Jones, P.A.; Takai, D. The role of DNA methylation in mammalian epigenetics. *Science* **2001**, *293*, 1068–1070. [CrossRef] [PubMed]
29. Reik, W.; Dean, W.; Walter, J. Epigenetic reprogramming in mammalian development. *Science* **2001**, *293*, 1089–1093. [CrossRef] [PubMed]
30. Reik, W.; Santos, F.; Mitsuya, K.; Morgan, H.P.; Dean, W. Epigenetic asymmetry in the mammalian zygote and early embryo: Relationship to lineage commitment? *Philos. Trans. R. Soc. B* **2003**, *358*, 1403–1409. [CrossRef] [PubMed]
31. Bird, A. DNA methylation patterns and epigenetic memory. *Genes Dev.* **2002**, *16*, 6–21. [CrossRef] [PubMed]
32. Bocchini, M.; Galla, G.; Pupilli, F. The vesicle traffcking regulator *PN_SCD1* is demethylated and overexpressed in florets of apomictic *Paspalum notatum* genotypes. *Sci. Rep.* **2018**, *8*, 3030. [CrossRef] [PubMed]

33. Marfil, C.F.; Camadro, E.L.; Masuelli, R.W. Phenotypic instability and epigenetic variability in a diploid potato of hybrid origin, *Solanum ruiz-lealii*. *BMC Plant Biol.* **2009**, *9*, 21. [CrossRef] [PubMed]
34. Ma, K.F.; Song, Y.P.; Jiang, X.B.; Zhang, Z.Y.; Li, B.L.; Zhang, D.Q. Photosynthetic response to genome methylation affects the growth of Chinese white poplar. *Tree Genet. Genomes* **2012**, *8*, 1407–1421. [CrossRef]
35. Cubas, P.; Vincent, C.; Coen, E. An epigenetic mutation responsible for natural variation in floral symmetry. *Nature* **1999**, *401*, 157–161. [CrossRef] [PubMed]
36. Telias, A.; Linwang, K.; Stevenson, D.E.; Cooney, J.M.; Hellens, R.P.; Allan, A.C.; Hoover, E.; Bradeen, J.M. Apple skin patterning is associated with differential expression of *MYB10*. *BMC Plant Biol.* **2011**, *11*, 93. [CrossRef] [PubMed]
37. Wang, Z.G.; Meng, D.; Wang, A.D.; Li, T.L.; Jiang, S.L.; Cong, P.H.; Li, T.Z. The methylation of the *PcMYB10* promoter is associated with green-skinned sport in Max red bartlett pear. *Plant Physiol.* **2013**, *162*, 885–896. [CrossRef] [PubMed]
38. Bai, S.L.; Tuan, P.A.; Saito, T.; Honda, C.; Hatsuyama, Y.; Ito, A.; Moriguchi, T. Epigenetic regulation of *MdMYB1* is associated with paper bagging-induced red pigmentation of apples. *Planta* **2016**, *244*, 573–586. [CrossRef] [PubMed]
39. Hoekenga, O.A.; Muszynski, M.G.; Cone, K.C. Developmental patterns of chromatin structure and DNA methylation responsible for epigenetic expression of a maize regulatory gene. *Genetics* **2000**, *155*, 1889–1902. [PubMed]
40. Kankel, M.W.; Ramsey, D.E.; Stokes, T.L.; Flowers, S.K.; Haag, J.R.; Jeddeloh, J.A.; Riddle, N.C.; Verbsky, M.L.; Richards, E.J. *Arabidopsis MET1* cytosine methyltransferase mutants. *Genetics* **2003**, *163*, 1109–1122. [PubMed]
41. Rudenko, G.N.; Ono, A.; Walbot, V. Initiation of silencing of maize *MuDR/Mu* transposable elements. *Plant J.* **2003**, *33*, 1013–1025. [CrossRef] [PubMed]
42. Chan, S.W.; Henderson, I.R.; Jacobsen, S.E. Gardening the genome: DNA methylation in *Arabidopsis thaliana*. *Nat. Rev. Genet.* **2005**, *6*, 351–360. [CrossRef] [PubMed]
43. Zhong, S.L.; Fei, Z.J.; Chen, Y.R.; Zheng, Y.; Huang, M.Y.; Vrebalov, J.; Mcquinn, R.; Gapper, N.E.; Liu, B.; Xiang, J.; et al. Single-base resolution methylomes of tomato fruit development reveal epigenome modifications associated with ripening. *Nat. Biotechnol.* **2013**, *31*, 154–159. [CrossRef] [PubMed]
44. An, Y.C.; Goettel, W.; Han, Q.; Bartels, A.; Liu, Z.R.; Xiao, W.Y. Dynamic changes of genome-wide DNA methylation during soybean seed development. *Sci. Rep.* **2017**, *7*, 12263. [CrossRef] [PubMed]
45. Song, Q.; Zhang, T.; Stelly, D.M.; Chen, Z.J. Epigenomic and functional analyses reveal roles for epialleles in the loss of photoperiod sensitivity during domestication of allotetraploid cottons. *Genome Biol.* **2017**, *18*, 99. [CrossRef] [PubMed]
46. Wang, W.S.; Qin, Q.; Sun, F.; Wang, Y.X.; Xu, D.D.; Li, Z.K.; Fu, B.Y. Genome-wide differences in DNA methylation changes in two contrasting rice genotypes in response to drought conditions. *Front. Plant Sci.* **2016**, *7*, 1675. [CrossRef] [PubMed]
47. Lu, X.K.; Wang, X.G.; Chen, X.G.; Shu, N.; Wang, J.J.; Wang, D.L.; Wang, S.; Fan, W.L.; Guo, L.X.; Guo, X.N.; et al. Single-base resolution methylomes of upland cotton (*Gossypium hirsutum* L.) reveal epigenome modifications in response to drought stress. *BMC Genom.* **2017**, *18*, 297. [CrossRef] [PubMed]
48. Xu, J.D.; Zhou, S.S.; Gong, X.Q.; Song, Y.; Van Nocker, S.; Ma, F.W.; Guan, Q.M. Single-base methylome analysis reveals dynamic epigenomic differences associated with water deficit in apple. *Plant Biotechnol. J.* **2018**, *16*, 672–687. [CrossRef] [PubMed]
49. Bock, C. Analysing and interpreting DNA methylation data. *Nat. Rev. Genet.* **2012**, *13*, 705–719. [CrossRef] [PubMed]
50. Meissner, A.; Mikkelsen, T.S.; Gu, H.C.; Wernig, M.; Hanna, J.; Sivachenko, A.; Zhang, X.L.; Bernstein, B.E.; Nusbaum, C.; Jaffe, D.B.; et al. Genome-scale DNA methylation maps of pluripotent and differentiated cells. *Nature* **2008**, *454*, 766–770. [CrossRef] [PubMed]
51. Doi, A.; Park, I.; Wen, B.; Murakami, P.; Aryee, M.J.; Irizarry, R.; Herb, B.R.; Ladd-Acosta, C.; Rho, J.; Loewer, S.; et al. Differential methylation of tissue- and cancer-specific CpG island shores distinguishes human induced pluripotent stem cells, embryonic stem cells and fibroblasts. *Nat. Genet.* **2009**, *41*, 1350–1353. [CrossRef] [PubMed]
52. Zhou, Y.; Wu, X.; Zhang, Z.; Gao, Z. Comparative proteomic analysis of floral color variegation in peach. *Biochem. Biophys. Res. Commun.* **2015**, *464*, 1101–1106. [CrossRef] [PubMed]

53. Suzuki, K.; Suzuki, T.; Nakatsuka, T.; Dohra, H.; Yamagishi, M.; Matsuyama, K.; Matsuura, H. RNA-seq-based evaluation of bicolor tepal pigmentation in Asiatic hybrid lilies (*Lilium* spp.). *BMC Genom.* **2016**, *17*, 611. [CrossRef] [PubMed]
54. Han, Y.; Vimolmangkang, S.; Soria-Guerra, R.E.; Korban, S.S. Introduction of apple *ANR* genes into tobacco inhibits expression of both *CHI* and *DFR* genes in flowers, leading to loss of anthocyanin. *J. Exp. Bot.* **2012**, *63*, 2437–2447. [CrossRef] [PubMed]
55. Sun, Y.; Li, H.; Huang, J. *Arabidopsis* TT19 functions as a carrier to transport anthocyanin from the cytosol to tonoplasts. *Mol. Plant* **2012**, *5*, 387–400. [CrossRef] [PubMed]
56. Zhou, H.; Peng, Q.; Zhao, J.B.; Owiti, A.; Ren, F.; Liao, L.; Wang, L.; Deng, X.B.; Jiang, Q.; Han, Y.P. Multiple R2R3-MYB transcription factors involved in the regulation of anthocyanin accumulation in peach flower. *Front. Plant Sci.* **2016**, *7*, 1557. [CrossRef] [PubMed]
57. Zhou, Y.; Zhou, H.; Linwang, K.; Vimolmangkang, S.; Espley, R.V.; Wang, L.; Allan, A.C.; Han, Y.P. Transcriptome analysis and transient transformation suggest an ancient duplicated MYB transcription factor as a candidate gene for leaf red coloration in peach. *BMC Plant Biol.* **2014**, *14*, 388. [CrossRef] [PubMed]
58. Qi, T.C.; Song, S.S.; Ren, Q.C.; Wu, D.W.; Huang, H.; Chen, Y.; Fan, M.; Peng, W.; Ren, C.M.; Xie, D.X. The Jasmonate-ZIM-Domain proteins interact with the WD-Repeat/bHLH/MYB complexes to regulate jasmonate-mediated anthocyanin accumulation and trichome initiation in *Arabidopsis thaliana*. *Plant Cell* **2011**, *23*, 1795–1814. [CrossRef] [PubMed]
59. Zilberman, D.; Gehring, M.; Tran, R.K.; Ballinger, T.; Henikoff, S. Genome-wide analysis of *Arabidopsis thaliana* DNA methylation uncovers an interdependence between methylation and transcription. *Nat. Genet.* **2007**, *39*, 61–69. [CrossRef] [PubMed]
60. Zhang, X.Y.; Yazaki, J.; Sundaresan, A.; Cokus, S.J.; Chan, S.W.L.; Chen, H.M.; Henderson, I.R.; Shinn, P.; Pellegrini, M.; Jacobsen, S. Genome-wide high-resolution mapping and functional analysis of DNA methylation in *Arabidopsis*. *Cell* **2006**, *126*, 1189–1201. [CrossRef] [PubMed]
61. Madlung, A.; Comai, L. The effect of stress on genome regulation and structure. *Ann. Bot.* **2004**, *94*, 481–495. [CrossRef] [PubMed]
62. Song, Q.; Decato, B.; Hong, E.E.; Zhou, M.; Fang, F.; Qu, J.H.; Garvin, T.; Kessler, M.; Zhou, J.; Smith, A.D. A reference methylome database and analysis pipeline to facilitate integrative and comparative epigenomics. *PLoS ONE* **2013**, *8*, e81148. [CrossRef] [PubMed]
63. Cokus, S.J.; Feng, S.H.; Zhang, X.Y.; Chen, Z.G.; Merriman, B.; Haudenschild, C.D.; Pradhan, S.; Nelson, S.F.; Pellegrini, M.; Jacobsen, S.E. Shotgun bisulphite sequencing of the *Arabidopsis* genome reveals DNA methylation patterning. *Nature* **2008**, *452*, 215–219. [CrossRef] [PubMed]
64. Feng, S.H.; Cokus, S.J.; Zhang, X.Y.; Chen, P.Y.; Bostick, M.; Goll, M.G.; Hetzel, J.; Jain, J.; Strauss, S.H.; Halpern, M.E.; et al. Conservation and divergence of methylation patterning in plants and animals. *Proc. Natl. Acad. Sci. USA* **2010**, *107*, 8689–8694. [CrossRef] [PubMed]
65. Chodavarapu, R.K.; Feng, S.H.; Ding, B.; Simon, S.A.; Lopez, D.; Jia, Y.L.; Wang, G.L.; Meyers, B.C.; Jacobsen, S.E.; Pellegrini, M. Transcriptome and methylome interactions in rice hybrids. *Proc. Natl. Acad. Sci. USA* **2012**, *109*, 12040–12045. [CrossRef] [PubMed]
66. Gent, J.I.; Ellis, N.A.; Guo, L.; Harkess, A.; Yao, Y.Y.; Zhang, X.Y.; Dawe, R.K. CHH islands: De novo DNA methylation in near-gene chromatin regulation in maize. *Genome Res.* **2013**, *23*, 628–637. [CrossRef] [PubMed]
67. Schmitz, R.J.; He, Y.P.; Valdeslopez, O.; Khan, S.M.; Joshi, T.; Urich, M.A.; Nery, J.R.; Diers, B.W.; Xu, D.; Stacey, G.; et al. Epigenome-wide inheritance of cytosine methylation variants in a recombinant inbred population. *Genome Res.* **2013**, *23*, 1663–1674. [CrossRef] [PubMed]
68. Kim, K.D.; Baidouri, M.E.; Abernathy, B.; Iwataotsubo, A.; Chavarro, C.; Gonzales, M.; Libault, M.; Grimwood, J.; Jackson, S.A. A comparative epigenomic analysis of polyploidy-derived genes in soybean and common bean. *Plant Physiol.* **2015**, *168*, 1433–1447. [CrossRef] [PubMed]
69. Li, X.Y.; Wang, X.F.; He, K.; Ma, Y.Q.; Su, N.; He, H.; Stolc, V.; Tongprasit, W.; Jin, W.W.; Jiang, J.M.; et al. High-resolution mapping of epigenetic modifications of the rice genome uncovers interplay between DNA methylation, histone methylation, and gene expression. *Plant Cell* **2008**, *20*, 259–276. [CrossRef] [PubMed]
70. Miura, A.; Nakamura, M.; Inagaki, S.; Kobayashi, A.; Saze, H.; Kakutani, T. An *Arabidopsis*jmjC domain protein protects transcribed genes from DNA methylation at CHG sites. *EMBO J.* **2009**, *28*, 1078–1086. [CrossRef] [PubMed]

71. Wang, J.; Marowsky, N.C.; Fan, C. Divergence of gene body DNA methylation and evolution of plant duplicate genes. *PLoS ONE* **2014**, *9*, e110357. [CrossRef] [PubMed]
72. Song, Q.X.; Lu, X.; Li, Q.T.; Chen, H.; Hu, X.Y.; Ma, B.; Zhang, W.K.; Chen, S.Y.; Zhang, J. Genome-wide analysis of DNA methylation in soybean. *Mol. Plant* **2013**, *6*, 1961–1974. [CrossRef] [PubMed]
73. Chopra, S.; Brendel, V.; Zhang, J.; Axtell, J.D.; Peterson, T. Molecular characterization of a mutable pigmentation phenotype and isolation of the first active transposable element from *Sorghum bicolor*. *Proc. Natl. Acad. Sci. USA* **1999**, *96*, 15330–15335. [CrossRef] [PubMed]
74. Feschotte, C.; Jiang, N.; Wessler, S.R. Plant transposable elements: Where genetics meets genomics. *Nat. Rev. Genet.* **2002**, *3*, 329–341. [CrossRef] [PubMed]
75. Li, X.; Chen, L.; Hong, M.Y.; Zhang, Y.; Zu, F.; Wen, J.; Yi, B.; Ma, C.Z.; Shen, J.X.; Tu, J.X.; et al. A large insertion in bHLH transcription factor BrTT8 resulting in yellow seed coat in *Brassica rapa*. *PLoS ONE* **2012**, *7*, e44145. [CrossRef] [PubMed]
76. Hong, L.L.; Qian, Q.; Tang, D.; Wang, K.J.; Li, M.; Cheng, Z.K. A mutation in the rice chalcone isomerase gene causes the golden hull and internode 1 phenotype. *Planta* **2012**, *236*, 141–151. [CrossRef] [PubMed]
77. Clegg, M.T.; Durbin, M.L. Tracing floral adaptations from ecology to molecules. *Nat. Rev. Genet.* **2003**, *4*, 206–215. [CrossRef] [PubMed]
78. Zabala, G.; Vodkin, L. The *wp* mutation of *Glycine max* carries a gene-fragment-rich transposon of the CACTA superfamily. *Plant Cell* **2005**, *17*, 2619–2632. [CrossRef] [PubMed]
79. Choi, J.; Hoshino, A.; Park, K.; Park, I.; Iida, S. Spontaneous mutations caused by a *Helitron* transposon, *Hel-It1*, in morning glory, *Ipomoea tricolor*. *Plant J.* **2007**, *49*, 924–934. [CrossRef] [PubMed]
80. Park, K.; Ishikawa, N.; Morita, Y.; Choi, J.; Hoshino, A.; Iida, S. A *bHLH* regulatory gene in the common morning glory, *Ipomoea purpurea*, controls anthocyanin biosynthesis in flowers, proanthocyanidin and phytomelanin pigmentation in seeds, and seed trichome formation. *Plant J.* **2007**, *49*, 641–654. [CrossRef] [PubMed]
81. Mirouze, M.; Vitte, C. Transposable elements, a treasure trove to decipher epigenetic variation: Insights from *Arabidopsis* and crop epigenomes. *J. Exp. Bot.* **2014**, *65*, 2801–2812. [CrossRef] [PubMed]
82. Kim, K.D.; Baidouri, M.E.; Jackson, S.A. Accessing epigenetic variation in the plant methylome. *Brief. Funct. Genom.* **2014**, *13*, 318–327. [CrossRef] [PubMed]
83. Fujimoto, R.; Kinoshita, Y.; Kawabe, A.; Kinoshita, T.; Takashima, K.; Nordborg, M.; Nasrallah, M.E.; Shimizu, K.K.; Kudoh, H.; Kakutani, T. Evolution and control of imprinted *FWA* Genes in the genus *Arabidopsis*. *PLoS Genet.* **2008**, *4*, e1000048. [CrossRef] [PubMed]
84. Tsugane, K.; Maekawa, M.; Takagi, K.; Takahara, H.; Qian, Q.; Eun, C.H.; Iida, S. An active DNA transposon *nDart* causing leaf variegation and mutable dwarfism and its related elements in rice. *Plant J.* **2006**, *45*, 46–57. [CrossRef] [PubMed]
85. Hashimoto, F.; Tanaka, M.; Maeda, H.; Shimizu, K.; Sakata, Y. Characterization of cyanic flower color of *Delphinium* cultivars. *J. Jpn. Soc. Hortic. Sci.* **2000**, *69*, 428–434. [CrossRef]
86. Zhang, J.; Wang, L.S.; Gao, J.M.; Xu, Y.J.; Li, L.F.; Li, C.H. Rapid separation and identification of anthocyanins from flowers of *Viola yedoensis* and *V. prionantha* by high-performance liquid chromatography-photodiode array detection-electrospray ionisationmass spectrometry. *Phytochem. Anal.* **2012**, *23*, 16–22. [CrossRef] [PubMed]
87. Bolger, A.M.; Lohse, M.; Usadel, B. Trimmomatic: A flexible trimmer for Illumina sequence data. *Bioinformatics* **2014**, *30*, 2114–2120. [CrossRef] [PubMed]
88. Krueger, F.; Andrews, S. Bismark: A flexible aligner and methylation caller for Bisulfite-Seq applications. *Bioinformatics* **2011**, *27*, 1571–1572. [CrossRef] [PubMed]
89. Langmead, B.; Salzberg, S.L. Fast gapped-read alignment with Bowtie. *Nat. Methods* **2012**, *9*, 357–359. [CrossRef] [PubMed]
90. Wang, L.; Zhang, J.; Duan, J.L.; Gao, X.X.; Zhu, W.; Lu, X.Y.; Yang, L.; Zhang, J.; Li, G.Q.; Ci, W.M.; et al. Programming and inheritance of parental DNA methylomes in mammals. *Cell* **2014**, *157*, 979–991. [CrossRef] [PubMed]
91. Lister, R.; Mukamel, E.A.; Nery, J.R.; Urich, M.A.; Puddifoot, C.A.; Johnson, N.D.; Lucero, J.; Huang, Y.; Dwork, A.J.; Schultz, M.D.; et al. Global epigenomic reconfiguration during mammalian brain development. *Science* **2013**, *341*, 1237905. [CrossRef] [PubMed]

92. Krzywinski, M.; Schein, J.E.; Birol, I.; Connors, J.M.; Gascoyne, R.D.; Horsman, D.; Jones, S.J.M.; Marra, M.A. Circos: An information aesthetic for comparative genomics. *Genome Res.* **2009**, *19*, 1639–1645. [CrossRef] [PubMed]
93. Lister, R.; Pelizzola, M.; Dowen, R.H.; Hawkins, R.D.; Hon, G.; Tontifilippini, J.; Nery, J.R.; Lee, L.K.; Ye, Z.; Ngo, Q.; et al. Human DNA methylomes at base resolution show widespread epigenomic differences. *Nature* **2009**, *462*, 315–322. [CrossRef] [PubMed]
94. Smallwood, S.A.; Lee, H.J.; Angermueller, C.; Krueger, F.; Saadeh, H.; Peat, J.R.; Andrews, S.; Stegle, O.; Reik, W.; Kelsey, G. Single-cell genome-wide bisulfite sequencing for assessing epigenetic heterogeneity. *Nat. Methods* **2014**, *11*, 817–820. [CrossRef] [PubMed]
95. Feng, H.; Conneely, K.N.; Wu, H.A. Bayesian hierarchical model to detect differentially methylated loci from single nucleotide resolution sequencing data. *Nucleic Acids Res.* **2014**, *42*, e69. [CrossRef] [PubMed]
96. Wu, H.; Xu, T.L.; Feng, H.; Chen, L.; Li, B.; Yao, B.; Qin, Z.H.; Jin, P.; Conneely, K.N. Detection of differentially methylated regions from whole-genome bisulfite sequencing data without replicates. *Nucleic Acids Res.* **2015**, *43*, e141. [CrossRef] [PubMed]
97. Park, Y.; Wu, H. Differential methylation analysis for BS-seq data under general experimental design. *Bioinformatics* **2016**, *32*, 1446–1453. [CrossRef] [PubMed]
98. Young, M.D.; Wakefield, M.J.; Smyth, G.K.; Oshlack, A. Gene ontology analysis for RNA-seq: Accounting for selection bias. *Genome Biol.* **2010**, *11*, 1–12. [CrossRef] [PubMed]
99. Kanehisa, M.; Araki, M.; Goto, S.; Hattori, M.; Hirakawa, M.; Itoh, M.; Katayama, T.; Kawashima, S.; Okuda, S.; Tokimatsu, T.; et al. KEGG for linking genomes to life and the environment. *Nucleic Acids Res.* **2007**, *36*, 480–484. [CrossRef] [PubMed]
100. Mao, X.; Cai, T.; Olyarchuk, J.G.; Wei, L. Automated genome annotation and pathway identification using the KEGG Orthology (KO) as a controlled vocabulary. *Bioinformatics* **2005**, *21*, 3787–3793. [CrossRef] [PubMed]
101. Kim, D.; Langmead, B.; Salzberg, S.L. HISAT: A fast spliced aligner with low memory requirements. *Nat. Methods* **2015**, *12*, 357–360. [CrossRef] [PubMed]
102. Trapnell, C.; Williams, B.A.; Pertea, G.; Mortazavi, A.; Kwan, G.; Van Baren, M.J.; Salzberg, S.L.; Wold, B.J.; Pachter, L. Transcript assembly and quantification by RNA-Seq reveals unannotated transcripts and isoform switching during cell differentiation. *Nat. Biotechnol.* **2010**, *28*, 511–515. [CrossRef] [PubMed]
103. Anders, S.; Huber, W. Differential expression analysis for sequence count data. *Genome Biol.* **2010**, *11*, 1–12. [CrossRef] [PubMed]
104. Anders, S.; Huber, W. *Differential Expression of RNA-Seq Data at the Gene Level-the DESeq Package*; European Molecular Biology Laboratory (EMBL): Heidelberg, Germany, 2012.
105. Pérez-Rodríguez, P.; Rianopachon, D.M.; Correa, L.G.G.; Rensing, S.A.; Kersten, B.; Muellerroeber, B. PlnTFDB: Updated content and new features of the plant transcription factor database. *Nucleic Acids Res.* **2010**, *38*, 822–827. [CrossRef] [PubMed]
106. Jin, J.; Zhang, H.; Kong, L.; Gao, G.; Luo, J. PlantTFDB 3.0: A portal for the functional and evolutionary study of plant transcription factors. *Nucleic Acids Res.* **2014**, *42*, 1182–1187. [CrossRef] [PubMed]
107. Zheng, Y.; Jiao, C.; Sun, H.; Rosli, H.G.; Pombo, M.A.; Zhang, P.; Banf, M.; Dai, X.; Martin, G.B.; Giovannoni, J.J.; et al. iTAK: A program for genome-wide prediction and classification of plant transcription factors, transcriptional regulators, and protein kinases. *Mol. Plant* **2016**, *9*, 1667–1670. [CrossRef] [PubMed]
108. Wang, M.J.; Yuan, D.J.; Tu, L.L.; Gao, W.H.; He, Y.H.; Hu, H.Y.; Wang, P.C.; Liu, N.; Lindsey, K.; Zhang, X.L. Long noncoding RNAs and their proposed functions in fibre development of cotton (*Gossypium* spp.). *New Phytol.* **2015**, *207*, 1181–1197. [CrossRef] [PubMed]
109. Ng, C.W.; Yildirim, F.; Yap, Y.S.; Dalin, S.; Matthews, B.J.; Velez, P.J.; Labadorf, A.; Housman, D.E.; Fraenkel, E. Extensive changes in DNA methylation are associated with expression of mutant huntingtin. *Proc. Natl. Acad. Sci. USA* **2013**, *110*, 2354–2359. [CrossRef] [PubMed]
110. Zhou, H.R.; Zhang, F.F.; Ma, Z.Y.; Huang, H.W.; Jiang, L.; Cai, T.; Zhu, J.K.; Zhang, C.Y.; He, X.J. Folate polyglutamylation is involved in chromatin silencing by maintaining global DNA methylation and histone H3K9 dimethylation in *Arabidopsis*. *Plant Cell* **2013**, *25*, 2545–2559. [CrossRef] [PubMed]

111. Yang, I.V.; Pedersen, B.S.; Rabinovich, E.I.; Hennessy, C.E.; Davidson, E.J.; Murphy, E.; Guardela, B.J.; Tedrow, J.; Zhang, Y.Z.; Singh, M.K.; et al. Relationship of DNA methylation and gene expression in idiopathic pulmonary fibrosis. *Am. J. Respir. Crit. Care Med.* **2014**, *190*, 1263–1272. [CrossRef] [PubMed]
112. Kretzmer, H.; Bernhart, S.H.; Wang, W.; Haake, A.; Weniger, M.A.; Bergmann, A.K.; Betts, M.J.; Carrillodesantapau, E.; Doose, G.; Gutwein, J.; et al. DNA methylome analysis in Burkitt and follicular lymphomas identifies differentially methylated regions linked to somatic mutation and transcriptional control. *Nat. Genet.* **2015**, *47*, 1316–1325. [CrossRef] [PubMed]

© 2018 by the authors. Licensee MDPI, Basel, Switzerland. This article is an open access article distributed under the terms and conditions of the Creative Commons Attribution (CC BY) license (http://creativecommons.org/licenses/by/4.0/).

International Journal of
Molecular Sciences

MDPI

Article

Classification and Genome-Wide Analysis of Chitin-Binding Proteins Gene Family in Pepper (*Capsicum annuum* L.) and Transcriptional Regulation to *Phytophthora capsici*, Abiotic Stresses and Hormonal Applications

Muhammad Ali [1], De-Xu Luo [2], Abid Khan [1], Saeed ul Haq [1], Wen-Xian Gai [1], Huai-Xia Zhang [1], Guo-Xin Cheng [1], Izhar Muhammad [3] and Zhen-Hui Gong [1,*]

1 College of Horticulture, Northwest A&F University, Yangling 712100, China; alinhorti@yahoo.com (M.A.); abidagriculturist@gmail.com (A.K.); saeed_ulhaq@nwafu.edu.cn (S.u.H.); gaiwenxian@163.com (W.-X.G.); 2016060124@nwsuaf.edu.cn (H.-X.Z.); lvge2011@126.com (G.-X.C.)

2 Xuhuai Region Huaiyin Institute of Agricultural Sciences, Huaian 223001, China; loudex2002@163.com

3 State Key Laboratory of Crop Stress Biology in Arid Areas, College of Life Sciences, Northwest A&F University, Yangling 712100, China; izeyaar@gmail.com

* Correspondence: zhgong@nwsuaf.edu.cn; Tel.: +86-029-8708-2102; Fax: +86-029-8708-2613

Received: 29 June 2018; Accepted: 26 July 2018; Published: 29 July 2018

Abstract: Chitin-binding proteins are pathogenesis-related gene family, which play a key role in the defense response of plants. However, thus far, little is known about the chitin-binding family genes in pepper (*Capsicum annuum* L.). In current study, 16 putative chitin genes (CaChi) were retrieved from the latest pepper genome database, and were classified into four distinct classes (I, III, IV and VI) based on their sequence structure and domain architectures. Furthermore, the structure of gene, genome location, gene duplication and phylogenetic relationship were examined to clarify a comprehensive background of the CaChi genes in pepper. The tissue-specific expression analysis of the CaChi showed the highest transcript levels in seed followed by stem, flower, leaf and root, whereas the lowest transcript levels were noted in red-fruit. *Phytophthora capsici* post inoculation, most of the CaChi (*CaChiI3*, *CaChiIII1*, *CaChiIII2*, *CaChiIII4*, *CaChiIII6*, *CaChiIII7*, *CaChiIV1*, *CaChiVI1* and *CaChiVI2*) were induced by both strains (PC and HX-9). Under abiotic and exogenous hormonal treatments, the *CaChiIII2*, *CaChiIII7*, *CaChiVI1* and *CaChiVI2* were upregulated by abiotic stress, while *CaChiI1*, *CaChiIII7*, *CaChiIV1* and *CaChiIV2* responded to hormonal treatments. Furthermore, *CaChiIV1*-silenced plants display weakened defense by reducing (60%) root activity and increase susceptibility to NaCl stress. Gene ontology (GO) enrichment analysis revealed that CaChi genes primarily contribute in response to biotic, abiotic stresses and metabolic/catabolic process within the biological process category. These results exposed that CaChi genes are involved in defense response and signal transduction, suggesting their vital roles in growth regulation as well as response to stresses in pepper plant. In conclusion, these finding provide basic insights for functional validation of the CaChi genes in different biotic and abiotic stresses.

Keywords: chitin-binding protein; chitinase; pepper; expression; biotic stress; abiotic stress

1. Introduction

Plants being sessile organisms are exposed to a number of stresses. External environmental fluctuations, different insect pest and pathogen considerably affect the growth, development, yield and quality [1]. To safe guard themselves against these threats, plants have evolved some sophisticated defense mechanisms. The inducible defense responses of plants include synthesis of signaling molecules, such as methyl jasmonate (MeJA), salicylic acid (SA) and ethylene (ET), which work in a complex network interaction that in turn regulates the expression of defense related genes (PR) and molecules such as reactive oxygen species (ROS), phytoalexins, proline, phenylpropanoids and pathogenesis-related genes [2,3]. Earlier studies revealed the significant role of these (PR) proteins in plant defense system [4,5]. During the biotic threat, plant defense mechanism consists of two typical interconnecting layers to develop plant immune system designated as effector-triggered immunity (ETI) and pattern-triggered immunity (PTI), thus participating in signal transduction [6,7]. A set of pathogenesis related (*PR-2* and *PR-5*) genes are involved in PTI and ETI, depending on the magnitude and time of the interacting signaling components [8,9].

Chitin-binding proteins (CBP), encoded by chitin-gene family, are PR proteins, which enhance resistance to different stresses in several crop plants [10–14]. These CBP proteins consist of one or several chitin-binding domains with high affinity and have a range of numerous complex glycoconjugates covering GlcNAc or *N*-acetyl-D-neuraminic acid (NeuNAc) as building blocks. Thus far, the chitinase responsible genes have been classified into seven different classes (classes I–VII) as they belongs to the glycoside_hydrolase_families, thus signifying that the chitinase isozymes were encoded by a family of multi-genes [4,15]. Some members of class I chitinases are localized in the vacuole, whereas other chitinases, such as the class III chitinases are positioned outside the cell [4]. Plant chitinases are responsible for the catalysis of chitin, the second most abundant polysaccharide after cellulose. Chitin is the part of the cell walls in most of fungi as well as in plants. Plant chitinases also have shown resistance to several pathogens, such as bacteria, viruses, and some abiotic stresses [16]. Certain chitinases are reported to take part in various physiological processes of plants, such as ethylene synthesis and embryogenesis [17]. CBPs are constitutively present in plant leaves, stems, seeds, flowers, and tubers. They are developmentally and tissue-specifically regulated [18,19]. Up to date, chitin genes have been cloned and characterized in numerous plants species, including *Arabidopsis thaliana* [20], *Triticum aestivum* [15], *Oryza sativa* [21], *Zea mays* [4] and *Sorghum bicolor* [22]. A class I chitin-binding proteins was isolated from *Hordeum vulgare* and has been shown antifungal activity [23]. The pathogen-inducible acidic class III chitinase proteins were isolated from *Nicotiana tobacum* after the infection of tobacco mosaic virus (TMV) [5] while endo-chitinase from *Trichoderma harzianum* showed higher resistance against phytopathogenic fungi in tobacco and apple [24,25].

Pepper (*Capsicum annuum* L.) is one of the essential Solanaceous vegetable crop possessing great economic value throughout the world. Its growth, yield and quality are reduced by numerous biotic factors such as bacterial wilt, *Phytophthora* blight, viral infections, insect pests and abiotic stresses (extreme temperatures, drought, salinity, and heavy metals) [26,27]. These stresses adversely affect the quality and yield of pepper plants. In response, plants have evolved some sophisticated defense mechanisms including oxidative burst and calluses into the cell wall and regulation of signaling networks to combat these stresses [7,28,29]. It has been reported that *Phytophthora capsici* infests pepper, eggplant, tomato, all cucurbits, and more recently snap and lima beans [30,31]. To control the attack of pathogen invasion in the host tissues, inducible biochemical reactions create a protective physiological condition [14].

The chitin-binding proteins are very important as they can enhance resistance against biotic and abiotic stress as well as in plant growth and development. The molecular function of chitin-binding protein genes in pepper plant are un-known. In the current study, sixteen chitin genes (CaChi) in pepper were mined through bioinformatics and their response to biotic and abiotic stresses and hormonal treatment were examined. Subsequently, the gene architecture, conserved domains, exon–intron structure, chromosomal location, gene duplication, gene ontology (GO) characterization, *cis*-acting regulatory elements in the promoter regions and phylogenetic relationships of the pepper chitin-binding protein were elucidated. This study provides a base for future research regarding pepper chitin-binding protein. Furthermore, differential expression was recorded against biotic (*Phytophthora capsici* two strains PC, HX-9) and abiotic (cold, drought, and salt) stress and hormonal treatment (SA,

MeJA, and ABA) along with tissue specific expression in different plant parts. This study provides a foundation for further characterization of CaChi members in pepper and valuable information regarding function of this significant gene family in other important crops as well.

2. Results

2.1. Identification, Classification and Annotation of Chitin Genes in Pepper

To comprehensively investigate and analyze the chitin-genes in pepper, the Hidden Markov Model (HMM) profile of the chitin-binding protein (Accession no. PF00187.17) was blast-searched in the pepper genome. As a result, 21 and 17 chitin genes were retrieved form CM334 and Zunla-1 databases, respectively. The gene sequences were aligned to avoid repetition and alternative splicing, and the longest sequences among them were chosen for further analysis. Among those 21 and 17 genes mined from the CM334 and Zunla databases, the genes having similar sequences with each other, were considered as a single gene. Consequently, we designed primer pairs (Table S1) for the amplification and confirmation of the doubtful gene sequences through cloning and sequencing. Finally, 16 predicted gene sequences were confirmed and then blast searched in NCBI. Nomenclature for the 16 CaChi was assigned based on their domains and chromosomal locations (Table 1 and Figure 1). The SMART results show that Chitin Binding Domain (CBD) was found in all 16 members while additional functional domains such as glycoside hydrolase_19_super family (*CaChiI1*, *CaChiI2*, and *CaChiI3*), chitinase glycoside_hydrolase_19 (*CaChiIV1* and *CaChiIV2*) and Barwin (*CaChiVI2* and *CaChi3*) were also found in this gene family (Figure 1 and Table S2). In addition, the characteristics of gene structure and protein size were quite different in CaChi gene family. The CDS of CaChi genes ranged from 258 bp (*CaChiVI1*) to 996 bp (*CaChiI1*), whereas the deduced proteins had 85–331 amino acids. The predicted *p*I values ranged from 4.71 (*CaChiIII7*) to 9.19 (*CaChiI2*), MW ranged from 9.06 (*CaChiVI1*) to 35.49 (*CaChiI1*) kDa and the instability index varied from 18.45 (*CaChiIV1*) to 68.89 (*CaChiIII4*) (Table 1 and Figure 1). The molecular formula shows that *CaChiIII3* contains the most (36) sulfur elements while *CaChiVI1* has the fewest (9) sulfur elements. All deduced proteins are shown in Table S2.

Table 1. List of Chitin-binding protein family genes identified in pepper and their sequence characteristics. Chr: chromosome; CDS: codding sequence; MW: molecular weight (kDa). the proteomic information was obtained from ExPASy (Available online: http://web.expasy.org/protparam/).

Name	Gene Locus ID	Chr	Position	CDS (bp)	MW	Instability Index	Introns
CaChiI1	*Capana07g001653*	7	195532737–195534304	996	35.49	38.08	2
CaChiI2	*Capana10g001143*	10	114626443–114627269	564	20.41	39.74	1
CaChiI3	*CA10g09850*	10	146386994–146387462	468	16.45	40.99	0
CaChiIII1	*Capana03g000778*	3	11663598–11664075	477	17.09	51.20	0
CaChiIII2	*Capana03g000780*	3	11754372–11754972	600	21.30	58.21	1
CaChiIII3	*CA03g30170*	3	245663113–245663923	810	28.93	55.61	0
CaChiIII4	*CA03g30180*	3	245700488–245700986	498	17.80	68.89	0
CaChiIII5	*CA03g30190*	3	245839016–245839764	618	22.45	35.35	1
CaChiIII6	*Capana07g001180*	7	161369376–161369835	459	16.15	47.70	0
CaChiIII7	*Capana07g001181*	7	161402785–161403394	609	21.38	64.04	0
CaChiIV1	*CA00g54030*	4	193080176–193081152	834	30.06	18.45	1
CaChiIV2	*Capana06g002084*	6	87334514–87335721	759	27.93	30.88	1
CaChiVI1	*CA07g09480*	7	147837320–147837578	258	9.06	22.23	0
CaChiVI2	*Capana08g001237*	8	126990120–126991323	606	21.31	21.57	1
CaChiVI3	*CA08g10220*	8	128199078–128200438	627	22.37	26.39	1
CaChiVI4	*CA12g08860*	12	49994036–49994360	324	11.94	34.02	0

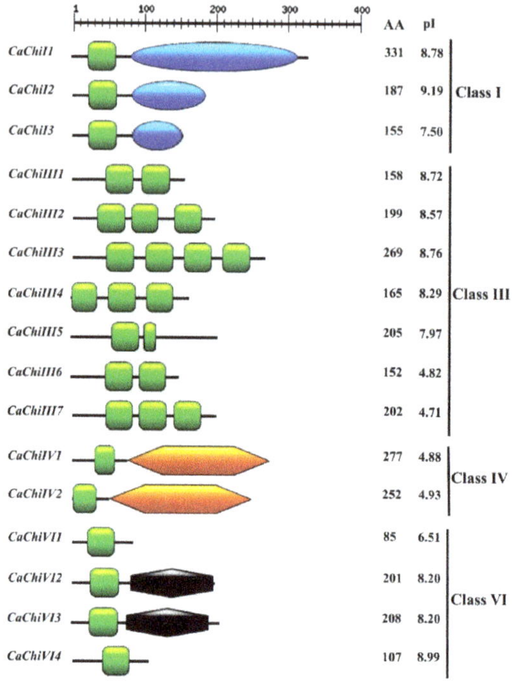

Figure 1. Domain architecture of CaChi classes I–VII in pepper and other plant species. The logos of domain organization were generated by Pfam database (Available online: http://pfam.xfam.org/search#tabview=tab0), and then further amendments were made with PhotoScape X. the aa: the number of amino acids; *p*I: isoelectric point; green ■: chitin binding domain (CBD); blue ■: glycoside hydrolase 19 super family; orange ■: chitinase glycoside hydrolase 19; and ■: Barwin.

2.2. Construction of Phylogenetic Tree, Exon/Intron Structure and Conserved Motif Analysis

To better understand the similarities and differences among the pepper and other plants chitin-binding protein genes, an unrooted phylogenetic tree was created using 162 chitin genes protein sequences from various plant species (Figure 2). These sequences used in the construction of phylogenetic tree were mainly from *Aegilops tauschii, Arabidopsis thaliana, Artemisia annua, Brassica napus, Brassica rapa, Bromus inermis, Bupleurum kaoi, Capsicum annuum, Capsicum baccatum, Capsicum chinense, Carica papaya, Cryptomeria japonica, Dionaea muscipula, Drosera rotundifolia, Euonymus europaeus, Gossypium barbadense, Gossypium hirsutum, Glycine max, Hevea brasiliensis, Hippophae rhamnoides, Hordeum vulgare, Limonium bicolor, Linum usitatissimum, Lupinus albus, Malus domestica, Mikania micrantha, Momordica charantia, Nepenthes khasiana, Nepenthes maxima, Nicotiana attenuate, Nicotiana benthamiana, Olea europaea, Oryza sativa, Persea americana, Pisum sativum, Psophocarpus tetragonolobus, Rehmannia glutinosa, Saccharum officinarum, Secale cereal, Sesamum indicum, Solanum tuberosum, Sorghum bicolor, Triphyophyllum peltatum, Triticum aestivum, Urtica dioica, Vitis vinifera, Zea diploperennis*, and *Zea mays*. The analysis shows that 16 CaChi were clearly classified into four distinct classes according to their sequence relatedness with previous research. Three CaChi (*CaChiI1, CaChiI2* and *CaChiI3*) were clustered in class I, seven CaChi (*CaChiIII1, CaChiIII2, CaCHiIII3, CaChiIII4, CaChiIII5, CaChiIII6* and *CaChiIII7*) in were clustered class III, two CaChi-genes (*CaChiIV1* and *CaChiIV2*) were clustered in class IV, and four CaChi (*CaChiVI1, CaChiVI2, CaChiVI3* and *CaChiVI4*) were clustered in class VI (Figure 2). Each class is highlighted with a different color following the previous chitin-binding protein genes classification [32,33].

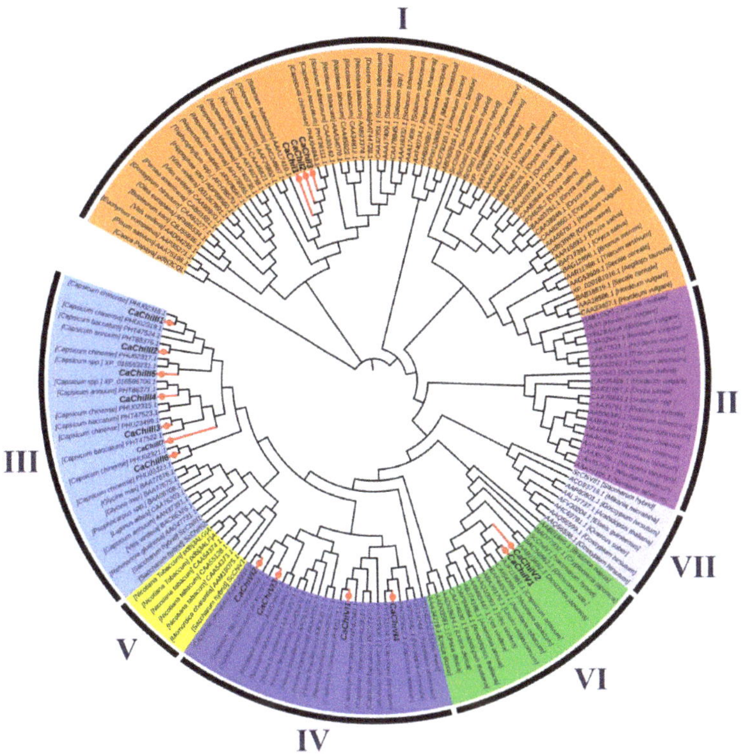

Figure 2. The phylogenic tree of chitin-binding protein family genes in pepper and other plant species. The phylogenetic tree was built using the neighbor-joining method and diagram was drawn using online iTOL (Available online: https://itol.embl.de/). The number of chitin-binding protein family genes were divided in I–VII well conserved groups.

The members within certain class exhibited higher identity percentage of the amino acids sequences (Figure 3a). The exon/intron structure analysis showed that out of 16 CaChi, 8 CaChi (50%) had no introns while 8 CaChi (50%) contained only one intron (Figure 3c). The conserved motifs of CaChi proteins were identified by online MEME server (Available online:). A sum of ten putative different motifs were obtained (Table S3). Motif 1 was found in all CaChi while motif 5 was found in most CaChi (except *CaChiIV1*, *CaChiIV2*, *CaChiIII2*, and *CaChiIII4*). Motif 6 was present in 50% of the CaChi (*CaChiIII1*, *CaChiIII2*, *CaChiIII3*, *CaChiIII4*, *CaChiIII6*, *CaChiIII7*, *CaChiIV1* and *CaChiIV2*). Motifs 2 and 10 were present in five and four CaChi, respectively. Motifs 3, 4 and 9 each were found in three different sequence of CaChi while motifs 7 and 8 each existed in two CaChi (Figure 3b and Table S3).

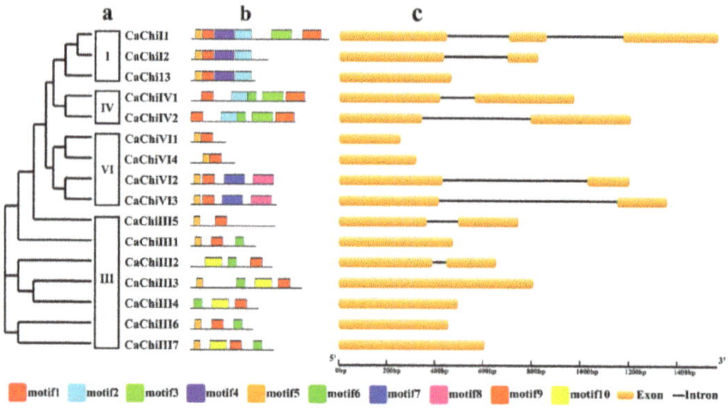

Figure 3. Phylogenetic relationship, domain organization and conserved motifs analysis of chitin-binding proteins family genes in pepper. (**a**) Phylogenetic analysis and classification of pepper genes. The phylogenetic tree was constructed via online iTOL (Available online: https://itol.embl.de/). (**b**) Motif analysis of pepper CaChi proteins. Motifs, numbered 1–10, were identified using MEME 4.11.2 software and are illustrated by different colors. Amino acid sequence of each motif is shown in Table S3. (**c**) Exon/intron structures of pepper chitins genes. Yellow boxes represent exons and introns are represented by black lines between two exons.

2.3. Chromosomal Location and Genes Duplication

According to the chromosomal location of the chitin-binding protein genes in pepper, the 16 CaChi were distributed across 7 out of 12 chromosomes of the pepper. Intriguingly, all CaChi members are random and non-randomly distributed across the chromosomes (Figure 4). The results showed that chromosome 3 had the highest number of CaChi (31.25%) as compared to other chromosomes. There were four genes (25%) on chromosome 7, while chromosomes 8 and 10 each have two genes. The remaining chromosomes (4, 6 and 12) each contained one gene. The duplication analysis showed that *CaChiIII1* have segmental duplication with *CaChiIII6* which occurred on chromosomes 3 and 7, respectively (Figure 4). *CaChiIII7* has two segmental duplication events with *CaChiIII2* and *CaChiIII5*. *CaChiIV1* also exhibited two segmental duplication events with *CaChiI1* and *CaChiIV2*. Moreover, one tandem duplication event was observed between *CaChiVI2* and *CaChiVI3*, which occurred on chromosome 8. Taken together, our findings suggest that, in the expansion of pepper CaChi genes, tandem and segmental duplication have an important contribution.

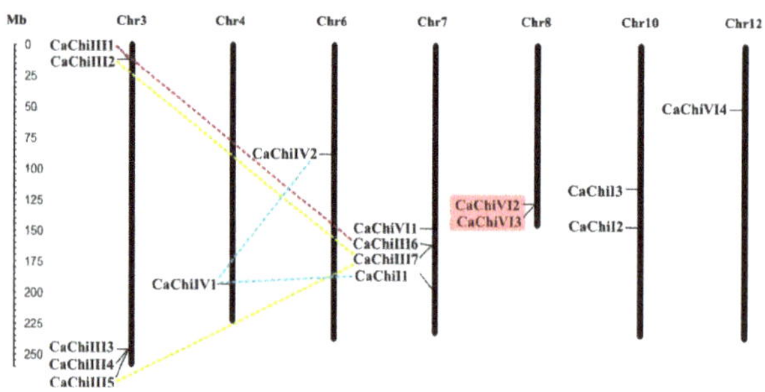

Figure 4. Chromosomal localization of CaChi of pepper plant, where the red shading box represents the tandem duplicated region. While the red, green and blue lines connection displaying segmentally duplicated genes.

2.4. Cis-Acting Elements and Gene Ontology (GO) Analysis of CaChi

To examine the possible *cis*-acting elements involvement in the stimulation of defense-related genes, the 1.5 kb upstream region from the start codon of all the CaChi genes were analyzed with Plant CARE online server. The silico analysis revealed that *Cis*-elements conferring responsiveness to plant hormones, biotic and abiotic stresses were found in the promoters of the CaChi. As shown in Figure 5 and Table S4, the heat stress elements (HSE) were identified in the promoters of all CaChi (except *CaChiI2*, *CaChiI3* and *CaChiVI3*), in which the HSE in the promoters of the *CaChiIV1* and *CaChiVI2* were highest (4) followed by *CaChiIII2*, *CaChiIII6* and *CaChiIV1* (each have 2). MeJA-responsiveness elements (CGTCA-motif) were found in the promoter region of 10 CaChi, where *CaChiIII1* had the highest number (5) of elements followed by *CaChiIII4* (3), while *CaChiIII5* and *CaChiIV2* each have two elements. *Cis*-acting elements involved in abscisic acid (ABA) responsiveness elements (ABRE), salicylic acid responsiveness (TCA-element) and ethylene-responsive element (ERE) were found in the promoter regions of six CaChi. The MYB binding site involved in drought-inducibility (MBS), resistance and stress responsiveness (TC-rich repeats), GA-responsive element (GARE-motif) and fungal elicitor-responsive element (W-box) were found in the promotor regions of 10, 8, 6 and 7 CaChi genes, respectively. The *cis*-acting element involved in low temperature sensitivity (LTR) was found in the promoter region of four CaChi. In addition, GA-responsive element (P box) and auxin-responsive elements (TGA-element and AuxRR-core) were also found in some of the CaChi promoter regions. All of the anticipated *cis*-elements were involved in response to signaling molecules and stresses.

Gene ontology (GO) enrichment analysis of CaChi were predicted by the gene ontology slim analysis using Blast2GO tool. The analysis comprised three categories, i.e., biological process, molecular function, and cellular component same as mentioned by Di et al. (2018) [34]. Our results showed that chitin catabolic processes and cell wall macromolecule catabolic processes, defense response to fungus, bacterium and response to stress were the highly regulated functions having role in biological process which support the function of the CaChi in the cell. In addition, prediction of molecular functions of CaChi proteins indicated that they had mostly involved in chitin binding capacity, chitinase activity and small molecules binding while cellular component analysis revealed that CaChi mostly localized in extracellular region. Furthermore, they can accumulate in subcellular parts of the cell such as vacuole, chloroplast, vacuole and plasma membrane (Figure 6).

Gene name	AuxRR-core	CGTCA-motif	ERE	HSE	TC-rich repeats	ABRE	Box-W1	WUN-motif	MBS	TGA-element	LTR	GARE-motif	P-box	TCA-elements
CaChiI1	0	0	0	1	0	0	0	0	0	0	0	0	0	0
CaChiI2	0	1	0	0	0	0	0	0	2	0	0	0	0	0
CaChiI3	0	0	0	0	0	0	2	0	0	0	0	0	0	0
CaChiIII1	0	5	6	1	0	0	0	0	5	0	1	2	0	0
CaChiIII2	0	1	0	2	3	0	0	0	0	0	0	1	2	1
CaChiIII3	1	1	2	1	1	0	0	0	0	0	0	0	0	0
CaChiIII4	1	3	0	1	2	1	1	0	5	1	0	0	0	0
CaChiIII5	0	2	0	1	0	0	0	1	0	1	2	0	0	0
CaChiIII6	0	0	0	2	2	2	1	0	1	0	0	0	0	0
CaChiIII7	0	1	1	1	0	0	1	0	1	1	1	1	0	2
CaChiIV1	1	2	0	4	0	0	2	0	2	0	0	1	0	1
CaChiIV2	0	1	0	1	4	0	0	0	1	1	0	1	0	2
CaChiVI1	0	1	1	2	1	2	0	0	1	0	0	0	0	0
CaChiVI2	0	0	1	4	1	1	0	0	1	0	0	0	1	1
CaChiVI3	0	0	0	0	0	3	1	0	0	0	0	1	0	1
CaChiVI4	0	0	0	1	3	0	1	0	1	0	1	0	0	0

Figure 5. *Cis*-acting regulatory elements in the promoter regions of CaChi genes. The *cis*-element positions in the individual CaChi promoter region was inferred from the Plant CARE website (Available online: http://bioinformatics.psb.ugent.be/webtools/plantcare/html/). The different number of *cis* regulatory elements represent in different colors.

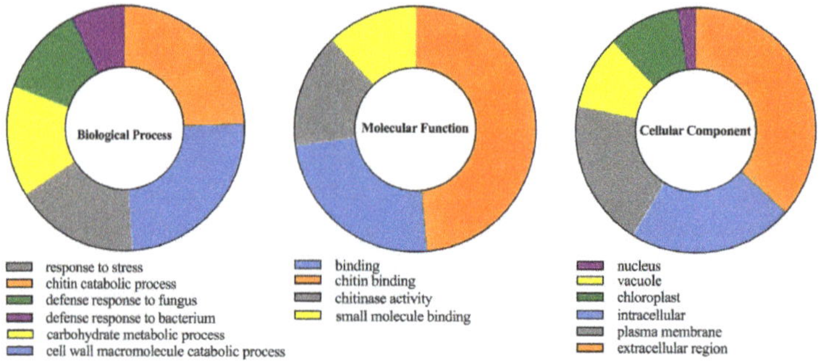

Figure 6. Gene ontology analysis of CaChi proteins in three categories (Biological processes, molecular functions and cellular component) using Blast2Go program. Different colors which are indicated near the graphics show different biological process, molecular functions and cellular component of pepper chitin-binding protein family genes.

2.5. *Expression Analysis of CaChi under* Phytophthora capsici *Strains Inoculation*

To examine the transcription levels of CaChi against the virulent (HX-9) and avirulent (PC) strain of *Phytophthora capsici*, pepper plants were inoculated with *P. capsici* via root drenching and their expression levels were analyzed by qRT-PCR. The results exposed that, post *P. capsici* inoculation, the CaChi were differentially expressed (Figure 7). Twelve CaChi (75%) were upregulated on different time points, and three members (*CaChiI1*, *CaChiIII3* and *CaChiVI4*) (18.75%) were downregulated to both strains on maximum time points. Initially, *CaChiI1* and *CaChiVI4* exhibited downregulation after inoculation with virulent strain (HX-9), then *CaChiI1* upregulated on 48 hpi and *CaChiVI4* on 72 hpi, which were 1.18 and 1.45, respectively. However, *CaChiIII2*, *CaChiIII6* and *CaChiVI1* exhibited progressive upregulation at all the time points in both strains, but *CaChiIII2* was reached to peak (29.39) at 48 hpi in HX-9, while *CaChiIII6* showed the highest transcription level after PC post inoculation (48 h), i.e., 26.17. Whereas *CaChiVI2* peaked at 12 hpi in virulent (29.38) and avirulent (29.75), *CaChiIV1* showed the highest expression compared with other CaChi, reaching a maximum at 38.52 (PC) and 45.51 (HX-9) and then slightly downregulated. Meanwhile, in the event of avirulent strain inoculation, *CaChiI2* (2.25), *CaChiIII1* (5.80) *CaChiIII3* (2.30) and *CaChiIII5* (2.59) were not predominantly upregulated but, at 72 hpi, showed slight expression. However, *CaChiIII4* and *CaChiVI2* were upregulated following the same pattern and reached a maximum 19.39 and 29.75, respectively, at 12 hpi. *CaChiI3* and *CaChiIV2* exhibited significant expression only to virulent (HX-9) strain. *CaChiIII7* revealed upregulation for both virulent and avirulent strains and reached a peak (31.10 and 37.13 respectively) at 6 hpi, then downregulated, and subsequently upregulated. Six hours post inoculation, *CaChiVI3* exhibited the highest transcription in PC strain and then later it was downregulated at every other time point; however, for HX-9 strain, its transcriptional level was raised.

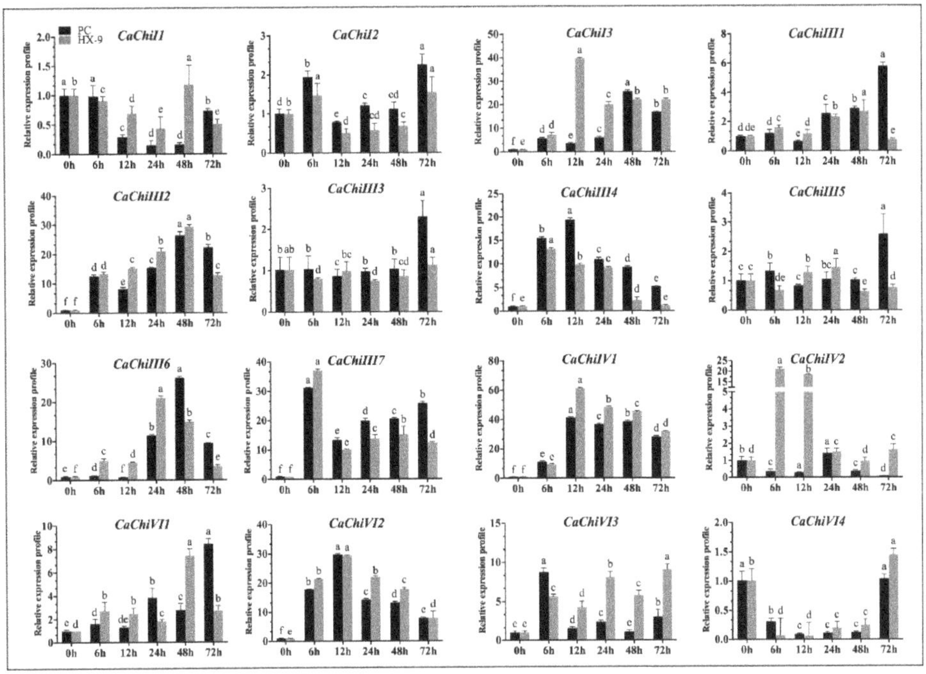

Figure 7. Expression profiles of CaChi in response to different strains of *Phytophthora capsici* (PC and HX-9). The samples were collected at different time points (0, 6, 12, 24, 48 and 72 hpt) and were analyzed by qRT-PCR. Mean values and SDs for three replicates are shown. Small letters (a–e) represent significant differences ($p < 0.05$).

2.6. Expression Profile of CaChi in Response to Abiotic Stresses

To examine the expression levels of the CaChi in response to abiotic stresses, eight representative genes (*CaChiI3, CaChiIII2, CaChiIII4, CaChiIII6, CaChiIII7, CaChiIV1, CaChiIV2* and *CaChiVI2*) were selected from the CaChi, in which at least one gene was selected from each class on the basis of their *cis*-acting elements response and expression to *P. capsici*. Then, they were subjected to NaCl, mannitol and cold stresses (Figure 8). *CaChiI3* showed no response to cold and NaCl stress while in response to mannitol it was gradually upregulated, reaching a maximum at 24 hpt (7.78), and then downregulated. *CaChiIII2* showed a slight upregulation at 6 hpt in response to NaCl while exhibited concomitant up- and downregulation in response of mannitol stress. In the case of cold stress, no expression was recorded. In response to NaCl, *CaChiIII4* initially exhibited no response, then upregulated at 6 hpt, reached a maximum at 12 hpt (16.84) and then showed a slight downregulation at 24 and 48 hpt, whereas no significant response was observed in response to cold and mannitol (Figure 8). *CaChiIII6* was not regulated by mannitol stress, whereas abrupt changes were observed to NaCl stress; however, highest expression was noticed at 6 hpt (21.32) in response to cold stress, where the expression was reduced in later hours. *CaChiIII7* was gradually upregulated in response to NaCl, reached a maximum at 6 hpt (5.53) and then downregulated. In response to mannitol, a slight up regulation was noted at 12 hpt and then downregulated similarly. In cold stress, significant expression (7.32) was observed in all time points. The transcript level of *CaChiIV1* was highly induced by mannitol, cold and NaCl at 12, 24 and 48 hpt, which were more than 12, 16 and 15 folds, respectively. *CaChiIV2* was initially downregulated by NaCl and mannitol stress and then exhibited an abrupt upregulation in response to NaCl at 12 hpt (3.34) and again downregulated. In response to mannitol stress, no significant expression occurred. In cold stress, it shown initial abrupt upregulation and then smoothly downregulated (Figure 8). Responding to NaCl stress, *CaChiVI2* was gradually upregulated, peaked at 6 hpt (9.47) then smoothly downregulated. In response to cold and mannitol, it was upregulated and reached a maximum at 6 hpt (5.86) and 48 hpt (5.40), respectively.

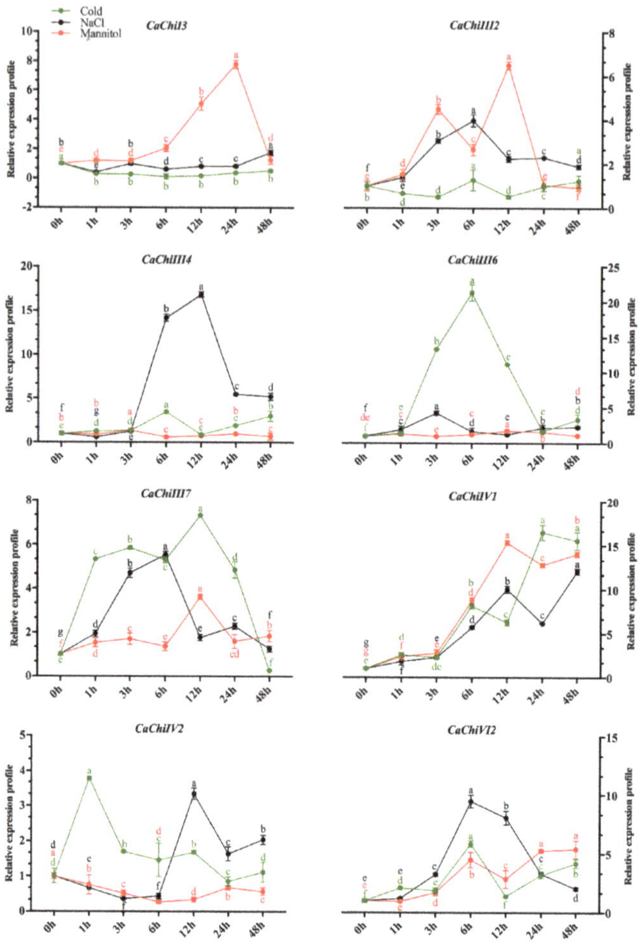

Figure 8. Expression profiles of CaChi genes in response abiotic stresses. The inducible expression patterns performed by qRT-PCR under Sodium chloride (NaCl) and Mannitol. Mean values and SDs for three replicates are shown. Small letters (a–e) represent significant differences ($p < 0.05$).

2.7. Expression Profile of CaChi in Response to Hormonal Treatments

Phyto-hormones and plant signaling molecules, such as MeJA, SA and ABA, are involved in various stress signaling pathways [35–38]. The above selected eight CaChi genes were also exposed to exogenous hormonal (MeJA, SA and ABA) treatments to explore the response of these target genes. We investigated the expression profiles of CaChi-genes in AA3 leaves. As shown in Figure 9, post SA and ABA treatment, five CaChi (*CaChiI3*, *CaChiIII2*, *CaChiIII6*, *CaChiIV1* and *CaChiVI2*) were significantly upregulated (>10, 11, 18, 6, 8 folds against SA and >6, 24, 52, 10 and 14 folds against ABA, respectively) at different time points, while *CaChiIV1* and *CaChiIV2* were gradually upregulated over time and maximum expressions at 48 (10 folds) and 12 hpt (two folds) were recorded in response to MeJA treatment. The *CaChiIII4* was initially upregulated 1 hpt after ABA treatment (>5 folds), and then irregular changes were noticed, while, in response to MeJA and SA, the expression level was very low or even not obvious, except for SA where an abrupt upregulation (2.29) was noted at all given time points. *CaChiIII7* responded to MeJA and reached a peak at 48 hpt (10.60), whereas the response to ABA was antagonistic-like initially where an upregulation and then smooth decline were observed. Expression was not induced by SA treatment. The transcript levels of *CaChiIV1* gene were induced by MeJA, steadily increased and reached a peak (10.90) at 48 hpt. In the case of SA treatment, it was gradually upregulated, reached a peak (6.82) at 6 hpt, downregulated at 12 and 24 hpt, and again upregulated at 48 hpt. After ABA treatment, transcription level was high at 1 (7 folds) and 12 hpt (10 folds). The *CaChiIV2* was upregulated by SA, reached to peak at 48 hpt (8.76). After ABA application, it abruptly upregulated at 1 and 12 hpt (>9 and 16 folds, respectively), while it showed no significant response to MeJA.

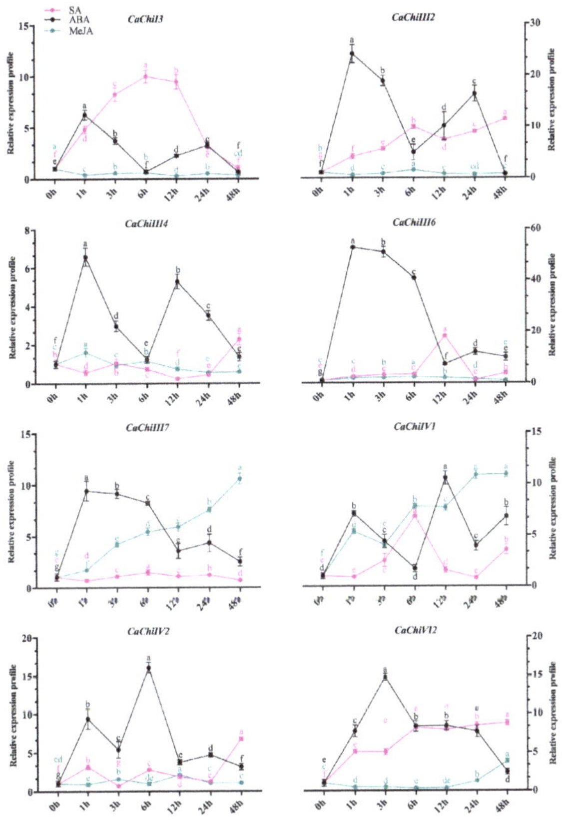

Figure 9. Expression profiles of CaChi in response hormones application. The inducible expression patterns performed by qRT-PCR under Salicylic acid (SA) and methyl-jasmonate (MeJA). Mean values and SDs for three replicates are shown. Small letters (a–f) represent significant differences ($p < 0.05$).

2.8. Expression Patterns of CaChi-Genes in Different Tissues

To further elucidate the expression characteristics of CaChi genes in various vegetative and reproductive tissues (leaf, stem, root, seven developmental phases of placenta and pericarp), we carried out in silico analysis using public transcriptomic database of pepper [39,40]. The different expression patterns of CaChi in pepper exhibited higher variance in distinctive tissues and stages, as demonstrated in heat map where from green to red display the index of expression (Figure 10A). Moreover, some CaChi had highest expression levels in all stages of plant growth and development, such as *CaChiI1*, *CaChiIII4*, *CaChiIII5*, *CaChiIV2* and *CaChiVI3*, while some of them had very low or no expression in all the tested tissues, i.e., *CaChiI3*, *CaChiIII1*, *CaChiIII6* and *CaChiVI1*. (Figure 10A). Some genes (*CaChiIV1* and *CaChi13*) are expressed only in particular tissue, mostly in seedling stage. Additionally, to further authenticate the CaChi expression level in various vegetative and reproductive tissues, we cultivated AA3 pepper variety in normal condition and different tissues were collected at different stages. Gene specific primers were used for qRT-PCR analysis (Table S5). As shown in Figure 10B, the expression pattern of CaChi dominantly expressed in seed except *CaChiVI2*, *CaChiVI3*, *CaChiVI4* and *CaChiIII5* expression were maximum in stem/flower, stem and stem, respectively. The lowest expression was recorded in red fruit comparing to other tissues except *CaChiVI1* (529.1), where the particular gene highly correspond to the development of red fruit, flower and leaf. Therefore, it can be assumed that the expression pattern implied by *CaChiVI1* may function importantly in red fruit development. Moreover, *CaChiI2*, *CaChiIII2*, *CaChiIII6*, *CaChiIII7* and *CaChiIV2* did not show highly significant expression to any tested tissues excluding seed but *CaChiIII6* having same results in silico. The integrated investigation of publicly available dataset revealed the ubiquitous expression of these genes and a number of CaChi exhibited a certain degree of tissue specificity. The noticeable differences in both expression level might be related to variation in biological materials, regulation of transcript, interpretation methodology and environmental fluctuation.

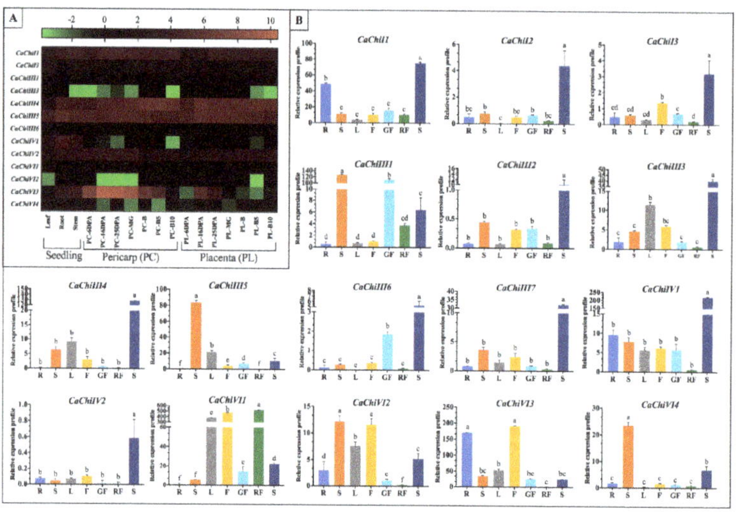

Figure 10. Developmental expression profile of chitin-binding protein family gene in pepper. (**A**) The expression pattern retrieved from the database of pepper (CM334), indicating different expression levels of CaChi genes in dissimilar organs. The results were log2 transformed before generating heat maps in leaf, root, stem, 6, 16, 25 days post-anthesis (6DPA, 16DPA, and 25DPA), mature green (MG), breaker (B), 5 and 10 days post-breaker (B5 and B10) of pericarp (PC) and placenta (PL). Three genes (*CaChiI2*, *CaChiIII2* and *CaChiIII7*) were from the Zunla-1 database, so they were not mentioned in the figure. (**B**) The graphs indicate tissue specific expression levels of chitin-binding protein family genes in pepper plant. The samples were collected from different parts root (R), stem (S), leaf (L), flower (F), green fruit (GF), red fruit (RF) and seed (S) analyzed by qRT-PCR. Data are the means of three independent qRT-PCR amplifications. Small letters (a–f) represent significant differences ($p < 0.05$).

2.9. Reduced Tolerance of CaChiIV1-Silenced Pepper Plants to NaCl

To evaluate the role of *CaChiIV1* under NaCl stress, the empty vector (used as control) and *CaChiIV1*-silenced plants were treated with NaCl (300 mM) solution. As shown in Figure 11A, the silencing of *CaChiIV1* significantly compromised resistance to NaCl stress. A greater transcript level of *CaChiIV1* was noted in the control (empty vector) plants than in the *CaChiIV1*-silenced plants in all time points, which is >3 folds. In addition, the transcript levels of other defense-related genes were also studied to see whether the silencing of *CaChiIV1* changes their expression. It was noted that, with the passage of time after NaCl treatment, the expression of *CaDEF1* (defensin) [41] and *CaSAR8.2A* (systemic acquired resistance) [42] were changed, but their rise in the control plants (empty vector) were greater compared to *CaChiIV1*-silenced plants (Figure 11B,C). Additionally, root activity was also studied, and the results revealed a significant decrease in the root activity after NaCl stress. At 24 h post NaCl stress, the root activity of the *CaChiIV1*-silenced plants (1.1) was less than TRV2:00 (2.7) (Figure 11D).

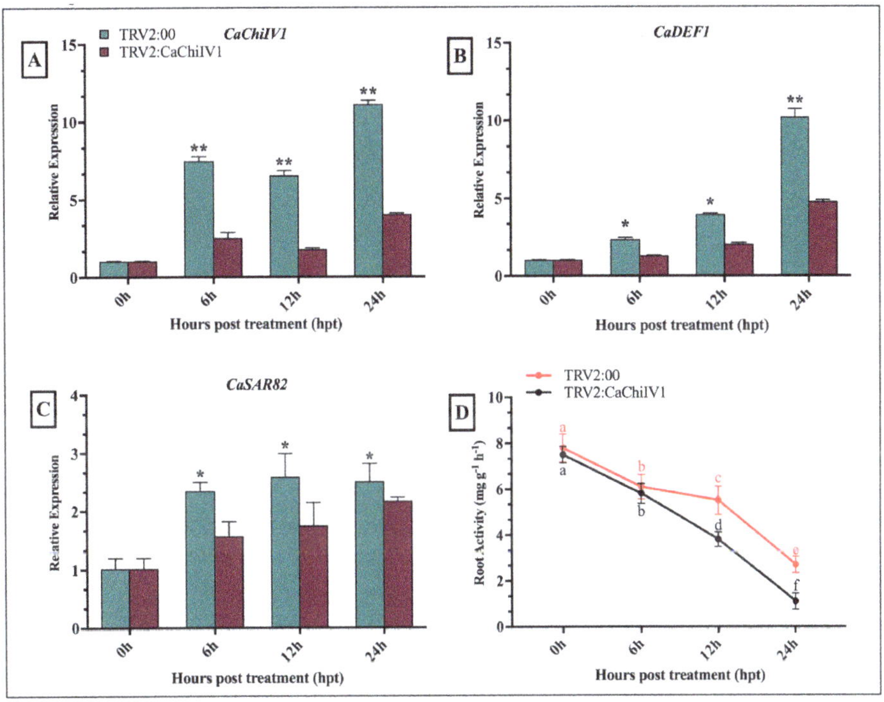

Figure 11. The *CaChiIV1*-silenced pepper plants exhibit reduce resistance to NaCl stress: (**A**) The transcript level of *CaChiIV1*; (**B**) transcript level of *CaDEF1*; (**C**) transcript level of *CaSAR82*; and (**D**) root activity of the control and *CaChiIV1*-silenced plants. Values are the means ± SD from three separate experiments. Small letters (a–f) and asterisk (*significant and **highly significant) denote significant variation ($p < 0.05$).

3. Discussion

Chitin-binding protein (CBP) is an important biotic and abiotic resistance responsive multigene family in plant [20,32,33,43], which play an important role to enhance resistance against stresses in different crops [15]. Chitin-binding proteins are well characterized class of PR proteins [44] which are speculated to be involved in the production of proline due to proline and glycine-rich region [20].

The number of chitin-binding protein genes varies in different plant species. Formerly, 24 chitins in *Arabidopsis thaliana*, 37 in *Oryza sativa* and 17 chitin genes in *Saccharum officinarum* were reported [20,32].

In the past, no comprehensive study has been conducted on genome-wide identification and characterization of chitin-binding proteins in pepper. Therefore, in the current study, we retrieved 16 CaChi genes from "CM334" and "Zunla-1" databases of pepper genome. Previously chitins were also found in different plant species [15,20,32,45]. The structural analysis revealed that out of 16 CaChi, eight (50%) CaChi contained introns in which seven genes (43.75%) had only one intron while *CaChiIV1* contain two introns (Figure 3c). Consequently, previously studies on the chitin genes in brassica and banana showed contrasting results [33,46]. The reason may be due to expansion in CaChi family in pepper plant and it may be concluded that the CaChi have undergone diverse gene structure changes during evolution process. The ORFs analysis revealed that the amino acid (aa) sequences ranged from 85 aa (*CaChiVI1*) to 331 aa (*CaChiI1*). The subcellular location of all the CaChi in pepper were predicted, and found that they exist in chloroplast, extracellular region, nucleus, cytoplasmic and vacuolar locations (Table 1). Nishizawa et al. (1999) and Collingel et al. (1993) [47,48] also identified chitins genes in *Oryza sativa*, *Pisum sativum* and *Hordeum vulgare* and described their subcellular localization in vacuole and extracellular.

Previous studies revealed that the nomenclature of the chitins gene family had seven (I–VII) distinct classes [4,15,33]. Therefore, in our study, we also classified CaChi genes with previous criteria into four classes (classes I, III, IV and VI). Parallel, results were also obtained by Backiyarani et al. (2015) [46]. Thus, there is strong possibility that the classification may be due to the homology of the protein sequences and the presence of core conserved domain and probably also in function. The phylogenetic analysis showed that CaChi genes can be divided into four distinct classes: *CaChiI*, *CaChiIII*, *CaChiIV* and *CaChiVI*. It was noticed that sequences contain the glyco_hydro_19 super family, chitinase_glyco_hydro_19 and barwin domains, and they were classified into classes I, IV and VI, respectively (Figure 2). Typical CaChi exhibited higher similarity in the sequence of their conserved domains but an obvious diversity in gene structure and protein size (Figure 1 and Table S2), implying an evolutionary relationship between CaChi genes. It shows that CaChi share a common ancestor and some similar biological functions [32,49].

The chromosomal locations exposed that CaChi were detected on seven diverse chromosomes of *Capsicum annuum*, the highest number of CaChi genes (5) were found on chromosome 3 followed by 7 which had four CaChi. In the duplication and transposition analysis of CaChi, we obtained five clusters of segmental duplication and two genes tandemly duplicated on chromosome 8 (Figure 4). Similarly, our findings are also supported by Cannon et al. (2004) and Backiyarani et al. (2015) [46,50], who also found that tandem and segmental duplication events in *Musa* spp. Therefore, the expansion of a gene family might be segmental and tandem duplication or transposition is the core evolutionary tools [50]. Thus, comparing with tandem duplication and transposition, segmental duplication happens more frequently because of polyploidy in utmost plants, which conserve numerous duplicated chromosomal blocks in their genomes [50]. These conclusions suggest that the pepper CaChi genes endured a complicated evolutionary history during the gene expansion and functional divergence.

Tissue specific expression is a common characteristic of the genes of a certain protein family in plants, which often reflects the functional collaborative and/or differences of the family members [51]. The preceding studies also shed light on developmentally regulated plant chitinases, clarifying their role in the specific physiological processes [52]. To further clarify the possible functions of the CaChi in the growth and development of *Capsicum annuum*, the transcription profiles of CaChi were studied through qRT-qPCR in different tissues. The results showed that mostly CaChi genes exhibited the highest expression during seed formation, followed by the stem, flower, leaf, root and green-fruit (Figure 10B), while the lowest expression was detected in red-fruit. This is similar to previous studies on *Brassica*, where comparatively higher expression levels were observed in flower followed by stem, leaves and roots [33]. Furthermore, Su et al. (2015) [53] also detected chitin gene expression in sugarcane and found high expression in stem pit compared with leaf, and stem epidermis. The transcriptomic analysis as shown in Figure 10A also showed the same expressions in most of the tested tissues but in some case, the transcriptional level is changes may be because of different cultivar was used and some other environmental factors may be involved. The pepper CaChi were expressed in an organ-specific way, suggesting the probable functions in distinct biological processes. CaChi genes suggesting their specific roles in heading stage implying its vital roles in growth regulation and stress response in pepper plant.

Plant diseases activated by fungal pathogens are one of the main concerns and promote defense to a plant pathogen is a difficult mechanism, which includes the triggering of several immune responses [1]. Several resistance genes, including pathogenesis related proteins, have been isolated and were used to improve the defense to different disease in plants. The pathogenesis related proteins are present during hypersensitive response to pathogens of bacteria, virus, fungi and are responsible for the induced resistance in plants. Although it is found that PR proteins not only develop resistance in plants but also play role against pathogen as well as abiotic stresses in susceptible condition [54,55]. In nature, the PR proteins are widely found in the form of chitins genes and play crucial role in plant defense system against the pathogen attack. Multiple antifungal chitinases (CH1, CH2 and CH3) have been reported in various crops such as, from *Sorghum bicolor* [15,56]. Recently, wheat class VII chitinases showed broad-spectrum antifungal activity against *Alternaria* sp., *Sarocladium oryzae, Fusarium sp., C. falcatum, Pestalotia theae* and *Rhizoctonia solani*. The partial mRNA sequences of the chitins *ScChiB1* were amplified from both cultivars (red rot-compatible and incompatible) of sugarcane [57]. In addition, chitin family genes were found to be involved in proline synthesis and primarily associated with defense and resistance against pathogens [32,49], and Western blotting study revealed greater and faster accumulation of chitin genes in a red rot-resistant cultivar [58].

The cellular predicted function of CaChi proteins indicated that they had mainly involved in defense response to fungi, bacterium infection, cell wall macromolecule catabolic process, chitin catabolic process and also in other stresses. The molecular functions of chitinase activities and chitin binding were predicted during GO analysis (Figure 6). Our results correlate with the results of Kovacs et al. (2013) and Rahul et al. (2013) [21,57], who find the functions of chitins genes in *Musa* spp. and *Saccharum officinarum*, respectively. In the current research work, during *Phytophthora capsici* inoculation (0–7 hpi), the transcription level of at least 11 CaChi (*CaChiI2, CaChiI3, CaChiIII1, CaChiIII2, CaChiIII4, CaChiIII6, CaChiIII7, CaChiIV1, CaChiVI1, CaChiVI2* and *CaChiVI3*) were highly induced by both strains (Figure 7). Moreover, the transcript levels of *CaChiI2, CaChiIII4* and *CaChiIII7* were higher in response to the PC-strain, whereas *CaChiI3, CaChiIII2, CaChiIV1, CaChiIV2, CaChiVI2* and *CaChiVI3* showed higher expression after inoculation with the HX-9 strain versus the PC-strain. These results suggest that CaChi genes are pathogen-inducible and perhaps also participate in immune system of pepper plant, thus helping in disease resistance. It has been found that CaChi are responsive to several biotic stresses and their transcripts were greatly increased (Figure 7). An earlier study also showed that chitins genes were induced by a fungus via *Phytophthora capsici* [49], whereas a class III sugarcane chitinase gene (*ScChi*) was exposed to be induced by *S. scitamineum* [59]. The expression of the *CABPR1* gene in pepper was higher in the virulent strain interaction versus the avirulent interaction [60]. However, in contrast to our previous studies [61], it was found that the expression of most *CaSBPs* was comparatively greater in the avirulent interaction than in the compatible interaction. Some other studies have revealed that the expression of oxysterol-binding protein gene (*CanOBP*) and a novel peroxidase gene (*CanPOD*) were higher in the incompatible interaction [62,63] and the defense-related genes such as b-1, 3-glucanase gene (*CABGLU*), disease-associated protein gene (*CABPR1*), and peroxidase gene (*CAPO1*) were expressed in a similar pattern, after the inoculation of virulent and avirulent strains of *Phytophthora capsici* on the roots of pepper as reported by Wang et al. (2013) [62]. The differences in the expression patterns of CaChi and other defense-related genes might be because of dissimilarities in the inoculation of *Phytophthora capsici* strains, duration of infection, cultivar or the variation in their compatibility systems.

Earlier studies have shown that chitins genes were expressed in different patterns in response to various abiotic stresses [16,59,64]. Based on GO analysis, CaChi genes imply a role in different stresses, in support of our study. Kumar et al. (2017) [65] also investigated the *OsWRKY71* gene enhanced cold tolerance and mainly involved in metabolic as well as in regulatory pathways in rice, while Eroglu and Aksoy (2017) [66] studied COP9 response under Fe deficiency in Arabidopsis. Hence, we extended our study to investigate the expression analysis of eight representative CaChi after salt (NaCl), cold (6 °C) and drought (mannitol) stresses (Figure 8). The results revealed that *CaChiI3*, *CaChiIII2*, and *CaChiIVI* genes showed significant response to mannitol, while *CaChiIII4*, *CaChiIII7*, *CaChiIVI CaChiIV2* and *CaChiV12* exhibited greater response to cold stress. Therefore, we assumed that chilling induced intracellular Ca^{2+} overload may enhance the ROS production which is key component response to chilling stress. The *CaChiIII6*, *CaChiIII7* and *CaChiIVI* genes expression was maximum at 6 hpt salt stress treatment. These results are supported by Yin et al. (2014) [67] who studied *CaAQP* gene in pepper under salt stress and conform the highest expression at 4 h salt stress but in latter hours the expression was dramatically reduced. The reason for downregulation of permeability of membrane is it results in limitation of water loss from vacuole. These effects propose that the dissimilar pepper chitin-binding protein have separate functions in response to numerous environmental stresses. However, as discussed above, different CaChi exhibit strong spatiotemporal and tissue-specific expressions, demonstrating an obvious collaborative and/or divergence in both biological roles and evolutionary relationship of the chitin-binding protein family genes in pepper.

Chitin-binding protein in plants are responsive for the certain level of abiotic (low temperature, drought, heavy metals and salt) stresses and plant hormones [1,58,59]. Previous reports show that MeJA. SA and ethylene are involved in signal compounds inducing two kinds of defense such as for induced systemic resistance (ISR) and systemic acquired resistance (SAR) [68]. The basic defense to biotrophic pathogens is mediated by SA [3]. In plant responses to environmental cues, MeJA plays the fastest role in resistance reaction via signal molecule reaction center and the genes which are related to MeJA showed upregulation, causing hyper accumulation of MeJA under biotic and abiotic stresses [69]. Several hormonal responsive elements were located in the promoter regions of CaChi, e.g. for MeJA (CGTCA-motif) [70] and SA (TCA-element) [71]. In light of this evidence, pepper plants were exposed to SA and MeJA stresses and their effects on the expression levels were investigated. The expression pattern of CaChi could be differentially regulated by MeJA, ABA and SA (Figure 9). External application of SA lead in an increased accumulation of *CaChiI1*, *CaChiIII2*, *CaChiIII6*, *CaChiIV1*, *CaChiIV2* and *CaChiIV2* expression, *CaChiIII7*, *CaChiIV1* and *CaChiVI2* showed maximum expression level against MeJA and *CaChiIII2*, *CaChiIII6*, *CaChiIII7* and *CaChiV12* exhibit increased transcription by ABA application. It should be noticed that the expression profiles of the members of CaChi have distinct characteristics approach in response to these hormone treatments. For example, Guo et al. (2013) [72] reported that the ABA responsive genes in pepper reduce cold stress injuries and help plants to combat unfavorable environment. Similarly, the evaluated level of these hormones may enhance the antioxidant activity also reduce the accumulation of reactive oxygen species (ROS) and thus play a crucial role in signaling pathways and ultimately in plant defense system.

For functional characterization of *CaChiIV1* gene initially, we searched the dataset of *cis*-regulatory elements in the promoter region and then we successfully knocked them down in pepper plant. Further, we performed an expression analysis under salt stress condition (Figure 11A). The results displayed significant response to salt stress as well as the defense related gene when compared to control plants. Similar to our results, rice *OsDIRs* exhibit greater response to salt stress as compared with mock-treated control seedlings [73]. Wu et al. (2009) [74] also reported similar finding during their studies on dirigent protein gene from the resurrection plant *Boea hygrometrica*. In current study, we also performed the detached leaf assay of the *CaChiIV1*-silenced gene with control plants treated with salt stress. The silenced plants showed the reduction in chlorophyll content, suggesting that due to knock down of *CaChiIV1* gene the pepper leaves are more susceptible to the NaCl stress (Figure 12). In pepper, the dehydrin *CaDHN1* silenced plants showed decrease in chlorophyll contents after three days of salt treatments [75]. Thus, the degree of leaf senescence in *CaChiIV1*-silenced plants are greater than control. Furthermore, the TTC reductase activity in the roots of pepper plant was measured after NaCl stress in both silenced and control plant with a duration of 0–24 h. the significant reduction in root activity was seen in silenced than control plants (Figure 11D). The root activity was significantly reduced of the silenced plants in avirulent strain of *P. capsica* than virulent strain [62]. Moreover, the *CaPTI1* gene in pepper also showed the significant differences in root activity [76]. The reason plants are susceptible to salt stress is that it causes severe injury in root tips and may lead to reduction in root activity. In conclusion, the *CaChiIV1* gene demonstrated a crucial role in biological processes and functionally involved in NaCl stress and defense response in pepper plant.

Figure 12. The *CaChiIV1*-silenced pepper plants reduced resistance to NaCl stress. Leaf discs phenotypes (0.5 cm in diameter) of the *TRV2:CaChiIV1* and TRV:00 plants in response to 0 mM, 100 mM and 300 mM NaCl stress after 48 h.

4. Materials and Methods

4.1. Identification and Sequence Analysis of CaChi Genes Family in Pepper

To identify the CaChi family members based on the conserved domain, accession No. "PF00187.17" was collected from Pfam database (http://pfam.xfam.org/) as described in our previous study [61,77]. To further authenticate the CaChi family members, the CaChi were aligned with DNAMAN to cross check in both CM334 (Available online: http://peppergenome.snu.ac.kr/download.php) [40] and Zunla-1 (Available online: http://peppersequence.genomics.cn/) [78] databases of pepper, while the obtained sequences were from the latest versions, i.e., v1.55 and v2.0, respectively. Furthermore, gene-specific primer pairs (Table S1) were designed by using Primer Premier 6.0 (Premier Biosoft International, Redwood City, CA, USA) to amplify the different target regions.

4.2. Phylogenetic Relationships, Sequence Alignment and Physio-Chemical Properties of CaChi Genes

Multiple sequence alignment of the pepper chitin-binding proteins were performed by ClustalW according to previous studies on plant chitinases [15,32,33]. The phylogenetic tree was built with iTOL (Available online: https://itol.embl.de/) [79] using neighbor-joining (NJ) method with 1000 bootstrap replicates. Nomenclature of the putative CaChi genes were assigned based on their class and chromosomal order. To compute the molecular formula, total number of items, instability index, molecular weight (MW), molecular formula (MF) and theoretical isoelectric point (*p*I), the amino acid sequences were blast in Expasy ProtoParam (Available online: http://web.expasy.org/

protparam/) [80] and WoLF 32 PSORT II (Available online: http://www.genscript.com/wolf-psort.html) [81]. The TargetP online tool (Available online: http://www.cbs.dtu.dk/services/TargetP/) [82] was used to predict the subcellular locations.

4.3. Exon–Intron Structure Analysis, Conserved Motifs and Domain Architecture

The gene structures (exon–intron) of CaChi were obtained by aligning the CDS sequences with their corresponding genomic sequences, and their structures were shaped by using online Gene-Structure-Display Server 2.0 as described by Kang et al. (2016) [83]. Conserved motifs of the CaChi genes were recognized using MEME tool (4.12.0) (Available online: http://meme-suite.org/tools/meme) as described by Guo et al. (2016) [84] with maximum number of motifs = 10. To increase the confidence level, all candidate protein sequences were further scrutinized for presence of the functional domains by the online tools, Conserved Domain Database (CDD) (Available online: http://www.ncbi.nlm.nih.gov/cdd/), SMART (Available online: http://smart.embl-heidelberg.de/), and EMBL-EBI (Available online: https://www.ebi.ac.uk/interpro/), and their diagrams were generated using online EXPASY server (Available online: https://prosite.expasy.org/mydomains/).

4.4. Analysis of Cis-Regulatory Elements and Gene Ontology of CaChi

To analyze the *cis*-acting elements, 1500 bp upstream from the start codon (ATG) of the CaChi were obtained from the pepper genome database (PGD) and queried against the PlantCARE (Available online: http://bioinformatics.psb.ugent.be/webtools/plantcare/html/) [85] and Neural Network Promoter Prediction (Available online: http://promotor.biosino.org/) online servers. The gene ontology analysis of CaChi protein sequences were obtained from Blast2GO (Available online: http://www.blast2go.com) program [86].

The amino acid sequences were blast in Blast2GO program and three groups of GO classification (molecular functions, biological process and cellular component) were recovered.

4.5. Chromosomal Location and Genes-Duplication Analysis of CaChi Genes

Information about the chromosomal location of CaChi genes were obtained from Pepper Genome Platform (PGP) (Available online: http://peppergenome.snu.ac.kr/) as described by Zhang et al. (2016) [61], and the genes were mapped on chromosomes using MapDraw [87]. Duplication analysis within the pepper genome was carried out with the criteria depicted by Gu et al. (2002) [88] further detail as: (1) the FASTA-alignable region among the two proteins should be more than 80% of the longer protein sequence; and (2) the identity between the two protein sequences (I) should be $I \geq 30\%$ if the alignable region is longer than 150 amino acid and $I \geq 0.01n + 4.8L^{-0.32(1 + \exp(-L/1000))}$ otherwise, where $n = 6$ and L is the alignable length between the two protein sequences of the gene [88,89].

4.6. Transcriptomic Data Analysis of the CaChi in Different Tissues

Publicly available transcriptomic data of root, stem, leaf, and for both pericarp and placenta at mature green (MG), breaker (B), 5 days post-breaker (5B), 10 days post-breaker, 6 days post anthesis (6DPA), 16 days post anthesis (16DPA) and 25 days post anthesis (25DPA) for pepper cultivar CM334 were retrieved from online server (Available online: http://peppergenome.snu.ac.kr/) which have been generated previously by Kim et al. (2014) [40]. The data were based on the Reads Per Kilobase per Million mapped reads (RPKM) analysis indicating the transcriptomic level of CaChi members and the results were presented using a heat map.

4.7. Plant Materials and Inoculation with Phytophthora capsici Strains

Pepper cultivar AA3, as well as *P. capsici* strains were obtained from the Laboratory of Vegetable Plant Biotechnology and Germplasm Innovation, Northwest A&F University-China. The method used for *P. capsici* inoculation was same as described by Khan et al. (2018) and Zhang et al. (2016) [61,77].

The root samples were amassed 0, 6, 12, 24, 48 and 72 h post inoculation (hpi) of *P. capsici*. For tissue specific expression, samples were collected from roots, stems, leaves, flower, green fruits, red fruits and seeds of untreated pepper plant for RNA extraction and qRT-PCR analysis [76,77].

4.8. Hormonal Applications and Abiotic Stresses Treatments

For hormonal treatments the plantlets were treated with 50 μM methyl jasmonate (MeJA), 5 mM salicylic acid (SA) and 0.57 mM ABA solution [90]. Plantlets were kept at 28 °C in condition of 16 h light/8 h dark photoperiod and samples were collected at 0, 1, 3, 6, 12, 24 and 48 h post treatment (hpt). For abiotic stresses, some pepper seedlings were exposed to low temperature (6 °C) in chamber and others separately treated with 300 mM both NaCl and mannitol for 0, 1, 3, 6, 12, 24 and 48 hpt [72]. The samples were directly frozen in liquid nitrogen after harvesting and saved at −80 °C for RNA extraction. The experiments were carried out in three biological replicates.

4.9. RNA Extraction and qRT-PCR Analysis

Total-RNA was extracted from different samples using Trizol reagent (Invitrogen, Carlsbad, CA, USA) following the instruction of manufacturer's protocol. Further treatment of RNA was achieved with RNase-free DNaseI to eliminate DNA contamination. The cDNA was synthesis by using the Prime-Script TM RT Reagent Kit (TaKaRa, Dalian, China). The quality of cDNA was checked by nanodrop (Thermo Scientific NanoDrop 2000C, Wilmington, DE, USA) and the required volume was calculated and adjusted the concentration up to 50 ng/μL. For qRT-PCR analysis, the gene-specific primers (Table S5) were designed by using Primer Premier 6.0 software package (Available online: http://www.premierbiosoft.com/primerdesign/index.html). The specificities of the primers were further confirmed through NCBI Primer BLAST (Available online: https://www.ncbi.nlm.nih.gov/tools/primer-blast/). The pepper ubiquitin-conjugating protein gene (*CaUbi3*) was used as internal control [91], with a little modification of annealing temperature (60 °C for 30 s). The relative expression levels of all the CaChi genes were calculated using the $2^{-\Delta\Delta Ct}$ method [92].

4.10. VIGS Assay of CaChiIV1

The VIGS approach was used for the knock-down of the *CaChiIV1* gene in the pepper plant cultivar AA3, and the VIGS assay was performed as described by Liu et al. (2016) [93]. For *pTRV2:CaChiIV1*, a 232-bp cDNA part of *CaChiIV1* gene was PCR-amplified. Briefly, the *CaChiIV1* gene was cloned into a pTRV2 vector to construct the recombinant plasmid *pTRV2:CaChiIV1*, which was further used in the subsequent research to confirm the exact silencing of *CaChiIV1* (primer pairs used for vector construction are given in Table S5. Afterwards, the freeze–thaw method was used to transform pTRV1, pTRV2 (negative control), and *pTRV2:CaPDS* (positive control) along with the combined vector *pTRV2:CaChiIV1* into an *Agrobacterium tumefaciens* strain (GV3101). *A. tumefaciens* harboring pTRV1 was mixed at a 1:1 ratio with pTRV2, pTRV2-CaPDS and pTRV2-*CaChiIV1*. The agrobacterium inocula suspensions harboring pTRV1, pTRV2:00, *pTRV2:CaPDS* or *pTRV2:CaChiIV1* (OD600 = 1.0) were infiltrated into the full extended cotyledons leaves of pepper plants using a 1.0 mL clean needleless syringe [94]. Then, these infiltrated plants were conserved at 18–22 °C in a plant growth chamber with a 16/8 h light/dark period as defined by Wang et al. (2013) [62,95]. Forty-five days post-infiltration, leaf samples from the control and *CaChiIV1*-silenced plants were collected to measure the silencing efficiency by RT-PCR. The triphenyltetrazolium chloride (TTC) method was used to measure the root activity [62,95]. Before the TTC test, root tips (approximately 0.2 g) from the control (TRV:00) and *CaChiIV1*-silenced (*pTRV2:CaChiIV1*) plants were collected at various time points after NaCl stress as described by Khan et al. (2018) [77]. These experiments were executed with three biological repeats.

4.11. Statistical analysis

The results were subjected to an analysis of variance (ANOVA) using SPSS software (SPSS version 23.0, SPSS Inc., Chicago, IL, USA), and the analyzed data were expressed as means ± standard deviation (SD) of three replications in all measured parameters. The least significant difference (LSD; $p < 0.05$) test was used to measure the significant differences among the given treatments. Data were presented in graphs and designed using GraphPad Prism 7.0 (GraphPad Software, Inc., LA Jolla, CA, USA).

5. Conclusions

The findings of our research work briefly explain that the pepper chitin-binding protein gene family performs a vital role in the complex signaling networks system in response to numerous biotic and abiotic stresses and exogenous hormone applications. Every member of CaChi has its specific expression pattern and functional preference. Meanwhile, the particular molecular mechanism and function of pepper chitin-binding protein are still unclear. Additionally, the chitins–genes connections are also weakly discussed. The various transcription pattern of pepper chitin genes has been observed in tissues due to fluctuations in environmental circumstances, for example salt, drought and hormones, suggesting the diverse roles and inimitable transcription levels of chitin genes in the pepper plant growth, development and response to different stresses. Accordingly, different changes in transcription level in the same pepper chitin gene against several biotic and ambient alterations confirm the differences in their mechanism of regulation. These outcomes clarify the background for further experiments and provide the basic knowledge to explore the role and the possible cross-talk between pepper chitin-binding proteins in plants.

Supplementary Materials: Supplementary materials can be found at http://www.mdpi.com/1422-0067/19/8/2216/s1.

Author Contributions: Conceptualization, M.A. and Z.-H.G., Methodology, M.A. and W.-X.G., Software, G.-X.C., Validation, M.A., I.M. and A.K., Formal Analysis, H.-X.Z.; Investigation, Z.-H.G., Resources, Z.-H.G., Data Curation, S.u.H., Writing-Original Draft Preparation, M.A., Writing-Review & Editing, I.M. and A.K., Supervision, Z.-H.G., Project Administration, Z.-H.G. and D.-X.L., Funding Acquisition, Z.-H.G. and D.-X.L.

Funding: This work was supported through funding from the Independent Innovation Fund Project of Agricultural Science and Technology in Jiangsu (No. CX (17) 3040), and the National Natural Science Foundation of China (No. 31272163 and No. U1603102).

Conflicts of Interest: The authors declare no conflict of interest.

References

1. Walters, D.; Walsh, D.; Newton, A.; Lyon, G. Induced resistance for plant disease control: maximizing the efficacy of resistance elicitors. *Phytopathology* **2005**, *95*, 1368–1373. [CrossRef] [PubMed]
2. Hahn, M.G. Microbial elicitors and their receptors in plants. *Annu. Rev. Phytopathol.* **1996**, *34*, 387–412. [CrossRef] [PubMed]
3. Thomma, B.P.H.J.; Penninckx, I.A.M.A.; Broekaert, W.F.; Cammue, B.P.A. The complexity of disease signaling in Arabidopsis. *Curr. Opin. Immunol.* **2001**, *13*, 63–68. [CrossRef]
4. Neuhaus, J.M.; Fritig, B.; Linthorst, H.J.M.; Meins, F.; Mikkelsen, J.D.; Ryals, J. A revised nomenclature for chitinase genes. *Plant Mol. Biol. Report.* **1996**, *14*, 102–104. [CrossRef]
5. Lawton, K.; Ward, E.; Payne, G.; Moyer, M.; Ryals, J. Acidic and basic class III chitinase mRNA accumulation in response to TMV infection of tobacco. *Plant Mol. Biol.* **1992**, *19*, 735–743. [CrossRef] [PubMed]
6. Hein, I.; Gilroy, E.M.; Armstrong, M.R.; Birch, P.R.J. The zig-zag-zig in oomycete-plant interactions. *Mol. Plant Pathol.* **2009**, *10*, 547–562. [CrossRef] [PubMed]
7. Jones, J.D.G.; Dangl, L. The plant immune system. *Nature* **2006**, *444*, 323–329. [CrossRef] [PubMed]
8. Thomma, B.P.H.J.; Nürnberger, T.; Joosten, M.H.A.J. Of PAMPs and effectors: the blurred PTI-ETI dichotomy. *Plant Cell* **2011**, *23*, 4–15. [CrossRef] [PubMed]
9. Tsuda, K.; Katagiri, F. Comparing signaling mechanisms engaged in pattern-triggered and effector-triggered immunity. *Curr. Opin. Plant Biol.* **2010**, *13*, 459–465. [CrossRef] [PubMed]
10. Ravi Kumar, M.N.V. A review of chitin and chitosan applications. *React. Funct. Polym.* **2000**, *46*, 1–27. [CrossRef]

11. Nakahara, K.S.; Masuta, C. Interaction between viral RNA silencing suppressors and host factors in plant immunity. *Curr. Opin. Plant Biol.* **2014**, *20*, 88–95. [CrossRef] [PubMed]
12. Seki, M.; Kamei, A.; Yamaguchi-Shinozaki, K.; Shinozaki, K. Molecular responses to drought, salinity and frost: Common and different paths for plant protection. *Curr. Opin. Biotechnol.* **2003**, *14*, 194–199. [CrossRef]
13. Ahmed, N.U.; Park, J.I.; Jung, H.J.; Kang, K.K.; Hur, Y.; Lim, Y.P.; Nou, I.S. Molecular characterization of stress resistance-related chitinase genes of Brassica rapa. *Plant Physiol. Biochem.* **2012**, *58*, 106–115. [CrossRef] [PubMed]
14. Thamil Arasan, S.K.; Park, J.I.; Ahmed, N.U.; Jung, H.J.; Hur, Y.; Kang, K.K.; Lim, Y.P.; Nou, I.S. Characterization and expression analysis of dirigent family genes related to stresses in Brassica. *Plant Physiol. Biochem.* **2013**, *67*, 144–153. [CrossRef] [PubMed]
15. Singh, A.; Isaac Kirubakaran, S.; Sakthivel, N. Heterologous expression of new antifungal chitinase from wheat. *Protein Expr. Purif.* **2007**, *56*, 100–109. [CrossRef] [PubMed]
16. Liu, J.-J.; Ekramoddoullah, A.K.M.; Zamani, A. A Class IV Chitinase Is Up-Regulated by Fungal Infection and Abiotic Stresses and Associated with Slow-Canker-Growth Resistance to *Cronartium ribicola* in Western White Pine (*Pinus monticola*). *Phytopathology* **2005**, *95*, 284–291. [CrossRef] [PubMed]
17. Hamid, R.; Khan, M.A.; Ahmad, M.; Ahmad, M.M.; Abdin, M.Z.; Musarrat, J.; Javed, S. Chitinases: An update. *J. Pharm. Bioallied Sci.* **2013**, *5*, 21–29. [PubMed]
18. Abeles, F.B.; Bosshart, R.P.; Forrence, L.E.; Habig, W.H. Preparation and purification of glucanase and chitinase from bean leaves. *Plant Physiol.* **1971**, *47*, 129–134. [CrossRef] [PubMed]
19. PEGG, G.F. Chitinase from Verticillium albo-atrum. *METHODS Enzymol.* **1988**, *161*, 474–479.
20. Xu, F.; Fan, C.; He, Y. Chitinases in Oryza sativa ssp. japonica and Arabidopsis thaliana. *J. Genet. Genom.* **2007**, *34*, 138–150. [CrossRef]
21. Kovács, G.; Sági, L.; Jacon, G.; Arinaitwe, G.; Busogoro, J.P.; Thiry, E.; Strosse, H.; Swennen, R.; Remy, S. Expression of a rice chitinase gene in transgenic banana ("Gros Michel", AAA genome group) confers resistance to black leaf streak disease. *Transgenic Res.* **2013**, *22*, 117–130. [CrossRef] [PubMed]
22. Mincoff, P.C.; Garcia Cortez, D.A.; Ueda-Nakamura, T.; Nakamura, C.V.; Dias Filho, B.P. Isolation and characterization of a 30 kD antifungal protein from seeds of Sorghum bicolor. *Res. Microbiol.* **2006**, *157*, 326–332. [CrossRef] [PubMed]
23. Kirubakaran, S.I.; Sakthivel, N. Cloning and overexpression of antifungal barley chitinase gene in Escherichia coli. *Protein Expr. Purif.* **2007**, *52*, 159–166. [CrossRef] [PubMed]
24. Lorito, M.; Woo, S.L.; Garcia, I.; Colucci, G.; Harman, G.E.; Pintor-Toro, J.a.; Filippone, E.; Muccifora, S.; Lawrence, C.B.; Zoina, A.; Tuzun, S.; Scala, F. Genes from mycoparasitic fungi as a source for improving plant resistance to fungal pathogens. *Proc. Natl. Acad. Sci. USA* **1998**, *95*, 7860–7865. [CrossRef] [PubMed]
25. Bolar, J.P.; Norelli, J.L.; Wong, K.W.; Hayes, C.K.; Harman, G.E.; Aldwinckle, H.S. Expression of Endochitinase from Trichoderma harzianum in Transgenic Apple Increases Resistance to Apple Scab and Reduces Vigor. *Phytopathology* **2000**, *90*, 72–77. [CrossRef] [PubMed]
26. Mittler, R. Abiotic stress, the field environment and stress combination. *Trends Plant Sci.* **2006**, *11*, 15–19. [CrossRef] [PubMed]
27. Wang, W.; Vinocur, B.; Altman, A. Plant responses to drought, salinity and extreme temperatures: Towards genetic engineering for stress tolerance. *Planta* **2003**, *218*, 1–14. [CrossRef] [PubMed]
28. Kunkel, B.N.; Brooks, D.M. Cross talk between signaling pathways in pathogen defense. *Curr. Opin. Plant Biol.* **2002**, *5*, 325–331. [CrossRef]
29. Singh, A.; Jha, S.K.; Bagri, J.; Pandey, G.K. ABA inducible rice protein phosphatase 2C confers ABA insensitivity and abiotic stress tolerance in arabidopsis. *PLoS One* **2015**, *10*, 1–24. [CrossRef] [PubMed]
30. Hausbeck, M.K.; Lamour, K.H. Research Progress and Management Challenges Phytophthora capsici on Vegetable Crops. *Plant Dis.* **2004**, *88*, 1292–1303. [CrossRef]
31. Lamour, K.H.; Stam, R.; Jupe, J.; Huitema, E. The oomycete broad-host-range pathogen Phytophthora capsici. *Mol. Plant Pathol.* **2012**, *13*, 329–337. [CrossRef] [PubMed]
32. Su, Y.; Xu, L.; Wang, S.; Wang, Z.; Yang, Y.; Chen, Y.; Que, Y. Identification, phylogeny, and transcript of chitinase family genes in sugarcane. *Sci. Rep.* **2015**, *5*, 10708. [CrossRef] [PubMed]
33. Ahmed, N.U.; Park, J.I.; Seo, M.S.; Kumar, T.S.; Lee, I.H.; Park, B.S.; Nou, I.S. Identification and expression analysis of chitinase genes related to biotic stress resistance in Brassica. *Mol. Biol. Rep.* **2012**, *39*, 3649–3657. [CrossRef] [PubMed]

34. Di, F.; Jian, H.; Wang, T.; Chen, X.; Ding, Y.; Du, H.; Lu, K.; Li, J.; Liu, L. Genome-wide analysis of the PYL gene family and identification of PYL genes that respond to abiotic stress in Brassica napus. *Genes (Basel)* **2018**, *9*, 156. [CrossRef] [PubMed]
35. Guo, M.; Zhai, Y.; Lu, J.; Chai, L.; Chai, W.; Gong, Z. Characterization of CaHsp70-1, a Pepper Heat-Shock Protein Gene in Response to Heat Stress and Some Regulation Exogenous Substances in *Capsicum annuum* L. *Int. J. Mol. Sci.* **2014**, 19741–19759. [CrossRef] [PubMed]
36. Fujita, M.; Fujita, Y.; Noutoshi, Y.; Takahashi, F.; Narusaka, Y.; Yamaguchi-Shinozaki, K.; Shinozaki, K. Crosstalk between abiotic and biotic stress responses: A current view from the points of convergence in the stress signaling networks. *Curr. Opin. Plant Biol.* **2006**, *9*, 436–442. [CrossRef] [PubMed]
37. Liu, Y.; Jiang, H.; Zhao, Z.; An, L. Abscisic acid is involved in brassinosteroids-induced chilling tolerance in the suspension cultured cells from Chorispora bungeana. *J. Plant Physiol.* **2011**, *168*, 853–862. [CrossRef] [PubMed]
38. Ma, X.; Ma, F.; Mi, Y.; Ma, Y.; Shu, H. Morphological and physiological responses of two contrasting Malus species to exogenous abscisic acid application. *Plant Growth Regul.* **2008**, *56*, 77–87. [CrossRef]
39. Guo, M.; Liu, J.-H.; Lu, J.-P.; Zhai, Y.-F.; Wang, H.; Gong, Z.-H.; Wang, S.-B.; Lu, M.-H. Genome-wide analysis of the CaHsp20 gene family in pepper: comprehensive sequence and expression profile analysis under heat stress. *Front. Plant Sci.* **2015**, *6*, 806. [CrossRef] [PubMed]
40. Kim, S.; Park, M.; Yeom, S.I.; Kim, Y.M.; Lee, J.M.; Lee, H.A.; Seo, E.; Choi, J.; Cheong, K.; Kim, K.T.; et al. Genome sequence of the hot pepper provides insights into the evolution of pungency in Capsicum species. *Nat. Genet.* **2014**, *46*, 270–278. [CrossRef] [PubMed]
41. Do, H.M.; Lee, S.C.; Jung, H.W.; Sohn, K.H.; Hwang, B.K. Differential expression and in situ localization of a pepper defensin (CADEF1) gene in response to pathogen infection, abiotic elicitors and environmental stresses in *Capsicum annuum*. *Plant Sci.* **2004**, *166*, 1297–1305. [CrossRef]
42. Lee, S.C.; Hwang, B.K. Identification of the pepper SAR8.2 gene as a molecular marker for pathogen infection, abiotic elicitors and environmental stresses in *Capsicum annuum*. *Planta* **2003**, *216*, 387–396. [CrossRef] [PubMed]
43. Boller, T.; Gehri, A.; Mauch, F.; Vögeli, U. Chitinase in bean leaves: induction by ethylene, purification, properties, and possible function. *Planta* **1983**, *157*, 22–31. [CrossRef] [PubMed]
44. Punja, Z.K.; Punja, Z.K. Genetic engineering of plants to enhance resistance to fungal pathogens—a review of progress and future prospects. *J. Plant Pathol* **2001**, *23*, 216–235. [CrossRef]
45. Xiao, Y.-H.; Li, X.-B.; Yang, X.-Y.; Luo, M.; Hou, L.; Guo, S.-H.; Luo, X.-Y.; Pei, Y. Cloning and characterization of a balsam pear class I chitinase gene (Mcchit1) and its ectopic expression enhances fungal resistance in transgenic plants. *Biosci. Biotechnol. Biochem.* **2007**, *71*, 1211–1219. [CrossRef] [PubMed]
46. Backiyarani, S.; Uma, S.; Nithya, S.; Chandrasekar, A.; Saraswathi, M.S.; Thangavelu, R.; Mayilvaganan, M.; Sundararaju, P.; Singh, N.K. Genome-Wide Analysis and Differential Expression of Chitinases in Banana Against Root Lesion Nematode (Pratylenchus coffeae) and Eumusa Leaf Spot (Mycosphaerella eumusae) Pathogens. *Appl. Biochem. Biotechnol.* **2015**, *175*, 3585–3598. [CrossRef] [PubMed]
47. Nishizawa, Y.; Nishio, Z.; Nakazono, K.; Soma, M.; Nakajima, E.; Ogaki, M.; Hibi, T. Enhanced resistance to blast (Magnaporthe grisea) in transgenic japonica rice by constitutive expression of rice chitinase. *Theor. Appl. Genet.* **1999**, *99*, 383–390. [CrossRef] [PubMed]
48. Collingel, D.B.; Kragh, K.M.; Mikkelsen, J.D.; Nielsen, K.K.; Rasrnssen', U.; Vad, K. Plant chitinases. *Plant J.* **1993**, *3*, 31–40. [CrossRef]
49. Liu, Z.; Shi, L.; Yang, S.; Lin, Y.; Weng, Y.; Li, X.; Hussain, A.; Noman, A.; He, S. Functional and Promoter Analysis of ChiIV3, a Chitinase of Pepper Plant, in Response to Phytophthora capsici Infection. *Int. J. Mol. Sci.* **2017**, *18*, 1661. [CrossRef] [PubMed]
50. Cannon, S.B.; Mitra, A.; Baumgarten, A.; Young, N.D.; May, G. The roles of segmental and tandem gene duplication in the evolution of large gene families in Arabidopsis thaliana. *BMC Plant Biol.* **2004**, *4*, 1–21. [CrossRef] [PubMed]
51. Muhammad, I.; Jing, X.Q.; Shalmani, A.; Ali, M.; Yi, S.; Gan, P.F.; Li, W.Q.; Liu, W.T.; Chen, K.M. Comparative in silico analysis of Ferric Reduction Oxidase (FRO) genes expression patterns in response to abiotic stresses, metal and hormone applications. *Molecules* **2018**, *23*, 1163. [CrossRef] [PubMed]
52. Passarinho, P.A.; Van Hengel, A.J.; Fransz, P.F.; De Vries, S.C. Expression pattern of the Arabidopsis thaliana AtEP3/AtchitIV endochitinase gene. *Planta* **2001**, *212*, 556–567. [CrossRef] [PubMed]

53. Su, Y.; Xu, L.; Wang, S.; Wang, Z.; Yang, Y.; Chen, Y.; Que, Y. Identification, Phylogeny, and Transcript of Chitinase Family Genes in Sugarcane. *Sci. Rep.* **2015**, *5*, 10708. [CrossRef] [PubMed]
54. Van Pelt-Heerschap, H.; Smit-Bakker, O. Analysis of defense-related proteins in stem tissue of carnation inoculated with a virulent and avirulent race of Fusarium oxysporum f.sp. dianthi. *Eur. J. Plant Pathol.* **1999**, *105*, 681–691. [CrossRef]
55. Broglie, K.; Chet, I.; Holliday, M.; Cressman, R.; Biddle, P.; Knowlton, S.; Mauvis, C.J.; Broglie, R. Transgenic Plants with Enhanced Resistance to the Fungal Pathogen Rhizoctonia solani. *Science* **1991**, *254*, 1194–1197. [CrossRef] [PubMed]
56. Krishnaveni, S.; Liang, G.H.; Muthukrishnan, S.; Manickam, A. Purification and partial characterization of chitinases from sorghum seeds. *Plant Sci.* **1999**, *144*, 1–7. [CrossRef]
57. Rahul, P.R.; Kumar, V.G.; Sathyabhama, M.; Viswanathan, R.; Sundar, A.R.; Malathi, P. Characterization and 3D structure prediction of chitinase induced in sugarcane during pathogenesis of Colletotrichum falcatum. *J. Plant Biochem. Biotechnol.* **2013**, *24*. [CrossRef]
58. Viswanathan, R.; Malathi, P.; Sundar, A.R.; Aarthi, S.; Premkumari, S.M.; Padmanaban, P. Differential induction of chitinases and thaumatin-like proteins in sugarcane in response to infection by Colletotrichum falcatum causing red rot disease Differenzielle. *J. Plant Dis. Prot.* **2005**, *112*, 417–425.
59. Su, Y.; Xu, L.; Fu, Z.; Yang, Y.; Guo, J.; Wang, S.; Que, Y. ScChi, encoding an acidic class III chitinase of sugarcane, confers positive responses to biotic and abiotic stresses in sugarcane. *Int. J. Mol. Sci.* **2014**, *15*, 2738–2760. [CrossRef] [PubMed]
60. Kim, Y.J.; Hwang, B.K. Pepper gene encoding a basic pathogenesis-related 1 protein is pathogen and ethylene inducible. *Physiol. Plant* **2000**, *118*, 51–60.
61. Zhang, H.-X.; Jin, J.-H.; He, Y.-M.; Lu, B.-Y.; Li, D.-W.; Chai, W.-G.; Khan, A.; Gong, Z.-H. Genome-Wide Identification and Analysis of the SBP-Box Family Genes under Phytophthora capsici Stress in Pepper (*Capsicum annuum* L.). *Front. Plant Sci.* **2016**, *7*, 1–14. [CrossRef] [PubMed]
62. Wang, J.E.; Liu, K.K.; Li, D.W.; Zhang, Y.L.; Zhao, Q.; He, Y.M.; Gong, Z.H. A novel peroxidase CanPOD gene of pepper is involved in defense responses to Phytophtora capsici infection as well as abiotic stress tolerance. *Int. J. Mol. Sci.* **2013**, *14*, 3158–3177. [CrossRef] [PubMed]
63. Liu, X.; Li-Ling, J.; Hou, L.; Li, Q.; Ma, F. Identification and characterization of a chitinase-coding gene from Lamprey (Lampetra japonica) with a role in gonadal development and innate immunity. *Dev. Comp. Immunol.* **2009**, *33*, 257–263. [CrossRef] [PubMed]
64. Fan, J.; Wang, H.; Feng, D.; Liu, B.; Liu, H.; Wang, J. Molecular characterization of plantain class I chitinase gene and its expression in response to infection by Gloeosporium musarum Cke and Massee and other abiotic stimuli. *J. Biochem.* **2007**, *142*, 561–570. [CrossRef] [PubMed]
65. Kumar, M.; Gho, Y.; Jung, K.; Kim, S.; Kim, S. Genome-Wide Identification and Analysis of Genes, Conserved between japonica and indica Rice Cultivars, that Respond to Low-Temperature Stress at the Vegetative Growth Stage. *Front. Plant Sci.* **2017**, *8*, 1–20. [CrossRef] [PubMed]
66. Eroglu, S.; Aksoy, E. Genome-wide analysis of gene expression profiling revealed that COP9 signalosome is essential for correct expression of Fe homeostasis genes in Arabidopsis. *BioMetals* **2017**, *30*, 685–698. [CrossRef] [PubMed]
67. Yin, Y.X.; Guo, W.L.; Zhang, Y.L.; Ji, J.J.; Xiao, H.J.; Yan, F.; Zhao, Y.Y.; Zhu, W.C.; Chen, R.G.; Chai, W.G.; Gong, Z.H. Cloning and characterisation of a pepper aquaporin, CaAQP, which reduces chilling stress in transgenic tobacco plants. *Plant Cell. Tissue Organ Cult.* **2014**, *118*, 431–444. [CrossRef]
68. Liu, B.; Xue, X.; Cui, S.; Zhang, X.; Han, Q.; Zhu, L.; Liang, X.; Wang, X.; Huang, L.; Chen, X.; Kang, Z. Cloning and characterization of a wheat b-1,3-glucanase gene induced by the stripe rust pathogen Puccinia striiformis f. sp. tritici. *Mol. Biol. Rep.* **2010**, *37*, 1045–1052. [CrossRef] [PubMed]
69. Wasternack, C. Jasmonates: An update on biosynthesis, signal transduction and action in plant stress response, growth and development. *Ann. Bot.* **2007**, *100*, 681–697. [CrossRef] [PubMed]
70. Rouster, J.; Leah, R.; Mundy, J.; Cameron-Mills, V. Identification of a methyl jasmonate-responsive region in the promoter of a lipoxygenase 1 gene expressed in barley grain. *Plant J.* **1997**, *11*, 513–523. [CrossRef] [PubMed]
71. Merkouropoulos, G.; Barnett, D.C.; Shirsat, A.H. The Arabidopsis extensin gene is developmentally regulated, is induced by wounding, methyl jasmonate, abscisic and salicylic acid, and codes for a protein with unusual motifs. *Planta* **1999**, *208*, 212–219. [CrossRef] [PubMed]

72. Guo, W.L.; Chen, R.G.; Gong, Z.H.; Yin, Y.X.; Li, D.W. Suppression Subtractive Hybridization Analysis of Genes Regulated by Application of Exogenous Abscisic Acid in Pepper Plant (*Capsicum annuum* L.) Leaves under Chilling Stress. *PLoS One* **2013**, *8*. [CrossRef] [PubMed]
73. Liao, Y.; Liu, S.; Jiang, Y.; Hu, C.; Zhang, X.; Cao, X.; Xu, Z.; Gao, X.; Li, L.; Zhu, J.; Chen, R. Genome-wide analysis and environmental response profiling of dirigent family genes in rice (Oryza sativa). *Genes Genomics* **2017**, *39*, 47–62. [CrossRef]
74. Wu, R.; Wang, L.; Wang, Z.; Shang, H.; Liu, X.; Zhu, Y.; Qi, D.; Deng, X. Cloning and expression analysis of a dirigent protein gene from the resurrection plant Boea hygrometrica. *Prog. Nat. Sci.* **2009**, *19*, 347–352. [CrossRef]
75. Chen, R.G.; Jing, H.; Guo, W.L.; Wang, S.B.; Ma, F.; Pan, B.G.; Gong, Z.H. Silencing of dehydrin CaDHN1 diminishes tolerance to multiple abiotic stresses in *Capsicum annuum* L. *Plant Cell Rep.* **2015**, *34*, 2189–2200. [CrossRef] [PubMed]
76. Jin, J.-H.; Zhang, H.-X.; Tan, J.-Y.; Yan, M.-J.; Li, D.-W.; Khan, A.; Gong, Z.-H. A New Ethylene-Responsive Factor CaPTI1 Gene of Pepper (*Capsicum annuum* L.) Involved in the Regulation of Defense Response to Phytophthora capsici. *Front. Plant Sci.* **2016**, *6*, 1–12. [CrossRef] [PubMed]
77. Khan, A.; Li, R.-J.; Sun, J.-T.; Ma, F.; Zhang, H.-X.; Jin, J.-H.; Ali, M.; ul Haq, S.; Wang, J.-E.; Gong, Z.-H. Genome-wide analysis of dirigent gene family in pepper (*Capsicum annuum* L.) and characterization of CaDIR7 in biotic and abiotic stresses. *Sci. Rep.* **2018**, *8*, 5500. [CrossRef] [PubMed]
78. Qin, C.; Yu, C.; Shen, Y.; Fang, X.; Chen, L.; Min, J.; Cheng, J.; Zhao, S.; Xu, M.; Luo, Y.; et al. Whole-genome sequencing of cultivated and wild peppers provides insights into Capsicum domestication and specialization. *Proc. Natl. Acad. Sci. USA* **2014**, *111*, 5135–5140. [CrossRef] [PubMed]
79. Letunic, I.; Bork, P. Interactive tree of life (iTOL) v3: an online tool for the display and annotation of phylogenetic and other trees. *Nucleic Acids Res.* **2016**, *44*, W242–W245. [CrossRef] [PubMed]
80. Gasteiger, E.; Gattiker, A.; Hoogland, C.; Ivanyi, I.; Appel, R.D.; Bairoch, A. ExPASy: The proteomics server for in-depth protein knowledge and analysis. *Nucleic Acids Res.* **2003**, *31*, 3784–3788. [CrossRef] [PubMed]
81. Horton, P.; Park, K.J.; Obayashi, T.; Fujita, N.; Harada, H.; Adams-Collier, C.J.; Nakai, K. WoLF PSORT: Protein localization predictor. *Nucleic Acids Res.* **2007**, *35*, 585–587. [CrossRef] [PubMed]
82. Emanuelsson, O.; Nielsen, H.; Brunak, S.; Von Heijne, G. Predicting subcellular localization of proteins based on their N-terminal amino acid sequence. *J. Mol. Biol.* **2000**, *300*, 1005–1016. [CrossRef] [PubMed]
83. Kang, W.H.; Kim, S.; Lee, H.A.; Choi, D.; Yeom, S.I. Genome-wide analysis of Dof transcription factors reveals functional characteristics during development and response to biotic stresses in pepper. *Sci. Rep.* **2016**, *6*, 1–12. [CrossRef] [PubMed]
84. Guo, M.; Liu, J.H.; Ma, X.; Zhai, Y.F.; Gong, Z.H.; Lu, M.H. Genome-wide analysis of the Hsp70 family genes in pepper (*Capsicum annuum* L.) and functional identification of CaHsp70-2 involvement in heat stress. *Plant Sci.* **2016**, *252*, 246–256. [CrossRef] [PubMed]
85. Lescot, M. PlantCARE, a database of plant cis-acting regulatory elements and a portal to tools for in silico analysis of promoter sequences. *Nucleic Acids Res.* **2002**, *30*, 325–327. [CrossRef] [PubMed]
86. Conesa, A.; Götz, S. Blast2GO: A comprehensive suite for functional analysis in plant genomics. *Int. J. Plant Genomics* **2008**, *2008*. [CrossRef] [PubMed]
87. Liu, R.; Meng, J. MapDraw: A microsoft Excel Macro for drawing genetic linkage maps based on given genetic linkage data. *Hereditas* **2003**, *25*, 317–321. [PubMed]
88. Gu, Z.; Cavalcanti, A.; Chen, F.; Bouman, P.; Li, W. Extent of Gene Duplication in the Genomes of Drosophila, Nematode, and Yeast. *Mol. Biol. Evol.* **2002**, *19*, 256–262. [CrossRef] [PubMed]
89. Rost, B. Twilight zone of protein sequence alignments. *Protein Eng. Des. Sel.* **1999**, *12*, 85–94. [CrossRef]
90. Guo, W.L.; Chen, R.G.; Gong, Z.H.; Yin, Y.X.; Ahmed, S.S.; He, Y.M. Exogenous abscisic acid increases antioxidant enzymes and related gene expression in pepper (*Capsicum annuum*) leaves subjected to chilling stress. *Genet. Mol. Res.* **2012**, *11*, 4063–4080. [CrossRef] [PubMed]
91. Wan, H.; Yuan, W.; Ruan, M.; Ye, Q.; Wang, R.; Li, Z.; Zhou, G.; Yao, Z.; Zhao, J.; Liu, S.; Yang, Y. Identification of reference genes for reverse transcription quantitative real-time PCR normalization in pepper (Capsicum annuum L.). *Biochem. Biophys. Res. Commun.* **2011**, *416*, 24–30. [CrossRef] [PubMed]
92. Schmittgen, T.D.; Livak, K.J. Analyzing real-time PCR data by the comparative CT method. *Nat. Protoc.* **2008**, *3*, 1101–1108. [CrossRef] [PubMed]

93. Liu, Z.Q.; Liu, Y.Y.; Shi, L.P.; Yang, S.; Shen, L.; Yu, H.X.; Wang, R.Z.; Wen, J.Y.; Tang, Q.; Hussain, A.; Khan, M.I.; Hu, J.; Liu, C.L.; Zhang, Y.W.; Cheng, W.; He, S.L. SGT1 is required in PcINF1/SRC2-1 induced pepper defense response by interacting with SRC2-1. *Sci. Rep.* **2016**, *6*, 1–16. [CrossRef] [PubMed]
94. Jing, H.; Li, C.; Ma, F.; Ma, J.-H.; Khan, A.; Wang, X.; Zhao, L.-Y.; Gong, Z.-H.; Chen, R.-G. Genome-Wide Identification, Expression Diversication of Dehydrin Gene Family and Characterization of CaDHN3 in Pepper (*Capsicum annuum* L.). *PLoS One* **2016**, *11*, e0161073. [CrossRef] [PubMed]
95. Ou, L.J.; Dai, X.Z.; Zhang, Z.Q.; Zou, X.X. Responses of pepper to waterlogging stress. *Photosynthetica* **2011**, *49*, 339–345. [CrossRef]

© 2018 by the authors. Licensee MDPI, Basel, Switzerland. This article is an open access article distributed under the terms and conditions of the Creative Commons Attribution (CC BY) license (http://creativecommons.org/licenses/by/4.0/).

International Journal of
Molecular Sciences

MDPI

Article

Identification of *WRKY* Gene Family from *Dimocarpus longan* and Its Expression Analysis during Flower Induction and Abiotic Stress Responses

Dengwei Jue [1,†], Xuelian Sang [1,†], Liqin Liu [1], Bo Shu [1], Yicheng Wang [1], Chengming Liu [2], Jianghui Xie [1,*] and Shengyou Shi [1,*]

1 Key Laboratory of Tropical Fruit Biology (Ministry of Agriculture), South Subtropical Crops Research Institute, Chinese Academy of Tropical Agricultural Sciences, Zhanjiang524091, China; juedengwei@126.com (D.J.); anyue1220@126.com (X.S.); lolitallq@163.com (L.L.); bshbest@163.com (B.S.); ychw08@163.com (Y.W.)

2 College of Horticulture, South China Agricultural University, Guangzhou 510642, China; cmliu@scau.edu.cn

* Correspondence: xiejianghui@21cn.com (J.X.); ssy7299@sohu.com (S.S.); Tel.: +86-075-9285-9112 (J.X.); +86-188-2063-2013 (S.S.)

† These authors contributed equally to this work.

Received: 6 June 2018; Accepted: 17 July 2018; Published: 25 July 2018

Abstract: Longan is an important fruit tree in the subtropical region of Southeast Asia and Australia. However, its blooming and its yield are susceptible to stresses such as droughts, high salinity, and high and low temperature. To date, the molecular mechanisms of abiotic stress tolerance and flower induction in longan have not been elucidated. WRKY transcription factors (TFs), which have been studied in various plant species, play important regulatory roles in plant growth, development, and responses to stresses. However, there is no report about WRKYs in longan. In this study, we identified 55 *WRKY* genes with the conserved WRKY domain and zinc finger motif in the longan genome. Based on the structural features of WRKY proteins and topology of the phylogenetic tree, the longan WRKY (DlWRKY) family was classified into three major groups (I–III) and five subgroups (IIa–IIe) in group II. Tissue expression analysis showed that 25 *DlWRKYs* were highly expressed in almost all organs, suggesting that these genes may be important for plant growth and organ development in longan. Comparative RNA-seq and qRT-PCR-based gene expression analysis revealed that 18 *DlWRKY* genes showed a specific expression during three stages of flower induction in "Sijimi" ("SJ"), which exhibited the "perpetual flowering" (PF) habit, indicating that these 18 *DlWRKY* genes may be involved in the flower induction and the genetic control of the perpetual flowering trait in longan. Furthermore, the RT-qPCR analysis illustrated the significant variation of 27, 18, 15, 17, 27, and 23 *DlWRKY* genes under SA (Salicylic acid), MeJA (Methyl Jasmonate), heat, cold, drought, or high salinity treatment, respectively, implicating that they might be stress- or hormone-responsive genes. In summary, we systematically and comprehensively analyzed the structure, evolution, and expression pattern of the *DlWRKY* genes. The results presented here increase our understanding of the WRKY family in fruit trees and provide a basis for the further elucidation of the biological function of *DlWRKY* genes in longan.

Keywords: longan; WRKY; expression analysis; flower induction; abiotic stress

1. Introduction

Longan (*Dimocarpuslongan* Lour.) is an important subtropical fruit tree in the family Sapindaceae, which is grown in many subtropical and tropical countries with most of the production in Southeast

Asia and Australia [1]. Biennial bearing is the most serious problem that affects longan fruit products. Among the factors that affect *D. longan* fruit yield, the difficulty and unstableness to blossom is one of the most challenging problems [2]. Floral bud induction of *D. longan* requires favorable conditions such as a period of low temperature (vernalization), suitable salinity, and dry conditions. To obtain a stable high yield, off-season flowering in longan is achieved by chemical treatment with potassium chlorate ($KClO_3$) application [3,4]. Nevertheless, the induction effect varies in different regions and varieties. Therefore, the study of the molecular regulatory mechanisms of flower induction and abiotic stress tolerance in longan is particularly important for understanding and solving the problems associated with fruit yield. However, due to the long generation time and lack of genome information, knowledge of the molecular regulatory mechanisms of flower induction and abiotic stress tolerance in longan is scarce.

As an important developmental process in the plant life cycle, flowering is directly linked to production whenever seeds or fruits are harvested [5]. The molecular and genetic bases of flowering have been well studied in *Arabidopsis thaliana* [6–8]. There are at least five major flowering pathways in *Arabidopsis*, including the photoperiod, autonomous, vernalization, gibberellin (GA), and aging pathways [9]. These pathways activate or inhibit floral transformation through a series of flower integron genes, such as the flowering locus T (*FT*), flowering locus C (*FLC*), and constans (*CO*) [10]. In addition, several transcription factors (TFs), such as MADS-domain TFs [11], NACs [12], MYBs [13], and DREBs [14], participate in the signaling of flowering regulation. As the seventh largest TF family in flowering plants, many WRKY genes are also involved in the determination of flowering time [15]. For example, in *A. thaliana*, the lines over-express *GsWRKY20*, *MlWRKY12*, and *WRKY71* in the flowers earlier than in the wild-type [16–18]. A recent research study found that two WRKY proteins (AtWRKY12 and AtWRKY13) played opposite functions in controlling the flowering time under short-day conditions in *A. thaliana* partly through mediating the effect of GA_3. The *wrky*12 mutant exhibits late flowering and the *wrky*13 mutant shows earlier flowering than that of the wild-type [19].

Abiotic stresses such as drought, heat, salt, and cold are the major causes of declined crop productivity worldwide. At the molecular level, several TFs, such as AP2/EREBP, NAC, WRKY, bZIP, MYB, and bHLH play a vital role in regulating downstream genes to protect plants from these stresses [20]. As one of the largest TF families in plants, the WRKY TFs also play pivotal roles in regulating many abiotic stress reactions [15]. In *Arabidopsis*, some of the *AtWRKYs* respond strongly to various abiotic stresses, such as salinity, drought, and cold [21–24]. In rice, 11 *OsWRKY* genes showed variable responses to salt, polyethylene glycol (PEG), and cold or heat stresses [25]. Overexpression of *OsWRKY47* increased both the drought tolerance and yield compared with wild-type plants [26]. In mulberry, *Morus013217* and *Morus002784* show high accumulation in response to cold and salt stresses. *Morus005757* shows significant up-regulation in response to dehydration stress, salinity stress, and SA and ABA (Abscisic acid) treatments [27]. Similar results were also found in wheat, common bean [28], grape [29], pineapple [30], soybean [31], moso bamboo [32], *Caragana intermedia* [33], peanut [34], and broomcorn millet [35]. These observations suggest that studying the WRKY gene families may provide valuable insights into the mechanism underlying abiotic stress tolerance in plants. As perennials growing in the subtropical and tropical area, some abiotic stresses, such as drought, heat, salt, and cold often have an adverse effect on the growth and yield of longan. However, given the lack of genome information, the identified and functions of *WRKY* genes in longan are still unknown.

In the present study, we performed a genome-wide identification of WRKY TFs in longan and analyzed their gene structures, conserved motifs, and expression patterns in nine different tissues. This work also determined the expression profiles of longan WRKY (*DlWRKY*) in three flowering stages of two longan cultivars and measured their transcript abundance in response to different phytohormone treatments and various abiotic stresses. This study provides a basis for future studies on *DlWRKY* gene family evolution and function.

2. Results

2.1. Identification of WRKY Gene Family in Longan

To extensively identify the WRKY genes in longan, whole-genome scanning was used to identify the genes which contain the particular domain by both the hidden Markov model (HMM) and Blastn search methods. In total, 59 candidate WRKY genes were identified (Table S1). After the WRKY domain scanning and sequence alignment, three genes (*Dlo_007676.1*, *Dlo_032703.1*, and *Dlo_028398.1*) without a complete predicted WRKY domain and one redundant gene (*Dlo_037584.1*) were removed. Finally, 55 *DlWRKY* genes were determined in the longan genome (Table 1). According to their chromosome locations, the 55 *DlWRKY* genes were designated *DlWRKY1*–*DlWRKY55*. In addition, the basic properties of *DlWRKY* genes, including the length of the full-length sequence, open reading frame (ORF), protein sequence, molecular weight (MW), and PI, were systematically evaluated (Table 1). The average length of these *DlWRKY* genes was 2417 bp and the length mainly centered on the range of 892 bp (*DlWRKY12*) to 5385 bp (*DlWRKY36*). Meanwhile, the length of the ORF was mainly distributed from 480 bp (*DlWRKY12* and *DlWRKY34*) to 3813 bp (*DlWRKY36*), with an average of 1237 bp. The length of the protein sequences ranged from 160 AA (*DlWRKY12* and *DlWRKY34*) to 1271 AA (*DlWRKY36*), with an average of 411 AA. The protein MW ranged from 18.10 kDa (*DlWRKY34*) to 143.77 kDa (*DlWRKY36*), with an average of 44.73 kDa. The predicted isoelectric point of the DlWRKY proteins varied from 4.62 (*DlWRKY22*) to 9.77 (*DlWRKY13*), with an average of 7.11.

Table 1. The information of the *DlWRKY* gene family.

Gene Name	Gene Locus ID	Location	ORF (bp)	Size (aa)	PI	MW (KDa)	Intron	Full Length
DlWRKY1	Dlo_000299.1	scaffold1:3145979:3147233	1071	356	9.63	38.76	2	1255
DlWRKY2	Dlo_026119.1	scaffold6:875263:878308	894	297	6.26	32.31	2	3046
DlWRKY3	Dlo_026149.1	scaffold6:1127159:1130416	1596	532	7.26	57.64	3	3258
DlWRKY4	Dlo_026267.1	scaffold6:2195842:2200859	1815	605	6.66	66.08	4	5018
DlWRKY5	Dlo_030713.1	scaffold8:184175:186167	1059	353	5.63	39.46	2	1993
DlWRKY6	Dlo_002181.1	scaffold11:1861336:1864296	1767	589	7.23	64.37	4	2961
DlWRKY7	Dlo_012455.1	scaffold23:1107291:1111499	1914	638	6.75	69.01	5	4209
DlWRKY8	Dlo_013053.2	scaffold24:1070557:1072933	1668	556	6.52	61.46	4	2377
DlWRKY9	Dlo_015501.2	scaffold29:1782026:1783019	762	254	8.99	28.30	4	994
DlWRKY10	dlo_037126.1	scaffold29:1793158:1794294	684	228	9.02	25.58	2	1146
DlWRKY11	Dlo_016404.1	scaffold31:1522675:1524675	1026	342	5.60	38.86	2	2001
DlWRKY12	Dlo_019125.1	scaffold38:1882835:1883726	480	160	5.16	18.38	2	892
DlWRKY13	Dlo_023965.1	scaffold53:1206068:1207919	1035	345	9.77	38.51	2	1852
DlWRKY14	Dlo_028963.1	scaffold71:878665:881164	1613	471	8.87	51.80	3	2500
DlWRKY15	Dlo_031097.1	scaffold81:147303:148636	972	324	6.33	35.30	2	1334
DlWRKY16	Dlo_033905.1	scaffold98:272537:275137	1419	473	5.82	51.20	2	2601
DlWRKY17	Dlo_001368.1	scaffold105:274029:278833	1425	475	6.10	52.14	4	4805
DlWRKY18	Dlo_003898.1	scaffold124:605265:607367	1053	351	9.04	39.34	1	2103
DlWRKY19	Dlo_003928.1	scaffold124:1058067:1061659	633	211	6.37	23.26	2	3593
DlWRKY20	Dlo_004435.1	scaffold129:429868:432959	1644	548	7.41	59.78	5	3092
DlWRKY21	Dlo_008095.1	scaffold167:682922:683969	714	238	5.14	26.59	2	1048
DlWRKY22	Dlo_008126.1	scaffold168:307774:310141	1245	415	4.62	44.95	1	2368
DlWRKY23	Dlo_008610.1	scaffold176:75022:76492	1023	341	8.62	38.08	4	1471
DlWRKY24	Dlo_009865.1	scaffold192:233555:234849	1071	357	5.50	39.25	2	1295
DlWRKY25	Dlo_011410.1	scaffold213:248908:250725	1038	346	5.93	38.65	2	1818
DlWRKY26	Dlo_011411.1	scaffold213:253855:257080	1122	374	6.00	40.22	2	3226
DlWRKY27	Dlo_012276.1	scaffold229:13116:15182	1005	335	7.16	37.09	2	2067
DlWRKY28	Dlo_012878.1	scaffold238:352167:354143	1527	509	5.89	55.49	3	1977
DlWRKY29	Dlo_013340.1	scaffold245:258019:261130	696	232	8.95	26.57	2	3112
DlWRKY30	Dlo_013413.1	scaffold247:267246:270528	2238	746	5.59	80.60	4	3283
DlWRKY31	Dlo_014324.1	scaffold266:341214:343700	663	221	7.71	25.38	3	2487
DlWRKY32	Dlo_015139.1	scaffold286:162902:164294	1059	353	6.32	38.46	2	1393
DlWRKY33	Dlo_015144.1	scaffold286:195837:198196	615	205	9.03	23.13	1	2360

Table 1. *Cont.*

Gene Name	Gene Locus ID	Location	ORF (bp)	Size (aa)	PI	MW (KDa)	Intron	Full Length
DlWRKY34	Dlo_015224.1	scaffold287:217068:218497	480	160	9.54	18.10	1	1430
DlWRKY35	Dlo_016828.1	scaffold322:63655:67562	1326	442	9.62	48.27	4	3908
DlWRKY36	Dlo_022548.1	scaffold487:170363:175747	3813	1271	5.15	143.77	5	5385
DlWRKY37	Dlo_023098.1	scaffold502:191885:193351	1056	352	9.46	38.46	2	1467
DlWRKY38	Dlo_023764.1	scaffold524:170088:173717	1533	511	8.66	55.75	3	3630
DlWRKY39	Dlo_025188.1	scaffold568:191129:193577	1530	510	8.26	55.75	5	2449
DlWRKY40	Dlo_025974.1	scaffold597:89062:90386	1110	370	5.07	40.99	2	1325
DlWRKY41	Dlo_026484.1	scaffold607:21585:23785	1218	406	6.06	45.38	4	2201
DlWRKY42	Dlo_027244.2	scaffold640:85638:89083	2298	766	5.15	83.59	4	3446
DlWRKY43	Dlo_027361.1	scaffold648:191661:193182	969	323	9.14	36.57	2	1522
DlWRKY44	Dlo_027614.1	scaffold657:107511:111179	1521	507	5.55	54.88	4	3669
DlWRKY45	Dlo_029034.1	scaffold711:179562:181398	1539	513	8.27	55.15	2	1837
DlWRKY46	Dlo_029939.1	scaffold757:33093:37889	1710	570	6.38	61.42	5	4797
DlWRKY47	Dlo_031466.1	scaffold829:42224:45497	1023	341	7.20	7.71	4	3274
DlWRKY48	Dlo_031469.1	scaffold829:58277:59797	990	330	9.06	36.21	3	1521
DlWRKY49	Dlo_031936.1	scaffold858:266912:269300	588	196	9.46	22.05	1	2389
DlWRKY50	Dlo_032595.1	scaffold896:87649:89238	1185	395	6.67	43.04	2	1590
DlWRKY51	Dlo_033966.1	scaffold980:88739:90132	933	311	5.14	34.86	2	1394
DlWRKY52	Dlo_001658.1	scaffold1077:66972:68290	918	306	6.26	33.96	4	1319
DlWRKY53	Dlo_002663.1	scaffold1135:95286:97669	1929	643	5.73	70.07	4	2384
DlWRKY54	Dlo_004749.1	scaffold1314:73982:75144	795	265	5.24	30.27	2	1163
DlWRKY55	Dlo_010873.1	scaffold2042:2013:3910	1023	341	9.42	37.97	3	1898

2.2. Phylogenetic Analysis of DlWRKY

A phylogenetic tree was constructed using the maximum likelihood (ML) method and based on multiple alignments of longan, grape, and *Arabidopsis* WRKY domain aa sequences. As shown in Figure 1, the phylogenetic results revealed that all the DlWRKY proteins could be categorized into three groups (I, II, and III). Eleven DlWRKY proteins were considered to be group I, which included two WRKY domains and a C_2H_2 (C–X_4–C–X_{22-23}–HXH) zinc finger motif. A total of 35 DlWRKY proteins contained one WRKY domain and a C_2H_2 (C–X_{4-5}–C–X_{23}–HXH) zinc-binding motif, which were classified as group II. The nine remaining genes were assigned to Group III, which consisted of a single WRKY domain and a C_2CH (C–X_7–C–X_{23}–HXC) zinc-binding motif. According to the WRKY subgroup classification of *Arabidopsis*, the DlWRKYs in Group II were further subdivided into five subgroups, including groups IIa (3), IIb (7) IIc (13), IId (6), and IIe (6).

2.3. Multiple Sequence Alignment and Structure Analysis

The WRKYGQK sequence is a considerably conservative motif of WRKY proteins and several variants of this conserved WRKY motif have been reported in plants [36], including WRKYGEK, WRKYGKK, WSKYEQK, and WRKYSEK. In the present study, this motif was observed in all longan WRKY proteins and three variants of this motif were also found. The majority of DlWRKY proteins contained the WRKYGQK motif, and WRKYGKK and WKKYRQK were observed in DlWRKY19 and DlWRKY47, respectively. The other remarkably conservative motif was a zinc finger structure which contained two types of zinc finger motifs: C–X_{4-5}–C–X_{22-23}–HXH and C–X_7–C–X_{23}–HXC. A total of 46 DlWRKY proteins contained C–X_{4-5}–C–X_{22-23}–HXH, and nine DlWRKY proteins contained C–X_7–C–X_{23}–HXC, which all belonged to Group III (Table S1).

According to the Gene Structure Display Server (GSDS) website, the number of introns was in the range of 1–5 in all the longan WRKY gene families, with most of *DlWRKY* genes containing 2–4 introns (n = 81.0%). The average number of introns was 2.82. In addition, the phylogenetic analysis of the *DlWRKY* gene family showed that the genes within the same group generally exhibited a similar exon/intron structure. For example, subgroup IIe contained two introns (Figure 2).

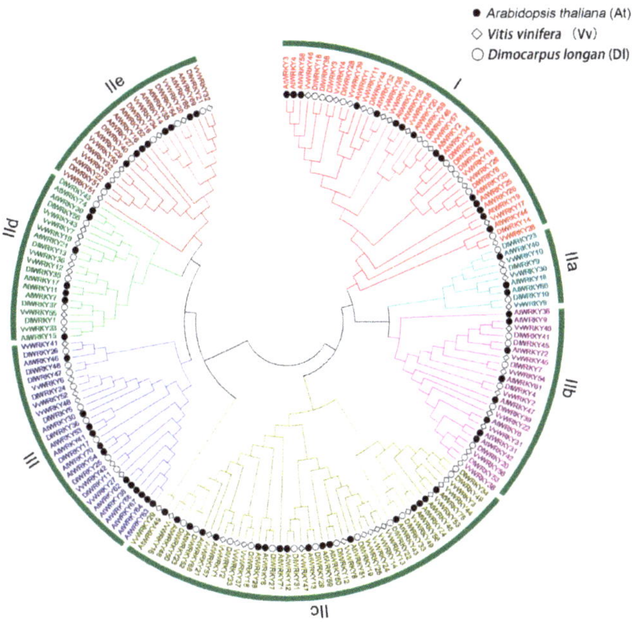

Figure 1. The phylogenetic analysis of the longan WRKY proteins with orthologous members from grape and *Arabidopsis.* The maximum likelihood phylogenetic tree was constructed by MEGA 6.0. Different groups of DlWRKY proteins are indicated by a circle and the different colors.

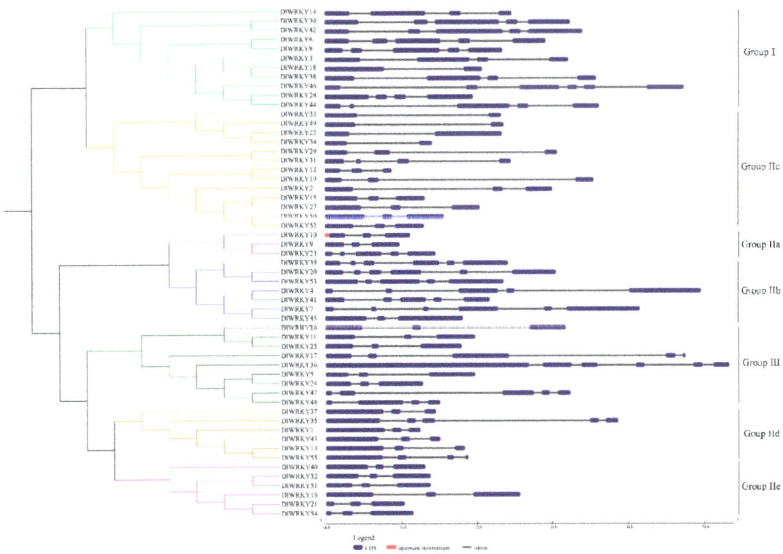

Figure 2. The unrooted phylogenetic tree (**left**) and gene structure (**right**) of 55 DlWRKY proteins. The phylogenetic tree was constructed by MEGA 6.0. The red color indicates the untranslated 5′- and 3′-regions; the blue color indicates exons; and the gray color indicates introns.

To further understand the similarity and diversity of motif composition among different DlWRKY proteins, a phylogenetic tree based on the full-length DlWRKY proteins was constructed (Figure 3). The motifs in the DlWRKY protein sequences were also predicted using MEME (http://meme.sdsc.edu/meme/cgi-bin/meme.cgi) (Figure 3 and Table S2). A total of 15 motifs were identified to illustrate the WRKY protein structure in longan. The results showed that the number of motifs in DlWRKYs ranged from 2 to 6, and the length of motifs ranged from 21 to 50 amino acids. Among the 15 identified motifs, motifs 1 and 2, characterized as WRKY domains, were broadly distributed across the DlWRKYs.

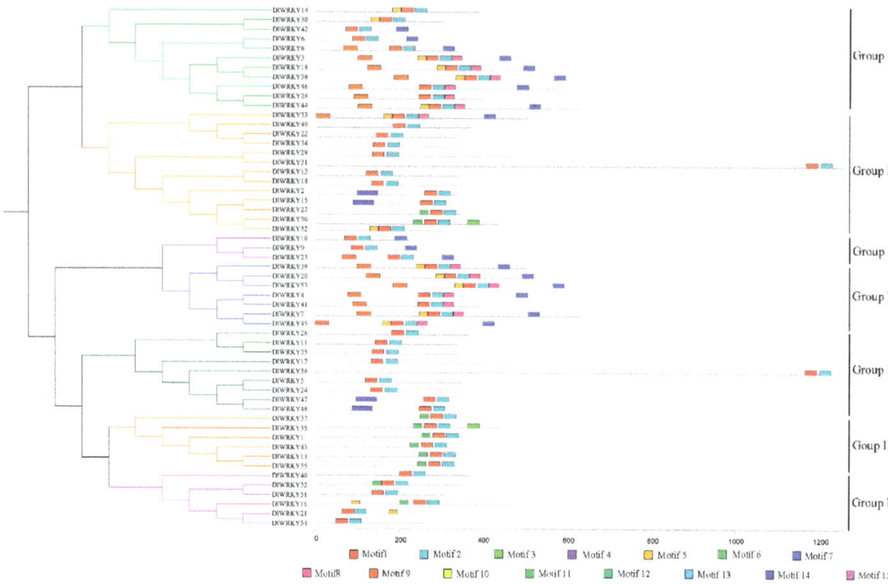

Figure 3. The unrooted phylogenetic tree (**left**) and conserved motifs (**right**) of 55 DlWRKY proteins. The phylogenetic tree was constructed using the same method used in Figure 2. Different colors represent various groups. MEME was used to predict motifs, and these motifs are represented by boxes.

2.4. Tissue-Specific Expression Patterns of DlWRKY

To generate expression profiles of *DlWRKY* genes under normal conditions, the expression levels of the 55 *DlWRKY* genes in the root, stem, leaf, seed, young fruit, pulp, pericarp, flower, and flower bud were investigated by the RNA-seq analysis. The $\log_{10}$ (FPKM + 0.01) values of the transcripts were clustered hierarchically and displayed in a heat map (Figure 4 and Table S3). The results showed that 96.36% (53 of 55) of *DlWRKYs* were expressed in young fruits and 94.55% were expressed in the pericarp, stems, and flower bud. A total of 90.91%, 89.09%, and 81.82% of *DlWRKYs* were expressed in the flower, leaf, root, and seed, respectively. Only a few *DlWRKY* genes were detected in pulps (67.27%). Approximately 60% (33 of 55) of the *DlWRKY* genes were expressed in each tested tissue, in which 25 *DlWRKY* genes (*DlWRKY*1, 2, 3, 5, 6, 8, 9, 13, 14, 23, 24, 28, 30, 32, 35, 37, 38, 39, 44, 49, 50, 52, 53, and 54) were highly expressed in at least six longan tissues. In contrast, 12 *DlWRKY* genes (*DlWRKY*10, 12, 18, 22, 26, 36, 40, 41, 42, 45, 47, and 48) were expressed at low levels in all tested tissues. Furthermore, *DlWRKY*22 only displayed a significantly low expression in the flower bud. *DlWRKY*10, 22, 41, 47, and 48 were preferential accumulation in two or three tissues.

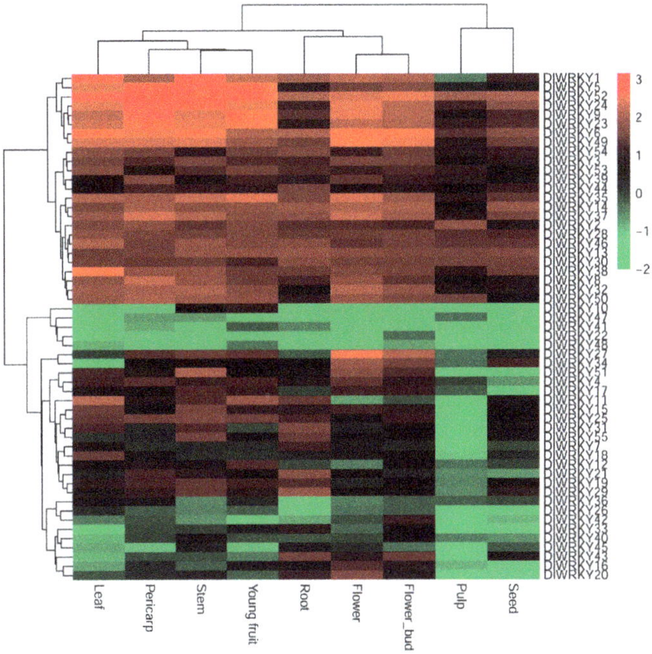

Figure 4. The heat map of the *DlWRKY* gene expression profiles in different tissues. The color scale represents the $\log_{10}$ expression values; the red and green colors indicate the higher or lower transcript abundances compared to the relevant control, respectively.

2.5. Comparative Expression Profiles of Two Longan Species during the Flowering Process

Although the involvement of many WRKY genes has been examined in the control of flowering time [15], the expression of *DlWRKY* genes during flower induction has not been studied extensively. In the present study, we also analyzed the expression patterns of 55 *DlWRKY* genes in two longan species during the three flowering stages by RNA-seq analysis (Table S4). Heat maps were constructed based on the $\log_{10}$ (FPKM + 0.01) values for the 55 *DlWRKY* genes (Figure 5a). Based on the criteria for p-values <0.05 and fold changes $\geq$2, the *DlWRKY* genes that were differentially expressed during the three flowering stages of the two longan species were identified. Interestingly, the results showed that all 55 *DlWRKY* genes were constructively expressed in the three test flowering stages of the "SX" longan, while 18 *DlWRKY* genes showed a specific expression in the "SJ" longan. Among the 18 *DlWRKY* genes, 12 (*DlWRKY*5, 7, 8, 9, 15, 21, 23, 24, 25, 39, 52, and 54) showed a continuously down-regulated expression through the three flowering stages, and four genes (*DlWRKY*16, 17, 41, and 42) showed an up-regulated expression. Moreover, two genes (*DlWRKY*10 and 48) showed a transient up-regulation at the second stage and a down-regulation at the third stage.

To validate the expression levels obtained from the RNA-seq data, twelve *DlWRKY* genes (*DlWRKY*1, 5, 9, 15, 16, 17,18,24,39,42, 48, and 50) were selected from the six different longan WRKY groups for the quantitative real-time reverse transcription polymerase chain reaction (qRT-PCR) analysis. Consistent with the result of the RNA-seq analysis, the transcript levels of all twelve *DlWRKY* genes did not exhibit any significant differences in the "SX" longan between the three flowering stages (Figure 5b). In addition, the relative expression level of *DlWRKY*1, *DlWRKY*18, and *DlWRKY*50 did not exhibit any significant differences in '""SJ" during the three flowering stages. The expression levels of *DlWRKY*16, 17, 42, and 48were up-regulated in the second and third stage. The transcript level of *DlWRKY*5, 9, 15, 24, and *DlWRKY*39 was down-regulated in the second and third stages (Figure 5b).

In general, the expression levels obtained by qRT-PCR for these genes are similar to the results obtained from the RNA-seq data.

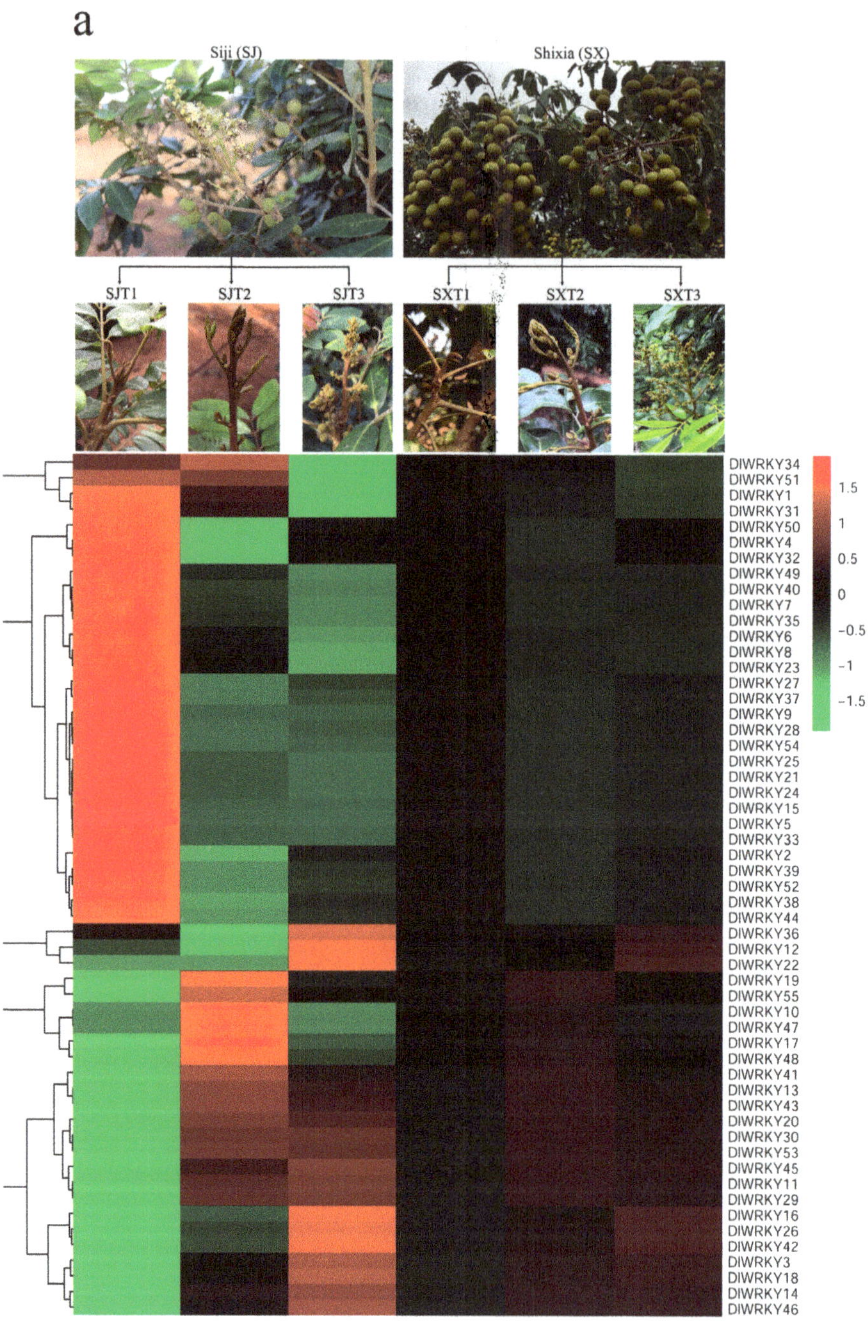

Figure 5. *Cont.*

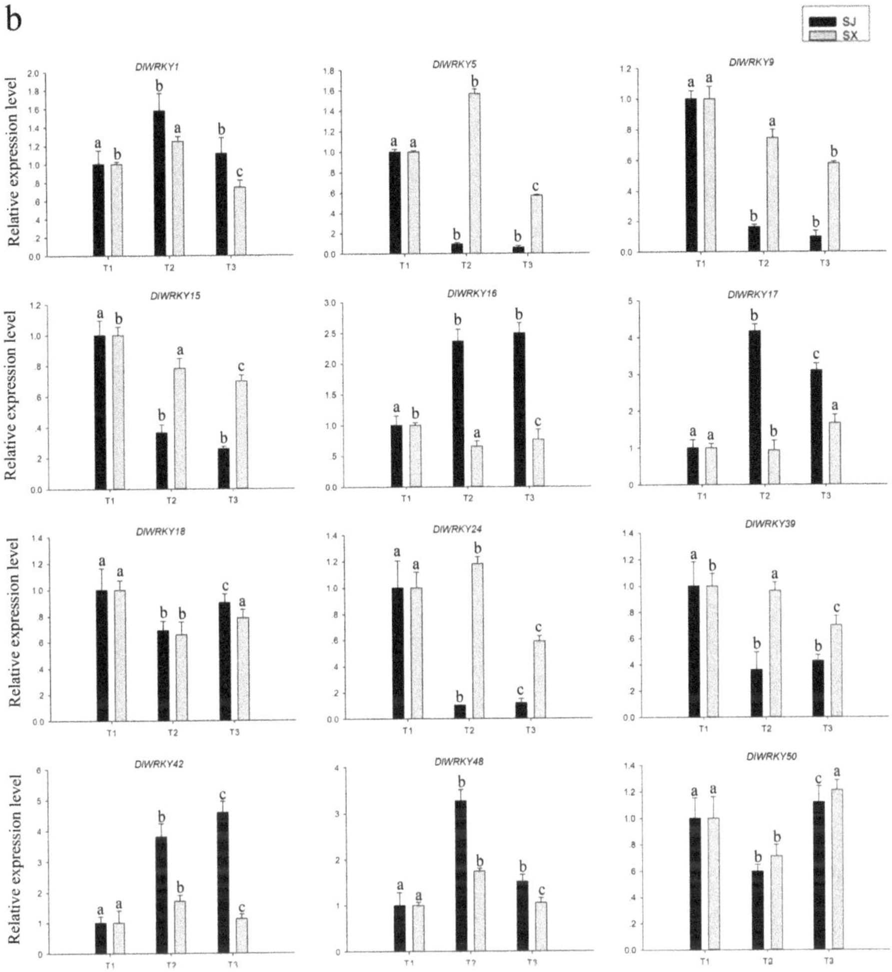

Figure 5. The expression profiles of *DlWRKY* in two longan species during the floral induction process. (**a**) A heat map showing the comparative expression level of the *WRKY* genes in the three flowering stages of"SJ" and "SX". The color scale represents the $\log_{10}$ expression values. Genes with comparatively low expression values are shown using shades of green, and high expression values are represented using shades of red. The three flowering stages of SJ are indicated by SJT1, SJT2, and SJT3. The three flowering stages of SX are indicated by SXT1, SXT2, and SXT3. (**b**) Relative expression levels of the twelve *DlWRKYs* during the three flowering stages of the two longan species by qRT-PCR. For each gene, the relative expression level in T1 (dormant apical bud) was set as one, and the longan *actin* gene was used as the internal expression control. The data represent the mean ± SD of the three replicates. Values with the same letter were not significantly different when assessed using Duncan's multiple range test ($p < 0.05$, $n = 3$).

2.6. Differential Expression of *DlWRKY* Genes in Response to Stress and Hormonal Treatments

The expression patterns of 55 *DlWRKY* genes were investigated in response to hormonal and various stresses by using qRT-PCR. As shown in Figure 6 and Figure S1, the majority of the *DlWRKY* genes (44 of 55) were up-regulated or down-regulated by >2-flod under at least one tested treatment, while eleven genes (*DlWRKY*2, 6, 12, 13, 14, 19, 29, 33, 34, 40, and 50) showed no significant differential expression in response to the given treatments. The SA treatment induced the expression of the 22 *DlWRKY* genes (*DlWRKY*1, 3, 5, 8, 10, 15, 16, 18, 23, 26, 27, 32, 36, 38, 42, 43, 45, 46, 48, 51, 54, and 55) but reduced the expression of five *DlWRKY* genes (*DlWRKY*9, 20, 24, 25, and 41). Fifteen *DlWRKY* genes (*DlWRKY*1, 3, 4, 8, 10, 16, 21, 32, 38, 39, 44, 45, 51, 53, and 55) were up-regulated, and three (*DlWRKY*20, 25, and 41) were down-regulated by MeJA treatment. For heat treatment, 11 (*DlWRKY*4, 9, 20, 27, 28, 35, 37, 39, 44, 49, and 52) and 4 (*DlWRKY*5, 8, 16, and 51) genes were down-regulated or up-regulated, respectively. A total of 17 *DlWRKY* (*DlWRKY*5, 7, 9, 17, 18, 20, 23, 25, 26, 31, 37, 39, 41, 42, 47, 51, and 54) genes showed up-regulated expressions, and no genes were down-regulated by cold treatment. Under the drought treatment, 20 (*DlWRKY*1, 4, 5, 8, 10, 11, 15, 16, 17, 21, 22, 25, 26, 27, 28, 30, 36, 45, 48, and 51) and 7 *DlWRKY* genes (*DlWRKY*9, 35, 37, 41, 44, 49, and 54) were up-regulated or down-regulated, respectively. Eighteen (*DlWRKY*1, 4, 5, 8, 10, 11, 15, 16, 18, 21, 23, 32, 36, 38, 42, 45, 48, and 51) and five *DlWRKY* genes (*DlWRKY*9, 20, 24, 37, and 41) were up-regulated or down-regulated, respectively, under high salinity treatment.

2.7. Analysis Related Cis-Elements in the Candidate *DlWRKY* Genes

To analyze the potential function of *DlWRKY* genes in response to various responses, the *cis*-elements in the promoter region of the *DlWRKY* genes were further analyzed. Among these 55 genes, 54 genes could perform *cis*-elements analysis except *DlWRKY*45, which only contain 270 promoter bases. All the *DlWRKY* genes shared the light-responsive boxes and stress-responsive boxes in their promoter. Hormone-related *cis*-elements, such as AuxRR-core, TCA-element, CGTCA-motif, GARE-motif, P-box, and ERE (Ethylene-responsive element), existed in the promoter of all *DlWRKY* genes except *DlWRKY*11, *DlWRKY*41, and *DlWRKY*52. Additionally, circadian-related *cis*-elements were found in the promoter of 39 *DlWRKY* genes and Meristem-related *cis*-elements were only presented in the promoter of 20 *DlWRKY* genes (Figure 7, Tables S5 and S6).

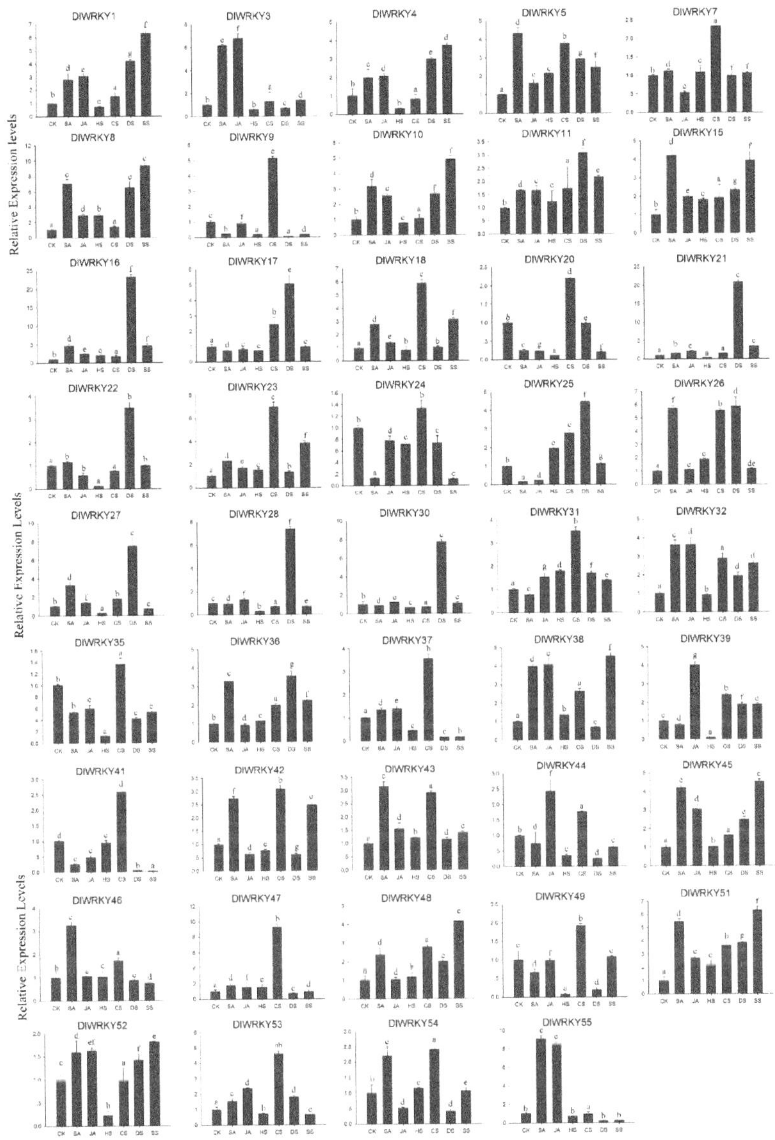

Figure 6. The expression patterns of the selected *DlWRKY* genes under various hormonal and abiotic stresses. The x-axis indicates various treatments and the y-axis indicates the relative expression level. Error bars were obtained from three independent biological replicates. Values with the same letter were not significantly different when assessed using Duncan's multiple range test ($p < 0.05$, $n = 3$). SA represents salicylic acid, JA represents jasmonic acid, HS represents heat stress, CS represents cold stress, DS represents drought stress, and SS represents salinity stress.

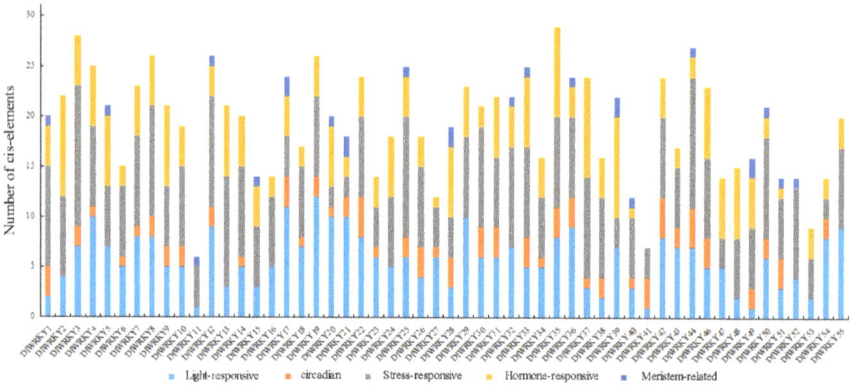

Figure 7. The predicted *cis*-elements in the promoter of the *DlWRKY* genes. The 1.5 kb sequences of 55 *DlWRKY* genes were analyzed with the PlantCARE software.

3. Discussion

The WRKY proteins, an important transcription factor superfamily which is involved in plant development and stress responses, have been widely detected in various organisms from single-celled green algae to monocots and dicots [15]. Recently, the successful genome sequencing of longan makes it possible to analyze WRKY TFs at the whole-genome level [37]. The present study is the first to identify and characterize WRKY proteins from whole-genome sequences of longan.

In this study, we identified 59 candidate WRKY genes in the longan genome (471.88 Mb) using the HMM and Blastn search methods. These genes included 58 *DlWRKYs*, which were also found by Lin et al. [37], and one gene *Dlo_*022548.1 (*DlWRKY*36) found in our study. Finally, after the WRKY domain scanning and sequence alignment, 55 *DlWRKY* genes were determined in the longan genome (Table 1). The number of *WRKY* genes in longan was similar to those found in grape (59 *VvWRKYs*), whose genome size is 487 Mb, which is similar to that of the longan genome [29]. However, the size of the WRKY family in longan is smaller than that in *A. thaliana* (72), *Oryza sativa* ssp. *Indica* (102), and the common bean (88), although their genome sizes are similar (*O. sativa* ssp. *Indica*, 466 Mb; common bean, 587 Mb) or even smaller (*A. thaliana*, 119 Mb) than the longan genome size (Table S7) [28,38,39]. Therefore, the number of WRKY family members is not necessarily correlated with the genome size. Previous studies showed that the only group I *WRKYs* are present in green algae and all *WRKY* genes originated from the group I C-terminal WRKY domains, whereas group II members were evolved in the common ancestor of land plants, and Group III members emerged in the common ancestor of seed plants [15]. In addition, as a newly defined and the most dynamic group with many duplication events, the differences in the number of *WRKY* genes in Group III are the primary cause of the sizes of *WRKY* gene families [40]. In the present study, the differences in the number of *WRKY* genes between longan and *Arabidopsis* mainly existed in groups IIc and III, indicating that the group IIc and III *WRKY* genes may play important roles in the functional evolution of *DlWRKYs*.

According to the classification scheme for the WRKY family of Eulgem et al. [41], the DlWRKY proteins were divided into three distinct clusters: groups I, II, and III. Group II proteins were further divided into five distinct groups: a–e (Figure 1 and Table 1). In addition, subgroup IIc contained the largest number of WRKY proteins. These results were consistent with the results observed in other species [28,29,42–44]. The WRKY motif was fairly conserved in longan WRKY proteins, and three variants of this motif were observed. All the DlWRKYs, except DlWRKY19 and DlWRKY47, possessed WRKYGQK. DlWRKY19, which belonged to subgroup IIc, possessed WRKYGKK. DlWRKY19, which belonged to subgroup III, possessed WKKYRQK. In the common bean, the variants WRKYGKK, WRKYGEK, WKKYEDK, and WKKYCEDK are mainly observed in subgroup IIc [28]; in mulberry,

WRKYGKK is detected in subgroup IIb [27]. Moreover, in rice, nine variants, most of which belong to groups III and IIc, are observed [45]. Previous studies showed that these variations of the WRKYGQK motif might change the DNA binding specificities of downstream target genes, and WRKY genes with the variations of the WRKYGQK motif may recognize binding sequences other than the W-box element ((C/T)TGAC(C/T)) [15]. Hence, the result suggested that DlWRKY19 and DlWRKY47 may possess different binding specificities and functions from those of other DlWRKY proteins.

WRKY family genes play important roles in diverse plant development and shown a tissue-specific expression in many plant species [15,40]. For example, *AtWRKY75* exerts a negative effect on root hair development [46]. *SUSIBA2* [47] and *MINISEED3* [48] play roles in the regulation of seed development. In grape, nearly half of the 59 *VvWRKY* genes show no significant organ/tissue-related differences in expression, and some clear spatial differences are noted [29]. In mulberry, 13 *WRKY* genes exhibit the highest expression in the *Morus notabilis* root tissue. A maximum of 25 *WRKYs* show the highest expression in the bark tissue, and 10 WRKY genes display the highest expression in other stages [27]. In the present study, the expression profiles of 55 longan *WRKY* genes in nine longan tissues were ascertained by RNA-seq analysis (Figure 4). The results demonstrated variation in the expression pattern of *DlWRKY* genes. In total, 25 *DlWRKY* genes (*DlWRKY*1, 2, 3, 5, 6, 8, 9, 13, 14, 23, 24, 28, 30, 32, 35, 37, 38, 39, 44, 49, 50, 52, 53, and 54) were highly expressed in at least six longan tissues. As highly expressed genes usually play important roles in plant development [44], we concluded that the 25 highly expressed *DlWRKY* genes might be important regulatory factors in longan development. It was found that group I and group IId *WRKY* genes are ancestral to other *WRKY* genes in plants or algae and are more likely to be constitutively expressed in different tissues [15,40]. For instance, most of the highly expressed *SiWRKY* genes belonged to group I and IId [40]. Consistent with these studies, in the present study, most of the members of groups I (9 of 11) and IId (4 of 6) were the highly expressed gene. In contrast, 12 *DlWRKY* genes were expressed at low levels in all tested tissues and these minimally expressed *DlWRKY* genes were distributed in almost all the *WRKY* gene subgroups except for IId. Meanwhile, six *DlWRKY* genes were preferential accumulation in no more than three tissues, implying that these genes might play crucial roles during the development of specific organs. Additionally, these specifically or minimally expressed *DlWRKY* genes could be induced under environment stimuli. For example, *DlWRKY*10, 22, 41, and 47 were not detected in leaves under normal conditions, but they were induced by different abiotic stresses (Figure 6). Similar results were also found in other studies [15,40,49].

Perpetual flowering is a crucial trait for fruit trees as it enlarges the production period [50]. To date, the genetic control of PF has been deciphered in several model plants. For example, In *Arabidopsis*, the PF trait is controlled by *PERPETUAL FLOWERING 1* (*PEP1*), an orthologue of the FLC floral repressor [51]. In the diploid strawberry and rose, the PF trait is due to a mutation in the orthologue of the *TERMINAL FLOWER* 1 (*TFL*1) floral repressor [50,52]. Recent studies showed that the PF trait of some cultivated strawberries is genetically controlled by the major *FaPFRU* locus, which is non-orthologous to *TFL*1 [53,54]. However, the multi-year delay in the onset of flowering and the long juvenile phase hampers the research of PF traits in perennials, such as longan. Although WRKY TFs regulate various plant developments, only a few data are available on whether WRKY TFs are involved in the flowering time regulation. Meanwhile, as a kind of TF, WRKY genes regulated plant flowering by being directly active or inhibiting the downstream target gene. For example, promoter sequences of *FT*, *LFY*, and *AP1* harbor W-boxes (TTTGACT/C); *AtWRKY*71 affects the flowering time of plants by directly regulating these genes [16]. In our study, all the 55 *DlWRKY* genes were constructively expressed in the three test flower induction process of the "SX" longan, while 18 *DlWRKY* genes showed a specific expression in the "SJ" longan (Figure 5a). This result indicated that these 18 *DlWRKY* genes may specifically be involved in the flower induction of "SJ". In summary, we proposed that these 18 *DlWRKY* genes may participate in the forming of the longan PF habit, which further studies are required to verify the function of these genes.

WRKY genes play crucial roles in the response to abiotic and biotic stress-induced defense signaling pathways [15]. Numerous studies have demonstrated that WRKY genes are expressed strongly and rapidly in response to particular abiotic stresses [15,22,29,40,52]. Consistent with these previous studies, our study showed that 44 *DlWRKY* genes (80%) showed up- or down-regulated expression in at least one tested treatment (Figure 6 and Figure S1), thereby highlighting the extensive involvement of *WRKY* genes in environmental adaptation. SA, JA, and Eth play important roles in biotic and abiotic stresses [55]. Many *WRKYs*, such as *AtWRKY28*, *AtWRKY46*, *AtWRKY70*, and *AtWRKY54*, play an important role in SA- and JA-dependent defense signaling pathways [53,56,57]. In the present study, 27 and 18 *DlWRKY* genes were up- or down-regulated by SA and MeJA treatment, respectively. For example, *DlWRKY25*, the orthologue of *AtWRKY70* and *AtWRKY54*, was regulated by the SA and JA treatments. *AtWRKY25* and *AtWRKY33* regulate plant adaptation to salinity stress through an interaction with their upstream or downstream target genes [58]; their orthologue *DlWRKY8* in longan was regulated by SA, JA, heat, drought, and salinity. In grape [29], *VvWRKY42* and its orthologue *DlWRKY11* in our study were up-regulated by salt treatment. Furthermore, we observed same orthologous genes with different expression patterns under stress treatment. *DlWRKY44* was down-regulated under drought, and its orthologous gene *VvWRKY35* was up-regulated under this stress treatment. *DlWRKY19* showed no significant differential expression in response to salinity, and its orthologous gene *VvWRKY25* was up-regulated [29]. We speculate that these orthologous genes may be involved in the different signaling pathways in different species. Additionally, only one gene (*DlWRKY52*) was significantly highly expressed under all abiotic stresses. These results indicated that the different *DlWRKYs* played different roles in regulating stress response and that further investigation of the functions of these *DlWRKY* genes is necessary. Differential responses of several *WRKYs* are regulated by the presence of *cis*-elements in their promoter region [27,40,49]. For example, *Morus*013217, which contains three LTREs in its promoter regions showed a strong response to cold stress [27]. Similar results were also found in our study. For instance, four HSEs were found in the promoter regions of *DlWRKY2*, which showed a strong response to heat stress. *DlWRKY36*, *DlWRKY46*, and *DlWRKY48* showed responsiveness to SA treatment and their expressions were all up-regulated, and more than two TCA-elements were found in their promoters. While the *DlWRKY11* and *DlWRKY52* hormone-related *cis*-elements existed in their promoter, they showed no response to the SA or MeJA treatments (Figure 6 and Table S6). Thus, these *cis*-elements could provide more evidence of the DlWRKY genes in response to different stresses or hormonal signaling.

4. Materials

4.1. Identification of Longan WRKY Genes

Longan whole-genome sequences, transcript data, and proteins were downloaded from the NCBI Sequence Read Archive (SRA315202) or ftp://climb.genomics.cn/pub/10.5524/100001_101000/100276/ [37]. The HMM profile of the WRKY DNA binding domain (PF03106) which was extracted from the Pfam database (http://pfam.sanger.ac.uk/) was used to obtain the potential members of the longan WRKY genes [59] and used to search the putative WRKY genes from the longan genome with HMMER 3.0 (http://hmmer.janelia.org/) with the default parameters and 0.01 as the cutoff value. Then, all non-redundant longan WRKY protein sequences were selected and the domain was conserved using Simple Modular Architecture Research Tool (http://smart.emblheidelberg.de/) [60].

4.2. Sequence Alignment, Phylogenetic Analysis, and Cis-Elements in the Promoters

The 72 *Arabidopsis* and 59 grape WRKY proteins described previously [29,38] were obtained from TAIR (http://www.arabidopsis.org/) and NCBI (http://www.ncbi.nlm.nih.gov/), respectively. By using Clustal X version 1.83, the WRKY protein sequences of *Arabidopsis* and longan were aligned for phylogenetic analysis. Based on this alignment, a bootstrapped ML (Maximum Likelihood) tree was constructed using MEGA (version 6.0) with the bootstrap test replicated 1000 times [61]. To assess the phylogenetic relationships among the members of the longan *WRKY* gene family, a phylogenetic tree was prepared according to the alignment of only the longan proteins. All DlWRKY transcription factors were classified into subgroups based on their structural features and evolutionary relationships. The 1500-bp sequences upstream of the start codon of the candidate *DlWRKY* genes were extracted from the longan genome sequences. The PlantCARE software (http://bioinformatics.psb.ugent.be/webtools/plantcare/html/) was used for searching the cis-acting elements [62].

4.3. Protein Feature Analysis

The ExPASy online tools (http://expasy.org/tools/) [63] were used to calculate the MW, the number of amino acids, the ORF, ORF length, and isoelectric point (pI) of DlWRKY proteins. The arrangements and the intron and exon junctions of the *DlWRKY* genes were analyzed by the GSDS, version 2.0 [64]. MEME (http://meme.sdsc.edu/meme/cgi-bin/meme.cgi) [65] was used to analyze the conserved motifs of the DlWRKY proteins with the following optimized parameters: any number of repetitions; maximum number of motifs: 15; and the optimum width of each motif: between 6 and 50 residues.

4.4. Expression Analysis of Longan WRKY Genes in Various Tissues and Different Flowering Stages

The RNA-seq data for analyzing the expression patterns of *WRKY* genes in different longan tissues were downloaded from the NCBI Sequence Read Archive (GSE84467). Three pairs of nine-year-old "SJ" and "SX" *D. longan* trees which displayed opposite flowering phenotype were used for comparative expression analysis of *DlWRKY* during floral induction. All those trees were grown at an experimental orchard in the South Subtropical Crops Research Institute of the Chinese Academy of Tropical Agricultural Science in Zhanjiang (110°16′ E, 21°10′ N), China. Three different kinds of apical buds, including the dormant stage (T1), the emergence of floral primordia stage (T2), and the floral organ formation stage (T3) of "SJ" and "SX", were used in this study. The samples obtained for the T1, T2, and T3 in "SJ" and "SX" were collected on 20 November 2016, 24 December 2016, and 1 January 2017, respectively. For each sample, we used three biological replicates from three different trees. Each biological replicate contained the mixed buds which were collected from the four cardinal directions of each tree. All samples were collected from 10:00 am to 12:00 am and were frozen immediately in liquid nitrogen and stored at −80 °C. According to the manufacturer's instructions, the total RNA was extracted by using the quick RNA Isolation Kit (Hua Yue Yang Bio Co., Ltd., Beijing, China) and the genomic DNA residues were removed during RNA extraction. We used an Agilent 2100 Bioanalyzer (Agilent, Santa Clara, CA, USA) to test the RNA concentration and the quality of each sample. The RNA quality was also confirmed by RNase free agarose gel electrophoresis. The RNA-seq experiment was performed as described by our previous study [66]. The RNA-seq data were uploaded to the NCBI Sequence Read Archive (SRS2241241, SRS2241242, SRS2241243, SRS2241244, SRS2241245, SRS2241246, SRS2241247, SRS2241248, SRS2241249, SRS2241250, SRS2241251, SRS2241252, SRS2241253, SRS2241254, SRS2241255, SRS2241256, SRS2241257, and SRS2241258). The fragments per kilobase of the exon model per million mapped values (FPKM) were $\log_{10}$-transformed, and heat maps with hierarchical clustering were exhibited using the software Mev4.9.0 [67].

4.5. Stress and Hormonal Treatments and Expression Profiling Using qRT-PCR

Twenty-seven one-year-old uniform grafted seedlings of "SJ", obtained from the South Subtropical Crops Research Institute of the Chinese Academy of Tropical Agricultural Science in Zhanjiang (110°16′E, 21°10′N) were used for stress and hormonal treatments. For hormone treatments, three seedlings were treated with methyl jasmonate (MeJA) or SA solution (100 μM) for 4 h at 28 °C, respectively. Meanwhile, three seedlings sprayed with water were used as a control. For heat and cold stresses, three samples were grown at 42 or 0 °C for 4 h, respectively, and three samples grown at 28 °C were used as a control. All the treatments were performed in a greenhouse. Six leaves were collected from each seedling and all samples were immediately frozen in liquid nitrogen and stored at −80 °C for expression analysis.

According to the manufacturer's instructions, the total RNA was obtained by using the SuperFast RNA extraction kit (Hua Yue Yang Bio Co.). The first-strand cDNA was synthesized by reverse transcription of the total RNA (500 ng) using PrimeScriptRTase (TaKaRa Biotechnology, Dalian, China). Gene-specific primers were designed according to the *DlWRKY* gene sequences using Primer Premier 5.0 and checked using Blastn in NCBI (Table S8). In addition, the longan *Actin*1 gene (Dlo_028674) was used as an internal control for normalization. qRT-PCR was conducted using the LightCycler® 480 Real-Time PCR System (Roche, Germany) and SYBR Green II PCR Master Mix (Takara, Dalian, China). The amplification program was as follows: 95 °C for 5 min, followed by 40 cycles of 95 °C for 15 s, and 60 °C for 1 min. Each reaction was performed in three replicates. The relative expression levels of the candidate genes were calculated by the $2^{-\Delta\Delta Ct}$ method. The analysis included cDNA from the three biological samples for each tissue, and all the reactions were run in triplicates. In the comparative expression analysis of the *DlWRKY* genes, genes that were up- or down-regulated by at least two-fold were considered differentially expressed.

5. Conclusions

It is essential to systematically analyze the function of transcription factors (TFs), since these genes can regulate the expression of many others, resulting in deep physiological modifications. Although *WRKY* genes have been identified in many other species, the information of longan *WRKY* is still unknown. In the present study, we conducted a genome-wide identification and analysis of the *WRKY* genes in longan. A total of 55 *DlWRKY* genes were identified in the longan genome. Phylogenetic analysis indicated that these 55 *DlWRKYs* could be divided into seven groups. An RNA-seq-based analysis showed that several of the identified *WRKY* genes may play various roles in the development of longan tissues. In addition, comparative expression analysis revealed that 18 *DlWRKY* genes might have participated in the regulation of longan flowering. Our RNA-seq, qRT-PCR, and promoter analyses revealed the gene expression profiles and implied that the response to different stress or hormonal signaling of some *DlWRKY* may be due to the cis-elements in their promoters. In summary, our results will facilitate further studies into the role of *DlWRKY* genes in response to abiotic stresses and the development of molecular breeding programs to enhance abiotic stress tolerance and increase yield in longans.

Supplementary Materials: Supplementary materials can be found at http://www.mdpi.com/1422-0067/19/8/2169/s1. The following are available online. Table S1. The candidate *WRKY* genes and their protein structure found in the longan genome. Table S2. The information for each motif of DlWRKYs. Table S3. FPKM values of *DlWRKY* genes in nine tissues of longan. Table S4. FPKM values of *DlWRKY* genes in the three flower induction stages of "SJ" and "SX" longan species. The red color indicates the genes which showed down-regulated expression; the blue color indicates the genes which showed up-regulated expression; and the green color indicates the genes that showed an up-regulated expression in the first two stages and a down-regulated expression in the third stage. Table S5. Details of the *cis*-elements identified in this study. Table S6. Predicted *cis*-elements in the promoter of the *DlWRKY* genes. Table S7. The WRKY gene number and genome size of different species. Table S8. Primers used in quantitative RT-PCR of *DlWRKY* genes. Figure S1. Expression patterns of selected *DlWRKY* genes which have no significant difference under various hormonal and abiotic stresses. The x-axis indicates various treatments, and the y-axis indicates the relative expression level. Error bars were obtained from three independent biological replicates.

Author Contributions: D.J., X.S., and S.S. conceived the experiments and D.J. performed the experiments. J.X. Additionally, C.L. analyzed the data, D.J. Additionally, S.S. contributed to the writing of the manuscript, L.L. provided the value comments and revised the grammar of the manuscript. B.S. provided help in the analysis of qRT-PCR. Y.W. prepared samples for RNA sequencing.

Funding: This work was supported by the Natural Science Foundation of China (31572087), the China Litchi and Longan Industry Technology Research System (CARS-32-02), the Central Public-interest Scientific Institution Basal Research Fund for Chinese Academy of Tropical Agricultural Sciences (No. 1630062018011) and the Natural Science Foundation of Hainan Province (20163111 and 317243).

Conflicts of Interest: The authors declare no conflict of interest.

Abbreviation

HMM	Hidden Markov model
NJ	Neighbor-joining
GSDS	Gene Structure Display Server
MW	The molecular weight
ORF	Open reading frame
pI	Isoelectric point
NCBI	National Center of Biotechnology Information
qRT-PCR	Quantitative real-time reverse transcription polymerase chain reaction
RNA-seq	RNA sequencing
SA	Salicylic acid
JA	Jasmonic acid

References

1. Matsumoto, T.K. Genes uniquely expressed in vegetative and potassium chlorate induced floral buds of *Dimocarpus longan*. *Plant Sci.* **2006**, *170*, 500–510. [CrossRef]
2. You, X.; Wang, L.; Liang, W.; Gai, Y.; Wang, X.; Chen, W. Floral reversion mechanism in longan (*Dimocarpus longan* lour.) revealed by proteomic and anatomic analyses. *J. Proteom.* **2012**, *75*, 1099–1118. [CrossRef] [PubMed]
3. Jia, T.; Wei, D.; Meng, S.; Allan, A.C.; Zeng, L. Identification of regulatory genes implicated in continuous flowering of longan (*Dimocarpus longan* L.). *PLoS ONE* **2014**, *9*, e114568. [CrossRef] [PubMed]
4. Zhang, H.N.; Shi, S.Y.; Li, W.C.; Shu, B.; Liu, L.Q.; Xie, J.H.; Wei, Y.Z. Transcriptome analysis of 'sijihua' longan (*Dimocarpus longan* L.) based on next-generation sequencing technology. *J. Hortic. Sci. Biotechnol.* **2016**, *91*, 180–188. [CrossRef]
5. Shabala, S.; Bose, J.; Hedrich, R. Salt bladders: Do they matter? *Trends Plant Sci.* **2014**, *19*, 687–691. [CrossRef] [PubMed]
6. Bluemel, M.; Dally, N.; Jung, C. Flowering time regulation in crops-what did we learn from arabidopsis? *Curr. Opin. Biotechnol.* **2015**, *32*, 121–129.
7. Dally, N.; Xiao, K.; Holtgräwe, D.; Jung, C. The b2 flowering time locus of beet encodes a zinc finger transcription factor. *Proc. Natl. Acad. Sci. USA* **2014**, *111*, 10365–10370. [CrossRef] [PubMed]
8. Andrés, F.; Coupland, G. The genetic basis of flowering responses to seasonal cues. *Nat. Rev. Genet.* **2012**, *13*, 627–639. [CrossRef] [PubMed]
9. Turnbull, C. Long-distance regulation of flowering time. *J. Exp. Bot.* **2011**, *62*, 4399–4413. [CrossRef] [PubMed]
10. Srikanth, A.; Schmid, M. Regulation of flowering time: All roads lead to Rome. *Cell. Mol. Life Sci.* **2011**, *68*, 2013–2037. [CrossRef] [PubMed]
11. Smaczniak, C.; Immink, R.G.; Muiño, J.M.; Blanvillain, R.; Busscher, M.; Busscher-Lange, J.; Dinh, Q.P.; Liu, S.; Westphal, A.H.; Boeren, S. Characterization of mads-domain transcription factor complexes in arabidopsis flower development. *Proc. Natl. Acad. Sci. USA* **2012**, *109*, 1560–1565. [CrossRef] [PubMed]
12. Yoo, S.Y.; Kim, Y.; Kim, S.Y.; Lee, J.S.; Ahn, J.H. Control of flowering time and cold response by a nac-domain protein in arabidopsis. *PLoS ONE* **2007**, *2*, e642. [CrossRef] [PubMed]
13. Shin, B.; Choi, G.; Yi, H.; Yang, S.; Cho, I.; Kim, J.; Lee, S.; Paek, N.C.; Kim, J.H.; Song, P.S. Atmyb21, a gene encoding a flower-specific transcription factor, is regulated by cop1. *Plant J.* **2002**, *30*, 23–32. [CrossRef] [PubMed]
14. Tong, Z.; Hong, B.; Yang, Y.; Li, Q.; Ma, N.; Ma, C.; Gao, J. Overexpression of two chrysanthemum dgdreb1 group genes causing delayed flowering or dwarfism in arabidopsis. *Plant Mol. Biol.* **2009**, *71*, 115–129. [CrossRef] [PubMed]
15. Chen, F.; Hu, Y.; Vannozzi, A.; Wu, K.; Cai, H.; Qin, Y.; Mullis, A.; Lin, Z.; Zhang, L. The wrky transcription factor family in model plants and crops. *Crit. Rev. Plant Sci.* **2017**, *36*, 311–335. [CrossRef]
16. Yu, Y.; Liu, Z.; Wang, L.; Kim, S.G.; Seo, P.J.; Qiao, M.; Wang, N.; Li, S.; Cao, X.; Park, C.M. Wrky71 accelerates flowering via the direct activation of flowering locus t and leafy in arabidopsis thaliana. *Plant J.* **2015**, *85*, 96–106. [CrossRef] [PubMed]

17. Cai, Y.; Chen, X.; Xie, K.; Xing, Q.; Wu, Y.; Li, J.; Du, C.; Sun, Z.; Guo, Z. Dlf1, a wrky transcription factor, is involved in the control of flowering time and plant height in rice. *PLoS ONE* **2014**, *9*, e102529. [CrossRef] [PubMed]
18. Yu, Y.; Hu, R.; Wang, H.; Cao, Y.; He, G.; Fu, C.; Zhou, G. Mlwrky12, a novel miscanthus transcription factor, participates in pith secondary cell wall formation and promotes flowering. *Plant Sci.* **2013**, *212*, 1–9. [CrossRef] [PubMed]
19. Li, W.; Wang, H.; Yu, D. The arabidopsis wrky transcription factors wrky12 and wrky13 oppositely regulate flowering under short-day conditions. *Mol. Plant* **2016**, *9*, 1492–1503. [CrossRef] [PubMed]
20. Kiranmai, K.; Lokanadha Rao, G.; Pandurangaiah, M.; Nareshkumar, A.; Amaranatha Reddy, V.; Lokesh, U.; Venkatesh, B.; Anthony Johnson, A.M.; Sudhakar, C. A novel wrky transcription factor, muwrky3 (*Macrotyloma uniflorum* lam. Verdc.) enhances drought stress tolerance in transgenic groundnut (*Arachis hypogaea* L.) plants. *Front. Plant Sci.* **2018**, *9*, 346. [CrossRef] [PubMed]
21. Li, S.; Fu, Q.; Chen, L.; Huang, W.; Yu, D. Arabidopsis thaliana wrky25, wrky26, and wrky33 coordinate induction of plant thermotolerance. *Planta* **2011**, *233*, 1237–1252. [CrossRef] [PubMed]
22. Han, C.; Lai, Z.; Shi, J.; Yong, X.; Chen, Z.; Xu, X. Roles of arabidopsis wrky18, wrky40 and wrky60 transcription factors in plant responses to abscisic acid and abiotic stress. *BMC Plant Biol.* **2010**, *10*, 281.
23. Kilian, J.; Whitehead, D.J.; Wanke, D.; Weinl, S.; Batistic, O.; D'Angelo, C.; Bornberg-Bauer, E.; Kudla, J.; Harter, K. The atgenexpress global stress expression data set: Protocols, evaluation and model data analysis of uv-b light, drought and cold stress responses. *Plant J.* **2007**, *50*, 347–363. [CrossRef] [PubMed]
24. Seki, M.; Narusaka, M.; Ishida, J.; Nanjo, T.; Fujita, M.; Oono, Y.; Kamiya, A.; Nakajima, M.; Enju, A.; Sakurai, T. Monitoring the expression profiles of 7000 arabidopsis genes under drought, cold and high-salinity stresses using a full-length cdna microarray. *Plant J.* **2002**, *31*, 279–292. [CrossRef] [PubMed]
25. Qiu, Y.; Jing, S.; Fu, J.; Li, L.; Yu, D. Cloning and analysis of expression profile of 13 wrky genes in rice. *Sci. Bull.* **2004**, *49*, 2159–2168. [CrossRef]
26. Raineri, J.; Wang, S.; Peleg, Z.; Blumwald, E.; Chan, R.L. The rice transcription factor oswrky47 is a positive regulator of the response to water deficit stress. *Plant Mol. Biol.* **2015**, *88*, 401–413. [CrossRef] [PubMed]
27. Baranwal, V.K.; Negi, N.; Khurana, P. Genome-wide identification and structural, functional and evolutionary analysis of wrky components of mulberry. *Sci. Rep.* **2016**, *6*, 30794. [CrossRef] [PubMed]
28. Wu, J.; Chen, J.; Wang, L.; Wang, S. Genome-wide investigation of wrky transcription factors involved in terminal drought stress response in common bean. *Front. Plant Sci.* **2017**, *8*, 380. [CrossRef] [PubMed]
29. Guo, C.; Guo, R.; Xu, X.; Gao, M.; Li, X.; Song, J.; Zheng, Y.; Wang, X. Evolution and expression analysis of the grape (*Vitis vinifera* L.) wrky gene family. *J. Exp. Bot.* **2014**, *65*, 1513–1528. [CrossRef] [PubMed]
30. Xie, T.; Chen, C.; Li, C.; Liu, J.; Liu, C.; He, Y. Genome-wide investigation of wrky gene family in pineapple: Evolution and expression profiles during development and stress. *BMC Genom.* **2018**, *19*, 490. [CrossRef] [PubMed]
31. Yang, Y.; Zhou, Y.; Chi, Y.; Fan, B.; Chen, Z. Characterization of soybean wrky gene family and identification of soybean wrky genes that promote resistance to soybean cyst nematode. *Sci. Rep.* **2017**, *7*, 17804. [CrossRef] [PubMed]
32. Li, L.; Mu, S.; Cheng, Z.; Cheng, Y.; Zhang, Y.; Miao, Y.; Hou, C.; Li, X.; Gao, J. Characterization and expression analysis of the wrky gene family in moso bamboo. *Sci. Rep.* **2017**, *7*, 6675. [CrossRef] [PubMed]
33. Wan, Y.; Mao, M.; Wan, D.; Yang, Q.; Yang, F.; Mandlaa; Li, G.; Wang, R. Identification of thewrkygene family and functional analysis of two genes in caragana intermedia. *BMC Plant Biol.* **2018**, *18*, 31. [CrossRef] [PubMed]
34. Song, H.; Wang, P.; Lin, J.Y.; Zhao, C.; Bi, Y.; Wang, X. Genome-wide identification and characterization ofwrkygene family in peanut. *Front. Plant Sci.* **2016**, *7*, 534. [CrossRef] [PubMed]
35. Yue, H.; Wang, M.; Liu, S.; Du, X.; Song, W.; Nie, X. Transcriptome-wide identification and expression profiles of the wrky transcription factor family in broomcorn millet (*Panicum miliaceum* L.). *BMC Genom.* **2016**, *17*, 343. [CrossRef] [PubMed]
36. Mohanta, T.K.; Park, Y.H.; Bae, H. Novel genomic and evolutionary insight of wrky transcription factors in plant lineage. *Sci. Rep.* **2016**, *6*, 37309. [CrossRef] [PubMed]
37. Lin, Y.; Min, J.; Lai, R.; Wu, Z.; Chen, Y.; Yu, L.; Cheng, C.; Jin, Y.; Tian, Q.; Liu, Q. Genome-wide sequencing of longan (*Dimocarpus longan* lour.) provides insights into molecular basis of its polyphenol-rich characteristics. *Gigascience* **2017**, *6*, 1–14. [CrossRef] [PubMed]

38. Gu, X.; Mao, Z.; Yu, H.; Zhang, Y.; Jiang, W.; Ling, J.; Huang, S.; Xie, B. Genome-wide analysis of wrky gene family in cucumis sativus. *BMC Genom.* **2011**, *12*, 471.
39. Yu, J.; Yang, H. A draft sequence of the rice genome (*Oryza sativa* L. ssp. Indica). *Science* **2002**, *296*, 1937–1942. [CrossRef] [PubMed]
40. Li, D.; Liu, P.; Yu, J.; Wang, L.; Dossa, K.; Zhang, Y.; Zhou, R.; Wei, X.; Zhang, X. Genome-wide analysis of wrky gene family in the sesame genome and identification of the wrky genes involved in responses to abiotic stresses. *BMC Plant Biol.* **2017**, *17*, 152. [CrossRef] [PubMed]
41. Eulgem, T.; Rushton, P.J.; Robatzek, S.; Somssich, I.E. The wrky superfamily of plant transcription factors. *Trends Plant Sci.* **2000**, *5*, 199–206. [CrossRef]
42. Shuai, L.; Luo, C.; Zhu, L.; Sha, R.; Qu, S.; Cai, B.; Wang, S. Identification and expression analysis of wrky transcription factor genes in response to fungal pathogen and hormone treatments in apple (*Malus domestica*). *J. Plant Biol.* **2017**, *60*, 215–230.
43. Zhi, Z.; Yang, L.; Wang, D.; Huang, Q.; Mo, Y.; Xie, G. Gene structures, evolution and transcriptional profiling of the wrky gene family in castor bean (*Ricinus communis* L.). *PLoS ONE* **2016**, *11*, e0148243.
44. Cheng, Y.; Jalalahammed, G.; Yu, J.; Yao, Z.; Ruan, M.; Ye, Q.; Li, Z.; Wang, R.; Feng, K.; Zhou, G. Putative wrkys associated with regulation of fruit ripening revealed by detailed expression analysis of the wrky gene family in pepper. *Sci. Rep.* **2016**, *6*, 39000. [CrossRef] [PubMed]
45. Zhang, Y.; Wang, L. The wrky transcription factor superfamily: Its origin in eukaryotes and expansion in plants. *BMC Evolut. Biol.* **2005**, *5*, 1.
46. Rishmawi, L.; Hülskamp, M. Non-cell-autonomous regulation of root hair patterning genes by wrky75 in arabidopsis thaliana. *Plant Physiol.* **2014**, *165*, 186. [CrossRef] [PubMed]
47. Sun, C.; Palmqvist, S.; Olsson, H.; Borén, M.; Ahlandsberg, S.; Jansson, C. A novel wrky transcription factor, susiba2, participates in sugar signaling in barley by binding to the sugar-responsive elements of the iso1 promoter. *Plant Cell* **2003**, *15*, 2076–2092. [CrossRef] [PubMed]
48. Luo, M.; Dennis, E.S.; Berger, F.; Peacock, W.J.; Chaudhury, A. Miniseed3 (mini3), a wrky family gene, and haiku2 (iku2), a leucine-rich repeat (lrr) kinase gene, are regulators of seed size in arabidopsis. *Proc. Natl. Acad. Sci. USA* **2005**, *102*, 17531–17536. [CrossRef] [PubMed]
49. Yang, X.; Li, H.; Yang, Y.; Wang, Y.; Mo, Y.; Zhang, R.; Zhang, Y.; Ma, J.; Wei, C.; Zhang, X. Identification and expression analyses ofwrkygenes reveal their involvement in growth and abiotic stress response in watermelon (*Citrullus lanatus*). *PLoS ONE* **2018**, *13*, e0191308.
50. Iwata, H.; Gaston, A.; Remay, A.; Thouroude, T.; Jeauffre, J.; Kawamura, K.; Oyant, L.H.S.; Araki, T.; Denoyes, B.; Foucher, F. The tfl1 homologue ksn is a regulator of continuous flowering in rose and strawberry. *Plant J. Cell Mol. Biol.* **2012**, *69*, 116–125. [CrossRef] [PubMed]
51. Vincent, C. Pep1 regulates perennial flowering in *Arabis alpina*. *Nature* **2009**, *459*, 423–427.
52. Koskela, E.A.; Hytönen, T. Mutation in terminal flower1 reverses the photoperiodic requirement for flowering in the wild strawberry fragaria vesca. *Plant Physiol.* **2012**, *159*, 1043–1054. [CrossRef] [PubMed]
53. Perrotte, J.; Gaston, A.; Potier, A.; Petit, A.; Rothan, C.; Denoyes, B. Narrowing down the single homoeologous fapfru locus controlling flowering in cultivated octoploid strawberry using a selective mapping strategy. *Plant Biotechnol. J.* **2016**, *14*, 2176–2189. [CrossRef] [PubMed]
54. Gaston, A.; Perrotte, J.; Lerceteauköhler, E.; Rousseaugueutin, M.; Petit, A.; Hernould, M.; Rothan, C.; Denoyes, B. Pfru, a single dominant locus regulates the balance between sexual and asexual plant reproduction in cultivated strawberry. *J. Exp. Bot.* **2013**, *64*, 1837–1848. [CrossRef] [PubMed]
55. Fujita, M.; Fujita, Y.; Noutoshi, Y.; Takahashi, F.; Narusaka, Y.; Yamaguchi-Shinozaki, K.; Shinozaki, K. Crosstalk between abiotic and biotic stress responses: A current view from the points of convergence in the stress signaling networks. *Curr. Opin. Plant Biol.* **2006**, *9*, 436–442. [CrossRef] [PubMed]
56. Besseau, S.; Li, J.; Palva, E.T. Wrky54 and wrky70 co-operate as negative regulators of leaf senescence in *Arabidopsis thaliana*. *J. Exp. Bot.* **2012**, *63*, 2667–2679. [CrossRef] [PubMed]
57. Li, J.; Brader, G.; Palva, E.T. The wrky70 transcription factor: A node of convergence for jasmonate-mediated and salicylate-mediated signals in plant defense. *Plant Cell* **2004**, *16*, 319–331. [CrossRef] [PubMed]
58. Jiang, Y.; Deyholos, M.K. Functional characterization of arabidopsis nacl-inducible wrky25 and wrky33 transcription factors in abiotic stresses. *Plant Mol. Biol.* **2009**, *69*, 91–105. [CrossRef] [PubMed]

59. Finn, R.D.; Mistry, J.; Tate, J.; Coggill, P.; Heger, A.; Pollington, J.E.; Gavin, O.L.; Gunasekaran, P.; Ceric, G.; Forslund, K. The pfam protein families database. *Nucleic Acids Res.* **2010**, *38*, D211–D222. [CrossRef] [PubMed]
60. Letunic, I.; Copley, R.R.; Schmidt, S.; Ciccarelli, F.D.; Doerks, T.; Schultz, J.; Ponting, C.P.; Bork, P. Smart 4.0: Towards genomic data integration. *Nucleic Acids Res.* **2004**, *32*, D142–D144. [CrossRef] [PubMed]
61. Tamura, K.; Stecher, G.; Peterson, D.; Filipski, A.; Kumar, S. Mega6: Molecular evolutionary genetics analysis version 6.0. *Mol. Biol. Evolut.* **2013**, *30*, 2725–2729. [CrossRef] [PubMed]
62. Lescot, M.; Déhais, P.; Thijs, G.; Marchal, K.; Moreau, Y.; Van de Peer, Y.; Rouzé, P.; Rombauts, S. Plantcare, a database of plant cis-acting regulatory elements and a portal to tools for in silico analysis of promoter sequences. *Nucleic Acids Res.* **2002**, *30*, 325–327. [CrossRef] [PubMed]
63. Gasteiger, E.; Gattiker, A.; Hoogland, C.; Ivanyi, I.; Appel, R.D.; Bairoch, A. Expasy: The proteomics server for in-depth protein knowledge and analysis. *Nucleic Acids Res.* **2003**, *31*, 3784–3788. [CrossRef] [PubMed]
64. Guo, A.Y.; Zhu, Q.H.; Chen, X.; Luo, J.C. Gsds: A gene structure display server. *Hereditas* **2007**, *29*, 1023–1026. [CrossRef] [PubMed]
65. Bailey, T.L.; Boden, M.; Buske, F.A.; Frith, M.; Grant, C.E.; Clementi, L.; Ren, J.; Li, W.W.; Noble, W.S. Meme suite: Tools for motif discovery and searching. *Nucleic Acids Res.* **2009**, *37*, 202–208. [CrossRef] [PubMed]
66. Jue, D.; Sang, X.; Liu, L.; Shu, B.; Wang, Y.; Xie, J.; Liu, C.; Shi, S. The ubiquitin-conjugating enzyme gene family in longan (*Dimocarpus longan* lour.): Genome-wide identification and gene expression during flower induction and abiotic stress responses. *Molecules* **2018**, *23*, 662. [CrossRef] [PubMed]
67. Saeed, A.; Sharov, V.; White, J.; Li, J.; Liang, W.; Bhagabati, N.; Braisted, J.; Klapa, M.; Currier, T.; Thiagarajan, M. Tm4: A free, open-source system for microarray data management and analysis. *Biotechniques* **2003**, *34*, 374–378. [PubMed]

© 2018 by the authors. Licensee MDPI, Basel, Switzerland. This article is an open access article distributed under the terms and conditions of the Creative Commons Attribution (CC BY) license (http://creativecommons.org/licenses/by/4.0/).

International Journal of
Molecular Sciences

Article

Whole-Transcriptome Sequence Analysis of *Verbena bonariensis* in Response to Drought Stress

Bei Wang [1], Xue-Qi Lv [1], Ling He [1], Qian Zhao [1], Mao-Sheng Xu [1], Lei Zhang [1], Yin Jia [1], Fan Zhang [1], Feng-Luan Liu [2,*] and Qing-Lin Liu [1,*]

1 Department of Ornamental Horticulture, Sichuan Agricultural University, 211 Huimin Road, Wenjiang District, Chengdu 611130, Sichuan, China; s20167108@stu.sicau.edu.cn (B.W.); 20150853@stu.sicau.edu.cn (X.-Q.L.); s20162113@stu.sicau.edu.cn (L.H.); s20167109@stu.sicau.edu.cn (Q.Z.); 20150782@stu.sicau.edu.cn (M.-S.X.); 14069@sicau.edu.cn (L.Z.); 13864@sicau.edu.cn (Y.J.); 13305@sicau.edu.cn (F.Z.)

2 Shanghai Key Laboratory of Plant Functional Genomics and Resources, Shanghai Chenshan Plant Science Research Center, The Chinese Academy of Science, Shanghai Chenshan Botanical Garden, 3888 Huagong Road, Songjiang District, Shanghai 201602, China

* Correspondence: liufengluan@csnbgsh.cn (F.-L.L.); 13854@sicau.edu.cn (Q.-L.L.); Tel.: +86-21-3779-2288 (ext. 911) (F.-L.L.); +86-28-8629-0881 (Q.-L.L.)

Received: 20 April 2018; Accepted: 8 June 2018; Published: 13 June 2018

Abstract: Drought is an important abiotic factor that threatens the growth and development of plants. *Verbena bonariensis* is a widely used landscape plant with a very high ornamental value. We found that Verbena has drought tolerance in production practice, so in order to delve into its mechanism of drought resistance and screen out its drought-resistance genes, we used the RNA-Seq platform to perform a de novo transcriptome assembly to analyze Verbena transcription response to drought stress. By high-throughput sequencing with Illumina Hiseq Xten, a total of 44.59 Gb clean data was obtained from T01 (control group) and T02 (drought experiment group). After assembly, 111,313 unigenes were obtained, and 53,757 of them were annotated by compared databases. In this study, 4829 differentially expressed genes were obtained, of which 4165 were annotated. We performed GO (Gene Ontology) and KEGG (Kyoto Encyclopedia of Genes and Genomes) pathway enrichment analyses, and explored a lot of differently expressed genes related to plant energy production, hormone synthesis, cell signal transduction, and metabolism to understand the stress response of Verbena in drought stress. In addition, we also found that a series of TFs related to drought-resistance of Verbena and provide excellent genetic resources for improving the drought tolerance of crops.

Keywords: *Verbena bonariensis*; drought stress; transcriptome sequencing; differentially expressed genes

1. Introduction

Adverse environmental factors such as low temperature and salt stress, together with drought, prevent plants from realizing their full genetic potential and have been the main problems facing agriculture [1]. Drought stress affects plant growth and development by affecting the plant respiration, growth, photosynthesis, assimilate partitioning, moisture and nutrient relationships, and drought-induced crop yield losses may outweigh losses from all other causes [2]. A series of physiological and biochemical reactions of the plant under drought conditions are altered through gene regulation, such as activation of respiration, repression of cell growth and photosynthesis, and stomatal closure [3]. Accelerating the pace of revealing drought-tolerance mechanisms will greatly help traditional breeding efforts and the application of modern genetic methods in improving the drought tolerance of crops [4].

Many of Verbenaceae's plants have important medicinal values, such as *Aloysia triphylla* [5] and *Cordia verbenace* [6]. Verbena (*Verbena bonariensis* L.) is an excellent ornamental landscape plant with extensive management. It has a flourishing and long flowering period, such that it plays an extremely important role in landscape layout. In addition to its ornamental value, Verbena has shown good performance under drought stress in production practice, which provides new thinking for our study on plants' responses to abiotic stress. However, the understanding of its drought-resistance mechanism is still in the early stage. Presently, high-throughput sequencing technology has been widely used to reveal plants' intrinsic physiological mechanism at the molecular level, such as model plants *Nicotiana tabacum* L. [7–9], *Oryza sativa* L. [10–12], and *A. thaliana* L. [13], as well as others, like *Glycine max* (Linn.) Merr. [14], *Brassica napus* L. [15], and *Cucumis sativus* L. [16], for its high accuracy and sensitivity of gene discovery. However, in non-model plants, the progress of drought-resistance research has been slow and unevenly developed, and is thus in need of a lot of effort. In this study, we analyzed the transcriptome and differently expressed genes of Verbena by using high-throughput sequencing technology, ultimately analyzing its mechanism of drought resistance and providing potential drought resistance gene information for resistance breeding work.

2. Results

2.1. Phenotypic and Physiological Indicators of Verbena under Drought Stress

In this study, the morphology of Verbena plants was not significantly affected by drought stress (Figure 1a,b). Different from the morphological indicators, physiological indicators of Verbena had undergone significant changes. The chlorophyll content of leaves first decreased rapidly and then increased by a small margin (Figure 2a), the content of proline (Pro) and soluble protein showed a trend of increasing (Figure 2b,c), the content of superoxide dismutase (SOD) reached the highest level on the 10th day, both catalase (CAT) and peroxidase (POD) gradually increased (Figures 2d and 3a,b), malonaldehyde (MDA) content also increased and relative water content (RWC) did not significantly decrease (Figure 3c,d).

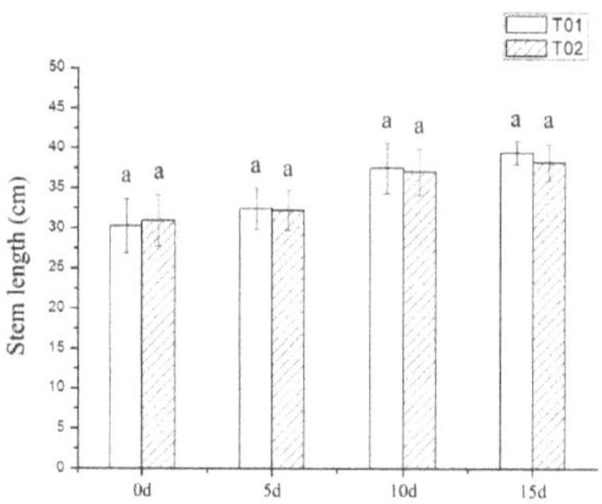

(a)

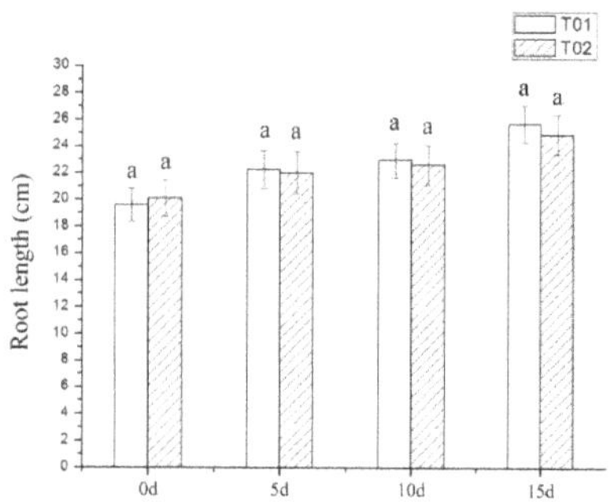

(b)

Figure 1. Morphological indexes of Verbena under drought stress. T01 is the control group and T02 is the drought experiment group: (**a**) The determination of stem length; (**b**) The determination of root length.

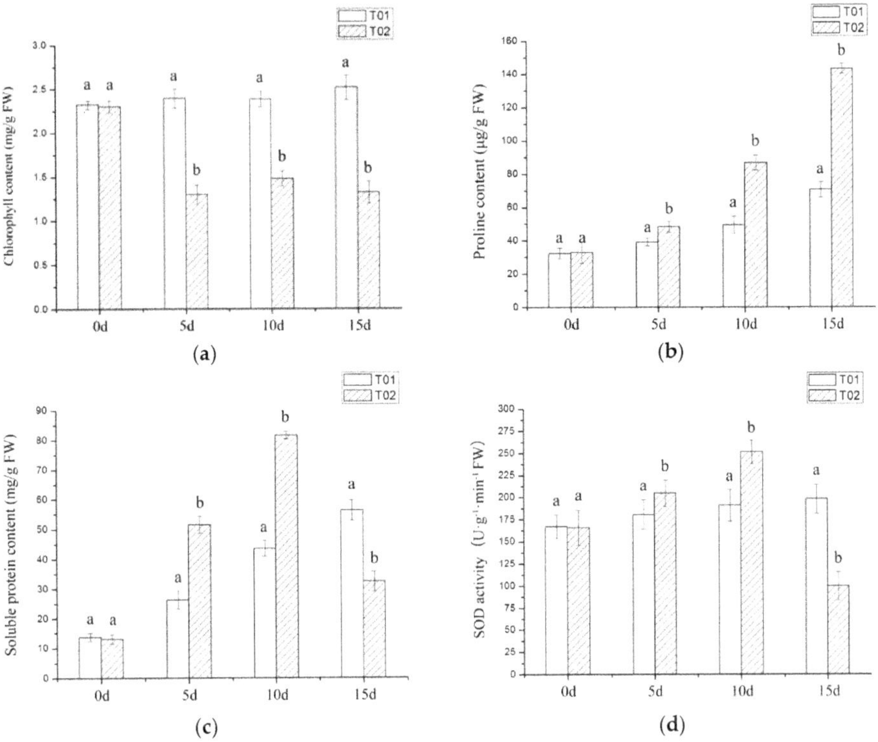

Figure 2. Physiological indexes of Verbena under drought stress. T01 is the control group and T02 is the drought experiment group: (**a**) The content of chlorophyll; (**b**) The content of Pro; (**c**) The content of soluble protein; (**d**) The content of SOD.

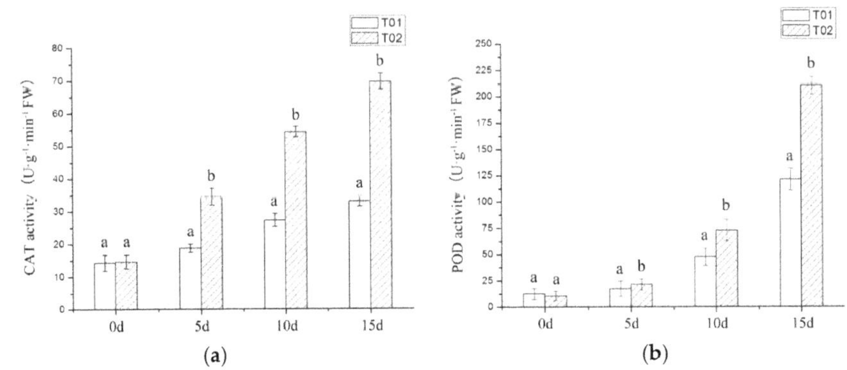

Figure 3. *Cont.*

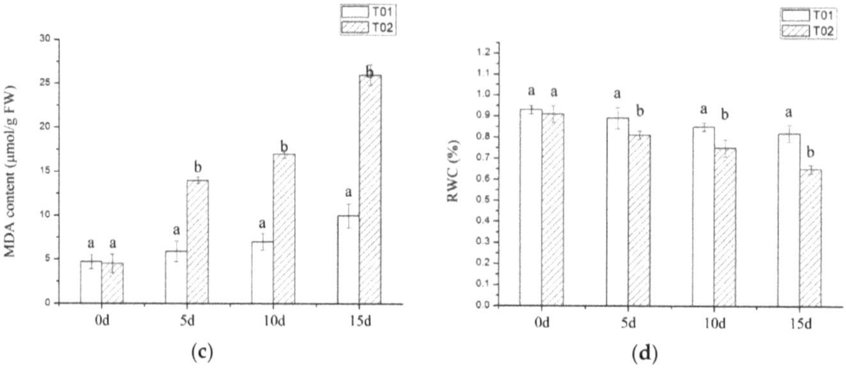

Figure 3. Physiological indexes of Verbena under drought stress. T01 is the control group and T02 is the drought experiment group: (**a**) The activity of CAT; (**b**) The activity of POD; (**c**) The activity of MDA; (**d**) The relative water content of leaves.

2.2. Sequencing and Annotation of Transcription and Unigenes

Based on Sequencing By Synthesis (SBS), six transcriptomes were sequenced by Illumina Hiseq Xten (Illumina, CA, USA). We obtained a total of 44.59 Gb clean data, and in each sample, the Q30 base was not less than 92.87%, the CG (guanine and cytosine basic groups) content was not less than 44.41% (Table SA1). The Pearson's Correlation Coeffiencient r between T1–T3 and T4–T6 was listed in Figure SA1. Using the de novo assembly program Trinity [17] to assemble short-reads, a total of 258,326 transcripts with an average length of 1139.68 bp were obtained. After continuing to cluster and assemble the transcripts for analysis and a total of 111,313 unigenes with an average length of 697.08 bp and N50 of 1223 bp were obtained, among which 40,340 (36.24%) were over 500 bp in length (Table SA2 and Figure SA2).

To predict and analyze the function of the Verbena unigenes, we used BLAST software to compare the amino acid sequence with NR (NCBI non-redundant), KOG (euKaryotic Orthologous Groups), COG (Clusters of Orthologous Groups), GO (Gene Ontology), KEGG (Kyoto Encyclopedia of Genes and Genomes), Pfam (Protein family) and Swissprot-Annotation (a manually annotated and reviewed protein sequence database) database, setting BLAST parameters E-value $\leq 1 \times 10^{-5}$ and HMMER parameters E-value $\leq 1 \times 10^{-10}$ as standard, a total of 53,757 unigenes was obtained, accounting for 48.29% (111,313) of the total. There were 51,352 (95.53%), 28,994 (53.54%), 14,836 (27.60%), 16,938 (31.51%), 20,988 (39.04%), 32,735 (60.89%) and 27,990 (54.53%), unigenes assigned to these databases, respectively (Table SA3). Only 48.29% of unigenes can be matched to known genes, which may be caused by the current lack of studies on Verbena.

A total of 18,104 (35.27%) of unigenes were annotated to *Sesamum indicum,* which means that *Sesamum indicum* had the highest level of homology with *Verbena*, followed by *Erythranthe guttata* (8.60%) and *Erysiphe necator* (7.56%). In addition, *Verbena* and *Sesamum indicum* are quite similar in morphology because they have spike inflorenscence, which is morphological proof of their high homology (Figure SA3).

There are 28,994 unigenes assigned to KOG database and 14,836 to COG database. In these two databases, the proportion of "Signal transduction mechanisms" related to plant resistance respectively occupied 9.85% and 9.45%. In addition, the amount of unigenes assigned to the classes related to plants' response to stress, such as "Defense mechanisms", "Secondary metabolites biosynthesis, transport and catabolism" and "Inorganic ion transport and metabolism", was 7.40% and 10.50% in the two databases, respectively (Figure SA4).

20,988 unigenes were annotated and classified into 3 categories of GO: cell component (CC), molecular function (MF), and biological processes (BP). Most of the genes were assigned to the biological process (59.16%), followed by the molecular function (23.87%) and cellular component (16.97%) (Figure SA5). Among these, the first three categories of BP were "metabolic process" (2.85%), "cellular process" (2.35%) and "single-organism process" (1.84%).

KEGG is a suite of databases and associated software for understanding and simulating higher-order functional behaviors of cells or the organisms based on their genome information. There were 16,938 (31.51%) unigenes allocated to 129 pathways of KEGG, and the pathways assigned with the most genes are "ribosome" (922), "carbon metabolism" (708) and "biosynthesis of amino acids" (622).

2.3. Analysis of Differentially Expressed Genes (DEGs)

Using Bowtie [18], the clean reads were compared to the unigene library. Genes' expression levels were estimated by RSEM (RNA-Seq by Expectation Maximization) [19], and according to the results, the expression abundance of a single gene was expressed as the value of FPKM (transcript fragment per million fragments). The volcano map showed that there were 4829 DEGs, where 3841 (79.54%) were up-regulated and 998 (20.46%) were down-regulated (Figure SA6). Finally, 4165 (85.72%) DEGs were annotated, most of which were annotated to NR (4155) and eggNOG (3957), followed by Pfam (3419), Swissprot (2577), KOG (2426), GO (1756), COG (1554) databases and KEGG (1521) (Table SA3). Clustering results of all DEGs are shown in Figure SA7.

Among the DEGs assigned to KOG and COG, the number of genes in these classes related to plant-resistance of abiotic stress such as "inorganic ion transport and metabolism", "secondary metabolites biosynthesis, transport and catabolism" and "defense mechanisms" were, respectively, 206 (8.50%) and 171 (11.00%), higher than it was for unigenes (Figure 4).

We found that BP, MF and CC accounted for 58.48%, 24.64% and 16.88% respectively by the analysis of 1756 DEGs in GO database. In the entire genetic background and DEGs, the enrichment of genes for each node in the GO database is listed in Figure 5. The top three classes in BP are "DNA integration", "RNA-dependent DNA replication", and "photosynthesis, light harvesting", the most notably of which is the enrichment of "photosynthesis, light harvesting", and the most noticeable of which in MF are the "acyl-CoA dehydrogenase activity" and "oxidoreductase activity, acting on paired donors, with oxidation of a pair of donors resulting in the reduction of molecular"; the *p*-values obtained by the KS test were 0.00024 and 0.00016, respectively. "Ribosome", "fungal-type vacuole membrane" and "nucleus" are the top three in CC, followed up by "photosystem II" and "photosystem I", with *p*-values of 8.3×10^{-5} and 1.4×10^{-4}. The top five significantly enriched nodes of BP, MF and CC are listed in Table SA4.

The DEGs were distributed to 126 lower pathways of KEGG. The first three pathways with the highest enrichment factor were "photosynthesis antenna proteins", "betalain biosynthesis" and "flavone and flavonol biosynthesis", with factors of 6.84, 6.36 and 4.71, respectively. The top ten most significantly enriched pathways of DEGs in KEGG are listed in Table SA5. In the "Photosynthesis antenna proteins" pathway, the class with the highest enrichment factor had a total of 20 DEGs, which were all annotated to the homologous gene annotations of "Light-harvesting Chlorophyll a/b Binding (LHC) Proteins"—the apoproteins of the light-harvesting complex of photosystem II (PSII) [20].

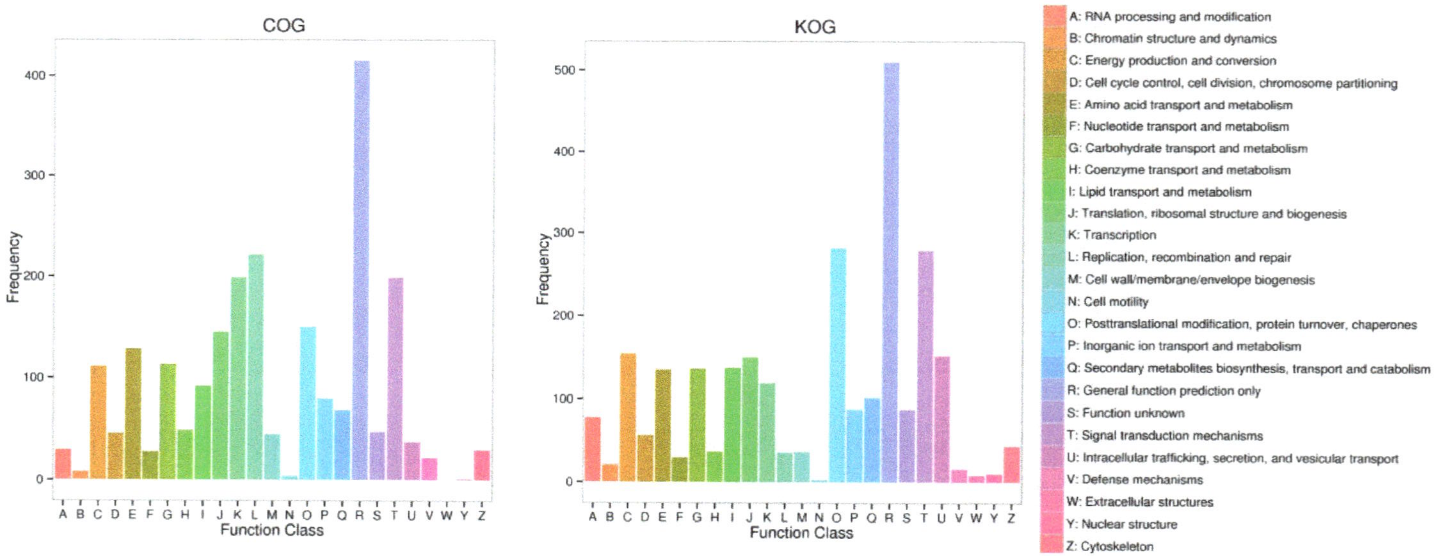

Figure 4. The bar chart of unigenes' functional classification annotated in COG and KOG databases. The abscissa is the function classifications of COG and KOG databases and the ordinate is the number of DEGs annotated in it.

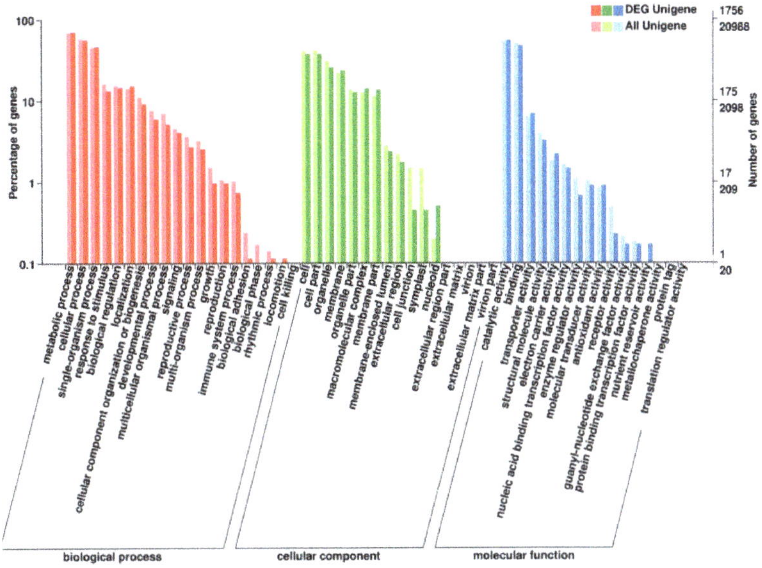

Figure 5. The bar chart of DEGs annotated in the GO classification. The ordinate at the left represents the percentage of the number of genes, the right ordinate represents the number of genes. The above of two ordinates is the number of DEGs, the following is the number of all genes. The abscissa is the classification of GO. The dark bar represents the number and proportion of DEGs that are enriched in GO function, and the light bar represents the number and proportion of genes that are enriched for each GO function.

2.4. DEGs of Transcription Factors (TFs) under Drought Stress

In this study, a total of 2146 (44.44%) transcription factor DEGs were identified, of which 1656 (77.17%) were up-regulated and 490 (22.83%) were down-regulated. The TFs mainly focused on bHLH, MYB, ERF, NAC and C2H2 families, with 228 (10.62%), 207 (9.65%), 176 (8.20%), 153 (7.13%) and 102 (4.75%) DEGs (Figure 6), respectively. The number of up-regulated genes of the ten were 171 (75%), 157 (75.85%), 137 (77.84%), 122 (79.74%) and 76 (74.51), respectively

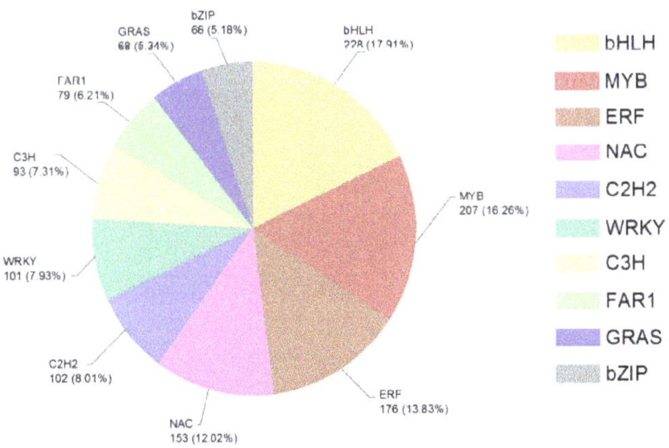

Figure 6. The sector diagram of TFs' classification and number in DEGs in response to drought stress.

2.5. Expression Level of DEGs' Changes and Verification Using qRT-PCR

We selected ten genes involved in different important biological processes to perform qRT-PCR, including genes related to "energy production and conversion", "transcription factors", "LHCB2 proteins", "glutamine synthetase", "protein kinases", "lipid transport and metabolism", "carbohydrate transport and metabolism" and "nitrogen metabolism". During the experiment, similar to indexes of physiological, the expression level of these DEGs were all up-regulated with the increase of stress time (Figure 7a,b). To further verify the expression level of genes obtained from Illumina Hiseq Xten, we compared the data obtained from the 15th day with sequencing, and the results showed a strong correlation between the two (Figure 8).

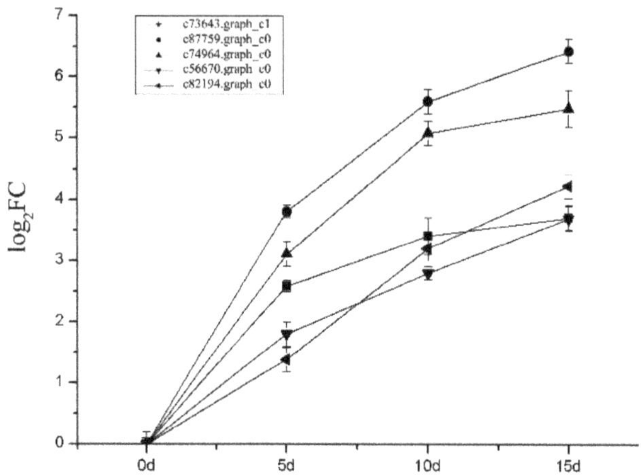

(**a**)

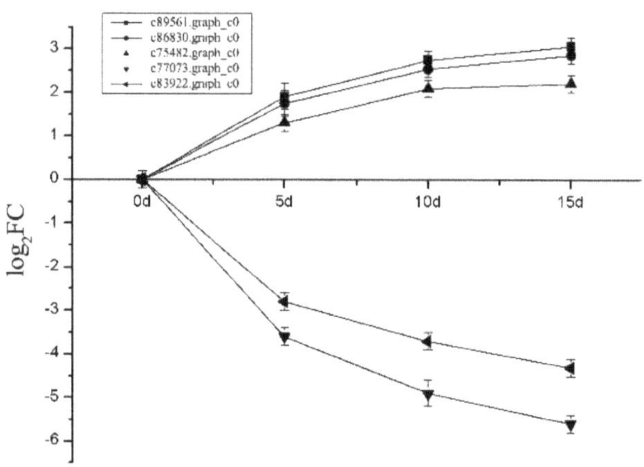

(**b**)

Figure 7. The line chart of the expression level of the 10 DEGs varied with the degree of drought during the experiment. (**a**) Five DEGs varied with the degree of drought during the experiment; (**b**) another five DEGs varied with the degree of drought during the experiment.

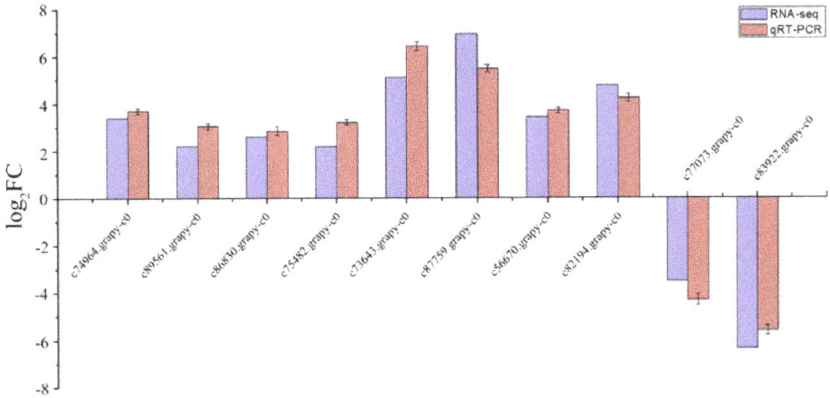

Figure 8. The bar chart of results of qRT-PCR in 15th day. The relative expression level of ten DEGs identified in the comparison between RNA-Seq and qRT-PCR. The genes relative expression level were determined by $2^{-\Delta\Delta Ct}$ method.

3. Discussion

3.1. Morphological and Physiological Index Analysis

Among the ten most abundant nodes in the BP of GO database, the most notable one is "photosynthesis, light harvesting". Many studies [21,22] have shown that water stress can change the plants' chlorophyll content, which can indicate the sensitivity of plants to water stress and directly affect the photosynthetic yield. Some studies have shown that drought increases the chlorophyll content of plants' leaves [23], while some other studies believe it would be gradually decreased [24]. In this study, the chlorophyll content decreased rapidly and then increased by a small margin. Although it is unclear what the mechanism of water stress on chlorophyll content is, the fact is that increasing the chlorophyll content will enhance plants' endurance to survive in adversity, and it is a kind of ability by which plants can adapt to drought stress, indicating that the Verbena has a certain drought tolerance.

In this experiment, the Pro content, soluble protein content, antioxidant enzyme activity and the MDA content were all increased. Pro and soluble proteins are the most common osmotic pressure regulators in drought-stressed plants. Plants that overproduce Pro and soluble protein might acquire the ability to tolerate environmental stresses such as drought and high salinity [25]. There are a large number of antioxidants that can prevent or repair the damage caused by reactive oxygen species and regulate redox-sensitive signaling pathways, such as POD, SOD, CAT and so on [26]. The degree of membrane damage on the leaves can be demonstrated by the content of MDA, which gradually increased in this study. Under the circumstances of drought force, the more drought-tolerant the plants are, the slower the water content of the leaves decreases. Figure 3d shows that on day 15, the RWC of T02 (stress group) was about 16% lower than that of T01 (control group), which performed better than *Hordeum vulgare* L. [27], *Zea mays* L. [28] and *Glycine max* (Linn.) Merr. [29], indicating that the leaves of Verbena have a certain water retaining capacity.

3.2. The Enrichment and Pathway Analysis of DEGs in GO and KEGG Databases

The chlorophyll content decreased sharply in the early stage of drought stress, and eventually increased to a lower level than T01. The 16 DEGs related to "porphyrin and chlorophyll metabolism" are listed in Table 1, and the enzymes of porphyrin metabolism were almost all up-regulated, while the enzymes of chlorophyll metabolism were almost all down-regulated. Analyzing the genes involved in the regulation of stomatal closure by ABA (abscisic acid), we found that some of the genes' expression levels of PYL and PP2C were significantly up-regulated (Table 2), and according to the analysis of the "carbon fixation in photosynthetic organisms" pathway, we found that a number of genes related to "C4-Dicarboxylic acid cycle and carbon fixation pathways in prokaryotes" process showed an upward trend. This indicates that under drought stress, the photosynthesis of Verbena may be mainly inhibited by stomatal closure, and at the same time it may provide an adequate carbon source for photosynthesis by enhancing the biological carbon sequestration pathway. The DEGs in "photosystem II" and "photosystem I" of CC also showed a high degree of enrichment, indicating that drought had a great impact on the photosynthesis of Verbena, which is the direction we should focus on. In MF, "acyl-CoA dehydrogenase" and "oxidoreductase" are obviously enriched. Previous study has shown that deficiencies of acyl-CoA dehydrogenases can lead to disorders of fatty acid oxidation, leading to life-threatening metabolic disorders [30]. It is known that drought stress will produce a large amount of reactive oxygen species (ROS) in plants and start the massive production of oxidoreductase such as SOD. Therefore, the increase of "acyl-CoA dehydrogenase" and "oxidoreductase" is of great significance for plants in responding to drought stress.

Table 1. DEGs (analysis of differentially expressed genes) in "Porphyrin and chlorophyll metabolism" pathway of KEGG (Kyoto Encyclopedia of Genes and Genomes).

Term	Gene ID	log2FC	Gene Description	FDR
UROD	c76376.graph_c0	2.33	uroporphyrinogen decarboxylase chloroplast precursor	1.44×10^{-5}
COX15	c59080.graph_c0	2.40	uroporphyrinogen decarboxylase chloroplast precursor	2.71×10^{-8}
FECH	c69481.graph_c0	2.58	protoporphyrin/coproporphyrin ferrochelatase	7.56×10^{-5}
	c86128.graph_c2	3.02	chloroplastic isoform X2	1.01×10^{-7}
EARS	c69469.graph_c0	2.43	glutamyl-tRNA reductase	1.14×10^{-6}
	c72758.graph_c0	2.31	Porphyrin and chlorophyll metabolism	2.72×10^{-3}
hemA	c85183.graph_c0	4.04	glutamyl-tRNA reductase 1, chloroplastic-like	1.98×10^{-11}
	c77400.graph_c0	−2.69	hypothetical protein	9.98×10^{-38}
	c77400.graph_c1	−2.65	glutamyl-tRNA reductase 1, chloroplastic	1.84×10^{-31}
	c77400.graph_c2	−2.68	glutamyl-tRNA reductase 1, chloroplastic	1.67×10^{-23}
chlH	c88820.graph_c1	−2.50	magnesium chelatase subunit H	1.91×10^{-60}
chlE	c77176.graph_c0	−2.24	magnesium-protoporphyrin IX monomethyl ester (oxidative) cyclase	7.54×10^{-37}
por	c85861.graph_c0	−3.48	protochlorophyllide reductase	1.36×10^{-11}
chlP	c80298.graph_c1	−2.94	geranylgeranyl diphosphate/geranylgeranyl-bacteriochlorophyllide a reductase	1.13×10^{-47}

Table 2. DEGs related to hormone synthesis in response to drought stress.

Term		Gene ID	log2FC	Gene Description	FDR
ABA	PYL/PYR	c72499.graph_c2	5.61	abscisic acid receptor PYR/PYL family (A)	5.09×10^{-14}
		c64811.graph_c0	−2.31	abscisic acid receptor PYR/PYL family (A)	2.69×10^{-58}
		c73702.graph_c1	−2.28	K14496 abscisic acid receptor PYR/PYL family (A)	0.00000743
	PP2C	c86830.graph_c0	2.60	probable protein phosphatase 2C 51	1.62×10^{-19}
SA	PR1	c31398.graph_c0	4.97	basic form of pathogenesis-related protein 1-like	2.11×10^{-159}
JA	JAZ	c75424.graph_c0	2.43	protein TIFY 10B-like	4.90×10^{-58}
		c75566.graph_c0	2.14	jasmonate ZIM domain-containing protein (A)	8.12×10^{-44}
		c77115.graph_c1	3.17	jasmonate ZIM domain-containing protein (A)	2.58×10^{-74}
		c77115.graph_c2	3.22	Protein TIFY 10B	2.80×10^{-87}
		c88229.graph_c0	2.13	protein TIFY 9-like	0.000018
	MYC2	c88848.graph_c1	2.14	transcription factor MYC2-like	1.13×10^{-24}
Auxin	GH3	c78593.graph_c1	2.24	auxin responsive GH3 gene family (A)	5.93×10^{-6}
		c83994.graph_c0	3.57	auxin responsive GH3 gene family (A)	3.63×10^{-62}
	SAUR	c76579.graph_c0	2.39	uncharacterized protein	3.12×10^{-16}
		c80406.graph_c5	2.71	hypothetical protein MIMGU_mgv1a0212152mg	4.51×10^{-26}
		c63583.graph_c0	−4.29	auxin-induced protein 10A5	1.74×10^{-17}
		c64412.graph_c0	−2.18	SAUR family protein (A)	5.04×10^{-9}
		c65963.graph_c0	−3.66	indole-3-acetic acid-induced protein ARG7-like	1.28×10^{-13}
		c84555.graph_c1	−4.21	SAUR family protein (A)	4.31×10^{-11}
Ethyle-ne	MPK6	c75482.graph_c0	2.19	mitogen-activated protein kinase 8	1.2624×10^{-3}
	EBF1/2	c70061.graph_c0	2.28	EIN3-binding F-box protein (A)	4.3002×10^{-3}

In the KEGG database, the first three pathways with the highest DEGs enrichment factors are "photosynthesis antenna proteins", "betalain biosynthesis" and "flavone and flavonol biosynthesis". After comparison, the homologous genes with almost all the down-regulated DEGs (Table 3) in "photosynthesis antenna proteins" pathway are related to LHC proteins, which may be involved in the drought resistance of plants and play an important role in crop environmental adaptability and yield [31]. Alberte R.S. et al. [32] showed that LHC proteins are a target easily attacked by water stress, and the loss of chlorophyll, the increase of chlorophyll a/b ratio and the decrease of photosynthetic unit under water stress are all caused by the decrease of LHC proteins. Vappaavuovi E. et al. [33] also confirmed that water stress reduced the LHC protein complex. In addition, some studies have showed that LHC proteins may be involved in the partial regulation of ABA signaling and play an active role in guarding cell signaling—these proteins may be induced by ABA and positive regulates ABA to inhibit stomatal opening [34,35]. Therefore, the decrease of LHC proteins will inhibit the production of ROS. However, due to the insufficiency research of LHC proteins, further research on its response to abiotic stress is needed. "Betalain biosynthesis" pathway has three distinctly up-regulated genes (Table 4). Betaine is the trimethyl derivative of the amino acid glycine, an efficient methyl donor that promotes fat metabolism and protein synthesis and the increasing of betaine biosynthesis has been shown to play an important role in osmoregulation of plants, which can help plants to withstand drought stress [36]. The "Flavone and flavonol biosynthesis" pathway has four distinctly up-regulated genes with an enrichment factor of 4.71, and studies have shown that the flavonoid substance may contribute to antioxidant functions in response to drought stress [37]. At present, there is still a lack of research on the changes in the content of secondary metabolites that have economic or medicinal value in Verbena under abiotic stress, but according to the schematic pathway figures of betalain and flavone (Figure 9a,b), we can preliminarily speculate that the content of flavonoids and betaine were up-regulated under drought conditions, while other plants also have similar research conclusions. Xing W. et al. [38] have shown that under drought stress, the endogenous leaf glycine betaine level of *A. thaliana* L. increased about 18-fold over that in the control plants, and similar results were also found in Pyrus bretschneideri Rehd [39] and *Hordeum vulgare* L. [40]. From the pathway of "flavone and flavonol biosynthese", we can see that all DEGs are up-regulated (Table 5) and there are two kinds of useful downstream production—luteolin and quercetin. Studies have shown that drought increased the accumulation of luteolin in *Ligustrum lucidum* Ait. [37], and quercetin also showed a positive impact on *Vigna unguiculata* L. Walp. [41], but in *Cabernet Sauvignon* [42], quercetin showed no obvious change under water stress. Therefore, further experiments are needed to investigate the effects of abiotic stress on secondary metabolite content in Verbena.

Table 3. DEGs in "Photosynthesis-antenna Proteins" pathway of KEGG.

Term	Gene ID	log2FC	Gene Description	FDR
LHCA1	c75167.graph_c0	−2.30	chlorophyll a-b binding protein 6, chloroplastic	1.13×10^{-47}
LHCA2	c57238.graph_c0	−2.30	chlorophyll a-b binding protein, chloroplastic	1.52×10^{-37}
LHCA3	c71085.graph_c0	−2.07	chlorophyll a-b binding protein 8, chloroplastic-like	3.11×10^{-14}
	c85515.graph_c0	−2.33	chlorophyll a-b binding protein 8, chloroplastic	2.41×10^{-138}
	c71085.graph_c1	−2.20	chlorophyll a-b binding protein 8, chloroplastic-like	5.63×10^{-5}
LHCA4	c57961.graph_c0	−3.78	chlorophyll a-b binding protein 4, chloroplastic	3.42×10^{-11}
	c81195.graph_c1	−3.42	agamous-like MADS-box protein AGL21 isoform X3	1.69×10^{-77}
	c31746.graph_c0	−3.92	chlorophyll a-b binding protein P4, chloroplastic-like	1.20×10^{-120}
LHCB1	c85665.graph_c1	−3.25	chlorophyll a/b-binding protein PS II-Type I	6.78×10^{-29}
	c85665.graph_c2	−3.68	chlorophyll a-b binding protein 21, chloroplastic-like	6.40×10^{-63}
	c83506.graph_c0	−3.53	chlorophyll a/b-binding protein, partial	1.02×10^{-12}
LHCB2	c31726.graph_c0	−2.58	chlorophyll a-b binding protein 5, chloroplastic	1.26×10^{-46}
	c77073.graph_c0	−3.51	chlorophyll A/B binding protein, putative	7.67×10^{-58}
LHCB3	c82382.graph_c0	−2.60	chlorophyll a-b binding protein 13, chloroplastic	5.22×10^{-47}
	c82382.graph_c1	−2.74	chlorophyll a-b binding protein 13, chloroplastic	5.88×10^{-32}
	c84778.graph_c0	−2.35	chlorophyll a-b binding protein 13, chloroplastic	1.14×10^{-20}
LHCB4	c57394.graph_c0	−3.66	chlorophyll a-b binding protein CP29.1, chloroplastic	4.63×10^{-60}
LHCB5	c72073.graph_c1	−2.03	chlorophyll a-b binding protein CP26, chloroplastic	4.96×10^{-43}
	c72073.graph_c0	−2.25	chlorophyll a-b binding protein CP26, chloroplastic	7.71×10^{-40}
LHCB6	c76630.graph_c1	−2.38	hypothetical protein MIMGU_mgv1a012260mg	2.07×10^{-23}

Table 4. DEGs in "Betalain biosynthesis" pathway of KEGG.

Term	Gene ID	log2FC	Gene Description	FDR
TYR	c83086.graph_c0	2.69	Tyrosinase	2.09×10^{-5}
COMT	c26366.graph_c0	3.02	catechol *O*-methyltransferase	7.64×10^{-5}
DOPA	c75132.graph_c0	3.38	PREDICTED: 4,5-DOPA dioxygenase extradiol-like	1.87×10^{-89}

Table 5. DEGs in "Flavone and flavonol biosynthese" pathway of KEGG.

Term	Gene ID	log2FC	Gene Description	FDR
E1.14.13.21	c32062.graph_c0	2.11	benzoate 4-monooxygenase cytochrome P450	1.06×10^{-5}
AOMT	c57467.graph_c0	4.20	PREDICTED: flavonoid 3' 5' -methyltransferase-like	1.59×10^{-13}
	c67675.graph_c0	3.17	PREDICTED: flavonoid 3' 5' -methyltransferase-like	2.84×10^{-22}
C12RT1	c69454.graph_c0	4.05	hypothetical protein MIMGU_mgv1a022315mg	1.81×10^{-17}

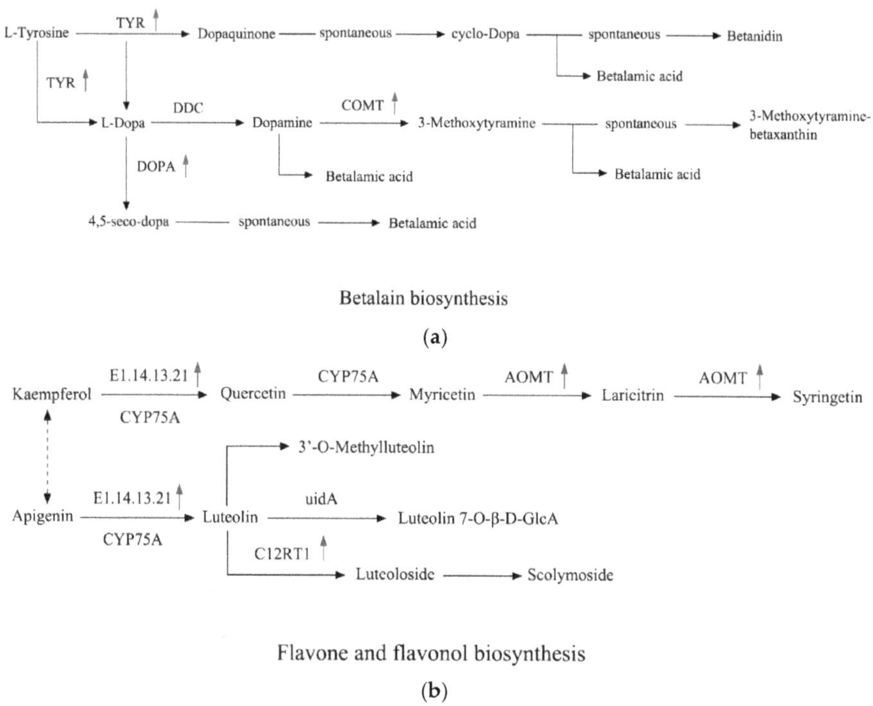

Figure 9. Two schematic pathway figures of "Betalain biosynthesis" and "Flavone and flavonol biosynthesis": (**a**) Effect of drought stress on the expression of genes associated with betaine. (**b**) Luteolin and quercetin metabolism.

3.3. Biological Mechanism of Verbena in Response to Drought Stress

The protective mechanisms of plants in response to stress are regulated by alterations in the expression level of stress-responsive genes. Among the DEGs assigned to KOG and COG, genes related to plants' stress response accounted for quite a high proportion; 8.50% and 11.00%, respectively. We performed GO and KEGG pathway enrichment analysis and excavated a group of important drought-responsive genes related to multiple biological mechanisms of plant energy production, hormone synthesis, cell signaling, and metabolism. In this study, there are 12, 25 and 24 genes differentially expressed in "glyoxylate and dicarboxylate metabolism", "citrate cycle (TCA cycle)" and "glycolysis/gluconeogenesis" pathway, respectively, and all of them were up-regulated. This indicates that the energy production of Verbena changed a lot, and the main method of providing ATP was changed from the photosynthetic phosphorylation to the oxidative phosphorylation.

ABA is one of the most important factors in abiotic stress response, and is involved in almost all plant activities, such as photosynthesis, ionic homeostasis, and antioxidant defense [43]. The quantities of gene expression controlling the key enzymes, such as PYL and PP2C in the ABA signaling pathway, have changed a lot. The vicissitudinous DEGs of PYL activate the high expression of PP2C and furthermore promote stomatal closure and decrease transpiration to reduce water loss. Under the "plant hormone signal transduction" pathway, in addition to the enzymes of ABA, the level of genes that control PR1 in the "salicylic acid (SA)", JAZ and MYC2 in the "jasmonic acid (JA)", GH3 and SAUR in the "auxin", and MPK6 together with EBF1_2 in "Ethylene" pathways, were almost all up-regulated (Table 2). This is in line with previous findings that the changes of hormonal synthesis and signal transduction are the conservative mechanisms by which plants response to adverse circumstances. It has been demonstrated that NO (nitric oxide) can act as a signaling molecule to activate ROS-scavenging enzymes in drought stress [44]. Under the pathway of "Nitrogen metabolism", we observed that the gene expression levels of NR and NirA in the assimilatory nitrate reduction pathway were down-regulated, and the expression of GLT1 and GLUL in the glutamate synthase pathway were up-regulated (Table 6). The glutamate synthase pathway leads to the process of glutamate metabolism and the production of ammonia, and a sufficient amount of oxidized glutamic acid will ensure that there is sufficient carbon skeleton for the tricarboxylic acid cycle to function effectively [45]. Therefore, we can speculate that improving the level of nitrogen metabolism is a very efficient method to help plants resist abiotic stress.

Carbohydrate metabolism is the center of the entire biological metabolism and involves the protein metabolism, lipid metabolism, nucleic acid metabolism and secondary metabolites production. Studies have shown that in rice [46], sugar can be used as a signal to induce the expression of genes associated with abiotic stress. Garg A.K. et al. [47] found that the accumulation of trehalose in rice enabled the transgenic rice to exhibit high salt tolerance, drought tolerance, and low-temperature stress. In this study, there were 28 and 12 DEGs in "Starch and sucrose metabolism" and "Pentose phosphate pathway", respectively, most of which were up-regulated, which indicates the importance of sugar for Verbena to cope with drought stress. Furthermore, we found a series of pathways for the up-regulation of DEGs expression: protein metabolism such as "Biosynthesis of amino acids", lipid metabolism such as "Fatty acid metabolism" and "Glycerophospholipid metabolism", nucleic acid metabolism such as "Pyrimidine metabolism" and "Purine metabolism", and metabolism of secondary products such as "alpha-Linolenic acid metabolism" and "Flavone and flavonol biosynthesis". All of these show that carbohydrate metabolism is very important for Verbena to improve its tolerance to adverse environmental conditions.

Moreover, "ubiquitin mediated proteolysis", an energy-consuming, highly efficient and highly directional protein degradation process, is also noteworthy. It plays an important role in many aspects, such as modulation of the immune and inflammatory responses, the regulation of cell cycle, control of signal transduction pathways, development and differentiation etc. [48,49]. In the present study, there are 16 DEGs in this pathway (Table 7), all of which up-regulate. Half are involved in the regulation of ubiquitin conjugating enzyme, and the others are involved in the regulation of ubiquitin ligase. At present, the understanding of this process is still very limited. The genes participated in this process during drought stress, and the mechanism of the enzymes which were regulated by these genes has yet to be studied.

Table 6. DEGs in "Nitrogen metabolism" pathway of KEGG.

Term	Gene ID	log2FC	Gene Description	FDR
NR	c88329.graph_c0	−2.65	Nitrate reductase 2	1.32×10^{-97}
NirA	c85021.graph_c0	−2.64	Ferredoxin–nitrite reductase	4.29×10^{-84}
GLUL	c89561.graph_c0	2.21	glutamine synthetase4	2.24×10^{-7}
GLT1	c85092.graph_c2	2.29	glutamate synthase (NADPH/NADH)	1.53×10^{-5}

Table 7. DEGs in "Ubiquitin mediated proteolysis" pathway of KEGG.

Term		Gene ID	log2FC	Gene Description	FDR
E2	UBE2A	c56569.graph_c0	2.73	ubiquitin-conjugating enzyme E2 A	9.53×10^{-7}
	UBE2O	c78080.graph_c1	2.32	ubiquitin-conjugating enzyme E2 O; A orthologs to drought gene GmMYB177	1.03×10^{-6}
	UBE2W	c43734.graph_c0	3.58	ubiquitin-conjugating enzyme E2 W	3.64×10^{-6}
	UBE2N	c61661.graph_c0	2.30	ubiquitin-conjugating enzyme E2 N	3.35×10^{-6}
	UBE2D-E	c89609.graph_c0	2.54	ubiquitin-conjugating enzyme E2 D/E	4.15×10^{-6}
	UBE2I	c46599.graph_c0	2.71	ubiquitin-conjugating enzyme E2 I; A orthologs to drought gene GmMYB177	3.46×10^{-8}
	UBE2G1	c26287.graph_c0	2.14	ubiquitin-conjugating enzyme E2 G1	2.449×10^{-3}
		c25902.graph_c0	3.11	ubiquitin-conjugating enzyme E2 G1	1.23×10^{-12}
E3	ARF-BP1	c84837.graph_c1	2.40	E3 ubiquitin-protein ligase HUWE1	2.58614×10^{-4}
	UBE4B	c71025.graph_c0	2.32	ubiquitin conjugation factor E4 B	1.45681×10^{-4}
	CYC4	c75600.graph_c0	2.46	peptidyl-prolyl cis-trans isomerase-like 2	1.46×10^{-5}
	PRP19	c60805.graph_c0	2.06	pre-mRNA-processing factor 19	2.56×10^{-5}
	Cul3	c63726.graph_c0	2.00	cullin 3 (A)	2.73×10^{-5}
	CYC4	c75600.graph_c0	2.46	peptidyl-prolyl cis-trans isomerase-like 2	1.46×10^{-5}
	SYVN	c79541.graph_c0	2.01	ubiquitin-protein ligase synoviolin	1.091839×10^{-3}
	Cdh1	c47817.graph_c0	2.93	cell division cycle 20-like protein 1, cofactor of APC complex (A)	1.53×10^{-5}
	TRIP12	c82561.graph_c0	2.52	E3 ubiquitin-protein ligase TRIP12	2.08×10^{-9}

We found evidence that Verbena responds to drought stress by altering energy synthesis pathways, decreasing transpiration, resetting hormone secretion levels, and increasing cell osmotic pressure and glucose metabolism. In general, Verbena's defensive response, like most plants under stress conditions, is a process of rebuilding physiologically, biochemically and metabolically, from the growth-oriented to the defensively based one. The genes mentioned in Section 3 are listed in Tables 1–7 and after counting the TFs to which these genes belong, we constructed the MYB TF phylogenetic tree using the reported abiotic stress genes to prepare for the next work (Figure SA8).

3.4. DEGs of Transcription Factors (TFs) under Drought Stress

There are 228 (10.62%) DEGs that were assigned to bHLH TF. The basic helix-loop-helix proteins are one of the largest transcription factor families and are widely distributed in eukaryotes [50]. Many important drought-tolerant genes have been discovered in the bHLH family. Over-expression of OsbHLH148 in the transgenic rice make the plant more drought-tolerant by regulating the jasmonate signal transduction pathway [51]. Up-regulation of bHLH122 can significantly increase ABA levels in cells and it is a positive regulator of drought, NaCl and osmotic signaling [52]. AtbHLH112 is a nuclear-localized protein induced by salt, drought and ABA, and it can increase proline levels and improve ROS scavenging ability to enhance stress tolerance [53]. At the same time, there are also a number of other TFs that play an important role in plants' resistance to abiotic stresses, such as MYB (9.65%), ERF (8.20%) and NAC (7.13%). In this study, most of the genes that enhance plants' stress resistance were achieved by overexpression, so we will pay more attention to the up-regulated genes and dig out the role of specific genes in the processes of stress resistance by transgenic and gene-silencing technology.

4. Materials and Methods

4.1. Plant Materials and Drought Treatments

Verbena bonariensis L. with the age of one month were used in this study. The experimental materials were incubated in greenhouse (25 °C, 16 h photoperiod, 50% RH), soil texture for the medium loam, the maximum soil moisture content of 80%. A total of 90 strains of Verbena were randomly divided into two groups T01 (control group) and T02 (drought experiment group), 1 seeding in each pot and 15 in one duplicated group, each of the T01 and T02 contained 3 duplicated groups named T1–T3 and T4–T6, respectively. Determination of soil moisture content was measured by the oven drying method [54]. The soil water content of T01 was kept at 80% of saturated soil moisture all the time, while watering of the T02 was stopped until the soil water content had reduced to 25% of the soil saturated water content. After that, the soil moisture content was measured every 2 days to replenish the amount of deficiency and maintained this drought status for 15 days.

At the 5th, 10th and 15th day, the mature leaves (3rd to 8th functional leaf) were selected randomly from the plants of T01 and T02 for the determination of physiological indexes. After the drought had been maintained for 15 days, samples were collected, rapidly frozen with liquid nitrogen, stored at −80 °C, and finally sent to Biomarker Technologies Co., Ltd. (Beijing, China) for whole-transcriptome sequencing.

4.2. Determination of Morphological and Physiological Characters

Vernier calipers were used to measure root length (whichever is longer) and stem length. Chlorophyll content was determined by NanoPhotometer® spectrophotometer (Implen, CA, USA) [55]; Pro content was measured using acidic-ninhydrin method [56]; estimation of soluble protein by Bradford method [57]; nitroblue tetrazolium blue (NBT) reduction method was used to determine the activity of SOD [58]; CAT activity was determined using UV absorption method [59]; POD activity was determined by guaiacol method [60]; determination of MDA content by thiobarbituric acid test (TBA) [61]. Relative water content (RWC) of leaves was measured according to the method of Tambussi E.A. et al. [62] using the formula: RWC = (FW − DW)/(turgid weight − DW) × 100%, where FW is the leaf fresh weight, DW is the leaf dry weight at 85 °C for 3 d, and SW is the turgid weight of leaves after soaking in water for 4 h at room temperature (approximately 25 °C). Physiological measurements were set up for three replicates to reduce the error.

4.3. Extraction of RNA, Library Preparation for Transcriptome Sequencing

To ensure the qualified samples were obtained for transcriptome sequencing, total RNA was extracted with Trizol kit (Invitrogen, Carlsbad, CA, USA) and its purity concentration and integrity detected by the Nanodrop, Qubit 2.0, Agilent 2100 method. Then, a total amount of 3 μg qualified RNA per sample was used as input material for the RNA sample preparations. According to manufacturer's recommendations, sequencing libraries were generated using NEBNext®Ultra™ RNA Library Prep Kit for Illumina® (New England Biolabs, Ipswich, MA, USA). The messenger RNAs (mRNAs) were separated from the total RNA by Oligo (dT) and were cleaved into short fragments at random. The first strand cDNA was synthesized by random hexamer primer, then the buffer, dNTPs, DNA polymerase I and RNase H were used to synthesize the second strand cDNAs. Lastly, the cDNAs were purified with AMPure XP beads and after end-repair and single nucleotide A (adenine) addition, the qualified cDNA libraries were constructed by PCR enrichment. After the cDNA libraries were constructed, Qubit 2.0 was used for preliminary quantification, and then the Agilent 2100 was used to detect the insert size of the libraries. After that, the Q-PCR method was used to accurately quantify the effective concentration of the libraries (effective library concentration > 2 nM) to ensure library quality. After passing the screening, high-throughput sequencing was performed with Illumina Hiseq Xten. The raw sequencing data have been submitted to the National Center for Biotechnology Information Search database (NCBI) Sequence Read Archive database with accession number SRP132610.

4.4. Transcriptome Assembly and Gene Functional Annotation

A total of 44.59 Gb of clean data was removed from the original data, including low-quality data. Then Trinity software was used to break reads into shorter K-mers, extend them to obtain Contigs, gather a collection of Contig clusters, and finally obtain the transcript sequence by using de Bruijin graphs algorithm method and reads.

Using BLAST software, the sequencing of the Unigene sequence was compared with NR, KOG, COG, GO, KEGG, Pfam and Swissprot-Annotation databases. After the prediction of the amino acid sequence of the unigene, HMMER software [63] was used to compare it with the Pfam database [64] to obtain functional annotation information of all unigenes.

4.5. Differential Expression Analysis

Differential expression analysis of the sample groups was performed using DESeq [65] to obtain a set of DEGs between the drought group and the control group, and the False Discovery Rate (FDR) was used to correct the p-value of the multiple-hypothesis test. In this study, FDR < 0.01 and FC (Fold Change) ≥ 4 were set as the threshold for significantly differential expression, and could be used to prepare for the follow-up study on the drought stress of Verbena.

4.6. Quantitative Real-Time PCR Analysis

In order to clarify the stress response variation tendency of Verbena at the transcriptional level on different stages of drought stress, we selected 10 DEGs to participate in different important biological processes for qRT-PCR with SsoFast™ EvaGreen® Supermix every 5 days. A 20 μL fluorescent quantitative reaction system contains 10 μL of stain, 2 μL of cDNA template and 300 nM of primers. The PCR settings are as follows:

Temperature	Time	Cycle
95 °C	30 s	
95 °C	15 s	40 cycles
60 °C	30 s	

Relative expression levels were calculated by the $2^{-\Delta\Delta Ct}$ method, and a β-actin gene of *Verbena bonariensis* (Forward primer: GAAAGATGGCTGGAAGAGGG, Reverse primer: GCTATGAA CTCCCTGATGGTC) was used as the reference for quantitative expression analysis. The expression pattern of DEGs was analyzed by melting furnace curve. The qRT-PCR primers are shown in Table SA6.

5. Conclusions

The development of plant genomics plays an important role in the effective use of modern molecular biology methods for the genetic improvement of species. After sequencing the transcriptome of drought-stressed Verbena materials by high-throughput sequencing and verifying the results by qRT-PCR, we finally identified DEGs and TFs related to stress resistance and analyzed their biomodulation mechanism, and this proved Verbena's commendable drought tolerance at physiological and molecular levels. These data will provide excellent genetic resources for improving the drought tolerance of crops and lay a good foundation for the follow-up study of Verbena.

Supplementary Materials: The following are available online at http://www.mdpi.com/1422-0067/19/6/1751/s1.

Author Contributions: B.W., X.-Q.L., and Q.-L.L. conceived and designed the experiments. B.W., X.-Q.L., M.-S.X. and Q.-L.L. performed the experiments. B.W., F.-L.L., Q.Z., L.H., Q.-L.L., L.Z., Y.J. and F.Z. analyzed the data. Q.-L.L. contributed reagents/ materials/ analysis tools. B.W. wrote the paper.

Funding: This research was funded by National Natural Science Foundation of China, grant number (31770742) and Innovative Training Program of Sichuan Agricultural Uniersity, grant number (201710626086).

Acknowledgments: This study was supported by Sichuan Agricultural University Ornamental Horticulture lab. We would like to acknowledge the contribution of Qing-Lin Liu for the provision of experimental materials and instruments.

Conflicts of Interest: The authors declare no conflict of interest.

References

1. Zhu, J.K. Salt and drought stress signal transduction in plants. *Annu. Rev. Plant Biol.* **2002**, *53*, 247–273. [CrossRef] [PubMed]
2. Farooq, M.; Wahid, A.; Kobayashi, N.; Fujita, D.; Basra, S.M.A. Plant drought stress: Effects, mechanisms and management. *Agron. Sustain. Dev.* **2009**, *29*, 185–212. [CrossRef]
3. Shinozaki, K.; Yamaguchi-Shinozaki, K. Gene networks involved in drought stress response and tolerance. *J. Exp. Bot.* **2007**, *58*, 221–227. [CrossRef] [PubMed]
4. Xiong, L.; Wang, R.; Mao, G.; Koczan, J.M. Identification of drought tolerance determinants by genetic analysis of root response to drought stress and abscisic acid. *Plant Physiol.* **2006**, *142*, 1065–1074. [CrossRef] [PubMed]
5. Mengistu, F.; Solomon, T.; Tsion, T.; Diriba, G.; Helen, G. In vitro protocol optimization for micropropagation of elite lemmon verbena (aloysia triphylla). *Afr. J. Plant Sci.* **2017**, *11*, 369–376. [CrossRef]
6. Sertié, J.A.; Basile, A.C.; Panizza, S.; Matida, A.K.; Zelnik, R. Pharmacological assay of cordia verbenacea; part 1. anti-inflammatory activity and toxicity of the crude extract of the leaves. *Planta Med.* **1988**, *54*, 7–10. [CrossRef] [PubMed]
7. Ni, L.; Ren, X.; Xiang, Z.; Wan, W.; Yang, D. Sequencing and characterization of leaf transcriptomes of six diploid Nicotiana species. *J. Biol. Res.* **2016**, *23*, 1–12. [CrossRef]
8. Xing, X.; Li, X.; Zhang, M.; Wang, Y.; Liu, B.; Xi, Q.; Zhao, K.; Wu, Y.; Yang, T. Transcriptome analysis of resistant and susceptible tobacco (Nicotiana tabacum) in response to root-knot nematode Meloidogyne incognita infection. *Biochem. Biophys. Res. Commun.* **2016**, *482*, 1114–1121. [CrossRef] [PubMed]
9. Bokvaj, P.; Hafidh, S.; Honys, D. Transcriptome profiling of male gametophyte development in Nicotiana tabacum. *Genom. Data* **2014**, *3*, 106–111. [CrossRef] [PubMed]
10. Kim, S.; Park, J.; Lee, J.; Shin, D.; Park, D.S.; Lim, J.S.; Choi, I.Y.; Seo, Y.S. Understanding pathogenic Burkholderia glumae metabolic and signaling pathways within rice tissues through in vivo transcriptome analyses. *Gene* **2014**, *547*, 77–85. [CrossRef] [PubMed]
11. Mohanty, B.; Kitazumi, A.; Cheung, C.Y.M.; Lakshmanan, M.; de Los Reyes, B.G.; Jang, I.C.; Lee, D.Y. Identification of candidate network hubs involved in metabolic adjustments of rice under drought stress by integrating transcriptome data and genome-scale metabolic network. *Plant Sci.* **2016**, *242*, 224–239. [CrossRef] [PubMed]
12. Li, X.; He, Y.; Yang, J.; Jia, Y.H.; Zeng, H.L. Gene mapping and transcriptome profiling of a practical photo-thermo-sensitive rice male sterile line with seedling-specific green-revertible albino leaf. *Plant Sci.* **2018**, *266*, 37–45. [CrossRef] [PubMed]
13. Boudichevskaia, A.; Heckwolf, M.; Althaus, L.; Kaldenhoff, R. Transcriptome analysis of the aquaporin AtPIP1;2 deficient line in Arabidopsis thaliana. *Genom. Data* **2015**, *4*, 162–164. [CrossRef] [PubMed]
14. Prince, S.J.; Joshi, T.; Mutava, R.N.; Syed, N.; Joao Vitor, M.S.; Patil, G.; Song, L.; Wang, J.; Lin, L.; Chen, W.; et al. Comparative analysis of the drought-responsive transcriptome in soybean lines contrasting for canopy wilting. *Plant Sci.* **2015**, *240*, 65–78. [CrossRef] [PubMed]
15. Kubala, S.; Garnczarska, M.; Wojtyla, Ł.; Clippe, A.; Kosmala, A.; Żmieńko, A.; Lutts, S.; Quinet, M. Deciphering priming-induced improvement of rapeseed (*Brassica napus* L.) germination through an integrated transcriptomic and proteomic approach. *Plant Sci.* **2015**, *231*, 94–113. [CrossRef] [PubMed]
16. Pawełkowicz, M.; Zieliński, K.; Zielińska, D.; Pląder, W.; Yagi, K.; Wojcieszek, M.; Siedlecka, E.; Bartoszewski, G.; Skarzyńska, A.; Przybecki, Z. Next generation sequencing and omics in cucumber (*Cucumis sativus* L.) breeding directed research. *Plant Sci.* **2015**, *242*, 77–88. [CrossRef]
17. Grabherr, M.G.; Grabherr, M.G.; Haas, B.J.; Yassour, M.; Levin, J.Z.; Thompson, D.A.; Amit, I.; Adiconis, X.; Fan, L.; Raychowdhury, R.; et al. Full length transcriptome assembly from RNA Seq data without a reference genome. *Nat. Biotechnol.* **2011**, *29*, 644–652. [CrossRef] [PubMed]

18. Langmead, B.; Trapnell, C.; Pop, M.; Salzberg, S.L. Ultrafast and memory-efficient alignment of short DNA sequences to the human genome. *Genome Biol.* **2009**, *10*, R25. [CrossRef] [PubMed]
19. Li, B.; Colin, N.D. RSEM: Accurate transcript quantification from RNA Seq data with or without a reference genome. *BMC Bioinf.* **2011**, *12*, 323. [CrossRef] [PubMed]
20. Liu, R.; Xu, Y.H.; Jiang, S.C.; Lu, K.; Lu, Y.F.; Feng, X.J.; Wu, Z.; Liang, S.; Yu, Y.T.; Wang, X.F.; et al. Light-harvesting chlorophyll a/b-binding proteins, positively involved in abscisic acid signalling, require a transcription repressor, WRKY40, to balance their function. *J. Exp. Bot.* **2013**, *64*, 5443–5456. [CrossRef] [PubMed]
21. Brix, H. The Effect of Water Stress on the Rates of Photosynthesis and Respiration in Tomato Plants and Loblolly Pine Seedlings. *Physiol. Plant.* **1962**, *15*, 10–20. [CrossRef]
22. Pelleschi, S.; Rocher, J.P.; Prioul, J.L. Effect of water restriction on carbohydrate metabolism and photosynthesis in mature maize leaves. *Plant Cell Environ.* **1997**, *20*, 493–503. [CrossRef]
23. Jackson, R.D.; Idso, S.B.; Reginato, R.J.; Pinter, P.J. Canopy temperature as a crop water stress indicator. *Water Resour. Res.* **1981**, *17*, 1133–1138. [CrossRef]
24. Huang, B.; Fry, J.; Wang, B. Water relations and canopy characteristics of tall fescue cultivars during and after drought stress. *HortScience* **1998**, *33*, 245–256.
25. Wang, B.M.; Chen, J.J.; Chen, L.S.; Wang, X.N.; Wang, R.; Ma, L.; Peng, S.F.; Luo, J.; Chen, Y.Z. Combined drought and heat stress in camellia oleifera, cultivars: Leaf characteristics, soluble sugar and protein contents, and rubisco gene expression. *Trees* **2015**, *29*, 1483–1492. [CrossRef]
26. Weydert, C.J.; Cullen, J.J. Measurement of superoxide dismutase, catalase and glutathione peroxidase in cultured cells and tissue. *Nat. Protoc.* **2010**, *5*, 51–66. [CrossRef] [PubMed]
27. Teulat, B.; Monneveux, P.; Wery, J.; Borries, C.; Souyris, I.; Charrier, A.; This, D. Relationships between relative water content and growth parameters under water stress in barley: A QTL study. *New Phytol.* **1997**, *137*, 99–107. [CrossRef]
28. Zygielbaum, A.I.; Gitelson, A.A.; Arkebauer, T.J.; Rundquist, D.C. Non-destructive detection of water stress and estimation of relative water content in maize. *Geophys. Res. Lett.* **2009**, *36*, 91–100. [CrossRef]
29. Singh, B.B.; Gupta, D.P. Proline accumulation and relative water content in soya bean (glycine max) varieties under water stress. *Ann. Bot.* **1983**, *52*, 109–110. [CrossRef]
30. Thorpe, C.; Kim, J.J. Structure and mechanism of action of the acyl-CoA dehydrogenases. *FASEB J.* **1995**, *9*, 718–725. [CrossRef] [PubMed]
31. Ganeteg, U.; Külheim, C.; Andersson, J.; Jansson, S. Is each light-harvesting complex protein important for plant fitness? *Plant Physiol.* **2004**, *134*, 502–509. [CrossRef] [PubMed]
32. Alberte, R.S.; Thornber, J.P. Water stress effects on the content and organization of chlorophyll in Mesphyll and bundle sheath chloroplasts of maize. *Plant Physiol.* **1997**, *59*, 351–353. [CrossRef]
33. Vapaavuori, E.; Nurmi, A. Chlorophyll-protein complexes in Salix sp. "aquatica gigantean" under strong and weak light. II. Effect of water stress on the chlorophyll-protein complexes and chloroplast ultrastructure. *Plant Cell Physiol.* **1982**, *23*, 791–801. [CrossRef]
34. Zhang, X.; Wang, H.; Takemiya, A.; Song, C.P.; Kinoshita, T.; Shimazaki, K. Inhibition of blue light-dependent H^+ pumping by abscisic acid through hydrogen peroxide-induced dephosphorylation of the plasma membrane H^+-ATPase in guard cell protoplasts. *Plant Physiol.* **2004**, *136*, 4150–4158. [CrossRef] [PubMed]
35. Xu, Y.H.; Liu, R.; Yan, L.; Liu, Z.Q.; Jiang, S.C.; Shen, Y.Y.; Wang, X.F.; Zhang, D.P. Light-harvesting chlorophyll a/b-binding proteins are required for stomatal response to abscisic acid inArabidopsis. *J. Exp. Bot.* **2012**, *63*, 1095–1106. [CrossRef] [PubMed]
36. Ratriyanto, A.; Mosenthin, R.; Bauer, E.; Eklund, M. Metabolic, osmoregulatory and nutritional functions of betaine in monogastric animals. *Asian-Aust. J. Anim. Sci.* **2009**, *22*, 1461–1476. [CrossRef]
37. Tattini, M.; Galardi, C.; Pinelli, P.; Massai, R.; Remorini, D.; Agati, G. Differential accumulation of flavonoids and hydroxycinnamates in leaves of Ligustrum vulgare under excess light and drought stress. *New Phytol.* **2004**, *163*, 547–561. [CrossRef]
38. Xing, W.; Rajashekar, C.B. Glycine betaine involvement in freezing tolerance and water stress in arabidopsis thaliana. *Environ. Exp. Bot.* **2001**, *46*, 21–28. [CrossRef]
39. Gao, X.P.; Yan, J.Y.; Liu, E.K.; Shen, Y.Y.; Lu, Y.F.; Zhang, D.P. Water stress induces in pear leaves the rise of betaine level that is associated with drought tolerance in pear. *J. Hortic. Sci. Biotechnol.* **2004**, *79*, 114–118. [CrossRef]

40. Hitz, W.D.; Ladyman, J.A.R.; Hanson, A.D. Betaine Synthesis and Accumulation in Barley during Field Water-Stress. *Crop Sci.* **1982**, *22*, 47–54. [CrossRef]
41. Goufo, P.; Moutinhopereira, J.M.; Jorge, T.F.; Correia, C.M.; Oliveira, M.R.; Eas, R.; António, C.; Trindade, H. Cowpea (*Vigna Unguiculatal.* walp.) metabolomics: Osmoprotection as a physiological strategy for drought stress resistance and improved yield. *Front. Plant Sci.* **2017**, *8*, 586. [CrossRef] [PubMed]
42. Quiroga, A.M. Water stress and abscisic acid exogenous supply produce differential enhancements in the concentration of selected phenolic compounds in cabernet sauvignon. *J. Berry Res.* **2015**, *2*, 33–44. [CrossRef]
43. Cutler, S.R.; Rodriguez, P.L.; Finkelstein, R.R.; Abrams, S.R. Abscisic acid: Emergence of a core signaling network. *Annu. Rev. Plant Biol.* **2010**, *61*, 651–679. [CrossRef] [PubMed]
44. Qiao, W.; Li, C.; Fan, L.M. Cross-talk between nitric oxide and hydrogen peroxide in plant responses to abiotic stresses. *Environ. Exp. Bot.* **2014**, *100*, 84–93. [CrossRef]
45. Robinson, S.A.; Slade, A.P.; Fox, G.G.; Phillips, R.; Ratcliffe, R.G.; Stewart, G.R. The role of glutamate dehydrogenase in plant nitrogen metabolism. *Plant Physiol.* **1991**, *95*, 509–516. [CrossRef] [PubMed]
46. Joo, J.; Lee, Y.H.; Kim, Y.; Nahm, B.H.; Song, S.I. Abiotic stress responsive rice ASR1 and ASR3 exhibit different tissue-dependent sugar and hormone-sensitivities. *Mol. Cells* **2013**, *35*, 421–435. [CrossRef] [PubMed]
47. Garg, A.K.; Kim, J.K.; Owens, T.J.; Ranwala, A.P.; Choi, Y.D.; Kochian, L.V.; Wu, R.J. Trehalose accumulation in rice plants confers high tolerance levels to different abiotic stresses. *Proc. Natl. Acad. Sci. USA* **2002**, *99*, 15898–15903. [CrossRef] [PubMed]
48. Ciechanover, A.; Orian, A.; Schwartz, A.L. Ubiquitin-mediated proteolysis: Biological regulation via destruction. *BioEssays* **2000**, *22*, 442–451. [CrossRef]
49. Olson, B.L.; Hock, M.B.; Ekholm, R.S.; Wohlschlegel, J.A.; Dev, K.K.; Kralli, A.; Reed, S.I. Scfcdc4 acts antagonistically to the pgc-1alpha transcriptional coactivator by targeting it for ubiquitin-mediated proteolysis. *Genes Dev.* **2008**, *22*, 252–264. [CrossRef] [PubMed]
50. Ji, X.; Nie, X.; Liu, Y.; Zheng, L.; Zhao, H.; Zhang, B.; Huo, L.; Wang, Y. A bHLH gene from Tamarix hispida improves abiotic stress tolerance by enhancing osmotic potential and decreasing reactive oxygen species accumulation. *Tree Physiol.* **2016**, *36*, 193–207. [CrossRef] [PubMed]
51. Seo, J.S.; Joo, J.; Kim, M.J.; Kim, Y.K.; Nahm, B.H.; Song, S.I.; Cheong, J.J.; Lee, J.S.; Kim, J.K.; Choi, Y.D. OsbHLH148, a basic helix-loop-helix protein, interacts with OsJAZ proteins in a jasmonate signaling pathway leading to drought tolerance in rice. *Plant J. Cell Mol. Biol.* **2011**, *65*, 907–921. [CrossRef] [PubMed]
52. Liu, W.; Tai, H.; Li, S.; Gao, W.; Zhao, M.; Xie, C.; Li, W.X. bHLH122, is important for drought and osmotic stress resistance in Arabidopsis, and in the repression of ABA catabolism. *New Phytol.* **2014**, *201*, 1192–1204. [CrossRef] [PubMed]
53. Liu, Y.; Ji, X.; Nie, X.; Qu, M.; Zheng, L.; Tan, Z.; Zhao, H.; Huo, L.; Liu, S.; Zhang, B.; et al. Arabidopsis AtbHLH112 regulates the expression of genes involved in abiotic stress tolerance by binding to their E box and GCG-box motifs. *New Phytol.* **2015**, *207*, 692–709. [CrossRef] [PubMed]
54. O'Kelly, B.C. Accurate Determination of Moisture Content of Organic Soils Using the Oven Drying Method. *Dry. Technol.* **2004**, *22*, 1767–1776. [CrossRef]
55. Chen, Y.E.; Liu, W.J.; Su, Y.Q.; Cui, J.M.; Zhang, Z.W.; Yuan, M.; Zhang, H.Y.; Yuan, S. Different response of photosystem II to short and long-term drought stress in *Arabidopsis* thaliana. *Physiol. Plant.* **2016**, *158*, 225–235. [CrossRef] [PubMed]
56. Zhi, M.; Li, X. Improvement on the method for measuring proline content. *Plant Physiol. Commun.* **2005**, *41*, 355–357.
57. Bradford, M.M. A rapid and sensitive method for the quantitation of microgram quantities of protein utilizing the principle of protein-dye binding. *Anal. Biochem.* **1976**, *72*, 248–254. [CrossRef]
58. Durak, I.; Yurtarslanl, Z.; Canbolat, O.; Akyol, O. A methodological approach to superoxide dismutase (SOD) activity assay based on inhibition of nitroblue tetrazolium (NBT) reduction. *Clin. Chim. Acta* **1993**, *214*, 103–104. [CrossRef]
59. Zhang, L.; Zhang, L.; Xi, D.; Luo, L.; Meng, F.; Li, Y.; Wu, C.A.; Guo, X. Cotton GhMPK2 is involved in multiple signaling pathways and mediates defense responses to pathogen infection and oxidative stress. *FEBS J.* **2011**, *278*, 1367–1378. [CrossRef] [PubMed]
60. Ranieri, A.; Petacco, F.; Castagna, A.; Soldatini, G.F. edox state and peroxidase system in sunflower plants exposed to ozone. *Plant Sci.* **2009**, *159*, 159–167. [CrossRef]

61. Ohya, T. Reactivity of alkanals towards malondialdehyde (MDA) and the effect of alkanals on MDA determination with a thiobarbituric acid test. *Biol. Pharm. Bull.* **1993**, *16*, 1078–1082. [CrossRef] [PubMed]
62. Tambussi, E.A.; Nogués, S.; Araus, J.L. Ear of durum wheat under water stress: Water relations and photosynthetic metabolism. *Planta* **2005**, *221*, 446–458. [CrossRef] [PubMed]
63. Eddy, S.R. Profile hidden Markov models. *Bioinformatics* **1998**, *14*, 755–763. [CrossRef] [PubMed]
64. Finn, R.D.; Bateman, A.; Clements, J.; Coggill, P.; Eberhardt, R.Y.; Eddy, S.R.; Heger, A.; Hetherington, K.; Holm, L.; Mistry, J. Pfam: The protein families database. *Nucl. Acids Res.* **2014**, *42*, D222. [CrossRef] [PubMed]
65. Anders, S.; Huber, W. Differential expression analysis for sequence count data. *Genome Biol.* **2010**, *11*, R106. [CrossRef] [PubMed]

© 2018 by the authors. Licensee MDPI, Basel, Switzerland. This article is an open access article distributed under the terms and conditions of the Creative Commons Attribution (CC BY) license (http://creativecommons.org/licenses/by/4.0/).

International Journal of
Molecular Sciences

MDPI

Article

Analysis of the Coding and Non-Coding RNA Transcriptomes in Response to Bell Pepper Chilling

Jinhua Zuo [1,2,3,4,5,*], Yunxiang Wang [6], Benzhong Zhu [7], Yunbo Luo [7], Qing Wang [1,2,3,4,*] and Lipu Gao [1,2,3,4,*]

1 Key Laboratory of Vegetable Postharvest Processing, Ministry of Agriculture, Beijing Vegetable Research Center, Beijing Academy of Agriculture and Forestry Sciences, Beijing 100097, China
2 Beijing Key Laboratory of Fruits and Vegetable Storage and Processing, Beijing Vegetable Research Center, Beijing Academy of Agriculture and Forestry Sciences, Beijing 100097, China
3 Key Laboratory of Biology and Genetic Improvement of Horticultural Crops (North China) of Ministry of Agriculture, Beijing Vegetable Research Center, Beijing Academy of Agriculture and Forestry Sciences, Beijing 100097, China
4 Key Laboratory of Urban Agriculture (North) of Ministry of Agriculture, Beijing Vegetable Research Center, Beijing Academy of Agriculture and Forestry Sciences, Beijing 100097, China
5 Boyce Thompson Institute for Plant Research, Cornell University Campus, Ithaca, NY 14853, USA
6 Beijing Academy of Forestry and Pomology Sciences, Beijing Academy of Agriculture and Forestry Sciences, Beijing 100093, China; yunxiangjkkl@126.com
7 Laboratory of Postharvest Molecular Biology of Fruits and vegetables, Department of Food Biotechnology, College of Food Science and Nutritional Engineering, China Agricultural University, Beijing 100083, China; zbz@cau.edu.cn (B.Z.); lyb@cau.edu.cn (Y.L.)
* Correspondence: zuojinhua@126.com (J.Z.); wangqing@nercv.org (Q.W.); gaolipu@nercv.org (L.G.)

Received: 21 May 2018; Accepted: 27 June 2018; Published: 9 July 2018

Abstract: Increasing evidence suggests that long non-coding RNAs (lncRNAs), circular RNAs (circRNAs), and microRNAs (miRNAs) have roles during biotic and abiotic stress, though their exact contributions remain unclear. To explore their biological functions in response to chilling in bell pepper, we examined their accumulation profiles by deep sequencing and identified 380 lncRNAs, 36 circRNAs, 18 miRNAs, and 4128 differentially expressed mRNAs in the chilled versus the non-chilled fruit. Gene ontology (GO) and Kyoto Encyclopedia of Genes and Genomes (KEGG) analyses revealed differentially expressed genes and putative ncRNA targets, including transcription factors of multiple classes, such as myeloblastosis (MYB), basic helix-loop-helix (bHLH), and ethylene response factor (ERF) transcription factors (TFs), enzymes involved in bio-oxidation and oxidative phosphorylation (serine/threonine-protein kinase, polyphenol oxidase, catalase, peroxidase, lipoxygenase, and ATPase), and cell wall metabolism-related enzymes (beta-galactosidase, pectate lyase, pectinesterase, and polygalacturonase). On the basis of the accumulation profiles, a network of putatively interacting RNAs associated with bell pepper chilling was developed, which pointed to ncRNAs that could provide the foundation for further developing a more refined understanding of the molecular response to chilling injury.

Keywords: analysis; non-coding RNA; transcriptomes; bell pepper; chilling injury

1. Introduction

Bell pepper (*Capsicum annuum*) is important from both nutritional and commercial standpoints because of its high vitamin C content and its widespread production throughout tropical, sub-tropical, and temperate regions [1–3]. To maintain the fruit quality, the pepper fruit must be cooled as quickly as possible after harvest [4]. However, pepper fruits are highly sensitive to cold and susceptible to chilling injury (CI) when transported or stored below 7 °C [5]. The main symptoms of chilling injury damage include deterioration of the calyx, sunken lesions, seed browning, and surface pitting [6,7]. CI limits the storage life and leads to a significant degradation of the postharvest nutritional quality and product value. However, cold storage is generally the most effective technology to maintain the quality of postharvest horticultural crops. Thus, it is important to overcome the chilling stress in commercially important chilling-sensitive crops [5,8].

With the development of deep-sequencing technology, numerous non-coding RNAs (ncRNAs) have been discovered in recent years [9]. The ncRNAs can be classified according to their length and function [10,11]. For instance, small ncRNAs of 20–30 nt are mostly microRNAs (miRNAs) and small interfering RNAs (siRNAs), usually associated with transcriptional and translational effects [12]. Medium ncRNAs of 50–200 nt and long ncRNAs (lncRNAs) over 200 nt are associated with splicing, gene inactivation, and translation [13,14]. Unlike linear mRNAs, circRNAs form covalently closed loop structures which originate from tRNAs, exons, introns, or combinations of these molecules to form stable circular RNAs [15–21]. Recently, both lncRNAs and circRNAs have been suggested to have properties as "miRNA sponges", whereby they contribute to the regulation of gene expression by operating as competing RNA (ceRNA), influencing a number of distinct biological processes [22,23].

Previously, in a study focused on pepper miRNAs, a comprehensive bioinformatics analysis revealed 11 miRNAs and 54 putative target genes [24]. Via later deep sequencing, 59 known miRNAs and 310 novel miRNAs were found in hot and black pepper [25,26]. The targets of the miRNAs were analyzed and, in some cases, identified as factors associated with fruit development, quality, and stress response [26,27]. In another study, using strand-specific RNA-sequencing, 2505 putative lncRNAs were identified, and many were associated with functions involved in fruit development and quality in hot pepper [28]. To better understand the molecular mechanisms involved in preventing CI, transcriptome profiling analyses of peppers treated with methyl jasmonate (MeJA) and Brassinosteroids (BRs) were performed [5,29]. However, little effort has been focused on the regulation of miRNAs, circRNAs, and lncRNAs in conjunction with mRNA expression during bell pepper chilling, and, as such, the broader non-coding RNA network involved in chilling response remains unclear.

In this study, high-throughput sequencing was employed to explore the regulation of ncRNAs during bell pepper chilling. We identified 380 lncRNAs, 36 circRNAs, 18 miRNAs, and 4128 differentially expressed mRNAs in response to chilling in pepper fruit. In addition, gene ontology (GO) and Kyoto encyclopedia of genes and genomes (KEGG) analyses revealed that several ncRNAs were involved in the chilling response, such as the WRKY and bHLH transcription factors, key enzymes, including polyphenol oxidase, catalase, peroxidase, and lipoxygenase involved in redox reaction, and cell wall metabolism-related enzymes, such as beta-galactosidase, pectate lyase, and polygalacturonase. Furthermore, the competing endogenous RNAs (ceRNAs) network of lncRNAs, circRNAs, mRNAs, and miRNAs was assessed by examining gene annotation to uncover influenced pathways and processes.

2. Results

2.1. Identification of Differential Expressed (DE) and Novel Non-Coding RNAs (ncRNAs)

In our results, 9848 lncRNAs were found: 84 were known lncRNAs, and 9764 were novel lncRNAs found in the control and chilling samples (Table S1). Among them, most of the lncRNAs were lincRNAs (8022, 81.5%), followed by antisense-lncRNAs (919, 9.3%), sense lncRNAs (682, 6.9%), and intronic-lncRNAs (225, 2.3%) (Figure 1A). In addition, 213 novel circRNAs were found, with many emanating from chromosome 8 (Table S1). The majority of circRNAs were over 3000 nt and from intergenic regions, while additional circRNAs were between 400 to 800 nt and derived from exons (Figure 1B, Table S1). In total, 281 miRNAs were found in our libraries with 120 known and 161 novel miRNAs. Most of the novel miRNAs were between 21 and 24 nt. The miRNAs nucleotide bias was also analyzed in our results, and, intriguingly, we found that the first nucleic acid bases were U and A, while the last was G (Table S1).

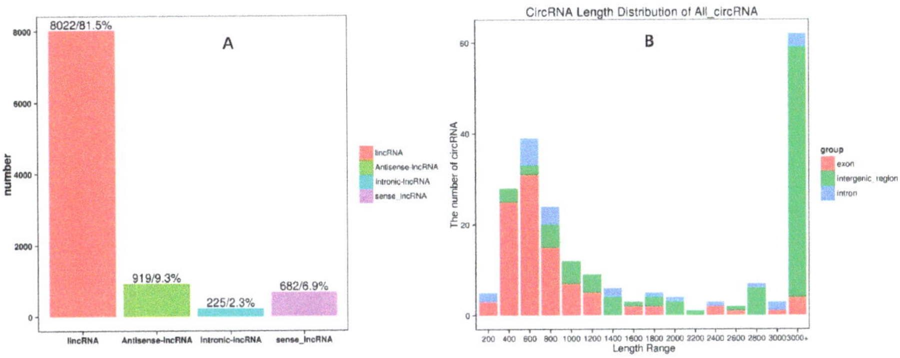

Figure 1. The four kinds of long non-coding RNAs (lncRNAs) were: long intergenic noncoding RNAs (lincRNAs) (8022, 81.5%), antisense-lncRNAs (919, 9.3%), sense lncRNAs (682, 6.9%), and intronic-lncRNAs (225, 2.3%) (**A**).

We compared the expression profiles of lncRNAs, circRNAs, miRNAs, and mRNAs between the control and chilling groups, and found that 380 lncRNAs, 36 circRNAs, 18 miRNAs, and 4128 mRNAs were differentially expressed (Figure 2, Table S2). Among them, 198 lncRNAs, 28 circRNAs, 3 miRNAs, and 2833 mRNAs were upregulated, whereas 182 lncRNAs, 8 circRNAs, 15 miRNAs, and 1295 mRNAs were downregulated in the chilling sample compared with the control. The differentially expressed non-coding RNAs are listed in Supplementary Table S2. The differentially expressed lncRNAs and mRNAs were widely distributed on the autosomal chromosomes, while the differentially expressed circRNAs were not found in chromosomes 5, 9, and 11, and their number was the largest in chromosome 1.

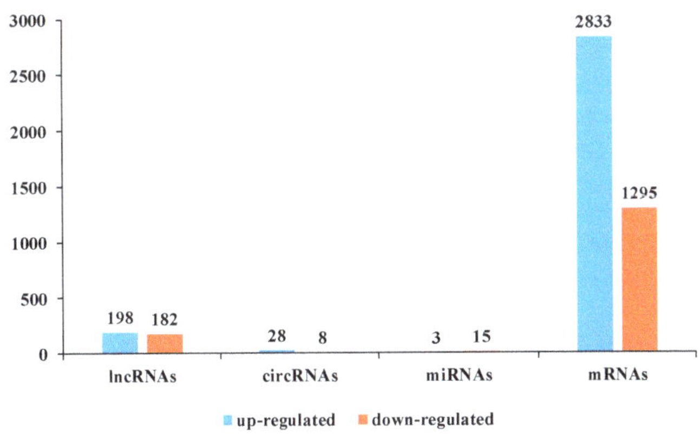

Figure 2. Differentially expressed (DE) ncRNAs: 380 lncRNAs, 36 circRNAs, 18 miRNAs, and 4128 mRNAs were found differentially expressed between the control and chilling groups.

2.2. GO and KEGG Pathway Analyses of ncRNAs

To explore the potential functions of the differential expressed non-coding RNAs, we performed GO and KEGG analyses. To our knowledge, the lncRNAs could regulate the expression of neighboring and overlapping coding genes; hence, lncRNAs likely regulate related mRNA genes [30]. The cis and trans targets of the differentially expressed lncRNAs were both analyzed, and the most relevant GO terms associated with biological processes and molecular functions contained many important response regulators and key enzymes involved in chilling injury. (Figure 3A, Table S3). The KEGG analysis results showed that the most frequently predicted pathways were involved in oxidative phosphorylation, carbon metabolism, ubiquitin-mediated proteolysis for the *cis*-acting targets of lncRNAs; in contrast, the trans-acting targets of lncRNAs were mainly involved in glyoxylate and dicarboxylate metabolism, carbon fixation in photosynthetic organisms, and carbon metabolism, indicating their specific regulation functions in chilling injury in bell pepper (Tables S3 and S4).

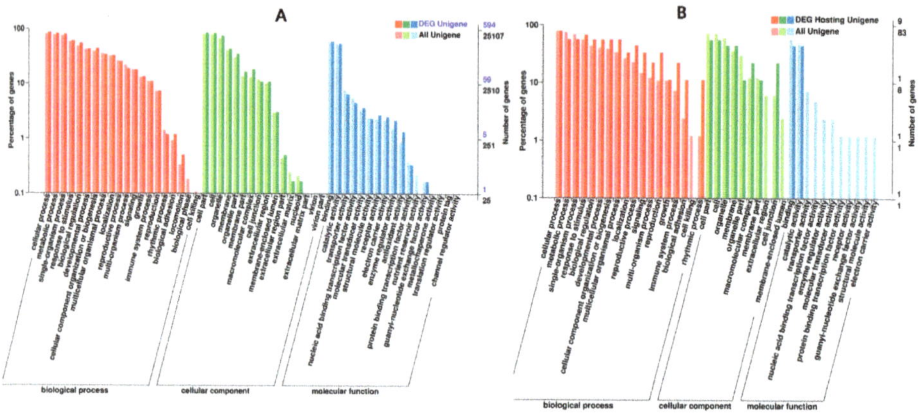

Figure 3. The gene ontology (GO) and Kyoto encyclopedia of genes and genomes (KEGG) analysis of the ncRNAs targets indicated that several important enzymes were involved in the chilling injury (CI).

The function of the differentially expressed miRNAs was also parsed. The most enriched GO terms were related to biological processes, such as response to oxygen-containing compound, response to abiotic stimulus, signal transduction, and hormone-mediated signaling pathway. However, the most relevant GO terms associated with molecular functions were protein binding, protein kinase activity, signaling receptor activity, and ATPase activity (Table S3). KEGG pathway analysis suggested that mRNAs were remarkably enriched in the pathways involved in RNA degradation and transport, peroxisome, mRNA surveillance pathway, plant hormone signal transduction, phenylpropanoid biosynthesis, and folate biosynthesis (Table S4).

So far, the majority of circRNAs has not been functionally annotated [31]. To explore the potential function of differentially expressed circRNAs, GO and KEGG pathway analyses of circRNAs were performed. Our data showed that the most relevant GO terms associated with biological processes were response to temperature stimulus, response to stress, signaling, cell communication, regulation of cellular process, regulation of primary metabolic process, and signal transduction (Figure 3B, Table S3). However, the KEGG pathways only included mRNA surveillance pathway and spliceosome (Table S4).

2.3. Comparative Parsing of LncRNAs and mRNAs and Function Analysis of DE mRNAs

There are many differences between lncRNAs and mRNAs, including their lengths, exon numbers, open reading frames, and expression levels. The number of lncRNAs decreased with the increase of the length (<3000 nt), whereas, the length of mRNAs was distributed from 400 to ≥3000 nt and had two peaks at 400 nt and ≥3000 nt. The number of the corresponding exons of lncRNAs was much less than that of mRNAs and mainly below 10; the mRNAs contained many exons, from 1 to > 30. The length of the corresponding open reading frames of lncRNAs was mainly between 50 and 300 nt, while the length of the corresponding open reading frames of mRNAs was mainly between 100 and 1100 nt (Table S5). An interactive analysis of the expression of lncRNAs and mRNAs was also conducted, and their distribution on the different chromosomes was described (Figure 4).

Figure 4. Distribution of the expression of lncRNAs and mRNAs on the different chromosomes.

For differentially expressed mRNAs, the most relevant GO terms associated with biological processes were defense response signaling pathway, response to auxin, response to abiotic stimulus, response to cold, and so on. However, the most relevant GO terms associated with molecular functions were protein kinase activity, transmembrane receptor protein kinase activity, protein serine/threonine kinase activity, signal transducer activity, and so on. KEGG pathway analysis indicated that the most frequently predicted pathways were involved in plant hormone signal transduction, phenylalanine metabolism, carbon metabolism, galactose metabolism, and so on. We found that many differentially expressed mRNAs were involved in chilling-related processes and could be divided into four different groups. The first group was transcription factors, including WRKY, MYB, bHLH, ERF, and NAC transcription factors, the second group was enzymes involved in bio-oxidation and oxidative phosphorylation, such as serine/threonine-protein kinase, polyphenol oxidase, catalase, peroxidase, lipoxygenase, and ATPase, the third group was cell wall metabolism, such as β-galactosidase, cellulose synthase, chitinase, pectate lyase, pectinesterase, and polygalacturonase, the fourth group was plant hormone-related processes, such as ethylene synthesis-related 1-aminocyclopropane-1-carboxlic acid synthase (ACS) and 1-aminocyclopropane-1-carboxlic acid oxygenase (ACO), abscisic acid receptor, gibberellin 2-β-dioxygenase, IAA-amino acid hydrolase, and salicylic acid-binding protein (Table S6).

2.4. Construction of the Competing Endogenous RNAs (ceRNAs) Network

It is reported that both lncRNAs and circRNAs can interact with miRNAs through microRNA response elements (MREs) within the ceRNA network [32,33]. We developed candidate ceRNA relationships through the miRNA target relationship and obtained 2972 pairs of ceRNA relationships. Then, we extracted three comprehensive ceRNA networks from the ceRNA relationship pairs, including 162 mRNAs, 81 lncRNAs, and 4 circRNAs (Figure 5, Table S7). More importantly, several important enzymes and transcription factors involved in chilling injury, such as ATPase, serine/threonine protein kinase, β-galactosidase, heat shock protein, ethylene-responsive transcription factor, were found in the ceRNA network in our results, indicating their specific cooperative regulation roles in chilling stress (Table S7). In addition, the functions of the key genes were annotated, and the first few most significantly enriched pathways were selected to extract the relationships between genes in multiple pathways and to integrate them into a pathway network. In the network, the key genes in the pathway are involved in lipid transport and metabolism which is important in the chilling stress process (Figure 6, Table S8).

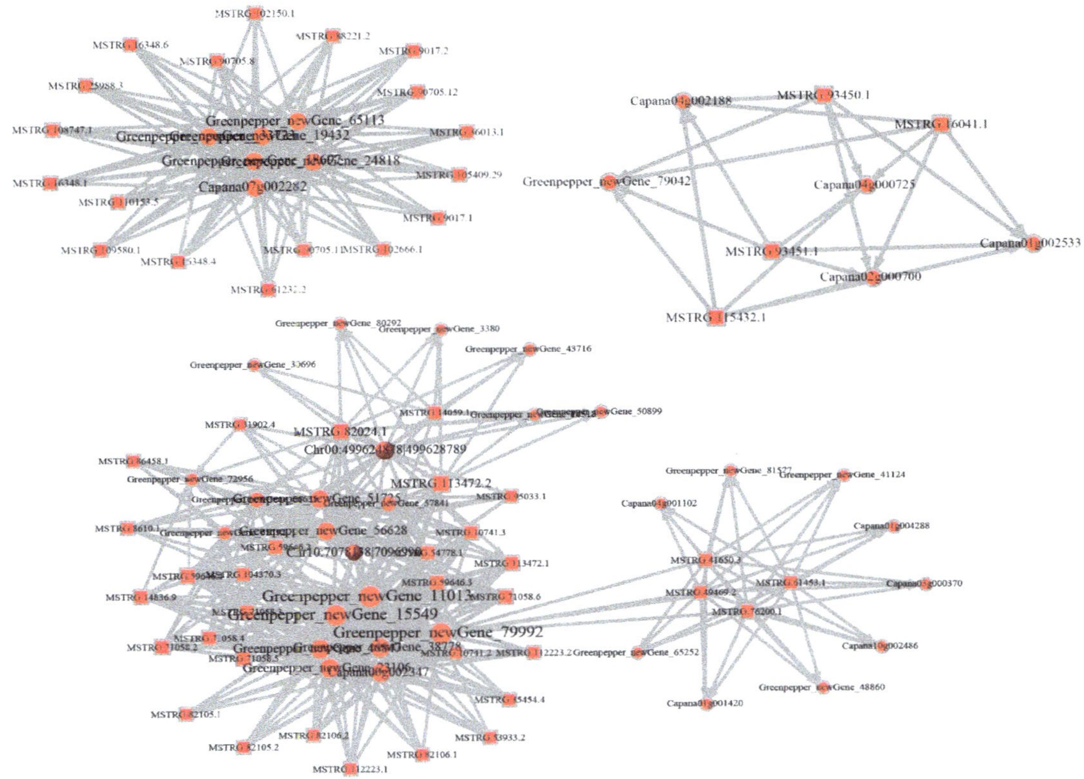

Figure 5. The ceRNA networks of mRNAs, lncRNAs, and circRNAs were parsed, and ceRNA relationship pairs were obtained, including 162 mRNAs, 81 lncRNAs, and 4 circRNAs.

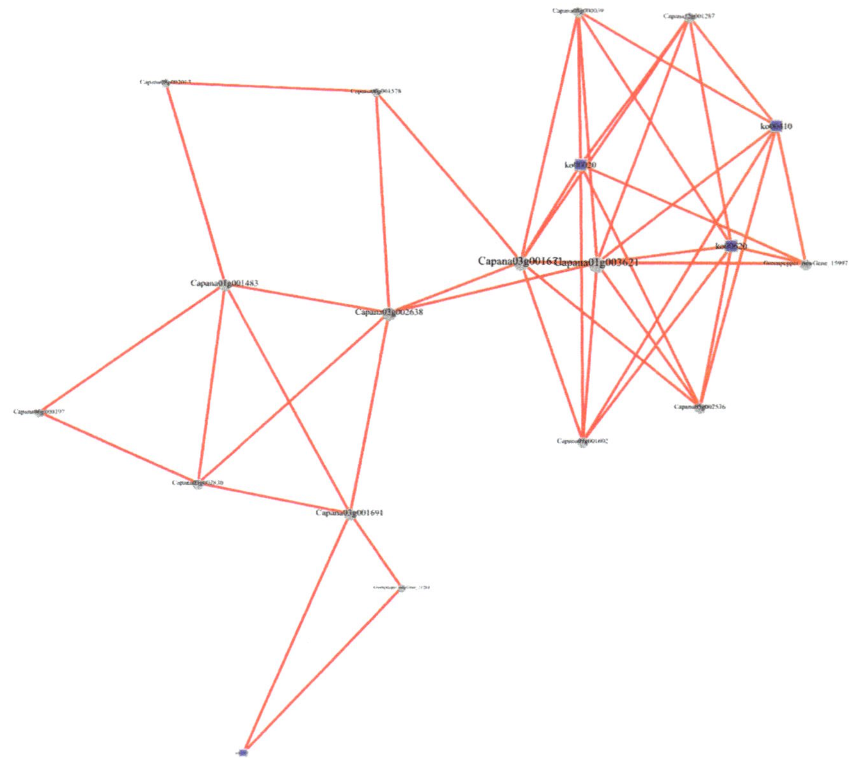

Figure 6. Pathway network from the annotation of the key genes. The key genes in the pathway are involved in lipid transport and metabolism, which play an important role in the chilling stress process.

3. Discussion

Emerging evidence shows that ncRNAs play important roles in cellular functions and especially in biotic and abiotic stresses [12]. Among all the ncRNAs, miRNAs, which perform their functions by mRNA slicing or inhibition at the post-transcriptional level, were most intensively studied [34,35]. Unlike miRNAs, the regulatory function of lncRNAs is difficult to understand because of its complexity, since lncRNAs can fold into secondary or higher orders of structure that make them more flexible in targeting proteins or gene sites [36]. Although thousands of circRNAs have been identified, their functions are largely unknown, but their spatio-temporal expression and tissue specificity indicate their potential biological roles in plants [37,38]. The cross-talk among mRNAs, lncRNA, and circRNA mediated by MREs, regulates biological processes and produces mass regulatory networks [35]. To explore the regulatory functions and complex interactions of ncRNAs in chilling injury, deep sequencing and bioinformatics technology were employed. In total, 380 lncRNAs, 36 circRNAs, 18 miRNAs, and 4128 differentially expressed mRNAs were identified, and three comprehensive ceRNA networks were found, which indicated their specific regulatory roles in chilling injury in bell pepper.

The study of ncRNAs in bell pepper is presently scanty. Few studies were focused on the regulation of lncRNAs and miRNAs in fruit development and quality in hot and black pepper [26,28,39]. At the present time, studies of ncRNA regulation in chilling injury in bell pepper are limited to the field of mRNAs [29]. This is the first report on the differential expression of lncRNA, mRNA, circRNA, and miRNA in chilling injury in bell pepper. In addition, we finely identified 9764 novel lncRNAs, 213 novel circRNAs, and 161 novel miRNAs which were enriched the ncRNAs library. Furthermore, 380 differentially expressed lncRNAs, 36 circRNAs, 18 miRNAs, and 4128 mRNAs were identified between the control and the chilling groups, which indicated their specific regulatory roles played in chilling injury.

In order to explore the potential regulatory functions of the ncRNAs differentially expressed between control and chilling injury groups, GO analysis was performed to further annotate the biological functions of the differentially expressed ncRNAs and their target genes. We noticed that a significant amount of GO terms of the differentially expressed ncRNAs genes was related to response to abiotic stimulus, signal transduction, hormone-mediated signaling pathway, and response to cold, and the molecular functions included protein kinase activity, ATPase, and protein serine/threonine kinase activity. This phenomenon is very intriguing, revealing the vital roles that ncRNAs play in chilling injury. In accordance with the results of the GO analysis, KEGG pathway analysis also revealed pathways related to RNA degradation, peroxisome, plant hormone signal transduction, and carbon metabolism, which indicated their specific functions in the chilling response. In addition, for the differentially expressed mRNAs, numerous mRNAs which encode key enzymes, including superoxide dismutase (SOD), polyphenol oxidase (PPO), and peroxidase (POD,) involved in the protection again oxidative damage by reactive oxygen species (ROS), were found in our results. SOD converts superoxide anion (O_2^-) to hydrogen peroxide (H_2O_2), which in turn is converted to water by Catalase (CAT) and POD [40]. In our results, *SOD*, *PPO*, and *POD* were significantly upregulated, consistently with previous results [5]. Furthermore, in this study, numerous transcription factors, such as the *ERFs*, *MYB*, *NAC*, and *WRKY*, were significantly upregulated by chilling stress, which was consistent with previous results [5].

Recently, circRNAs were proposed to harbor miRNAs and were discovered to be enriched with functional miRNA-binding sites [41]. So far, there has been no report on ceRNAs in bell pepper fruit. Here, we constructed a lncRNA–circRNA–mRNA ceRNA network for bell pepper chilling stress based on our deep-sequencing data for the first time. In total, 162 mRNAs, 81 lncRNAs and 4 circRNAs were included in the ceRNA network. Several targets of the non-coding RNAs in the network were key enzymes in chilling injury, such as ATPase, which is an important enzyme in energy metabolism in bell pepper [42], serine/threonine protein kinase, and β-galactosidase, which are important in signaling and plant defense reaction and cell wall metabolism, respectively [37,43]. In addition, several transcription factors, such as ethylene-responsive transcription factor and heat shock factors, which play specific regulatory roles in the chilling response, were identified [44,45]. In addition, a pathway network was also constructed with the key genes of the KEGG analysis, revealing that the most important pathway was involved in lipid transport and metabolism, which are important in the chilling stress process [46,47]. These findings provide a theoretical basis for deciphering novel mechanisms of chilling injury and for the functional characterization of ceRNA networks in the future studies.

4. Materials and Methods

4.1. Sample Collection and Preparation

Green bell peppers (*C. annuum* L. cv. Jingtian) were harvested from a green house in the "Xiaotangshan" and quickly transported to the lab. The control fruits were stored at 10 °C, whereas the chilled fruits were stored at 1 °C for 72 h. Bell pepper fruit pericarp samples were collected, frozen in liquid nitrogen, and stored at −80 °C for the subsequent experiments.

4.2. Methods of RNA Extraction and Detection

The RNA samples were extracted with RNA Extraction Kit (RN40, Aidlab Biotechnologies, Beijing, China). RNA integrity was assessed using the RNA Nano 6000 Assay Kit of the Agilent Bioanalyzer 2100 system (Agilent Technologies, Santa Clara, CA, USA), to ensure the use of qualified samples for sequencing.

Library preparation for sRNA sequencing: A total amount of 2.5 ng RNA per sample was used as input material for the RNA sample preparations. Sequencing libraries were generated using NEB Next Ultra small RNA Sample Library Prep Kit for Illumina (NEB, Ipswich, MA, USA), following the manufacturer's recommendations, and index codes were added to attribute sequences to each sample. First of all, the 3′SR Adaptor was ligated and mixed for Illumina. The RNA and nuclease-free water were mixed after incubation for 2 min at 70 °C in a preheated thermal cycler, which was then transferred to ice. A 3′Ligation Reaction Buffer (2×) was then added and mixed with the 3′Ligation Enzyme Mix, after which, the 3′SR Adaptor was ligated and incubated for 1 h at 25 °C in a thermal cycler. To prevent adaptor–dimer formation, the SR RT Primer hybridizes to the excess of 3′SR Adaptor (that remains free after the 3′ligation reaction) and transforms the single-stranded DNA adaptor into a double-stranded DNA molecule (dsDNAs) that is not a substrate for ligation. Subsequently, the 5′SR Adaptor was ligated. Then, reverse transcription produced the synthetic first chain. Last, PCR amplification and Size Selection were performed. A polyacrylamide gel electrophoresis (PAGE) gel was used for fragment screening, rubber cutting recycling as the pieces get small RNA libraries. At last, the PCR products were purified (AMPure XP system, Beckman Coulter, Beverly, MA, USA), and the library quality was assessed on the Agilent Bioanalyzer 2100 system (Agilent Technologies, Santa Clara, CA, USA).

Library preparation for lncRNAs and circRNAs sequencing: A total amount of 1.5 μg RNA (for circRNA it was 2.0 μg) per sample was used as input material for rRNA removal using the Ribo-Zero rRNA Removal Kit (Epicentre, Madison, WI, USA). Sequencing libraries were generated using NEBNext® Ultra™ Directional RNA Library Prep Kit for Illumina® (NEB, USA), following the manufacturer's recommendations, and index codes were added to attribute sequences to each sample. Briefly, fragmentation was carried out using divalent cations under an elevated temperature in NEBNext First-Strand Synthesis Reaction Buffer (5×). First-strand cDNA was synthesized using random hexamer primers and Reverse Transcriptase. Second-strand cDNA synthesis was subsequently performed using DNA Polymerase I and RNase H. The remaining overhangs were converted into blunt ends via exonuclease and polymerase activities. After adenylation of the 3′ ends of the DNA fragments, NEBNext Adaptor with a hairpin loop structure was ligated to prepare for hybridization. In order to select insert fragments of preferentially 150–200 bp (for circRNA it was 150–250 bp) in length, the library fragments were purified with AMPure XP Beads (Beckman Coulter). Then, 3 μL

USER Enzyme (NEB, USA) was used with size-selected, adaptor-ligated cDNA at 37 °C for 15 min before PCR. Then PCR was performed with Phusion High-Fidelity DNA polymerase, Universal PCR primers and Index(X) Primer. At last, the PCR products were purified (AMPure XP system), and the library quality was assessed on the Agilent Bioanalyzer 2100 and qPCR.

4.3. Clustering, Sequencing, and Quality Control

The clustering of the index-coded samples was performed on a cBot Cluster Generation System using TruSeq PE Cluster Kit v4-cBot-HS (Illumina) according to the manufacturer's instructions. After cluster generation, the library preparations were sequenced on an Illumina Hiseq platform, and paired-end reads were generated. The raw data (raw reads) of fastq format were firstly processed through in-house perl scripts. In this step, clean data (clean reads) were obtained by removing reads containing adapters, reads containing ploy-N, and low-quality reads from the raw data. At the same time, Q20, Q30, GC content, and sequence duplication level of the clean data were calculated. All the downstream analyses were based on clean data with high quality (Supplied by BioMarker, Beijing, China).

4.4. NcRNAs Identity

The transcriptome was assembled using the StringTie (https://ccb.jhu.edu/software/stringtie/index.shtml) [48] based on the reads mapped to the reference genome. The assembled transcripts were annotated using the gff compare program (Cuffcompare 2.2.1, http://cole-trapnell-lab.github.io/cufflinks/manual/). The unknown transcripts were used to screen for putative lncRNAs. Three computational approaches, namely, CPC (0.9-r2, http://cpc.cbi.pku.edu.cn/)/CNCI(v2, http://www.ncbi.nlm.nih.gov/pubmed/23892401)/Pfam(v1.5, http://pfam.xfam.org/)/CPAT(v1.2.2, http://lilab.research.bcm.edu/cpat/) [49–52], were combined to sort non-protein-coding RNA candidates from putative protein-coding RNAs in the unknown transcripts. Putative protein-coding RNAs were filtered out using a minimum length and exon number threshold. Transcripts with lengths over 200 nt and with more than two exons were selected as lncRNA candidates and further screened using CPC/CNCI/Pfam/CPAT that have the power to distinguish protein-coding genes from non-coding genes. The different types of lncRNAs, including long intergenic noncoding RNAs (lincRNAs), intronic lncRNAs, anti-sense lncRNAs, sense lncRNAs were selected using cuff compare (Cuffcompare 2.2.1, http://cole-trapnell-lab.github.io/cufflinks/manual/)(Supplied by BioMarker).

We used CIRI (CircRNA Identifier, v2.0.5) [53] tools to identify circRNA; it scans SAM files twice and collects sufficient information to identify and characterize circRNAs. Briefly, during the first scanning of SAM alignment, CIRI detects junction reads with PCC signals that reflect a circRNA candidate. Preliminary filtering is implemented using paired-end mapping (PEM) and GT–AG splicing signals for the junctions. After clustering the junction reads and recording each circRNA candidate, CIRI scans the SAM alignment again to detect additional junction reads and, meanwhile, performs further filtering to eliminate false-positive candidates resulting from incorrectly mapped reads of homologous genes or repetitive sequences. Finally, the identified circRNAs are output with annotation information.

Using Bowtie software, the clean reads were analyzed respectively with Silva database, GtRNAdb database, Rfam database, and Repbase database sequence alignment, to filter ribosomal RNA (rRNA), transfer RNA (tRNA), small nuclear RNA (snRNA), small nucleolar RNA (snoRNA), and other ncRNA and repeats. The remaining reads were used to detect known miRNA and novel miRNA, predicted by comparing with known miRNAs from the miRBase. Randfold tools soft (v2.1.7) was used for novel miRNA secondary structure prediction (Supplied by BioMarker, Beijing, China).

4.5. Differential Expression Analysis

Differential expression analysis of two conditions or groups was performed using the DESeq R package (1.18.0, http://www.bioconductor.org/packages/release/bioc/html/DESeq.html) [54].

DESeq provides statistical routines for determining differential expression in digital gene expression, lncRNAs, circRNAs, and miRNAs expression data, using a model based on the negative binomial distribution. The resulting P values were adjusted using the Benjamini and Hochberg's approach for controlling the false discovery rate. Genes, lncRNAs, and circRNAs with an adjusted p-value < 0.01 and an absolute value of log2 (Fold change) > 1 found by DESeq were assigned as differentially expressed. miRNAs with an adjusted $p < 0.05$ found by DESeq were assigned as differentially expressed (Supplied by BioMarker, Beijing, China).

4.6. Gene Function Annotation

Gene function was annotated on the basis of the following databases: Nr (NCBI non-redundant protein sequences; ftp://ftp.ncbi.nih.gov/blast/db/FASTA/); Pfam (Protein family; http://pfam.xfam.org/); KOG/COG (Clusters of Orthologous Groups of proteins; http://www.ncbi.nlm.nih.gov/KOG); Swiss-Prot (A manually annotated and reviewed protein sequence database; http://www.uniprot.org/); KEGG (Kyoto Encyclopedia of Genes and Genomes; http://www.genome.jp/kegg/); GO (Gene Ontology; http://www.geneontology.org/).

4.7. GO and KEGG Pathway Enrichment Analysis

GO enrichment analysis of the differentially expressed genes (DEGs) was implemented by the GOseq R packages based on Wallenius non-central hyper-geometric distribution. We used KOBAS software to test the statistical enrichment of differentially expressed genes in KEGG pathways [55].

4.8. CeRNAs Network Analysis of ncRNAs

A hypergeometric test was executed for each ceRNA pair separately, which was defined by four parameters: (i) N was the total number of miRNAs used to predict targets; (ii) K was the number of miRNAs that interact with the chosen gene of interest; (iii) n was the number of miRNAs that interact with the candidate ceRNA of the chosen gene; (iv) c was the common miRNA number between these two genes. The test calculates the P-value by using the following formula:

$$P = \sum_{i=c}^{\min(K,n)} \frac{\binom{K}{i}\binom{N-K}{n-i}}{\binom{N}{n}}$$

All p-values were subject to false discovery rate (FDR) correction. The following features were necessary for ceRNAs: (i) number of miRNAs that interact with the candidate ceRNA ≥ 5; (ii) FDR < 0.05 [56].

Supplementary Materials: Supplementary materials can be found at http://www.mdpi.com/1422-0067/19/7/2001/s1.

Author Contributions: J.Z. conceived and designed the experiments and wrote the manuscript; Y.W., B.Z., Y.L., Q.W., and L.G. participated in the related experiments and analyzed the data.

Funding: This work was supported by the National Natural Science Foundation of China (31772022), the Natural Science Foundation of Beijing (6182016), the National Key Research and Development Program of China (2016YFD0400901), the China Agriculture Research System Project (CARS-23), Special innovation ability construction fund of Beijing Academy of Agricultural and Forestry Sciences (20180404), the Young Investigator Fund of Beijing Academy of Agricultural and Forestry Sciences (201709), Beijing Academy of Agriculture and Forestry fruit and vegetable preservation and processing innovation team (201602), the International Cooperation Fund Project of Beijing Academy of Agricultural and Forestry Sciences.

Acknowledgments: The authors acknowledge James Giovannoni (USDA/Cornell) and Lance Courtney (Cornell) for useful discussions during the preparation of this manuscript.

Conflicts of Interest: The authors declare no conflict of interest.

References

1. Pickersgill, B. Genetic resources and breeding of Capsicum spp. *Euphytica* **1997**, *96*, 129–133. [CrossRef]
2. Edirisinghe, M.; Ali, A.; Maqbool, M.; Alderson, P.G. Chitosan controls postharvest anthracnose in bell pepper by activating defense-related enzymes. *J. Food Sci. Technol.* **2014**, *51*, 4078–4083. [CrossRef] [PubMed]
3. Wang, Q.; Ding, T.; Zuo, J.; Gao, L.; Fan, L. Amelioration of postharvest chilling injury in sweet pepper by glycine betaine. *Postharvest Biol. Technol.* **2016**, *112*, 114–120. [CrossRef]
4. Liu, L.; Wei, Y.; Shi, F.; Liu, C.; Liu, X.; Ji, S. Intermittent warming improves postharvest quality of bell peppers and reduces chilling injury. *Postharvest Biol. Technol.* **2015**, *101*, 18–25. [CrossRef]
5. Shin, S.Y.; Park, M.H.; Choi, J.W.; Kim, J.G. Gene network underlying the response of harvested pepper to chilling stress. *J. Plant Physiol.* **2017**, *219*, 112–122. [CrossRef] [PubMed]
6. Özden, Ç.; Bayindirli, L. Effects of combinational use of controlled atmosphere, cold storage and edible coating applications on shelf life and quality attributes of green peppers. *Eur. Food Res. Technol.* **2002**, *214*, 320–326. [CrossRef]
7. Lim, C.S.; Kang, S.M.; Cho, J.L. Bell pepper (*Caprigum amtuum* L.) fruits are susceptible to chilling injury at the breaker stage of ripeness. *HortScience* **2007**, *42*, 1659–1664.
8. Cuadra-Crespo, P.; del Amor, F.M. Effects of postharvest treatments on fruit quality of sweet pepper at low temperature. *J. Sci. Food Agric.* **2010**, *90*, 2716–2722. [CrossRef] [PubMed]
9. Liu, X.; Hao, L.; Li, D.; Zhu, L.; Hu, S. Long non-coding RNAs and their biological roles in plants. *Genom. Proteom. Bioinform.* **2015**, *13*, 137–147. [CrossRef] [PubMed]
10. Costa, F.F. Non-coding RNAs: New players in eukaryotic biology. *Gene* **2005**, *357*, 83–94. [CrossRef] [PubMed]
11. Jin, J.; Liu, J.; Wang, H.; Wong, L.; Chua, N.H. PLncDB: Plant long non-coding RNA database. *Bioinformatics* **2013**, *29*, 1068–1071. [CrossRef] [PubMed]
12. Gomes, A.Q.; Nolasco, S.; Soares, H. Non-coding RNAs: Multi-tasking molecules in the cell. *Int. J. Mol. Sci.* **2013**, *14*, 16010–16039. [CrossRef] [PubMed]
13. Ma, L.; Bajic, V.B.; Zhang, Z. On the classification of long non-coding RNAs. *RNA Biol.* **2013**, *10*, 925–933. [CrossRef] [PubMed]
14. Liu, T.T.; Zhu, D.; Chen, W.; Deng, W.; He, H.; He, G. A global identification and analysis of small nucleolar RNAs and possible intermediate-sized non-coding RNAs in Oryza sativa. *Mol. Plant* **2013**, *6*, 830–846. [CrossRef] [PubMed]
15. Meng, X.; Li, X.; Zhang, P.; Wang, J.; Zhou, Y.; Chen, M. Circular RNA: An emerging key player in RNA world. *Brief. Bioinform.* **2017**, *18*, 547–557. [CrossRef] [PubMed]
16. Jeck, W.R.; Sorrentino, J.A.; Wang, K.; Slevin, M.K.; Burd, C.E.; Liu, J.; Marzluff, W.F.; Sharpless, N.E. Circular RNAs are abundant, conserved, and associated with ALU repeats. *RNA* **2013**, *19*, 141–157. [CrossRef] [PubMed]
17. Zhang, Y.; Zhang, X.O.; Chen, T.; Xiang, J.F.; Yin, Q.F.; Xing, Y.H.; Zhu, S.; Yang, L.; Chen, L.L. Circular intronic long non-coding RNAs. *Mol. Cell* **2013**, *51*, 792–806. [CrossRef] [PubMed]
18. Talhouarne, G.J.; Gall, J.G. Lariat intronic RNAs in the cytoplasm of Xenopus tropicalis oocytes. *RNA* **2014**, *20*, 1476–1487. [CrossRef] [PubMed]
19. Chen, L.L.; Yang, L. Regulation of circRNA biogenesis. *RNA Biol.* **2015**, *12*, 381–388. [CrossRef] [PubMed]
20. Li, Z.; Huang, C.; Bao, C.; Chen, L.; Lin, M.; Wang, X.; Zhong, G.; Yu, B.; Hu, W.; Dai, L.; et al. Exon-intron circular RNAs regulate transcription in the nucleus. *Nat. Struct. Mol. Biol.* **2015**, *22*, 256–264. [CrossRef] [PubMed]
21. Lu, Z.; Filonov, G.S.; Noto, J.J.; Schmidt, C.A.; Hatkevich, T.L.; Wen, Y.; Jaffrey, S.R.; Matera, A.G. Metazoan tRNA introns generate stable circular RNAs in vivo. *RNA* **2015**, *21*, 1554–1565. [CrossRef] [PubMed]
22. Li, X.; Ao, J.; Wu, J. Systematic identification and comparison of expressed profiles of lncRNAs and circRNAs with associated co-expression and ceRNA networks in mouse germline stem cells. *Oncotarget* **2017**, *8*, 26573–26590. [CrossRef] [PubMed]
23. Wang, Y.; Wang, Q.; Gao, L.; Zhu, B.; Luo, Y.; Deng, Z.; Zuo, J. Integrative analysis of circRNAs acting as ceRNAs involved in ethylene pathway in tomato. *Physiol. Plant.* **2017**, *161*, 311–321. [CrossRef] [PubMed]
24. Kim, H.J.; Baek, K.H.; Lee, B.W.; Choi, D.; Hur, C.G. In silico identification and characterization of microRNAs and their putative target genes in Solanaceae plants. *Genome* **2011**, *54*, 91–98. [CrossRef] [PubMed]

25. Hwang, D.G.; Park, J.H.; Lim, J.Y.; Kim, D.; Choi, Y.; Kim, S.; Reeves, G.; Yeom, S.I.; Lee, J.S.; Park, M.; et al. The hot pepper (*Capsicum annuum*) microRNA transcriptome reveals novel and conserved targets: A foundation for understanding MicroRNA functional roles in hot pepper. *PLoS ONE* **2013**, *8*, e64238. [CrossRef] [PubMed]
26. Liu, Z.; Zhang, Y.; Ou, L.; Kang, L.; Liu, Y.; Lv, J.; Wei, G.; Yang, B.; Yang, S.; Chen, W.; et al. Identification and characterization of novel microRNAs for fruit development and quality in hot pepper (*Capsicum annuum* L.). *Gene* **2017**, *608*, 66–72. [CrossRef] [PubMed]
27. Joy, N.; Soniya, E.V. Identification of a miRNA candidate reflects the possible significance of transcribed microsatellites in the hairpin precursors of black pepper. *Funct. Integr. Genom.* **2012**, *12*, 387–395. [CrossRef] [PubMed]
28. Ou, L.; Liu, Z.; Zhang, Z.; Wei, G.; Kang, L.; Yang, B.; Yang, S.; Lv, J.; Liu, Y.; Chen, W.; et al. Noncoding and coding transcriptome analysis reveals the regulation roles of long noncoding RNAs in fruit development of hot pepper (*Capsicum annuum* L.). *Plant Growth Regul.* **2017**, *83*, 141–156. [CrossRef]
29. Li, J.; Yang, P.; Kang, J.; Gan, Y.; Yu, J.; Calderón-Urrea, A.; Lyu, J.; Zhang, G.; Feng, Z.; Xie, J. Transcriptome Analysis of Pepper (*Capsicum annuum*) Revealed a Role of 24-Epibrassinolide in Response to Chilling. *Front. Plant Sci.* **2016**, *7*, 1281. [CrossRef] [PubMed]
30. Huang, M.; Zhong, Z.; Lv, M.; Shu, J.; Tian, Q.; Chen, J. Comprehensive analysis of differentially expressed profiles of lncRNAs and circRNAs with associated co-expression and ceRNA networks in bladder carcinoma. *Oncotarget* **2016**, *7*, 47186–47200. [CrossRef] [PubMed]
31. Salzman, J. Circular RNA Expression: Its Potential Regulation and Function. *Trends Genet.* **2016**, *32*, 309–316. [CrossRef] [PubMed]
32. Salmena, L.; Poliseno, L.; Tay, Y.; Kats, L.; Pandolfi, P.P. A ceRNA hypothesis: The rosetta stone of a hidden RNA language? *Cell* **2011**, *146*, 353–358. [CrossRef] [PubMed]
33. Hansen, T.B.; Jensen, T.I.; Clausen, B.H.; Bramsen, J.B.; Finsen, B.; Damgaard, C.K.; Kjems, J. Natural RNA circles function as efficient microRNA sponges. *Nature* **2013**, *495*, 384–388. [CrossRef] [PubMed]
34. Ghildiyal, M.; Zamore, P.D. Small silencing RNAs: An expanding universe. *Nat. Rev. Genet.* **2009**, *10*, 94–108. [CrossRef] [PubMed]
35. Dou, C.; Cao, Z.; Yang, B.; Ding, N.; Hou, T.; Luo, F.; Kang, F.; Li, J.; Yang, X.; Jiang, H.; et al. Changing expression profiles of lncRNAs, mRNAs, circRNAs and miRNAs during osteoclastogenesis. *Sci. Rep.* **2016**, *6*, 21499. [CrossRef] [PubMed]
36. Guttman, M.; Rinn, J.L. Modular regulatory principles of large non-coding RNAs. *Nature* **2012**, *482*, 339–346. [CrossRef] [PubMed]
37. Zuo, J.; Wang, Q.; Zhu, B.; Luo, Y.; Gao, L. Deciphering the roles of circRNAs on chilling injury in tomato. *Biochem. Biophys. Res. Commun.* **2016**, *479*, 132–138. [CrossRef] [PubMed]
38. Wang, H.; Zhao, Y.; Chen, M.; Cui, J. Identification of Novel Long Non-coding and Circular RNAs in Human Papillomavirus-Mediated Cervical Cancer. *Front. Microbiol.* **2017**, *8*, 1720. [CrossRef] [PubMed]
39. Asha, S.; Sreekumar, S.; Soniya, E.V. Unraveling the complexity of microRNA-mediated gene regulation in black pepper (*Piper nigrum* L.) using high-throughput small RNA profiling. *Plant Cell Rep.* **2016**, *35*, 53–63. [CrossRef] [PubMed]
40. Gill, S.S.; Tuteja, N. Polyamines and abiotic stress tolerance in plants. *Plant Signal. Behav.* **2010**, *5*, 26–33. [CrossRef] [PubMed]
41. Li, L.J.; Zhao, W.; Tao, S.S.; Leng, R.X.; Fan, Y.G.; Pan, H.F.; Ye, D.Q. Competitive endogenous RNA network: Potential implication for systemiclupus erythematosus. *Expert Opin. Ther. Targets* **2017**, *21*, 639–648. [CrossRef] [PubMed]
42. Lurie, S.; Lipsker, R.Z.; Aloni, B. Effects of paclobutrazol and chilling temperatures on lipids, antioxidants and ATPase activity of plasma membrane isolated from green bell pepper fruits. *Physiol. Plant.* **1994**, *91*, 593–598. [CrossRef]
43. Afzal, A.J.; Wood, A.J.; Lightfoot, D.A. Plant receptor-like serine threonine kinases: Roles in signaling and plant defense. *Mol. Plant-Microbe Interact.* **2008**, *21*, 507–517. [CrossRef] [PubMed]
44. Li, H.Y.; Chang, C.S.; Lu, L.S.; Liu, C.A.; Chan, M.T.; Chang, Y.Y. Over-expression of Arabidopsis thaliana heat shock factor gene (*AtHsfA1b*) enhances chilling tolerance in transgenic tomato. *Bot. Bull. Acad. Sin.* **2003**, *44*, 129–140.

45. Phukan, U.J.; Jeena, G.S.; Tripathi, V.; Shukla, R.K. Regulation of Apetala2/Ethylene Response Factors in Plants. *Front. Plant Sci.* **2017**, *8*, 150. [CrossRef] [PubMed]
46. Leisso, R.S.; Gapper, N.E.; Mattheis, J.P.; Sullivan, N.L.; Watkins, C.B.; Giovannoni, J.J.; Schaffer, R.J.; Johnston, J.W.; Hanrahan, I.; Hertog, M.L.; et al. Gene expression and metabolism preceding soft scald, a chilling injury of 'Honeycrisp' apple fruit. *BMC Genom.* **2016**, *17*, 798. [CrossRef] [PubMed]
47. Zuo, J.; Wang, Q.; Han, C.; Ju, Z.; Cao, D.; Zhu, B.; Luo, Y.; Gao, L. SRNAome and degradome sequencing analysis reveals specific regulation of sRNA in response to chilling injury in tomato fruit. *Physiol. Plant.* **2017**, *160*, 142–154. [CrossRef] [PubMed]
48. Pertea, M.; Kim, D.; Pertea, G.M.; Leek, J.T.; Salzberg, S.L. Transcript-level expression analysis of RNA-seq experiments with HISAT, StringTie and Ballgown. *Nat. Protoc.* **2016**, *11*, 1650–1667. [CrossRef] [PubMed]
49. Kong, L.; Zhang, Y.; Ye, Z.Q.; Liu, X.Q.; Zhao, S.Q.; Wei, L.; Gao, G. CPC: Assess the protein-coding potential of transcripts using sequence features and support vector machine. *Nucleic Acids Res.* **2007**, *35*, 345–349. [CrossRef] [PubMed]
50. Sun, L.; Luo, H.; Bu, D.; Zhao, G.; Yu, K.; Zhang, C.; Liu, Y.; Chen, R.; Zhao, Y. Utilizing sequence intrinsic composition to classify protein-coding and long non-coding transcripts. *Nucleic Acids Res.* **2013**, *41*, e166. [CrossRef] [PubMed]
51. Finn, R.D.; Bateman, A.; Clements, J.; Coggill, P.; Eberhardt, R.Y.; Eddy, S.R.; Heger, A.; Hetherington, K.; Holm, L.; Mistry, J.; et al. Pfam: The protein families database. *Nucleic Acids Res.* **2014**, *42*, 222–230. [CrossRef] [PubMed]
52. Wang, L.; Park, H.J.; Dasari, S.; Wang, S.; Kocher, J.P.; Li, W. CPAT: Coding-Potential Assessment Tool using an alignment-free logistic regression model. *Nucleic Acids Res.* **2013**, *41*, e74. [CrossRef] [PubMed]
53. Gao, Y.; Wang, J.; Zhao, F. CIRI: An efficient and unbiased algorithm for de novo circular RNA identification. *Genome Biol.* **2015**, *16*, 4. [CrossRef] [PubMed]
54. Anders, S.; Huber, W. Differential expression analysis for sequence count data. *Genome Biol.* **2010**, *11*, 106. [CrossRef] [PubMed]
55. Mao, X.; Cai, T.; Olyarchuk, J.G.; Wei, L. Automated genome annotation and pathway identification using the KEGG Orthology (KO) as a controlled vocabulary. *Bioinform* **2005**, *21*, 3787–3793. [CrossRef] [PubMed]
56. Li, J.H.; Liu, S.; Zhou, H.; Qu, L.H.; Yang, J.H. starBase v2.0: Decoding miRNA-ceRNA, miRNA-ncRNA and protein-RNA interaction networks from large-scale CLIP-Seq data. *Nucleic Acids Res.* **2014**, *42*, 92–97. [CrossRef] [PubMed]

© 2018 by the authors. Licensee MDPI, Basel, Switzerland. This article is an open access article distributed under the terms and conditions of the Creative Commons Attribution (CC BY) license (http://creativecommons.org/licenses/by/4.0/).

International Journal of
Molecular Sciences

MDPI

Article

Genome-Wide Bioinformatics Analysis of *MAPK* Gene Family in Kiwifruit (*Actinidia Chinensis*)

Gang Wang, Tao Wang, Zhan-Hui Jia, Ji-Ping Xuan, De-Lin Pan, Zhong-Ren Guo and Ji-Yu Zhang *

Institute of Botany, Jiangsu Province and Chinese Academy of Sciences, Nanjing 210014, China; wg20092011@163.com (G.W.); immmorer@163.com (T.W.); 13915954315@163.com (Z.-H.J.); xuanjiping@cnbg.net (J.-P.X.); PPxsperfect@163.com (D.-L.P.); zhongrenguo@cnbg.net (Z.-R.G.)

* Correspondence: maxzhangjy@163.com; Tel.: +86-025-8434-7033

Received: 16 July 2018; Accepted: 20 August 2018; Published: 24 August 2018

Abstract: Mitogen activated protein kinase (MAPK) cascades are universal signal transduction modules that play crucial roles in various biotic and abiotic stresses, hormones, cell division, and developmental processes in plants. Mitogen activated protein kinase (MAPK/MPK), being a part of this cascade, performs an important function for further appropriate cellular responses. Although MAPKs have been investigated in several model plants, no systematic analysis has been conducted in kiwifruit (*Actinidia chinensis*). In the present study, we identified 18 putative MAPKs in the kiwifruit genome. This gene family was analyzed bioinformatically in terms of their chromosome locations, sequence alignment, gene structures, and phylogenetic and conserved motifs. All members possess fully canonical motif structures of MAPK. Phylogenetic analysis indicated that *AcMAPK*s could be classified into five subfamilies, and these gene motifs in the same group showed high similarity. Gene structure analysis demonstrated that the number of exons in *AcMAPK* genes ranged from 2 to 29, suggesting large variation among kiwifruit *MAPK* genes. The expression profiles of these *AcMAPK* genes were further investigated using quantitative real-time polymerase chain reaction (qRT-PCR), which demonstrated that *AcMAPKs* were induced or repressed by various biotic and abiotic stresses and hormone treatments, suggesting their potential roles in the biotic and abiotic stress response and various hormone signal transduction pathways in kiwifruit. The results of this study provide valuable insight into the putative physiological and biochemical functions of *MAPK* genes in kiwifruit.

Keywords: mitogen-activated protein kinase (MAPK); kiwifruit; phylogenetic relationships; gene expression; biotic and abiotic stresses

---

1. Introduction

Plants are often challenged by different biotic and abiotic stresses in nature, including pathogen infection, cold, drought, salt, and oxidative stresses; thus, they have developed some sophisticated signaling networks to sense and transmit environmental stimuli at the molecular or cellular levels [1]. A series of highly elaborate signaling networks are composed of some stress-activated molecular pathways [2]. Mitogen-activated protein kinase (MAPK) cascades play an important role in protein phosphorylation of signal transduction events and are one of the major mechanisms in controlling intracellular response to extra cellular signals in plants [3,4].

MAPK cascades are involved in the protein phosphorylation of signal transduction events that contribute to signaling [5], and MAPK cascades are classically composed of three protein kinases: MAPK (MAPK/MPK), MAPK kinase (MAPKK/MKK), and MAPK kinase kinase (MAPKKK/MAP3K/MEKK), but sometimes contain a MAPK kinase kinase kinase

(MAPKKKK/MAP4K) that phosphorylates the corresponding downstream substrates [6–8]. MAPK can catalyze the phosphorylation of a substrate protein by chemically adding phosphate groups from adenosine triphosphate (ATP) [9]. MAP3Ks are the first component of this phosphorelay cascade, which phosphorylates two serine/threonine residues in a conserved S/T-$X_{3\text{-}5}$-S/T motif of the MKK activation loop. Then, MKKs are dual-specificity kinases that activate the downstream MAPK through TDY or TEY phosphorylation motif in the activation loop (T-loop) [3,4,10]. The activated MAPK ultimately phosphorylates various downstream substrates, including transcription factors and other signaling components that regulate the expression of downstream genes [11]. MAPK proteins contain 11 evolutionary conserved kinase domains that may be involved in substrate specificity or protein–protein interaction [1,12].

Compared with MAPKKs and MAP3Ks, MAPKs act at the bottom of MAPK cascades in much greater numbers and show more complexity and sequence diversity. MAPK cascade proteins have TEY or TDY phosphorylation motifs in their activation loops between kinase domains VII and VIII, which provide protein-binding domains for the activation of MAPKs [3,6]. Plant MAPKs can be separated into four groups (A, B, C, and D) based on the phylogenetic relationships of the amino acid sequence and the phosphorylation motif. Members of the A, B, and C subfamily have the TEY motif at its phosphorylation site, and members of the D subfamily possess the TDY motif [3,4].

The MAPK proteins belong to a complex gene family in plants [13]. The identification and characterization of different members of the MAPK cascades have been revealed by genome sequencing projects in various plant species. The model plants that have been most studied are *Arabidopsis thaliana* and rice; there are 20 MAPKs in the *A. thaliana* genome [3], whereas the rice genome contains 17 MAPKs [14]. Recent research has reported that a total of 16, 19, 16, 14, 12, 17, 10, and 15 homologs in *MAPK* family genes have been identified from tomato (*Solanum lycopersicum*) [15], maize (*Zea mays*) [16], purple false brome (*Brachypodium distachyon*) [17], grapevine [13,18] and strawberry (*Fragaria vesca*) [19], tobacco (*Nicotiana tabacum*) [20], mulberry (*Moraceae morus*) [21], wheat (*Triticum aestivum*) [22] genomes, and as many as 21, 26, and 25 putative *MAPK* genes were identified in poplar (*Populus trichocarpa*) [23], apple (*Malus domestica*) [24], and banana (*Musa acuminata*), respectively.

In plants, MAPKs are involved in cellular responses to the regulation of the cell cycle, plant growth and development, hormones, and responses to biotic and abiotic stresses [7,25]. To date, several plant MAPK signaling cascades have been characterized in detail. The MEKK1-MKK4/5-MPK3/6 cascade was the first characterized signaling module in Arabidopsis, which up-regulated the expression of the transcription factors of WRKY22/29 and then increased resistance to both fungal and bacterial pathogens [25,26]. In addition, *AtMPK3* and *AtMPK6* are involved in the anther, embryo, inflorescence development, and stomatal distribution on the leaf surface [27,28]. The MEKK1-MKK1/2-MPK4 cascade was shown to positively regulate defense responses against necrotrophic fungi while negatively regulating defenses against biotrophic pathogens [29,30], also shown to be activated by drought, cold, and salt stresses [31]. *MAPK* genes in other important crops have also attracted considerable attention. For example, *OsMAPK3* and *OsMAPK6* are induced by a chitin elicitor in rice [32], *OsMPK5* is activated by pathogens and abiotic stresses [1], and overexpression of *OsMAPK33* enhances sensitivity to salt stress in rice through unfavorable ion homeostasis as negative regulators [33]. *ZmMPK3*, *ZmMPK5*, and *ZmMPK17* genes in maize are involved in signal transduction pathways associated with different environmental stresses [34–36]. Overexpression of *BnMAPK4* enhances resistance to *Sclerotinia sclerotiorum* in transgenic *Brassica napus* [37]. *GhMPK7* (*Gossypium hirsutum*) is induced by pathogen infection, and may be an important regulator in broad spectrum disease resistance and plant growth and development [38]. The expression of *VvMAPK3* and *VvMAPK6* genes were induced by salinity and drought [18].

Kiwifruit (*Actinidia chinensis*) is a nutritionally and commercially important and valuable fruit, well known for its remarkably high vitamin C content. For example, the Hongyang kiwifruit, which is derived from *A. chinensis* var. *chinensis* [39], is becoming a favorite of consumers, growers, and breeders due to its unique phenotype and high premium price at market. To date, systematic investigations and functional analyses of the MAPK gene family have not been reported for *A. chinensis*, despite the importance of MAPK proteins in multiple biological processes. Recently, the genome of a heterozygous kiwifruit cultivar "Hongyang" (*A. chinensis* var. *chinensis*) was sequenced [40], suggesting that kiwifruit has potential as a model organism for fruit trees. As such, it has become an imperative to compare the functions of gene families, particularly those having vital functions with the gene families characterized from *Arabidopsis* [41], which provides an opportunity for systematic analysis of *MAPK* in the kiwifruit species. With the rapid development of molecular biology and bioinformatics, the mining and positioning of functional genes in plant genome-wide data have become research hotspots. Due to the importance of MAPKs in diverse biological and physiological processes as well as their potential application to the development of improved stress tolerant transgenic plants, we performed the classification and phylogeny of the *MAPK* gene family of kiwifruit through bioinformatics analysis. Additionally, we conducted a comprehensive analysis of all the identified *AcMAPK* genes to determine which of these genes contribute to stress and hormone responses using quantitative real-time polymerase chain reaction (qRT-PCR) analysis. These data further provide information about the relationship between MAPK function and growth and development, disease resistance, and stress response of kiwifruit. The results of our identification and comprehensive investigation of the MAPK gene family in kiwifruit provide a theoretical basis for future gene cloning and expression, especially for the genetic improvement in the breeding of kiwifruit.

2. Results

2.1. Identification of MAPK Family Genes in Kiwifruit

To identify MAPK family genes from the *A. chinensis* genome, both Hidden Markov Model (HMM) and BLAST searches were performed using *Arabidopsis* and *Vitis. vinifera* MAPK proteins as query sequences. The comparison of the sequence of candidate proteins from BLAST and HMM hits were completed and 25 AcMAPK proteins were identified with top hits for *AtMAPK* and *VvMAPK* orthologs with an e-value cutoff of $1 \times e^{-50}$. Then, some sequences were excluded because they encode very short polypeptides of amino acids, or did not contain the known conserved motifs of the MAPK family proteins by phylogenetic and conserved domains analysis. After multiple steps of screening and validation of the conserved domains, we finally identified 18 putative *AcMAPK* genes and the AcMAPK proteins were named according to the Gene ID number from the *A. chinensis* genome, designated as AcMAPK1–AcMAPK18 (Table 1), which was further supported by multiple sequence alignment analyses (Figure S1). The sequence data of all above *MAPK* genes are shown in Supplementary Material File 1. These putative *AcMAPK* genes were predicted to encode 336 (AcMAPK9) to 1056 (AcMAPK6) amino acids in length, with putative molecular weights (Mw) ranging from 38.82 kDa (AcMAPK12) to 119.46 kDa (AcMAPK6), and protein isoelectric points (pIs) ranging from 4.52 (AcMAPK11) to 9.82 (AcMAPK13). The subcellular localization was predicated and the putative AcMAPKs were located in the cytoplasm, nucleus, and chloroplast, except for AcMAPK9 and AcMAPK11, which were present in the peroxisome and vacuolar, respectively (Table 1).

Table 1. The characteristics of putative *MAPK* genes in kiwifruit.

Name	Gene ID	Chromosome	Length of Protein in AA (Amino Acid)	CDS (Coding Sequences) Length in bp	MW (Molecular Weights) (kDa)	PI (Protein Isoelectric Points)	Number of Exons	T-Loop	Subcellular Location
AcMAPK1	Achn005721	15	374	1125	42.93	6.44	6	TEY	Chloroplast
AcMAPK2	Achn025711	11	379	1140	43.65	7.12	3	TEY	Nuclear, Cytoplasm
AcMAPK3	Achn060571	28	340	1023	38.99	4.89	6	TEY	Cytoplasm
AcMAPK4	Achn074341	Un	376	1131	43.36	6.51	6	TEY	Cytoplasm
AcMAPK5	Achn082251	23	344	1035	39.74	5.41	6	TEY	Cytoplasm
AcMAPK6	Achn098501	20	1056	3171	119.46	9.14	29	TEY	Cytoplasm
AcMAPK7	Achn131961	Un	371	1116	42.56	6.89	2	TEY	Cytoplasm
AcMAPK8	Achn132381	25	434	1305	49.42	7.39	7	TEY	Cytoplasm
AcMAPK9	Achn135551	1	336	1011	39.11	7.90	3	TEY	Peroxisome
AcMAPK10	Achn146591	13	601	1806	67.76	9.17	9	TDY	Chloroplast
AcMAPK11	Achn195331	13	425	1278	48.30	4.52	8	TEY	Vacuolar
AcMAPK12	Achn209161	1	340	1023	38.82	6.25	5	TEY	Cytoplasm
AcMAPK13	Achn228801	25	451	1356	51.08	9.82	16	TEY	Cytoplasm, Nuclear
AcMAPK14	Achn237151	29	862	2589	97.02	9.25	10	TDY	Nuclear, Chloroplast
AcMAPK15	Achn248791	Un	475	1428	54.89	5.15	7	TEY	Cytoplasm
AcMAPK16	Achn252431	2	403	1212	46.22	5.69	6	TEY	Cytoplasm
AcMAPK17	Achn296271	1	702	2109	79.02	9.03	11	TDY	Chloroplast
AcMAPK18	Achn377281	15	409	1230	46.34	5.40	16	TDY	Nuclear

The multiple sequence alignment data showed that all the putative AcMAPKs contain the classical TXY motif (Figure S1), which is located in the activation loop [6]. Moreover, MAPKs have a common docking (CD) domain that was observed in the extended C-terminal region, which is defined as (LH [D/E] XX [D/E] EPXC) and functions as a docking site for MAPKKs. Altogether, we identified 18 *MAPK* genes in kiwifruit.

To determine the chromosomal distribution of the identified AcMAPKs, the physical locations of the sequences of the 18 *AcMAPK* genes on the kiwifruit chromosomes were investigated. As shown in the location image (Figure S2 and Table 1), 18 genes were mapped on 11 chromosomes (including unknown chromosomes). Chromosomes 2, 11, 20, 23, 28, and 29 only contained one gene: *AcMAPK16*, *AcMAPK2*, *AcMAPK6*, *AcMAPK5*, *AcMAPK3*, and *AcMAPK14*, respectively. Chromosomes 13, 15, and 25 contained two genes, Chromosomes 1, and unknown contained three genes, respectively.

2.2. Phylogenetic Relationship Analysis of MAPK Gene in Kiwifruit

In order to evaluate the evolutionary relationships among the MAPK proteins, a phylogenetic tree was constructed with amino acid sequences of 18 putative *AcMAPKs* from kiwifruit, 20 *AtMAPKs* from Arabidopsis, and 14 *VvMAPKs* from grapevine. In plants, MAPK proteins have diverged into four major subfamilies (A, B, C, and D) [3], as shown in Figure 1. The phylogenetic analysis showed that the 18 putative *AcMAPKs* could be divided into five distinct groups (groups A, B, C, D, and E) together with their MAPK orthologs in *Arabidopsis* and grapevine, which are more groups than identified in previous reports [42]. *AcMAPKs* belonging to the A, B, C, and E subfamilies all possess a TEY motif, except for *AcMAPK18*, which harbors a TDY motif, whereas the D subfamily possesses a TDY motif at the activation site (Table 1).

AcMAPK5, *AcMAPK12*, *AcMAPK15* and *AcMAPK16* genes are clustered in Group A, which contains well-characterized *MAPK* genes including *AtMPK3*, *AtMPK6*, *VvMPK12*, and *VvMPK14* genes. *AcMAPK1*, *AcMAPK3*, *AcMAPK4*, *AcMAPK8*, and *AcMAPK11* genes belong to Group B, which ncludes *AtMPK4*, *AtMPK5*, *AtMPK11*, *AtMPK12*, *VvMPK9*, and *VvMPK11* genes. Group C contained three genes: *AcMAPK2*, *AcMAPK7*, and *AcMAPK9* genes. Group D includes *AcMAPK10*, *AcMAPK14*, and *AcMAPK17* genes of the kiwifruit MAPKs (Figure 1), which have a TDY motif, consistently found in members of the other MAPK subfamily. *AcMAPK6*, *AcMAPK13*, and *AcMAPK18*, genes belonging to group E, were separated from other groups (Figure 1).

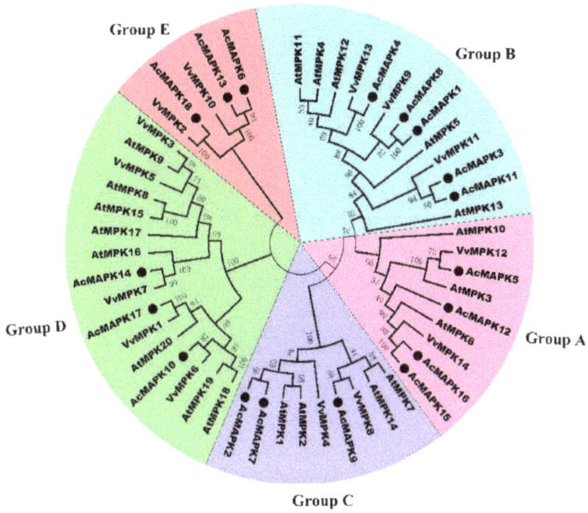

Figure 1. Phylogenetic relationship of putative *MAPK* genes in *Arabidopsis chinensis*, *V. vinifera*, and *A. thaliana*. The phylogenetic tree was created using MEGA5.0 program with the neighbor-joining (NJ) method. Bootstrap values for 1000 replicates are indicated at each branch. Letters A–E indicate different groups of MAPKs.

2.3. Gene Structure Analysis of MAPK Gene in Kiwifruit

The identification of exon-intron structures for each *AcMAPK* gene was determined by aligning corresponding genomic DNA sequences. The exon/intron structures of putative *AcMAPK* genes could also be divided into five subgroups based on their phylogenetic relationship (Figure 2). We found that *AcMAPK* genes in different groups have strikingly different exon/intron structures, but that the gene structures of putative *AcMAPK* members in the same group were highly conserved in kiwifruit (Figure 2). The putative *AcMAPK* members were composed of exons varying from five to seven in Group A. Group B contains exons varying from six to eight, whereas those of Group C only had two or three exons. Nine to 11 exons were present in the *AcMAPK* genes in Group D, and Group E had a larger number of exons with variable exon lengths than other groups; in this group, the number of exons varied from 16 to 29 (Figure 2).

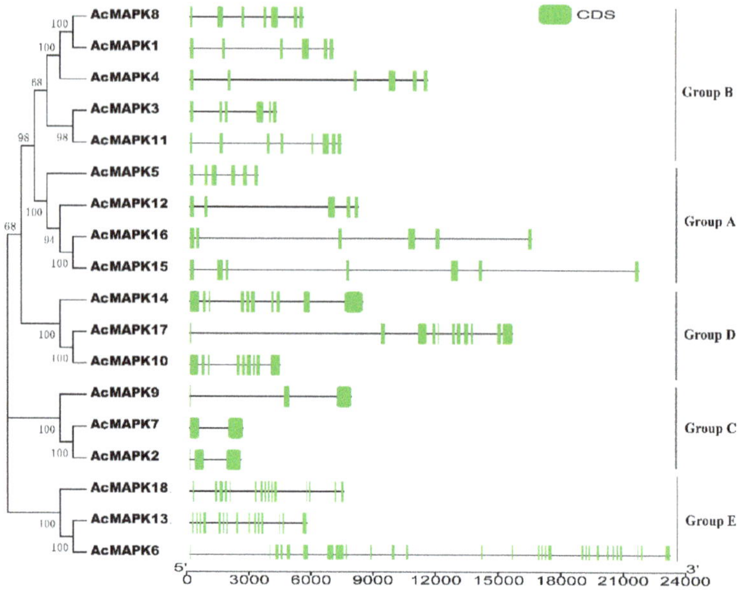

Figure 2. The phylogenetic analysis and intron/exon structures of putative *MAPK* genes in *A. chinensis*. The phylogenetic tree (left panel) was created using MEGA5.0 program with the neighbor-joining (NJ) method. Exon/intron structures of the *MAPK* genes are shown in the right panel. The green boxes indicate the exons, whereas the single lines indicate introns. Gene models were drawn to scale as indicated on bottom.

2.4. The Conserved Motifs Domain and Promoter Regions Analysis of MAPK Gene in Kiwifruit

To explore the structural diversity of the *AcMAPK* genes, we submitted the 18 putative AcMAPK protein sequences to the online MEME program to search for conserved motifs (Figure 3, Supplementary Material File 2) [43]. As shown in Figure 3, 20 conserved motifs were identified. Specifically, all the identified AcMAPKs contained motifs 1, 3 (contained the TXY signature motif), and 5 (Figure 3), indicating that all the kiwifruit MAPKs were typical of the MAPK family. Additionally, the majority of AcMAPKs contained the 13 protein kinase motifs (motifs 1, 2, 3, 4, 5, 6, 7, 8, 9, 10, 12, 14, and 15) (Figure 3). We found all the members identified in the same subfamily shared similar conserved motifs. For instance, along with all the conserved motifs, most MAPK proteins in Groups A and B had specific motif 11 at the N-terminal region, whereas 18 motifs only existed in most MAPKs in Group C. MAPKs in group D contained specific motif 19 at the N-terminal region as well as motif 16 at the C-terminal region, and motifs 13 and 17 only existed in Group E of the MAPK proteins (Figure 3).

To further investigate the potential functions and transcriptional regulation of these putative *AcMAPK* genes, we identified the *cis*-regulatory elements by the transcriptional start site (ATG) using 1500 bp upstream regions. We found a large amount of pathogen-related, stress-related, and hormone-related *cis*-elements in the putative promoter regions of the putative *AcMAPK* genes in kiwifruit. Some genes contain more *cis*-elements, and some genes contain less (Figure S3, Supplementary Material File 3).

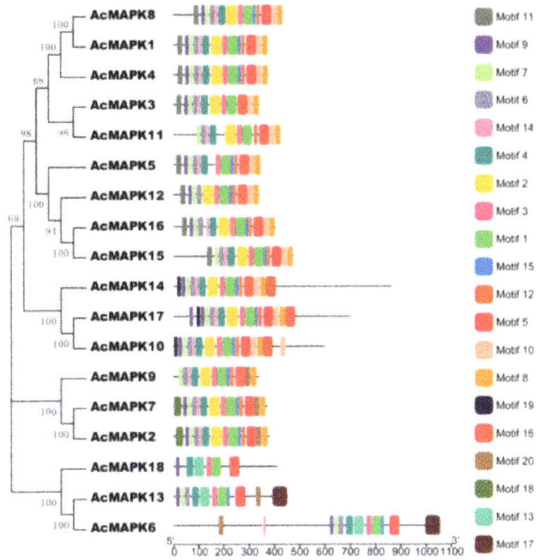

Figure 3. The conserved motifs of kiwifruit putative MAPKs according to the phylogenetic relationship. All motifs were identified online with the MEME program with the complete amino acid sequences of the 18 MAPKs. Different colors of the boxes represent different motifs in the corresponding position of each AcMAPK proteins. Detailed information of the 20 motifs is provided in Supplementary Material File 2.

2.5. Expression Profiles of AcMAPK Genes in Response to Hormone Treatments

To investigate the contribution of *AcMAPK* to various hormone treatments, we subjected four-week-old seedlings of Jinkui (*A. chinensis* var. *deliciosa*) to examine the expression patterns of 18 *AcMAPK* genes using quantitative real-time PCR. In order to obtain a comprehensive view and compare the effects of different treatments on a given gene, the produced heat-map graphic of the expression profiles for all genes and all hormone treatments is provided in Figure 4. It was interesting that the transcript levels of almost all genes were down-regulated in response to hormone treatments (Figure 4, Figure S4, and Figure S5). In our work, the transcript levels of all *AcMAPK* genes were down-regulated after abscisic acid (ABA) treatment (Figure 4 and Figure S4A). *AcMAPK5*, *AcMAPK 9*, *AcMAPK15*, and *AcMAPK16* genes were up-regulated at four hours. *AcMAPK17* showed obvious up-regulation at 4 and 48 h after 1-aminocyclopropanecarboxylic acid (ACC) treatment (Figure 4 and Figure S4B). These genes (*AcMAPK1*, *AcMAPK5*, and *AcMAPK9*) were significantly up-regulated at 12 h after salicylic acid (SA) treatment, and *AcMAPK5*, *AcMAPK9*, *AcMAPK15*, *AcMAPK16*, and *AcMAPK17* were induced by jasmonic acid (JA) treatment (Figure 4 and Figure S5). In these genes, *AcMAPK5* demonstrated significantly higher induction after the hormone treatments than other genes.

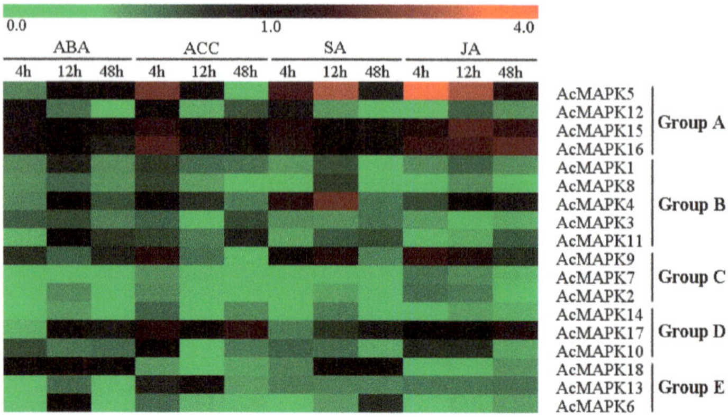

Figure 4. Hierarchical clustering of the expression profiles of *AcMAPK* genes in response to different hormones treatments in kiwifruit leaves. ABA: treatments with abscisic acid, ACC: treatments with 1-Aminocyclopropanecarboxylic Acid, SA: treatments with salicylic acid; JA: treatments with jasmonic acid, details of the treatments are reported in Materials and Methods. The heat-map demonstrates the relative fold-change expression for all *AcMAPK* genes in response to the different hormone treatments in comparison to their respective controls. Red and green colors represent increased or decreased expression levels, respectively, in comparison to controls, as reported by the scale. Genes were clustered according to phylogenetic relationships in expression profiles. Relative expression values for each gene and each treatment are provided in Figures S4 and S5.

2.6. Expression Patterns of AcMAPK Genes under Abiotic and Biotic Stresses

We also investigated the expression of *AcMAPK* genes in response to various abiotic and biotic stress responses with different hormone treatments (Figure 5, Figure S6, and Figure S7). In response to cold stress, the expression of five *AcMAPK* genes (*AcMAPK5*, *AcMAPK9*, *AcMAPK10*, *AcMAPK11*, and *AcMAPK12*) were significantly up-regulated throughout the treatment process, and *AcMAPK4* was up-regulated at 48 h of treatment; whereas *AcMAPK2*, *AcMAPK6*, *AcMAPK7*, *AcMAPK13*, and *AcMAPK18* genes were down-regulated at all treated time points (Figure 5 and Figure S6A). After heat treatment, nine *AcMAPK* genes (*AcMAPK1*, *AcMAPK5*, *AcMAPK10*, *AcMAPK11*, *AcMAPK14*, *AcMAPK15*, *AcMAPK16*, *AcMAPK17* and *AcMAPK18*) were up-regulated after four hours of heat stress treatment at 48 °C. The *AcMAPK11* gene was significantly up-regulated (Figure 5 and Figure S6B). With salt treatment, the expression of *AcMAPK4*, *AcMAPK5*, *AcMAPK9*, and *AcMAPK12* genes were significantly up-regulated at all treatment time points, and *AcMAPK10*, *AcMAPK13* and *AcMAPK17* genes were up-regulated at several treated time points, whereas the remaining genes were almost down-regulated under salt treatment (Figure 5 and Figure S7A). Almost all the *AcMAPKs* genes (except *AcMAPK2*, *AcMAPK3* and *AcMAPK9*) were down-regulated after *Pseudomonas syringae* pv. *actinidiae* (Psa) treatment (Figure 5 and Figure S7B).

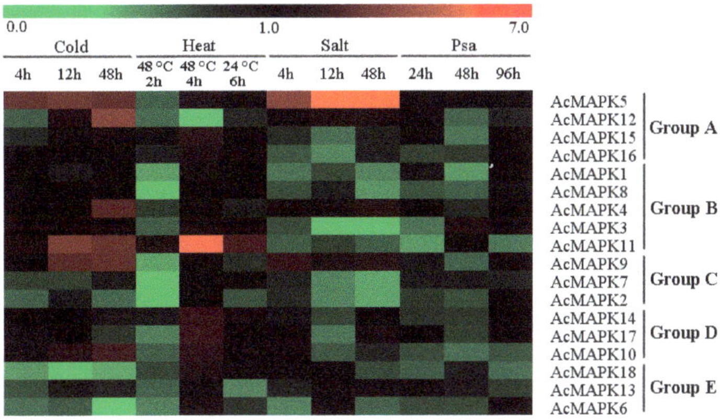

Figure 5. Hierarchical clustering of the expression profiles of *AcMAPK* genes in response to various biotic and abiotic stresses. Cold: treatment at 4 °C; Heat: treatment at 48 °C and 24 °C; Salt: treatment with NaCl; Psa: *Pseudomonas syringae* pv. *actinidiae* infection. Details of the treatments are reported in Materials and Methods. The heat-map depicts the fold-change of the relative expression of all *AcMAPK* genes in response to the various treatments in comparison to their respective controls. Red and green colors represent increased or decreased expression levels, respectively, in comparison to controls, as reported by the scale. Genes were clustered according to phylogenetic relationships in expression profiles. Relative expression values for each gene and each treatment are provided in Figures S6 and S7.

3. Discussion

In recent years, the characterization of gene families has been useful for studying their function [44]. The accuracy and reliability of gene family evolutionary characterization analysis depend on the genomic sequences. The availability of the complete kiwifruit genome sequence has made it possible to identify all the MAPK family members in this plant species for the first time. In this study, we identified 18 putative *MAPK* genes in the *A. chinensis* genome. The numbers are comparable to those in *A. thaliana* genome, where 20 MAPK members have been identified [3], but the genome size of *A. chinensis* (~616.1 Mb) is approximately four times that of the *A. thaliana* genome (~125 Mb). The 18 members in kiwifruit is a larger number than found in grapevine (14 members) and strawberry (12 members), but smaller than in apple (26 members) and banana (25 members) in fruit. The full-length sequences of putative *AcMAPK* ranged from 336 to 1056 amino acids. Variation in the length of the entire *MAPK* gene is usually due to differences in the length of the MAPK domain or the number of introns [18]. We found most members in the same group share a similar exon/intron structure, which was similar to other plants, including Arabidopsis, tomato, and poplar [1,7,15]. So, the exon/intron structures of each gene cluster originated from tandem or segmental duplication events in the *MAPK* gene family and tended to share similar structure organizations, except for tiny differences. The results were consistent with those of domain and phylogenetic analyses performed.

In plants, *MAPK* genes have diverged into four subfamilies based on the conserved residues of the TEY/TDY motifs in the activation loop region (T-loop) [3]. However, phylogenetic analysis showed that the 18 putative *AcMAPKs* were divided into five distinct groups (A, B, C, D, and E), together with their MAPK orthologs in *Arabidopsis* and grapevine, which is more than previously reports [42]. *AcMAPK5*, *AcMAPK12*, *AcMAPK15*, and *AcMAPK16* belong to Group A, which contains *AtMPK3* and *AtMPK6* (Figure 1). It has been well-characterized that *AtMPK3* is activated in response to pathogens and abiotic stresses, and *AtMPK6* can be activated by various abiotic and biotic stresses [1]. *AcMAPK1*, *AcMAPK3*, *AcMAPK4*, *AcMAPK8*, and *AcMAPK11* belong to Group B, which includes *AtMPK4*, *AtMPK5*, *AtMPK11*, *AtMPK12*, *VvMPK9*, and *VvMPK11* (Figure 1). The MAPKs in Group B are involved in both abiotic stress responses and cell division in Arabidopsis. *AtMPK4* and its upstream

AtMKK2 can be activated by biotic and abiotic stresses [31]. Group C contained three genes: *AcMAPK2*, *AcMAPK7*, and *AcMAPK9* (Figure 1). Members from this group in other plant species are known to be regulated by both biotic and abiotic stresses. For example, *AtMPK1* in Group C is regulated by salt stress treatment [4], and *AtMPK1* and *AtMPK2* are activated by ABA [45]. In addition, the rice *BWMK1* and alfalfa *TDY1* genes in Group C are activated by wounding and pathogens [46]. Group D includes *AcMAPK10*, *AcMAPK14*, and *AcMAPK17* of the kiwifruit MAPKs (Figure 1), which have the TDY motif in their T-loop, which are consistently found in members of the other MAPK groups. We found that Group D is the largest group of MAPKs in most plant species. *AcMAPK6*, *AcMAPK13*, and *AcMAPK18*, belonging to Group E, were separated from other groups (Figure 1). The *AcMAPKs* genes of Group E are found only in the grapevine genome among other plant species; there were no orthologs of *AtMAPK* in *A. thaliana*.

The result of our examination of the conserved motifs domain found all the identified *AcMAPKs* contained motifs 1, 3 (contained the TXY signature motif), and 5 (Figure 3), indicating that all the kiwifruit MAPKs were typical of the MAPK family. Above, we stated that all members identified in the same subfamily shared similar conserved motifs. For instance, along with all the conserved motifs, most MAPK proteins in Groups A and B had specific motif 11 at the N-terminal region, whereas 18 motifs only existed in most MAPKs in Group C. The MAPKs of Group D contained the specific motif 19 at the N-terminal region as well as motif 16 at the C-terminal region, and motifs 13 and 17 only existed in Group E of the MAPK proteins (Figure 3). This suggests functional consistency among the MAPK members in the same group. Moreover, motifs in each group were diverse, in accordance with the intron/exon structure of each group. Thus, the composition and the sequential order of these motifs in the same group showed high similarity. A large amount of stress-, pathogen-, and hormone-related *cis*-elements were found in the putative promoter regions of the *AcMAPK* genes in kiwifruit as shown by *cis*-regulatory elements analysis. The existence of these *cis*-elements suggested that these *AcMAPK* genes might have potential functions in various stress signaling pathways. Similar *cis*-elements were found in *MAPK* genes of tomato [15] and *B. distachyon* [17].

A large number of reports demonstrated the involvement of *MAPK* genes in response to various biotic and abiotic stresses and hormone signaling [4]. AtMAPK3 and AtMAPK6 of Group A are the most prominent kinases, which have been widely studied and have been strongly associated with various environmental stresses in *Arabidopsis* [11]. In this study, he transcription level of the *AcMAPK5* gene, which is the kiwifruit orthologue of the *AtMAPK3* gene, showed an obvious up-regulation response to cold, heat, salt, ACC, SA, and JA treatments, which indicates that *AcMAPK5* might be an important regulator in response to abiotic stresses and hormone signaling molecules. Notably, the gene expression down-regulation of *AcMAPK12* observed in any hormone treatment was induced transcriptionally by cold and salt stress, suggesting that activation of AcMAPK12 protein kinase activities might be not correlated with their transcript levels, similar to AtMAPK6. However, the expression of *AcMAP15* and *AcMAP16* genes was repressed by most hormone (except for JA) and heat treatments; these results are similar to previous reports in which the MKK3/MPK6 module was proposed to participate in JA signaling [47]. The MAPKs of Group B (*AtMAPK4* and *AtMAPK11*) have been implicated in pathogen defense and abiotic stress responses [48]. The relationship of MAPK signaling pathways and SA in plant abiotic stress responses was recently characterized [49]. The *AcMAPK11* gene showed an obvious up-regulation response to cold and heat stress. The *AcMAPK4* gene was induced by cold, salt, SA, and ABA treatments, suggesting the involvement of these genes in abiotic stress tolerance and hormone signal transduction in kiwifruit. The expression of *AcMAPK* genes from Group B was repressed by most stresses and hormone treatments, suggesting that *AcMAPK* genes of Group B may function in an early stage of stress signaling transduction as negative regulators in kiwifruit. The MAPKs of Group C in *Arabidopsis* are activated by ABA, providing evidence for a role in an ABA-induced MAPK pathway in plant stress signaling [50]. However, the transcription level of *AcMAPK2*, *AcMAPK7*, and *AcMAPK9* from Group C were down-regulated by ABA treatment in this study, which suggests that the involvement of these genes in ABA signaling might be regulated at the level of translation. The *AcMAPK9* gene was induced by cold, heat, salt, *P. syringae*, ACC, SA, and JA treatments, which suggests that this gene might also have important functions in abiotic stress and hormone signaling. *AtMPK7* was significantly up-regulated in response to cold stress [17]. *AcMAPK9*, which showed the highest homology to *AtMPK7*, showed strong activation by cold stress, suggesting a similar function. The *MAPK* genes of Group D have not been as well studied as those of Groups A and B. *AcMAPK10* and *AcMAPK17* genes in Group D were induced by cold, heat, salt, and ACC treatments. It is interesting that expression of *AcMAPK14* was up-regulated by all biotic and abiotic stresses, whereas down-regulation induced by all hormone treatments. Together, these results indicate possible roles of the *MAPK* genes of Group D in abiotic stress responses and hormone signaling. The *AcMAPK*s genes of Group E are found only in the grapevine genome among other plant species; there were no orthologs of *AtMAPK* in *Arabidopsis*. The expression of Group E gene members was repressed by most biotic and abiotic stresses, similar to their response to hormone treatments, except for heat treatment. However, more research is needed to determine the specific functions of the MAPK family of genes by additional experiments.

4. Materials and Methods

4.1. Genome-Wide Identification of MAPK Genes in Kiwifruit

For identification of the *MAPK* gene family, the sequences of *Arabidopsis* MAPK cascade proteins were obtained from TAİR (https://www.arabidopsis.org/). The MAPK protein sequences of grapevine were obtained from the *V. vinifera* proteome 12× database (http://www.genoscope.cns.fr/externe/GenomeBrowser/Vitis/). Kiwifruit (*A. chinensis*) assembly and annotation were downloaded from kiwifruit genome database (http://bioinfo.bti.cornell.edu/cgi-bin/kiwi/download.cgi). These sequences were used as queries to search against the kiwifruit protein databases by the BLASTP program with an e-value of $1 \times e^{-50}$ as the threshold. The local Hidden Markov Model-based searches (HMMER: http://hmmer.janelia.org/), built from all the known MAPK protein sequences from *Arabidopsis* and grapevine, were used to identify the *MAPK* genes in kiwifruit. To identify predicted *AcMAPK* genes accurately from genome sequences, the unique sequences obtained from the above-mentioned programs were further filtered based on the typical structural features of plant MAPK proteins as previously reported [1,13]. The *AcMAPK* genes were accepted only if they contained the essential TDY or TEY signature motif and the 11 conserved subdomains.

4.2. Sequence Alignment, Phylogenetic Analysis, Chromosomal Location, and Gene Structure Construction

The protein theoretical molecular weight and isoelectric point were predicted using compute pI/MW (http://au.expasy.org/tools). Multiple alignments of the nucleotide and amino acid sequences were performed using ClustalW [51]. The phylogenetic analysis was constructed based on the sequences of MAPK proteins from *Arabidopsis*, *V. vinifera*, and kiwifruit using a neighbor-joining (NJ) method with 1000 bootstrap replicates and visualized with MEGA5 software [52]. The chromosomal distribution of all *AcMAPK* genes was determined based on the results of identification, and subsequently the location images of *AcMAPK* genes were drawn with MapInspect software (http://www.softsea.com/review/MapInspect.html). The exon/intron structure analysis of the *AcMAPK* genes was conducted and displayed by comparing CDSs and their corresponding gene sequences from genomic using the Gene Structure Display Serve [53]. The MEME program was used to statistically identify conserved motifs in the complete amino acid sequences of AcMAPK proteins [43].

4.3. Cis-Element Analysis of Putative Promoter Regions

To investigate *cis*-elements in the promoter regions of the identified genes, we downloaded the genomic DNA sequences upstream from the kiwifruit database to search the initiation codon (ATG) of each gene [54]. The putative *cis*-regulatory elements in the promoter regions sequences were analyzed via the PLACE database (http://www.dna.affrc.go.jp/PLACE/).

4.4. Plant Materials and Treatments

The kiwifruit cultivar "Jinkui" (*Actinidia chinensis* var. *deliciosa*) were maintained in vitro on Murashige and Skoog (MS) medium supplied with 6-benzylaminopurine (6-BA, 3.0 mg/L, Sigma-Aldrich, St. Louis, MO, USA), and naphthalene acetic acid (NAA, 0.2 $mg \cdot L^{-1}$, Sigma) under a 16/8 h photoperiod (100 $\mu mol\ m^{-2} \cdot s^{-1}$) at 25 °C in a growth chamber. Four-week-old plants were used for hormones, freezing (4 °C), and heat stress treatments. Shoots with good growth vigor were collected from Jinkui kiwifruit trees and cultured in MS medium, and maintained in growth chambers, then used for *Pseudomonas syringae* pv. *actinidiae* (Psa) treatments. The conditions included a temperature of 25 °C and 12/12 h light/dark cycles. Two-year-old Jinkui cutting seedlings, which were used for salt treatment, were grown in nutrient soil in a greenhouse at a temperature of 25–28 °C during the day and 20–25 °C during the night.

Several of the stress treatments were performed in kiwifruit as described previously [55]. For treatments with abscisic acid (ABA), 1-aminocyclopropanecarboxylic acid (ACC), salicylic acid (SA) and jasmonic acid (JA), plants with eight fully expanded leaves per tissue-culture container (240 mL) were sprayed with 0.01 mM ABA, 0.01 mM ACC, 0.1 mM SA, and 0.02 mM JA. All the chemicals were purchased from Sigma-Aldrich and dissolved in sterile distilled water. The leaves were harvested at 0, 4, 12, and 48 h post-treatment. For cold stress, seedlings were grown at 4 °C for 0, 4, 12, and 48 h. For heat stress, seedlings were grown at 48 °C for 0, 2, and 4 h, and then at 24 °C for another 6 h. For salt stress, the cutting seedlings were soaked at high salinity (200 mM NaCl) for 0, 4, 12, and 48 h. The seedling leaves and seedling cuttings from both treated and control plants were harvested in the above treatments. For *Pseudomonas syringae* pv. *actinidiae* (Psa) bacterial infection, bacterial cells were suspended in distilled water and adjusted to an OD_{600} = 0.2, and injected into the seedling stems, which were carved with a knife. Only carved seedling stems were used as the control (CK), inoculated with Psa, and sampled at 24, 48, and 96 h. Every treated sample had a corresponding regularly-watered control. Three biological replicates were collected per time point, each comprising five independent plants. All samples were immediately frozen in liquid nitrogen and stored at −80 °C.

4.5. Total RNA Isolation and qRT-PCR Expression Analysis

Total RNA was extracted from the collected samples as described previously with some modifications [5]. Reverse transcription of mRNA was synthesized with a Prime Script™ RT Reagent Kit (Perfect Real Time, TaKaRa, Ostu, Japan) with 1 μg total RNA. The cDNA samples were diluted 1:10 with sterile double-distilled water and stored at −20 °C before being used.

The expressions of *AcMAPKs* were examined by qRT-PCR using a SYBR Green method on an ABI 7300 Real-time PCR System (Applied Biosystems, Waltham, MA, USA). The primer sequences used were designed based on gene sequences and the Beacon designer software (NJ, USA), as shown in Supplementary Material File 4 in this study. Kiwifruit actin was used as the housekeeping gene to monitor cDNA abundance [56]. qRT-PCR was carried out as described previously [55]. The relative gene expression level was calculated according to the $2^{-\Delta\Delta C_t}$ method, where $\Delta\Delta C_t = (C_{t\ target\ gene} - C_{t\ actin})_{treatment} - (C_{t\ target\ gene} - C_{t\ actin})_{ck}$ [5,57]. To visualize the relative expression levels data, 0 h at each treatment was normalized as "1", which are presented as the mean fold changes between treated and control samples at each time point ± standard deviations (SDs). The expression data of the 18 *AcMAPK* genes were transformed in $\log_2$ values and used for heat map generation. The heat map was created with MeV4.8 software (Boston, MA, USA) (http://www.tm4.org/mev/).

4.6. Statistical Analysis

Statistical analyses were performed using SPSS version 17.0 software (Chicago, MI, USA) and Excel. All results of expression data are indicated as means ± standard deviations (SDs), and the level of significance between different time points was set at $p < 0.05$.

5. Conclusions

Kiwifruit (*Actinidia chinensis*) has become an important commercial fruit due to its pleasant flavor and nutritional components that benefit human health [39]. However, research progress on kiwifruit has been relatively slow compared to grape and apple horticultural crops. MAPK cascade is one of the major pathways in plants, with MAPK as the downstream molecule of the MAPK cascade playing an important role in singling [6]. In this study, we identified 18 putative *MAPK* genes from the kiwifruit genome and established their classification and phylogenetic relationships, gene structure, conserved protein domains/motifs, and promoter regions. The phylogenetic relationship of MAPKs among kiwifruit demonstrated that the 18 *AcMAPK* genes were grouped into five subgroups (A, B, C, D, and E), and most genes within the same group generally share similar exon/intron patterns and conserved protein domains and motifs. Our analyses strongly supported the identity of each subgroup. The expression profiles of the MAPK cascade genes in various biotic and abiotic stresses and hormones treatments were discussed. The majority of the MAPK cascade genes could be induced by one or more specific treatments. In summary, our study provides an overview of the *MAPK* gene family in kiwifruit, which will be helpful in the biochemical functional characterization of the MAPK cascades in kiwifruit.

Supplementary Materials: Supplementary materials can be found at http://www.mdpi.com/1422-0067/19/9/2510/s1.

Author Contributions: G.W. and J.-Y.Z. designed and initiated this study. G.W. and Z.-H.J. carried out the bioinformatics analyses. G.W. and D.-L.P. performed the qRT-PCR experiments. G.W. and J.-Y.Z. wrote the manuscript. T.W., J.-P.X. and Z.-R.G. helped in discussions of the manuscript. All authors read and approved the final manuscript.

Acknowledgments: This research is supported by the Natural Science Foundation of Jiangsu Province of China (General Program, BK20171328).

Conflicts of Interest: The authors declare no conflict of interest.

References

1. Hamel, L.P.; Nicole, M.C.; Sritubtim, S.; Morency, M.J.; Ellis, M.; Ehlting, J.; Beaudoin, N.; Barbazuk, B.; Klessig, D. Ancient signals: Comparative genomics of plant MAPK and MAPKK gene families. *Trends Plant Sci.* **2006**, *11*, 192–198. [CrossRef] [PubMed]
2. Romeis, T. Protein kinases in the plant defence response. *Curr. Opin. Plant Biol.* **2001**, *4*, 407–414. [CrossRef]
3. Group, M. Mitogen-activated protein kinase cascades in plants: A new nomenclature. *Trends Plant Sci.* **2002**, *7*, 301 308.
4. Rodriguez, M.C.; Petersen, M.; Mundy, J. Mitogen-activated protein kinase signaling in plants. *Annu. Rev. Plant Biol.* **2010**, *61*, 621–649. [CrossRef] [PubMed]
5. Wang, G.; Lovato, A.; Polverari, A.; Wang, M.; Liang, Y.H.; Ma, Y.C.; Cheng, Z.M. Genome-wide identification and analysis of mitogen activated protein kinase kinase kinase gene family in grapevine (*Vitis vinifera*) *BMC Plant Biol.* **2014**, *14*, 219. [CrossRef] [PubMed]
6. Jonak, C.; Okresz, L.; Bogre, L.; Hirt, H. Complexity, cross talk and integration of plant MAP kinase signalling. *Curr. Opin. Plant Biol.* **2002**, *5*, 415–424. [CrossRef]
7. Pitzschke, A.; Schikora, A.; Hirt, H. MAPK cascade signalling networks in plant defence. *Curr. Opin. Plant Biol.* **2009**, *12*, 421–426. [CrossRef] [PubMed]
8. Champion, A.; Picaud, A.; Henry, Y. Reassessing the MAP3K and MAP4K relationships. *Trends Plant Sci.* **2004**, *9*, 123–129. [CrossRef] [PubMed]
9. Mohanta, T.K.; Arora, P.K.; Mohanta, N.; Parida, P.; Bae, H. Identification of new members of the MAPK gene family in plants shows diverse conserved domains and novel activation loop variants. *BMC Genom.* **2015**, *16*, 58. [CrossRef] [PubMed]
10. Doczi, R.; Okresz, L.; Romero, A.E.; Paccanaro, A.; Bogre, L. Exploring the evolutionary path of plant MAPK networks. *Trends Plant Sci.* **2012**, *17*, 518–525. [CrossRef] [PubMed]

11. Colcombet, J.; Hirt, H. Arabidopsis MAPKs: A complex signalling network involved in multiple biological processes. *Biochem. J.* **2008**, *413*, 217–226. [CrossRef] [PubMed]
12. Huang, X.S.; Luo, T.; Fu, X.Z.; Fan, Q.J.; Liu, J.H. Cloning and molecular characterization of a mitogen-activated protein kinase gene from Poncirus trifoliata whose ectopic expression confers dehydration/drought tolerance in transgenic tobacco. *J. Exp. Bot.* **2011**, *62*, 5191–5206. [CrossRef] [PubMed]
13. Wang, G.; Lovato, A.; Liang, Y.H.; Wang, M.; Chen, F.; Tornielli, G.B.; Polverari, A.; Pezzotti, M.; Cheng, Z.M. Validation by isolation and expression analyses of MAPK gene family in grapevine (*Vitis vinifera* L.). *Aust. J. Grape Wine Res.* **2014**, *2*, 255–262. [CrossRef]
14. Reyna, N.S.; Yang, Y. Molecular analysis of the rice MAP kinase gene family in relation to Magnaporthe grisea infection. *Mol. Plant Microbe Interact.* **2006**, *19*, 530–540. [CrossRef] [PubMed]
15. Kong, F.; Wang, J.; Cheng, L.; Liu, S.; Wu, J.; Peng, Z.; Lu, G. Genome-wide analysis of the mitogen-activated protein kinase gene family in Solanum lycopersicum. *Gene* **2012**, *499*, 108–120. [CrossRef] [PubMed]
16. Kong, X.; Pan, J.; Zhang, D.; Jiang, S.; Cai, G.; Wang, L.; Li, D. Identification of mitogen-activated protein kinase kinase gene family and MKK-MAPK interaction network in maize. *Biochem. Biophys. Res. Commun.* **2013**, *441*, 964–969. [CrossRef] [PubMed]
17. Chen, L.; Hu, W.; Tan, S.; Wang, M.; Ma, Z.; Zhou, S.; Deng, X.; Zhang, Y.; Huang, C.; Yang, G. Genome-wide identification and analysis of MAPK and MAPKK gene families in Brachypodium distachyon. *PLoS ONE* **2012**, *7*. [CrossRef] [PubMed]
18. Cakir, B.; Kilickaya, O. Mitogen-activated protein kinase cascades in *Vitis vinifera*. *Front. Plant Sci.* **2015**, *6*, 556. [CrossRef] [PubMed]
19. Zhou, H.; Ren, S.; Han, Y.; Zhang, Q.; Qin, L.; Xing, Y. Identification and Analysis of Mitogen-Activated Protein Kinase (MAPK) Cascades in Fragaria vesca. *Int. J. Mol. Sci.* **2017**, *18*, 1766. [CrossRef] [PubMed]
20. Zhang, X.; Cheng, T.; Wang, G.; Yan, Y.; Xia, Q. Cloning and evolutionary analysis of tobacco MAPK gene family. *Mol. Biol. Rep.* **2013**, *40*, 1407–1415. [CrossRef] [PubMed]
21. Wei, C.; Liu, X.; Long, D.; Guo, Q.; Fang, Y.; Bian, C.; Zhang, D.; Zeng, Q.; Xiang, Z.; Zhao, A. Molecular cloning and expression analysis of mulberry MAPK gene family. *Plant Physiol. Biochem.* **2014**, *77*, 108–116. [CrossRef] [PubMed]
22. Lu, K.; Guo, W.; Lu, J.; Yu, H.; Qu, C.; Tang, Z.; Li, J.; Chai, Y.; Liang, Y. Genome-Wide Survey and Expression Profile Analysis of the Mitogen-Activated Protein Kinase (MAPK) Gene Family in *Brassica rapa*. *PLoS ONE* **2015**, *10*. [CrossRef] [PubMed]
23. Nicole, M.C.; Hamel, L.P.; Morency, M.J.; Beaudoin, N.; Ellis, B.E.; Seguin, A. MAP-ping genomic organization and organ-specific expression profiles of poplar MAP kinases and MAP kinase kinases. *BMC Genom.* **2006**, *7*, 223. [CrossRef] [PubMed]
24. Zhang, S.; Xu, R.; Luo, X.; Jiang, Z.; Shu, H. Genome-wide identification and expression analysis of MAPK and MAPKK gene family in Malus domestica. *Gene* **2013**, *531*, 377–387. [CrossRef] [PubMed]
25. Asai, T.; Tena, G.; Plotnikova, J.; Willmann, M.R.; Chiu, W.L.; Gomez-Gomez, L.; Boller, T.; Ausubel, F.M.; Sheen, J. MAP kinase signalling cascade in *Arabidopsis* innate immunity. *Nature* **2002**, *415*, 977–983. [CrossRef] [PubMed]
26. Galletti, R.; Ferrari, S.; De Lorenzo, G. Arabidopsis MPK3 and MPK6 play different roles in basal and oligogalacturonide- or flagellin-induced resistance against Botrytis cinerea. *Plant Physiol.* **2011**, *157*, 804–814. [CrossRef] [PubMed]
27. Bush, S.M.; Krysan, P.J. Mutational evidence that the *Arabidopsis* MAP kinase MPK6 is involved in anther, inflorescence, and embryo development. *J. Exp. Bot.* **2007**, *58*, 2181–2191. [CrossRef] [PubMed]
28. Gray, J.E.; Hetherington, A.M. Plant development: YODA the stomatal switch. *Curr. Biol.* **2004**, *14*, 488–490. [CrossRef] [PubMed]
29. Petersen, M.; Brodersen, P.; Naested, H.; Andreasson, E.; Lindhart, U.; Johansen, B.; Nielsen, H.B.; Lacy, M.; Austin, M.J.; Parker, J.E. *Arabidopsis* map kinase 4 negatively regulates systemic acquired resistance. *Cell* **2000**, *103*, 1111–1120. [CrossRef]
30. Qiu, J.L.; Zhou, L.; Yun, B.W.; Nielsen, H.B.; Fiil, B.K.; Petersen, K.; Mackinlay, J.; Loake, G.J.; Mundy, J.; Morris, P.C. *Arabidopsis* mitogen-activated protein kinase kinases MKK1 and MKK2 have overlapping functions in defense signaling mediated by MEKK1, MPK4, and MKS1. *Plant Physiol.* **2008**, *148*, 212–222. [CrossRef] [PubMed]

31. Teige, M.; Scheikl, E.; Eulgem, T.; Doczi, R.; Ichimura, K.; Shinozaki, K.; Dangl, J.L.; Hirt, H. The MKK2 pathway mediates cold and salt stress signaling in *Arabidopsis*. *Mol. Cell* **2004**, *15*, 141–152. [CrossRef] [PubMed]
32. Kishi-Kaboshi, M.; Okada, K.; Kurimoto, L.; Murakami, S.; Umezawa, T.; Shibuya, N.; Yamane, H.; Miyao, A.; Takatsuji, H.; Takahashi, A.; et al. A rice fungal MAMP-responsive MAPK cascade regulates metabolic flow to antimicrobial metabolite synthesis. *Plant J.* **2010**, *63*, 599–612. [CrossRef] [PubMed]
33. Lee, S.K.; Kim, B.G.; Kwon, T.R.; Jeong, M.J.; Park, S.R.; Lee, J.W.; Byun, M.O.; Kwon, H.B.; Matthews, B.F.; Hong, C.B.; et al. Overexpression of the mitogen-activated protein kinase gene OsMAPK33 enhances sensitivity to salt stress in rice (*Oryza sativa* L.). *J. Biosci.* **2011**, *36*, 139–151. [CrossRef] [PubMed]
34. Wang, J.; Ding, H.; Zhang, A.; Ma, F.; Cao, J.; Jiang, M. A novel mitogen-activated protein kinase gene in maize (*Zea mays*), ZmMPK3, is involved in response to diverse environmental cues. *J. Integr. Plant Biol.* **2010**, *52*, 442–452. [CrossRef] [PubMed]
35. Zhang, A.; Zhang, J.; Ye, N.; Cao, J.; Tan, M.; Zhang, J.; Jiang, M. ZmMPK5 is required for the NADPH oxidase-mediated self-propagation of apoplastic H_2O_2 in brassinosteroid-induced antioxidant defence in leaves of maize. *J. Exp. Bot.* **2010**, *61*, 4399–4411. [CrossRef] [PubMed]
36. Pan, J.; Zhang, M.; Kong, X.; Xing, X.; Liu, Y.; Zhou, Y.; Liu, Y.; Sun, L.; Li, D. ZmMPK17, a novel maize group D MAP kinase gene, is involved in multiple stress responses. *Planta* **2012**, *235*, 661–676. [CrossRef] [PubMed]
37. Wang, Z.; Mao, H.; Dong, C.; Ji, R.; Cai, L.; Fu, H.; Liu, S. Overexpression of *Brassica napus MPK4* enhances resistance to *Sclerotinia sclerotiorum* in oilseed rape. *Mol. Plant Microbe Interact.* **2009**, *22*, 235–244. [CrossRef] [PubMed]
38. Shi, J.; An, H.L.; Zhang, L.; Gao, Z.; Guo, X.Q. GhMPK7, a novel multiple stress-responsive cotton group C MAPK gene, has a role in broad spectrum disease resistance and plant development. *Plant Mol. Biol.* **2010**, *74*, 1–17. [CrossRef] [PubMed]
39. Li, W.; Ding, Z.; Ruan, M.; Yu, X.; Peng, M.; Liu, Y. Kiwifruit R2R3-MYB transcription factors and contribution of the novel AcMYB75 to red kiwifruit anthocyanin biosynthesis. *Sci. Rep.* **2017**, *7*, 16861. [CrossRef] [PubMed]
40. Huang, S.; Ding, J.; Deng, D.; Tang, W.; Sun, H.; Liu, D.; Zhang, L.; Niu, X.; Zhang, X.; Meng, M.; et al. Draft genome of the kiwifruit *Actinidia chinensis*. *Nat. Commun.* **2013**, *4*, 2640. [CrossRef] [PubMed]
41. Nonis, A.; Ruperti, B.; Pierasco, A.; Canaguier, A.; Adam-Blondon, A.F.; Di Gaspero, G.; Vizzotto, G. Neutral invertases in grapevine and comparative analysis with *Arabidopsis*, poplar and rice. *Planta* **2008**, *229*, 129–142. [CrossRef] [PubMed]
42. Kumar, K.R.; Kirti, P.B. A mitogen-activated protein kinase, AhMPK6 from peanut localizes to the nucleus and also induces defense responses upon transient expression in tobacco. *Plant Physiol. Biochem.* **2010**, *48*, 481–486. [CrossRef] [PubMed]
43. Bailey, T.L.; Williams, N.; Misleh, C.; Li, W.W. MEME: Discovering and analyzing DNA and protein sequence motifs. *Nucleic Acids Res.* **2006**, *34*, 369–373. [CrossRef] [PubMed]
44. Tohge, T.; Fernie, A.R. Combining genetic diversity, informatics and metabolomics to facilitate annotation of plant gene function. *Nat. Protoc.* **2010**, *5*, 1210–1227. [CrossRef] [PubMed]
45. Ortiz-Masia, D.; Perez-Amador, M.A.; Carbonell, J.; Marcote, M.J. Diverse stress signals activate the C1 subgroup MAP kinases of *Arabidopsis*. *FEBS Lett.* **2007**, *581*, 1834–1840. [CrossRef] [PubMed]
46. Lynch, M.; O'Hely, M.; Walsh, B.; Force, A. The probability of preservation of a newly arisen gene duplicate. *Genetics* **2001**, *159*, 1789–1804. [PubMed]
47. Takahashi, F.; Yoshida, R.; Ichimura, K.; Mizoguchi, T.; Seo, S.; Yonezawa, M.; Maruyama, K.; Yamaguchi-Shinozaki, K.; Shinozaki, K. The mitogen-activated protein kinase cascade MKK3-MPK6 is an important part of the jasmonate signal transduction pathway in Arabidopsis. *Plant Cell* **2007**, *19*, 805–818. [CrossRef] [PubMed]
48. Meng, X.; Zhang, S. MAPK cascades in plant disease resistance signaling. *Annu. Rev. Phytopathol.* **2013**, *51*, 245–266. [CrossRef] [PubMed]
49. Meldau, S.; Ullman-Zeunert, L.; Govind, G.; Bartram, S.; Baldwin, I.T. MAPK-dependent JA and SA signalling in Nicotiana attenuata affects plant growth and fitness during competition with conspecifics. *BMC Plant Biol.* **2012**, *12*, 213. [CrossRef] [PubMed]

50. Doczi, R.; Brader, G.; Pettko-Szandtner, A.; Rajh, I.; Djamei, A.; Pitzschke, A.; Teige, M.; Hirt, H. The Arabidopsis mitogen-activated protein kinase kinase MKK3 is upstream of group C mitogen-activated protein kinases and participates in pathogen signaling. *Plant Cell* **2007**, *19*, 3266–3279. [CrossRef] [PubMed]
51. Hung, J.H.; Weng, Z. Sequence Alignment and Homology Search with BLAST and ClustalW. *Cold Spring Harb. Protoc.* **2016**, *2016*. [CrossRef] [PubMed]
52. Tamura, K.; Peterson, D.; Peterson, N.; Stecher, G.; Nei, M.; Kumar, S. MEGA5: Molecular evolutionary genetics analysis using maximum likelihood, evolutionary distance, and maximum parsimony methods. *Mol. Biol. Evol.* **2011**, *28*, 2731–2739. [CrossRef] [PubMed]
53. Guo, A.Y.; Zhu, Q.H.; Chen, X.; Luo, J.C. GSDS: A gene structure display server. *Yi Chuan* **2007**, *29*, 1023–1026. [CrossRef] [PubMed]
54. Wang, J.; Pan, C.; Wang, Y.; Ye, L.; Wu, J.; Chen, L.; Zou, T.; Lu, G. Genome-wide identification of MAPK, MAPKK, and MAPKKK gene families and transcriptional profiling analysis during development and stress response in cucumber. *BMC Genom.* **2015**, *16*, 386. [CrossRef]
55. Zhang, J.Y.; Huang, S.N.; Wang, G.; Xuan, J.P.; Guo, Z.R. Overexpression of Actinidia deliciosa pyruvate decarboxylase 1 gene enhances waterlogging stress in transgenic *Arabidopsis thaliana*. *Plant Physiol. Biochem.* **2016**, *106*, 244–252. [CrossRef] [PubMed]
56. Yin, X.R.; Allan, A.C.; Chen, K.S.; Ferguson, I.B. Kiwifruit EIL and ERF genes involved in regulating fruit ripening. *Plant Physiol.* **2010**, *153*, 1280–1292. [CrossRef] [PubMed]
57. Livak, K.J.; Schmittgen, T.D. Analysis of relative gene expression data using real-time quantitative PCR and the $2(-\Delta\Delta C_t)$ Method. *Methods* **2001**, *25*, 402–408. [CrossRef] [PubMed]

© 2018 by the authors. Licensee MDPI, Basel, Switzerland. This article is an open access article distributed under the terms and conditions of the Creative Commons Attribution (CC BY) license (http://creativecommons.org/licenses/by/4.0/).

International Journal of
Molecular Sciences

MDPI

Article

Genome-Wide Analysis and Cloning of the Apple Stress-Associated Protein Gene Family Reveals *MdSAP15*, Which Confers Tolerance to Drought and Osmotic Stresses in Transgenic *Arabidopsis*

Qinglong Dong †, Dingyue Duan †, Shuang Zhao, Bingyao Xu, Jiawei Luo, Qian Wang, Dong Huang, Changhai Liu, Chao Li, Xiaoqing Gong, Ke Mao * and Fengwang Ma *

State Key Laboratory of Crop Stress Biology for Arid Areas/Shaanxi Key Laboratory of Apple, College of Horticulture, Northwest A & F University, Yangling 712100, China; dong19850412@163.com (Q.D.); duandingyue207@foxmail.com (D.D.); zhsh812972738@126.com (S.Z.); 18392610250@163.com (B.X.); m18391488428@163.com (J.L.); wangqian123@nwafu.edu.cn (Q.W.); Mrhaodee@126.com (D.H.); chliu@nwafu.edu.cn (C.L.); lc453@163.com (C.L.); gongxq0103@nwsuaf.edu.cn (X.G.)

* Correspondence: maoke2002@nwsuaf.edu.cn (K.M.); fwm64@nwsuaf.edu.cn or fwm64@sina.com (F.M.); Tel.: +86-029-8708-2613 (K.M.); +86-029-8708-2648 (F.M.)

† These two authors contributed equally to this work.

Received: 8 July 2018; Accepted: 13 August 2018; Published: 21 August 2018

Abstract: Stress-associated proteins (SAPs) are novel A20/AN1 zinc finger domain-containing proteins that are now favorable targets to improve abiotic stress tolerance in plants. However, the SAP gene family and their biological functions have not been identified in the important fruit crop apple (*Malus* × *domestica* Borkh.). We conducted a genome-wide analysis and cloning of this gene family in apple and determined that the overexpression of *MdSAP15* enhances drought tolerance in *Arabidopsis* plants. We identified 30 SAP genes in the apple genome. Phylogenetic analysis revealed two major groups within that family. Results from sequence alignments and analyses of 3D structures, phylogenetics, genomics structure, and conserved domains indicated that apple SAPs are highly and structurally conserved. Comprehensive qRT-PCR analysis found various expression patterns for *MdSAPs* in different tissues and in response to a water deficit. A transgenic analysis showed that the overexpression of *MdSAP15* in transgenic *Arabidopsis* plants markedly enhanced their tolerance to osmotic and drought stresses. Our results demonstrate that the SAP genes are highly conserved in plant species, and that *MdSAP15* can be used as a target gene in genetic engineering approaches to improve drought tolerance.

Keywords: apple; SAP gene family; expression analysis; function analysis; drought stress; osmotic stress

1. Introduction

The growth, development, and survival of plants is constantly challenged by a variety of biotic and abiotic environmental factors. Plants utilize complex molecular mechanisms that regulate patterns of gene expression to protect themselves against these stresses [1,2]. Some key modulators of stress responses have been characterized and have emerged as appropriate targets to enhance abiotic stress tolerance in many plants. They include NAC domain-containing transcription factors, DRE/CRT-binding transcription factors (DREBs/CBFs), mitogen-activated protein kinases (MAPKs), stress-associated proteins (SAPs), and heat shock factor/proteins (HSPs/HSF) [1,3–6]. Among these, the SAPs are a newly identified class of zinc finger proteins (ZFPs) that play crucial roles in various abiotic stress responses by numerous plants [1,2,7].

The SAP gene family members have two special ZF domains: the highly conserved A20 domain, which was first isolated in human umbilical vein endothelial cells with the characterization of a tumor necrosis factor (TNF)-α-inducible protein; and/or the AN1 domain, which is also highly conserved and first identified from *Xenopus laevis* animal hemisphere 1 (AN1) maternal RNA with the delineation of the ubiquitin-like protein [8,9]. The SAP proteins expressed in *Arabidopsis thaliana* (hereafter *Arabidopsis*), rice (*Oryza sativa*), tomato (*Solanum lycopersicum*), and cotton (*Gossypium hirsutum*) have been classified into five groups (I through V) based on results from their phylogenetic analyses [10,11]. One significant feature of plant SAPs is the very frequent occurrence of intronless genes [2]. For example, 11 rice SAP genes, 15 from desert poplar (*Populus euphratica*), and 30 from cotton lack introns and show a remarkably higher percentage of intronless genes [2,6,11].

The roles of SAP genes are increasingly being reported in plants. Transcriptional levels are induced by multiple stresses and provide a positive reinforcement of tolerance to abiotic stress. Rice A20/AN1 protein (OSISAP1/OsSAP1), the first identified plant SAP gene, is induced after different types of stress treatments are applied [1]. The overexpression of *OSISAP1* confers tolerance to dehydration, cold, and salt in transgenic seedlings of tobacco (*Nicotiana tabacum*) [1]. Furthermore, *ZFP177* (*OsSAP9*) and *AtSAP5* are induced by numerous challenges and have significant roles in improving abiotic stress tolerance [12,13]. Similar results have been described for SAPs from maize (*Zea mays*) [14], medicago (*Medicago truncatula*) [15], banana (*Musa* sp.) [16], the halophyte grass *Aeluropus littoralis* [17], and poplar (*Populus alba* × *P. glandulosa*) [18]. These genes also function in biotic stress responses. For example, Tyagi et al. [19] have analyzed the expression patterns of rice SAP gene family members in response to pathogen elicitors, and discovered that *OsSAP1*, *OsSAP8*, and *OsSAP11* are up-regulated. Transgenic tobacco overexpressing *OsiSAP1* shows significantly enhanced basal resistance against infection by the bacterial pathogen *Pseudomonas syringae* pv. *Tabaci* [19].

The SAP genes also help regulate signal transduction and phytohormone synthesis. In rice, the overexpression of *OsDOG* (*OsiSAP11*) [20] and *OsZFP185* (*OsiSAP4*) [21] results in dwarf phenotypes, a decrease in gibberellic acid (GA) contents, and deficient cell elongation. Furthermore, *OsZFP185* negatively regulates the expression of several genes related to abscisic acid (ABA) biosynthesis, and interferes with ABA-mediated tolerance to salt, drought, and cold [21]. Various SAPs can function as E3 ubiquitin ligases, redox sensors, and/or regulators of gene expression under stress [7,13,22,23]. Other novel biological functions for SAP genes will continue to be reported.

Based on their highly conserved A20/AN1 domains, members of the SAP gene family have been identified and characterized in *Arabidopsis* [24], rice [24], maize [14], tomato [10], cotton [11], desert poplar [6], and medicago [25]. Although extensive genomic analyses have provided considerable details about this family in several species, members in apple (*Malus* × *domestica* Borkh.) have not been as thoroughly investigated. Nevertheless, recent completion of the draft genome sequence for apple has enabled genome-wide analyses of its SAP genes [26–28]. Here, we identified SAP members in apple and examined their A20/AN1 domain, protein and gene structures, conserved domains, phylogenetic relationships, chromosomal locations, *cis*-acting elements, and expression patterns for *MdSAPs* cloned in response to water deficits. We also overexpressed *MdSAP10* in *Arabidopsis* and investigated its function. Our results will serve as a basis for exploring the molecular roles of SAPs. By facilitating further studies into their functions in abiotic stress responses, we can continue our efforts to introduce improved apple cultivars.

2. Results

2.1. Identification and Annotation of Apple SAP Genes

To identify the genes in the apple genome that encode SAP proteins, we conducted a BlastP of the apple genome database and identified 32 putative family members. We then used the Pfam and NCBI Conserved Domain Database (NCBI-CDD) databases for verification, searching for the A20/AN1 domain in the amino acid sequences encoded by all 32 genes. From this, we confirmed the identity of 30 typical apple SAP genes in the original dataset (Table 1).

We next cloned all of the full-length apple SAP genes based on predicted nucleotide sequences in the apple genome and in the NCBI nucleotide database. As shown in Table 1, this revealed that the full-length cDNAs of *MdSAP7*, *-8*, *-10*, *-12*, *-14*, *-15*, *-16*, *-19*, *-21*, *-23*, *-25*, *-28*, and *-29* had been isolated and confirmed by RT-PCR (Supplementary Sequence A1). Their corresponding 5′- and 3′-UTRs were then amplified.

2.2. Structures and Conserved Domains of Apple SAP Genes

To gain insights into the structural diversity of SAP genes in apple, we analyzed the phylogenetic tree, exon–intron organization, and conserved domains in the coding sequences. The *MdSAP* proteins were classified as groups I and II based on their phylogenetic relationships (Figure 1A). Gene structure analysis indicated that *MdSAP1* through *MdSAP4*, *MdSAP6* through *MdSAP21*, *MdSAP23*, *MdSAP24*, *MdSAP26*, *MdSAP29*, and *MdSAP30* contained no introns, whereas *MdSAP5*, *MdSAP22*, *MdSAP25*, *MdSAP27*, and *MdSAP28* had one each (Figure 1B). Conserved domain analysis revealed that all of the *MdSAP* proteins included A20 and/or AN1 domain(s). *MdSAP1* through *MdSAP4*, *MdSAP7* through *MdSAP20*, *MdSAP23*, *MdSAP24*, *MdSAP26*, *MdSAP29*, and *MdSAP30* contained an A20 domain and an AN1 domain; *MdSAP5*, *MdSAP6*, and *MdSAP22* had single AN1 domains; and *MdSAP21*, *MdSAP25*, *MdSAP27*, and *MdSAP28* each had two AN1 domains. In addition, *MdSAP25* contained a C2H2 domain at the C terminal (Figures 1C and 2).

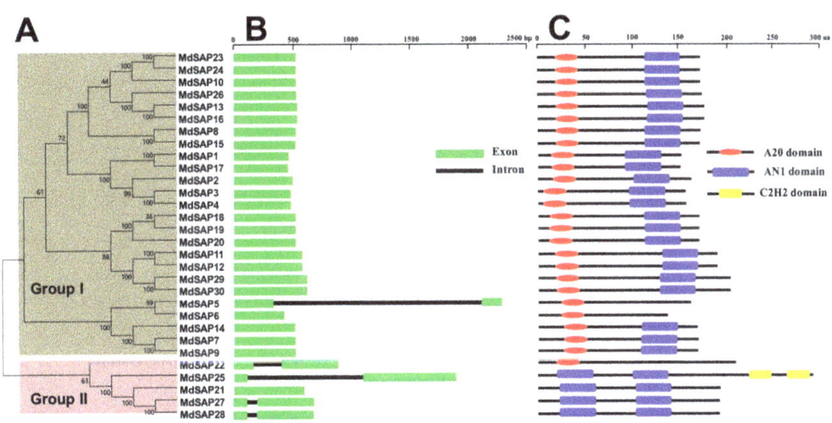

Figure 1. (**A**) Phylogenetic relationships; (**B**) Structures for 30 genes; and (**C**) Analysis of conserved domains for stress-associated protein (SAP) genes in apple. A phylogenetic tree for full-length amino acid sequences was constructed with MEGA software and the NJ method.

Table 1. Properties of SAPs identified from apple genome.

Gene Name	Gene ID [1]	Zinc Finger Domain	Protein Length (aa)	Molecular Weight (kDa)	Theoretical Isoeletrical Point	Chromosome Location
MdSAP1	MDP0000494946	A20-AN1	151	16.53	8.41	chr1:24227744..24228199
MdSAP2	MDP0000588934	A20-AN1	162	17.71	8.62	chr2:22718230..22718718
MdSAP3	MDP0000122842	A20-AN1	156	17.07	8.27	chr2:22745813..22746283
MdSAP4	MDP0000237812	A20-AN1	156	17.07	8.27	chr2:22753663..22754133
MdSAP5	MDP0000316313	A20	161	18.11	8.95	chr2:24190887..24193156
MdSAP6	MDP0000543745	A20	136	15.18	6.37	chr2:25865864..25866277
MdSAP7	MDP0000362676	A20-AN1	169	18.42	9.05	chr2:35848074..35848583
MdSAP8	MDP0000874708	A20-AN1	172	18.24	8.12	chr2:35871562..35872080
MdSAP9	MDP0000362677	A20-AN1	169	18.42	9.05	chr2:35878425..35878935
MdSAP10	MDP0000164222	A20-AN1	172	18.41	7.52	chr3:2250369..2250887
MdSAP11	MDP0000516205	A20-AN1	189	20.09	6.78	chr4:8801881..8802450
MdSAP12	MDP0000506127	A20-AN1	189	20.09	6.78	chr4:8807038..8807607
MdSAP13	MDP0000286185	A20-AN1	176	18.97	7.46	chr6:7080970..7081500
MdSAP14	MDP0000263150	A20-AN1	168	18.51	9.16	chr7:613109..613615
MdSAP15	MDP0000292844	A20-AN1	171	18.17	8.12	chr7:697510..698025
MdSAP16	MDP0000294781	A20-AN1	176	18.97	7.46	chr7:10610840..10611370
MdSAP17	MDP0000133254	A20-AN1	150	16.41	8.41	chr7:22721788..22722240
MdSAP18	MDP0000139359	A20-AN1	170	18.59	8.67	chr8:16795514..16796026
MdSAP19	MDP0000707978	A20-AN1	170	18.59	8.67	chr8:16801945..16802457
MdSAP20	MDP0000296953	A20-AN1	170	18.59	8.67	chr8:16804379..16804891
MdSAP21	MDP0000211516	AN1-AN1	193	21.36	8.67	chr9:2267267..2267848
MdSAP22	MDP0000165407	A20	209	23.09	8.29	chr9:27693011..27693884
MdSAP23	MDP0000231017	A20-AN1	172	18.18	7.51	chr11:2865175..2865693
MdSAP24	MDP0000683912	A20-AN1	172	18.18	7.51	chr11:2875969..2876487
MdSAP25	MDP0000652898	AN1-AN1-C2H2-C2H2	293	32.04	8.26	chr12:7214047..7215925
MdSAP26	MDP0000086327	A20-AN1	174	18.78	7.46	chr14:7829264..7829785
MdSAP27	MDP0000141121	AN1-AN1	192	21.39	8.72	chr17:2716099..2716758
MdSAP28	MDP0000853499	AN1-AN1	192	21.39	8.72	chr17:2716210..2716869
MdSAP29	MDP0000661416	A20-AN1	203	21.86	7.66	unanchored:14381067..14381678
MdSAP30	MDP0000284856	A20-AN1	203	21.86	7.66	unanchored:14408647..14409258

[1] Gene ID in apple genome (https://www.rosaceae.org/gb/gbrowse/malus_x_domestica/).

2.3. Multiple Sequence Alignments and Three-Dimensional Structure of Apple A20/AN1 Domains

Multiple alignments demonstrated that the A20/AN1 domains are conserved among the *MdSAP* proteins (Figure 2). We then produced sequence logos that further showed that these domains were highly conserved at each residue position (Figure 3A,C; Supplementary Sequences A2 and A3). Afterward, the SWISS-MODEL web server was used for modeling and analysis of homology among protein structures. For this, we built the A20 domain and AN1 domain homology models and evaluated them using the homologous templates 2KZY.pdb and 1WFP, respectively (Figure 3B,D). The 3D models indicated that, respectively, the A20 domain and the AN1 domain in the *MdSAP7* structure most closely matched the A20 domain of ubiquitin receptor ZNF216 and the zf-AN1 domain of the *Arabidopsis* F5O11.17 protein (PDB ID: 2KZY.1.A, 38% sequence identity for residues 18–50; 1WFP.1.A, 47% sequence identity for residues 96–148).

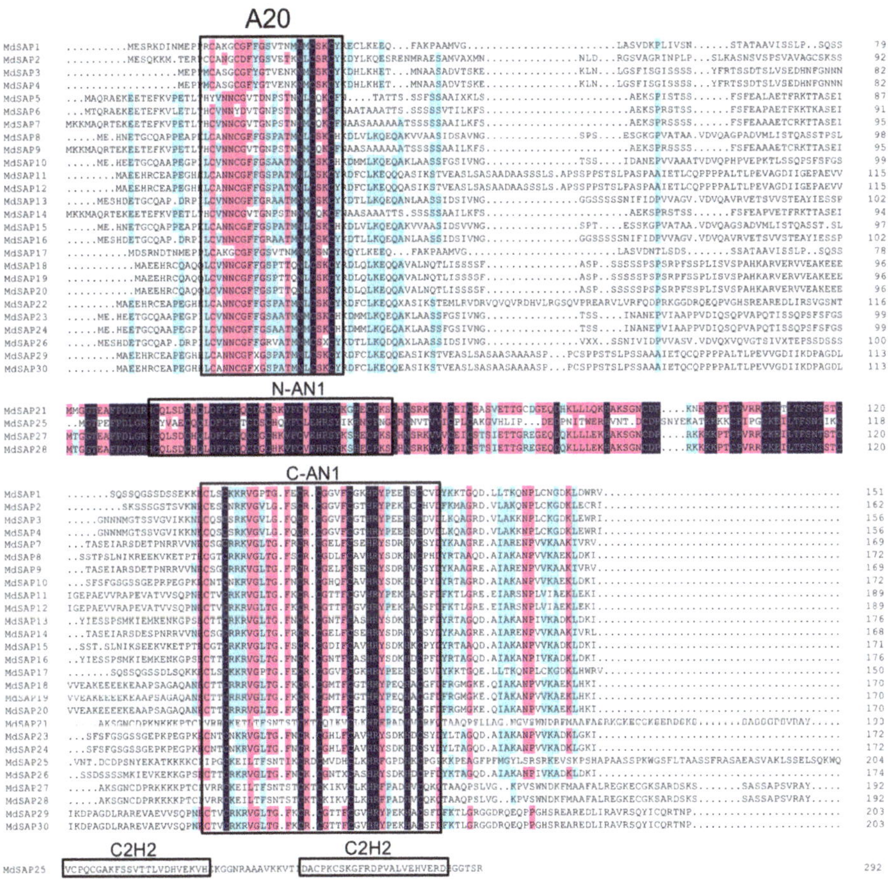

Figure 2. Multiple alignments of A20/AN1 domain and C2H2 amino acids in apple SAPs, using the DNAMAN program. Conserved domains are boxed, and identical amino acids are shown against a dark blue background (Similarity: dark blue = 100%; pink > 75%; cyan > 50%).

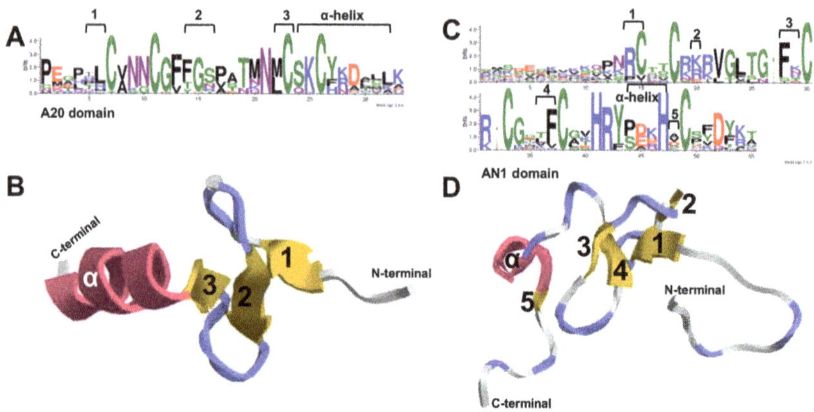

Figure 3. (**A**) Sequence logos for A20 domain in 26 *MdSAP* proteins, generated via WebLogo; (**B**) Three-dimensional tertiary structural model of A20 domain (PDB ID: 2KZY.1.A); (**C**) Sequence logos of AN1 domain in 28 *MdSAP* proteins; and (**D**) Three-dimensional tertiary structural model of AN1 domain (PDB ID: 1WFP.1.A). Within each stack, symbol height indicates the relative frequency of each amino acid at that position. Logos for the A20 domain and AN1 domain were obtained through multiple alignments of 26 and 28 *MdSAP* protein sequences, respectively. At the top of the corresponding amino acid sequences, arabic numbers (1–5) indicate β-sheets in A20 and AN1 domains. In (**B**,**D**), α-helices are red, β-sheets (arabic numbers 1–5) are yellow, and strands are blue/gray. Three-dimensional representations were generated with RasTop software.

2.4. Phylogenetic Analysis of SAP Proteins

To examine the evolutionary relationships among plant SAP proteins, we used MEGA 6 and constructed unrooted phylogenetic trees from full-length protein sequences encoded by 453 SAP genes in 32 species (Supplementary File B1). Two major groups were revealed: I, containing an A20 domain and an AN1 domain; and II, containing two AN1 domains. Members in Group I were further classified into four subgroups (Ia-Id), while Group II members were assigned to two subgroups: IIa, containing two AN1 domains and one or two C2H2 domain(s); and IIb, containing only two AN1 domains. Among these 30 *MdSAP* proteins, 25 (*MdSAP1-MdSAP20*, *MdSAP23*, *MdSAP24*, *MdSAP26*, *MdSAP29*, and *MdSAP30*) could be unambiguously classified as Group I, while four (*MdSAP21*, *MdSAP25*, *MdSAP27*, and *MdSAP28*) were assigned to Group II based on their relationship with the other SAP proteins (Figure 4 and Supplementary File B2). Further analysis revealed that the 25 Group-I apple proteins belonged to subgroups Ia (seven genes), Ib (five), Ic (five), and Id (eight). Subgroup IIa contained *MdSAP25*, while IIb contained *MdSAP21*, *MdSAP27*, and *MdSAP28*.

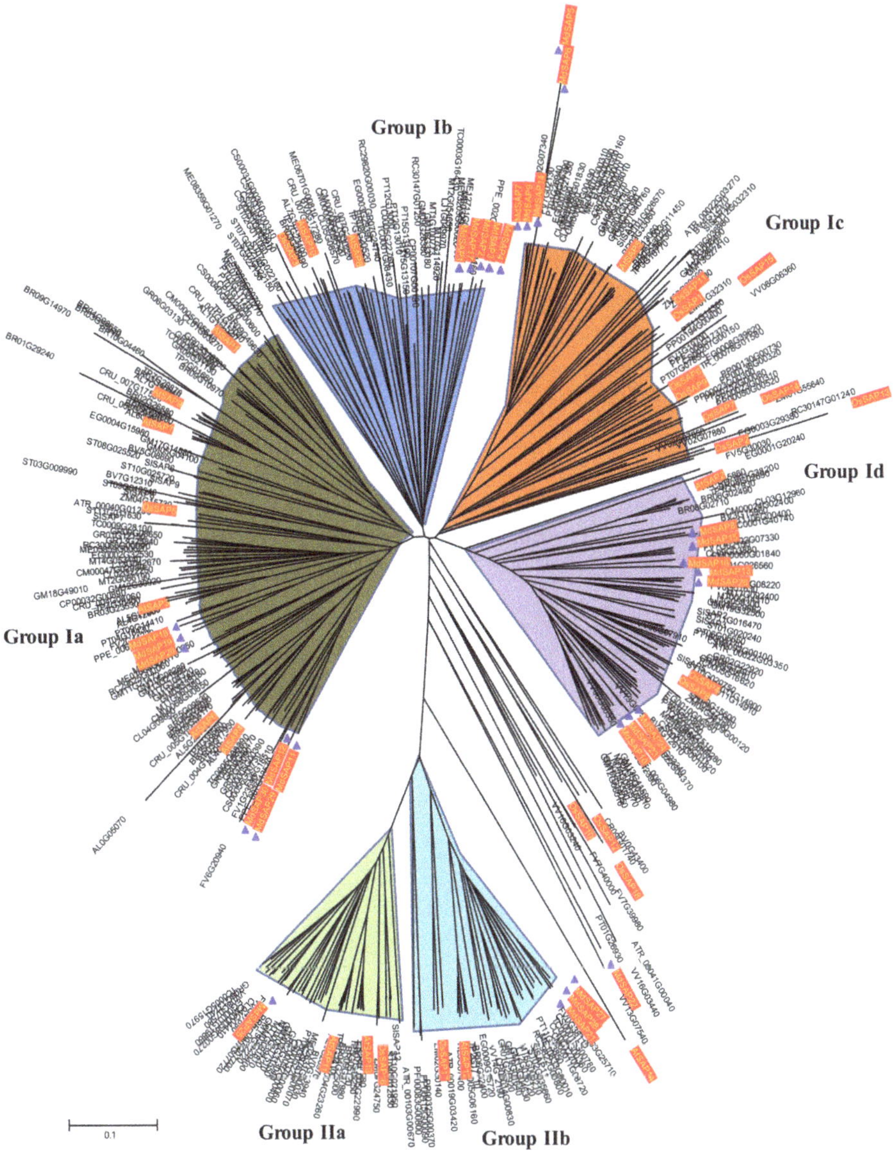

Figure 4. Phylogenetic analysis of 453 SAP proteins in 32 species. Unrooted NJ tree was constructed with MEGA 6 software, using full-length amino acid sequences. Tree comprises six subgroups (Ia–Id; IIa and IIb).

2.5. Genome Distribution of Apple SAP Genes

We determined the genomic locations of these apple SAP genes based on their mapping coordinates. In all, 28 of the 30 *MdSAP* genes were assigned to chromosomes 1–4, 6–9, 11, 12, 14, and 17 (Table 1; Figure 5). However, we could not conclusively map two genes (*MdSAP29* and *MdSAP30*) to any chromosome. The genes were unevenly distributed among the 12 chromosomes, with Chromosome 2 containing the most (eight genes), followed by Chromosome 7 (five genes), and one each for chromosomes 1, 3, 6, 12, and 14.

The apple gene family appears to have expanded during the process of genome evolution [26]. To uncover the mechanism underlying this expansion, we investigated gene duplication events, including tandem and segmental duplications, and found that many *MdSAP* genes (19/30, or 63.33%) were present in two or more copies (Figure 5). In all, 17 had undergone tandem duplication, while two were subjected to segment duplication. Those segment duplications produced many homologs of SAP genes on different chromosomes, while tandem duplications produced SAP gene clusters or hotspots (blue and red font in Figure 5). A relatively recent genome-wide duplication is that in the Pyreae tribe, which was thought to result in the transition of nine ancestral chromosomes to 17 chromosomes [26]. We noted here that multiple gene pairs were each linked to at least six potential chromosomal segmental duplications (Figure 5, pairs of bars in grey areas), e.g., large sections of chromosomes 9 and 17, 3 and 11, and 7 and 2.

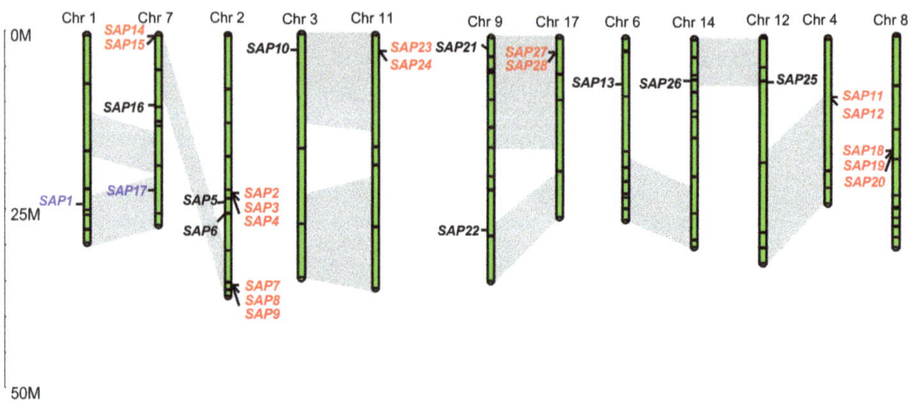

Figure 5. Chromosomal locations of 28 apple SAP genes. Scale is in megabases (Mb). Red font, tandem duplication; blue font, segmental duplication; grey area, genome-wide duplications.

2.6. Promoter Sequence Analysis of Apple SAP Genes

To investigate putative *cis*-acting elements in their promoter regions, we isolated approximately 1500-bp genomic sequences upstream of the start codon from our *MdSAPs*. Along with some *cis* elements involved in light-responsiveness (Table 2 and Supplementary File B3), we found that many were responsive to various stresses and correlative hormones. In total, 10 types of *cis* elements were discovered in the 13 promoters. They were associated with responses to hypoxia, heat, chilling, drought, pathogens, wounding, or hormones such as salicylic acid, methyl jasmonate, ABA, or ethylene. Therefore, we concluded that these *cis* elements play important roles in plant stress responses.

Sequences and functions for ABRE (ABA response element), ARE (anaerobic response element), CGTCA (MeJA-responsiveness), ERE (ethylene-responsive element), HSE (heat shock response element), LTR (low-temperature response element), MBS (MYB binding site involved in drought response), TCA (salicylic acid response element), TC-rich repeat (defense and stress responsiveness), and W-box (elicitation; wounding and pathogen responsiveness; binding site of WRKY type transcription factors) were obtained from the PlantCARE database (http://bioinformatics.psb.ugent.be/webtools/plantcare/html/). Digits represent the number of regulatory elements on plus/minus strand. Blank space indicates no corresponding cis-acting element in either strand of the promoter.

Table 2. The *cis*-acting elements of 13 promoters in apple SAP genes.

Cis-Acting Elements	ABRE	ARE	CGTCA	ERE	HSE	LTR	MBS	TCA	TC-Rich Repeat	W-Box
Stress to Response	**ABA**	**Hypoxia**	**MeJA**	**Ethylene**	**Heat**	**Chilling**	**Drought**	**SA**	**Defense**	**Pathogen**
MdSAP7	2/2	2/0	2/2		1/0	1/1		0/2	0/2	
MdSAP8		0/1			0/1		3/0	0/3		0/1
MdSAP10	2/0	0/2	0/1				1/0	0/2		
MdSAP12			1/0			1/0				
MdSAP14		2/0	1/1		2/0	1/0		1/0	1/0	0/1
MdSAP15	0/1	0/2			0/1		0/1	0/1	1/0	1/0
MdSAP16		2/2		2/0					1/0	3/0
MdSAP19		0/2		0/1		1/1		1/1	2/0	
MdSAP21		1/0	1/0		1/2				1/0	
MdSAP23	1/0	1/4	0/2			1/0				1/0
MdSAP25	1/1	1/1	1/1			0/2	1/1	0/1	0/1	
MdSAP28		0/4	1/2		1/1	1/1	1/1			
MdSAP29		0/2	1/2		1/1	1/1		1/0	2/0	

ABA: abscisic acid; ABRE: ABA response element; ARE: anaerobic response element; CGTCA: MeJA-responsiveness; ERE: ethylene-responsive element; HSE: heat shock response element; LTR: low-temperature response element; MBS: MYB binding site involved in drought response; TCA: salicylic acid response element; TC-Rich Repeat: defense and stress responsiveness; W-box: elicitation, wounding, and pathogen responsiveness/binding site of WRKY type transcription factors.

2.7. Expression Profiles of MdSAP Genes

Knowing the patterns of expression in various tissue types can help us understand gene functions. Firstly, we collected different tissues including young roots, stems, fully expanded leaves, flowers, and mature fruit (70 mm, red peel, 150 days after bloom), from apple plants that were five years old after bud grafting. The scion was *Malus domestica* "Golden Delicious", and the rootstock was *M. hupehensis*. Then, we isolated the full-length cDNA, 5′-UTR, and 3′-UTR sequences for 13 *MdSAP* genes and used specific primers for our qRT-PCR assays. These *MdSAP*s were constitutively expressed in the five tissues examined here, albeit at different levels of transcription (Figure 6A). For example, *MdSAP7*, *-8*, *-10*, *-15*, *-19*, and *-29* were most highly expressed in the leaves; *MdSAP12*, *-16*, *-21*, and *-28* were most highly expressed in the fruits; and *MdSAP14*, *-23*, and *-25* were most highly expressed in the roots. To induce a water deficit, irrigation was withheld up to 8 days, while the designated control plants continued to receive normally scheduled irrigation ("Golden Delicious" scions and *M. hupehensis* rootstocks). In response to drought stress, the expression of *MdSAP15*, *-25*, and *-28* was significantly induced from that detected in the non-stressed control plants (Figure 6B), transcripts of *MdSAP7* and *-21* mRNAs were significantly reduced, and expression of the other MdSAP genes remained constant.

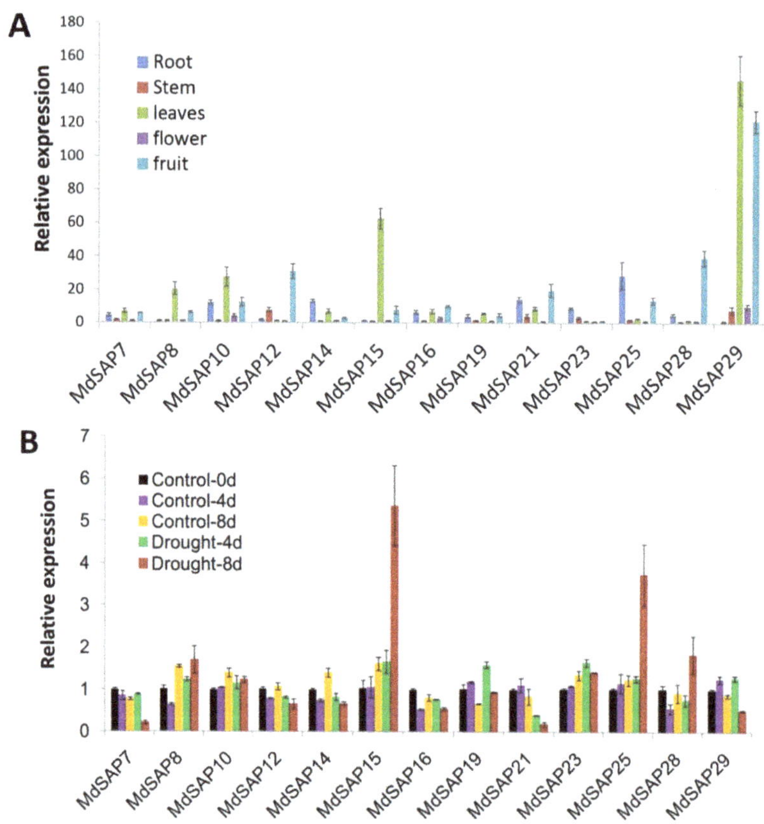

Figure 6. Tissue-specific expression and drought response of 13 *MdSAP* genes. (**A**) Expression patterns of 13 *MdSAP* genes in apple tissues. (**B**) Expression patterns of 13 *MdSAP* genes in response to drought stress, i.e., normally scheduled irrigation withheld for 0, 4, or 8 days. Three independent replicates were used for calculations. Error bars indicate standard deviation.

2.8. *MdSAP15* Overexpression Enhances Osmotic Stress Tolerance by *Arabidopsis* Seedlings

Since the transcription of *MdSAP15* mRNA was significantly accumulated under drought stress, we chose this gene for investigating its biological functions in *Arabidopsis*. After kanamycin-resistance screening and PCR detection using *Arabidopsis* genomic DNAs as templates, more than five transformants were identified and confirmed, with elevated levels of *MdSAP15* transcripts (Figure 7A). From these, we selected three transgenic lines (L1, 4, and 5) with high *MdSAP15* expression to evaluate its potential functioning in response to osmotic and drought stresses. For the osmotic stress assay, five-day-old seedlings grown on Murashige and Skoog (MS) agar plates were vertically plated on an MS agar medium supplemented with 0, 200, or 300 mM of mannitol. While the roots of "Col" and *MdSAP15* OE seedlings displayed similar growth characteristics on MS agar medium plates (Figure 7B), their growth was affected when treated with different concentrations of mannitol. For example, the primary roots and fresh weights of the transgenics were longer and heavier than those of the wild-type (WT) "Col" upon exposure to 200 or 300 mM of mannitol (Figure 7C,D).

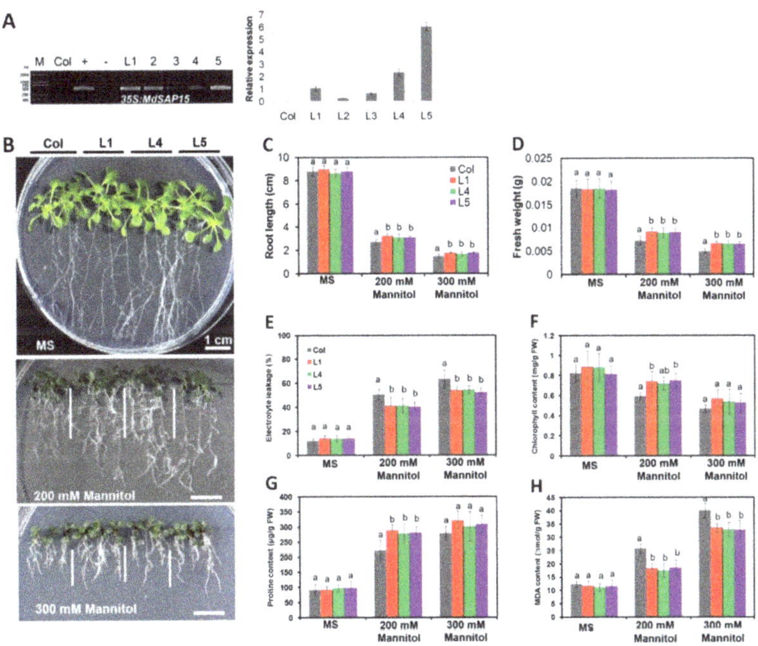

Figure 7. Overexpression of *MdSAP15* in *Arabidopsis* increased osmotic tolerance in response to mannitol treatment. (**A**) PCR identification and relative expression analysis of *MdSAP15* in Col-0 wild-type (Col) and transgenic *Arabidopsis* lines. M, DNA marker; −, negative control (H2O); +, positive control (plasmid DNA of *35S::MdSAP15* pRI101AN vector). Specific primers for *MdSAP15* were used to detect relative expression levels of Col and transgenic *Arabidopsis* lines. (**B**) Representative images of "Col" and transgenic seedlings at 5 days after seeds had been cultivated for 11 days on MS medium alone (0 mM mannitol) or MS medium supplemented with 200 or 300 mM of mannitol. Bars = 1 cm. (**C**) Primary root lengths and (**D**) fresh weights of "Col" and transgenic seedlings measured on Day 5 after plants had been exposed for 11 days to osmotic stress. (**E**) Relative electrolyte leakage, and levels of chlorophyll (**F**), proline (**G**), and malondialdehyde (MDA) (**H**) in "Col" and transgenic seedlings, measured on Day 5 following treatment on MS medium with 0 mM, 200 mM, or 300 mM of mannitol for 11 d. Error bars represent SD based on three independent replicates. For (**C**–**H**), bars not labeled with same letters in each panel indicate values are significantly different at $p < 0.05$, based on one-way ANOVA and Duncan's tests.

2.9. MdSAP15-Overexpressing Arabidopsis Seedlings Have Improved Physiological Traits Associated with Osmotic Stress Tolerance When Compared with "Col" Wild Type

For further investigation of this *MdSAP15*-mediated enhancement of tolerance to osmotic stress, we measured relative electrolyte leakage (REL) and concentrations of chlorophyll, proline, and malondialdehyde (MDA), all of which are important markers of such tolerance. Under normal growing conditions, we did not observe any obvious differences in REL. However, in response to mannitol exposure, REL values were significantly higher in "Col" seedlings than in the transgenics (Figure 7E). Levels of chlorophyll and proline were similar between "Col" and OE plants under control conditions, but were higher in the transgenics after they were treated with 200 mM of mannitol (Figure 7F,G). Furthermore, MDA concentrations did not differ among genotypes under normal conditions, but were lower in the overexpressed (OE) lines than in the WT following mannitol treatment (Figure 7H). These results suggested that the overexpression of *MdSAP15* in *Arabidopsis* seedlings leads to enhanced osmotic stress tolerance.

2.10. MdSAP15 Overexpression Enhances Drought Tolerance in Transgenic Arabidopsis Plants

To evaluate drought tolerance, we simultaneously germinated seeds of the "Col" and *MdSAP15*-overexpressing lines and grew the seedlings on MS plates for one week before transplanting them into soil for another three weeks of culture. Drought conditions were then imposed by withholding water. After 20 days of stress, all of the plants exhibited symptoms related to severe water loss, although those symptoms were milder for the OE transgenics than for the WT. When rehydration began (2 days of refreshment), most of the "Col" plants did not recover (Figure 8A), whereas the survival rate was significantly higher for the three transgenic lines (Figure 8B). These results again indicated that the overexpression of *MdSAP15* in *Arabidopsis* plants enhances their degree of drought tolerance.

Figure 8. Assessment of drought tolerance in "Col" and transgenic *Arabidopsis* plants. (**A**) Representative images of four-week-old plants; (**B**) Survival rates of WT and transgenic lines 2 days after rehydration began. Error bars represent SD based on three independent replicates. For (**B**), bars not labeled with same letters in each panel indicate values are significantly different at $p < 0.05$, based on one-way ANOVA and Duncan's tests.

3. Discussion

The proteins encoded by SAP genes comprise large families and are broadly distributed in higher plants [2]. Apple is an economically important woody plant and the most widely cultivated fruit crop in the world. Sequencing of its genome has provided a good platform for genome-wide analyses of all putative gene families in apple, including the DREB [29], MYB [30], MADS-box [31], and WRKY [32,33] families. However, genome-wide information about apple SAP genes has remained unknown, while members of that family have been identified in other plant species [6,10,11,14,24,25]. Moreover, the content of SAP genes varies substantially among species. For example, *Brassica rapa*, *Glycine max*, *Solanum tuberosum*, *Salix purpurea*, *Populus trichocarpa*, and cotton each have a relatively large number of SAP members, i.e., 28, 26, 19, 19, 19, and 19, respectively; while *Chlamydomonas reinhardtii*, *Lotus japonicus*, *Carica papaya*, and *Amborella trichopoda* have relatively few, i.e., 2, 6, 7, and 7, respectively (Table 3). Here, we determined that the apple genome contains 30 SAP genes, making this family much larger than in any other species.

Table 3. Numbers of SAP gene family members in various species.

Plant Species	A20-AN1	A20-A20-AN1	A20	AN1	AN1-AN1	AN1-AN1-C2H2	AN1-AN1-C2H2-C2H2	Total Number
Malus domestica	23	0	3	0	3	0	1	30
Arabidopsis thaliana	10	0	0	1	1	1	1	14
Oryza sativa	11	1	1	3	1	0	1	18
Populus trichocarpa	15	0	0	2	1	0	1	19
Solanum lycopersicum	9	0	0	1	2	0	1	13
Gossypium hirsutum	14	0	0	2	2	0	1	19
Populus euphratica	15	0	0	0	2	0	1	18
Arabidopsis lyrata	12	0	0	0	1	1	1	15
Amborella trichopoda	3	0	1	1	1	0	1	7
Brassica rapa	18	0	1	5	1	2	1	28
Beta vulgaris	6	0	0	1	0	0	1	8
Citrullus lanatus	7	0	0	2	1	0	1	11
Cucumis melo	10	0	0	0	1	0	1	12
Carica papaya	5	0	0	0	1	0	1	7
Chlamydomonas reinhardtii	1	0	0	1	0	0	0	2
Capsella rubella	10	0	0	0	1	1	1	13
Citrus sinensis	10	0	0	0	1	0	1	12
Eucalyptus grandis	8	1	0	1	1	0	0	11
Fragaria vesca	12	0	0	1	1	0	1	15
Glycine max	18	0	0	2	2	0	4	26
Gossypium raimondii	14	0	0	2	1	0	2	19
Lotus japonicus	4	0	0	1	1	0	0	6
Manihot esculenta	14	0	0	1	1	0	1	17
Medicago truncatula	11	0	0	2	1	0	2	16
Physcomitrella patens	6	0	0	1	2	0	1	10
Prunus persica	8	0	0	0	1	0	2	11
Ricinus communis	5	0	0	2	1	0	1	9
Solanum tuberosum	13	0	0	3	1	0	2	19
Theobroma cacao	10	0	0	0	1	0	1	12
Thellungiella parvula	11	0	0	0	1	1	1	14
Vitis vinifera	4	0	1	4	2	0	0	11
Zea mays	8	0	0	1	1	0	1	11

Segmental, tandem, and whole-genome duplications are critical for both the diversification of gene functions and the rearrangement and expansion of genomes [31–34]. Whole-genome duplication events have occurred in apple [26], and tandem, segmental, and whole-genome duplications have caused some apple gene families to expand, including the MYB [30], MADS-box [31], and WRKY [32] families. We learned here that two *MdSAP* genes have undergone segmental duplication, and 17 have undergone tandem duplication. In addition, multiple gene pairs have each been linked to six potential chromosomal segmental duplications (Figure 5). Similar results have been reported for the Medicago SAP gene family. Our findings suggest that transposition events and the whole-genome and chromosomal segmental duplications have led to the expansion of the apple SAP gene family, and might partially explain why more SAP genes are present in apple than in any other species.

In several distinct species, the zinc finger types of some family members have either disappeared or increased in number. For example, the A20-A20-AN1 zinc finger occurs only in rice and *Eucalyptus grandis*; the A20 type is found in apple, rice, *Amborella trichopoda*, *B. rapa*, and grape (*Vitis vinifera*); the

AN1-AN1-C2H2 zinc-finger exists in *Arabidopsis thaliana*, *A. lyrata*, *B. rapa*, *Capsella rubella*, *Thellungiella parvula*, and desert poplar (Table 3). We might speculate that the loss or the increase in zinc finger types of SAP genes in these genomes means that they are critical for the complicated enzymatic activity that is present in those species. In this study, we determined that apple SAPs are highly and structurally conserved based on analyses of gene structure, conserved domains, sequence alignments, 3D structures, and phylogenetics (Figures 1–4). Similar results have been reported for *Arabidopsis*, rice, maize, tomato, cotton, desert poplar, and medicago [6,10,11,14,24,25].

Although members of the SAP gene family in *Arabidopsis*, rice, tomato, and cotton are arranged into five groups [10,11,24], the SAP proteins expressed in *Arabidopsis*, desert poplar, *Populus trichocarpa*, *Salix purpurea*, and *S. suchowensis* are classified into two major groups: I (Ia–If) and II (IIa and IIb) [6]. Due to the small number of plant species that have been examined, we cannot yet accurately analyze the evolutionary relationships within that gene family. In this study, we were able to divide those members into two major groups (Ia–Id/IIa and IIb) by comparing the apple genome with SAP proteins from 31 other species (Figure 4). We believe that this result from our evolution analysis is convincing. Usually, exon–intron structural diversity can provide important evidence for phylogenetic relationships and play a valuable role in the evolution of gene families [31]. An intronless structure is typical of SAP genes in various species, and is a key characteristic of that family. However, one exception is the grape genome, for which only two VvSAP members lack introns, while 10 members each contain one intron. Furthermore, we discovered here that the intronless gene structure of SAP genes is the dominant arrangement among Group I members, whereas most of the Group II members contain at least one intron (Table 4 and Supplementary File B2). Thus, the prevalence of an intronless gene structure reflects the ancient origin of SAP genes, and links well with their rapid accumulation of transcripts due to reduced post-transcriptional processing [2,35].

Table 4. Statistics for numbers of intronless members within different groups of the SAP gene family.

Group	Ia	Ib	Ic	Id	IIa	IIb
Intronless number	105	43	63	61	1	5
Total number	123	53	74	78	39	34
Percentage (%)	85.36	81.13	85.13	78.2	2.56	14.7

Plant SAPs are quickly induced by multiple abiotic stresses [1,12–14,36–39]. They include rice *OsiSAP1*/*OsSAP1*, which responds to drought, salt, cold, submergence, mechanical wounding, and ABA [1]; and *ZFP177* (*OsSAP9*), which is also from rice, and shows enhanced expression in response to cold, heat, and PEG6000 [12]. The expression of *OsiSAP8* in tobacco and rice is enhanced by salt, cold, heat, desiccation, wounding, submergence, heavy metals, and ABA [36]. Similarly, SAP genes in *Aeluropus littoralis*, banana, *Arabidopsis*, and maize respond to salt, cold, drought, and osmotic stresses in a tissue and stress-specific manner [13,14,16,37–39]. In this study, we comprehensively analyzed the expression patterns of 13 cloned SAP genes under drought stress. Whereas the expression of *MdSAP15*, *-25*, and *-28* was significantly induced, transcripts levels for *MdSAP7* and *-21* mRNAs were significantly reduced (Figure 6B). Our results suggest that these genes have important roles in the response to water deficits.

The constitutive expression of SAP genes confers tolerance to multiple challenges. The overexpression of rice *OsiSAP1*/*OsSAP1*, *OsiSAP8*, and *AlSAP* in tobacco and rice increases their tolerance to numerous abiotic stresses [1,17,36,37,40]. Similar findings have been described for the overexpression of *AtSAP5* in cotton and *Arabidopsis* [13,41]. The overexpression of *AtSAP13* and *MusaSAP1* in *Arabidopsis* and banana leads to greater drought and salt tolerances [16,39]. The downregulation of *PagSAP1* improves salt tolerance in poplar and alters the regulation of genes involved in maintaining cellular ionic homeostasis [18]. We noted here that *MdSAP15* overexpression conferred increased tolerance to osmotic stress by increasing the root lengths and fresh weights of transgenic *Arabidopsis* seedlings when compared with the WT. This overexpression also influenced a range of parameters associated with abiotic stress responses, including REL and the concentrations of chlorophyll, proline, and MDA, all of which are often used to evaluate the degree of plant tolerance under stress conditions [42–44]. Measured values for all of them were favorably affected in our transgenic lines (Figure 7). Finally, our experiments with induced water deficits demonstrated that the transgenic *Arabidopsis* plants showed milder stress symptoms when compared with the WT, and they also had higher survival rates during the period of rehydration and recovery from drought treatment (Figure 8). Taken together, these results indicate that the overexpression of *MdSAP15* in *Arabidopsis* plants leads to enhanced drought tolerance. This work provides a basis for exploring the molecular roles of SAPs and facilitates further investigations into the functions of these genes in abiotic stress responses. Our data also lay a solid foundation for future efforts to introduce improved apple cultivars.

4. Materials and Methods

4.1. Identification of Apple SAP Genes

We downloaded the database of the *Arabidopsis* SAP family from the TAIR website (http://www.arabidopsis.org/) [24]. As query sequences for BlastP (http://www.rosaceae.org/tools/ncbi_blast) against predicted apple proteins, we used 14 *Arabidopsis* SAP proteins and the consensus protein sequences of the A20/AN1 domain hidden Markov model (HMM) profile (A20-like zinc finger, PF01754; AN1 zinc finger, PF01428) from the Pfam database (http://pfam.xfam.org/family/PF01754; http://pfam.xfam.org/family/PF01428). We then searched all of those SAP sequences against the apple genome database (https://www.rosaceae.org/gb/gbrowse/malus_x_domestica/) with HMMER v3.0 and BlastP [31,32]. Confirming the reliability of those protein sequences ensured that the A20 and/or AN1 domains were present in each candidate MdSAP protein. For this, we used the Pfam database (http://pfam.sanger.ac.uk/search) and NCBI Conserved Domain Database (NCBI-CDD; http://www.ncbi.nlm.nih.gov/Structure/cdd/wrpsb.cgi) [34].

4.2. Sequence Alignments and Phylogenetic Analysis

We performed multiple sequence alignments of 30 *MdSAP* protein sequences, using DNAMAN 6.0.3.99 with its default parameters [45]. A phylogenetic tree for the SAP gene family was constructed with MEGA 6.0 software (www.megasoftware.net) and the neighbor-joining (NJ) method, together with the full-length amino acid sequences of 453 SAPs from 32 plant species, including apple (Supplementary Sequence A2). Related sequences were downloaded from the resource Plaza 3.0 (http://bioinformatics.psb.ugent.be/plaza/). We used the following parameters in the NJ method: bootstrap (1000 replicates), complete deletion, and amino: *p*-distance [34].

4.3. Sequence Logos and Structure Model Analysis

Sequence logos for the A20 domain in 26 *MdSAP* genes and the AN1 domain in 28 *MdSAP* genes were generated by the application WebLogo (http://weblogo.threeplusone.com) (Supplementary Sequences A3 and A4). We used the web server SWISS-MODEL (http://swissmodel.expasy.org/) for modeling and predicting the homology of protein structures for those two domains [46]. The proposed 3D structure was modeled on the original NMR structure in PDB ID: 2KZY and 1WFP, and RasTop 2.2 software (http://www.geneinfinity.org/rastop/) was used to present that model [47].

4.4. Analyses of Intron–Exon Structure, Genome Distribution, and Gene Duplications

Genomic sequences (apple v1.0), gene distributions on chromosomes, and genome locations of SAPs in apple and 31 other species were downloaded from Plaza 3.0 (Supplementary Sequence A1) [48]. The structural features of these *MdSAP* genes, including exons/introns, numbers, and locations, were obtained and presented by using the gene structure display server (GSDS) web-based bioinformatics tool (http://gsds.cbi.pku.edu.cn/) [49]. The chromosomal positions of all of the *MdSAP* genes were located via MapInspect (www.plantbreeding.wur.nl/UK/software_mapinspect.html) [50]. Segmental and tandem-duplication events were investigated according to the method of Tian et al. [31].

4.5. Prediction of Cis-Acting Elements in Promoters

To examine the putative *cis*-acting elements in the promoters of apple SAP genes, we isolated sequences that were 1500 bp upstream of the translational start codon, using the contig sequences of that genome and PCR amplification. Details for the promoters used here are listed in Table 5 and Supplementary File B3. Possible *cis*-acting elements in those promoters were then predicted according to the Plant CARE database (http://bioinformatics.psb.ugent.be/webtools/plantcare/html/) [51].

Table 5. Application of primers and sequences. ORF: open reading frames.

Use	Primer Name	Forward Primer (5′–3′)	Reverse Primer (5′–3′)
Complete	MdSAP7	ATGAAAAAAATGGCACAGAGAA	TCAAACCCGGACGATCTTTGCGG
ORF	MdSAP8	ATGGAGCACAATGAGACAGGAT	TCAGATTTTATCCAGCTTTTCT
amplification	MdSAP10	ATGGAGCACGAGGAGACTGGATG	TTAGATTTTATCGAGCTTCTCA
	MdSAP12	ATGGCGGAAGAGCACAGATGCG	TCAAATCTTCTCGAGCTTCTCG
	MdSAP14	ATGAAAAAAATGGCACAGAGAA	TCAGAGCCGGACGATCTTCGCA
	MdSAP15	ATGGAGCACAATGAGACAGGATG	TCAGATTTTATCCAGCTTGTCTG
	MdSAP16	ATGGAATCTCATGATGAAACTG	CTAGATTTTGTCAAGTTTGTCTG
	MdSAP19	ATGGCGGAAGAGCATCGTTGCCA	TCAAATCTTATGCAGCTTCTCCG
	MdSAP21	ATGATGGGAGGAACAGAAGCTT	TCAATACGCTCGAACAGATGGCC
	MdSAP23	ATGGAGCACGAGGAGACTGGATG	TTAGATTTTACCAAGCTTGTCAG
	MdSAP25	ATGGGAACTCCGGAATTCCCAGA	CTATGCTCTTGAAGTACCGCCGT
	MdSAP28	ATGACGGGAGGAACAGAAGCTTT	TCAATAAGCTCGAACAGAAGGC
	MdSAP29	ATGGCGGAAGAGCACAGATGCGA	TCACGGATTTGTACGTTGGCAA
Promoter	MdSAP7	CAGATTTTGTTCAAATGTAGG	TGGGCGATGGAGGAGACAGAAAT
amplification	MdSAP8	TGTTTCAATTGCGTTCTTGAGG	CATTGTAATTCGCTTAAGTTCT
	MdSAP10	ACCTTTTCCAAAACCGTTATTAG	TGCGAAAACCAACAATTAATGG
	MdSAP12		
	MdSAP14	GTAAAGAGGTTAGTGGCCCTGAA	CAAATTCTGATCGATCGATCGAT
	MdSAP15	ATGCGCTTTACTGTTTTTTCAGT	CATTGTAATTCGCTAAGTCCTT
	MdSAP16	CACGAGGAGAGCACTAAAATGGA	CACCAAGAAAACCTCGCCGTTT
	MdSAP19	ACCTTTCTTTTGAGAAGTTTGT	TGCAATTCCAAAACAAATTATTC
	MdSAP21	ATGGATTCTAGTTTGATTTGGGC	GATTTTTCAGTTTGTTAAATTTT
	MdSAP23	ATATTTCCATCACATTGAATAA	CTACTCAGCTTACCTGCAAAGAG
	MdSAP25	GCAGGTAGAGTTTCAAAGTACG	AAATTTTGTATGTACAACACTA
	MdSAP28	ACAGGTCACCGTGGTGACTCCGG	GTCGGTCGGTCGGTCTGGGGTTG
	MdSAP29	GTGCTTTTTGTTGGAACACAAAG	CGATCGAGAGGACAAAAATATTA
qRT-PCR	MdSAP7	TCGTCCGGGTTTGATGATTT	TCCCCGGTCTCTGAATTTCG
	MdSAP8	GGGAAGCGGATAGGAACCAT	CTTGGGAGCTTCAGGAGGAG
	MdSAP10	GATTATCGCACTGCTGGACG	AGTGCTAAGATACCGCTGCA
	MdSAP12	GTTGGTCATAGCCGAGAAGC	ATCAGCTTAATTCCCACGCG
	MdSAP14	GCTCTGACCGGTTTGACAAT	TTGCTGATGATCTCCGGGAG
	MdSAP15	ATGATTACCGGACTGCTGCT	CCACATGGGTAGAAATGAGAGC
	MdSAP16	GCCAATCCTATCGTGAAGGC	GAGACCTATGCAGACAAGAAGC
	MdSAP19	CGATTTCAGAGGGATGGGGA	CAACCATCCCCTACCCCAAT
	MdSAP21	AGGGAAAGAATGCGGGAAGA	CGAAGAAACATGAAACTGCGG
	MdSAP23	GCCAACCCTGTCGTAAAAGC	TGCTAAGATACCGCTGCAGA
	MdSAP25	AATCCAATCCAAGCCTCGGA	TCCCATCCGAATTTTGCACG
	MdSAP28	TGCTTTGAGGGAAGGGAAAGA	ACATCGAATTGTGGAAGCAGA
	MdSAP29	TTCCTCCTCGCACAGATCAG	TCCGCCATGTCTACAGTCAA
	MdMDH	CGTGATTGGGTACTTGGAAC	TGGCAAGTGACTGGGAATGA
pRI-101AN	MdSAP15	TTGATACATATGCCCGTCGAC ATGGAGCACAAT	AGAGTTGTTGATTCAGGATC CTCAGATTTTATC
	35S	CGCACAATCCCACTATCCTT	
	qRT-MdSAP15	AGTCGTTGCAGCATCCATTG	GGAAGCCTGTGTTGAGATAAGC
	AtActin2	GTGAAGGCTGGATTTGCAGGA	AACCTCCGATCCAGACACTGT

4.6. Plant Materials, Growth Conditions, and Stress Treatments

Young roots, stems, and fully expanded leaves, as well as flowers and mature fruit (70 mm, red peel, 150 days after bloom), were collected from apple plants that were five years old after bud grafting.

The scion was *Malus domestica* "Golden Delicious", and the rootstock was *M. hupehensis*. Samples used for examining the effects of water deficits were harvested from plants three months after bud grafting was performed with "Golden Delicious" scions and *M. hupehensis* rootstocks. These grafted plants were grown in pots (height, 320 mm; diameter, 300 mm) in a greenhouse and treatments began when the plants were approximately 500-mm tall. To induce a water deficit, irrigation was withheld from certain plants for up to 8 days while the designated control plants continued to receive normally scheduled irrigation [52]. Our sampling schedule involved harvesting mature leaves at the middle nodes on days 0, 4, and 8 of the deficit period. All of the tissues were frozen immediately in liquid N_2 and stored at −80 °C.

Seedlings of *Arabidopsis thaliana* L. (Heyn), cv. Columbia ("Col"), were used for genetic transformations and assays of osmotic and drought tolerance. They were cultured in a growth chamber under a 16-h photoperiod at 23 °C. For the drought tolerance assay, water was withheld from four-week-old plants for 20 days before they were rewatered. Survival rates were scored 2 days after rewatering began. Well-watered plants were used as the negative control. For the osmotic stress assay, five-day-old seedlings grown on MS agar plates were vertically plated on an MS agar medium supplemented with 0, 200, or 300 mM of mannitol. Their root lengths, fresh weights, relative electrolyte leakage (REL), and concentrations of chlorophyll, malondialdehyde (MDA), and proline were measured 11 days after that transfer. All of the experiments were repeated three times.

4.7. Cloning of MdSAPs and qRT-PCR Analysis

We extracted total RNA from previously frozen apple tissues according to the CTAB method and from *Arabidopsis* leaves, using Trizol reagent (Thermo Fisher Scientific-CN; Shanghai, China; https://www.thermofisher.com/cn/zh/home.html) [53]. Two micrograms of total RNA were collected for synthesizing first-strand cDNA. For cloning *MdSAP*, complete open reading frames were obtained via RT-PCR from fully expanded leaves of "Golden Delicious" apple, using specific primers listed in Table 5. The 5′- and 3′-untranslated regions (UTRs) were obtained with a Rapid Amplification for cDNA Ends kit (TaKaRa, Dalian, China). For the qRT-PCR assays, reverse transcription was performed with 1 μg of total RNA from each sample, followed by PCR-amplification of 1 μL of the product. We conducted the qRT-PCR assays in 20-μL reaction mixtures that contained 10 μL of SYBR® Premix Ex Taq™ (TaKaRa; Beijing, China; http://www.takarabiomed.com.cn), and used an iQ5 instrument (Bio-Rad, Hercules, CA, USA) as described before [34]. Thermal cycling included an initial 3 min at 95 °C; then 40 cycles of 10 s at 95 °C, 30 s at 58 °C, and 15 s at 72 °C; followed by 3 min at 72 °C and then 81 cycles of 7 s each, increasing by an increment of 0.5 °C from 55 °C to 95 °C. Three biological replicates were tested in each assay, and ΔCt values were calculated by using *MdMDH* as our endogenous control [54]. Relative quantification was calculated according to the $2^{-\Delta\Delta Ct}$ method [55], and dissociation curve analysis was performed for determining the specificity of the amplifications.

4.8. Vector Construction and Plant Transformation

To construct the *MdSAP15* overexpression (OE) vectors, we performed RT-PCR to isolate the full-length cDNA of *MdSAP15* from fully expanded leaves of the "Golden Delicious" apple. The cDNA was cloned into pRI 101-AN plant transformation vectors that were driven by the cauliflower mosaic virus (CaMV) 35S promoter. For *Arabidopsis* transformation, the recombinant plasmid described above was introduced into the "Col-0" ecotype via the *Agrobacterium tumefaciens* GV3101-mediated floral dip method. Seeds of the transgenic plants were individually harvested and screened with kanamycin monosulfate. Homozygous transgenic lines were used for further investigations.

4.9. Measurements of Physiological Indices

Relative electrolytic leakage was examined as described by Tan et al. [56]. Chlorophyll concentrations were determined using the protocol of Liang et al. [57], while MDA levels were obtained as described by Wei et al. [42]. Proline concentrations were calculated according to the method described by Dong et al. [43].

4.10. Statistical Analysis

All of the data were analyzed with IBM SPSS Statistics v.20 software (https://www.ibm.com/support/knowledgecenter/SSLVMB_20.0.0/com.ibm.spss.statistics_20.kc.doc/pv_welcome.html). One-way ANOVA and Duncan's tests were used to compare the results. Differences between treatments were considered statistically significant at $p < 0.05$.

Supplementary Materials: Supplementary materials can be found at http://www.mdpi.com/1422-0067/19/9/2478/s1. Supplementary Sequence A1: Genomic information for cloned *MdSAPs*; Supplementary Sequence A2: Full-length amino acid sequences of 453 SAPs from 32 plant species; Supplementary Sequence A3: Sequence logo for the A20 domain in 26 *MdSAP* genes; Supplementary Sequence A4: Sequence logo for the AN1 domain in 28 *MdSAP* genes; Supplementary File B1: The IDs of SAP gene family members in various species; Supplementary File B2: The intron numbers and groups of SAP gene family members in various species; Supplementary File B3: Prediction of *cis*-acting elements in *MdSAP* promoters.

Author Contributions: F.M., K.M. and Q.D. collected the public dataset, performed bioinformatics analysis, and drafted the manuscript. D.D., S.Z., B.X., J.L., Q.W. and D.H. contributed to the bioinformatics analysis and preparation of all figures and tables. Q.D., D.D., C.L. (Changhai Liu), C.L. (Chao Li) and X.G. conducted the experiments. F.M. and Q.D. conceived this study and reviewed the manuscript. All authors have read and approved the final manuscript.

Funding: This work was supported by the State Key Program of the National Natural Science Foundation of China (31330068), the earmarked fund for the China Agriculture Research System (CARS-27), the National Natural Science Foundation of China (31401852 and 31701894), the China Postdoctoral Science Foundation (2017M620474), and the Fundamental Research Funds for the Central Universities (2452017064 and 2452016186).

Acknowledgments: The authors are grateful to Priscilla Licht for help in revising our English composition.

Conflicts of Interest: The authors declare no conflict of interest.

References

1. Mukhopadhyay, A.; Vij, S.; Tyagi, A.K. Overexpression of a zinc-finger protein gene from rice confers tolerance to cold, dehydration, and salt stress in transgenic tobacco. *Proc. Natl. Acad. Sci. USA* **2004**, *101*, 6309–6314. [CrossRef] [PubMed]
2. Giri, J.; Dansana, P.K.; Kothari, K.S.; Sharma, G.; Vij, S.; Tyaqi, A.K. SAPs as novel regulators of abiotic stress response in plants. *BioEssays* **2013**, *35*, 639–648. [CrossRef] [PubMed]
3. Mizoi, J.; Shinozaki, K.; Yamaguchi-Shinozaki, K. AP2/ERF family transcription factors in plant abiotic stress responses. *Biochim. Biophys. Acta* **2012**, *1819*, 86–96. [CrossRef] [PubMed]
4. Nakashima, K.; Takasaki, H.; Mizoi, J.; Shinozaki, K.; Shinozaki, K.; Yamaguchi-Shinozaki, K. NAC transcription factors in plant abiotic stress responses. *Biochim. Biophys. Acta* **2012**, *1819*, 97–103. [CrossRef] [PubMed]
5. Scharf, K.D.; Berberich, T.; Ebersberger, I.; Nover, L. The plant heat stress transcription factor (Hsf) family: Structure, function and evolution. *Biochim. Biophys. Acta* **2012**, *1819*, 104–109. [CrossRef] [PubMed]
6. Jia, H.; Li, J.; Zhang, J.; Ren, Y.; Hu, J.; Lu, M. Genome-wide survey and expression analysis of the stress-associated protein gene family in desert poplar, *Populus euphratica*. *Tree Genet. Genomes* **2016**, *12*, 78. [CrossRef]
7. Giri, J.; Vij, S.; Dansana, P.K.; Tyagi, A.K. Rice A20/AN1 zinc-finger containing stress-associated proteins (SAP1/11) and a receptor-like cytoplasmic kinase (OsRLCK253) interact via A20 zinc-finger and confer abiotic stress tolerance in transgenic *Arabidopsis* plants. *New Phytol.* **2011**, *191*, 721–732. [CrossRef] [PubMed]
8. Dixit, V.M.; Green, S.; Sarma, V.; Holzman, L.B.; Wolf, F.W.; O'Rourke, K.; Ward, P.A.; Prochownik, E.V.; Marks, R.M. Tumor necrosis factor-alpha induction of novel gene products in human endothelial cells including a macrophage-specifc chemotaxin. *J. Biol. Chem.* **1990**, *265*, 2973–2978. [PubMed]
9. Linnen, J.M.; Bailey, C.P.; Weeks, D.L. Two related localized mRNAs from *Xenopus laevis* encode ubiquitin-like fusion proteins. *Gene* **1993**, *128*, 181–188. [CrossRef]

10. Solanke, A.U.; Sharma, M.K.; Tyagi, A.K.; Sharma, A.K. Characterization and phylogenetic analysis of environmental stress-responsive SAP gene family encoding A20/AN1 zinc fnger proteins in tomato. *Mol. Genet. Genom.* **2009**, *282*, 153–164. [CrossRef] [PubMed]
11. Gao, W.; Long, L.; Tian, X.; Jin, J.; Liu, H.; Zhang, H.; Xu, F.; Song, C. Genome-wide identification and expression analysis of stress-associated proteins (SAPs) containing A20/AN1 zinc finger in cotton. *Mol. Genet. Genom.* **2016**, *291*, 2199–2213. [CrossRef] [PubMed]
12. Huang, J.; Wang, M.M.; Jiang, Y.; Bao, Y.M.; Huang, X.; Sun, H.; Xu, D.Q.; Lan, H.X.; Zhang, H.S. Expression analysis of rice A20/AN1-type zinc fnger genes and characterization of *ZFP177* that contributes to temperature stress tolerance. *Gene* **2008**, *420*, 135–144. [CrossRef] [PubMed]
13. Kang, M.; Fokar, M.; Abdelmageed, H.; Allen, R.D. *Arabidopsis* SAP5 functions as a positive regulator of stress responses and exhibits E3 ubiquitin ligase activity. *Plant Mol. Biol.* **2011**, *75*, 451–466. [CrossRef] [PubMed]
14. Xuan, N.; Jin, Y.; Zhang, H.; Xie, Y.; Liu, Y.; Wang, G. A putative maize zinc-finger protein gene, ZmAN13, participates in abiotic stress response. *Plant Cell Tissue Organ Cult.* **2011**, *107*, 101–112. [CrossRef]
15. Gimeno-Gilles, C.; Gervais, M.L.; Planchet, E.; Satour, P.; Limami, A.M.; Lelievre, E. A stress-associated protein containing A20/AN1 zing-finger domains expressed in *Medicago truncatula* seeds. *Plant Physiol. Biochem.* **2011**, *49*, 303–310. [CrossRef] [PubMed]
16. Sreedharan, S.; Shekhawat, U.K.; Ganapathi, T.R. *MusaSAP1*, a A20/AN1 zinc finger gene from banana functions as a positive regulator indifferent stress responses. *Plant Mol. Biol.* **2012**, *80*, 503–517. [CrossRef] [PubMed]
17. Ben Saad, R.; Fabre, D.; Mieulet, D.; Meynard, D.; Dingkuhn, M.; Al-Doss, A.; Guiderdoni, E.; Hassairi, A. Expression of the *Aeluropus littoralis* AlSAP gene in rice confers broad tolerance to abiotic stresses through maintenance of photosynthesis. *Plant Cell Environ.* **2012**, *35*, 626–643. [CrossRef] [PubMed]
18. Yoon, S.K.; Bae, E.K.; Lee, H.; Choi, Y.I.; Han, M.; Choi, H.; Kang, K.S.; Park, E.J. Downregulation of stress-associated protein 1 (PagSAP1) increases salt stress tolerance in poplar (*Populus alba* × *P. glandulosa*). *Trees* **2018**, *32*, 823–833. [CrossRef]
19. Tyagi, H.; Jha, S.; Sharma, M.; Giri, J.; Tyagi, A.K. Rice SAPs are responsive to multiple biotic stresses and overexpression of OsSAP1, an A20/AN1 zinc-fnger protein, enhances the basal resistance against pathogen infection in tobacco. *Plant Sci.* **2014**, *225*, 68–76. [CrossRef] [PubMed]
20. Liu, Y.; Xu, Y.; Xiao, J.; Ma, Q.; Li, D.; Xue, Z.; Chong, K. OsDOG, a gibberellin-induced A20/AN1 zinc-fnger protein, negatively regulates gibberellin-mediated cell elongation in rice. *J. Plant Physiol.* **2011**, *168*, 1098–1105. [CrossRef] [PubMed]
21. Zhang, Y.; Lan, H.; Shao, Q.; Wang, R.; Chen, H.; Tang, H.; Zhang, H.; Huang, J. An A20/AN1-type zinc finger protein modulates gibberellins and abscisic acid contents and increases sensitivity to abiotic stress in rice (*Oryza sativa*). *J. Exp. Bot.* **2016**, *67*, 315–326. [CrossRef] [PubMed]
22. Ströher, E.; Wang, X.J.; Roloff, N.; Klein, P.; Husemann, A.; Dietz, K.J. Redox-dependent regulation of the stress-induced zinc-finger protein SAP12 in *Arabidopsis thaliana*. *Mol. Plant* **2009**, *2*, 357–367. [CrossRef] [PubMed]
23. Kothari, K.S.; Dansana, P.K.; Giri, J.; Tyagi, A.K. Rice stress associated protein 1 (OsSAP1) interacts with aminotransferase (OsAMTR1) and pathogenesis-related 1a protein (OsSCP) and regulates abiotic stress responses. *Front. Plant Sci.* **2016**, *7*, 1057. [CrossRef] [PubMed]
24. Vij, S.; Tyagi, A.K. Genome-wide analysis of the stress associated protein (SAP) gene family containing A20/AN1 zinc-fnger(s) in rice and their phylogenetic relationship with *Arabidopsis*. *Mol. Genet. Genom.* **2006**, *276*, 565–575. [CrossRef] [PubMed]
25. Zhou, Y.; Zeng, L.; Chen, R.; Wang, Y.; Song, J. Genome-wide identification and characterization of stress-associated protein (SAP) gene family encoding A20/AN1 zinc-finger proteins in *Medicago truncatula*. *Arch. Biol. Sci.* **2018**, *70*, 87–98. [CrossRef]
26. Velasco, R.; Zharkikh, A.; Affourtit, J.; Dhingra, A.; Cestaro, A.; Kalyanaraman, A.; Fontana, P.; Bhatnagar, S.K.; Troggio, M.; Pruss, D.; et al. The genome of the domesticated apple (*Malus* × *domestica* Borkh.). *Nat. Genet.* **2010**, *42*, 833–839. [CrossRef] [PubMed]
27. Li, X.; Ling, K.; Zhang, J.; Xie, Y.; Wang, L.; Yan, Y.; Wang, N.; Xu, J.; Li, J.; Li, C.; et al. Improved hybridde novogenome assembly of domesticated apple (*Malus x domestica*). *Gigascience* **2016**, *5*, 35. [CrossRef] [PubMed]

28. Daccord, N.; Celton, J.M.; Linsmith, G.; Becker, C.; Choisne, N.; Schijlen, E.; van de Geest, H.; Bianco, L.; Micheletti, D.; Velasco, R.; et al. High-quality de novo assembly of the apple genome and methylome dynamics of early fruit development. *Nat. Genet.* **2017**, *49*, 1099–1106. [CrossRef] [PubMed]
29. Zhao, T.; Liang, D.; Wang, P.; Liu, J.; Ma, F. Genome-wide analysis and expression profiling of the DREB transcription factor gene family in Malus under abiotic stress. *Mol. Genet. Genom.* **2012**, *287*, 423–436. [CrossRef] [PubMed]
30. Cao, Z.; Zhang, S.; Wang, R.; Zhang, R.; Hao, Y. Genome wide analysis of the apple MYB transcription factor family allows the identification of MdoMYB121 gene conferring abiotic stress tolerance in plants. *PLoS ONE* **2013**, *8*, e69955.
31. Tian, Y.; Dong, Q.; Ji, Z.; Chi, F.; Cong, P.; Zhou, Z. Genome-wide identification and analysis of the MADS-box gene family in apple. *Gene* **2015**, *555*, 277–290. [CrossRef] [PubMed]
32. Gu, Y.; Ji, Z.; Chi, F.; Qiao, Z.; Xu, C.; Zhang, J.; Dong, Q.; Zhou, Z. Bioinformatics and expression analysis of the WRKY gene family in apple. *Sci. Agric. Sin.* **2015**, *48*, 3221–3238.
33. Meng, D.; Li, Y.; Bai, Y.; Li, M.; Cheng, L. Genome-wide identification and characterization of WRKY transcriptional factor family in apple and analysis of their responses to waterlogging and drought stress. *Plant Physiol. Biochem.* **2016**, *103*, 71–83. [CrossRef] [PubMed]
34. Dong, Q.; Zhao, S.; Duan, D.; Tian, Y.; Wang, Y.; Mao, K.; Zhou, Z.; Ma, F. Structural and functional analyses of genes encoding VQ proteins in apple. *Plant Sci.* **2018**, *272*, 208–219. [CrossRef] [PubMed]
35. Jeffares, D.C.; Penkett, C.J.; Bahler, J. Rapidly regulated genes are intron poor. *Trends Genet.* **2008**, *24*, 375–378. [CrossRef] [PubMed]
36. Kanneganti, V.; Gupta, A.K. Overexpression of *OsiSAP8*, a member of stress associated protein (SAP) gene family of rice confers tolerance to salt, drought and cold stress in transgenic tobacco and rice. *Plant Mol. Biol.* **2008**, *66*, 445–462. [CrossRef] [PubMed]
37. Ben Saad, R.; Zouari, N.; Ben Ramdhan, W.; Azaza, J.; Meynard, D.; Guiderdoni, E.; Hassairi, A. Improved drought and salt stress tolerance in transgenic tobacco overexpressing a novel A20/AN1 zinc-finger "*AlSAP*" gene isolated from the halophyte grass *Aeluropus littoralis*. *Plant Mol. Biol.* **2010**, *72*, 171–190. [CrossRef] [PubMed]
38. Kang, M.; Lee, S.; Abdelmageed, H.; Reichert, A.; Lee, H.K.; Fokar, M.; Mysore, K.S.; Allen, R.D. *Arabidopsis* stress associated protein 9 mediates biotic and abiotic stress responsive ABA signaling via the proteasome pathway. *Plant Cell Environ.* **2017**, *40*, 702–716. [CrossRef] [PubMed]
39. Dixit, A.; Tomar, P.; Vaine, E.; Abdullah, H.; Hazen, S.; Dhankher, O.P. A stress-associated protein, AtSAP13, from *Arabidopsis thaliana* provides tolerance to multiple abiotic stresses. *Plant Cell Environ.* **2018**, *41*, 1171–1185. [CrossRef] [PubMed]
40. Dansana, P.K.; Kothari, K.S.; Vij, S.; Tyagi, A.K. *OsiSAP1* overexpression improves water-deficit stress tolerance in transgenic rice by affecting expression of endogenous stress-related genes. *Plant Cell Rep.* **2014**, *33*, 1425–1440. [CrossRef] [PubMed]
41. Hozain, M.; Abdelmageed, H.; Lee, J.; Kang, M.; Fokar, M.; Allen, R.D.; Holaday, A.S. Expression of *AtSAP5* in cotton up-regulates putative stress-responsive genes and improves the tolerance to rapidly developing water deficit and moderate heat stress. *J. Plant Physiol.* **2012**, *169*, 1261–1270. [CrossRef] [PubMed]
42. Wei, Z.; Gao, T.; Liang, B.; Zhao, Q.; Ma, F.; Li, C. Effects of exogenous melatonin on methyl viologen-mediated oxidative stress in apple leaf. *Int. J. Mol. Sci.* **2018**, *19*, 316.
43. Dong, Q.L.; Liu, D.D.; An, X.H.; Hu, D.G.; Yao, Y.X.; Hao, Y.J. *MdVHP1* encodes an apple vacuolar H^+-PPase and enhances stress tolerance in transgenic apple callus and tomato. *J. Plant Physiol.* **2011**, *168*, 2124–2133. [CrossRef] [PubMed]
44. Tu, M.; Wang, X.; Feng, T.; Sun, X.; Wang, Y.; Huang, L.; Gao, M.; Wang, Y.; Wang, X. Expression of a grape (*Vitis vinifera*) bZIP transcription factor, VlbZIP36, in *Arabidopsis thaliana*, confers tolerance of drought stress during seed germination and seedling establishment. *Plant Sci.* **2016**, *252*, 311–323. [CrossRef] [PubMed]
45. Zhou, K.; Hu, L.; Li, P.; Gong, X.; Ma, F. Genome-wide identification of glycosyltransferases converting phloretin to phloridzin in malus species. *Plant Sci.* **2017**, *265*, 131–145. [CrossRef] [PubMed]
46. Liu, Y.; Guan, X.; Liu, S.; Yang, M.; Ren, J.; Guo, M.; Huang, Z.; Zhang, Y. Genome-wide identification and analysis of tcp transcription factors involved in the formation of leafy head in Chinese cabbage. *Int. J. Mol. Sci.* **2018**, *19*, 847. [CrossRef] [PubMed]

47. Gu, Y.B.; Ji, Z.R.; Chi, F.M.; Qiao, Z.; Xu, C.N.; Zhang, J.X.; Zhou, Z.S.; Dong, Q.L. Genome-wide identification and expression analysis of the WRKY gene family in peach. *Hereditas* **2016**, *38*, 254–270. [PubMed]
48. Proost, S.; Van Bel, M.; Vaneechoutte, D.; Van de Peer, Y.; Inzé, D.; Mueller-Roeber, B.; Vandepoele, K. PLAZA 3.0: An access point for plant comparative genomics. *Nucleic Acids Res.* **2015**, *43*, 974–981. [CrossRef] [PubMed]
49. Liu, Q.; Dang, H.; Chen, Z.; Wu, J.; Chen, Y.; Chen, S.; Luo, L. Genome-wide identification, expression, and functional analysis of the sugar transporter gene family in cassava (manihot esculenta). *Int. J. Mol. Sci.* **2018**, *19*, 987. [CrossRef] [PubMed]
50. Mao, K.; Dong, Q.; Li, C.; Liu, C.; Ma, F. Genome wide identification and characterization of apple bHLH transcription factors and expression analysis in response to drought and salt stress. *Front. Plant Sci.* **2017**, *8*, 480. [CrossRef] [PubMed]
51. Wang, P.; Sun, X.; Jia, X.; Ma, F. Apple autophagy-related protein MdATG3s afford tolerance to multiple abiotic stresses. *Plant Sci.* **2017**, *256*, 53–64. [CrossRef] [PubMed]
52. Shao, Y.; Qin, Y.; Zou, Y.; Ma, F. Genome-wide identification and expression profiling of the SnRK2 gene family in *Malus prunifolia*. *Gene* **2014**, *552*, 87–97. [CrossRef] [PubMed]
53. Wang, N.; Guo, T.; Sun, X.; Jia, X.; Wang, P.; Shao, Y.; Liang, B.; Gong, X.; Ma, F. Functions of two *Malus hupehensis* (Pamp.) Rehd. YTPs (MhYTP1 and MhYTP2) in biotic- and abiotic-stress responses. *Plant Sci.* **2017**, *261*, 18–27. [PubMed]
54. Perini, P.; Pasquali, G.; Margis-Pinheiro, M.; de Oliviera, P.R.D.; Revers, L.F. Reference genes for transcriptional analysis of flowering and fruit ripening stages in apple (*Malus* × *domestica* Borkh.). *Mol. Breed.* **2014**, *34*, 829–842. [CrossRef]
55. Livak, K.J.; Schmittgen, T.D. Analysis of relative gene expression data using real-time quantitative PCR and the $2^{-\Delta\Delta CT}$ method. *Methods* **2001**, *25*, 402–408. [CrossRef] [PubMed]
56. Tan, Y.; Li, M.; Yang, Y.; Sun, X.; Wang, N.; Liang, B.; Ma, F. Overexpression of *MpCYS4*, a phytocystatin gene from *Malus prunifolia* (Willd.) Borkh., enhances stomatal closure to confer drought tolerance in transgenic *Arabidopsis* and apple. *Front. Plant Sci.* **2017**, *8*, 33. [CrossRef] [PubMed]
57. Liang, B.; Li, C.; Ma, C.; Wei, Z.; Wang, Q.; Huang, D.; Chen, Q.; Li, C.; Ma, F. Dopamine alleviates nutrient deficiency-induced stress in *malus hupehensis*. *Plant Physiol. Biochem.* **2017**, *119*, 346–359. [CrossRef] [PubMed]

© 2018 by the authors. Licensee MDPI, Basel, Switzerland. This article is an open access article distributed under the terms and conditions of the Creative Commons Attribution (CC BY) license (http://creativecommons.org/licenses/by/4.0/).

International Journal of
Molecular Sciences

MDPI

Article

Genome-Wide Identification and Characterization of Warming-Related Genes in *Brassica rapa* ssp. *pekinensis*

Hayoung Song [1,†], Xiangshu Dong [2,†], Hankuil Yi [1], Ju Young Ahn [1], Keunho Yun [1], Myungchul Song [3], Ching-Tack Han [3] and Yoonkang Hur [1,*]

1 Department of Biological Sciences, Chungnam National University, Daejeon 34141, Korea; hysong@cnu.ac.kr (H.S.); hankuil.yi@cnu.ac.kr (H.Y.); wnduds357@naver.com (J.Y.A.); keunho0307@gmail.com (K.Y.)
2 School of Agriculture, Yunnan University, Kunming 650091, China; dongxiangshu_123@163.com
3 Department of Life Science, Sogang University, Seoul 04107, Korea; s000692@sogang.ac.kr (M.S.); cthan@ccs.sogang.ac.kr (C.-T.H.)
* Correspondence: ykhur@cnu.ac.kr; Tel.: +82-42-821-6279
† These authors contributed equally to this work.

Received: 11 May 2018; Accepted: 6 June 2018; Published: 11 June 2018

Abstract: For sustainable crop cultivation in the face of global warming, it is important to unravel the genetic mechanisms underlying plant adaptation to a warming climate and apply this information to breeding. Thermomorphogenesis and ambient temperature signaling pathways have been well studied in model plants, but little information is available for vegetable crops. Here, we investigated genes responsive to warming conditions from two *Brassica rapa* inbred lines with different geographic origins: subtropical (Kenshin) and temperate (Chiifu). Genes in Gene Ontology categories "response to heat", "heat acclimation", "response to light intensity", "response to oxidative stress", and "response to temperature stimulus" were upregulated under warming treatment in both lines, but genes involved in "response to auxin stimulus" were upregulated only in Kenshin under both warming and minor-warming conditions. We identified 16 putative high temperature (HT) adaptation-related genes, including 10 heat-shock response genes, 2 transcription factor genes, 1 splicing factor gene, and 3 others. *BrPIF4*, *BrROF2*, and *BrMPSR1* are candidate genes that might function in HT adaptation. Auxin response, alternative splicing of *BrHSFA2*, and heat shock memory appear to be indispensable for HT adaptation in *B. rapa*. These results lay the foundation for molecular breeding and marker development to improve warming tolerance in *B. rapa*.

Keywords: warming; BrHSFA2; BrHSP18.2s; transcriptome; alternative splicing; Kenshin

1. Introduction

Global warming poses a serious threat to agriculture, as it threatens crop productivity and food safety worldwide [1,2]. Various strategies have been utilized to facilitate the breeding or engineering of thermotolerant crops, including regulating the expression of heat shock (HS) transcription factor (HSF) and heat shock response (HSR) genes, as well as the use of molecular markers [1,3,4].

In a model plant *Arabidopsis thaliana*, three types of tolerance responses to heat exposure have been identified: basal thermotolerance, acquired thermotolerance, and warming tolerance [5–7]. Plants with basal thermotolerance can survive when grown at 21–22 °C (normal growth condition), exposed to 42–45 °C for 0.5–1 h, and examined after 5–7 days. Plants with acquired thermotolerance can survive when grown under normal conditions, transferred to 36–38 °C (moderate heat stress) for 1.5 h (referred to as "priming"), recovered at 21–22 °C for 2 h, subjected to over 45 °C, and examined after 5–7 days.

Plants with warming tolerance (as opposed to heat-stress tolerance) survive when grown under normal conditions, subjected to 12 °C for 2 days, and treated with warming conditions (27 °C) for 3 h. The long-term adaptation of plants to warmer growth conditions is thought to induce developmental reprograming. Identifying and applying warming-related genes in *Brassica* crop species poses a major challenge for crop breeding for improved tolerance to global warming.

Several marker genes for thermotolerance responses are currently available [8]. Genes encoding an exportin family protein (*XPO1A*, AT5G17020) and heat shock protein (HSP) 101 (*HSP101*, AT1G74310) are markers for basal thermotolerance. Acquired thermotolerance by priming is divided into two categories: short-term and long-term acquired thermotolerance. *HSP101* expression represents short-term acquired thermotolerance, while long-term acquired thermotolerance is characterized by the expression of several marker genes, including those encoding Rotamase FKBP1/FK506-binding protein 62 (*ROF1/FKBP62*, AT3G25230), ROF2/FKBP65 (AT5G48570), heat HSF factor A2 (*HSFA2*, AT2G26150), HSP101, and heat stress-associated 32 kD (*Hsa32*, AT4G21320). During the long-term acquired thermotolerance response, the expression levels of small HSP genes (*sHSPs*), *HSP70s*, *ROS* genes, and ascorbate peroxidase (*APX*) also increase [7].

Warming treatment does not trigger the expression of HSR genes, but other genes, such as *HSP70* (AT3g12580) [6] and *Phytochrome-Interacting Factor4* (*PIF4*, AT2G43010) [9–11], are core components in this process. PIF4 controls morphological acclimation to high temperatures (HT) via auxin [9,12]. Phytochrome B (PhyB) was recently shown to function upstream of PIF4 [13]. Changes in ambient temperatures induce alternative splicing of a large number of genes [14]. Due to increases in global temperatures, the mechanism used by plants to sense small variations in ambient temperatures is becoming an increasing focus of study. It is important to elucidate whether crops that have long been cultivated in regions with different climates, such as Chinese cabbage, have developed similar responses to warming to those found in *Arabidopsis*.

Two inbred Chinese cabbage lines, Chiifu and Kenshin, have different geographic origins: Chiifu originated in temperate regions, whereas Kenshin originated in subtropical and tropical regions. Kenshin has long been used as a breeding stock to develop heat-tolerant *Brassica* species [15,16]. In addition, these two inbred lines show different electrolyte leakage rates in response to HT exposure and different expression of many genes [17]. The long history and intensive breeding of these two Chinese cabbage lines make them promising targets for transcriptome analysis after warming treatment to identify warming-related genes in this crop. These genes could then be used to develop molecular markers and to generate climate-change-resilient *Brassica* crops under global warming conditions. In the current study, we used the Br135K microarray (Version 3) to identify differentially expressed genes (DEGs) upon warming treatment in Chiifu and Kenshin Chinese cabbage, confirmed their expression patterns by qRT-PCR, and further characterized the expression of patterns of several candidate warming-relating genes. The results of this study lay the foundation for breeding Chinese cabbage lines with improved tolerance to warming conditions.

2. Results

2.1. Transcriptome Analysis of Plants under Warming, Minor-Warming, and Low-Temperature (LT) Conditions Using the Br135K Microarray

To identify putative warming-related (or HT adaptation-related) genes, we carried out Br135K microarray analysis of samples from two inbred lines, Chiifu and Kenshin, under three conditions (22 °C, 12 °C, and 12 → 28 °C) (Figure 1). The experiments were repeated twice; the mean values are summarized in Table S2. The microarray data have been deposited in "NCBI (https://www.ncbi.nlm.nih.gov/)" with [geo] GSE113637. Among the 41,173 genes deposited on the Br135K microarray, 14,222 (35%) showed probe intensity (PI) values <500 in all samples, whereas 26,951 (65%) showed PI values >500 in at least one sample. Of these 26,951 genes, 2104 had no *Arabidopsis* counterpart, i.e., NA (nonannotated genes). We subjected these 26,951 genes to further analysis because genes with a PI value of 500 (cutoff value) can easily be examined using standard RT-PCR. Responsive genes (i.e., DEGs) were defined as having at least a 2-fold change (cutoff value) in expression between comparative conditions. As shown in Figure 1, three types of comparisons were made: warming (samples treated for 2 days at 12 °C vs. samples after 3 h exposure to 28 °C), minor-warming (samples grown at 22 °C [control] vs. samples after 3 h exposure to 28 °C), and low-temperature (LT) treatment (samples grown at 22 °C vs. samples treated for 2 days at 12 °C).

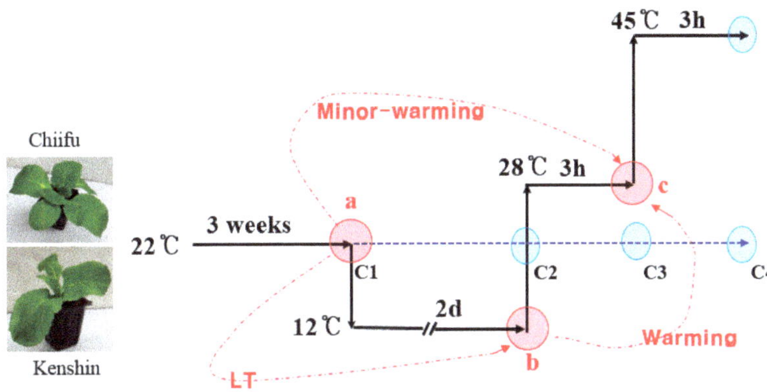

Figure 1. Temperature treatment and sampling schedule. Collection times are indicated by circles. Red circles (**a**–**c**) represent sampling times for the microarray experiments as well as qRT-PCR analysis. Blue circles indicate that the collected samples were only used for qRT-PCR. Shoots from five individual plants were sampled and frozen in liquid nitrogen. Treatments were as follows: (**a**) (22 °C) to (**b**) (12 °C); low-temperature (LT) conditions; (**a**) (22 °C) to (**c**) (28 °C), minor-warming conditions; and (**b**) (12 °C) to (**c**) (28 °C), warming conditions.

2.1.1. Warming-Responsive Genes

We identified 6862 warming-responsive genes. Similar numbers of these genes were upregulated in Chiifu and Kenshin, but more were downregulated in Chiifu (Figure 2A; Table S3). Genes in the Gene Ontology (GO) biological process categories "response to heat", "heat acclimation", "response to light intensity", "response to oxidative stress", and "response to temperature stimulus" were enriched among upregulated genes in both lines (group (c) in Figure 2B,C). The categories "response to hormone stimulus", "plasma membrane", "defense response", and "lipid biosynthetic process" were enriched among genes that were specifically upregulated in Kenshin (group (b) in Figure 2B,C). Genes in the categories "response to water deprivation", "response to osmotic stress", "response to sucrose stimulus", and "response to cold" were enriched among downregulated genes in both lines (group (f) in Figure 2B,C). Genes in two categories, "transcription regulatory activity" and "response to auxin stimulus", were upregulated in Kenshin but downregulated in Chiifu under warming conditions (group (h) in Figure 2B,C).

2.1.2. Minor Warming-Responsive Genes

We identified 6596 minor warming-responsive genes; more of these genes were upregulated in Kenshin and downregulated in Chiifu (Figure 3A; Table S4). Upregulated genes in both lines were enriched in similar categories to those of warming-responsive genes, such as "response to heat", "heat acclimation", "response to light intensity", "response to oxidative stress", and "response to temperature stimulus" (group (c) in Figure 3B,C). Genes upregulated specifically in Kenshin were enriched in the categories "response to stimulus", "response to stress", "transcription factor activity", and "response to water deprivation" (group (b) in Figure 3B,C). Genes in three categories, "response to stimulus", "response to light stimulus", and "response to auxin stimulus", were upregulated in Kenshin but downregulated in Chiifu in response to minor-warming conditions (group (h) in Figure 3B,C). The category "response to auxin stimulus" was enriched among genes upregulated in Kenshin under both warming and minor-warming conditions, suggesting that the auxin response might contribute to heat tolerance in Kenshin.

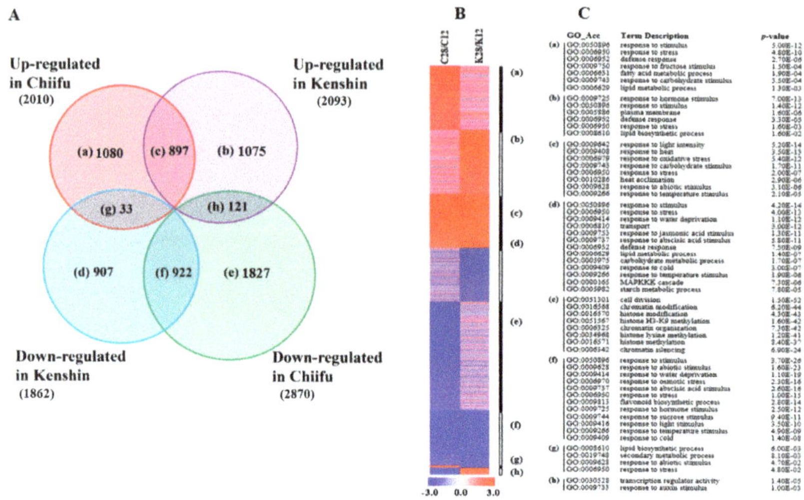

Figure 2. Analysis of warming-responsive genes from two contrasting inbred lines, Chiifu and Kenshin. (**A**) Venn diagram of DEGs with over 2-fold differences in expression; a–h indicate the groups of genes in each category; (**B**) Heatmap of DEGs in each group based on fold change; (**C**) Gene Ontology (GO) enrichment analysis of DEGs in each group, with p values obtained using the agriGO tool (http://bioinfo.cau.edu.cn/agriGO/index.php). C and K of Figure 2B indicate Chiifu and Kenshin, respectively.

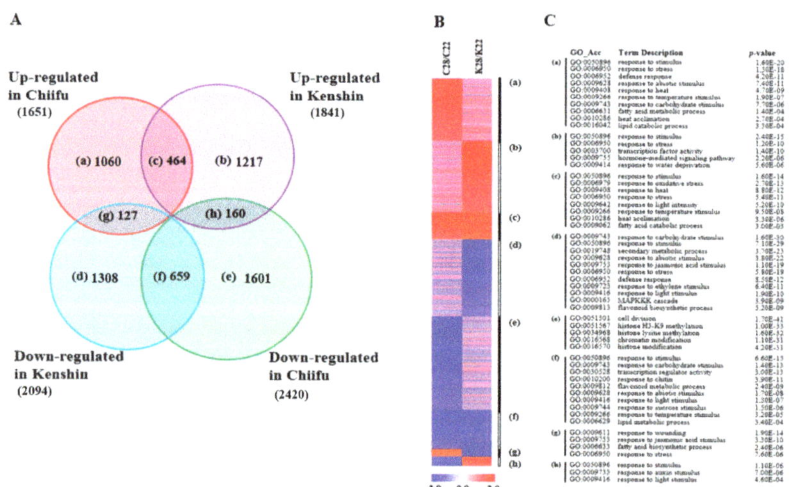

Figure 3. Analysis of minor warming-responsive genes from two contrasting inbred lines, Chiifu and Kenshin. (**A**) Venn diagram of DEGs with over 2-fold differences in expression; a–h indicate the groups of genes in each category; (**B**) Heatmap of DEGs in each group based on fold change; (**C**) GO enrichment analysis of DEGs in each group, with p values obtained using the agriGO tool (http://bioinfo.cau.edu.cn/agriGO/index.php). C and K of Figure 3B indicate Chiifu and Kenshin, respectively.

2.1.3. LT-Responsive Genes

The identification of LT-responsive genes was not the main objective of this study, but associations between cold- and heat-stress signaling through calcium signaling, ROS signaling, and protein degradation have been reported [18–20], and changes in DNA methylation were also shown to be involved in heat tolerance [21]. In addition, we wanted to examine the possible association between LT treatment and the warming response. We identified 5205 LT-responsive genes, for which more genes were upregulated in Kenshin and downregulated in Chiifu (Figure 4; Table S5), implying that Kenshin is more sensitive to LT exposure than Chiifu. Many GO categories were identified as enriched among upregulated genes in both lines, including "response to cold", "response to stress", and "lipid transport and metabolic process". Although similar GO categories were enriched among LT-responsive genes and genes in the two categories mentioned above (warming and minor warming), the expression of warming- and minor warming-responsive genes might not be influenced by LT treatment (Table S5).

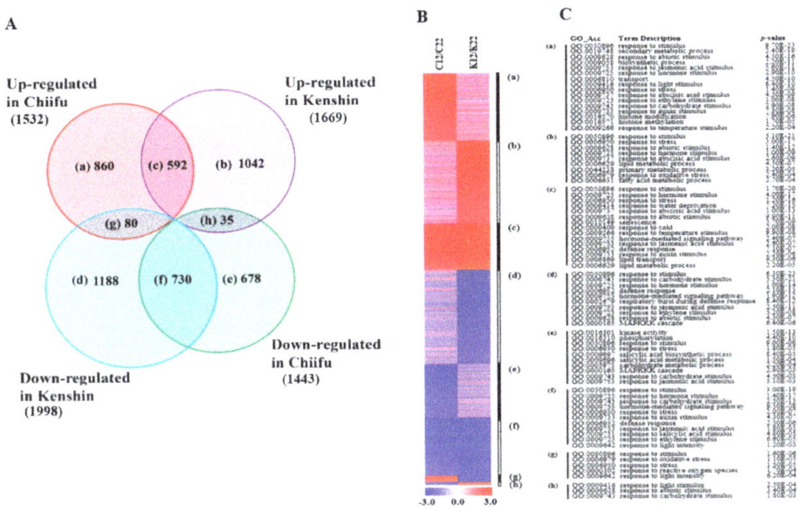

Figure 4. Analysis of LT-responsive genes from two contrasting inbred lines, Chiifu and Kenshin. (**A**) Venn diagram of DEGs with over 2-fold differences in expression; a–h indicate the groups of genes in each category; (**B**) Heatmap of DEGs in each group based on fold change; (**C**) GO enrichment analysis of DEGs in each group, with *p* values obtained using the agriGO tool (http://bioinfo.cau.edu.cn/agriGO/index.php). C and K of Figure 4B indicate Chiifu and Kenshin, respectively.

2.1.4. Genes Upregulated by Both Warming and Minor-Warming Treatment

Among DEGs, we attempted to identify genes that were upregulated by warming, minor warming, and both treatments (Table 1; Tables S6–S14). The numbers of genes upregulated under each condition were the same as those listed in Figures 1–3, but 759 and 726 genes were upregulated by both conditions in Chiifu and Kenshin, respectively (Table S6). We reasoned that these genes might be involved in acquired thermotolerance and/or long-term adaptation to HT in both Chiifu and Kenshin. To specifically identify genes that might be involved in long-term adaptation to HT in the subtropical species Kenshin, we extracted specifically expressed genes (SEGs) from our data set (Tables S8–S13). We identified 85 genes that were specifically upregulated over 2-fold by both minor-warming and warming conditions in Chiifu but not in Kenshin (Table S10). In addition, 86 genes were specifically upregulated over 2-fold by both minor-warming and warming conditions in Kenshin but not in Chiifu (Table S13). Among these, 27 genes were nonannotated and unknown genes. Genes commonly upregulated in both Chiifu and Kenshin under both minor-warming and warming conditions included

15 *HSPs* and chaperone genes (Table S14). We subjected the genes listed in Tables S12–S14 to further GO enrichment analysis (Tables 2–4).

Table 1. Summary of genes upregulated by various treatments. SEGs (specifically expressed genes) represent genes showing an over 2-fold change in expression under the indicated condition but no change or downregulation under the other conditions.

Inbred Line (Treatment)	Comparison	No. of Genes (Table S5)	SEGs (Tables S7–S12)	Comparison	No. of Genes (Table S6)
Chiifu (22 °C → 12 °C → 28 °C)	28 °C/22 °C	1651	121	12 °C/22 °C	1532
	28 °C/12 °C	2010	49	28 °C/12 °C	2010
	Both	759	85	Both	40
Kenshin (22 °C → 12 °C → 28 °C)	28 °C/22 °C	1841	193	12 °C/22 °C	1669
	28 °C/12 °C	2093	146	28 °C/12 °C	2093
	Both	726	86	Both	59

12 °C/22 °C: LT conditions; 28 °C/12 °C: warming conditions; 28 °C/22 °C: minor-warming conditions.

Table 2. Functional classification of genes specifically expressed in response to HT acclimation and/or adaption conditions. The table was constructed based on Tables S11–S13 using agriGO (http://bioinfo.cau.edu.cn/agriGO/) based on GO information for *Arabidopsis* homologs. W, warming, MW, minor warming.

Classification	Kenshin (W)	Kenshin (W + MW)	Kenshin/Chiifu (W + MW)
Heat acclimation	-	1	6
Response to heat	-	-	13
Response to stress	8	6	8
Transcription factor activity	9	9	9
Transferase activity	9	5	12
Transport	7	6	6
Carbohydrate metabolic process	8	4	4
Ligase activity	-	5	1
Lipid biosynthetic process	14	4	3
Oxidation reduction	-	1	2
Response to auxin stimulus	-	3	1
Response to oxidative stress	-	3	2
Response to salicylic acid stimulus	-	3	3
Chromosome organization	-	-	5
Response to hormone stimulus	22	-	-
Intracellular membrane bound organelle	23	-	-
Signal transduction	6	-	-
Primary metabolic process	7	-	-
Catalytic activity	7	-	-
Plasma membrane	3	-	-
Ion binding	5	-	-
Unclassified	4	15	34
Unknown protein	6	5	14
Not annotated	8	16	34
Total	132	86	157

Table 3. Genes associated with the adaptation of Kenshin to HT based on microarray analysis. The selection criteria were (1) intrinsic levels of expression in Kenshin at 22 °C over 2-fold higher than those in Chiifu; and (2) expression levels under both warming and minor-warming conditions at least 2-fold higher in Kenshin than in Chiifu. HSR, heat shock response; TF, transcription factor; SF, splicing factor.

Classification	*At*_Locus	Gene Description	*Br*_SEQ_ID	Expression Level (Probe Intensity)						Fold Change			
				Chiifu (C)			Kenshin (K)			Intrinsic	Warming	Warming	Minor-W
				22 °C	12 °C	28 °C	22 °C	12 °C	28 °C	K22/C22	K28/K12	C28/C12	K28/K22
HSR	AT2G26150	Heat shock transcription factor A2 (HSFA2)	*Bra000557* *	68	223	1256	256	542	1786	3.8	3.3	5.6	7.0
	AT5G62020	Heat shock transcription factor B2A (HSFB2A)	*Bra029292* *	385	374	3164	875	587	3814	2.3	6.5	8.5	4.4
	AT4G25200	Mitochondrion-localized small heat shock protein 23.6 (HSP23.6-MITO)	*Bra013872* *	333	323	1308	984	265	3551	3.0	13.4	4.0	3.6
	AT4G10250	HSP20-like chaperones superfamily protein (HSP22.0)	*Bra027999* *	27	71	378	85	37	677	3.2	18.1	5.3	7.9
	AT1G54050	HSP20-like chaperones superfamily protein	*Bra030910* *	827	945	4041	2406	2649	9204	2.9	3.5	4.3	3.8
	AT2G29500	HSP20-like chaperones superfamily protein	*Bra018383* *	1870	2774	11,166	5211	2428	19,026	2.8	7.8	4.0	3.7
			Bra018384	2575	5095	6789	5906	5362	14,692	2.3	2.7	1.3	2.5
			Bra031725	1996	4409	5909	6292	5275	13,450	3.2	2.5	1.3	2.1
	AT5G51440	HSP20-like chaperones superfamily protein	*Bra029174* *	406	376	5772	1657	292	7627	4.1	26.1	15.3	4.6
	AT5G59720	Heat shock protein 18.2 (HSP18.2)	*Bra002539* *	73	259	747	206	58	5114	2.8	88.5	2.9	24.8
	AT5G12020	17.6 kDa class II heat shock protein (HSP17.6II)	*Bra006137* *	52	271	2513	909	310	2036	17.6	6.6	9.3	2.2
	AT5G48570	FKBP-type peptidyl-prolyl *cis-trans* isomerase family protein (ROF2)	*Bra037477* *	309	283	4967	1632	352	6607	5.3	18.8	17.6	4.0
	AT1G72660	P-loop containing nucleoside triphosphate hydrolases superfamily protein	*Bra016043* *	212	185	2034	967	889	2723	4.6	3.1	11.0	2.8
	AT5G47830	Unknown protein	*Bra020728* *	523	544	4910	1310	1076	4854	2.5	4.5	9.0	3.7
	AT3G14200	Chaperone DnaJ-domain superfamily protein	*Bra027363* *	1333	2129	5365	2857	2634	7477	2.1	2.8	2.5	2.6
TF	AT2G23690	HTH-type transcriptional regulator	*Bra039208*	1122	1383	3449	2567	1072	6076	2.3	5.7	2.5	2.4
	AT5G56840	MYB-like transcription factor family protein	*Bra002790* *	89	154	589	251	144	767	2.8	5.3	3.8	3.0
	AT5G52600	MYB domain protein 82 (MYB82)	*Bra029113*	64	334	132	205	371	1579	3.2	4.3	0.4	7.7
	AT2G24645	Transcriptional factor B3 family protein	*Bra032079*	43	85	77	182	244	1013	4.3	4.2	0.9	5.6
	AT1G70270	Transcription factor	*Bra007905*	246	797	368	1127	952	3529	4.6	3.7	0.5	3.1
	AT5G15150	Homeobox 3 (HB3)	*Bra023506*	197	389	165	389	372	1184	2.0	3.2	0.4	3.0
	AT5G66940	Dof-type zinc finger DNA-binding family protein	*Bra012119*	488	1459	209	1159	983	2746	2.4	2.8	0.1	2.4

Table 3. *Cont.*

Classification	*At*_Locus	Gene Description	*Br*_SEQ_ID	Expression Level (Probe Intensity)						Fold Change			
				Chiifu (C)			Kenshin (K)			Intrinsic	Warming	Warming	Minor-W
				22 °C	12 °C	28 °C	22 °C	12 °C	28 °C	K22/C22	K28/K12	C28/C12	K28/K22
TF	AT3G62090	Phytochrome interacting factor 3-like 2 (PIL2/PIF6)	*Bra007660* *	218	239	537	426	684	1883	2.0	2.8	2.2	4.4
	AT5G10970	C2H2 and C2HC zinc fingers superfamily protein	*Bra009000*	303	776	668	823	878	2180	2.7	2.5	0.9	2.6
	AT1G23380	KNOTTED1-like homeobox gene 6 (KNAT6)	*Bra016348*	140	970	475	273	1004	2202	2.0	2.2	0.5	8.1
	AT4G18610	Light-dependent short hypocotyl 9 (LSH9)	*Bra021000*	83	85	77	189	112	605	2.3	5.4	0.9	3.2
	AT2G42610	Light-dependent short hypocotyl 10 (LSH10)	*Bra016865*	422	617	197	1008	1013	2079	2.4	2.1	0.3	2.1
SF	AT1G65660	Pre-mRNA splicing Prp18-interacting factor (SMP1)	*Bra023741*	141	232	1190	664	589	1712	4.7	2.9	5.1	2.6
Others	AT4G36430	Peroxidase superfamily protein	*Bra017761*	125	381	182	543	686	2278	4.4	3.3	0.5	4.2
	AT1G16530	ASYMMETRIC LEAVES 2-like 9 (LBD3/ASL9)	*Bra026042*	171	393	272	522	824	1729	3.1	2.1	0.7	3.3
			Bra026716	209	325	291	454	534	1063	2.2	2.0	0.9	2.3
	AT5G59670	Leucine-rich repeat protein kinase family protein	*Bra020300*	129	93	80	288	147	622	2.2	4.2	0.9	2.2
	AT4G19530	Disease resistance protein (TIR-NBS-LRR class) family	*Bra027594*	158	206	202	532	536	1591	3.4	3.0	1.0	3.0
	AT2G32660	Receptor like protein 22 (RLP22)	*Bra021803*	163	72	201	960	963	2013	5.9	2.1	2.8	2.1
	AT1G51860	Leucine-rich repeat protein kinase family protein	*Bra030411*	92	75	41	874	1051	1992	9.5	1.9	0.5	2.3
	AT1G53350	Disease resistance protein (CC-NBS-LRR class) family	*Bra037453* *	185	155	420	539	706	1425	2.9	2.0	2.7	2.6
	AT4G08570	Heavy metal transport/detoxification superfamily protein	*Bra037865*	37	45	505	255	67	2044	6.9	30.5	11.1	8.0
	AT5G66110	Heavy metal transport/detoxification superfamily protein (HIPP27)	*Bra009662*	115	329	449	1102	1195	7086	9.5	5.9	1.4	6.4
	AT1G79360	Organic cation/carnitine transporter 2 (OCT2)	*Bra035111*	53	37	23	135	111	568	2.5	5.1	0.6	4.2
	AT2G04100	MATE efflux family protein	*Bra015133*	78	181	255	350	510	1175	4.5	2.3	1.4	3.4
	AT2G35460	Late embryogenesis abundant (LEA) hydroxyproline-rich glycoprotein family	*Bra028562*	215	78	879	575	123	2595	2.7	21.1	11.3	4.5
	AT2G25450	2-oxoglutarate (2OG) and Fe(II)-dependent oxygenase superfamily protein	*Bra021671*	246	68	240	530	644	5146	2.2	8.0	3.5	9.7
	AT1G28030	2-oxoglutarate (2OG) and Fe(II)-dependent oxygenase superfamily protein	*Bra021552*	59	73	121	129	169	1079	2.2	6.4	1.7	8.4
	AT2G39310	Jacalin-related lectin 22 (JAL22)	*Bra005053*	16	99	45	36	109	826	2.3	7.6	0.5	23.1

Table 3. *Cont.*

Classification	*At*_Locus	Gene Description	*Br*_SEQ_ID	Expression Level (Probe Intensity)						Fold Change			
				Chiifu (C)			Kenshin (K)			Intrinsic	Warming	Warming	Minor-W
				22 °C	12 °C	28 °C	22 °C	12 °C	28 °C	K22/C22	K28/K12	C28/C12	K28/K22
Others	AT3G16900	LURP-one-like protein	*Bra021211*	217	205	358	1035	524	2674	4.8	5.1	1.7	2.6
	AT4G36380	Cytochrome P450 superfamily protein (ROT3)	*Bra011678* *	52	139	350	395	205	842	7.6	4.1	2.5	2.1
	AT4G24110	NADP-specific glutamate dehydrogenase	*Bra013763*	92	79	129	204	176	637	2.2	3.6	1.6	3.1
	AT3G51000	Alpha/beta-Hydrolases superfamily protein	*Bra036841*	69	21	23	676	873	3074	9.8	3.5	1.1	4.5
	AT4G22460	Bifunctional inhibitor/lipid-transfer protein/seed storage 2S albumin superfamily protein	*Bra013619*	152	251	304	541	555	1932	3.6	3.5	1.2	3.6
	AT4G13410	Nucleotide-diphospho-sugar transferases superfamily protein (CSLA15)	*Bra008638*	276	385	600	656	523	1384	2.4	2.6	1.6	2.1
	AT5G40650	Succinate dehydrogenase 2-2 (SDH2-2)	*Bra028469*	468	531	545	2517	2395	6064	5.4	2.5	1.0	2.4
	AT5G05390	Laccase 12 (LAC12)	*Bra009111*	279	1072	1201	1118	1176	2906	4.0	2.5	1.1	2.6
	AT3G17820	Glutamine synthetase 1.3 (GLN1.3)	*Bra021276* *	1197	1053	2579	2798	3121	7326	2.3	2.3	2.4	2.6
	AT5G61260	Plant calmodulin-binding protein-related	*Bra029324*	256	194	163	811	1078	2480	3.2	2.3	0.8	3.1
	AT5G64870	SPFH/Band 7/FHB domain-containing membrane-associated protein family	*Bra024333*	270	569	701	559	484	1113	2.1	2.3	1.2	2.0
	AT1G20575	Nucleotide-diphospho-sugar transferases superfamily protein	*Bra025828*	1293	888	2710	3467	3248	7310	2.7	2.3	3.1	2.1
	AT4G39140	RING/U-box superfamily protein	*Bra025860*	27	27	59	2422	2404	5394	88.5	2.2	2.2	2.2
	AT3G09260	Glycosyl hydrolase superfamily protein (BGLU23)	*Bra034060*	278	412	393	1034	995	2106	3.7	2.1	1.0	2.0
	AT3G06550	O-acetyltransferase family protein	*Bra040276*	250	250	310	492	762	1570	2.0	2.1	1.2	3.2
	AT1G29590	Eukaryotic initiation factor 4E protein (eIF4E3)	*Bra032325*	299	445	486	1304	1365	2571	4.4	1.9	1.1	2.0
	AT4G19430	Unknown protein	*Bra013396*	103	16	42	252	17	726	2.5	41.6	2.7	2.9
	NA	NA	*Bra012220*	183	112	97	5190	4062	11,440	28.4	2.8	0.9	2.2
	NA	NA	*Bra025861*	53	59	147	1979	1928	4522	37.3	2.3	2.5	2.3
	NA	NA	*Bra010352*	596	604	1785	4135	4187	9370	6.9	2.2	3.0	2.3

* Gene in Table S13 (genes upregulated in both Chiifu and Kenshin under both minor-warming and warming conditions).

Table 4. Summary of the expression levels of *B. rapa* genes shown to be HT responsive in *Arabidopsis*.

Marker	At_Locus	Gene Description	*Br*_SEQ_ID	Expression Level (Probe Intensity)						Fold Change			
				Chiifu			Kenshin			Chiifu		Kenshin	
				22 °C	12 °C	28 °C	22 °C	12 °C	28 °C	28/22 °C	28/12 °C	28/22 °C	28/12 °C
Basal thermotolerance	AT1G74310	Heat shock protein 101 (HSP101)	*Bra003807*	1457	1411	3457	1696	2238	3594	2.4	2.5	2.1	1.6
			Bra015922	2725	3674	6053	5409	5601	8821	2.2	1.6	1.6	1.6
	AT5G17020	Exportin 1A (XPO1A)	*Bra006382*	9010	8711	9368	8345	9747	11,271	1.0	1.1	1.4	1.2
			Bra008580	10,093	10,241	10,261	9834	8047	12,608	1.0	1.0	1.3	1.6
			Bra023593	6942	7056	7948	6939	6511	7561	1.1	1.1	1.1	1.2
Acquired thermotolerance	AT3G25230	Rotamase FKBP 1 (ROF1/FKBP62)	*Bra013224*	11	18	42	15	59	40	3.8	2.4	2.7	0.7
	AT5G48570	Rotamase FKBP 2 (ROF2/FKBP65)	*Bra037477* *	309	283	4967	1632	352	6607	16.1	17.6	4.0	18.8
	AT2G26150	Heat shock transcription factor A2 (HSFA2)	*Bra000557* *	68	223	1256	256	542	1786	18.5	5.6	7.0	3.3
Warming	AT2G18790	Phytochrome B (PHYB)	*Bra001650*	717	770	829	402	1083	527	1.2	1.1	1.3	0.5
			Bra022192	13,874	10,184	7989	16,575	11,087	13,685	0.6	0.8	0.8	1.2
	AT3G12580	Heat shock protein 70 (HSP70)	*Bra001457*	1560	2512	8099	9758	6732	13,032	5.2	3.2	1.3	1.9
			Bra038734	1049	1159	7871	6488	4093	7597	7.5	6.8	1.2	1.9
	AT2G43010	Phytochrome interacting factor 4 (PIF4)	*Bra000283* *	19,533	12,261	15,413	13,610	9359	24,186	0.8	1.3	1.8	2.6
			Bra037742	3718	4145	3350	1988	3218	4296	0.9	0.8	2.2	1.3
Other HSPs	AT2G25140	Casein lytic proteinase B4 (HSP98.7/CLPB4)	*Bra007816*	645	702	1353	1332	1269	2425	2.1	1.9	1.8	1.9
	AT4G16660	HSP 70 family protein	*Bra038496*	1160	1868	2286	1503	2091	1505	2.0	1.2	1.0	0.7
	AT4G25200	Mitochondrion-localized small heat shock protein 23.6 (HSP23.6-MITO)	*Bra013872* *	333	323	1308	984	265	3551	3.9	4.0	3.6	13.4
	AT4G10250	HSP20-like chaperones superfamily protein (HSP22.0-L)	*Bra000703* *	1174	1055	1638	1352	2898	6069	1.4	1.6	4.5	2.1
			Bra027999 *	27	71	378	85	37	677	14.1	5.3	7.9	18.1
	AT4G27670	Heat shock protein 21 (HSP21)	*Bra026317*	45	166	729	310	54	603	16.1	4.4	1.9	11.2
	AT5G47590	Heat shock protein HSP20/alpha crystallin family	*Bra022051*	1119	1560	1914	1045	3295	2588	1.7	1.2	2.5	0.8
			Bra022079	4462	6564	7455	4309	9320	12,845	1.7	1.1	3.0	1.4
			Bra022083	1201	1103	1815	939	1902	1852	1.5	1.6	2.0	1.0
			Bra022084	1042	1120	1737	760	1635	1540	1.7	1.6	2.0	0.9
	AT1G53540	HSP20-like chaperones superfamily protein	*Bra018216* *	10,773	3454	10,773	7716	3159	16,904	1.0	3.1	2.2	5.4
	AT1G54050	HSP20-like chaperones superfamily protein	*Bra012949*	5473	4945	10796	1568	1799	1843	2.0	2.2	1.2	1.0
			Bra030910 *	827	945	4041	2406	2649	9204	4.9	4.3	3.8	3.5
	AT2G29500	HSP20-like chaperones superfamily protein	*Bra018383* *	1870	2774	11,166	5211	2428	19,026	6.0	4.0	3.7	7.8
			Bra018384	2575	5095	6789	5906	5362	14,692	2.6	1.3	2.5	2.7
			Bra031725	1996	4409	5909	6292	5275	13,450	3.0	1.3	2.1	2.5
	AT4G27890	HSP20-like chaperones superfamily protein	*Bra040837*	77	29	232	128	199	637	3.0	8.1	5.0	3.2
	AT5G51440	HSP20-like chaperones superfamily protein	*Bra029174*	406	376	5772	1657	292	7627	14.2	15.3	4.6	26.1
	AT5G59720	Heat shock protein 18.2 (HSP18.2)	*Bra002539* *	73	259	747	206	58	5114	10.2	2.9	24.8	88.5
			Bra006697 *	1129	779	2431	765	426	5739	2.2	3.1	7.5	13.5
			Bra020295 *	1659	383	1167	2436	197	5383	0.7	3.0	2.2	27.3
	AT5G12020	17.6 kDa class II heat shock protein (HSP17.6II)	*Bra006137* *	52	271	2513	909	310	2036	48.6	9.3	2.2	6.6
			Bra008920	3065	1599	5017	1267	1426	3693	1.6	3.1	2.9	2.6

* Notable genes possibly related to the HT response in *B. rapa*.

2.1.5. Genes Upregulated by Both LT and Warming Treatment

Since both LT (transfer from 22 to 12 °C) and warming (transfer from 12 to 28 °C) are considered to be temperature-stress treatments, we analyzed genes responsive to LT and warming conditions (Table 1; Tables S7 and S15). We expected that genes upregulated by LT would also be upregulated by warming, but only a small number of genes in these categories overlapped: 40 and 59 genes in Chiifu and Kenshin, respectively (Table 1; Table S15). Interestingly, only one gene, *BrHSFA2* (Bra000557, an ortholog of AT2G26150), was upregulated under both conditions as well as in both inbred lines. Therefore, BrHSFA2 represents a candidate temperature-specific regulator in Chinese cabbage.

In contrast to the single upregulated gene, many genes were downregulated by both LT and warming (Table S16). Whereas no gene was upregulated under both LT and warming conditions in Chiifu, 28 genes were downregulated under both conditions in Kenshin, implying that Kenshin is more sensitive to changes in temperature. The expression levels of these 28 genes were highest under normal growth conditions and lowest at 28 °C. Well-known genes in this category include genes encoding HY5-homolog (*HYH*; AT3G17609; Bra022225, Bra021258), gibberellin 2-oxidase 1 (*GA2OX1*; AT1G78440; Bra008362), CONSTANS-like 1 (*COL1*; AT5G15850; Bra023541), and CONSTANS-like 2 (*COL2*; AT3G02380; Bra021464, Bra001043).

2.2. GO Analysis of Warming- and Minor Warming-Responsive Genes

To identify genes associated with acclimation or adaptation to HT, we functionally classified upregulated genes in Kenshin under warming conditions, in Kenshin under both warming and minor-warming conditions, and in both lines under warming and minor-warming conditions (Tables S12–S14) via GO analysis (Table 2). Unexpectedly, no gene was identified in the "heat acclimation" or "response to heat" category among genes only expressed in Kenshin in response to warming conditions, but these categories were enriched among upregulated genes in both lines. Only one "heat acclimation"-related gene (*BrCYP71B2*) was identified in Kenshin under both minor-warming and warming conditions. However, several genes in putative HT adaptation-related gene categories were identified, such as 6 "heat acclimation", 13 "response to heat", and "chromosome organization" genes, among genes upregulated by warming and minor warming in both lines. Genes in the "heat acclimation" category included *BrHSP18.2*, *BrROF2* (FKBP-type peptidyl-prolyl *cis-trans* isomerase family protein), *BrHSP20-L*, *BrHSFA2*, *BrMge1* (*Mitochondrial GrpE2*), and an unknown gene. Genes in the "response to heat" category included most sHSPs, and "chromosome organization" genes included *BrHON4*, *BrSMP1* (*Swellmap 1*), *BrSWIB*, *BrENTG/VHS* family protein, and *BrSWC6* (SWR1 complex subunit 6) (Table S14). Categories that were specifically enriched in Kenshin under warming conditions included "lipid biosynthetic process", "response to hormone stimulus" (most were auxin responsive), "intracellular membrane bound organelle" (most of unknown function), and "signal transduction" (most were defense related) (Table S12).

2.3. Identification of Genes Associated with HT Adaptation in Kenshin

To identify HT adaptation-related genes, we adopted several selection criteria based on the results shown in Table 2. The selection criteria were: (1) expression levels in Kenshin at 22 °C over 2-fold higher than those in Chiifu (we hypothesized that genes essential for long-term adaptation to HT would exhibit high basal levels of expression); and (2) expression levels under both warming and minor-warming conditions at least 2-fold higher in Kenshin than in Chiifu. Sixty-four genes were identified (Table 3), including 16 genes that were upregulated in both lines under minor-warming and warming conditions (asterisks in Table 3): 10 HSR genes, 2 TF (transcription factor) genes, 1 SF (splicing factor) gene, and 3 other genes. The two transcription factor genes were *BrPIF6* (Bra007660; phytochrome-interacting factor 3-like 2; PIL2/PIF6) and *Bra006853* (MYB-like transcription factor family protein). Genes in other categories included *Bra037453* (disease resistance protein (CC-NBS-LRR class) family) for "response to stress", *BrGLN1.3* (Bra021276; glutamine synthetase 1.3) for "ligase activity", and *BrROT3* (Bra011678; cytochrome P450 superfamily protein (ROT3)) for "lipid biosynthesis process". These 16 genes might play important roles in HT adaptation in *B. rapa*. We subjected three of these genes to further analysis: *BrHSFA2*, *BrHSP18.2s*, and *BrSMP1* (Bra023741).

2.4. Comparison of HT-Related Gene Expression between B. rapa and Arabidopsis

Genes involved in thermotolerance (basal thermotolerance, acquired thermotolerance, and warming tolerance) and the associated marker genes are well known in the model plant *Arabidopsis*. To determine whether the same set of genes functions in *B. rapa*, we compared the expression patterns of these genes and other *HSP* genes with our microarray data (Table 4). Warming genes (*PHYB*, *HSP70*, and *PIF4*) identified in *Arabidopsis* were highly expressed in all *B. rapa* samples, with no notable increase in expression upon warming treatment, implying that genes responsible for long-term HT adaptation in *B. rapa* are different from *Arabidopsis* warming genes. In other cases, we assumed differences among samples, such as two contrasting lines in *B. rapa* vs. an ecotype of *Arabidopsis*. *BrPIF4* (Bra000283) expression appeared to be somewhat related to HT adaptation in *B. rapa*. Two Chinese cabbage genes homologous to acquired thermotolerance-related genes in *Arabidopsis*, *BrROF2* and *BrHSFA2*, appear to be critical for warming adaptation in *B. rapa*. The expression levels of several *HSP* genes were also consistent with warming treatment, pointing to their possible involvement in adaptation to HT.

2.5. Confirmation of Microarray Data via qRT-PCR

To confirm the expression levels of the genes detected by microarray analysis, we performed RT-PCR analysis of several selected genes (Figure 5). Although RT-PCR appears to be less sensitive than microarray analysis, RT-PCR results are often used to support microarray data. The expression levels of most of these genes increased upon warming treatment (28 °C), with maximum levels detected at 45 °C. These genes included three heat-acclimation-related genes (*BrHSFA2*, *BrHSFB2A*, and *BrROF2*), various *HSP* genes (especially *sHSPs*), peroxidase family genes, and others. Several genes showed high basal expression levels that further increased upon warming conditions: *BrHSP98.7*, *BrHSP70*, *BrHSP21*, three *BrHSP20Ls* (Bra30910, Bra01883, Bra01884), and *BrMPSR1*. As expected, all of these genes showed higher basal expression levels in Kenshin than in Chiifu.

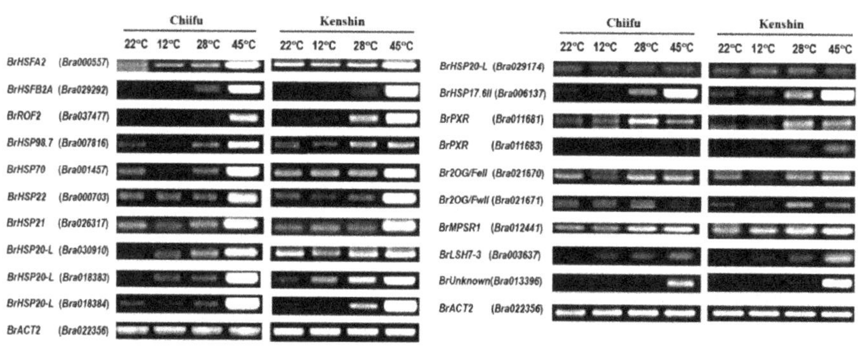

Figure 5. RT-PCR analysis of selected genes identified by microarray analysis. The expression levels of these genes obtained by microarray analysis are summarized in Table S16.

2.6. Expression of BrHSFA2 and BrHSP18.2

Based on our data (Tables 3 and 4) and previous reports [22–24], we selected *BrHSFA2* and *BrHSP18.2A-C* for further analysis of their possible involvement in response to warming and HT conditions. To examine whether *BrHSFA2* undergoes alternative splicing upon HT exposure, as does *Arabidopsis HSFA2*, and whether the intron sequences of this gene are the same in Kenshin and Chiifu, we cloned and sequenced at least 10 clones of the *BrHSFA2* intron region including part of

Exon 1 and Exon 2 from both lines. These 548 bp fragments, including the 337 bp intron sequence, were 100% identical between Chiifu and Kenshin (NCBI accession MH310901, MH310902). We then compared *BrHSFA2* with the homologous sequence from *Arabidopsis* to investigate whether *BrHSFA2* also undergoes alternative splicing (Figure S1). As shown in Figure S1B, *BrHSFA2* might contain a mini-exon with a TAG stop codon, which produces a truncated version of the BrHSFA2 polypeptide via alternative splicing. The truncated version of BrHSFA2 has different C-terminal amino acids from *Arabidopsis* HSFA2 (Figure S1C).

To confirm that alternative splicing occurs in *BrHSFA2*, we carried out qRT-PCR with a reverse primer consisting of possible mini-exon-derived mRNA (Table S1; Figure 6). The levels of an alternatively spliced form of the transcript were higher in Kenshin than in Chiifu upon warming conditions (Figure 6D), while the total transcript levels (full-length + alternatively spliced form) were higher in Chiifu (Figure 6C). The level of full-length mRNA was higher in Kenshin than in Chiifu (Figure 6B). These results indicate that alternative splicing occurs in the intron of *BrHSFA2*, that this process facilitates the expression of full-length *BrHSFA2* as in *Arabidopsis*, and that the levels of the alternatively spliced form of this gene are higher in Kenshin (adapted to HT) than in Chiifu. To examine any association of the alternative splicing of *BrHSFA2* with *BrSMP1*, encoding a spliceosome component and a candidate gene involved in HT adaptation in Kenshin (Table 3), we examined the expression of *BrSMP1* under the same conditions (Figure S3). The expression pattern of *BrSMP1* upon warming and HS was proportional to *BrHSFA2* expression, suggesting the possible involvement of BrSMP1 in alternative splicing of *BrHSFA2* upon HT treatment.

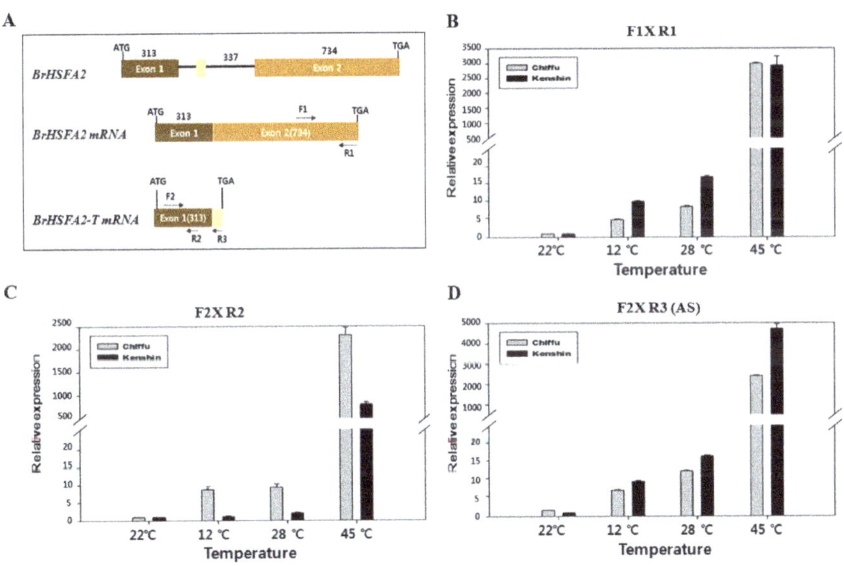

Figure 6. *BrHSFA2* expression in Chiifu and Kenshin during warming and heat shock treatments. qRT-PCR was performed with primer sets described in Table S1 and data analysis was carried out using qPCR value of three replicates. (**A**) Genomic organization of *BrHSFA2* and possible mRNAs with primer positions indicated; (**B**) Full-length *BrHSFA2* mRNA levels; (**C**) *BrHSFA2* mRNA containing both full-length and truncated (alternatively spliced) forms; (**D**) Truncated form of *BrHSFA2* mRNA.

HSFA2 is responsible for maintaining HS memory up to two days in *Arabidopsis* by maintaining histone methylation, thereby enabling the quick induction of HSR genes upon recurring HS [23,24]. HSFA2 has the most pronounced effect on *Arabidopsis HSP18.2* (*Hsp18.1-CI*/ AT5G59720) [23]. *B. rapa* possesses three homologs corresponding to *AtHSP18.2*, *BrHSP18.2A*

(Bra002539), *BrHSP18.2B* (Bra020295), and *BrHSP18.2C* (Bra006697), as listed in order from the highest to lowest identity with *AtHSP18.2*. These three genes encode highly identical polypeptides (93–96% identity) (Figure S2), but we successfully generated primer sets to distinguish each gene (Table S1). A gradual increase in temperature (by 5 °C every 2 h [HS]) strongly upregulated all three genes at 37 °C, with the greatest increase observed for *BrHSP18.2A* and *BrHSP18.2B* in Kenshin (Table 5). However, under warming conditions, compared with 27 °C during HS treatment, tremendous increases were observed in the expression of *BrHSP18.2A* and *BrHSP18.2B* in Kenshin, but there was a several-fold higher increase in *BrHSP18.2C* expression in Chiifu than in Kenshin. These results appear to reflect the different responses of *B. rapa* to HT acclimation compared with *Arabidopsis*, as well as differences between Kenshin and Chiifu.

Table 5. Expression of *BrHSP18.2* family genes in Chiifu and Kenshin under various temperature conditions. Expression level (fold change) was calculated based on qRT-PCR values of three replicates using *BrACT2* as a standard. Heat shock treatment was performed by increasing the temperature 5 °C every 2 h.

Gene	Line	Expression (Fold Change)								
		Heat Shock					Warming			
		22 °C	27 °C	32 °C	37 °C	42 °C	22 °C	12 °C	28 °C	45 °C
BrHSP18.2A	Chiifu	1.0	2.2	10.3	3220.4	4132.9	1.0	0.5	8.4	7154.1
	Kenshin	1.0	0.8	5.5	4921.5	49,617.7	1.0	0.5	106.7	34,142.0
BrHSP18.2B	Chiifu	1.0	0.5	8.7	74.4	357.9	1.0	1.2	1.3	241.4
	Kenshin	1.0	3.9	5.7	160.5	1738.6	1.0	1.8	42.3	31,296.8
BrHSP18.2C	Chiifu	1.0	1.4	13.8	140.0	646.8	1.0	1.8	7.8	7540.8
	Kenshin	1.0	2.3	4.0	84.1	2960.7	1.0	1.5	3.1	733.5

2.7. *BrHSP18.2* Promoter Analysis

The differential expression levels of the three *BrHSP18.2*s upon warming treatment prompted us to analyze cis-elements in their promoters. We designed a common forward primer based on a comparison of known *B. oleracea* genes, *B. napus* genes, three *B. rapa* genes (*BrHSP18.2A*: *Bra002539*, *BrHSP18.2B*: *Bra020295*, *BrHSP18.2C*: *Bra006697*), and AT5G59720, as well as specific reverse primers for each gene (Table S1). We obtained promoter sequences of different sizes (from the ATG start codon): 1257 bp for *BrHSP18.2A*, 969 bp for *BrHSP18.2B*, and 1199 and 1236 bp for *BrHSP18.2C*_Chiifu and Kenshin, respectively (NCBI accession MH310903-8). The promoter sequences of *BrHSP18.2A* and *B* were identical between Chiifu and Kenshin, but the promoter sequences of *BrHSP18.2C* were not similar between the two inbred lines. However, the promoters of the three genes were different from each other but with partially conserved regions.

Despite sequence difference among the *BrHSP18.2A*, *B*, and *C* promoters, four HSE-binding modules [25] were present between −53 and −194 upstream of the ATG start codon: two head-to-head (nGAAnnTTCn) and two tail-to-tail (nTTCnnGAAn) modules (Figure S4). Two modules at −194 to −180 were overlapping. In *Arabidopsis*, eight HSEs (a(g,t,c)GAAn, a(g,t,c)GnAn, or a(g,t,c)Gann) have been detected between −97 and −53 bp in *HSP18.2*, and six HSE deletions were detected, leading to a loss of promoter activity [26]. In *BrHSP18.2*s, seven HSEs were present. The finding that *BrHSP18.2A*, *B*, and *C* possess sufficient numbers of HSEs for HSF binding, as well as possessing identical HSEs, points to the importance of having sufficient numbers of HSFs (or other elements) to control *HSP* expression levels.

3. Discussion

Plants exposed to HT exhibit reduced growth and development, as well as changes in signaling cascades or gene expression, representing adaptive responses to HT [27]. Global climate change has prompted breeders to develop thermotolerant crop varieties, including vegetable crops. Identifying genes that regulate plant responses to warming (or HT adaptation) and elucidating the mechanisms underlying their functions will be crucial for coping with the effects of global warming on agriculture. To identify and characterize warming-associated genes from Chinese cabbage, we subjected Chiifu and Kenshin, two contrasting inbred lines with respect to geographic origin and temperature responsiveness, to transcriptome analysis with a newly developed 3′-tiling microarray (135 K) covering the whole *Brassica rapa* genome using newly designed temperature treatments. Although the warming condition is different from the heat stress, some of HSR genes like *sHSPs* showed similar upregulated patterns as described in previous work [17]. However, this analysis yielded several novel findings, including the discovery of putative HT-adaptive genes, alternative splicing of *BrHSFA2*, and the expression patterns of its target genes.

3.1. Transcriptome Analysis

Of the three treatments, both warming and minor-warming conditions upregulated genes enriched in five biological process categories in both inbred lines, including "response to heat", "heat acclimation", and "response to temperature stimulus" (Figures 2 and 3), indicating that the general responses of both lines are similar. However, genes involved in "response to auxin stimulus" were upregulated by both warming and minor warming in Kenshin, but not in Chiifu. This category of genes might be important for HT adaptation in Chinese cabbage. There is increasing evidence for an association between HT and auxin responses—HT reduces auxin biosynthesis [28], and HSPs such as sHSP22 [29] and cytosolic HSP90 [30] regulate auxin responses.

Our transcriptome analysis indicated that *BrHSFA2* was upregulated or induced in both lines under all three treatment conditions (LT, warming, and minor warming). *Arabidopsis HSFA2* is induced by HT and plays a role in the maintenance of HS memory [26,31,32]. *Arabidopsis HSFA2* expression is also associated with H_2O_2/ROS stress [33] and salt/osmotic stress [34], suggesting that it plays an important role in responses to various stress conditions. The induction of *BrHSFA2* expression by LT treatment suggests that BrHSFA2 targets LT-responsive genes besides HSR genes upon exposure to HT.

3.2. Functional Classification of Putative Warming Genes

As shown in Table 2 and Tables S12–S14, most genes belonging to the GO categories "heat acclimation", "response to heat", and "chromosome organization" were upregulated in both inbred lines, but their levels were higher in Kenshin than in Chiifu. These results suggest that the minor differences in expression of these genes in *B. rapa* could result from long-term adaptation to HS or that the genes identified in *Arabidopsis* could be associated with short-term HS adaptation. *BrCYP71B2* (involved in "heat acclimation") was upregulated in Kenshin under both minor-warming and warming conditions; its homolog is also upregulated upon HS in *Arabidopsis* [35], indicating its possible involvement in the HSR in *B. rapa*. Some putative HT adaptation-related genes have well-known functions in *Arabidopsis*. For example, *Mge1* is induced by heat (under the control of HsfA1) and confers thermotolerance under priming conditions [36]. *Arabidopsis* SMP1 and SWC6 are associated with splicing [37] and chromatin remodeling [38], respectively. These findings suggest that these genes might play a role in HT adaptation in Kenshin. Kenshin-specific genes were classified into four categories ("lipid biosynthetic process", "response to hormone stimulus", "intracellular membrane bound organelle", and "signal transduction") (Table 2; Table S12), but none were found to be involved in HT adaptation, except for auxin-responsive genes. Further studies are needed to investigate the roles of these genes in *B. rapa*.

3.3. Putative HT Adaptation-Related Genes in Kenshin

To identify and characterize putative HT adaptation-related genes in Kenshin, we applied new cutoff criteria (Table 3) compared with known genes from *Arabidopsis* (Table 4) and confirmed their expression by RT-PCR (Figure 5). Unexpectedly, *B. rapa* warming genes appeared to differ from those of *Arabidopsis*, suggesting that different warming adaptation mechanisms or different sets of genes might function in *B. rapa* upon warming conditions. *Arabidopsis* warming genes such as *PHYB*, *HSP70*, and *PIF4* were highly expressed in all *B. rapa* samples, with no notable increase upon warming treatment. Only the expression pattern of *BrPIF4* (Bra000283) appeared to be somewhat related to HT adaptation, and two homologs of acquired thermotolerance-related genes in *Arabidopsis*, *BrROF2* and *BrHSFA2*, appeared to be responsive to warming in *B. rapa*. PIF4, a basic helix-loop-helix (bHLH) transcription factor, is a central regulator of ambient temperature signaling in *Arabidopsis* [9]. PIF4-mediated thermomorphogenesis is associated with the circadian clock [39,40], auxin [9,41,42], other phytohormones [43,44], and epigenetic modification [6]. Quint et al. (2016) [11] indicated that PIF controls thermomorphogenesis via three molecular circuitries: (1) transcriptional regulation of circadian clock genes; (2) post-translation regulation by phosphorylation and degradation; and (3) phytohormonal control through interactions at various levels. These findings and our expression data suggest that BrPIF4 plays a role in the adaptation of Kenshin to HT.

Most warming-responsive genes in Kenshin are orthologs of *Arabidopsis* genes involved in acquired thermotolerance: HSR genes, s*HSPs*, peroxidase family genes, and disease-resistance genes (Table 3; Figure 5). The expression of other *HSP* genes and *BrMPSR1* also increased in response to warming conditions, suggesting their possible involvement in HT adaptation (Figure 5). The roles of a few s*HSPs* in heat tolerance have been examined, including genes encoding HSP 21 (*HSP21*; Bra026317, AT4G27670) [45] and 17.6 kDa class II HSP (*HSP17.6II*; AT5G12020; Bra006137, Bra008920) [29]. However, many s*HSPs*, such as *HSP21*, *HSP22.0*, *HSP18.2*, and *ASCORBATE PEROXIDASE 2* (*APX2*), are HS memory-related genes [22] and are targeted by HSFA2 to help maintain HS memory [31,32]. Class III peroxidases (PRXs) are plant-specific enzymes encoded by multigene families that are involved in lignification, cell elongation, stress responses, and seed germination [46]. Ascorbate peroxidase (APX), a key antioxidant enzyme, participates in various abiotic stress responses and in maintaining cellular homeostasis [47]. In *Arabidopsis*, MPSR1 (Misfolded Protein Sensing RING E3 Ligase 1) is involved in the rapid degradation of misfolded proteins due to protein-damaging stress, thereby controlling proteotoxic stress in the cytoplasm [48]. In *B. rapa*, two *MPSR1* genes (*BrMPSR1-1* (*Bra012441*) and *BrMPSR1-2*(*Bra016290*)) appear to be regulated at the transcriptional level or regulated in evolutionarily divergent ways. These genes might also participate in HT tolerance in Kenshin. Together, these findings support the notion that these genes play a role in long-term adaptation to HT in Kenshin.

3.4. *BrHSFA2 and Its Target BrHSP18.2s*

The expression of *Arabidopsis HSFA2* is dependent on HS (the expression of which is amplified by the production of its alternatively spliced form), increases the expression of target HSR genes such as *HSP18.2*, and confers acquired thermotolerance or HS memory. In the current study, *BrHSFA2* expression and splicing, and the expression of its target gene, *BrHSP18.2s* (Figure 6; Table 5), followed a similar pattern to that of *Arabidopsis* under warming conditions, implying that the warming response of *B. rapa* is similar to acquired thermotolerance in *Arabidopsis*.

Acquired thermotolerance by exposure to moderate HS confers tolerance to normally lethal HT [7]; this thermotolerance is maintained as HS memory for several days [32,49,50]. Three HS memory maintenance-related genes have been identified in *Arabidopsis*, which maintain this memory for several days after the plant returns to nonstress temperature conditions: two days for *HSFA2* [23,24,28], three days for heat stress-associated 32 kD protein gene (*Hsa32*) [49], and three days for *miR156* [22]. The maintenance of HS memory results from the induced hypermethylation of target genes (*HSR* genes), although not all target genes are hypermethylated [51].

Arabidopsis HSFA2 is a key regulator of responses to various types of stress including heat, high light, and ROS stress and is required for extending acquired thermotolerance by maintaining the expression of *HSP* genes [26,28,31]. HSFA2 is a regulatory component responsive to the accumulation of misfolded proteins in the cytosol [52]. HSFA2 also induces abscisic-acid-mediated heat tolerance by upregulating *HSPs* in both a monocot (fescue) and a dicot (*Arabidopsis*) [53]. HSFA2 is responsible for maintaining HS memory up to two days by maintaining histone methylation, thereby inducing HSR gene expression upon recurring HS [23,24]. Many small *HSP* genes, such as *HSP21*, *HSP22.0*, *HSP18.2*, and *ASCORBATE PEROXIDASE 2* (*APX2*), are HS memory-related genes in *Arabidopsis* [22,24], whereas *Hsp70* (AT3G12580) and *Hsp101* (AT1G74310) are non-HS memory-related genes [24]. *Arabidopsis HSFA2* produces an alternatively spliced form (truncated form) upon HS, and this truncated form in turn increases *HSFA2* transcription levels [54]. This scenario appears to operate in *B. rapa* as well, where *BrHSFA2* undergoes alternative splicing, is upregulated, and induces/maintains target gene expression/memory. Alternative splicing by a spliceosome complex is an important mechanism in the sensing of (and adaptation to) small variations in ambient temperature [14], as well as acquired thermotolerance conditions [48], in *Arabidopsis*. These findings imply that temperature changes, including HS and priming, lead to alternative splicing. Alternative splicing might contribute to long-term adaptation to HT in Chinese cabbage, in which thermotolerance does not appear to be due to morphological and architectural changes caused by high ambient temperatures, as found in other plants (thermomorphogenesis) [11].

3.5. *BrHSP18.2s* Promoters and Their Possible Control

The induction of *Arabidopsis HSP18.2* (*Hsp18.1-CI*/AT5G59720) expression by HSFA2, a major thermotolerance HSF [24,26,31], is related to the role of HSFA2 in sustaining H3K4 methylation [24]. The promoter activity of *ArabidopsisHSP18.2* is highest at 35 °C [55]. This promoter contains eight HSE modules between −97 and −53 bp; the deletion of two modules maintains promoter activity, but a deletion of six modules causes a dramatic reduction in promoter activity [26]. All *BrHSP18.2* promoters contain seven HSE modules, implying that there is no difference in BrHSFA2 binding among the three *BrHSP18.2* promoters. In other organisms, at least two nGAAn units arranged head-to-head (nGAAnnTTCn) or tail-to-tail (nTTCnnGAAn) are required in the promoters of *HSPs* for efficient HSF binding [25]. Four HSE units were found in all *BrHSP18.2s* promoters, one of which is overlapping (Figure S4), indicating that *BrHSP18.2A* to *C* contain sufficient numbers of HSEs for TF binding. The binding of HSFA2 with several target genes including *HSP18.2* has been assessed [26], showing that two modules of a TATA-proximal HSE (nGAAnnTTCn) are essential for transcriptional activation by HSFA2. These modules are also conserved in all *BrHSP18.2s* promoters, suggesting that the expression differences among *BrHSP18.2s* upon HT exposure or warming conditions might be due to the presence of different numbers of HSFs such as *BrHSFA2* (which is controlled by alternative splicing) and/or other factor(s).

4. Materials and Methods

4.1. Plant Materials

Seeds of two Chinese cabbage (*Brassica rapa* ssp. *pekinensis*) inbred lines, Chiifu and Kenshin, were kindly provided by Woori Seed Co., Sejong City, Korea. The seeds were sown in a 32-hole tray (6 × 6 × 6 cm × 32 holes) and grown for approximately 3 weeks in a growth chamber at 22 °C under a 16 h light/8 h dark photoperiod with a photon flux density of 140 μmol m^{-2} s^{-1}. For warming treatment, the plants were subjected to 12 °C for 2 days under the same photoperiod and transferred to 28 °C for 3 h. For extreme HS treatment, warming-exposed plants were further incubated at 45 °C in a growth chamber for 3 h. The samples were collected at the end of each treatment (Figure 1). Humidity of growth chambers was set to 70 ± 10%. Shoots from five individual plants were sampled and quickly frozen in liquid nitrogen. To prepare the other experimental samples, plants grown for 3 weeks were subjected to various temperature treatments.

4.2. Br135K Microarray Analysis

The Br135K microarray (Brapa_V3_microarray, 3′-Tiling microarray) is a high-density DNA array prepared with Maskless Array Synthesizer (MAS) technology by NimbleGen (http://www.nimblegen.com/) [55]. Probes were designed from 41,173 genes from *Brassica rapa* accession Chiifu-401-42 (http://brassicadb.org/brad/). All three probes were 60 mers with 30 bp overlaps in 120 bp regions (60 bp of coding sequence plus 60 bp of the 3′UTR for each gene), representing 123,647 features. Fifty features from five markers (*GUS*, *GFP*, *Bar*, *Kan*, *Hyg*) were also included. Total and polysomal RNA were extracted using an RNeasy Mini kit (Qiagen, GmbH, Hilden, Germany) and the RNA protect Reagent (Qiagen), and contaminating DNA was removed by on-column DNase digestion with RNase-free DNase (Promega, Madison, WI, USA). Labeling, data processing, and background correction were performed as described previously [56]. To assess the reproducibility of the microarray analysis, the experiment was repeated using independently prepared total RNA samples from two biological replicates. To obtain insights regarding the putative biological functions and biochemical pathways of the DEGs, enrichment analysis was carried out by searching the GO [57], agriGO [58], and Kyoto Encyclopedia of Genes and Genomes [59] databases.

4.3. RNA Extraction, RT-PCR, and qRT-PCR

Total RNA was extracted from the plant samples using an RNeasy Mini kit (Qiagen). The RNA was treated with RNase-free DNase (Promega) to remove genomic DNA contamination. RT-PCR was performed using an Avian Myeloblastosis Virus (AMV) One-step RT-PCR kit (Takara, Kusatsu, Shiga, Japan). The gene-specific primers used to analyze the selected genes are listed in Table S1. For qRT-PCR, the RNA was subjected to first-strand cDNA synthesis using an Ace-α kit with Oligo-dT primers (Toyobo, Osaka, Japan). The primer sequences were designed according to sequences from the *Brassica* database (BRAD, http://brassicadb.org/brad/). PCR was performed using SYBR® Green Realtime PCR Master Mix-Plus (Toyobo, Japan) under the following cycling conditions: 30 s at 95 °C followed by 30 cycles of 95 °C for 5 s, 58 °C for 10 s, and 72 °C for 15 s.

4.4. Gene Cloning and Sequence Analysis

To analyze the intron sequence of *BrHSFA2* and promoter sequences of *B. rapa* small *HSP18.2* genes (*BrHSP18.2*s), genomic DNA was cloned and analyzed. Genomic DNA was isolated from Chiifu and Kenshin leaves using a DNeasy Plant Mini kit (Qiagen GmbH, Hilden, Germany). Primers were designed based on sequences listed in the BRAD website (Table S1). Genomic PCR was performed under the following conditions: denaturation (5 min at 94 °C), 30 cycles of amplification (30 s at 94 °C, 30 s at 52 °C, and 3 min at 72 °C), and a final extension (7 min at 72 °C). The PCR products were purified using a MEGA-Spin Gel Extraction kit (Intron Biotech. Inc., Sungnam, Korea) and cloned into the TA-vector using a T&A Cloning kit (RBC Bioscience Corp., New Taipei City, Taiwan). *Escherichia coli* (DH5α) cells were transformed with plasmid DNA carrying the desired insert. Plasmid DNA was purified using DNA-Spin (Intron Biotech. Inc., Sungnam, Korea) prior to sequencing (Macrogen, Seoul, Korea). To eliminate PCR and sequencing errors, at least 10 clones per gene were sequenced and analyzed. Any possible PCR and/or sequencing errors were eliminated by aligning independent sequences (http://www.genome.jp/tools-bin/clustalw).

5. Conclusions

This is the first report of the effects of long-term adaptation to warmer growth conditions or HT-adaptation of the gene expression profile in crops. We were able to derive the following conclusions from this study. Many DEGs were overlapping between minor-warming and warming conditions and in both lines examined, Chiifu and Kenshin. Most HT adaptation-associated genes in Chinese cabbage are homologous to acquired thermotolerance-related genes in *Arabidopsis*. Sixteen putative HT adaptation-related genes were identified: 10 HSR genes (including *BrHSFA2* and *sHSPs*), 2 TF genes

(*BrPIF6* and a *BrMyB*), 1 SF gene (Pre-mRNA splicing Prp18-interacting factor, *BrSMP1*), and 3 other genes. Three additional genes, *BrPIF4*, *BrROF2*, and *BrMPSR1*, were also identified as candidate genes involved in HT adaptation. The degree of expression of *HSR* genes such as *BrHSP18.2*s appears to be related to the levels of HSF protein such as BrHSFA2 rather than their own promoter activity. Adaptation to HT in Chinese cabbage appears to be due to changes in the auxin response, increases in the alternative splicing of *BrHSFA2* to amplify its expression, HS memory of HSR genes, and their increases in expression upon recurring HT. The genes identified in this study could be utilized in molecular breeding and marker development after further analysis.

Supplementary Materials: Supplementary materials can be found at http://www.mdpi.com/1422-0067/19/6/1727/s1.

Author Contributions: Conceptualization: H.Y., C.-T.H., Y.H. Data curation: J.Y.A., X.D., C.-T.H. Formal analysis: J.Y.A., M.S. Funding acquisition: Y.H. Investigation: J.Y.A., X.D. Methodology: H.S., M.S., K.Y. Software: X.D., J.Y.A. Supervision: Y.H. Validation: J.Y.A., X.D., H.Y., Y.H. Writing-original draft: X.D., J.Y.A. Writing-review & editing: H.Y., C.-T.H., Y.H.

Acknowledgments: This work was supported by a grant from the Research Fund of Chungnam National University (CNU), Daejeon, Korea, to Yoonkang Hur (2017-1828-01).

Conflicts of Interest: The authors declare no conflict of interest.

References

1. Bita, C.E.; Gerats, T. Plant tolerance to high temperature in a changing environment, scientific fundamentals and production of heat stress-tolerant crops. *Front. Plant Sci.* **2013**, *4*, 273. [CrossRef] [PubMed]
2. Driedonks, N.; Rieu, I.; Vrienzen, W.H. Breeding for plant heat tolerance at vegetative and reproductive stages. *Plant Reprod.* **2016**, *29*, 67–79. [CrossRef] [PubMed]
3. Fragkostefanakis, S.; Röth, S.; Schleiff, E.; Scharf, K.D. Prospects of engineering thermotolerance in crops through modulation of heat stress transcription factor and heat shock protein networks. *Plant Cell Environ.* **2015**, *38*, 1881–1895. [CrossRef] [PubMed]
4. Kole, C.; Muthamilarasan, M.; Henry, R.; Edwards, D.; Sharma, R.; Abberton, M.; Batley, J.; Bentley, A.; Blakeney, M.; Bryant, J.; et al. Application of genomics-assisted breeding for generation of climate resilient crops, progress and prospects. *Front. Plant Sci.* **2015**, *6*, 563. [CrossRef] [PubMed]
5. Larkindale, J.; Vierling, E. Core genome responses involved in acclimation to high temperature. *Plant Physiol.* **2008**, *146*, 748–761. [CrossRef] [PubMed]
6. Kumar, S.V.; Wigge, P.A. H2A.Z-containing nucleosomes mediate the thermosensory response in *Arabidopsis*. *Cell* **2010**, *140*, 136–147. [CrossRef] [PubMed]
7. Mittler, R.; Finka, A.; Goloubinoff, P. How do plants fee the heat? *Trends Biochem. Sci.* **2012**, *37*, 118–125. [CrossRef] [PubMed]
8. Yeh, C.H.; Kaplinsky, N.J.; Hu, C.; Charng, Y.Y. Some like it hot, some like it warm, phenotyping to explore thermotolerance diversity. *Plant Sci.* **2012**, *195*, 10–23. [CrossRef] [PubMed]
9. Koini, M.A.; Alvey, L.; Allen, T.; Tilley, C.A.; Harberd, N.P.; Whitelam, G.C.; Franklin, K.A. High temperature-mediated adaptations in plant architecture require the bHLH transcription factor PIF4. *Curr. Biol.* **2009**, *19*, 408–413. [CrossRef] [PubMed]
10. Kumar, S.V.; Lucyshyn, D.; Jaeger, K.E.; Alos, E.; Alvey, E.; Harberd, N.P.; Wigge, P.A. Transcription factor PIF4 controls the thermosensory activation of flowering. *Nature* **2012**, *484*, 242–245. [CrossRef] [PubMed]
11. Quint, M.; Delker, C.; Franklin, K.A.; Wigge, P.A.; Halliday, K.J.; van Zanten, M. Molecular and genetic control of plant thermomorphogenesis. *Nat. Plants* **2016**, *2*, 15190. [CrossRef] [PubMed]
12. Proveniers, M.C.; van Zanten, M. High temperature acclimation through PIF4 signaling. *Trends Plant Sci.* **2013**, *18*, 59–64. [CrossRef] [PubMed]
13. Jung, J.H.; Domijan, M.; Klose, C.; Biswas, S.; Ezer, D.; Gao, M.; Khattak, A.K.; Box, M.S.; Charoensawan, V.; Cortijo, S.; et al. Phytochromes function as thermosensors in *Arabidopsis*. *Science* **2016**, *354*, 886–889. [CrossRef] [PubMed]

14. Verhage, L.; Severing, E.I.; Bucher, J.; Lammers, M.; Busscher-Lange, J.; Bonnema, G.; Rodenburg, N.; Proveniers, M.C.; Angenent, G.C.; Immink, R.G. Splicing-related genes are alternatively spliced upon changes in ambient temperatures in plants. *PLoS ONE* **2017**, *12*, e0172950. [CrossRef] [PubMed]
15. Hossain, M.M.; Inden, H.; Asahira, T. Interspecific hybrids between *Brassica campestris* L. and *B. oleracea* L. through embryo and ovary culture. *Mem. Coll. Agric. Kyoto Univ.* **1989**, *135*, 21–30.
16. Yamagishi, H.; Hossain, M.M.; Yonezawa, K. Morphology, fertility and cross-compatibility of somatic hybrids between *Brassica oleracea* L. and *B. campestris* L. *Sci. Horticult.* **1994**, *58*, 283–288. [CrossRef]
17. Dong, X.; Yi, H.; Lee, J.; Nou, I.S.; Han, C.T.; Hur, Y. Global gene-expression analysis to identify differentially expressed genes critical for the heat stress response in *Brassica rapa*. *PLoS ONE* **2015**, *10*, e0130451. [CrossRef] [PubMed]
18. Chinnusamy, V.; Zhu, J.; Zhu, J.K. Cold stress regulation of gene expression in plants. *Trends Plant Sci.* **2007**, *12*, 444–451. [CrossRef] [PubMed]
19. Scharf, K.D.; Berberich, T.; Ebersberger, I.; Nover, L. The plant heat stress transcription factor (Hsf) family, structure, function and evolution. *Biochim. Biophys. Acta* **2012**, *1819*, 104–119. [CrossRef] [PubMed]
20. Zhu, J.K. Abiotic stress signaling and response in plants. *Cell* **2016**, *167*, 313–324. [CrossRef] [PubMed]
21. Liu, T.; Li, Y.; Daun, W.; Huang, F.; Hou, X. Cold acclimation alters DNA methylation patterns and confers tolerance to heat and increase growth rate in *Brassica rapa*. *J. Exp. Bot.* **2017**, *68*, 1213–1224. [CrossRef] [PubMed]
22. Stief, A.; Altmann, S.; Hoffmann, K.; Pant, B.D.; Scheible, W.R.; Bäurle, I. *Arabidopsis miR156* regulates tolerance to recurring environmental stress through SPL transcription factors. *Plant Cell* **2014**, *26*, 1792–1807. [CrossRef] [PubMed]
23. Lämke, J.; Brzezinka, K.; Bäurle, I. HSFA2 orchestrates transcriptional dynamics after heat stress in *Arabidopsis thaliana*. *Transcription* **2016**, *7*, 111–114. [CrossRef] [PubMed]
24. Lämke, J.; Brzezinka, K.; Altmann, S.; Bäurle, I. A hit-and-run heat shock factor governs sustained histone methylation and transcriptional stress memory. *EMBO J.* **2016**, *35*, 162–175. [CrossRef] [PubMed]
25. Perisic, O.; Xiao, H.; Lis, J.T. Stable binding of *Drosophia* heat shock factor to head-to-head and tail-to-tail repeats of conserved 5 bp recongnition unit. *Cell* **1989**, *59*, 797–806. [CrossRef]
26. Nishizawa-Yokoi, A.; Yoshida, E.; Yabuta, Y.; Shigeoka, S. Analysis of the regulation of target genes by an *Arabidopsis* heat shock transcription factor, HsfA2. *Biosci. Biotechnol. Biochem.* **2009**, *73*, 890–895. [CrossRef] [PubMed]
27. Hasanuzzaman, M.; Nahar, K.; Alam, M.M.; Roychowdhury, R.; Fujita, M. Physiological, biochemical, and molecular mechanisms of heat stress tolerance in plants. *Int. J. Mol. Sci.* **2013**, *14*, 9643–9684. [CrossRef] [PubMed]
28. Oshino, T.; Miura, S.; Kikuchi, S.; Hamada, K.; Yano, K.; Watanabe, M.; Higashitani, A. Auxin depletion in barley plants under high-temperature conditions represses DNA proliferation in organelles and nuclei via transcriptional alterations. *Plant Cell Environ.* **2011**, *34*, 284–290. [CrossRef] [PubMed]
29. Li, G.; Li, J.; Hao, R.; Guo, Y. Activation of catalase activity by a peroxisome-localized small heat shock protein Hsp17.6CII. *J. Genet. Genom.* **2017**, *44*, 395–404. [CrossRef] [PubMed]
30. Watanabe, E.; Mano, S.; Hara-Nishimura, I.; Nishimura, M.; Yamada, K. HSP90 stabilizes auxin receptor TIR1 and ensures plasticity of auxin responses. *Plant Signal. Behav.* **2017**, *12*, e1311439. [CrossRef] [PubMed]
31. Nishizawa, A.; Yabuta, Y.; Yoshida, E.; Maruta, T.; Yoshimura, K.; Shigeoka, S. *Arabidopsis* heat shock transcription factor A2 as a key regulator in response to several types of environmental stress. *Plant J.* **2006**, *48*, 535–547. [CrossRef] [PubMed]
32. Charng, Y.Y.; Liu, H.C.; Liu, N.Y.; Chi, W.T.; Wang, C.N.; Chang, S.H.; Wang, T.T. A heat-inducible transcription factor, HsfA2, is required for extension of acquired thermotolerance in *Arabidopsis*. *Plant Physiol.* **2007**, *143*, 251–262. [CrossRef] [PubMed]
33. Miller, G.; Mittler, R. Could heat shock transcription factors function as hydrogen peroxide sensors in plants? *Ann. Bot.* **2006**, *98*, 279–288. [CrossRef] [PubMed]
34. Ogawa, D.; Yamaguchi, K.; Nishiuchi, T. High-level overexpression of the *Arabidopsis HsfA2* gene confers not only increased themotolerance but also salt/osmotic stress tolerance and enhanced callus growth. *J. Exp. Bot.* **2007**, *58*, 3373–3383. [CrossRef] [PubMed]

35. Lim, C.J.; Yang, K.A.; Hong, J.K.; Choi, J.S.; Yun, D.J.; Hong, J.C.; Chung, W.S.; Lee, S.Y.; Cho, M.J.; Lim, C.O. Gene expression profiles during heat acclimation in *Arabidopsis thaliana* suspension-culture cell. *J. Plant Res.* **2006**, *119*, 373–383. [CrossRef] [PubMed]
36. Hu, C.; Lin, S.Y.; Chi, W.T.; Charng, Y.Y. Recent gene duplication and subfunctionalization produced a mitochondrial GrpE, the nucleotide exchange factor of the Hsp70 complex, specialized in thermotolerance to chronic heat stress in *Arabidopsis*. *Plant Physiol.* **2012**, *158*, 747–758. [CrossRef] [PubMed]
37. Clay, N.K.; Nelson, T. The recessive epigenetic *swellmap* mutation affects the expression of two stop II splicing factors required for the transcription of the cell proliferation gene *STRUWWELPETER* and for the the timing of cell cycle arrest in the *Arabidopsis* leaf. *Plant Cell* **2005**, *17*, 1994–2008. [CrossRef] [PubMed]
38. Lázaro, A.; Gómez-Zambrano, A.; López-González, L.; Piñeiro, M.; Jarillo, J.A. Mutations in the *Arabidopsis SWC6* gene, encoding a component of the SWR1 chromatin remodelling complex, accelerate flowering time and alter leaf and flower development. *J. Exp. Bot.* **2008**, *59*, 653–666. [CrossRef] [PubMed]
39. Nomoto, Y.; Kubozono, S.; Miyachi, M.; Yamashino, T.; Nakamichi, N.; Mizuno, T. A circadian clock- and PIF4-mediated double coincidence mechanism is implicated in the thermosensitive photoperiodic control of plant architectures in *Arabidopsis thaliana*. *Plant Cell Physiol.* **2012**, *53*, 1965–1973. [CrossRef] [PubMed]
40. Nomoto, Y.; Kubozono, S.; Miyachi, M.; Yamashino, T.; Nakamichi, N.; Mizuno, T. Circadian clock and PIF4-mediated external coincidence mechanism coordinately integrates both of the cues from seasonal changes in photoperiod and temperature to regulate plant growth in *Arabidopsis thaliana*. *Plant Signal. Behav.* **2013**, *8*, e22863. [CrossRef] [PubMed]
41. Franklin, K.A.; Lee, S.H.; Patel, D.; Kumar, S.V.; Spartz, A.K.; Gu, C.; Ye, S.; Yu, P.; Breen, G.; Cohen, J.D.; et al. Phytochrome-interacting factor 4 (PIF4) regulates auxin biosynthesis at high temperature. *Proc. Natl. Acad. Sci. USA* **2011**, *108*, 20231–20235. [CrossRef] [PubMed]
42. Sun, J.; Qi, L.; Li, Y.; Chu, J.; Li, C. PIF4-mediated activation of YUCCA8 expression integrates temperature into the auxin pathway in regulating *Arabidopsis* hypocotyl growth. *PLoS Genet.* **2012**, *8*, e1002594. [CrossRef] [PubMed]
43. Stavang, J.A.; Gallego-Bartolomé, J.; Gómez, M.D.; Yoshida, S.; Asami, T.; Olsen, J.E.; García-Martínez, J.L.; Alabadí, D.; Blázquez, M.A. Hormonal regulation of temperature-induced growth in *Arabidopsis*. *Plant J.* **2009**, *60*, 589–601. [CrossRef] [PubMed]
44. Oh, E.; Zhu, J.Y.; Wang, Z.Y. Interaction between BZR1 and PIF4 integrates brassinosteroid and environmental responses. *Nat. Cell Biol.* **2012**, *14*, 802–809. [CrossRef] [PubMed]
45. Bernfur, K.; Rutsdottir, G.; Emanuelsson, C. The chloroplast-localized small heat shock protein Hsp21 associates with the thylakoid membranes in heat-stressed plants. *Protein Sci.* **2017**, *26*, 1773–1784. [CrossRef] [PubMed]
46. Shigeto, J.; Tsutsumi, Y. Diverse functions and reactions of class III peroxidases. *New Phytol.* **2015**, *209*, 1395–1402. [CrossRef] [PubMed]
47. Pandey, S.; Fartyal, D.; Agarwal, A.; Shukla, T.; James, D.; Kaul, T.; Negi, Y.K.; Arora, S.; Reddy, M.K. Abiotic stress tolerance in plants: Myriad roles of ascorbate peroxidase. *Front. Plant Sci.* **2017**, *8*, 581. [CrossRef] [PubMed]
48. Kim, J.H.; Cho, S.K.; Oh, T.R.; Ryu, M.Y.; Yang, S.W.; Kim, W.T. MPSR1 is a cytoplasmic PQC E3 ligase for eliminating emergent misfolded proteins in *Arabidopsis thaliana*. *Proc. Natl. Acad. Sci. USA* **2017**, *114*, E10009. [CrossRef] [PubMed]
49. Charng, Y.Y.; Liu, H.C.; Liu, N.Y.; Hsu, F.C.; Ko, S.S. *Arabidopsis* Hsa32, a novel heat shock protein, is essential for acquired thermotolerance during long recovery after acclimation. *Plant Physiol.* **2006**, *140*, 1297–1305. [CrossRef] [PubMed]
50. Meiri, D.; Breiman, A. *Arabidopsis* ROF1 (FKBP62) modulates thermotolerance by interacting with HSP90.1 and affecting the accumulation of HsfA2-regulated sHSPs. *Plant J.* **2009**, *59*, 387–399. [CrossRef] [PubMed]
51. Lämke, J.; Bäurle, I. Epigenetic and chromatin-based mechanisms in environmental stress adaptation and stress memory in plants. *Genome Biol.* **2017**, *18*, 124. [CrossRef] [PubMed]
52. Sugio, A.; Dreos, R.; Apricio, F.; Maule, A.J. The cytosolic protein response as a subcomponent of the wider heat shock response in *Arabidopsis*. *Plant Cell* **2009**, *21*, 642–654. [CrossRef] [PubMed]
53. Wang, X.; Zhuang, L.; Shi, Y.; Huang, B. Up-regulation of *HSFA2c* and *HSPs* by ABA contributing to improved heat tolerance in tall fescue and *Arabidopsis*. *Int. J. Mol. Sci.* **2017**, *18*, 1981. [CrossRef] [PubMed]

54. Liu, J.; Sun, N.; Liu, M.; Liu, J.; Du, B.; Wang, X.; Qi, X. An autoregulatory loop controlling *Arabidopsis HsfA2* expression, role of heat shock-induced alternative splicing. *Plant Physiol.* **2013**, *162*, 512–521. [CrossRef] [PubMed]
55. Takahashi, T.; Naito, S.; Komeda, Y. The *Arabidopsis HSP18.2*promoter/*GUS* gene fusion in transgenic *Arabidopsis* plants, a powerful tool for the isolation of regulatory mutants of the heat-shock response. *Plant J.* **1992**, *2*, 751–761. [CrossRef]
56. Jung, H.J.; Dong, X.; Park, J.I.; Thamilarasan, S.K.; Lee, S.S.; Kim, Y.K.; Lim, Y.P.; Nou, I.S.; Hur, Y. Genome-wide transcriptome analysis of two contrasting *Brassica rapa* doubled haploid lines under cold-stresses using Br135K oligomeric chip. *PLoS ONE* **2014**, *9*, e106069. [CrossRef] [PubMed]
57. Ashburner, M.; Ball, C.A.; Blake, J.A.; Botstein, D.; Butler, H.; Cherry, J.M.; Davis, A.P.; Dolinski, K.; Dwight, S.S.; Eppig, J.T.; et al. Gene ontology, tool for the unification of biology. The Gene Ontology Consortium. *Nat. Genet.* **2000**, *25*, 25–29. [CrossRef] [PubMed]
58. Du, Z.; Zhou, X.; Ling, Y.; Zhang, Z.; Su, Z. agriGO, A GO analysis toolkit for the agricultural community. *Nucleic Acids Res.* **2010**, *38*, W64–W70. [CrossRef] [PubMed]
59. Kanehisa, M.; Araki, M.; Goto, S.; Hattori, M.; Hirakawa, M.; Itoh, M.; Katayama, T.; Kawashima, S.; Okuda, S.; Tokimatsu, T.; et al. KEGG for linking genomes to life and the environment. *Nucleic Acids Res.* **2008**, *36*, D480–D484. [CrossRef] [PubMed]

© 2018 by the authors. Licensee MDPI, Basel, Switzerland. This article is an open access article distributed under the terms and conditions of the Creative Commons Attribution (CC BY) license (http://creativecommons.org/licenses/by/4.0/).

International Journal of
Molecular Sciences

MDPI

Article

Genome-Wide Screening and Characterization of the *Dof* Gene Family in Physic Nut (*Jatropha curcas* L.)

Peipei Wang [1,2], Jing Li [1], Xiaoyang Gao [1], Di Zhang [1,2], Anlin Li [1,2] and Changning Liu [1,*]

1 Key Laboratory of Tropical Plant Resource and Sustainable Use, Xishuangbanna Tropical Botanical Garden, Chinese Academy of Sciences, Kunming 650223, China; wangpeipei@xtbg.ac.cn (P.W.); lijing3@xtbg.ac.cn (J.L.); gaoxiaoyang@xtbg.ac.cn (X.G.); zhangdi_net@foxmail.com (D.Z.); lianlin@xtbg.ac.cn (A.L.)

2 Faculty of Life Sciences, University of Chinese Academy of Sciences, Beijing 100049, China

* Correspondence: liuchangning@xtbg.ac.cn; Tel.: +86-691-8713009

Received: 12 April 2018; Accepted: 23 May 2018; Published: 29 May 2018

Abstract: Physic nut (*Jatropha curcas* L.) is a species of flowering plant with great potential for biofuel production and as an emerging model organism for functional genomic analysis, particularly in the Euphorbiaceae family. DNA binding with one finger (Dof) transcription factors play critical roles in numerous biological processes in plants. Nevertheless, the knowledge about members, and the evolutionary and functional characteristics of the *Dof* gene family in physic nut is insufficient. Therefore, we performed a genome-wide screening and characterization of the *Dof* gene family within the physic nut draft genome. In total, 24 *JcDof* genes (encoding 33 JcDof proteins) were identified. All the *JcDof* genes were divided into three major groups based on phylogenetic inference, which was further validated by the subsequent gene structure and motif analysis. Genome comparison revealed that segmental duplication may have played crucial roles in the expansion of the *JcDof* gene family, and gene expansion was mainly subjected to positive selection. The expression profile demonstrated the broad involvement of *JcDof* genes in response to various abiotic stresses, hormonal treatments and functional divergence. This study provides valuable information for better understanding the evolution of *JcDof* genes, and lays a foundation for future functional exploration of *JcDof* genes.

Keywords: *Jatropha curcas*; *Dof* gene family; transcription factor; phylogenetic analysis; gene expression analysis

1. Introduction

Physic nut (*Jatropha curcas* L.) is a perennial small tree in the spurge family, Euphorbiaceae, with a high seed-oil content (40–50%). It can grow easily in barren soil and endure drought and saline environments, thus, having a broad adaptability in various agro-climatic conditions. Given its great potential for biofuel production, nowadays, Physic nut is attracting much attention due to the gradual depletion and a cost increase of fossil energy resources [1,2]. However, there are still a number of challenges in physic nut industries. For example, most physic nut germplasms are monoecious with very low ratios of female to male flowers (approximately female:male = 1:13–29), which considerably reduces the seed yield in physic nut [3,4]. Another serious drawback to the use of physic nut is the presence of toxic components, such as lectin, trypsin inhibitor, and phorbol esters, in all parts of the plant [5]. Therefore, in-depth understanding of the structure and function of key gene families and metabolic pathways of physic nut is essential for improving its crop productivity and commercialization.

Additionally, physic nut is a potential model organism for functional genomic analysis, particularly in Euphorbiaceae. Physic nut is a diploid species ($2n = 22$) [6], with a relatively small genome size (approximately 416Mbp) compared to other members of the Euphorbiaceae [7,8].

According to the most recently updated version, the assembled genome has a total length of 320.5 Mbp consisting of 27,127 putative protein-coding genes, and most of the scaffolds have been anchored on the genetic linkage map [9]. In addition, high-throughput sequencing has mushroomed over the past decade, stimulating the transcriptome profiling analyses. Gene expression profiles of physic nut from different tissues, developmental stages, and biotic/abiotic stresses were constructed [10,11]. Additionally, more than 170 biosamples of physic nut are publicly available from NCBI (National Coalition Building Institute) (as of 5 February 2018). All these genomic and transcriptional data provided a valuable resource for basic and applied studies in physic nut.

Transcription factors (TFs), also known as trans-acting elements, are DNA-binding proteins that specifically bind cis-acting elements in the eukaryotic promoters to activate or inhibit transcriptional regulation [12,13]. As its important roles in the regulation of plant gene expression [14], the structure and function of TFs are now becoming research hotspots in plant molecular biology. The gene expression involves different classes of TFs, which have evolved to regulate a variety of plant-specific genes or signals [15,16]. The *Dof* gene family is a typical example of such TFs. The Dof proteins typically have 200–400 amino acids and contain two main functional domains. One is the Dof domains in the N-terminal domain includes a highly-conserved single zinc-finger structure formed by a CX2CX21CX2C motif, where one Zn^{2+} can covalently combine with four Cys residues. The other one is a regulatory C-terminal domain [17–19]. For instance, the transcription activation domain of the maize *ZmDof1* gene is formed by 44 amino acid residues located at the C-terminal domain [20]. In spite of high level homology in the Dof domain, the rest of the amino acid sequences in the proteins are divergent, coinciding with their expected diverse functions [19].

Previous studies disclosed a variety of roles of Dof proteins in gene expression regulation when associated with plant-specific phenomena, including metabolism, photoperiodic regulation, phytohormone responses, defense responses, and other aspects of plant development [21–23]. Maize *Dof1* (*ZmDof1*) is the first member of the *Dof* gene family identified in plants which may be involved in novel molecular mechanisms underlying tissue-specific and light-regulated gene expression in plants [24]. In addition, recent studies revealed that two closely-related *Arabidopsis thaliana* L. *Dof* genes, *AtDof3.7* (*DAG1*) and *AtDof2.5* (*DAG2*), are maternal genes involved in the control of seed germination, although their actions are opposite [25]. A rice Dof protein, OsDof3, affected the DNA binding of GAMYB to GARE, which is important for combinational regulation of the transcriptional response to Gibberellic acid (GA), and might be a mediator for GA signaling during germination [19,26]. Although none of roles has been confirmed conclusively, Dof proteins apparently participate in the regulation of various processes.

Until now, the *Dof* genes have been identified and characterized in different plant species, such as *A. thaliana*, *Oryza sativa* L. [27], *Glycine max* L. [28], *Triticuma estivum* L. [29], and *Ricinus communis* L. [30]. However, understanding of the *Dof* gene family members and their evolutionary and functional characteristics in physic nut is limited. Based on the genomic and transcriptional data, we focused on the identification, characterization and functional exploration of the *Dof* genes in physic nut. In total, 33 JcDof proteins (encoded by 24 *JcDof* genes) were identified and characterized. These *JcDof* genes were further divided into three major groups by comparison to their orthologs/paralogs in castor bean (*Ricinus communis* L.) and *A. thaliana*. The gene structure, motif, and phylogenetic analysis revealed that genes within each group exhibited similar gene structure and protein motif arrangements. Segmental duplication probably played crucial roles in the expansion of the *JcDof* gene family, and the gene expansion was mainly subjected to positive selection. The expression profile demonstrated the broad involvement of *JcDof* genes in response to various abiotic stresses and hormonal treatments and their possible functional divergence. Taken together, our results provide valuable information for understanding the *JcDof* genes' evolution, and lay a foundation for future functional analysis of the *JcDof* genes.

2. Results

2.1. Identification and Characterization of Dof Genes in Physic Nut

To extensively identify all the Dof candidate members in the physic nut genome, we used a whole-genome scanning to identify genes that encode proteins containing the Dof DNA-binding domain by both BLASTP and HMM profile search. Initially, the Dof protein sequences from *Arabidopsis thaliana* and their HMM profiles of the Dof domain were used as the BLASTP and HMMER query sequences to screen the physic nut genome. Subsequently, it was examined for the presence of the Dof domain using the SMART software and NCBI Conserved Domain database for all the Dof candidate sequences. Eventually, we identified 24 candidates of *Dof* genes in total, represented by 33 transcripts in physic nut (Table S1). Based on their gene loci, we designated each Dof protein uniquely as JcDof-1, and JcDof-2 to JcDof-24.

In addition, we systematically evaluated the basic properties of JcDof protein, including domain position, protein length, molecular weight (Mw), isoelectric point (pI), instability coefficient, and orthologous genes (Table 1). The average length of these Dof protein sequences was 339 amino acid residues and the length mainly centered on the range of 160–518 amino acid residues. Correspondingly, the molecular weights were mainly distributed from 18.2 kDa (JcDof-1) to 55.7 kDa (JcDof-6). The predicted isoelectric point of Dof proteins varied from 4.65 (JcDof-21) to 9.42 (JcDof-3). The instability coefficient of JcDof protein showed a variation from 39.4 (JcDof-17) to 61.74 (JcDof-7.3-5). The location of JcDof protein conserved domain was analyzed by SMART. It was found that the domain positions of JcDof proteins encoded by the same gene (i.e., JcDof proteins that are generated by alternative splicing of the same gene model) were similar, but quite different for those encoded by different genes.

Table 1. The information of the *JcDof* gene family.

Gene ID	Protein Name	Protein Model	Position of Dof Domain	Protein Length	Mw	pI	Instability Index	Orthologous
105649666	JcDof-1	XP_012091770.1	40–98	160	18,217	9.4	49.94	AT1G29160.1
105649561 *	JcDof-2.1	XP_012091631.1	67–125	331	36,503	8.74	58.25	AT1G28310.2
	JcDof-2.2	XP_012091630.1	67–125	365	40,036	8.72	60.68	AT1G28310.2
105649560 *	JcDof-3.1	XP_012091629.1	39–97	321	34,862	9.42	61.1	AT5G60850.1
	JcDof-3.2	XP_012091628.1	39–97	321	34,862	9.42	61.1	AT5G60850.1
105649506	JcDof-4	XP_012091561.1	48–106	283	31,234	8.87	58.45	AT1G28310.2
105645621	JcDof-5	XP_012086658.1	39–97	302	32,787	8.69	56.53	AT4G24060.1
105644078	JcDof-6	XP_012084716.1	148–206	518	55,717	5.88	43.82	AT5G62430.1
105642820 *	JcDof-7.1	XP_012083166.1	38–96	256	27,889	9.04	58.51	AT3G61850.4
	JcDof-7.2	XP_012083165.1	41–99	259	28,222	9.04	57.94	AT3G61850.4
	JcDof-7.3	XP_012083164.1	23–81	272	29,648	8.89	61.74	AT3G61850.4
	JcDof-7.4	XP_012083162.1	23–81	272	29,648	8.89	61.74	AT3G61850.4
	JcDof-7.5	XP_012083161.1	23–81	272	29,648	8.89	61.74	AT3G61850.4
	JcDof-7.6	XP_012083160.1	38–96	287	31,275	8.87	59.86	AT3G61850.4
	JcDof-7.7	XP_012083159.1	41–99	290	31,607	8.87	59.35	AT3G61850.4
105641716	JcDof-8	XP_012081705.1	32–90	344	36,852	8.97	52.37	AT5G65590.1
105640671	JcDof-9	XP_012080436.1	31–89	334	36,528	6.86	57.05	AT5G60850.1
105640546	JcDof-10	XP_012080278.1	121–179	471	51,491	6.61	54.99	AT5G39660.1
105640379	JcDof-11	XP_012080063.1	52–110	312	34,986	6.75	39.5	AT5G62940.1
105639962	JcDof-12	XP_012079559.1	26–84	236	24,419	9.17	48.6	AT3G50410.1
105639655	JcDof-13	XP_012079164.1	63–121	326	34,900	9.08	48.8	AT1G07640.3
105639642	JcDof-14	XP_012079148.1	26–84	246	25,139	8.72	41.9	AT3G50410.1
105636282 *	JcDof-15.1	XP_012074917.1	75–133	353	36,578	9.14	50.3	AT2G37590.1
	JcDof-15.2	XP_012074916.1	84–142	362	37,605	9.14	50.73	AT2G37590.1
105635894	JcDof-16	XP_012074418.1	10–68	282	30,892	5.14	48.76	AT3G52440.1
105633699	JcDof-17	XP_012071724.1	21–79	245	25,883	8.52	39.4	AT1G47655.1
105632564	JcDof-18	XP_012070363.1	101–159	497	53,629	7.78	43.27	AT3G47500.1

Table 1. *Cont.*

Gene ID	Protein Name	Protein Model	Position of Dof Domain	Protein Length	Mw	pI	Instability Index	Orthologous
105631489	JcDof-19	XP_012069011.1	18–76	249	26,497	8.26	47.16	AT3G21270.1
105630455	JcDof-20	XP_012067660.1	129–187	465	51,091	6.8	47.74	AT5G39660.1
105629142	JcDof-21	XP_012066060.1	28–86	287	32,762	4.65	51.26	AT1G21340.1
105628246	JcDof-22	XP_012065018.1	71–129	315	33,921	9.23	51.88	AT2G28810.1
105628152	JcDof-23	XP_012064896.1	36–94	290	32,430	6.65	41.41	AT2G28510.1
105647749	JcDof-24	XP_012089351.1	70–128	338	35,691	9.19	50.23	AT3G55370.3

* These genes are regulated by alternative splicing mechanisms. Mw: Molecular weight; pI: Isoelectric point.

2.2. DNA-Binding Domain Conservation Analysis of JcDof Protein

Dof protein usually has a DNA-binding domain of approximate 40–60 amino acid residues in the N-terminus. This domain contains a highly-conserved CX2CX21CX2C single zinc-finger structure, which is essential for the zinc finger configuration and loop stability. In this study, the conservation of DNA-binding domain of JcDof proteins was analyzed. Multiple protein sequence alignments against *Dof* DNA-binding domain of JcDof proteins revealed that all of them were highly conserved. Especially, we found 20 highly-conserved (100% identical in all 33 JcDof proteins) amino acids CPRC-S–TKFCY-NNY—QPR-FCK-C in the 29 amino acid-long region which corresponded to the CX2CX21CX2C single zinc-finger structure (Figure 1).

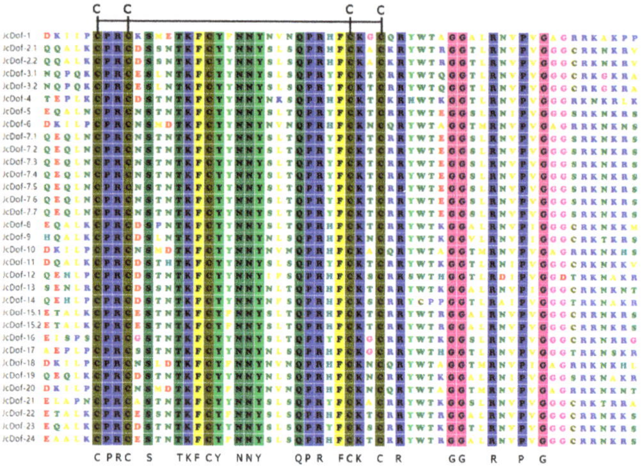

Figure 1. Multiple protein sequences alignments against Dof DNA-binding domain in *JcDof* genes. The identical amino acids are shown in bottom and the four cysteine residues are indicated on top.

2.3. Phylogenetic Analysis and Classification of JcDof Proteins

To explore the phylogenetic relationships of JcDof proteins, we carried out phylogenetic analysis on Dof proteins from physic nut and other two plant species, including *Ricinus communis*, also from the Euphorbiaceae family, and *A. thaliana*, as an outgroup (detailed information on all of the Dof proteins is listed in Supplementary Table S2). A phylogenetic tree was reconstructed including 24 physic nut, 21 *R. communis* and 36 *A. thaliana* Dof proteins (Figure 2). For each gene, we chose the longest protein formed by alternative splicing. The resulting phylogenetic tree was clustered into three major groups (A, B, and C), and they were considered to be evidentfor distinct phylogenetic lineages, which were supported by a bootstrap value over 80%. The two external nodes at the end of the same clades of phylogenetic tree were likely to represent the closest homologous gene pairs.

Of the three major groups, Group C was the first main clade, containing 19 physic nut Dof proteins, 17 *R. communis* Dof proteins, and 25 *A. thaliana* Dof proteins, which were further divided into two sub-groups, C1 and C2, supported by a bootstrap value over 40%. Group A was the second major clade with five physic nut Dof proteins, four *R. Communis* Dof proteins, and seven *A. Thaliana* Dof proteins. Group B was the minimal clade, with only four proteins. Distinguishingly, the Group B Dof proteins were only found in *Arabidopsis*, which could be explained by species/lineage-specific gene gain or loss events. We further checked the GO (Gene Ontology) annotations of these four *Arabidopsis Dof* genes, and found that comparing with the *Arabidopsis Dof* genes in other groups, two of these four genes (*At4g21030, At4g21050*) have some specific annotations, such as "cotyledon development", "mucilage metabolic process involved in seed coat development", "regulation of secondary shoot formation", and "fruit development", which implied the possible function divergence of Dof genes in group B (Supplementary Table S3 for detailed information). The phylogenetic tree showed that Dofs in the Group A and C were duplicated several times before the divergence of these three species, and were highly conserved among *J. curcas*, *R. communis*, and *A. thaliana*. In addition, the physic nut Dof proteins were more closely related, evolutionarily, to *R. communis* than to the *Arabidopsis* Dof proteins.

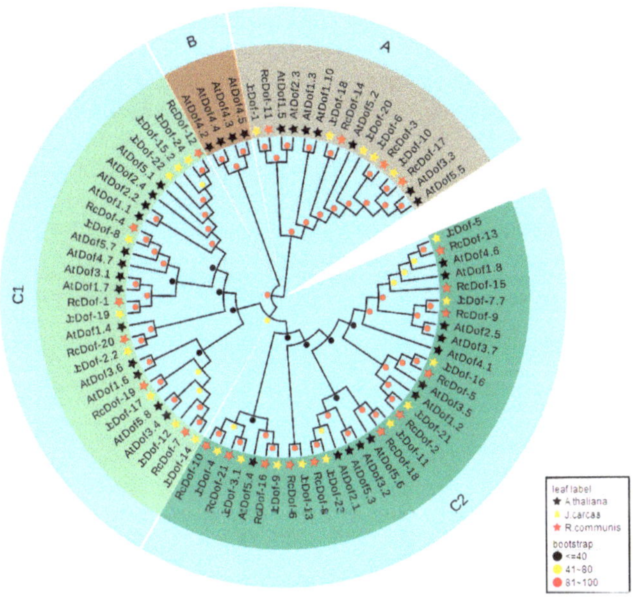

Figure 2. Phylogenetic relationships among *J. curcas, A. thaliana,* and *R. communis* Dof proteins. The neighbor-joining tree was created using the MEGA6.0 program (bootstrap value set at 1000). Thirty-six (36) AtDof proteins marked with black pentacle, 24 JcDof proteins marked with yellow pentacle, and 21RcDof proteins marked with red pentacle. The resulting phylogenetic tree was clustered into three major groups (A, B, and C), which were supported by a bootstrap value over 80%. The Dof proteins in Group C were further divided into two sub-groups, C1 and C2, supported by a bootstrap value over 40%. The detailed information of all the Dof proteins is listed in Supplementary Table S2.

2.4. *JcDof* Gene Structures and Conserved Motifs in *JcDof* Proteins

Introns and exons are the backbones of genes. Their numbers and distribution patterns are an evolutionary mark for a gene family. We, therefore, compared the intron-exon structure of each *JcDof* gene. The results revealed that the gene structure pattern was consistent with the phylogenetic analysis. Based on the exon-intron structures, the number of introns varied from one to three in *J. curcas* (Figure 3b). There are ten *JcDof* genes with one intron (41.7%), 12 *JcDof* genes with two introns (50%), and two *JcDof* genes with three introns (8.3%). All of the *JcDof* genes in subfamily A possessed two introns, while the number of introns of the *JcDof* gene in subfamily C varied from one to three.

Our classification of *Dof* genes was also verified by the conserved motif analysis. All of the Dof protein sequences were loaded into the MEME analysis tool to identify the conserved motifs. As a result, a total of ten conserved motifs were observed, which were statistically-significant with *E*-values less than 1×10^{-40} (Figure 3a, described in detail in Supplementary Figure S1 and Table S4). The motifs of Dof proteins identified by MEME were between 13–43 amino acids in length. Among them, Motif-1 is a common motif in all Dof proteins, corresponding to the CX2CX21CX2C single zinc-finger structure in the Dof domain, which was the highly-homologous core region of *Dof* family (Figure 3c). While all of the Group B proteins and many of the Group C1 and C2 proteins only contain Motif-1, some Dof proteins have extra specific motifs, which may be relevant to different functions. The Dof proteins from Group A had the most complicated motif patterns, and Motif-2, Motif-4, Motif-5, and Motif-9 were specific for them. While Group C members have relatively simple motif patterns compared with Group A, they also had group-specific motifs, such as Motif-6, Motif-8, and Motif-10, but not all the group members have these specific motifs. For further elucidation of the potential roles of the Group A specific motifs, we checked the GO annotations of the Group A genes in *Arabidopsis*. Interestingly, we found that comparing with the *Arabidopsis Dof* genes in other groups, most of the genes in Group A (5 out of 7) have some flower-development-related annotations, such as "flower development", "negative regulation of long-day photo periodism", "flowering", "negative regulation of short-day photo periodism", "regulation of timing of transition from vegetative to reproductive phase", and "vegetative to reproductive phase transition of meristem", which implied the possible function divergence of the *Dof* genes in group A (see Supplementary Table S3 for detailed information).

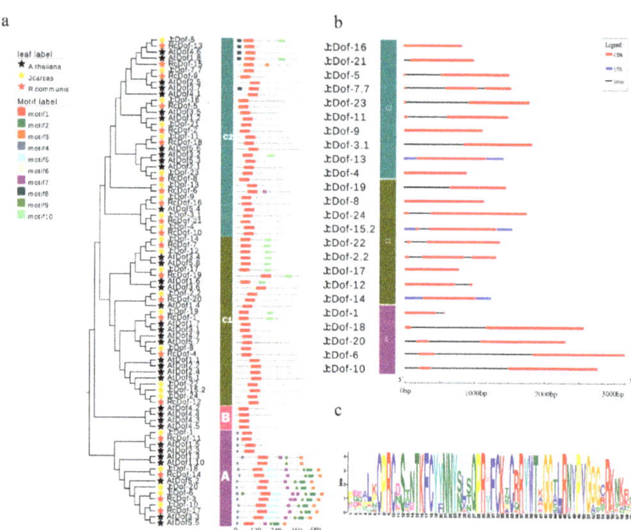

Figure 3. *JcDof* gene structures and conserved motifs in JcDof proteins. (**a**) The distribution of 10 conserved motifs in Dof proteins; (**b**) Gene structures of *JcDof* genes. CDS, UTR and introns were depicted by filled red boxes, blue boxes, and single black lines; and (**c**) Motif-1, corresponding to theCX2CX21CX2C single zinc-finger structure. The detailed motif's sequences are shown in Figure S1 and Table S4.

2.5. Chromosomal Locations and Gene Duplication Events of *JcDof* Genes

In order to explore the mechanism of evolution and amplification of *JcDof* gene, the chromosomal locations and gene duplication events of *JcDof* genes were further analyzed. The chromosomal distribution of *JcDof* genes was plotted using Map Inspect software (Figure 4). The duplication events of *JcDof* genes were also examined, and *Dof* gene-pairs arising from segmental and tandem duplication were marked with light blue line and dark blue rectangles, respectively. From Figure 4 we can find that some *Dof* genes, such as *JcDof-19*, have been duplicated several times to form more than one duplicated gene-pair with other genes; and some *JcDof* genes, such as *JcDof-15*, *JcDof-22*, and *JcDof-24*, are evolutionarily too close to resolve their gene duplication order (the duplication pairs are described in detail in Supplementary Table S5). The gene expansion of the *Dof* family in physic nut mainly resulted from segmental duplication, and tandem duplication also played a minor role. In total, 26 pairs of segmental duplicated *JcDof* genes (93% of all duplicated genes) and two pairs of tandem duplicated *JcDof* genes (7% of all duplicated genes) were found. For most of the duplicated gene pairs (22 out of 28), the pairwise *JcDof* genes often came from the same phylogenetic group, with very high sequence similarities. Specifically, tandem duplicated genes have higher sequence similarity than segmental duplicated genes (Table S5).

To further understand the evolutionary constraints acting on all of the duplicated *JcDof* genes, we calculated the non-synonymous substitution rate (*Ka*), synonymous substitution rate (*Ks*) and *Ka*/*Ks* for all of the 28 pairs duplicated genes (Figure 5 and Table S5). We found 23 pairs duplicated genes whose *Ka*/*Ks* were more than one (accounting for 82% of all the duplicated genes) and five pairs duplicated genes whose *Ka*/*Ks* ratio were less than one (accounting for 18% of all the duplicated gene pairs) (Table S5). This implied that most of the *Dof* duplicated gene pairs tended to be subjected to positive selection, which may play important roles in the origin of adaptive phenotypes and the possible function divergence in *JcDof* genes.

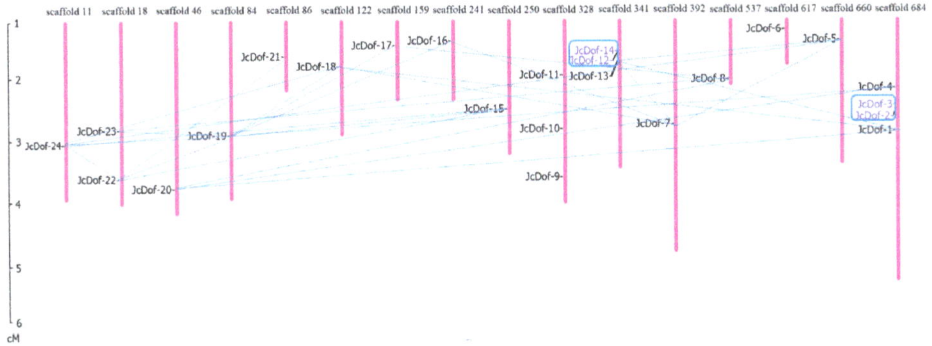

Figure 4. Chromosomal locations and gene duplication events of *JcDof*. Respective scaffold numbers are indicated at the top of each bar. The scale on the left is in centimorgan (cM). The *JcDof* gene pairs of segmental and tandem duplication are linked by pale blue lines and marked in dark blue rectangles, respectively. The detailed information of duplication pairs are described in Supplementary Table S5.

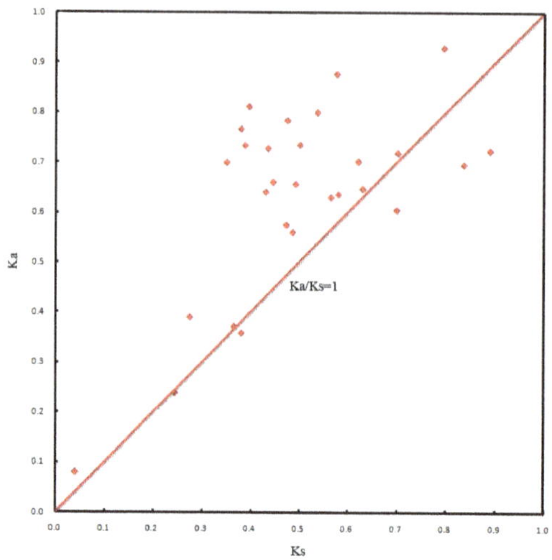

Figure 5. The *Ka/Ks* value of duplicated *JcDof* gene pairs. 28 pairs of duplicated *JcDof* genes and 23 pairs duplicated genes with *Ka/Ks* more than one. The detailed *Ka/Ks* information of duplication pairs are described in Supplementary Table S5.

2.6. Expression Patterns of JcDof Genesunder Different Abiotic Stress and Hormone Treatments

In order to further study the possible function divergence of *JcDof* genes, we investigated the expression level of *JcDof* genes under various abiotic stresses and hormonal treatments by using the public transcriptome data from NCBI SRA database (Supplementary Tables S6 and S7 for detailed information). We employed a heatmap to visualize a global transcription profile of the *JcDof* genes. As shown in Figure 6, *JcDof* genes showed diverse responses to various treatments, and significant differences were found in response to 6-Benzylaminopurine (BA), salt, and drought treatments (two-fold increases or decreases compared to controls).

In the BA treatment experiments (gene expression data collected from roots), compared with the negative control (mock), three genes (*JcDof-1*, *JcDof-8*, and *JcDof-10*) exhibited significant responses. Among them, *JcDof-1* and *JcDof-10* showed reduced expression when responding to BA treatment, with more than two-fold (*JcDof-1*) and nearly four-fold (*JcDof-10*) decreasing, respectively. Meanwhile, *JcDof-8* showed a significantly up-regulated expression with more than four-fold increase. We further checked the GO annotations of their Arabidopsis orthologs, and found they were annotated as "seed coat development" (*JcDof-1*, *AT1G29160.1*), "guard cell differentiation, positive regulation of transcription, regulation of cell wall pectin metabolic process, stomatal movement" (*JcDof-8*, *AT5G65590.1*), and "flower development" (*JcDof-10*, *AT5G39660.1*) respectively, which may imply the possible roles of these three genes (Supplementary Table S3 for detailed information).

We further analyzed the expression patterns of the *JcDof* genes in salt- and drought-stressed roots and leaves at different times: 2 h, 2 days, and 7 days (salt-stressed); 13 days, 49 days, and 52 days (drought-stressed). The fold changes of gene expression were calculated between abiotic stress treatments and controls. Many *JcDof* genes exhibited significant responses, and some of them showed significant up- or down-regulation in both roots and leaves, such as *JcDof-8*, *JcDof-17*, and *JcDof-20* in salt-stressed treatments, and *JcDof-6*, *JcDof-8*, *JcDof-10*, *JcDof-14*, *JcDof-17*, and *JcDof-21* in drought-stressed treatments. Most of these significantly up- or down-regulated genes (seven out of nine) tended to show similar expression changes (up- or down-regulation) in both roots and leaves. The only two exceptions were *JcDof-20* and *JcDof-14*. *JcDof-20* showed significantly reduced expression in leaves (from 2 h to 7 days) when responding to salt treatment, while *JcDof-20* expression in salt-treated roots first decreased (at 2 h), and then increased significantly (two days and seven days). Another gene, *JcDof-14*, showed significantly reduced expression in leaves (in 49 days) when responding to drought treatment, while *JcDof-14* expression in drought-treated roots first increased (in 13 days), and then decreased significantly (49 days and 52 days).

We have also checked the differential expression patterns of the duplicated *JcDof* gene pairs, and found that if *JcDof* genes differentially expressed in some stress treatments, and their duplicated counterparts were more likely not to show differential expression (27 pairs vs. 20 pairs, Supplementary Table S8 for detailed information). We think these results are consistent with our *Ka*/*Ks* results, that most of the duplicated *JcDof* genes tended to be subjected to positive selection, and implied the possible function divergence in *JcDof* genes.

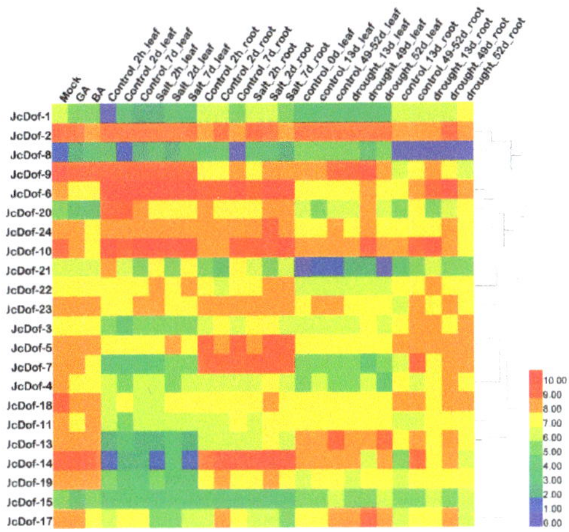

Figure 6. Expression patterns of *JcDof* genes under different treatments. The heatmap was generated by HemI software using the expression data of the *JcDof* genes, and normalized log_2 transformed values were used with hierarchical clustering represented by the color scale (0–10). Blue indicates low expression, and red indicates high expression. The samples were: roots and leaves (salt- and drought-stressed at different times), and roots (BA treatment). The detailed information of expression data are described in Supplementary Tables S6 and S7.

3. Discussion

The Euphorbiaceae family includes some of the most efficient biomass accumulators, such as physic nut, castor bean, cassava, and rubber tree [9,31]. Crop improvement in Euphorbiaceae for sustainable industrial raw materials and food production requires more extensive genome-wide studies on these species. Notably, physic nut has become an ideal model organism in Euphorbiaceae for further functional genomics analysis due to its sequenced genome, genetic linkage map, and abundance of high-throughput transcriptome data. Studies on physic nut will provide insights into the investigation of other Euphorbiaceae organisms.

Genome-wide gene family analysis is abasic and a key step to understanding the gene structure, function, and evolution [32]. The *Dof* gene family has been shown to play crucial roles in the regulatory network of plant defense, including responses to diverse biotic and abiotic stresses [22,23,33,34]. Until now, the *Dof* genes have been identified and characterized in different plant species, but not in the promising energy plant physic nut yet. Therefore, we conducted a comprehensive analysis of the *JcDof* family in physic nut, along with their homologs in *R. communis* and *A. thaliana*, to study their phylogenetic relationships and potential functions.

In total, we identified 24 *JcDof* genes in the physic nut genome. Compared with the number of *Dof* genes in *A. thaliana* (36 genes from TAIR), the size of physic nut *Dof* gene family is much smaller [35], although the assembled genome size of physic nut is approximately three times larger than the *A. thaliana* genome (320.5 Mbp vs. 125 Mbp) [9,36]. Correspondingly, we had discovered that the members from Group B, one of the major groups in the phylogenetic tree, all pertained to *AtDof* genes. In addition, Subgroup C1 contained 13 *AtDof* genes; while only nine *JcDof* genes were noted. Subgroup C2 had 12 *AtDof* genes and 10 *JcDof* genes. These results suggested that *JcDof* and *AtDof* genes should arise through different duplication events, and might have undergone species/lineage-specific gene gain or loss.

Both tandem duplication and segmental duplication contributed to the variation in gene family number and distribution [37,38]. In total, 26 gene-pairs from segmental duplication and two from tandem duplication were found in physic nut. We calculated the *Ka*/*Ks* ratios for these duplicated *JcDof* paralog genes, and found most of the duplicated genes pairs had *Ka*/*Ks* ratios over 1, implying that positive selection played an important role in the evolution of *JcDof* genes, and high-throughput expression data analysis further confirmed the functional diversity of *JcDof* genes. *JcDof* genes showed diverse responses to various treatments, and might participate in different stress/hormone-responding regulatory processes. This work provides valuable information for understanding the evolution of *JcDof* genes and lays a foundation for future functional analysis of *Dof* genes in the process of growth, development, and Dof-mediated regulation in physic nut.

4. Materials and Methods

4.1. Data Sources

The physic nut genomic and proteomic sequences were downloaded from the NCBI database (Available online: https://www.ncbi.nlm.nih.gov/, Assembly JatCur_1.0). The Dof protein sequences of *A. thaliana* were obtained from the Arabidopsis genome database (TAIR 9.0 release, Available online: http://www.arabidopsis.org/) [35]. The Dof protein sequences of castor bean were obtained from the PlantTFDB database (Available online: http://planttfdb.cbi.pku.edu.cn/) [16]. The physic nut gene expression data were collected from the SRA database (Available online: https://www.ncbi.nlm.nih.gov/) [39].

4.2. Dof Gene Identification and Characterization

To identify all the possible *Dof* genes in physic nut, both local BLASTP [40] and Hidden Markov model (HMM) searches were performed [41]. For BLASTP, the known Dof proteins from Arabidposis were taken as queries and the *E*-value was set to 1×10^{-10}. For the HMM search, the HMM profile of the Dof domain was used as query and the *E*-value was set to 1 [24]. All the retrieved sequences were further scanned and tested using SMART (Available online: http://smart.embl-heidelberg.de/) [42] and NCBI Conserved Domains database (Available online: http://www.ncbi.nih.gov/Structure/cdd/cdd.shtml) for authentication of the presence of Dof domain [43]. We manually removed redundant sequences that do not have Dof domain or have incomplete encoding frame. Parameters, such as protein length, molecular weight, isoelectric point, and instability coefficient of all the Dof proteins in physic nut were predicted using ExPASy Proteomics Server (Available online: http://prosite.expasy.org/) [44]. The orthologous genes of JcDof proteins in *A. thaliana* were predicted by BLASTP.

4.3. DNA-Binding Domain Conservation Analysis of JcDof Protein

The conserved regions of JcDof proteins were extracted by DNAMAN tool (version 2.6 Lynnon Biosoft, Quebec City, QC, Canada) [45]. We then identified highly-conserved Dof domain for all Dof proteins by multiple sequence alignment analysis using ClustalW MEGA integration software [46].

4.4. Phylogenetic Analysis

Physic nut, *A. thaliana*, and *R. communis* Dof protein sequences were pretreated by GUIDANCE2 online tool to remove unreliable columns [47]. The phylogenetic relationship among the Dof proteins was analyzed using ClustalW and the dendrogram was constructed using MEGA (v6.0, Tokyo Metropolitan University, Tokyo, Japan) by neighbor-joining method, with the following parameters: Poisson correction, pairwise deletion, and 1000-bootstrap replicates [48].

4.5. Gene Structure of Dof Proteins

Positional information for both the gene sequences and the corresponding coding sequences was loaded into the Gene Structure Display Server (GSDS v2.0, Available online: http://gsds.cbi.pku.edu.cn/) to obtain information on intron/CDS structure [49]. The coordinates of the Dof domain in each protein were recalculated into the coordinates in the corresponding gene sequence and featured in the gene structure.

4.6. Detection of Additional Conserved Motifs

To identify additional conserved motifs outside the Dof domain of physic nut Dof proteins, we used Multipel Expectation Maximization for Motif Elucidation (MEME v4.11.2, Available online: http://meme.nbcr.net/meme/) [50]. The limits on maximum width, minimum width, and maximum number of motifs were specified as 5, 150, and 10, respectively. The motifs were numbered serially according to their order in MEME. Those motifs common to genes in one of the three similarity groups were designated as the group-specific signatures.

4.7. Chromosomal Localization

According to the chromosomal positions of genes, we drew a map of the distribution of *Dof* genes throughout the physic nut genome using MapInspect software (Available online: http://mapinspect.software.informer.com/) [51]. The *Dof* gene pairs resulting from segmental or tandem duplication were linked by lines and marked in blue rectangle, respectively.

4.8. Detection of Gene Duplication Events and Estimation of Synonymous (Ks) and Nonsynonymous (Ka) Substitutions per Site and Their Ratio

Duplicated gene pairs derived from segmental or tandem duplication were identified in physic nut genome based on the method described in the Plant Genome Duplication Database [52,53]. An all-against-all BLASTP comparison (E-value $\leq 1 \times 10^{-20}$) provided the gene pairs for syntenic clustering determined by MCScanX (E-value $\leq 1 \times 10^{-20}$) [54]. Tandem duplication arrays were identified using BLASTP with a threshold of E-value $< 1\times 10^{-20}$, and one unrelated gene among cluster members was tolerated, as described for *A. thaliana*. Pairs from segmental and tandem duplications were used to estimate *Ka*, *Ks*, and their ratio. Coding sequences from segmentally and tandemly duplicated *Dof* gene pairs were aligned by PRANK [55] and trimmed by Gblocks. The software DnaSP (Available online: http://www.softpedia.com/get/Science-CAD/DnaSP.shtml) [56] was then used to compute *Ka* and *Ks* values for each pair following the YN model (a simple model of voting) [57]. If $Ka/Ks > 1$, there is positive selection pressure; if $Ka/Ks = 1$, there is neutral selection or natural selection pressure; if $Ka/Ks < 1$, there is a purification selection effect [58,59].

4.9. Expression Analysis of Physic Nut Dof Genes

The original expression data for *JcDof* genes under different treatments (including gibberellins [GA], 6-Benzylaminopurine[BA], high salt concentration and drought) were retrieved from NCBI SRA database (Available online: https://www.ncbi.nlm.nih.gov/). All the data were analyzed using Tuxedo suite (TopHat and Cufflinks, http://post.queensu.ca/~rc91/NGS/TuxedoTutorial.html) and then upper-quartile normalized and log transformed. Heat maps were generated by means of the HemI toolkit (Available online: http://hemi.biocuckoo.org/) with average linkage hierarchical clustering [60,61].

5. Conclusions

In conclusion, a total of 24 *Dof* genes were identified from physic nut, and these *Dof* genes were further divided into three major groups based on the phylogenetic inference. The gene structures, conserved motifs, gene duplicated events, selection pressures, and expression profiling of these *JcDof* genes were analyzed. A genome comparison discovered that the expansion of the *Dof* gene family in physic nut mainly resulted from segmental duplication, and this expansion was mainly subjected to positive selection. The expression profile demonstrated the broad involvement of *JcDof* genes in different hormonalor abiotic stressed treatments. Among them, three genes (*JcDof-1*, *JcDof-8*, and *JcDof-10*) exhibited significant responses to the BA treatment. Furthermore, many *JcDof* genes were significantly responsive to the salt and drought treatments. On the whole, this study provides an extensive resource for understanding the *Dof* genes in physic nut.

Supplementary Materials: The following are available online at http://www.mdpi.com/1422-0067/19/6/1598/s1.

Author Contributions: C.L. conceived and supervised this study. C.L. and P.W. designed the experiments. P.W. carried out the experiments, and analyzed and interpreted the data. J.L., X.G., D.Z. and A.L. participated in the discussion and provided valuable advice and practical contributions. C.L., P.W. and J.L. wrote the manuscript. All authors reviewed, edited, and approved the final manuscript.

Acknowledgments: This work was supported by the following grants: National Natural Science Foundation of China (grant nos. 31471220, 91440113), Start-up Fund from Xishuangbanna Tropical Botanical Garden, and 'Top Talents Program in Science and Technology' from Yunnan Province.

Conflicts of Interest: The authors declare no conflict of interest.

Abbreviations

TF	Transcription factor
HMM	Hidden Markov model
Mw	Molecular weight
pI	Isoelectric point
GA	Gibberellic acid
BA	6-Benzylaminopurine

References

1. Bhasanutra, R.; Sutiponpeibun, S. Jatropha curcas oil as a substitute for diesel engine oil. *Int. Energy J.* **2017**, *4*, 56–70.
2. Openshaw, K. A review of Jatropha curcas: An oil plant of unfulfilled promise. *Biomass Bioenergy* **2000**, *19*, 1–15. [CrossRef]
3. Chen, M.-S.; Pan, B.-Z.; Fu, Q.; Tao, Y.-B.; Martínez-Herrera, J.; Niu, L.; Ni, J.; Dong, Y.; Zhao, M.-L.; Xu, Z.-F. Comparative transcriptome analysis between gynoecious and monoecious plants identifies regulatory networks controlling sex determination in *Jatropha curcas*. *Front. Plant Sci.* **2017**, *7*, 1953. [CrossRef] [PubMed]
4. Pan, B.-Z.; Xu, Z.-F. Benzyladenine treatment significantly increases the seed yield of the biofuel plant *Jatropha curcas*. *J. Plant Growth Regul.* **2011**, *30*, 166–174. [CrossRef]
5. Makkar, H.; Becker, K.; Sporer, F.; Wink, M. Studies on nutritive potential and toxic constituents of different provenances of *Jatropha curcas*. *J. Agric. Food Chem.* **1997**, *45*, 3152–3157. [CrossRef]
6. Dehgan, B. Phylogenetic significance of interspecific hybridization in *Jatropha* (Euphorbiaceae). *Syst. Bot.* **1984**, *9*, 467–478. [CrossRef]
7. Carvalho, C.R.; Clarindo, W.R.; Praça, M.M.; Araújo, F.S.; Carels, N. Genome size, base composition and karyotype of *Jatropha curcas* L., an important biofuel plant. *Plant Sci.* **2008**, *174*, 613–617. [CrossRef]

8. Rahman, A.Y.A.; Usharraj, A.O.; Misra, B.B.; Thottathil, G.P.; Jayasekaran, K.; Feng, Y.; Hou, S.; Ong, S.Y.; Ng, F.L.; Lee, L.S. Draft genome sequence of the rubber tree *Hevea brasiliensis*. *BMC Genom.* **2013**, *14*, 75. [CrossRef] [PubMed]
9. Wu, P.; Zhou, C.; Cheng, S.; Wu, Z.; Lu, W.; Han, J.; Chen, Y.; Chen, Y.; Ni, P.; Wang, Y. Integrated genome sequence and linkage map of physic nut (*Jatropha curcas* L.), a biodiesel plant. *Plant J.* **2015**, *81*, 810–821. [CrossRef] [PubMed]
10. Zhang, L.; Zhang, C.; Wu, P.; Chen, Y.; Li, M.; Jiang, H.; Wu, G. Global analysis of gene expression profiles in physic nut (*Jatropha curcas* L.) seedlings exposed to salt stress. *PLoS ONE* **2014**, *9*, e97878. [CrossRef] [PubMed]
11. Zhang, C.; Zhang, L.; Zhang, S.; Zhu, S.; Wu, P.; Chen, Y.; Li, M.; Jiang, H.; Wu, G. Global analysis of gene expression profiles in physic nut (*Jatropha curcas* L.) seedlings exposed to drought stress. *BMC Plant Biol.* **2015**, *15*, 17. [CrossRef] [PubMed]
12. Takahashi, M.U.; Nakagawa, S. Transcription Factor Genes. In *Evolution of the Human Genome I*; Springer: Berlin, Germany, 2017; pp. 241–263.
13. Riaño-Pachón, D.M.; Ruzicic, S.; Dreyer, I.; Mueller-Roeber, B. PlnTFDB: An integrative plant transcription factor database. *BMC Bioinform.* **2007**, *8*, 42. [CrossRef] [PubMed]
14. Noguero, M.; Atif, R.M.; Ochatt, S.; Thompson, R.D. The role of the DNA-binding One Zinc Finger (DOF) transcription factor family in plants. *Plant Sci.* **2013**, *209*, 32–45. [CrossRef] [PubMed]
15. Franco-Zorrilla, J.M.; López-Vidriero, I.; Carrasco, J.L.; Godoy, M.; Vera, P.; Solano, R. DNA-binding specificities of plant transcription factors and their potential to define target genes. *Proc. Natl. Acad. Sci. USA* **2014**, *111*, 2367–2372. [CrossRef] [PubMed]
16. Jin, J.; Tian, F.; Yang, D.-C.; Meng, Y.-Q.; Kong, L.; Luo, J.; Gao, G. PlantTFDB 4.0: Toward a central hub for transcription factors and regulatory interactions in plants. *Nucleic Acids Res.* **2016**, *45*, D1040–D1045. [CrossRef] [PubMed]
17. Yanagisawa, S.; Schmidt, R.J. Diversity and similarity among recognition sequences of Dof transcription factors. *Plant J.* **1999**, *17*, 209–214. [CrossRef] [PubMed]
18. Yanagisawa, S. Dof domain proteins: Plant-specific transcription factors associated with diverse phenomena unique to plants. *Plant Cell Physiol.* **2004**, *45*, 386–391. [CrossRef] [PubMed]
19. Yanagisawa, S. The Dof family of plant transcription factors. *Trends Plant Sci.* **2002**, *7*, 555–560. [CrossRef]
20. Chen, X.; Wang, D.; Liu, C.; Wang, M.; Wang, T.; Zhao, Q.; Yu, J. Maize transcription factor Zmdof1 involves in the regulation of *Zm401* gene. *Plant Growth Regul.* **2012**, *66*, 271–284. [CrossRef]
21. Gupta, S.; Malviya, N.; Kushwaha, H.; Nasim, J.; Bisht, N.C.; Singh, V.; Yadav, D. Insights into structural and functional diversity of Dof (DNA binding with one finger) transcription factor. *Planta* **2015**, *241*, 549–562. [CrossRef] [PubMed]
22. Venkatesh, J.; Park, S.W. Genome-wide analysis and expression profiling of DNA-binding with one zinc finger (Dof) transcription factor family in potato. *Plant Physiol. Biochem.* **2015**, *94*, 73–85. [CrossRef] [PubMed]
23. Yanagisawa, S. Structure, Function, and Evolution of the Dof Transcription Factor Family. In *PlantTranscription Factors*; Elsevier: New York, NY, USA, 2015; pp. 183–197.
24. Shu, Y.; Song, L.; Zhang, J.; Liu, Y.; Guo, C. Genome-wide identification and characterization of the Dof gene family in Medicago truncatula. *Genet. Mol. Res.* **2015**, *14*, 10645–10657. [CrossRef] [PubMed]
25. Wang, T.; Yue, J.-J.; Wang, X.-J.; Xu, L.; Li, L.-B.; Gu, X.-P. Genome-wide identification and characterization of the Dof gene family in moso bamboo (*Phyllostachys heterocycla* var. pubescens). *Genes Genom.* **2016**, *38*, 733–745. [CrossRef]
26. Wu, Q.; Li, D.; Li, D.; Liu, X.; Zhao, X.; Li, X.; Li, S.; Zhu, L. Overexpression of OsDof12 affects plant architecture in rice (*Oryza sativa* L.). *Front. Plant Sci.* **2015**, *6*, 833. [CrossRef] [PubMed]
27. Lijavetzky, D.; Carbonero, P.; Vicente-Carbajosa, J. Genome-wide comparative phylogenetic analysis of the rice and Arabidopsis Dof gene families. *BMC Evol. Biol.* **2003**, *3*, 17. [CrossRef] [PubMed]
28. Guo, Y.; Qiu, L.-J. Retraction: Genome-Wide Analysis of the Dof Transcription Factor Gene Family Reveals Soybean-Specific Duplicable and Functional Characteristics. *PLoS ONE* **2016**, *11*, e0167019. [CrossRef] [PubMed]
29. Shaw, L.M.; McIntyre, C.L.; Gresshoff, P.M.; Xue, G.-P. Members of the Dof transcription factor family in Triticum aestivum are associated with light-mediated gene regulation. *Funct. Integr. Genom.* **2009**, *9*, 485. [CrossRef] [PubMed]

30. Jin, Z.; Chandrasekaran, U.; Liu, A. Genome-wide analysis of the Dof transcription factors in castor bean (*Ricinus communis* L.). *Genes Genom.* **2014**, *36*, 527–537. [CrossRef]
31. Chan, A.P.; Crabtree, J.; Zhao, Q.; Lorenzi, H.; Orvis, J.; Puiu, D.; Melake-Berhan, A.; Jones, K.M.; Redman, J.; Chen, G. Draft genome sequence of the oilseed species Ricinus communis. *Nat. Biotechnol.* **2010**, *28*, 951. [CrossRef] [PubMed]
32. Marquez, Y.; Höpfler, M.; Ayatollahi, Z.; Barta, A.; Kalyna, M. Unmasking alternative splicing inside protein-coding exons defines exitrons and their role in proteome plasticity. *Genome Res.* **2015**, *25*, 995–1007. [CrossRef] [PubMed]
33. Singh, K.B.; Foley, R.C.; Oñate-Sánchez, L. Transcription factors in plant defense and stress responses. *Curr. Opin. Plant Biol.* **2002**, *5*, 430–436. [CrossRef]
34. Ma, J.; Li, M.-Y.; Wang, F.; Tang, J.; Xiong, A.-S. Genome-wide analysis of Dof family transcription factors and their responses to abiotic stresses in Chinese cabbage. *BMC Genom.* **2015**, *16*, 33. [CrossRef] [PubMed]
35. Lamesch, P.; Berardini, T.Z.; Li, D.; Swarbreck, D.; Wilks, C.; Sasidharan, R.; Muller, R.; Dreher, K.; Alexander, D.L.; Garcia-Hernandez, M. The Arabidopsis Information Resource (TAIR): Improved gene annotation and new tools. *Nucleic Acids Res.* **2011**, *40*, D1202–D1210. [CrossRef] [PubMed]
36. Initiative, A.G. Analysis of the genome sequence of the flowering plant Arabidopsis thaliana. *Nature* **2000**, *408*, 796. [CrossRef] [PubMed]
37. Cannon, S.B.; Mitra, A.; Baumgarten, A.; Young, N.D.; May, G. The roles of segmental and tandem gene duplication in the evolution of large gene families in Arabidopsis thaliana. *BMC Plant Biol.* **2004**, *4*, 10. [CrossRef] [PubMed]
38. Leister, D. Tandem and segmental gene duplication and recombination in the evolution of plant disease resistance genes. *Trends Genet.* **2004**, *20*, 116–122. [CrossRef] [PubMed]
39. Leinonen, R.; Sugawara, H.; Shumway, M.; Collaboration, I.N.S.D. The sequence read archive. *Nucleic Acids Res.* **2010**, *39* (Suppl. 1), D19–D21. [CrossRef] [PubMed]
40. Mount, D.W. Using the basic local alignment search tool (BLAST). *Cold Spring Harb. Protoc.* **2007**, *2007*. [CrossRef] [PubMed]
41. Schuster-Böckler, B.; Bateman, A. An introduction to hidden Markov models. *Curr. Protoc. Bioinform.* **2007**, *18*, A.3A.1–A.3A.9.
42. Letunic, I.; Doerks, T.; Bork, P. SMART 7: Recent updates to the protein domain annotation resource. *Nucleic Acids Res.* **2011**, *40*, D302–D305. [CrossRef] [PubMed]
43. Marchler-Bauer, A.; Derbyshire, M.K.; Gonzales, N.R.; Lu, S.; Chitsaz, F.; Geer, L.Y.; Geer, R.C.; He, J.; Gwadz, M.; Hurwitz, D.I. CDD: NCBI's conserved domain database. *Nucleic Acids Res.* **2014**, *43*, D222–D226. [CrossRef] [PubMed]
44. Gasteiger, E.; Gattiker, A.; Hoogland, C.; Ivanyi, I.; Appel, R.D.; Bairoch, A. ExPASy: The proteomics server for in-depth protein knowledge and analysis. *Nucleic Acids Res.* **2003**, *31*, 3784–3788. [CrossRef] [PubMed]
45. Woffelman, C. *DNAMAN for Windows, Version 2.6*; Lynon Biosoft, Institute of Molecular Plant Sciences, Leiden University: Leiden, The Netherlands, 1994.
46. Thompson, J.D.; Gibson, T.; Higgins, D.G. Multiple sequence alignment using ClustalW and ClustalX. *Curr. Protoc. Bioinform.* **2002**. [CrossRef] [PubMed]
47. Sela, I.; Ashkenazy, H.; Katoh, K.; Pupko, T. GUIDANCE2: Accurate detection of unreliable alignment regions accounting for the uncertainty of multiple parameters. *Nucleic Acids Res.* **2015**, *43*, W7–W14. [CrossRef] [PubMed]
48. Tamura, K.; Stecher, G.; Peterson, D.; Filipski, A.; Kumar, S. MEGA6: Molecular evolutionary genetics analysis version 6.0. *Mol. Biol. Evol.* **2013**, *30*, 2725–2729. [CrossRef] [PubMed]
49. Hu, B.; Jin, J.; Guo, A.-Y.; Zhang, H.; Luo, J.; Gao, G. GSDS 2.0: An upgraded gene feature visualization server. *Bioinformatics* **2014**, *31*, 1296–1297. [CrossRef] [PubMed]
50. Bailey, T.L.; Boden, M.; Buske, F.A.; Frith, M.; Grant, C.E.; Clementi, L.; Ren, J.; Li, W.W.; Noble, W.S. MEME SUITE: Tools for motif discovery and searching. *Nucleic Acids Res.* **2009**, *37* (Suppl. 2), W202–W208. [CrossRef] [PubMed]
51. He, H.; Dong, Q.; Shao, Y.; Jiang, H.; Zhu, S.; Cheng, B.; Xiang, Y. Genome-wide survey and characterization of the WRKY gene family in Populus trichocarpa. *Plant Cell Rep.* **2012**, *31*, 1199–1217. [CrossRef] [PubMed]
52. Lee, T.-H.; Tang, H.; Wang, X.; Paterson, A.H. PGDD: A database of gene and genome duplication in plants. *Nucleic Acids Res.* **2012**, *41*, D1152–D1158. [CrossRef] [PubMed]

53. Kohler, A.; Rinaldi, C.; Duplessis, S.; Baucher, M.; Geelen, D.; Duchaussoy, F.; Meyers, B.C.; Boerjan, W.; Martin, F. Genome-wide identification of NBS resistance genes in Populus trichocarpa. *Plant Mol. Biol.* **2008**, *66*, 619–636. [CrossRef] [PubMed]
54. Wang, Y.; Tang, H.; DeBarry, J.D.; Tan, X.; Li, J.; Wang, X.; Lee, T.-H.; Jin, H.; Marler, B.; Guo, H. MCScanX: A toolkit for detection and evolutionary analysis of gene synteny and collinearity. *Nucleic Acids Res.* **2012**, *40*, e49. [CrossRef] [PubMed]
55. Löytynoja, A.; Goldman, N. webPRANK: A phylogeny-aware multiple sequence aligner with interactive alignment browser. *BMC Bioinform.* **2010**, *11*, 579. [CrossRef] [PubMed]
56. Librado, P.; Rozas, J. DnaSP v5: A software for comprehensive analysis of DNA polymorphism data. *Bioinformatics* **2009**, *25*, 1451–1452. [CrossRef] [PubMed]
57. Yang, Z.; Nielsen, R. Estimating synonymous and nonsynonymous substitution rates under realistic evolutionary models. *Mol. Biol. Evol.* **2000**, *17*, 32–43. [CrossRef] [PubMed]
58. Hurst, L.D. The Ka/Ks ratio: Diagnosing the form of sequence evolution. *Trends Genet.* **2002**, *18*, 486–487. [CrossRef]
59. Ito, T.M.; Trevizan, C.B.; dos Santos, T.B.; de Souza, S.G.H. Genome-Wide Identification and Characterization of the Dof Transcription Factor Gene Family in *Phaseolus vulgaris* L. *Am. J. Plant Sci.* **2017**, *8*, 3233. [CrossRef]
60. Deng, W.; Wang, Y.; Liu, Z.; Cheng, H.; Xue, Y. HemI: A toolkit for illustrating heatmaps. *PLoS ONE* **2014**, *9*, e111988. [CrossRef] [PubMed]
61. Song, S.; Zhou, H.; Sheng, S.; Cao, M.; Li, Y.; Pang, X. Genome-Wide Organization and Expression Profiling of the SBP-Box Gene Family in Chinese Jujube (*Ziziphus jujuba* Mill.). *Int. J. Mol. Sci.* **2017**, *18*, 1734. [CrossRef] [PubMed]

© 2018 by the authors. Licensee MDPI, Basel, Switzerland. This article is an open access article distributed under the terms and conditions of the Creative Commons Attribution (CC BY) license (http://creativecommons.org/licenses/by/4.0/).

International Journal of
Molecular Sciences

MDPI

Article

Genome-Wide Identification and Expression Analysis of the UGlcAE Gene Family in Tomato

Xing Ding, Jinhua Li, Yu Pan, Yue Zhang, Lei Ni, Yaling Wang and Xingguo Zhang *

Key Laboratory of Horticulture Science for Southern Mountainous Regions (Chinese Ministry of Education), College of Horticulture and Landscape Architecture, Southwest University, Chongqing 400715, China; dingdingxing@email.swu.edu.cn (X.D.); ljh502@swu.edu.cn (J.L.); pany1020@swu.edu.cn (Y.P.); zy18109037077@163.com (Y.Z.); m15090062251@163.com (L.N.); yalingwangx@163.com (Y.W.)
* Correspondence: zhangdupian@swu.edu.cn; Tel.: +86-23-68250974

Received: 12 April 2018; Accepted: 23 May 2018; Published: 27 May 2018

Abstract: The UGlcAE has the capability of interconverting UDP-D-galacturonic acid and UDP-D-glucuronic acid, and UDP-D-galacturonic acid is an activated precursor for the synthesis of pectins in plants. In this study, we identified nine *UGlcAE* protein-encoding genes in tomato. The nine *UGlcAE* genes that were distributed on eight chromosomes in tomato, and the corresponding proteins contained one or two trans-membrane domains. The phylogenetic analysis showed that *SlUGlcAE* genes could be divided into seven groups, designated *UGlcAE1* to *UGlcAE6*, of which the *UGlcAE2* were classified into two groups. Expression profile analysis revealed that the *SlUGlcAE* genes display diverse expression patterns in various tomato tissues. Selective pressure analysis indicated that all of the amino acid sites of SlUGlcAE proteins are undergoing purifying selection. Fifteen stress-, hormone-, and development-related elements were identified in the upstream regions (0.5 kb) of these *SlUGlcAE* genes. Furthermore, we investigated the expression patterns of *SlUGlcAE* genes in response to three hormones (indole-3-acetic acid (IAA), gibberellin (GA), and salicylic acid (SA)). We detected firmness, pectin contents, and expression levels of UGlcAE family genes during the development of tomato fruit. Here, we systematically summarize the general characteristics of the *SlUGlcAE* genes in tomato, which could provide a basis for further function studies of tomato *UGlcAE* genes.

Keywords: *Solanum lycopersicum*; *UGlcAE* gene family; identification; characterization; plant hormones; gene expression

1. Introduction

As a major component of the primary cell walls of plants [1], pectins are essential for remodeling cell wall and normal cell-cell adhesion during cellular growth [2–5]. D-galacturonic acid (GalA) is the constituent of the capsular polysaccharides and lipopolysaccharides of several bacterial species [6]. In plants, GalA residues, which are the precursor of pectin formation, are contained in the backbone of all pectin polymers [7]. UDP-D-galacturonic acid (UDP-GalA), which is the activated nucleotide sugar form of GalA, is required in the synthesis of GalA-containing polymers. UDP is the abbreviation of uridine diphosphate and it is a nucleotide diphosphate that is made up of a pyrophosphate group, a pentose ribose, and a nucleated base uracil. UDP-GalA is synthesized via 4-epimerization of UDP-D-glucuronic acid (UDP-GlcA), which is a nucleotide sugar that is formed by the reputed inositol oxygenation pathway [8] or by the dehydrogenation of UDP-D-glucose (UDP-Glc) in the upstream [9]. Therefore, enzymes that are related to the formation of UDP-GalA and UDP-GlcA are likely to play critical roles in pectin biosynthesis [7,10].

UDP-D-glucuronic acid 4-epimerase (UGlcAE) is capable of reversibly converting UDP-GlcA and UDP-GalA [6,11]. According to previous literatures [12–15], UDP-D-glucuronic acid 4-epimerase is another name of UDP-D-glucuronate 4-epimerase. Both GAE and UGlcAE are the abbreviations of UDP-D-glucuronic acid 4-epimerase. The abbreviation is unified as UGlcAE in this study.

The UGlcAE, which is a specific membrane-bound 4-epimerase [13], is considered to evolve from some chlamydial bacteria [15]. The UGlcAE is also recognized as a key enzyme in regulating pectin biosynthesis due to its function. In 1958, the isolation of the epimerase was firstly reported [16], and subsequently it was also isolated from *Cyanobacterium anabaena* flos-aquae [17,18] and plants [10,19–22].

Although its function is believed to interconvert UDP-GlcA and UDP-GalA, the UGlcAEs from different organisms have distinct biochemical properties. For example, some UGlcAEs were substrate specific [6,11–14], while the others displayed substrate promiscuity [23]. In addition, UGlcAEs in *Poaceae* species differed from homologs in *Arabidopsis* [13,14], and different UGlcAEs in the same species also shows biochemical properties that varied differentially. For example, UGlcAE3 in *Ornithogalum caudatum* could catalyse the reversible conversion of UDP-GalA and UDP-GlcA; however, OcUGlcAE1 and OcUGlcAE2 did not have this activity [24].

The evolutionary relationship of the *UGlcAE* gene family in plants is not clear, and we hardly know anything about this gene family in tomato (*Solanum lycopersicum*). Tomato, a berry fruit, is considered to be an important economically vegetable worldwide due to its good quality and high yield [25]. With the completion of the whole genome sequencing of tomato [26], it promotes the genome-wide identification of gene families and functional analysis in tomato [27]. Therefore, studies of the *UGlcAE* gene family in tomato could develop potential strategies for improving *Solanum*-related crops genetically and stimulate new research directions, and considering the potentially important functions of the UGlcAE proteins can expand our knowledge of tomato UGlcAE isoforms. In this study, we identified and characterized the tomato *UGlcAE* gene family on a genome-wide scale, and the tissue- and organ-specific expression of *UGlcAE* family under the normal conditions and in response to three hormone treatments were analyzed according to *cis*-acting elements analysis. Analysis of this family not only identifies its members and characteristics in tomato, but it also lays a foundation for future functional analyses of *UGlcAE* genes.

2. Results

2.1. The Identification of UGlcAE Gene Family in Tomato

To identify the *UGlcAE* genes in the tomato, we searched for sequences that contained the particular domain in the tomato protein database using the hidden Markov model (HMM) model of PF01370, and we found nine potential genes (Table 1). The open reading frame (ORF) lengths of *UGlcAE* that were identified in this study ranged from 1221 bp to 1359 bp, encoding peptides varied from 406 to 452 amino acids (aa). All nine *UGlcAE* genes had a single exon.

2.2. Phylogenetic Analysis of the UGlcAE Genes in Tomato and Other Species

To evaluate the classification of the *UGlcAE* genes in *S. lycopersicum*, we analyzed the sequence features in 10 different species, including *S. lycopersicum*, *C. sativus*, *C. annuum*, *S. tuberosum*, *A. thaliana*, *N. tabacum*, *P. trichocarpa*, *S. pennelli*, *Z. mays*, and *A. lyrata* subsp. *lyrata*, and we constructed a unrooted phylogenetic tree of the *UGlcAE* genes (Figure 1) using the N-J methods. The orthologous relationships were evident. Only the tree topology is shown, and the branch lengths do not represent the estimated numbers of amino acid replacements [28].

Table 1. Information about the nine isoforms of the tomato SlUGlcAE gene family.

No.	Gene Accession No.	NCBI Name	Chr.	Gene Name	Location	*Arabidopsis* Homologous	Size (AA)	ORF (bp)	Exon
1	Solyc07g006220	XP_010323223	7	SlUGlcAE1	ch07:1039601-1041900	AT4G30440	425	1278	1
2	Solyc12g010540	XP_004251783	12	SlUGlcAE1-like	ch12:3531001-3533400	AT4G30440	432	1299	1
3	Solyc03g083550	XP_004234942	3	SlUGlcAE2-like	ch03:53489301-53491700	AT1G02000	406	1221	1
4	Solyc01g091200	XP_004229804	1	SlUGlcAE3	ch01:84887601-84890000	AT4G00110	435	1308	1
5	Solyc05g050990	XP_004240046	5	SlUGlcAE3-like1	ch05:61200401-61202800	AT4G00110	435	1308	1
6	Solyc10g018260	XP_004248561	10	SlUGlcAE3-like2	ch10:7314201-7316600	AT4G00110	435	1308	1
7	Solyc08g079440	XP_004245716	8	SlUGlcAE5	ch08:62963401-62965900	AT4G12250	445	1338	1
8	Solyc09g092330	XP_004247883	9	SlUGlcAE6	ch09:71458501-71461000	AT3G23820	452	1359	1
9	Solyc05g053790	XP_004239852	5	SlUGlcAE6-like	ch05:63814001-63816400	AT3G23820	433	1302	1

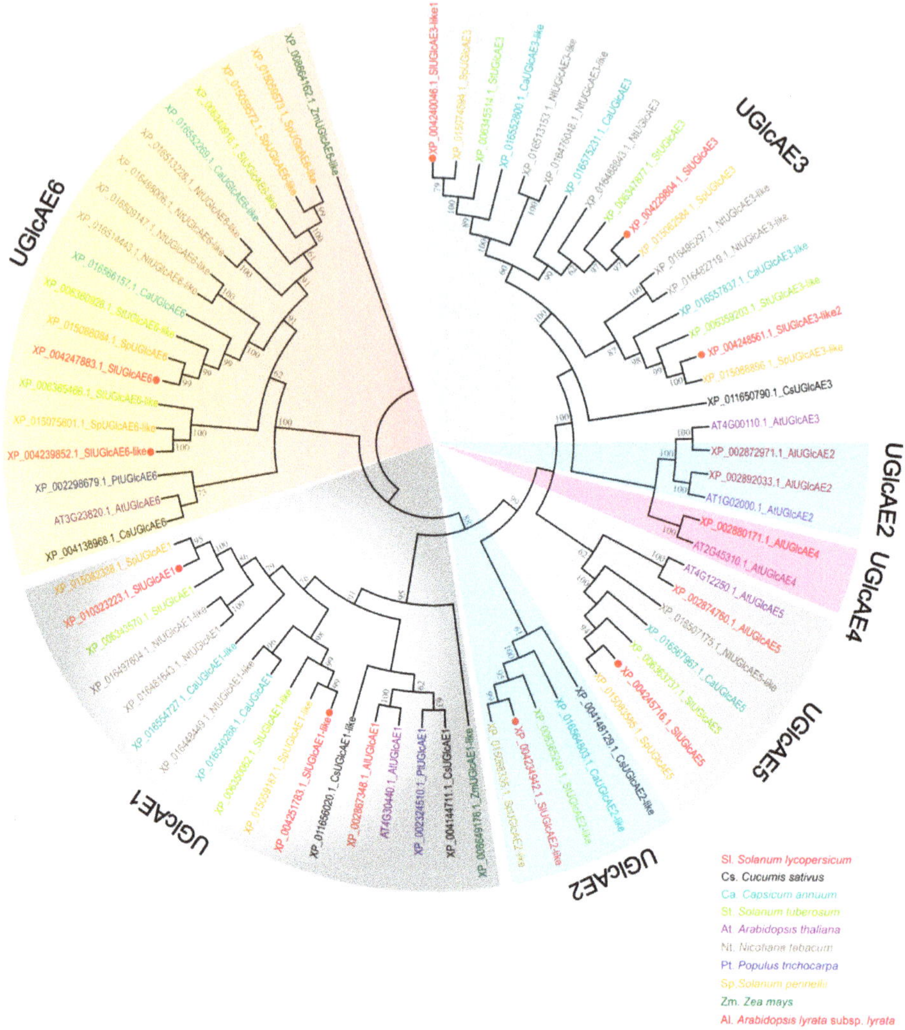

Figure 1. Phylogenetic analysis of the *UGlcAE* gene family based amino acids in tomato and other nine species. The unrooted neighbor-joining phylogenetic tree is generated by MEGA 5. The sequence names included three parts: the source numbers from NCBI, the abbreviation of species names, and their respective subfamilies. Red dots highlight the tomato *UGlcAE* genes.

Combing with the sequence similarity of the mature proteins, the employed *UGlcAE* genes are distributed into seven groups (Figure 1). In addition to *UGlcAE2* being divided into two clusters, the other five subfamilies are clustered separately.

Interestingly, the *UGlcAE4* subfamily is specifically present in *A. thaliana* and *A. lyrata* subsp. *lyrata*, whereas it is absent from other species in this study. This implies that *UGlcAE4* may be associated with distinctive functions. It is noteworthy that *UGlcAE2* genes were classified into two different groups based on their evolutionary relationship. This finding indicates that *UGlcAE2* genes may evolve into new features, which have not been known until today. Moreover, phylogenetic analyses showed that the *UGlcAE3* gene in *A. thaliana* was clustered together with *UGlcAE2* genes of *A. thaliana* and *A. lyrata* subsp. *lyrata*, suggesting that there may be some gene fusion among them.

2.3. Structures of the UGlcAE Genes in the Tomato and Arabidopsis Thaliana

Introns, especially UTR introns, in *UGlcAE* genes may influence the expression level [29]. To analyze the structural characteristics of the *UGlcAE* genes in tomato and *Arabidopsis thaliana*, their gene structures were mapped according to the genome sequences and corresponding coding sequences of *SlUGlcAE* and *AtUGlcAE* genes (Figure 2). We found that all of the *SlUGlcAE* and *AtUGlcAE* genes do not contain intron in their genomic sequences. In other words, nine *UGlcAE* genes in the tomato and six *UGlcAE* genes in *Arabidopsis thaliana* are single exon structures.

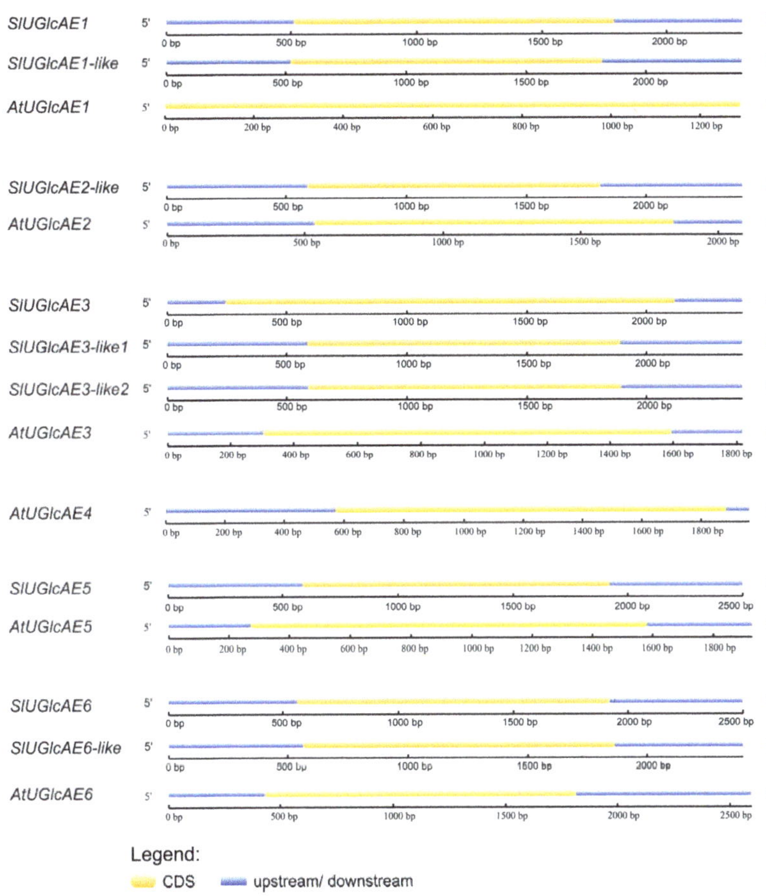

Figure 2. Exon-intron structures of nine tomato *UGlcAE* genes and six *Arabidopsis thaliana UGlcAE* genes. The yellow sections represent the exons, and the blue parts indicate upstream/downstream regions.

2.4. Chromosomal Distribution of the UGlcAE Genes in Tomato

To characterize the distribution of *UGlcAE* genes in the tomato genome, the physical locations of *UGlcAE* genes on the tomato chromosomes were obtained. According to the genomic sequences of *UGlcAE* genes, nine *SlUGlcAE* genes were mapped to eight chromosomes, including chromosome 1, 3, 5, 7, 8, 9, 10, and 12 without regularities of tandem duplication, whose positions were indicated by the black lines in the tomato chromosomes (Figure 3). Two *UGlcAE* genes (*SlUGlcAE3-like1* and *SlUGlcAE6-like*) were located on chromosome 5, and the other seven genes (*SlUGlcAE1*, *SlUGlcAE1-like*, *SlUGlcAE2-like*, *SlUGlcAE3*, *SlUGlcAE3-like2*, *SlUGlcAE5*, and *SlUGlcAE6*) were assigned to different chromosomes, but no gene was mapped to chromosome 2, 4, 6, and 11. There were no tandem duplication events among *UGlcAE* family members of tomato, suggesting that the functional differentiation may exist among the *SlUGlcAE* family members. Almost all of the *UGlcAE* genes in tomato are located near the ends of the chromosome.

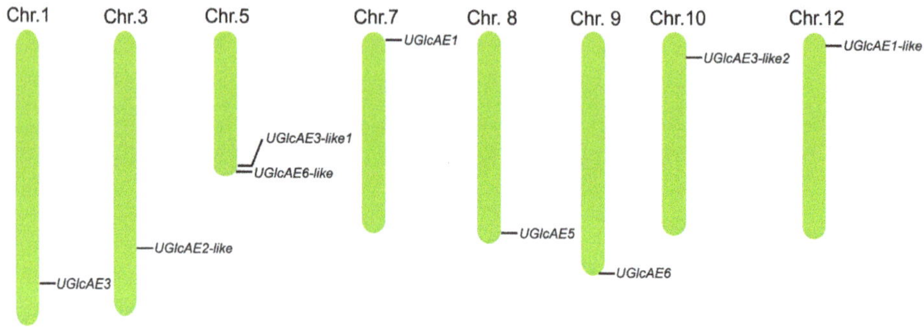

Figure 3. Chromosomal localization of the UGlcAE family genes in tomato. The numbers above each chromosome indicates the chromosome number.

2.5. Sequence Alignments and Hydrophilicity Analysis of SlUGlcAE Family

UGlcAE is one of the short-chain dehydrogenase/reductase (SDR) enzyme families, and therefore the amino acid sequences of SlUGlcAE and AtUGlcAE contained two conserved motifs that existed in SDR protein families [30,31]. As shown in Figure 4, the two motifs contain an N-terminal GxxGxxG (x represents any amino acid) sequence to bind to NAD $(P)^+$, and a motif (YxxxK), which play a catalytic role [32].

As indicated in Supplementary Materials Figure S1, all of the SlUGlcAE proteins were trans-membrane proteins. The result was consistent with previous reports [13,14,24]. Among the nine proteins, most of SlUGlcAEs had only one trans-membrane helice with more than 80% probability, except that SlUGlcAE1 and SlUGlcAE5 were more likely to contain two trans-membrane helices.

2.6. Spatiotemporal Expression Patterns Analysis of UGlcAE Genes in Tomato

To gain the expression patterns of *UGlcAE* genes in different tissues and organs of tomato, and in the developmental stages of the fruit, we performed expression patterns analysis of the *SlUGlcAE* genes (Figure 5) with the RNA-Seq database on the website of the functional genomics database of the tomato plant. The expression profiles of the nine tomato *UGlcAE* genes showed different patterns of temporal- and tissue-specific expression (Figure 5). The results of the tomato cultivar showed that three genes, including *SlUGlcAE1*, *SlUGlcAE1-like*, and *SlUGlcAE6*, were strongly expressed in the bud, flower, leaf, root, and most fruit ripening stages (Figure 5A). The same three genes (*SlUGlcAE1*, *SlUGlcAE1-like*, *SlUGlcAE6*) showed a similar expression characteristic in the cultivated tomato, with a lower expression at the breaker stage (Figure 5A). In addition, *SlUGlcAE1* and *SlUGlcAE6* had a high expression in leaf and most fruit development stages, and *SlUGlcAE1-like* and *SlUGlcAE3* exhibited a high expression in leaf of currant tomato (Figure 5B). Specifically, *SlUGlcAE6-like* exhibited a very low expression level in root, bud, leave, flower, and fruit of the cultivar tomato and wild relative *Solanum pimpinellifolium* plants (Figure 5A,B).

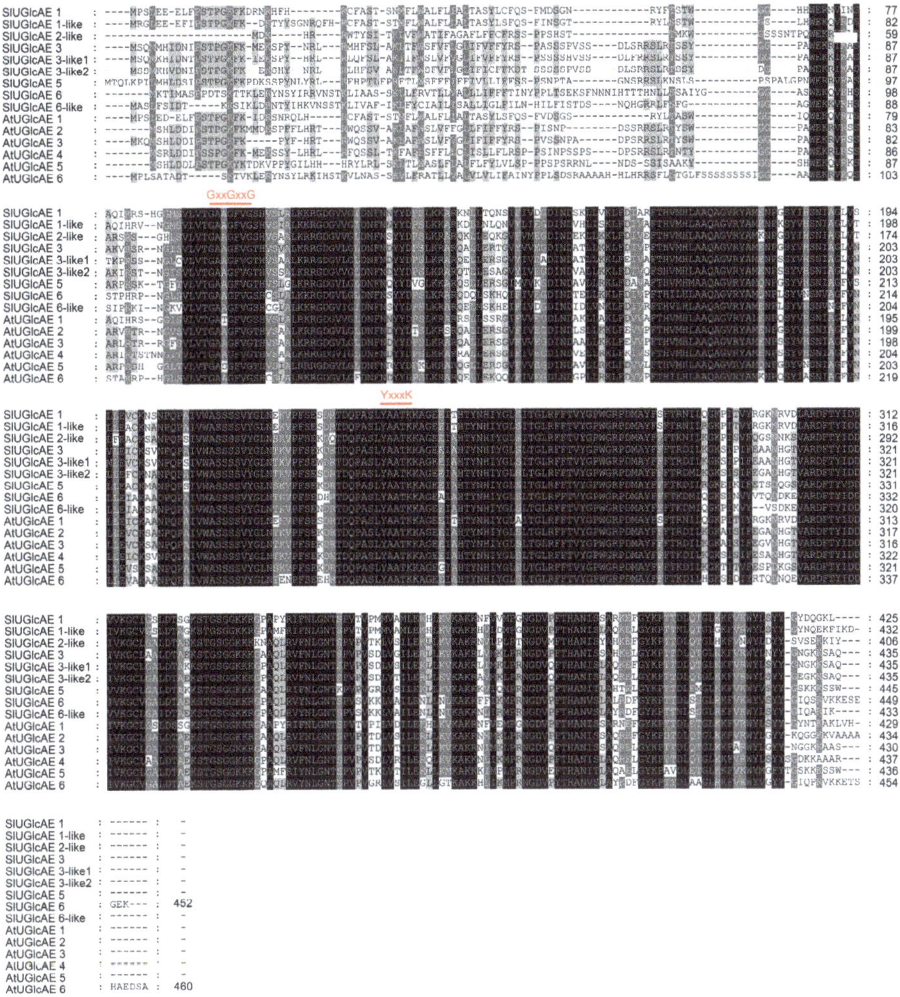

Figure 4. Amino acid sequence alignment of SlUGlcAE in tomato and AtUGlcAE in *Arabidopsis thaliana*. The black backgrounds indicate the strictly conserved residues, and the gray backgrounds indicate the similar amino acid residues. The GxxGxxG and YxxxK motifs are marked above the sequence alignment in red.

2.7. GO Analysis of the UGlcAE Genes in the Tomato

In order to compare the product functions of *UGlcAE* genes in tomato and *Arabidopsis*, we analyzed *SlUGlcAE* genes and its six orthologous genes in *Arabidopsis*. As it is shown in the gene ontology (GO) map (Figure 6), all of the *UGlcAE* genes of *A. thaliana* are involved in cellular components, molecular functions, and biological processes. However, the *SlUGlcAE* genes just have roles in molecular functions and biological processes.

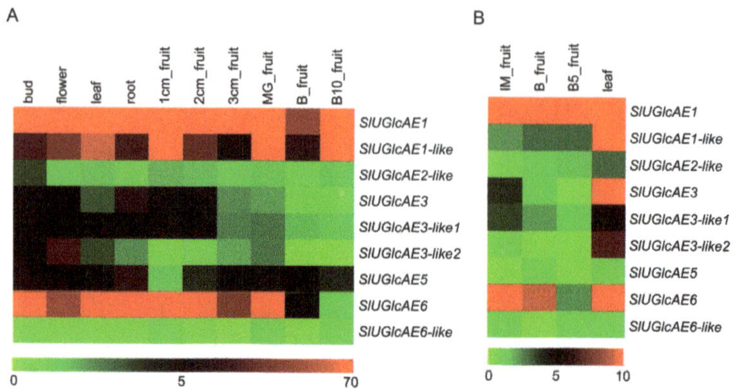

Figure 5. Heatmap analysis of UGlcAE family gene expression in various organs of tomato. (**A**) tomato cultivar *Solanum lycopersicum* (mature green stage (MG), breaker stage (B), ten days after breaker stage (B10)); (**B**) wild relative *Solanum pimpinellifolium* (immature green stage (IMG), breaker stage (B), five days after breaker stage (B5)). The expression data was gained from pubic RNA-seq data and shown as $\log_2$ as calculated by FPKM values (fragments per kilo base of exon model per million) mapped reads. The green boxes represent the lower expression level, whereas the red boxes represent the higher expression level.

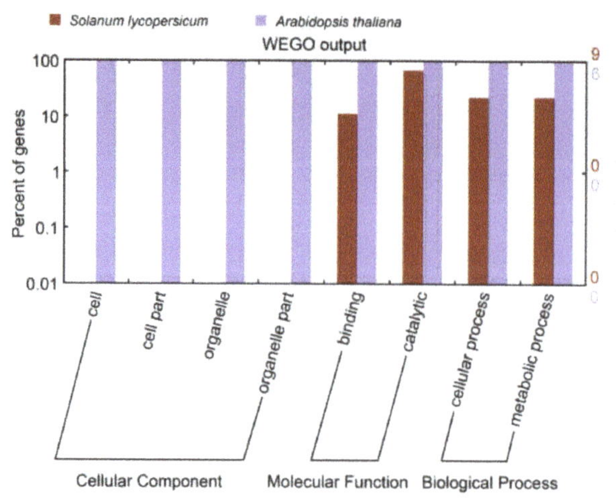

Figure 6. Assignment of Gene Ontology categories to *UGlcAE* genes in tomato. The lengths of the rectangular columns indicate the number of genes that participated in the corresponding classification. The purple rectangular columns mean gene functions of *UGlcAE* in *Arabidopsis*, the red rectangular columns represent gene functions of *UGlcAE* in tomato. All of the gene functions were classified into three categories, which were further divided into eight minor terms.

2.8. Selective Pressure on UGlcAE Proteins in the Tomato

To examine the evolutionary conservation of the UGlcAE proteins, the selective pressure on the UGlcAE was analyzed with SELECTON. We found that the domain of SlUGlcAE3-like1 protein was undergoing strong purifying selection (Figure 7). Selective pressure analyses of the other SlUGlcAE proteins were also analyzed and the results are shown in Supplementary Material Figure S2. These results confirm that the *SlUGlcAE* genes are undergoing strong purifying selection. The amino acids that are emphasized in yellow are under positive selection; however, no positive selection site was found in this selection analysis. These results confirm that these gene family members were very conservative in evolution, which imply them playing a pivotal function in pectin biosynthesis. Selection pressure in the promoter regions of *SlUGlcAE* genes indicated that they are also undergoing negative selection (Figure S3).

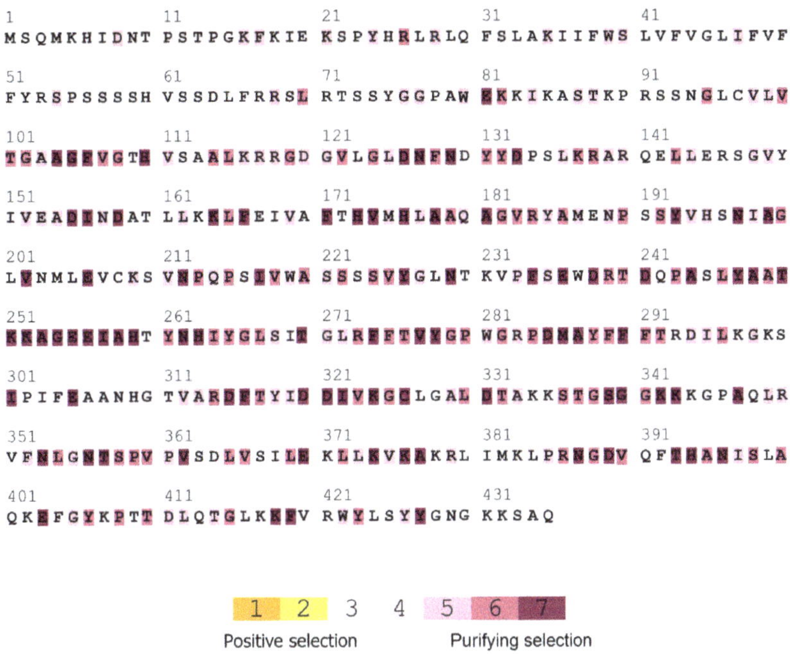

Figure 7. Selection pressure analysis of the UGlcAE proteins in tomato. The red shades represent ω < 1 (purifying selection). Amino acid sequence of SlUGlcAE3-like1 is shown, and the sequences of other SlUGlcAE proteins are presented in the Supplementary Materials Figure S2.

2.9. *Cis*-Acting Elements Analysis of the UGlcAE Genes in the Tomato

To explore the *cis*-acting elements of *SlUGlcAE* genes, we analyzed the 0.5 kb upstream sequences of nine *SlUGlcAE* genes using online software Plant CARE and the result was shown in Figure 8 and Supplementary Materials Table S1. The analysis result of 1.5 kb upstream genomic sequences of genes was shown in Supplementary Materials Table S2.

Kinds, numbers, and locations of *cis*-elements in the upstream of *SlUGlcAE* genes were shown in Figure 8A, and the functional descriptions of these stress-related, hormone-related, and development-related *cis*-elements were exhibited in Figure 8B. As shown in Figure 8A, there are three *cis*-acting elements that are related to hormone, including TGA-element, TATC-box and TCA-element, and 5 stress-related elements including HSE (heat stress-related element), TC-rich repeats (*cis*-acting element involved in defense and stress responsiveness), LTR (*cis*-acting element involved in low-temperature responsiveness), WUN-motif (wound-responsive element), and MBS (MYB Binding Site), and seven elements that are involved in development (Skn-1_motif, HD-Zip 1, HD-Zip 2, circadian, CAT-box, W box, and ELI-box3). Among them, the 0.5 kb upstream regions of four *SlUGlcAE* genes were found to be the presence of heat stress-related element (HSE), of which had two HSE elements in the 0.5 kb upstream region of *SlUGlcAE6* and 1 HSE elements in the 0.5 kb upstream regions of *SlUGlcAE1*, *SlUGlcAE3*, and *SlUGlcAE5*. Furthermore, defense- and stress-response element (TC-rich repeats) was identified in the 0.5 kb upstream regions of four *SlUGlcAE* genes (*SlUGlcAE1-like*, *SlUGlcAE3-like1*, *SlUGlcAE3-like2*, and *SlUGlcAE5*), and circadian element (circadian) was found in the 0.5 kb upstream regions of three *SlUGlcAE* genes (*SlUGlcAE1-like*, *SlUGlcAE3-like1*, and *SlUGlcAE3-like2*), and endosperm expression-related element (Skn-1_motif) was discovered in the 0.5 kb upstream regions of two *SlUGlcAE* genes (*SlUGlcAE1* and *SlUGlcAE6*), and other 11 elements are all present in the 0.5 kb upstream regions of only one *SlUGlcAE* gene. Four elements are located in 0.5 kb upstream region of *SlUGlcAE1*, and element numbers are diversified in the 0.5 kb upstream region of other genes, respectively (four in *SlUGlcAE1-like*, three in *SlUGlcAE2-like*, one in *SlUGlcAE3*, two in *SlUGlcAE3-like1*, four in *SlUGlcAE3-like2*, five in *SlUGlcAE5*, four in *SlUGlcAE6*, and zero in *SlUGlcAE6-like*).

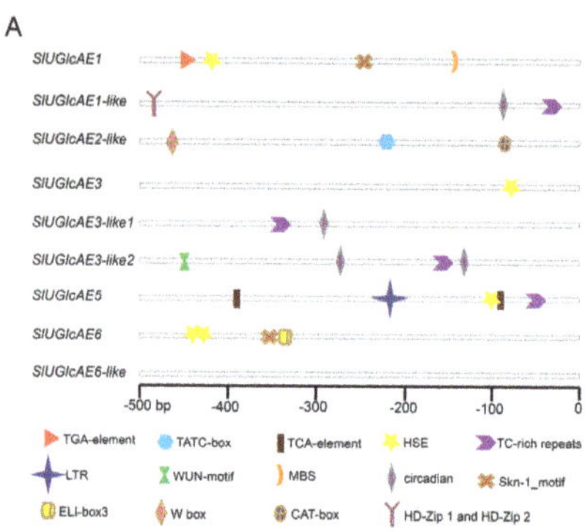

B

Element name	Function annotation
TGA-element	auxin-responsive element
TATC-box	cis-acting element involved in gibberellin-responsiveness
TCA-element	cis-acting element involved in salicylic acid responsiveness
HSE	cis-acting element involved in heat stress responsiveness
TC-rich repeats	cis-acting element involved in defense and stress responsiveness
LTR	cis-acting element involved in low-temperature responsiveness
WUN-motif	wound-responsive element
MBS	MYB Binding Site
circadian	cis-acting regulatory element involved in circadian control
Skn-1_motif	cis-acting regulatory element required for endosperm expression
ELI-box3	elicitor-responsive element
W box	No functional description
CAT-box	cis-acting regulatory element related to meristem expression
HD-Zip 1	element involved in differentiation of the palisade mesophyll cells
HD-Zip 2	element involved in the control of leaf morphology development

Figure 8. Kinds and numbers of stress-related, hormone-related, and development-related *cis*-elements in the upstream of *SlUGlcAE* genes. (**A**) Various symbols indicate different *cis*-acting elements; and (**B**) Element names and their functional descriptions.

No of stress-, hormone-, and development-related element was found in the 0.5 kb upstream region of *SlUGlcAE6-like* gene, and the previous spatial expression patterns showed *SlUGlcAE6-like* was lowly expressed at every stages, which are highly consistent. Thus, we infer that *SlUGlcAE6-like* may not be involved in the process of growth and development. In addition, HD-Zip 1 and HD-Zip 2 are located in the same position of *SlUGlcAE1* and *SlUGlcAE1-like*, implying that the two elements might be closely related and complementary to each other. Furthermore, other stress-, hormone-, and development-unrelated *cis*-acting elements have also been identified. For example, core promoter element (TATA-box) and common *cis*-acting element (CAAT-box) are present in the 0.5 kb upstream regions of all nine *UGlcAE* genes. Light responsive *cis*-acting regulatory elements (GATA-motif, chs-CMA1a, G-box) and enhancer (TA-rich region) could also be found.

2.10. Expression Patterns of SlUGlcAE Family Genes in Response to IAA, GA and SA

Plant hormones, such as IAA, GA, and SA are used as endogenous messengers in response to biotic and abiotic stresses in plants [33]. It has been reported that the treatments of plants by exogenous hormones often lead to transient and rapid transcriptional changes in the whole genome [34]. According to *cis*-acting elements analysis of the *SlUGlcAE* genes upstream, three *cis*-acting elements that are related to plant hormones (IAA, GA, and SA) are located in 0.5 kb upstream genomic sequences of *SlUGlcAE1*, *SlUGlcAE2-like*, and *SlUGlcAE5*, respectively (Figure 9A). Thus, we investigated the expression profiles of *SlUGlcAE1*, *SlUGlcAE2-like*, and *SlUGlcAE5* with IAA, GA, and SA treatments (Figure 9B–D).

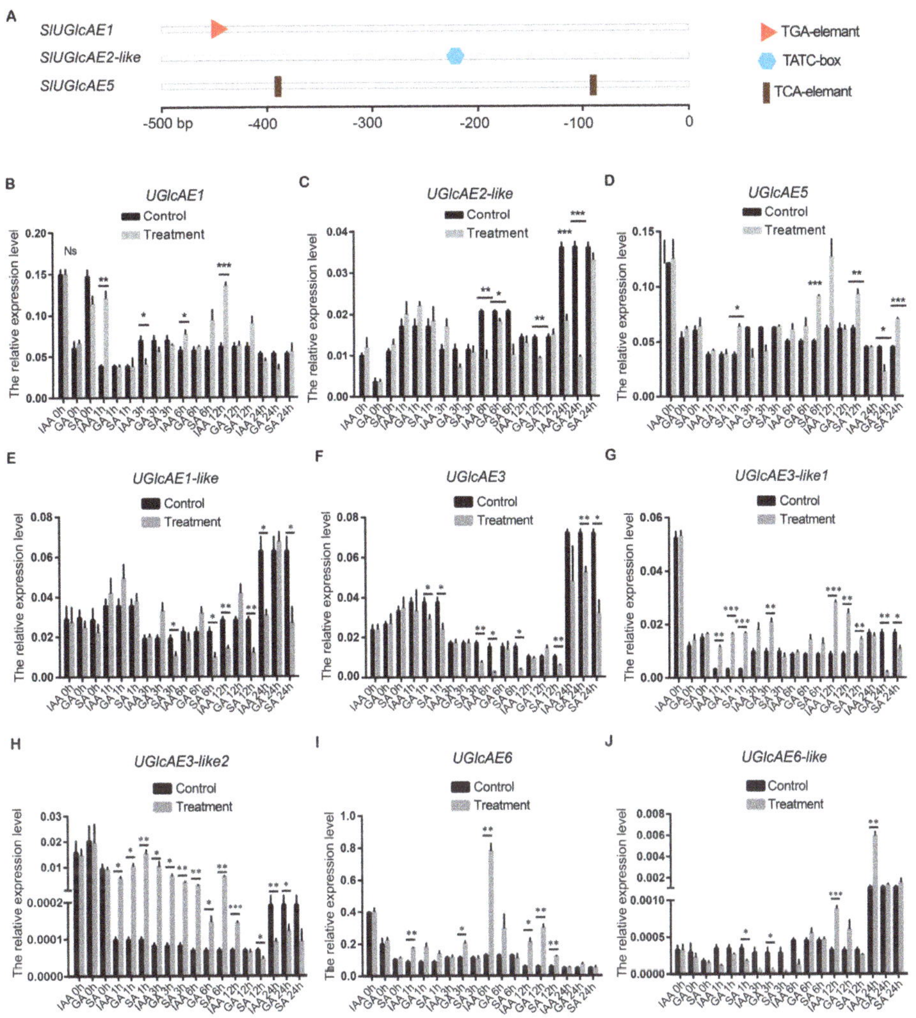

Figure 9. The detection of expressional level of *SlUGlcAE* genes after plant hormone treatments. (**A**) Hormone-related *cis*-elements prediction in the upstream of three *SlUGlcAE* genes (TGA-element: auxin-responsive element, TATC-box: gibberellin-responsive element, TCA-element: salicylic acid-responsive element); (**B**) The qPCR expression analysis of *SlUGlcAE1* response to plant hormone treatments; (**C**) Responses of *SlUGlcAE2-like* to plant hormones; (**D**) Expression profile analysis of *SlUGlcAE5* under plant hormone treatments; (**E**) Expression of *SlUGlcAE1-like* after hormone treatments; (**F**) Expression level of *SlUGlcAE3* after three hormones treatments; (G) Responses of *SlUGlcAE3-like1* under hormone treatments; (**H**) Expression profile analysis of *SlUGlcAE3-like2* to plant hormones; (**I**) The expression analysis of *SlUGlcAE6* under three hormones treatments; and, (**J**) Expression pattern of *SlUGlcAE6-like* response to plant hormone treatments. The error bars represent the SEM. * $p < 0.05$, ** $p < 0.01$, *** $p < 0.001$, Ns: No significant.

The expression of *SlUGlcAE1* in response to IAA was increased at 1 h, decreased at 3 h, and increased thereafter, including 6 and 12 h, and returned to background level until 24 h (Figure 9B). On the whole, the expressional level of *SlUGlcAE1* gene was up-regulated after IAA treatment, although accompanying an unclear reason of being down-regulated only at 3 h. The expressional level of *SlUGlcAE1* gene was not significant different when comparing to the control in response to GA and SA throughout the treated process (Figure 9B). Combining to the result analysis of *cis*-acting elements in *SlUGlcAE1* gene upstream, we found that *SlUGlcAE1* expressional regulation was consistent with TGA *cis*-acting elements existing in the upstream of this gene, which is an auxin-responsive element, while GA or SA responsive element is non-existent (Figure 9A). This indicated that the expression of *SlUGlcAE1* gene could be regulated by IAA, but it could not be regulated by GA and SA (Figure 9B). The *SlUGlcAE2-like* gene under IAA treatment was down-regulated at 6 and 24 h, and no obvious difference was found at other time points (Figure 9C). The change of *SlUGlcAE2-like* expression pattern was unobvious under SA treatment. *SlUGlcAE2-like* showed increasingly strong down-regulation in expression at 6, 12 and 24 h, while similar expression level to the control within the first 3 h under GA treatment (Figure 9C). These results indicated that the expression of *SlUGlcAE2-like* gene could be regulated by GA, and be regulated by IAA irregularly, but not be regulated by SA (Figure 9C). This is aligned with the previous *cis*-acting elements analysis result. That is the fact that there are a gibberellin-responsive element (TATC-box) within 0.5 kb genomic sequences of *SlUGlcAE2-like* gene upstream (Figure 9A) and another gibberellin-responsive element (GARE-motif) within 1.5 kb genomic sequences outside 0.5 kb genomic sequences of *SlUGlcAE2-like* gene upstream (Table S2). There is an auxin-responsive element (AuxRR-core) within 1.5 kb genomic sequences outside 0.5 kb genomic sequences of *SlUGlcAE2-like* gene upstream (Table S2). *SlUGlcAE5* after IAA treatment showed a similar expression level to the control at different time points (Figure 9D). The expression level of *SlUGlcAE5* after GA treatment was no significant change in the first 12 h, but it reduced at 24 h with unknown reasons). *SlUGlcAE5* was up-regulated at 1 h, not affected at 3 h, and up-regulated continuously at later time points, including 6, 12, and 24 h after SA treatment (Figure 9D). The results suggested the expression of *SlUGlcAE5* gene could indeed be regulated by SA. It keeps consistent with the result of *cis*-acting elements analysis, which has a salicylic acid-responsive element (TCA element) in 0.5 kb genomic sequences of *SlUGlcAE5* gene upstream (Figure 9A). Taken together, these data suggest that although the result of *cis*-acting elements analysis may contain a few false positives, the prediction of these three hormone response sites in our study is still relatively reliable.

In addition, we also examined the expression level of the other six *UGlcAE* genes after three hormones treatments. The results were shown in Figure 9E–J. Both *SlUGlcAE1-like* and *SlUGlcAE3* were down-regulated by SA (Figure 9E,F). *SlUGlcAE3-like2* was up-regulated by the three hormones within 12 h of hormone treatment. However, it was down-regulated by three hormones at 24 h (Figure 9H). *SlUGlcAE3-like1*, *SlUGlcAE6*, and *SlUGlcAE6-like* were affected by different hormones at different times after hormones treatments (Figure 9G,I,J).

2.11. The Firmness, Pectin Content and the Expression Level of UGlcAE Family Genes in Tomato Fruits at Different Development Stages

To further explore the change of firmness, pectin contents, and expression levels of *UGlcAE* genes, we investigated the firmness, pectin contents, and the expression levels of *UGlcAE* family genes in fruits at different stages of tomato development (Figure 10). As the fruit matured, the firmness of tomato fruit decreased gradually (Figure 10A). The content of water-soluble pectin (WSP) in the development of tomato fruit showed an increased trend, reached a maximum at the MG stage, and then decreased (Figure 10B). The increase of WSP content in the early may be because of the accumulation of pectin as the fruit grows. The content of WSP gradually decreased after the MG stage, probably due to the gradual degradation of partially WSP by some pectinases. As shown in Figure 10C, nine genes of the UGlcAE family have different expression patterns in the development of tomato fruit. Among them, four genes (*UGlcAE1*, *UGlcAE1-like*, *UGlcAE5*, and *UGlcAE6*) showed relatively high expression levels, and other five genes had lower expression levels. The expression levels of both *UGlcAE1* and *UGlcAE5* first increased, reached the maximum value at the MG stage, and then decreased. This is consistent with the trend of WSP content in tomato fruit development.

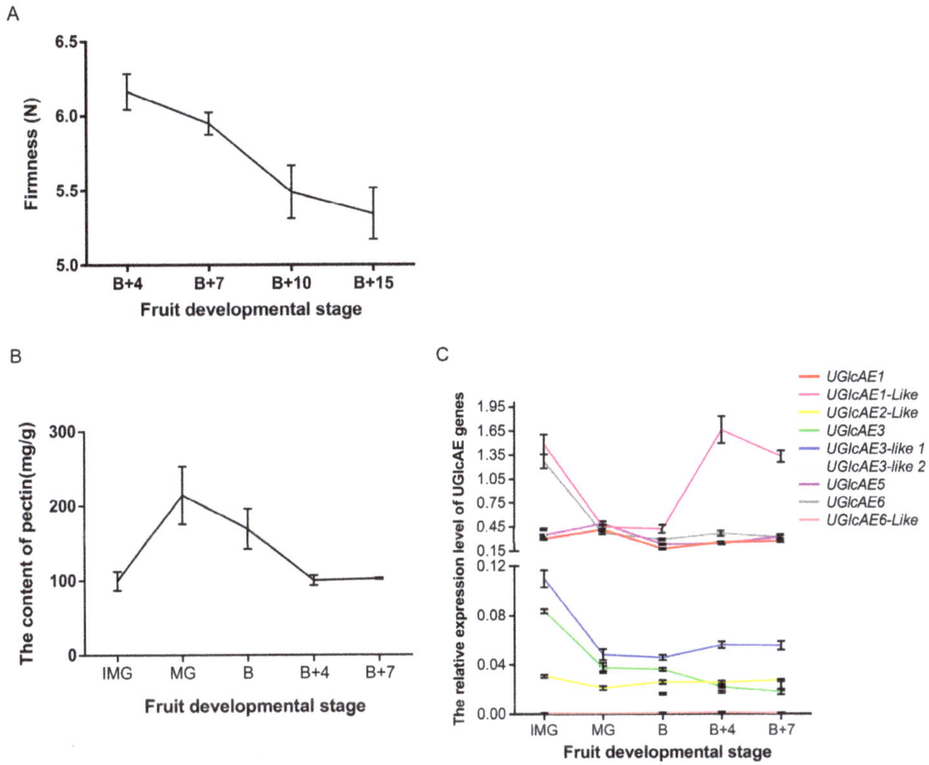

Figure 10. Firmness, pectin contents, and expression levels of UGlcAE family genes during the development of tomato fruit. (**A**) Firmness of tomato fruit in five different stages; (**B**) water-soluble pectin (WSP) contents of fruit in five development stages; and (**C**) Expression level of nine UGlcAE family genes in stages.

3. Discussion

UGlcAE is capable of reversibly interconverting UDP-GlcA and UDP-GalA, which plays an important role in pectin synthesis. It bring many new opportunities to study gene families in an evolutionary context with various plant genomes being sequenced [4]. To investigate the phylogenetic relationship of *UGlcAE* gene family members, we searched and collected the amino acid sequences of UGlcAE from 10 plant species. Allof the six subfamilies exist in *Arabidopsis thaliana*. However, in other plant species, the numbers of subfamilies of *UGlcAE* genes vary from three to five. Interestingly, the UGlcAE4 subfamily is specifically present in *Arabidopsis thaliana* and *Arabidopsis lyrata* subsp. *lyrata*, whereas it is absent from the other eight plant species. According to this analysis, we infer that the UGlcAE4 protein may have played specific roles in *Arabidopsis*.

It is mentionable that the members of the UGlcAE2 subfamily were not classified into the same cluster. This indicates that there are great differences in the sequences of different members within the UGlcAE2 subfamily. We further speculate that the UGlcAE2 subfamily may be dividing new functions. Moreover, some members of the UGlcAE2 subfamily grouped with UGlcAE3 in *Arabidopsis*. It indicates that there are similarities in the sequences of these UGlcAE2 members and AtUGlcAE3. Therefore, it is likely that the members of the UGlcAE2 subfamily and AtUGlcAE3 might have similar functions or undergo gene fusion.

Similar to the previous studies of the UGlcAE in *Arabidopsis* [12,13], two branches of the phylogenetic tree are trustworthily occupied by UGlcAE1 and UGlcAE6, respectively, meanwhile, UGlcAE2, UGlcAE3, UGlcAE4, and UGlcAE5 are located together in one branch of the phylogenetic tree. This result implies a more ancient role of UGlcAE1 and UGlcAE6, concurrently, the other UGlcAEs might have evolved later [12,13].

In all ten plant species, only *Arabidopsis* contains all of the six *UGlcAE* subfamilies and every subfamily has at least one member. This is comprehensive and regular, which is very congruent with its identity of the model plant.

The expression patterns of the nine genes differed in the different tissues and development stages of tomato. However, it is still possible to find a certain rule from Figure 5. As mentioned by Mølhøj in 2004 [13], the heatmap representation of all the expression patterns reveals that *UGlcAE1* and *UGlcAE6* subfamilies (except *UGlcAE6-like*) were strongly expressed in cultivar tomato, whereas *UGlcAE2*, *UGlcAE3*, and *UGlcAE5* subfamilies were lowly expressed isoforms. However, *UGlcAE6-like* showed considerably lower expression levels in tomato. This is consistent with the result of *cis*-element analysis of *UGlcAE* gene families in Figure 8, which is no significant (stress-, hormone- and development-related) *cis*-acting elements being found within the range of 0.5 kb in front of the *UGlcAE6-like* gene coding region.

The expression trends of *UGlcAE1* and *UGlcAE5* in tomato fruit development were consistent with those of WSP content, indicating that *UGlcAE1* and *UGlcAE5* may be more closely related to the formation of WSP during the fruit ripening when compared to other members of the *UGlcAE* gene family. The expression level of *UGlcAE5* was high in Figure 10C and low in Figure 5A, which indicate that *UGlcAE5* may be easily affected by some factors in the environment and cause its expression level to be unstable. In addition, other results of Figure 10C (high expression level of the three genes (*UGlcAE1*, *UGlcAE1-like*, and *UGlcAE6*) and low expression level of the five genes (*UGlcAE2-like*, *UGlcAE3*, *UGlcAE3-like1*, *UGlcAE3-like2* and *UGlcAE6-like*)) were basically consistent with the results of Figure 5A. This may suggest that the expression of these eight genes is relatively stable during tomato fruit development.

After three hormones treatments, the expression of *UGlcAE1* was more susceptible to IAA, and the expression of *UGlcAE5* was more susceptible to SA. These results suggest that the WSP content of tomato may be more susceptible to IAA and SA in fruit development. *UGlcAE6-like* exhibited the very low expressions in Figures 5, 9J and 10C, indicating that *UGlcAE6-like* is less likely to affect WSP content during the tomato fruit ripening.

Pectin degradation is a major effect on fruit softening [35]. The identifications of the family genes help to understand more about these genes and can better investigate the mechanisms of pectin production and degradation. An in-depth understanding of specific gene expression during ripening and maturation of tomato fruits [36] will enable the precise manipulation of expression of new associate genes to more precisely control the mechanisms of cell wall modification and softening. This is still an outstanding question so far [35].

4. Materials and Methods

4.1. Data Set Collection and Identification of SlUGlcAE Genes

The protein databases of all ten species were retrieved from the National Center for Biotechnology Information (NCBI) FTP site (available online: http://www.ncbi.nlm.nih.gov/Ftp/). The cDNA, CDS, and genome sequence data in tomato were downloaded from the Solanaceae Genomics Network (SGN) (available online: http://solgenomics.net) [37] and Tomato Functional Genomics Database (TFGD) (available online: http://ted.bti.cornell.edu) [38]. Other information and sequences of *Arabidopsis thaliana* UGlcAEs (AtUGlcAEs) were obtained from the *Arabidopsis* Information Resource (TAIR; available online: http://www.arabidopsis.org/) [39]. The UGlcAE proteins of tomato (SlUGlcAEs) were predicted depending on the UGlcAE hidden Markov model (HMM) profile from the Pfam database (available online: http://pfam.sanger.ac.uk/) [40], which was used to search the *S. lycopersicum* UGlcAE proteins sequences by the HMMSEARCH program from HMMER software (available online: http://hmmer.janelia.org) [41]. In the case of the uncompleted protein databases, all of the results were then used as queries in TBLASTN searches against the tomato genomic sequences. To further confirm UGlcAE proteins, the domains of candidate sequences were predicted with the Pfam online server (available online: http://pfam.sanger.ac.uk/) [40] and SMART online server (available online: http://smart.embl-heidelberg.de/) [42]. The tomato genomic sequences were also checked using BLASTP at the NCBI site (available online: http://blast.ncbi.nlm.nih.gov), retaining only those sequences with highly significant matches to annotated UGlcAE proteins. The same procedure was used to search UGlcAE family members in the protein databases of the following nine species: *Cucumis sativus*, *Capsicum annuum*, *Solanum tuberosum*, *Arabidopsis thaliana*, *Nicotiana tabacum*, *Populus trichocarpa*, *Solanum pennelli*, *Zea mays*, and *Arabidopsis lyrata* subsp. *lyrata*.

The tomato *UGlcAE* gene subfamilies were named according to the orthologous *UGlcAE* genes in the *A. thaliana* genome. The subfamilies of *UGlcAE* genes in the tomato were distinguished by Arabic numerals, and different members of a subfamily were designated with the numbers.

4.2. Phylogenetic Analysis

A phylogenetic tree of UGlcAE was constructed by analyzing full-length proteins from *S. lycopersicum*, *C. sativus*, *C. annuum*, *S. tuberosum*, *A. thaliana*, *N. tabacum*, *P. trichocarpa*, *S. pennelli*, *Z. mays*, and *A. lyrata* subsp. *lyrata* in the MEGA5 software (Center for Evolutionary Medicine and Informatics, Arizona State University, Tempe, AZ, USA) using the Neighbor-Joining method [43]. Bootstrap analysis was employed using 1000 replicates.

4.3. Selective Pressure Analysis on UGlcAE Proteins in the Tomato

The ratio of non-synonymous to synonymous substitutions (dN/dS; termed ω) at each codon site of each protein was identified, according to an empirical Bayesian method using the Server for the identification of site-specific positive selection and purifying selection (SELECTON version 2.4, Tel Aviv University, Tel Aviv, Israel [44,45]. Selection pressure analysis can be used to identify purifying or positive selection of specific areas in a sequence, and the sites that ω values significantly >1 or <1 suggest positive (Darwinian) or purifying, respectively [46]. The selection pressure acting on the coding sequences of the *SlUGlcAE* genes was recognized with the M8 model (extra category $\omega s \geq 1$, beta distribution, and positive selection allowed). In order to ensure the accuracy of the results, a likelihood ratio test was used to test the significance of the ω values [47], which compares two nested models: a null model that assumes no positive selection (M8a) and an alternative model that assumes positive selection (M8). Non-nested models, including M8a (extra category ωs set to 1) and MEC (positive selection allowing model), were also used in the pressure analysis.

4.4. Chromosomal Location

Locations of the *UGlcAE* genes on the tomato chromosomes were obtained using NCBI website (available online: http://www.ncbi.nlm.nih.gov/mapview/), according to their positions in the SGN (available online: http://solgenomics.net/) [37].

4.5. Gene Structure Analysis

To analyze gene structure, the exon, and intron structures of *SlUGlcAE* and *AtUGlcAE* genes were generated using the Gene Structure Display Server 2.0 (available online: http://gsds.cbi.pku.edu.cn) [48] by aligning the CDS sequences with the corresponding genomic DNA sequences from the SGN (available online: http://solgenomics.net/) [37].

4.6. Gene Ontology Analysis

For the gene ontology (GO) analysis, the *UGlcAE* gene family members in tomato and *Arabidopsis* were classified according to their each GO numbers from the SGN (available online: http://solgenomics.net/) [37] and the TAIR (available online: http://www.arabidopsis.org/) [39]. After being normalized by the online service CapitalBio Molecule Annotation System (MAS) 3.0 (available online: http://bioinfo.capitalbio.com/mas3/) [49], their GO numbers were identified and visualized with BGI WEGO (available online: http://wego.genomics.org.cn/cgi-bin/wego/index.pl) [50].

4.7. Analysis of Expression Profile of UGlcAE Genes in Tomato Various Tissues

The expression profile was obtained through analyzing microarray data. The microarray data were downloaded from the Tomato Functional Genomics Database (available online: http://ted.bti.cornell.edu/cgi-bin/TFGD/digital/home.cgi) [38], including the *UGlcAE* genes expression in 10 tissues (bud, flower, leaf, root, 1 cm fruit, 2 cm fruit, 3 cm fruit, mature green stage (MG) fruit, breaker stage (B) fruit, and ten days after breaker stage (B10) fruit) of the tomato cultivar (*Solanum lycopersicum*), and four tissues (immature green stage (IMG) fruit, breaker stage (B) fruit, five days after breaker stage (B5) fruit, and leaf) of the wild species (*Solanum pimpinellifolium*). Only genes with an at least five units average expression signal at one time point and the similar trend in different biological replicates were considered to be expressed at the time point. The expression patterns of the *SlUGlcAE* genes were estimated by intensity values and were visualized using MultiExperiment Viewer (Broad Institute of MIT and Harvard University, Boston, MA, USA [51].

4.8. Sequence Alignments and Prediction of Transmembrane Domains of SlUGlcAE Family

All nine SlUGlcAE and six AtUGlcAE protein sequences were aligned using the (version 5.0.6, North Carolina State University, Raleigh, NC, USA), and then the results were output by genedoc program. Next, the hydrophilicity of the SlUGlcAE protein sequences was predicted by the trans-membrane Hidden Markov model algorithm (available online: http://www.cbs.dtu.dk/services/TMHMM/) [52].

4.9. The Analysis of SlUGlcAE Family Protein Domains

The domains of UGlcAE family proteins in tomato were analyzed by the Pfam (available online: http://pfam.xfam.org/search) [40].

4.10. Cis-Elements in the Upstream of SlUGlcAE Genes

For identifying the *cis*-acting elements of UGlcAE genes upstream, we obtained the sequences of upstream regions (1.5 kb) of nine *SlUGlcAE* genes from NCBI (available online: https://www.ncbi.nlm.nih.gov/) and identified *cis*-acting motifs by PlantCARE (available online: http://bioinformatics.psb.ugent.be/webtools/plantcare/html/) [53].

4.11. Hormone Treatments

S. lycopersicum plants were grown at 25 ± 2 °C with a 12 h light/dark photoperiod. The humidity was maintained at approximately 60% to 70%, and the photosynthetic photon flux density was controlled at about 120 μmol photons/m^2/s. When the seedlings were six weeks old, the plants were treated with IAA (100 μM), GA (100 μM) and SA (100 μM), respectively [54]. Plant leaves were collected at 0, 1, 3, 6, 12, and 24 h after treatments, immediately frozen in liquid nitrogen, and then stored at −80 °C until use.

4.12. Plant Materials

Other *S. lycopersicum* seedlings were grown in the same conditions with the above mentioned seeds (see Section 4.11). Different fruits were harvested in the following five stages: immature green stage (IMG), mature green stage (MG), breaker stage (B), four days after breaker stage (B + 4), and seven days after breaker stage (B + 7). All of the plant samples were retrieved at the same time each day, and then frozen in liquid nitrogen and stored at −80 °C.

4.13. Real-Time PCR

Total RNA was extracted from the leaves using the Total RNA Kit (BioTeke Corporation, Beijing, China), following the manufacturer's instructions. Integrity of the RNA was verified by agarose gel electrophoresis. Synthesis of the cDNA was performed from the total RNA samples using the PrimeScript™ RT Reagent Kit, according to the protocol with gDNA Eraser (TaKaRa, Dalian, China). All of the primer sequences are shown in Table 2. *EF1α* gene was used as the internal control under abiotic stress [55], and the *SlCAC* gene was selected as an internal standard during tomato development [56] to quantitate the expression of *SlUGlcAE* genes. Real-time PCR was performed using CFX96 Touch™ real-time PCR system (Bio-Rad, Hercules, CA, USA) with a SYBR Premix Ex Taq™ II Kit (Bio-Rad). The reactions were carried out in the following conditions: denaturation at 94 °C for 4 min, 40 cycles of 5 s at 95 °C, 30 s at 60 °C, 15 s at 95 °C, 20 s at 60 °C, and 15 s at 95 °C. Three biological duplications were used. The $2^{-\Delta\Delta Ct}$ method was used to visualize and analyze the real-time PCR data [57,58].

Table 2. Primer sequences used for quantitative real-time PCR in the paper.

Primer Name	Sense Sequence (5′ → 3′)	Antisense Sequence (5′ → 3′)
EF1a	TACTGGTGGTTTTGAAGCTG	AACTTCCTTCACGATTTCATCATA
CAC	CCTCCGTTGTGATGTAACTGG	ATTGGTGGAAAGTAACATCATCG
SlUGlcAE1	TGTAAAATGGCTAATCCACAACCT	AAAAACCGCAATCCAGTAATCG
SlUGlcAE1-like	ACCGGTGTTTCGCTTCAACGAGT	AAGACTACCCCATGTGGAGGAGAG
SlUGlcAE2-like	GCGAGTCTATACGCTGCCACA	CGTCTTCTTACCACCACTTCCTG
SlUGlcAE3	CAACCCCAGGAAAGTTCAAGATGG	GACGAAGAAGCTGGAGATCTGTAG
SlUGlcAE3-like1	AGGCAGCTAATCATGGCACAGTC	AAGATCAGATACCGGGACAGGTG
SlUGlcAE3-like2	TCATGGGACTGTTGCTAGGGACT	CCTTGGCAACTTCATCACAGCTC
SlUGlcAE5	TGTAAAATGGCTAATCCACAACCT	AAAACCGCAATCCAGTAATCG
SlUGlcAE6	CCACCTGACACAAGCAAAACCAC	GGAGGATAGAAGGTTATGGGTAGTGG
SlUGlcAE6-like	GGACTGATCAACCAGCTAGTCTC	CGTAAACCTTGATCGGCTTCCCTTG

4.14. Fruit Firmness Measurement and Determination of Water-Soluble Pectin Content

As described by Wu and Abbott [59], the fruit firmness was quantified using a Firmness tester (GY-2). Fifteen unbroken tomatoes were taken from each group. The equator of the fruit was placed under a flat probe, and the maximum value was read after pressing down. Each fruit was measured at least three times. Test parameters: Probe pressure rate 1 mm/s, Pressing distance 3 mm.

The content of pectin in tomato fruit was detected by the water-soluble pectin content kit of Suzhou Keming Biotechnology Co., Ltd. (Suzhou, China The principle of determination is to use the acid solution to extract water-soluble pectin, and to determine the content of pectin by carbazole colorimetry. Pectin is hydrolyzed to galacturonic acid, which condenses with carbazole reagent in sulfuric acid solution. The resulting material has a maximum absorption peak at 530 nm.

5. Conclusions

The *Solanaceae* genus is one of the most morphologically various plant families, with more than 3000 described species being distributed worldwide [60]. Apart from being of economic value, tomato is also a model crop for fleshy fruit development [35,61]. In addition, tomato is still the first horticultural crop for which its genome has been sequenced [62]. In this study, we identified nine *SlUGlcAE* genes and analyzed the spatiotemporal expression patterns, the phylogenetic relationships, the selective pressure, the *cis*-acting elements, and so on. We also focused on the response patterns of nine *SlUGlcAE* genes to IAA, GA, and SA, according to the results of *cis*-acting elements analysis. Moreover, the firmness decreased gradually, and WSP showed an increased trend, reached a maximum at the MG stage, and then decreased in the development of tomato fruit. All of the results above have allowed for us to identify tomato orthologs that are related to known *UGlcAE* genes in *Arabidopsis* for in-depth studies. It would also accelerate for executing functional studies based genomics to elucidate their elaborate roles in tomato fruit development, and to be helpful for revealing the roles of other members in the *Solanaceae* genus.

Supplementary Materials: Supplementary materials can be found at http://www.mdpi.com/1422-0067/19/6/1583/s1.

Author Contributions: X.Z., X.D. conceived and designed the study; X.D. collected data, processed and analyzed the data, created the figures, wrote the manuscript and completed the submission; J.L. and Y.P. helped design the experiments, provided the research facility and helped write the manuscript; Y.Z., L.N. and Y.W. provided daily care to the plant and helped collect plant leaves at different times.

Acknowledgments: This work was supported by the National High Technology Research and Development Program of China (863 Program, Grant no. 2010AA10060705).

Conflicts of Interest: The authors declare no conflict of interest.

Abbreviations

UGlcAE/GAE	UDP-D-glucuronic acid 4-epimerase
IAA	indole-3-acetic acid
GA	gibberellin
SA	salicylic acid
GalA	D-galacturonic acid
UDP-GlcA	UDP-D-glucuronic acid
UDP-Glc	UDP-D-glucose
UDP-GalA	UDP-D-galacturonic acid
SDR	short-chain dehydrogenase/reductase
WSP	water-soluble pectin

References

1. Carpita, N.C.; Gibeaut, D.M. Structural models of primary cell walls in flowering plants: Consistency of molecular structure with the physical properties of the walls during growth. *Plant J.* **1993**, *3*, 1–30. [CrossRef] [PubMed]
2. Derbyshire, P.; McCann, M.C.; Roberts, K. Restricted cell elongation in arabidopsis hypocotyls is associated with a reduced average pectin esterification level. *BMC Plant Biol.* **2007**, *7*, 31. [CrossRef] [PubMed]
3. Krupková, E.; Immerzeel, P.; Pauly, M.; Schmülling, T. The tumorous shoot development2 gene of arabidopsis encoding a putative methyltransferase is required for cell adhesion and co-ordinated plant development. *Plant J.* **2007**, *50*, 735–750. [CrossRef] [PubMed]
4. Mccarthy, T.W.; Der, J.P.; Honaas, L.A.; Depamphilis, C.W.; Anderson, C.T. Phylogenetic analysis of pectin-related gene families in physcomitrella patens and nine other plant species yields evolutionary insights into cell walls. *BMC Plant Biol.* **2014**, *14*, 79. [CrossRef] [PubMed]
5. Mouille, G.; Ralet, M.C.; Cavelier, C.; Eland, C.; Effroy, D.; Hematy, K.; McCartney, L.; Truong, H.N.; Gaudon, V.; Thibault, J.F.; et al. Homogalacturonan synthesis in *Arabidopsis thaliana* requires a golgi-localized protein with a putative methyltransferase domain. *Plant J.* **2007**, *50*, 605–614. [CrossRef] [PubMed]
6. Gu, X.; Wages, C.J.; Davis, K.E.; Guyett, P.J.; Bar-Peled, M. Enzymatic characterization and comparison of various poaceae UDP-GlcA 4-epimerase isoforms. *J. Biochem.* **2009**, *146*, 527–534. [CrossRef] [PubMed]
7. Fry, S.C. *Pectins and Their Manipulation*; Seymour, G.B., Knox, J.P., Eds.; Blackwell Publishing: Oxford, UK, 2002; 262p, ISBN 1-841-27228-0.
8. Loewus, F.; Chen, M.S.; Loewus, M.W. The myo-inositol oxidation pathway to cell wall polysaccharides *. *Biogenes. Plant Cell Wall Polysacch.* **1973**, 1–27. [CrossRef]

9. Tenhaken, R.; Thulke, O. Cloning of an enzyme that synthesizes a key nucleotide-sugar precursor of hemicellulose biosynthesis from soybean: UDP-glucose dehydrogenase. *Plant Physiol.* **1996**, *112*, 1127–1134. [CrossRef] [PubMed]
10. Orellana, A.; Mohnen, D. Enzymatic synthesis and purification of [(3)h]uridine diphosphate galacturonic acid for use in studying golgi-localized transporters. *Anal. Biochem.* **1999**, *272*, 224–231. [CrossRef] [PubMed]
11. Broach, B.; Gu, X.; Bar-Peled, M. Biosynthesis of UDP-glucuronic acid and UDP-galacturonic acid in *Bacillus cereus* subsp. Cytotoxis NVH 391-98. *FEBS J.* **2012**, *279*, 100–112. [CrossRef] [PubMed]
12. Usadel, B.; Schlüter, U.; Mølhøj, M.; Gipmans, M.; Verma, R.; Kossmann, J.; Reiter, W.D.; Pauly, M. Identification and characterization of a UDP-D-glucuronate 4-epimerase in arabidopsis. *FEBS Lett.* **2004**, *569*, 327–331. [CrossRef] [PubMed]
13. Marx, M.; Schmandt, C. The biosynthesis of D-galacturonate in plants. Functional cloning and characterization of a membrane-anchored UDP-D-glucuronate 4-epimerase from arabidopsis. *Plant Physiol.* **2004**, *135*, 1221–1230.
14. Gu, X.; Barpeled, M. The biosynthesis of UDP-galacturonic acid in plants. Functional cloning and characterization of arabidopsis UDP-D-glucuronic acid 4-epimerase. *Plant Physiol.* **2004**, *136*, 4256–4264. [CrossRef] [PubMed]
15. Yin, Y.; Huang, J.; Gu, X.; Barpeled, M.; Xu, Y. Evolution of plant nucleotide-sugar interconversion enzymes. *PLoS ONE* **2011**, *6*, e27995. [CrossRef] [PubMed]
16. Feingold, D.S.; Neufeld, E.F.; Hassid, W.Z. Enzymic synthesis of uridine diphosphate glucuronic acid and uridine diphosphate galacturonic acid with extracts from phaseolus aureus seedlings. *Arch. Biochem. Biophys.* **1958**, *78*, 401–406. [CrossRef]
17. Gaunt, M.A.; Maitra, U.S.; Ankel, H. Uridine diphosphate galacturonate 4-epimerase from the blue-green alga anabaena flos-aquae. *J. Biol. Chem.* **1974**, *249*, 2366–2372. [PubMed]
18. Ankel, H.; Tischer, R.G. UDP-D-glucuronate 4-epimerase in blue-green algae. *Biochim. Biophys. Acta* **1969**, *178*, 415–419. [CrossRef]
19. Feingold, D.S.; Neufeld, E.F.; Hassid, W.Z. The 4-epimerization and decarboxylation of uridine diphosphate D-glucuronic acid by extracts from *Phaseolus aureus* seedlings. *J. Biol. Chem.* **1960**, *235*, 910. [PubMed]
20. Dalessandro, G.; Northcote, D.H. Possible control sites of polysaccharide synthesis during cell growth and wall expansion of pea seedlings (*Pisum sativum* L.). *Planta* **1977**, *134*, 39–44. [CrossRef] [PubMed]
21. Dalessandro, G.; Northcote, D.H. Changes in enzymic activities of nucleoside diphosphate sugar interconversions during differentiation of cambium to xylem in sycamore and poplar. *Biochem. J.* **1977**, *162*, 267–279. [CrossRef] [PubMed]
22. Liljebjelke, K.; Adolphson, R.; Baker, K.; Doong, R.L.; Mohnen, D. Enzymatic synthesis and purification of uridine diphosphate [^{14}C]galacturonic acid: A substrate for pectin biosynthesis. *Anal. Biochem.* **1995**, *225*, 296–301. [CrossRef] [PubMed]
23. Frirdich, E.; Whitfield, C. Characterization of glakp, a UDP-galacturonic acid C4-epimerase from *Klebsiella pneumoniae* with extended substrate specificity. *J. Bacteriol.* **2005**, *187*, 4104–4115. [CrossRef] [PubMed]
24. Yin, S.; Sun, Y.J.; Liu, M.; Li, L.N.; Kong, J.Q. cDNA isolation and functional characterization of UDP-D-glucuronic acid 4-epimerase family from *Ornithogalum caudatum*. *Molecules* **2016**, *21*, 1505. [CrossRef] [PubMed]
25. Seymour, G.B.; Østergaard, L.; Chapman, N.H.; Knapp, S.; Martin, C. Fruit development and ripening. *Ann. Rev. Plant Biol.* **2013**, *64*, 219–241. [CrossRef] [PubMed]
26. Consortium, T.G. The tomato genome sequence provides insights into fleshy fruit evolution. *Nature* **2012**, *485*, 635–641. [CrossRef] [PubMed]
27. Koh, A.; Yoshiyuki, O.; Kaori, I.; Kentaro, Y.; Hideki, N.; Eli, K.; Atsushi, T. Functional genomics of tomato in a post-genome-sequencing phase. *Breed. Sci.* **2013**, *63*, 14–20.
28. Frelin, O.; Agrimi, G.; Laera, V.L.; Castegna, A.; Richardson, L.G.L.; Mullen, R.T.; Lermaortiz, C.; Palmieri, F.; Hanson, A.D. Identification of mitochondrial thiamin diphosphate carriers from arabidopsis and maize. *Funct. Integr. Genom.* **2012**, *12*, 317–326. [CrossRef] [PubMed]
29. Chung, B.Y.; Simons, C.; Firth, A.E.; Brown, C.M.; Hellens, R.P. Effect of 5′UTR introns on gene expression in *Arabidopsis thaliana*. *BMC Genom.* **2006**, *7*, 120. [CrossRef] [PubMed]

30. Thoden, J.B.; Hegeman, A.D.; Wesenberg, G.; Chapeau, M.C.; Frey, P.A.; Holden, H.M. Structural analysis of UDP-sugar binding to UDP-galactose 4-epimerase from *Escherichia coli*. *Biochemistry* **1997**, *36*, 6294–6304. [CrossRef] [PubMed]
31. Oppermann, U.; Filling, C.; Hult, M.; Shafqat, N.; Wu, X.; Lindh, M.; Shafqat, J.; Nordling, E.; Kallberg, Y.; Persson, B.; et al. Short-chain dehydrogenases/reductases (SDR): The 2002 update. *Chem.-Biol. Interact.* **2003**, *143–144*, 247–253. [CrossRef]
32. Wierenga, R.K.; Terpstra, P.; Hol, W.G.J. Prediction of the occurrence of the ADP-binding βαβ-fold in proteins, using an amino acid sequence fingerprint. *J. Mol. Biol.* **1986**, *187*, 101–107. [CrossRef]
33. Wolters, H.; Jürgens, G. Survival of the flexible: Hormonal growth control and adaptation in plant development. *Nat. Rev. Genet.* **2009**, *10*, 305–317. [CrossRef] [PubMed]
34. Seif ElYazal, S.A.; Seif ElYazal, M.A.; Dwidar, E.F.; Rady, M.M. Phytohormone crosstalk research: Cytokinin and its crosstalk with other phytohormones. *Curr. Protein Pept. Sci.* **2015**, *16*, 395–405. [CrossRef]
35. Wang, D.; Yeats, T.H.; Uluisik, S.; Rose, J.; Seymour, G.B. Fruit softening: Revisiting the role of pectin. *Trends Plant Sci.* **2018**, *23*, 302–310. [CrossRef] [PubMed]
36. Fernandezpozo, N.; Zheng, Y.; Snyder, S.; Nicolas, P.; Shinozaki, Y.; Fei, Z.; Catala, C.; Giovannoni, J.J.; Rose, J.K.; Mueller, L.A. The tomato expression atlas. *Bioinformatics* **2017**, *33*, 2397–2398. [CrossRef] [PubMed]
37. Mueller, L.A.; Solow, T.H.; Taylor, N.; Skwarecki, B.; Buels, R.; Binns, J.; Lin, C.; Wright, M.H.; Ahrens, R.; Wang, Y. The sol genomics network: A comparative resource for solanaceae biology and beyond. *Plant Physiol.* **2005**, *138*, 1310–1317. [CrossRef] [PubMed]
38. Fei, Z.; Joung, J.G.; Tang, X.; Zheng, Y.; Huang, M.; Lee, J.M.; Mcquinn, R.; Tieman, D.M.; Alba, R.; Klee, H.J. Tomato functional genomics database: A comprehensive resource and analysis package for tomato functional genomics. *Nucleic Acids Res.* **2011**, *39*, D1156–D1163. [CrossRef] [PubMed]
39. Lamesch, P.; Berardini, T.Z.; Li, D.; Swarbreck, D.; Wilks, C.; Sasidharan, R.; Muller, R.; Dreher, K.; Alexander, D.L.; Garcia-Hernandez, M.; et al. The arabidopsis information resource (tair): Improved gene annotation and new tools. *Nucleic Acids Res.* **2012**, *40*, D1202–D1210. [CrossRef] [PubMed]
40. Finn, R.D.; Bateman, A.; Clements, J.; Coggill, P.; Eberhardt, R.Y.; Eddy, S.R.; Heger, A.; Hetherington, K.; Holm, L.; Mistry, J. Pfam: The protein families database. *Nucleic Acids Res.* **2014**, *42*, D222–D230. [CrossRef] [PubMed]
41. Finn, R.D.; Clements, J.; Eddy, S.R. Hmmer web server: Interactive sequence similarity searching. *Nucleic Acids Res.* **2011**, *39*, W29–W37. [CrossRef] [PubMed]
42. Schultz, J.; Milpetz, F.; Bork, P.; Ponting, C.P. Smart, a simple modular architecture research tool: Identification of signaling domains. *Proc. Natl. Acad. Sci. USA* **1998**, *95*, 5857–5864. [CrossRef] [PubMed]
43. Saitou, N.; Nei, M. The neighbor-joining method: A new method for reconstructing phylogenetic trees. *Mol. Biol. Evol.* **1987**, *4*, 406–425. [PubMed]
44. Doronfaigenboim, A.; Stern, A.; Mayrose, I.; Bacharach, E.; Pupko, T. Selecton: A server for detecting evolutionary forces at a single amino-acid site. *Bioinformatics* **2005**, *21*, 2101–2103. [CrossRef] [PubMed]
45. Stern, A.; Doronfaigenboim, A.; Erez, E.; Martz, E.; Bacharach, E.; Pupko, T. Selecton 2007: Advanced models for detecting positive and purifying selection using a bayesian inference approach. *Nucleic Acids Res.* **2007**, *35*, W506–W511. [CrossRef] [PubMed]
46. Yang, Z.; Nielsen, R. Codon-substitution models for detecting molecular adaptation at individual sites along specific lineages. *Mol. Biol. Evol.* **2002**, *19*, 908–917. [CrossRef] [PubMed]
47. Anisimova, M.; Bielawski, J.P.; Yang, Z.H. Accuracy and power of the likelihood ratio test in detecting adaptive molecular evolution. *Mol. Biol. Evol.* **2001**, *18*, 1585–1592. [CrossRef] [PubMed]
48. Hu, B.; Jin, J.; Guo, A.; Zhang, H.; Luo, J.; Gao, G. Gsds 2.0: An upgraded gene feature visualization server. *Bioinformatics* **2015**, *31*, 1296–1297. [CrossRef] [PubMed]
49. Song, J.; Li, Z.; Tong, X.; Chen, C.; Chen, M.; Meng, G.; Chen, P.; Li, C.; Xin, Y.; Gai, T. Genome-wide identification and characterization of fox genes in the silkworm, bombyx mori. *Funct. Integr. Genom.* **2015**, *15*, 511–522. [CrossRef] [PubMed]
50. Ye, J.; Fang, L.; Zheng, H.; Zhang, Y.; Chen, J.; Zhang, Z.; Wang, J.; Li, S.; Li, R.; Bolund, L. Wego: A web tool for plotting go annotations. *Nucleic Acids Res.* **2006**, *34*, W293–W297. [CrossRef] [PubMed]
51. Yeung, K.Y.; Fraley, C.; Murua, A.; Raftery, A.E.; Ruzzo, W.L. Model-based clustering and data transformations for gene expression data. *Bioinformatics* **2001**, *17*, 977–987. [CrossRef] [PubMed]

52. Krogh, A.; Larsson, B.; Von, H.G.; Sonnhammer, E.L. Predicting transmembrane protein topology with a hidden markov model: Application to complete genomes. *J. Mol. Biol.* **2001**, *305*, 567–580. [CrossRef] [PubMed]
53. Lescot, M.; Déhais, P.; Thijs, G.; Marchal, K.; Moreau, Y.; Van de Peer, Y.; Rouzé, P.; Rombauts, S. Plantcare, a database of plant cis-acting regulatory elements and a portal to tools for in silico analysis of promoter sequences. *Nucleic Acids Res.* **2002**, *30*, 325–327. [CrossRef] [PubMed]
54. Yu, C.; Cai, X.; Ye, Z.; Li, H. Genome-wide identification and expression profiling analysis of trihelix gene family in tomato. *Biochem. Biophys. Res. Commun.* **2015**, *468*, 653–659. [CrossRef] [PubMed]
55. Zhu, M.; Chen, G.; Zhang, J.; Zhang, Y.; Xie, Q.; Zhao, Z.; Pan, Y.; Hu, Z. The abiotic stress-responsive nac-type transcription factor SLNAC4 regulates salt and drought tolerance and stress-related genes in tomato (*Solanum lycopersicum*). *Plant Cell Rep.* **2014**, *33*, 1851–1863. [CrossRef] [PubMed]
56. Zhu, M.; Chen, G.; Zhou, S.; Tu, Y.; Wang, Y.; Dong, T.; Hu, Z. A new tomato nac (NAM/ATAF1/2/CUC2) transcription factor, SLNAC4, functions as a positive regulator of fruit ripening and carotenoid accumulation. *Plant Cell Physiol.* **2014**, *55*, 119–135. [CrossRef] [PubMed]
57. Li, Z.; Peng, R.; Tian, Y.; Han, H.; Xu, J.; Yao, Q. Genome-wide identification and analysis of the myb transcription factor superfamily in *Solanum lycopersicum*. *Plant Cell Physiol.* **2016**, *57*, 1657–1677. [CrossRef] [PubMed]
58. Li, Z.; Zhang, L.; Wang, A.; Xu, X.; Li, J. Ectopic overexpression of SLHsFA3, a heat stress transcription factor from tomato, confers increased thermotolerance and salt hypersensitivity in germination in transgenic arabidopsis. *PLoS ONE* **2013**, *8*, e54880. [CrossRef] [PubMed]
59. Wu, T.; Abbott, J.A. Firmness and force relaxation characteristics of tomato stored intact or as slices. *Postharvest Biol. Technol.* **2002**, *24*, 59–68. [CrossRef]
60. Sandra, K.; Lynn, B.; Michael, N.; Spooner, D.M. Solanaceae—A model for linking genomics with biodiversity. *Comp. Funct. Genom.* **2004**, *5*, 285–291.
61. Seymour, G.B.; Manning, K.; Eriksson, E.M.; Popovich, A.H.; King, G.J. Genetic identification and genomic organization of factors affecting fruit texture. *J. Exp. Bot.* **2002**, *53*, 2065–2071. [CrossRef] [PubMed]
62. Seymour, G.B. Genomics meets horticulture. *J. Hortic. Sci. Biotechnol.* **2006**, *81*, 173. [CrossRef]

© 2018 by the authors. Licensee MDPI, Basel, Switzerland. This article is an open access article distributed under the terms and conditions of the Creative Commons Attribution (CC BY) license (http://creativecommons.org/licenses/by/4.0/).

MDPI
St. Alban-Anlage 66
4052 Basel
Switzerland
Tel. +41 61 683 77 34
Fax +41 61 302 89 18
www.mdpi.com

International Journal of Molecular Sciences Editorial Office
E-mail: ijms@mdpi.com
www.mdpi.com/journal/ijms

www.ingramcontent.com/pod-product-compliance
Lightning Source LLC
Chambersburg PA
CBHW061458220926
51850CB00004B/123
* 9 7 8 3 0 3 6 5 7 2 2 7 7 *